nonassociative algebra and its applications

LECTURE NOTES IN PURE AND APPLIED MATHEMATICS

1. *N. Jacobson*, Exceptional Lie Algebras
2. *L.-Å. Lindahl and F. Poulsen*, Thin Sets in Harmonic Analysis
3. *I. Satake*, Classification Theory of Semi-Simple Algebraic Groups
4. *F. Hirzebruch et al.*, *Differentiable Manifolds and Quadratic Forms*
5. *I. Chavel*, Riemannian Symmetric Spaces of Rank One
6. *R. B. Burckel*, Characterization of C(X) Among Its Subalgebras
7. *B. R. McDonald et al.*, Ring Theory
8. *Y.-T. Siu*, Techniques of Extension on Analytic Objects
9. *S. R. Caradus et al.*, Calkin Algebras and Algebras of Operators on Banach Spaces
10. *E. O. Roxin et al.*, Differential Games and Control Theory
11. *M. Orzech and C. Small*, The Brauer Group of Commutative Rings
12. *S. Thomier*, Topology and Its Applications
13. *J. M. Lopez and K. A. Ross*, Sidon Sets
14. *W. W. Comfort and S. Negrepontis*, Continuous Pseudometrics
15. *K. McKennon and J. M. Robertson*, Locally Convex Spaces
16. *M. Carmeli and S. Malin*, Representations of the Rotation and Lorentz Groups
17. *G. B. Seligman*, Rational Methods in Lie Algebras
18. *D. G. de Figueiredo*, Functional Analysis
19. *L. Cesari et al.*, Nonlinear Functional Analysis and Differential Equations
20. *J. J. Schäffer*, Geometry of Spheres in Normed Spaces
21. *K. Yano and M. Kon*, Anti-Invariant Submanifolds
22. *W. V. Vasconcelos*, The Rings of Dimension Two
23. *R. E. Chandler*, Hausdorff Compactifications
24. *S. P. Franklin and B. V. S. Thomas*, Topology
25. *S. K. Jain*, Ring Theory
26. *B. R. McDonald and R. A. Morris*, Ring Theory II
27. *R. B. Mura and A. Rhemtulla*, Orderable Groups
28. *J. R. Graef*, Stability of Dynamical Systems
29. *H.-C. Wang*, Homogeneous Branch Algebras
30. *E. O. Roxin et al.*, Differential Games and Control Theory II
31. *R. D. Porter*, Introduction to Fibre Bundles
32. *M. Altman*, Contractors and Contractor Directions Theory and Applications
33. *J. S. Golan*, Decomposition and Dimension in Module Categories
34. *G. Fairweather*, Finite Element Galerkin Methods for Differential Equations
35. *J. D. Sally*, Numbers of Generators of Ideals in Local Rings
36. *S. S. Miller*, Complex Analysis
37. *R. Gordon*, Representation Theory of Algebras
38. *M. Goto and F. D. Grosshans*, Semisimple Lie Algebras
39. *A. I. Arruda et al.*, Mathematical Logic
40. *F. Van Oystaeyen*, Ring Theory
41. *F. Van Oystaeyen and A. Verschoren*, Reflectors and Localization
42. *M. Satyanarayana*, Positively Ordered Semigroups
43. *D. L Russell*, Mathematics of Finite-Dimensional Control Systems
44. *P.-T. Liu and E. Roxin*, Differential Games and Control Theory III
45. *A. Geramita and J. Seberry*, Orthogonal Designs
46. *J. Cigler, V. Losert, and P. Michor*, Banach Modules and Functors on Categories of Banach Spaces
47. *P.-T. Liu and J. G. Sutinen*, Control Theory in Mathematical Economics
48. *C. Byrnes*, Partial Differential Equations and Geometry
49. *G. Klambauer*, Problems and Propositions in Analysis
50. *J. Knopfmacher*, Analytic Arithmetic of Algebraic Function Fields
51. *F. Van Oystaeyen*, Ring Theory
52. *B. Kadem*, Binary Time Series
53. *J. Barros-Neto and R. A. Artino*, Hypoelliptic Boundary-Value Problems
54. *R. L. Sternberg et al.*, Nonlinear Partial Differential Equations in Engineering and Applied Science
55. *B. R. McDonald*, Ring Theory and Algebra III
56. *J. S. Golan*, Structure Sheaves Over a Noncommutative Ring
57. *T. V. Narayana et al.*, Combinatorics, Representation Theory and Statistical Methods in Groups
58. *T. A. Burton*, Modeling and Differential Equations in Biology
59. *K. H. Kim and F. W. Roush*, Introduction to Mathematical Consensus Theory

60. *J. Banas and K. Goebel*, Measures of Noncompactness in Banach Spaces
61. *O. A. Nielson*, Direct Integral Theory
62. *J. E. Smith et al.*, Ordered Groups
63. *J. Cronin*, Mathematics of Cell Electrophysiology
64. *J. W. Brewer*, Power Series Over Commutative Rings
65. *P. K. Kamthan and M. Gupta*, Sequence Spaces and Series
66. *T. G. McLaughlin*, Regressive Sets and the Theory of Isols
67. *T. L. Herdman et al.*, Integral and Functional Differential Equations
68. *R. Draper*, Commutative Algebra
69. *W. G. McKay and J. Patera*, Tables of Dimensions, Indices, and Branching Rules for Representations of Simple Lie Algebras
70. *R. L. Devaney and Z. H. Nitecki*, Classical Mechanics and Dynamical Systems
71. *J. Van Geel*, Places and Valuations in Noncommutative Ring Theory
72. *C. Faith*, Injective Modules and Injective Quotient Rings
73. *A. Fiacco*, Mathematical Programming with Data Perturbations I
74. *P. Schultz et al.*, Algebraic Structures and Applications
75. *L Bican et al.*, Rings, Modules, and Preradicals
76. *D. C. Kay and M. Breen*, Convexity and Related Combinatorial Geometry
77. *P. Fletcher and W. F. Lindgren*, Quasi-Uniform Spaces
78. *C.-C. Yang*, Factorization Theory of Meromorphic Functions
79. *O. Taussky*, Ternary Quadratic Forms and Norms
80. *S. P. Singh and J. H. Burry*, Nonlinear Analysis and Applications
81. *K. B. Hannsgen et al.*, Volterra and Functional Differential Equations
82. *N. L. Johnson et al.*, Finite Geometries
83. *G. I. Zapata*, Functional Analysis, Holomorphy, and Approximation Theory
84. *S. Greco and G. Valla*, Commutative Algebra
85. *A. V. Fiacco*, Mathematical Programming with Data Perturbations II
86. *J.-B. Hiriart-Urruty et al.*, Optimization
87. *A. Figa Talamanca and M. A. Picardello*, Harmonic Analysis on Free Groups
88. *M. Harada*, Factor Categories with Applications to Direct Decomposition of Modules
89. *V. I. Istrătescu*, Strict Convexity and Complex Strict Convexity
90. *V. Lakshmikantham*, Trends in Theory and Practice of Nonlinear Differential Equations
91. *H. L. Manocha and J. B. Srivastava*, Algebra and Its Applications
92. *D. V. Chudnovsky and G. V. Chudnovsky*, Classical and Quantum Models and Arithmetic Problems
93. *J. W. Longley*, Least Squares Computations Using Orthogonalization Methods
94. *L. P. de Alcantara*, Mathematical Logic and Formal Systems
95. *C. E. Aull*, Rings of Continuous Functions
96. *R. Chuaqui*, Analysis, Geometry, and Probability
97. *L. Fuchs and L. Salce*, Modules Over Valuation Domains
98. *P. Fischer and W. R. Smith*, Chaos, Fractals, and Dynamics
99. *W. B. Powell and C. Tsinakis*, Ordered Algebraic Structures
100. *G. M. Rassias and T. M. Rassias*, Differential Geometry, Calculus of Variations, and Their Applications
101. *R.-E. Hoffmann and K. H. Hofmann*, Continuous Lattices and Their Applications
102. *J. H. Lightbourne III and S. M. Rankin III*, Physical Mathematics and Nonlinear Partial Differential Equations
103. *C. A. Baker and L. M. Batten*, Finite Geometrics
104. *J. W. Brewer et al.*, Linear Systems Over Commutative Rings
105. *C. McCrory and T. Shifrin*, Geometry and Topology
106. *D. W. Kueke et al.*, Mathematical Logic and Theoretical Computer Science
107. *B.-L. Lin and S. Simons*, Nonlinear and Convex Analysis
108. *S. J. Lee*, Operator Methods for Optimal Control Problems
109. *V. Lakshmikantham*, Nonlinear Analysis and Applications
110. *S. F. McCormick*, Multigrid Methods
111. *M. C. Tangora*, Computers in Algebra
112. *D. V. Chudnovsky and G. V. Chudnovsky*, Search Theory
113. *D. V. Chudnovsky and R. D. Jenks*, Computer Algebra
114. *M. C. Tangora*, Computers in Geometry and Topology
115. *P. Nelson et al.*, Transport Theory, Invariant Imbedding, and Integral Equations
116. *P. Clément et al.*, Semigroup Theory and Applications
117. *J. Vinuesa*, Orthogonal Polynomials and Their Applications
118. *C. M. Dafermos et al.*, Differential Equations
119. *E. O. Roxin*, Modern Optimal Control
120. *J. C. Díaz*, Mathematics for Large Scale Computing

121. *P. S. Milojević* Nonlinear Functional Analysis
122. *C. Sadosky,* Analysis and Partial Differential Equations
123. *R. M. Shortt,* General Topology and Applications
124. *R. Wong,* Asymptotic and Computational Analysis
125. *D. V. Chudnovsky and R. D. Jenks,* Computers in Mathematics
126. *W. D. Wallis et al.*, Combinatorial Designs and Applications
127. *S. Elaydi,* Differential Equations
128. *G. Chen et al.*, Distributed Parameter Control Systems
129. *W. N. Everitt,* Inequalities
130. *H. G. Kaper and M. Garbey,* Asymptotic Analysis and the Numerical Solution of Partial Differential Equations
131. *O. Arino et al.*, Mathematical Population Dynamics
132. *S. Coen,* Geometry and Complex Variables
133. *J. A. Goldstein et al.*, Differential Equations with Applications in Biology, Physics, and Engineering
134. *S. J. Andima et al.*, General Topology and Applications
135. *P Clément et al.*, Semigroup Theory and Evolution Equations
136. *K. Jarosz*, Function Spaces
137. *J. M. Bayod et al.*, *p*-adic Functional Analysis
138. *G. A. Anastassiou*, Approximation Theory
139. *R. S. Rees*, Graphs, Matrices, and Designs
140. *G. Abrams et al.*, Methods in Module Theory
141. *G. L. Mullen and P. J.-S. Shiue*, Finite Fields, Coding Theory, and Advances in Communications and Computing
142. *M. C. Joshi and A. V. Balakrishnan*, Mathematical Theory of Control
143. *G. Komatsu and Y. Sakane*, Complex Geometry
144. *I. J. Bakelman*, Geometric Analysis and Nonlinear Partial Differential Equations
145. *T. Mabuchi and S. Mukai*, Einstein Metrics and Yang–Mills Connections
146. *L. Fuchs and R. Göbel*, Abelian Groups
147. *A. D. Pollington and W. Moran*, Number Theory with an Emphasis on the Markoff Spectrum
148. *G. Dore et al.*, Differential Equations in Banach Spaces
149. *T. West*, Continuum Theory and Dynamical Systems
150. *K. D. Bierstedt et al.*, Functional Analysis
151. *K. G. Fischer et al.*, Computational Algebra
152. *K. D. Elworthy et al.*, Differential Equations, Dynamical Systems, and Control Science
153. *P.-J. Cahen, et al.*, Commutative Ring Theory
154. *S. C. Cooper and W. J. Thron*, Continued Fractions and Orthogonal Functions
155. *P. Clément and G. Lumer*, Evolution Equations, Control Theory, and Biomathematics
156. *M. Gyllenberg and L. Persson*, Analysis, Algebra, and Computers in Mathematical Research
157. *W. O. Bray et al.*, Fourier Analysis
158. *J. Bergen and S. Montgomery*, Advances in Hopf Algebras
159. *A. R. Magid*, Rings, Extensions, and Cohomology
160. *N. H. Pavel*, Optimal Control of Differential Equations
161. *M. Ikawa*, Spectral and Scattering Theory
162. *X. Liu and D. Siegel*, Comparison Methods and Stability Theory
163. *J.-P. Zolésio*, Boundary Control and Variation
164. *M. Křížek et al.*, Finite Element Methods
165. *G. Da Prato and L. Tubaro*, Control of Partial Differential Equations
166. *E. Ballico*, Projective Geometry with Applications
167. *M. Costabel et al.*, Boundary Value Problems and Integral Equations in Nonsmooth Domains
168. *G. Ferreyra, G. R. Goldstein, and F. Neubrander*, Evolution Equations
169. *S. Huggett*, Twistor Theory
170. *H. Cook et al.*, Continua
171. *D. F. Anderson and D. E. Dobbs*, Zero-Dimensional Commutative Rings
172. *K. Jarosz*, Function Spaces
173. *V. Ancona et al.*, Complex Analysis and Geometry
174. *E. Casas*, Control of Partial Differential Equations and Applications
175. *N. Kalton et al.*, Interaction Between Functional Analysis, Harmonic Analysis, and Probability
176. *Z. Deng et al.*, Differential Equations and Control Theory
177. *P. Marcellini et al.* Partial Differential Equations and Applications
178. *A. Kartsatos*, Theory and Applications of Nonlinear Operators of Accretive and Monotone Type
179. *M. Maruyama*, Moduli of Vector Bundles
180. *A. Ursini and P. Aglianò*, Logic and Algebra
181. *X. H. Cao et al.*, Rings, Groups, and Algebras
182. *D. Arnold and R. M. Rangaswamy,* Abelian Groups and Modules
183. *S. R. Chakravarthy and A. S. Alfa*, Matrix-Analytic Methods in Stochastic Models

184. *J. E. Andersen et al.*, Geometry and Physics
185. *P.-J. Cahen et al.*, Commutative Ring Theory
186. *J. A. Goldstein et al.*, Stochastic Processes and Functional Analysis
187. *A. Sorbi*, Complexity, Logic, and Recursion Theory
188. *G. Da Prato and J.-P. Zolésio*, Partial Differential Equation Methods in Control and Shape Analysis
189. *D. D. Anderson*, Factorization in Integral Domains
190. *N. L. Johnson*, Mostly Finite Geometries
191. *D. Hinton and P. W. Schaefer*, Spectral Theory and Computational Methods of Sturm–Liouville Problems
192. *W. H. Schikhof et al.*, *p*-adic Functional Analysis
193. *S. Sertöz*, Algebraic Geometry
194. *G. Caristi and E. Mitidieri*, Reaction Diffusion Systems
195. *A. V. Fiacco*, Mathematical Programming with Data Perturbations
196. *M. Křížek et al.*, Finite Element Methods: Superconvergence, Post-Processing, and A Posteriori Estimates
197. *S. Caenepeel and A. Verschoren*, Rings, Hopf Algebras, and Brauer Groups
198. *V. Drensky et al.*, Methods in Ring Theory
199. *W. B. Jones and A. Sri Ranga*, Orthogonal Functions, Moment Theory, and Continued Fractions
200. *P. E. Newstead*, Algebraic Geometry
201. *D. Dikranjan and L. Salce*, Abelian Groups, Module Theory, and Topology
202. *Z. Chen et al.*, Advances in Computational Mathematics
203. *X. Caicedo and C. H. Montenegro*, Models, Algebras, and Proofs
204. *C. Y. Yıldırım and S. A. Stepanov*, Number Theory and Its Applications
205. *D. E. Dobbs et al.*, Advances in Commutative Ring Theory
206. *F. Van Oystaeyen*, Commutative Algebra and Algebraic Geometry
207. *J. Kakol et al.*, *p*-adic Functional Analysis
208. *M. Boulagouaz and J.-P. Tignol*, Algebra and Number Theory
209. *S. Caenepeel and F. Van Oystaeyen*, Hopf Algebras and Quantum Groups
210. *F. Van Oystaeyen and M. Saorin*, Interactions Between Ring Theory and Representations of Algebras
211. *R. Costa et al.*, Nonassociative Algebra and Its Applications

Additional Volumes in Preparation

nonassociative algebra and its applications

the fourth international conference

edited by

Roberto Costa
Alexander Grishkov
Henrique Guzzo, Jr.
Luiz A. Peresi
University of São Paulo
São Paulo, Brazil

CRC Press
Taylor & Francis Group
Boca Raton London New York

CRC Press is an imprint of the
Taylor & Francis Group, an **informa** business

CRC Press
Taylor & Francis Group
6000 Broken Sound Parkway NW, Suite 300
Boca Raton, FL 33487-2742

First issued in hardback 2019

ISBN-13: 978-0-8247-0406-3 (pbk)
ISBN-13: 978-1-138-40176-1 (hbk)

Library of Congress Cataloging-in-Publication Data

Nonassociative algebra and its applications : the fourth international conference / edited by Roberto Costa ... [et al.].
p. cm. — (Lecture notes in pure and applied mathematics ; v. 211)
ISBN 0-8247-0406-1 (alk. paper)
1. Nonassociative algebras—Congresses. I. Costa, Roberto. II. Series.

QA252 .N66 2000
512'.24—dc21 00-026394

Visit the Taylor & Francis Web site at
http://www.taylorandfrancis.com

and the CRC Press Web site at
http://www.crcpress.com

Preface

The Fourth International Conference on Nonassociative Algebra and Its Applications was held in São Paulo (Brazil). The three previous Conferences were in: Novosibirsk (Russia, 1988), Tashkent (Russia, 1990), and Oviedo (Spain, 1993).

The Organising Committee members were Roberto Costa, Alexander Grishkov, Henrique Guzzo, Jr., and Luiz Antonio Peresi, from the University of São Paulo. The Scientific Committee was made up of the following members:

Benkart, G. (University of Wisconsin, U.S.)
Costa, R. (USP, Brazil)
Elduque, A. (Universidad de La Rioja, Spain)
Ferreira, L. A. (UNESP, Brazil)
Forger, M. (USP, Brazil)
González, S. (Universidad de Oviedo, Spain)
Hentzel, I. R. (Iowa State University, U.S.)
Kamiya, N. (Shimane University, Japan)
Kaup, W. (Universität Tubingen, Germany)
Kleinfeld, E. (University of Iowa, U.S.)
Kostrikin, A. I. (Moscow State University, Russia)
Kuzmin, E. N. (Institute of Mathematics, Novosibirsk, Russia)
Lyubich, Y. (Technion University, Israel)
Mallol, C. (Universidad de La Frontera, Chile)
Martínez, C. (Universidad de Oviedo, Spain)
McCrimmon, K. (University of Virginia, U.S.)
Medvedev, Y. (York University, Canada)
Meyberg, K. (Technische Universität München, Germany)
Moody, R. (University of Alberta, Canada)
Myung, H. C. (KAIST, Taejon, South Korea)
Okubo, S. (University of Rochester, U.S.)
Osborn, J. M. (University of Wisconsin, U.S.)
Pchelintsev, S. (Lenin Moscow Pedagogical Institute, Russia)
Petersson, H. (Fern-Universität, Germany)
Polcino Milies, F. C. (USP, Brazil)
Racine, M. (University of Ottawa, Canada)
Schafer, R. D. (Massachusetts Institute of Technology, U.S.)
Shestakov, I. P.(Institute of Mathematics, Novosibirsk, Russia)
Slinko, A. (University of Auckland, New Zealand)
Taft, E. (Rutgers University, U.S.)
Zelmanov, E. (Yale University, U.S.)

Nearly 100 people came to the conference. Several topics in nonassociative algebra were presented, including applications to population genetics theory and physics.

During the five days of the conference, 14 plenary talks were delivered by members of the Scientific Committee and also some invited speakers. Twelve conferences on some special topics were presented, as well as 35 short communications.

The editors would like to thank Professor L. A. Ferreira, who was in charge of the conferences in the field of physics. We regret the death of Professor M. V. Saveliev some months after the meeting.

The editors want also to express thanks to the following institutions for their financial support: The Institute of Mathematics and Statistics of the University of São Paulo, Fundação de Amparo à Pesquisa do Estado de São Paulo (FAPESP), Sociedade Brasileira de Matemática (SBM), CAPES, and CCInt-USP.

The editors also thank the referees for their invaluable work of revision of the written version of the conferences and communications.

Roberto Costa
Alexander Grishkov
Henrique Guzzo, Jr.
Luiz A. Peresi

Contents

Contributors

Helena Albuquerque University of Coimbra, Coimbra, Portugal

Y. Bahturin Moscow State University, Moscow, Russia

H. Belich Centro Brasileiro de Pesquisas Fisicas, Rio de Janeiro, Brazil

Pilar Benito University of La Rioja, Logroño, Spain

A. Catalan University of La Frontera, Temuco, Chile

Roberto Costa Institute of Mathematics and Statistics, University of São Paulo, São Paulo, Brazil

João Carlos da Motta Ferreira Federal University of Mato Grosso do Sul, Campo Grande, Brazil

Cristina Draper University of La Rioja, Logroño, Spain

Alberto Elduque University of Zaragoza, Zaragoza, Spain

Mohammed El Houari Institute of Mathematics and Statistics, University of São Paulo, São Paulo, Brazil

Luiz A. Ferreira Institute of Theoretical Physics –IFT/UNESP, São Paulo-SP, Brazil

Alice Fialowski Eötvös Loránd University, Budapest, Hungary

M. A. Garcia-Muñiz University of Oviedo, Oviedo, Spain

J. F. Gomes Institute of Theoretical Physics – UNESP, São Paulo-SP, Brazil

Carlos Gomez-Ambrosi University of Zaragoza, Zaragoza, Spain

S. González University of Oviedo, Oviedo, Spain

Edgar G. Goodaire Memorial University of Newfoundland, St. John's, Newfoundland, Canada

Alexander Grishkov Institute of Mathematics and Statistics, University of São Paulo, São Paulo, Brazil

Henrique Guzzo, Jr. Institute of Mathematics and Statistics, University of São Paulo, São Paulo, Brazil

Irvin Roy Hentzel Iowa State University, Ames, Iowa

Noriaki Kamiya Shimane University, Japan

Bakhadyr Khoussainov University of Auckland, Auckland, New Zealand

Erwin Kleinfeld University of Iowa, Iowa City, Iowa

Margaret Kleinfeld University of Iowa, Iowa City, Iowa

Jesus Laliena University of La Rioja, Logroño, Spain

C. López-Díaz University of Oviedo, Oviedo, Spain

Yu. Lyubich Technion, Haifa, Israel

Shahn Majid University of Cambridge, Cambridge, England

C. Mallol University of La Frontera, Temuco, Chile

C. Martínez University of Oviedo, Oviedo, Spain

J. Martínez Moreno University of Granada, Granada, Spain

Martin Mathieu The Queen's University of Belfast, Belfast, Northern Ireland

F. E. Mendonça da Silveira Institute of Theoretical Physics – UNESP, São Paulo, Brazil

M. L. Merlini Giuliani Federal University of Santa Maria, Santa Maria, Brazil

Artibano Micali University of Montpellier II, Montpellier, France

Hyo Chul Myung Korea Institute for Advanced Study and KAIST, School of Mathematics, Seoul, Korea

Alexander I. Nesterov CUCEI, University of Guadalajara, Guadalajara, Jalisco, Mexico

Susumu Okubo University of Rochester, Rochester, New York

J. Marshall Osborn University of Wisconsin, Madison, Wisconsin

R. Paunov Centro Brasileiro de Pesquisas Fisicas, Rio de Janeiro, Brazil

Luiz Antonio Peresi University of São Paulo, São Paulo, Brazil

J. M. Perez-Izquierdo University of La Rioja, Logroño, Spain,

César Polcino Milies Institute of Mathematics and Statistics, University of São Paulo, São Paulo SP, Brazil

A. V. Razumov Institute for High Energy Physics, Moscow region, Russia

L. V. Sabinin University of Quintana Roo, Chetumal, Quintana Roo, Mexico

Liudmila Sabinina Faculty of Sciences, UAEM., Morelos, Mexico

M. V. Saveliev Institute for High Energy Physics, Moscow region, Russia

L. V. Sbitneva University of Quintana Roo, Chetumal, Quintana Roo, Mexico

J. Setó University of Oviedo, Oviedo, Spain

Ivan P. Shestakov Sobolev Institute of Mathematics, Novosibirsk, Russia

Ivan P. Shestakov University of São Paulo, São Paulo-SP, Brazil

Pavel Shumyatsky University of Brasilia, Brasilia – DF, Brazil

Arkadii Slinko University of Auckland, Auckland, New Zealand

G. M. Sotkov Departamento de Campos e Partículas – CBPF, Rio de Janeiro – RJ, Brazil

R. Varro University of Montpellier III, Montpellier, France

P. Vicente University of León, León, Spain

G. P. Wene The University of Texas at San Antonio, San Antonio, Texas

Richard Wiegandt Mathematical Institute, HAS, Budapest, Hungary

M. Zaicev Moscow State University, Moscow, Russia

Kaiming Zhao Institute of Systems Science, Academia Sinica, Beijing, China

A. H. Zimerman Institute of Theoretical Physics – UNESP, São Paulo–SP, Brazil

New approach to octonions and Cayley algebras

HELENA ALBUQUERQUE[1] Departamento de Matematica-Faculdade de Ciencias e Tecnologia, Universidade de Coimbra, Apartado 3008, 3000 Coimbra, Portugal

SHAHN MAJID[2] Department of Applied Mathematics and Theoretical Physics, University of Cambridge, Cambridge CB3 9EW, UK

ABSTRACT We announce a new approach to the octonions as *quasiassociative* algebras. We strip out the categorical and quasi-quantum group considerations in our longer paper and present here (without proof) some of the more algebraic conclusions.

1 INTRODUCTION

Usually one recognises the nonassociativity of the octonions by saying that they are instead alternative algebras. While this is true, the property of being alternative has a much weaker character than associativity and, as a result, many standard ideas and constructions for associative algebras do not go through in the alternative case. In our paper [1], to which this note is a short introduction, we have introduced a full solution to this problem based on modern ideas from category theory and Drinfeld's theory of quasiquantum groups.

Without going into any details (see [1]), the new formulation is that the octonions and other Cayley algebras live naturally as objects in a monoidal category[2][3]. For any three objects V, W, Z in such a category there is an associator isomorphism $\Phi_{V,W,Z} : (V \otimes W) \otimes Z \to V \otimes (W \otimes Z)$ which performs the rebracketting. Mac Lane's pentagon condition on Φ ensures that *we can do all constructions as if there*

[1] Supported by CMUC-JNICT and by Praxis 2/2.1/Mat7458/94

[2] Royal Society University Research Fellow and Fellow of Pembroke College, Cambridge

are no brackets (i.e. as if $\otimes$ is strictly associative). After writing any desired constructions as the composition of various maps, we simply insert Φ as needed for the compositions to make sense, and all different ways to do this will give the same result (this is Mac Lane's coherence theorem). So working in such a category is no harder than usual associative linear algebra. For example, an algebra A in such a category means

$$\bullet \circ (\bullet \otimes \mathrm{id}) = \bullet \circ (\mathrm{id} \otimes \bullet) \circ \Phi_{A,A,A}$$

for the product $\bullet$, where Φ is inserted for the bracketting to make sense. So, recognizing the octonions as such a *quasiassociative algebra* (or *quasialgebra* for short) makes them as good as associative in the precise sense explained above.

In [1] we introduce and study a class of such quasialgebras that contain composition algebras and more general algebras obtained by a generalised Cayley-Dickson process. All algebras are considered over a field k of characteristic different from 2. The required class of quasialgebras arises naturally by a certain 'Drinfeld twisting'[4][3] or deformation of classical group algebras, as follows.

2 QUASIALGEBRAS $k_F G$

First of all, we know that if we consider the set of complex numbers, the quaternions or the octonions, all these algebras have something in common: If we choose a suitable basis and remove the $\pm$ signs from the multiplication tables of these algebras, we have the tables of the additive groups $G = \mathbb{Z}_2$ (for complex numbers), $G = \mathbb{Z}_2 \times \mathbb{Z}_2$ (for quaternions) and $G = \mathbb{Z}_2 \times \mathbb{Z}_2 \times \mathbb{Z}_2$ (for octonions)[5]. We view the signs in the multiplication tables as an invertible 2-cochain $F : G \times G \to k$ (a nowhere-zero function which is 1 when either argument is the group identity $e \in G$). Writing $F(x, y) = (-1)^{f(x,y)}$, one has explicitly[1],

$$G = \mathbb{Z}_2, \quad f(x,y) = xy, \quad x, y \in \mathbb{Z}_2 \qquad \text{(Complex numbers)},$$

$$G = (\mathbb{Z}_2)^2, \quad f(\vec{x}, \vec{y}) = x_1 y_1 + (x_1 + x_2) y_2 \qquad \text{(Quaternions)},$$

where $\vec{x} = (x_1, x_2) \in G$ is a vector notation and the components x_1, x_2 are viewed in the field $\mathbb{Z}_2$.

$$G = (\mathbb{Z}_2)^3, \quad f(\vec{x}, \vec{y}) = \sum_{i \le j} x_i y_j + y_1 x_2 x_3 + x_1 y_2 x_3 + x_1 x_2 y_3 \quad \text{(Octonions)},$$

where $\vec{x} = (x_1, x_2, x_3) \in G$ is a vector notation. Similarly for higher Cayley algebras where $G = (\mathbb{Z}_2)^n$.

From the group G and the cochain F we recover the complex, quaternion and octonion algebras as the 'deformation' $k_F G$ of the group algebra of the appropriate G. This is the vector space with basis labeled by G and the product

$$x \bullet y = F(x, y) xy$$

for $x, y \in G$, where xy is the group product in G. So this is a kind of deformation of the usual group algebra of G. In quantum groups we do the deformation by introducing a parameter q such that when q tends to 1 we have the original algebras. Here we do the deformation by introducing a cochain F.

PROPOSITION 1 *[1] Let G be a group and F any invertible 2-cochain. Then $k_F G$ is a G-graded quasialgebra with associator Φ determined by the coboundary ϕ of F. Explicitly, it is quasiassociative in the sense*

$$(x \bullet y) \bullet z = \phi(x, y, z) x \bullet (y \bullet z)$$

for all $x, y, z \in G$, and

$$\phi(x, y, z) = \frac{F(x, y)F(xy, z)}{F(y, z)F(x, yz)}.$$

To explain the setting here, for G a group and $\phi : G \times G \times G \to k$ an invertible group 3-cocycle, the category of G-graded vector spaces becomes monoidal with the associator Φ determined by the 3-cocycle ϕ and the grading. A quasialgebra with this form of Φ is called a *G-graded quasialgebra*. It consists of an algebra A, a G-grading which respects the product and unit (so the degree of $1 \in A$ is $e \in G$, the group identity), and the quasiassociativity law

$$(a \bullet b) \bullet c = \phi(|a|, |b|, |c|) a \bullet (b \bullet c)$$

for all elements $a, b, c \in A$ of degree $|a|, |b|, |c|$. In our case, $k_F G$ is such a G-graded quasialgebra with ϕ built from F and with $|x| = x$ for $x \in G$.

For the complex number and the quaternion algebras, F is closed, i.e. ϕ is trivial and the algebras happen to be strictly associative. This is because f in these cases is quadratic. As soon as we introduce a cubic or higher 'interaction' term in f, as in the case of the octonions, ϕ typically becomes nontrivial and the algebra $k_F G$ nonassociative. In the case of the Octonions it is

$$\phi(\vec{x}, \vec{y}, \vec{z}) = (-1)^{(\vec{x} \times \vec{y}) \cdot \vec{z}}$$

(the vector cross product and vector dot product in the exponent, i.e. the determinant $|\vec{x}\ \vec{y}\ \vec{z}|$). On the other hand, if we simply drop the cubic or higher terms in the above family, we clearly obtain the corresponding Clifford algebra with negative signature[3] (the relations are immediate and the dimensions match) i.e. these are obtained as $k_F G$ with

$$G = (Z_2)^n, \quad f(\vec{x}, \vec{y}) = \sum_{i \le j} x_i y_j \qquad \text{(Clifford algebras)},$$

as the associative version of the octonion or Cayley algebra, which is another way to see the close relationship between these and Clifford algebras. The positive signature algebras are obtained similarly with $i < j$ in f.

[3] We would like to thank Tony Smith for asking us to clarify this point

Also, we are mainly interested in G Abelian and specialise to this case from now on. For ϕ of the coboundary form, the category of G-graded spaces is symmetric[2], i.e. for any two objects V, W there is a generalised transposition isomorphism $\Psi_{V,W}$: $V \otimes W \to W \otimes V$. A quasialgebra A is *quasicommutative* if $\bullet = \bullet \circ \Psi_{A,A}$. This is the case for all $k_F G$ with Ψ determined by a function $\mathcal{R}(x,y) = F(x,y)/F(y,x)$. Explicitly,

$$x \bullet y = \mathcal{R}(x,y) y \bullet x,$$

for all $x, y \in G$. For complex numbers, quaternions and octonions, etc., the function $\mathcal{R}$ has the simple form

$$\mathcal{R}(x,y) = \begin{cases} 1 & \text{if } x = e \text{ or } y = e \text{ or } x = y \\ -1 & \text{otherwise.} \end{cases}$$

We call this important case *altercommutative*. By contrast, for the Clifford algebras one has

$$\mathcal{R}(\vec{x}, \vec{y}) = (-1)^{\sum_{i \neq j} x_i y_j}$$

which is not in general altercommutative (for $n > 2$).

3 ALGEBRAIC PROPERTIES OF $k_F G$

We have just seen that the functions $\phi, \mathcal{R}$ built from F allow for the categorical setting of the algebras $k_F G$ as quasiassociative and quasicommutative. In particular, the algebra is associative iff $\phi = 1$ and commutative iff $\mathcal{R} = 1$. We now summarise less obvious results expressing the more conventional algebraic properties of $k_F G$ in terms of these functions $\phi, \mathcal{R}$. We refer to [1] for proofs and details.

PROPOSITION 2 *[1] $k_F G$ is an alternative algebra* iff

$$\phi^{-1}(y,x,z) + \mathcal{R}(x,y)\phi^{-1}(x,y,z) = 1 + \mathcal{R}(x,y)$$

$$\phi(x,y,z) + \mathcal{R}(z,y)\phi(x,z,y) = 1 + \mathcal{R}(z,y)$$

for all $x, y, z \in G$. In this case,

$$\phi(x,x,y) = \phi(x,y,y) = \phi(x,y,x) = 1$$

for all $x, y \in G$.

Next we consider involutions. Since we have a special basis of $k_F G$ it is natural to consider involutions diagonal in this basis.

PROPOSITION 3 *[1] $k_F G$ admits an involution which is diagonal in the basis G* iff $\mathcal{R}(x,y) = \frac{s(x)s(y)}{s(xy)}$ *for some 1-cochain $s : G \to k$ (a nowhere-zero function with $s(e) = 1$) obeying $s^2 = 1$. In this case, one has $\mathcal{R}(x,y) = \mathcal{R}(y,x)$ and $\phi(x,y,z) = \phi(z,y,x)^{-1}$ for all $x, y, z \in G$.*

The corresponding involution here is $\sigma(x) = s(x)x$ for all $x \in G$. Let A be a finite dimensional algebra with identity element 1 and let σ be an involution in A. We say that σ is a strong involution if $a + \sigma(a)$, $a \bullet \sigma(a) \in k1$ for all $a \in A$.

PROPOSITION 4 *[1] k_FG admits a diagonal strong involution σ* iff
i) $G \simeq (Z_2)^n$ for some n,
ii) $\sigma(e) = e$, $\sigma(x) = -x$ for all $x \neq e$,
iii) k_FG is altercommutative.

Given k_FG we have a natural function $s(x) = F(x,x)$ and consider now the possibility of defining a strong involution using this. For all statements of simplicity in the following we assume $|G| > 2$.

PROPOSITION 5 *[1] If $\sigma(x) = F(x,x)x$ for all $x \in G$ defines a strong involution, then the algebra k_FG is simple and the following are equivalent*
i) k_FG is an alternative algebra,
ii) k_FG is a composition algebra.

Finally, it is possible to characterize a natural class of k_FG algebras that are composition algebras,

PROPOSITION 6 *[1] Let k_FG admit a strong diagonal involution $\sigma(x) = s(x)x$. Then $q(x) = x \bullet \sigma(x)$ makes k_FG a composition algebra* iff
i) $s(xy)F(x,y)^2F(xy,xy) = s(x)s(y)F(x,x)F(y,y)$, for all $x, y \in G$.
ii) $F(x,xz)F(y,yz)F(z,z)s(z) + F(x,yz)F(y,xz)F(xyz,xyz)s(xyz) = 0$, for all $x, y \in G$ with $x \neq y$.

An important corollary of the last result is:

COROLLARY 7 *[1] If $G \cong (Z_2)^n$ then the Euclidean norm quadratic function defined by $q(x) = 1$ for all $x \in G$ makes k_FG a composition algebra* iff
i) $F^2(x,y) = 1$ for all $x, y \in G$
ii) $F(x,xz)F(y,yz) + F(x,yz)F(y,xz) = 0$, for all $x, y, z \in G$ with $x \neq y$.

In this case $\sigma(x) = F(x,x)x$ for all $x \in G$ is a strong involution and k_FG is simple and alternative.

4 CAYLEY-DICKSON PROCESS FOR k_FG

We have a generalisation of the Cayley-Dickson process as follows. Again, details are in [1].

DEFINITION 8 *Let G be an Abelian group F a 2-cochain on it. For any 1-cochain $s : G \to k$ and $\alpha \neq 0$ we define $\bar{G} = G \times Z_2$ and on it the 2-cochain $\bar{F}$ and 1-cochain $\bar{s}$,*

$$\bar{F}(x,y) = F(x,y), \quad \bar{F}(x,vy) = s(x)F(x,y), \quad \bar{F}(vx,y) = F(y,x),$$

$$\bar{F}(vx, vy) = \alpha s(x)F(y, x), \quad \bar{s}(x) = s(x), \quad \bar{s}(vx) = -1$$

for all $x, y \in G$. *Here* $x \equiv (x, e)$ *and* $vx \equiv (x, \nu)$ *denote elements of* $\bar{G}$, *where* $\boldsymbol{Z}_2 = \{e, \nu\}$ *with product* $\nu^2 = e$.

We say that $k_{\bar{F}}\bar{G}$ is the *generalised Cayley-Dickson extension* of $k_F G$ associated to s, α. The motivation is that if $\sigma(x) = s(x)x$ is a strong involution, then $k_{\bar{F}}\bar{G}$ is the usual Cayley-Dickson extension of $k_F G$ associated to σ, α. Note that since all unital composition algebras over k are obtained by repeated Cayley-Dickson extension[6], they are all of the form of a quasialgebra $k_F G$ in the last proposition of the preceding section with G a power of $\boldsymbol{Z}_2$.

The natural application for our generalised Cayley-Dickson process is when $k_F G$ admits a diagonal involution $\sigma(x) = s(x)x$ (but not necessarily strong). In this case we have:

PROPOSITION 9 *[1] The 3-cocycle* $\bar{\phi}$ *of* $k_{\bar{F}}\bar{G}$ *is given by*

$$\bar{\phi}(x, y, z) = \phi(x, y, z), \quad \bar{\phi}(vx, y, z) = \mathcal{R}(y, z)\phi(x, y, z),$$

$$\bar{\phi}(x, vy, z) = \mathcal{R}(y, z)\mathcal{R}(xy, z)\phi(x, y, z), \quad \bar{\phi}(x, y, vz) = \mathcal{R}(x, y)\phi(x, y, z),$$

$$\bar{\phi}(vx, vy, z) = \mathcal{R}(xy, z)\phi(x, y, z), \quad \bar{\phi}(vx, y, vz) = \mathcal{R}(y, z)\mathcal{R}(x, y)\phi(x, y, z),$$

$$\bar{\phi}(x, vy, vz) = \mathcal{R}(x, yz)\phi(x, y, z), \quad \bar{\phi}(vx, vy, vz) = \mathcal{R}(xy, z)\mathcal{R}(x, y)\phi(x, y, z),$$

for $x, y, z \in G$.

Using this calculation, one may show under the same assumptions:

PROPOSITION 10 *[1]*

i) $k_{\bar{F}}\bar{G}$ *is associative* iff $k_F G$ *is associative and commutative.*

ii) If $k_F G$ *has trivial centre then* $k_{\bar{F}}\bar{G}$ *is alternative* iff $k_F G$ *is associative and* $s(x) = -1$ *for all* $x \in G$ *and* $x \neq e$.

iii) If s *defines a strong diagonal involution then* $k_{\bar{F}}\bar{G}$ *is alternative* iff $k_F G$ *is associative.*

This extends to more general $k_F G$ some well-known considerations for octonions and higher Cayley algebras.

5 CONCLUDING REMARKS

We conclude with some remarks about other classes if quasialgebras. In fact, the input data for our $k_F G$ construction is clearly very general. If we denote the elements of the finite group G by $\{x_1 = e, x_2, \ldots, x_n\}$ then we can represent the cochain F by an $n \times n$ matrix with entries $F_{ij} = F(x_i, x_j)$. It has 1 in the first row and column and all entries non-zero. Conversely, any such matrix will do for a cochain and yield a quasialgebra. Therefore it is a wide-open question what other groups and cochains might be interesting; here we list just a few natural classes.

First of all, motivated by composition algebras for the Euclidean norm (isomorphic to complex, quaternions or octonions), where the cochain is represented by certain normalized Hadamard matrices, a natural more general class of examples is $k_F G$ where F is a general normalised Hadamard matrix. Some results for (and low-dimensional examples of) this kind of $k_F G$ are given in [1]. Hadamard matrices have even dimension but odd dimensional examples of $k_F G$ can be obtained by taking F with first column and row 1 and the rest an Hadamard matrix.

Another general choice of F, overlapping with the extended Hadamard case, is (for G any group),

$$F(x,y) = \begin{cases} 1 & \text{if } x = e \text{ or } y = e \text{ or } x \neq y \\ -1 & \text{otherwise} \end{cases},$$

where one may show that $k_F G$ is simple, commutative and in general nonassociative.

Other interesting examples of $k_F G$ quasialgebras come from the theory of finite fields, and include the generalisations of octonions based on Galois sequences in [5].

Finally, we would also like to recall that our approach answers such questions as what means a representation of the octonions. Following the method explained in the introduction, a representation of the quasialgebra $k_F G$ means a G-graded vector space V and a degree-preserving action $\triangleright$ obeying

$$(x \bullet y) \triangleright v = \phi(x, y, |v|) x \triangleright (y \triangleright v)$$

for all $v \in V$. This is explained in [1], where it is also shown that a representation is equivalent to an algebra map from $k_F G$ to a certain quasialgebra $\mathrm{End}_\phi(V)$ of *quasimatrices* associated to any V.

REFERENCES

[1] H.Albuquerque and S. Majid. Quasialgebra structure of the octonions. Preprint Damtp/97-138 and math.QA/9802116.

[2] S. Mac Lane. Categories for the Working Mathematician. Springer Verlag, 1974.

[3] S.Majid. Foundations of Quantum Group Theory. Cambridge University Press, 1995.

[4] V.G. Drinfeld. QuasiHopf algebras. Leningrad Math. J. 1:1419-1457, 1990.

[5] G.Dixon. Division Algebras: Octonions, Quaternions, Complex Numbers and the Algebraic Design of Physics. Kluwer, 1994.

[6] K.A. Zhevlakov, A.M. Slin'ko, I.P. Shestakov, and A.I.Shirshov. Rings That are Nearly Associative. Academic Press, 1982.

Identities of graded alternative algebras

Y. BAHTURIN and M. ZAICEV Department of Algebra,Faculty of Mechanics and Mathematics, Moscow State University, 119899 Moscow, RUSSIA e-mail: bahturin@mech.math.msu.su and zaicev@nw.math.msu.su

Dedicated to Alexei Ivanovich KOSTRIKIN on his 70th birthday

1 INTRODUCTION

In this paper we study identities of alternative algebras over a field.
The main results are Theorems 1, 2, 3. In Theorem 1 we prove that for any constant $a > 0$ and any alternative PI-algebra A the growth of identities of A is asymptotically less $\frac{n!}{a^n}$. The statement of this result is similar to a theorem about Lie algebras due to A. Grishkov [7], but we leave open the question about the possibility of bounding the growth from above by a function of the form b^n for a suitable b. In the class of associative algebras this estimate had been obtained more than 20 years ago (see [10]).
Our second result treats the identities of *graded* alternative algebras. It is well-known [6] that if $A = \oplus_{g \in G} A_g$ is an associative algebra graded by a finite semigroup G and any subalgebra A_e, where e is an idempotent element in G, satisfies a non-trivial identity then also A is a PI-algebra. It is known also that the degree of identity guaranteed for A does not depend on the algebra itself and is completely determined by the degrees of identities satisfied by A_e's and the order of G [2]. Later similar results had been obtained for Lie algebras, special Jordan algebras and some other types of algebras (see [4], [5]). In Theorem 2 we prove that alternative algebras graded also enjoy a similar property (G commutative).

This work was partially supported by Russian Foundation for Fundamental Research Grants # 6-01-00146 and 96-15-96050.

Our third main result, Theorem 3, shows that the matrix algebra over an alternative PI-algebra is a non-associative algebra with a non-trivial identity of a very specific form. This theorem can be viewed as a generalization of the classical result from [8] saying that any matrix algebra over an associative PI-algebra is a PI-algebra itself.

2 DEFINITIONS AND PRELIMINARY RESULTS

We recall some basic notions. Let A be an algebra over a field Φ. It is called *alternative* if the following identities hold:

$$(xx)y = x(xy) \quad , \qquad y(xx) = (yx)x \quad . \tag{1}$$

It follows from (1) that a sufficient condition for an algebra to be alternative is the associativity of any 2-generator subalgebra. By a well-known Artin's Theorem (see, for example, [13]) this condition is also necessary

We introduce the following notation: $x \circ y = xy + yx$. The linearization procedure applied to the second of identities (1) after a change of notation yields a relation of the form:

$$a(x \circ y) = (ax)y + (ay)x, \tag{2}$$

which will be repeatedly used in what follows. Following [13] we write also $\langle a_1 a_2 \ldots a_k \rangle$ for the left - normed product of the elements $a_1, \cdots, a_k$ of an alternative algebra. Specifically, $\langle a_1 a_2 \rangle = a_1 a_2$ and $\langle a_1 \cdots a_k \rangle = \langle a_1 \cdots a_{k-1} \rangle a_k$. We will also omit the brackets in the left - normed Jordan product and write $a_1 \circ a_2 \circ \ldots \circ a_k = (a_1 \circ a_2 \circ \ldots \circ a_{k-1}) \circ a_k$.

According to a conventional definition (see [13]) an algebra A is called an alternative *PI-algebra*, if it satisfies an *essential* identity

$$f(x_1, \ldots, x_n) \equiv 0.$$

A non-associative polynomial f from a free alternative algebra is called essential if after forgetting all brackets it becomes a non-trivial associative polynomial.

We need a result about the identities in alternative algebras. If char $F \neq 2$ this easily follows from Lemma 5.14 of [13].

LEMMA 1. *Any alternative PI - algebra satisfies an identity of the form*

$$\langle x_0 x_1 \ldots x_m \rangle \equiv \sum_{1 \neq \sigma \in S_n} \alpha_\sigma \langle x_0 x_{\sigma(1)} \cdots x_{\sigma(m)} \rangle. \tag{3}$$

Proof. Let A be an alternative PI-algebra and $V = \text{var}\, A$ the variety generated by A. We denote by B the relatively free 2-generator algebra from V. Then B is an the PI-algebra. Since the growth of identities of any associative variety is at most exponential B satisfies an identity of the form

$$f = \sum_{\sigma \in S_n} \beta_\sigma y \circ x_{\sigma(1)} \circ \ldots \circ x_{\sigma(n)} \equiv 0. \tag{4}$$

One can assume that the coefficient of the Jordan monomial $y \circ x_1 \circ \ldots \circ x_n$ in (4) is non-zero. We perform in (4) a substitution $\varphi(y) = x \circ y^{(t_0)}, \varphi(x_1) = x \circ y^{(t_1)}, \ldots, \varphi(x_n) = x \circ y^{(t_n)}$, where $t_0 < t_1 < \ldots < t_n$, and $x \circ y^{(k)}$ is a Jordan monomial $x \circ y \circ \ldots \circ y$ of degree k in y. If we introduce the lexicographic order on the monomials in x, y in the free associative algebra by setting $x > y$ then the leading monomial of the element $(x \circ y^{(t_0)}) \circ (x \circ y^{(t_1)}) \circ \ldots \circ (x \circ y^{(t_n)})$, written associatively, is equal to $a = xy^{t_0}xy^{t_1}\ldots xy^{t_n}$ since $t_0 < t_1, \ldots, t_n$. On the other hand, the leading term of any $(x \circ y^{(t_0)}) \circ (x \circ y^{(t_{\sigma(1)})}) \circ \ldots \circ (x \circ y^{(t_{\sigma(n)})})$ is strictly less than a, as soon as $\sigma \neq 1$. This means, that $\varphi(f) \equiv 0$ is a non-trivial identity in B, while $\varphi(f)$ is a Jordan polynomial in two variables. Hence, in A we have an identity $\varphi(f) \equiv 0$, and the also $z\varphi(f) \equiv 0$. After a linearization of this identity by x, y in the free alternative algebra and making a change of variables, considering (2) we obtain a relation

$$\sum_{\sigma \in S_m} \alpha_\sigma \langle x_0 x_{\sigma(1)} \ldots x_{\sigma(m)} \rangle \equiv 0. \tag{5}$$

The non-triviality of the coefficients in (5) follows since the associative polynomial

$$\sum_{\sigma \in S_m} \alpha_\sigma x_{\sigma(1)} \ldots x_{\sigma(m)}$$

is a linearization in the free associative algebra of a Jordan polynomial $\varphi(f)$. Since not all coefficients in (5) are zero, it easily implies (3), and Lemma 1 has been proven. □

3 MAIN RESULTS

We denote by Alt(Y) the free alternative algebra with the free generating set $Y = \{y_1, y_2, \ldots\}$. As before, let A be an alternative PI-algebra, I the ideal of identities of A in Alt(Y). We consider in Alt(Y) the subspace P_n of multilinear elements in $y_1, \ldots, y_n$.

THEOREM 1. *Let A satisfy an identity (3) of degree m. Then for any $c > 0$ there exists n, depending only on m and c, such that*

$$\dim \frac{P_n}{P_n \cap I} < \frac{n!}{c^n}.$$

Proof. We first consider the subspace $Q_n \subset P_{n+1}$ spanned by all left - normed monomials $\langle y_0 y_{i_1} \ldots y_{i_n} \rangle$. A monomial $\langle y_0 y_{i_1} \ldots y_{i_n} \rangle$ from Q_n is called *k-decomposable*, if there exist indices $1 \leq j_1 < \ldots < j_k \leq n$, such that $i_{j_s} > i_{1+j_s} > i_{2+j_s} > \ldots > i_{j_k}$ for all $s = 1, \ldots, k$. In other words, in our monomial we can find k variables, each of which greater than all the variables between it and the last one, and also greater then the last, k-th variable. All the rest monomials in Q_n we will call *k-indecomposable*. For the proof of Theorem 1 we need the following.

LEMMA 2. *Let A satisfy an identity of the form (3). Then modulo I the subspace Q_n coincides with the linear span V of m-indecomposable monomials.*

Proof. We order the variables $y_0, y_1, \ldots, y_n$ by setting $y_0 < y_1 < \ldots < y_n$ and extend this order to the lexicographic order on the monomials from Q_n, comparing the letters from the left to the right. Then there exists a counterexample which is minimal in the sense of this order, that is, an m-decomposable left-normed monomial a which is not in $I+V$. Let $a = \langle y_0 y_{i_1} \ldots y_{i_n} \rangle$ and $j_1 < \ldots < j_m$ be a set of indices determining an m-decomposition on a. We introduce the following notation. If $j_1 > 1$ then we set $a_0 = \langle y_0 y_{i_1} \cdots y_l \rangle$ where $l = i_{j_1 - 1}$. If $j_1 = 1$ then $a_0 = y_0$. Further, we denote by $b_r, 1 \leq r \leq m-1$ the left-normed Jordan product of the variables y_s with the indices running from i_{j_r} to $i_{j_{r+1}-1}$. For $r = m$ we set $b_m = y_l$, $l = i_{j_m}$. By a_{m+1} we will for the convenience denote the multiplication from the right by the remaining variables y_{i_s} with the indices $s = j_m + 1, \ldots, n$, if $j_m < n$. From (2) it follows that for any $c, z_1, \ldots, z_r$ from $\mathrm{Alt}(Y)$ we have an equation

$$c\langle z_1 \circ z_2 \circ \ldots \circ z_r \rangle - \langle c z_1 \ldots z_r \rangle = \sum_{1 \neq \sigma \in S_n} \langle c z_{\sigma(1)} \ldots z_{\sigma(r)} \rangle. \quad (6)$$

By the minimality of the counterexample selected, it follows from (6) that we have

$$a \equiv \langle a_0 b_1 \ldots b_m a_{m+1} \rangle (\mathrm{mod}(I+V)).$$

Now (3) implies:

$$a \equiv \sum_{1 \neq \sigma \in S_n} \alpha_\sigma \langle a_0 b_{\sigma(1)} \ldots b_{\sigma(m)} a_{m+1} \rangle (\mathrm{mod}(I+V)). \quad (7)$$

If we apply (2) to each summand on the right hand side of (7) we will obtain an expression for a modulo $I+V$ in the form of a linear combination of left-normed products $\langle y_0 y_{k_1} \ldots y_{k_n} \rangle$, each of which is lexicographically less than a and thus belongs to $I+V$. The contradiction thus obtained completes the proof of Lemma 2. □

We continue the proof of Theorem 1. By a Theorem of A. I. Shirshov (see [13]) P_n is the linear span of the products $\langle y_1 \ldots y_k \rangle$, where each y_i in its turn is a left-normed product of some of $x_1, \ldots, x_n$, $y_i = x_{\alpha_1} \ldots x_{\alpha_s}$, with $\alpha_1 < \ldots < \alpha_s$. From here it follows, that if we denote by $S(n,k)$ the Stirling number of the second kind, that is, the number of decompositions of an n-element set into the disjoint union of k subsets then

$$\dim \frac{P_n}{P_n \cap I} \leq \sum_{k=0}^{n} S(n,k) n \dim \frac{Q_{n-1}}{Q_{n-1} \cap I}. \quad (8)$$

Let $q_n = \dim \frac{Q_{n-1}}{Q_{n-1} \cap I}$. From Lemma 2 it follows that q_n does not exceed the number of m-indecomposable monomials in Q_{n-1}.It has been shown in [3] (see also [1]) that for any $a > 0$ there exists n_0, depending only on m and a, such that

$$q_n < \frac{n!}{a^n}$$

for all $n > n_0$. In order to obtain the upper bound for the sum on the right side of (8), we first consider the sum

$$\sum_{k=0}^{n} S(n,k)\frac{k!}{a^k}, \tag{9}$$

in which $a = 6c$, and c is the constant from the proof of our Theorem. It is well-known (see [12], Ch.1), that for any natural $t \geq n$ we have

$$t^n = \sum_{k=0}^{n} S(n,k)(t)_k,$$

in which $(t)_k = t(t-1)(t-2)\ldots(t-k+1)$. From here one can obtain an estimate for the value (9):

$$\sum_{k=0}^{n} S(n,k)\frac{k!}{a^k} < \frac{1}{a^n}\sum_{k=0}^{n} S(n,k)k! < \frac{1}{a^n}\sum_{k=0}^{n} S(n,k)(n)_k = \frac{n^n}{a^n}. \tag{10}$$

From Stirling's formula for $n!$ it is easy to derive $n^n < 3^n n!$ hence, from (10), it follows inequality

$$\sum_{k=0}^{n} S(n,k)\frac{k!}{a^k} < \frac{n!}{b^n}, \tag{11}$$

where $b = \frac{a}{3} = 2c$.

We now split the sum on the right side of (8) into two components

$$A = \sum_{k=0}^{n_0} S(n,k)nq_k; \quad B = \sum_{k=n_0+1}^{n} S(n,k)nq_k.$$

Then, as it follows from (11), $B < n\frac{n!}{b^n}$. In order to estimate A we notice, that the values $S(n,k)$ satisfy the inequalities: $S(n,2) \leq 2^n, S(n,k) \leq 2^n S(n,k-1)$. It follows then that $S(n,k) \leq t^n$, where $t = 2^{n_0}$, for all $k \leq n_0$. Considering an obvious inequality $q_k \leq (k-1)!$, we obtain

$$A \leq n(n_0)!n_0t^n = Dnt^n, \tag{12}$$

where D and t are some constants depending only on c and m. Now from (8), (11) and (12) we have the bound

$$\dim\frac{P_n}{P_n \cap I} \leq Dnt^n + B < Dnt^n + n\frac{n!}{b^n} = Dnt^n + n\frac{n!}{2^n c^n}.$$

For given c, D and t one can select $n_1 = n_1(D,t,c)$, such that $Dnt^n < \frac{1}{2}\frac{n!}{c^n}$ for all $n \geq n_1$. Since $\frac{n}{2^n} \leq \frac{1}{2}$ for all natural n,

$$\dim\frac{P_n}{P_n \cap I} < \frac{1}{2}\frac{n!}{c^n} + \frac{1}{2}\frac{n!}{c^n} = \frac{n!}{c^n},$$

and Theorem 1 has been proven. □

Let now G be a commutative semigroup and

$$A = \sum_{g \in G} A_g$$

a G-graded alternative algebra, that is, $A_g A_h \subset A_{gh}$. We will say that A is an algebra with a finite grading, if $A_g \neq 0$ only for a finite number of elements $g \in G$.

THEOREM 2. *Let $A = \sum_{g \in G} A_g$ be a alternative algebra over a field Φ with a finite grading. If for any idempotent $e \in G$ the subalgebra A_e is a PI-algebra then also A is an alternative PI-algebra. Moreover, the degree of the nontrivial identity in A is bounded from above by a function only on the degrees of non-trivial identities in A_e and on the number of elements $g \in G$, for which $A_g \neq 0$.*

Proof. We introduce some necessary notation. Let $H = \{g_1, \dots, g_t\}$ be a subset of all elements from G, for which $A_g \neq 0$. We consider a G-graded set of the variables $X = \cup_{g \in G} X_g$, where $X_g = \{x_1^g, x_2^g, \dots\}$, and generate by this the set a free alternative algebra $\mathrm{Alt}(X)$. By $\overline{\mathrm{Alt}}(Y)$ we denote the alternative subalgebra in $\mathrm{Alt}(X)$ generated by the elements $\{y_1, y_2, \dots\}$, where $y_i = x_i^{g_1} + \dots + x_i^{g_t}$. It is obvious that $\overline{\mathrm{Alt}}(Y)$ is a free alternative algebra and $\mathrm{Alt}(X)$ is endowed by a natural G-grading,

$$\mathrm{Alt}(X) = \bigoplus_{g \in G} Alt_g(X).$$

We denote by Ψ the set all homomorphisms $\psi : \overline{\mathrm{Alt}}(Y) \to A$, and by Ψ_0 the set all homomorphisms $\psi_0 : \mathrm{Alt}(X) \to A$, for which $\psi_0(x_i^g) \in A_g$. We set

$$I = \bigcap_{\psi \in \Psi} \mathrm{Ker}\ \psi \quad \triangleleft\ \overline{\mathrm{Alt}}(Y), \quad J = \bigcap_{\psi_0 \in \Psi_0} \mathrm{Ker}\ \psi_0 \quad \triangleleft\ \mathrm{Alt}(X).$$

The elements from I (resp., J) are the *non-graded* (resp., *graded*) identities of algebra A.

Let e be an idempotent of G. By a hypothesis of Theorem A_e is PI-algebra. Repeating verbatim the argument of Lemma 1, one can observe, that for any $g \in G$ and $x_0 \in A_g$ the elements $x_1, \dots, x_m \in A_e$ satisfy a left-normed identity of the form (3), and, moreover, from the proof of Lemma it follows, that the degree m of this identity depends only on the degree of the initial identity of A_e. Since A is a algebra with finite grading one can assume that for all e and g there exists m such that we have a graded identity

$$\langle x_0^g x_1^e \dots x_m^e \rangle \equiv \sum_{1 \neq \sigma \in S_n} \alpha_\sigma^{e,g} \langle x_0^g x_{\sigma(1)}^e \dots x_{\sigma(m)}^e \rangle,$$

that is, we have an inclusion

$$\langle x_0^g x_1^e \dots x_m^e \rangle - \sum_{1 \neq \sigma \in S_n} \alpha_\sigma^{e,g} \langle x_0^g x_{\sigma(1)}^e \dots x_{\sigma(m)}^e \rangle \in J. \tag{13}$$

We denote now by P_n subspace in $\overline{\mathrm{Alt}}(Y)$ of multilinear polynomials in $y_1,\dots,y_n$, and by $P_{n,q}$ a subspace of multilinear polynomials in $x_1^{q_1},\dots,x_n^{q_n}$ in $\mathrm{Alt}(X)$, where $q=(q_1,\dots q_n)\in G^n$. In order to prove the theorem it is sufficient for some n, bounded by a function of m and t, to obtain the inequality

$$\dim\frac{P_n}{P_n\cap I}<n!\,, \tag{14}$$

since if (14) holds the ideal I cannot be zero. .
We show first, that $P_n\cap I=P_n\cap J$ in $\mathrm{Alt}(X)$. Let $f=f(y_1,\dots,y_n)\in I$, ψ_0 is a homomorphism $\mathrm{Alt}(X)\to A$. We consider a homomorphism $\psi:\overline{\mathrm{Alt}}(Y)\to A$, for which $\psi(y_i)=\psi(x_i^{g_1}+\dots+x_i^{g_t})=\psi_0(x_i^{g_1})+\dots+\psi_0(x_i^{g_t})=\psi_0(y_i)$. As $f\in I$ we have $\psi_0(f)=\psi(f)=0$, so that $f\in\operatorname{Ker}\psi_0$. Hence, we have proved the inclusion $I\subset J$. Let now $f=f(y_1,\dots,y_n)$ be a multilinear polynomial in $y_1,\dots,y_n$, an element in J, and ψ is an arbitrary homomorphism from $\overline{\mathrm{Alt}}(Y)$ into A. Then $\psi(y_i)=b_i^{g_1}+\dots+b_i^{g_t}$, where $b_i^{g_j}\in A_{g_j}$

$$\psi(f)=\sum_{\alpha,\beta,\dots,\gamma}f(b_1^{g_\alpha},b_2^{g_\beta},\dots,b_n^{g_\gamma}).$$

But all summands on the right side equal zero hence $f\in I$ and then $P_n\cap I=P_n\cap J$ has been verified.
Now, considering the inclusion $P_n\subset\bigoplus_{q\in G^n}P_{n,q}$, we obtain the inequality

$$\dim\frac{P_n}{P_n\cap I}=\dim\frac{P_n}{P_n\cap J}=\dim\frac{P_n+J}{J}\le\sum_{q\in G^n}\dim\frac{P_{n,q}+J}{J}=$$

$$\sum_{q\in G^n}\dim\frac{P_{n,q}}{P_{n,q}\cap J}=\sum_{q\in H^n}\dim\frac{P_{n,q}}{P_{n,q}\cap J}. \tag{15}$$

The inequality (14) will follow from (15) if we prove, that for some n, depending only on m and t, for any $q\in G^n$ we have

$$\dim\frac{P_{n,q}}{P_{n,q}\cap J}<\frac{n!}{t^n}, \tag{16}$$

since the number of the summands on the right side (15) is equal to $|H|=t^n$.
The proof of (16) is similar to the proof of Theorem 1. Change n by $n+1$, fix a tuple $q=(q_1,\dots,q_{n+1})$ and write, for convenience,

$$x_1^{q_1}=x_0,\dots,x_{n+1}^{q_{n+1}}=x_n.$$

We introduce on $x_0,\dots,x_n$ an order by setting $x_0<x_1<\dots<x_n$. As before, we denote by $Q_{n,q}$ subspace in $P_{n,q}$, spanned by the monomials $\langle x_0x_{i_1}\dots x_{i_n}\rangle$ and consider k-indecomposable monomials, where the number k will be chosen later. We prove now an analogue of Lemma 2 for the graded case.

LEMMA 3. *Modulo J the subspace $Q_{n,q}$ coincides with the linear span V of all k - indecomposable monomials, where $k = \frac{t^N-1}{t-1}$, and $N = t(m-1)+1$.*

Proof. As in Lemma 2, consider the lexicographically minimal counterexample

$$a = \langle x_0 x_{i_1} \ldots x_{i_n} \rangle,$$

that exists if Lemma is wrong. Let $j_1 < \ldots < j_k$ be a set of indices defining a k-decomposition for a. We set $a_0 = \langle x_0 x_{i_1} \ldots x_l \rangle$, where $l = i_{j_1 - 1}$, We denote by $b_r, 1 \le r \le k-1$ a left-normed Jordan product of the variables x_s where $s = i_{j_r}, \ldots, i_{j_{r+1}-1}$. If $r = k$ we will assume $b_k = x_{i_s}$ where $s = j_k$. As before, for brevity by a_{k+1} we denote the right multiplication by all remaining variables x_{i_s}. As in Lemma 2,

$$a \equiv \langle a_0 b_1 \ldots b_k a_{k+1} \rangle (\mathrm{mod}(J+V)).$$

By construction, the elements $b_1, \ldots, b_k$ are homogeneous relative the G - grading. Let $B_i \in Alt_{p_i}(X), i = 1, \ldots, k$. We consider the product $p_1 p_2 \ldots p_k$ in G. By Lemma 1 from [5] for k selected by us there exist $1 \le t_0 < \ldots < t_m \le k$ such that

$$p_{t_0} p_{t_0+1} \cdots p_{t_1 - 1} = p_{t_1} p_{t_1+1} \cdots p_{t_2-1} = \ldots = p_{t_{m-1}} p_{t_{m-1}+1} \cdots p_{t_m} = e$$

$e^2 = e$ in G. Now we set $c_0 = \langle a_0 b_1 \ldots b_{t_0 - 1} \rangle, c_1 = b_{t_0} \circ b_{t_0+1} \circ \ldots \circ b_{t_1 - 1}, \ldots, c_m = b_{t_{m-1}} \circ b_{t_{m-1}+1} \circ \ldots \circ b_{t_m}$, and by c_{m+1} we denote the right multiplication by all remaining b_i and by the operator a_{k+1}. As before, by (2), the lexicographically leading term of $\langle c_0 c_1 \ldots c_{m+1} \rangle$ is equal to the leading term of $\langle a_0 b_1 \ldots b_k a_{k+1} \rangle$ and leading term of a hence

$$a \equiv \langle c_0 c_1 \ldots c_{m+1} \rangle (\mathrm{mod}(J+V))$$

by the minimality of the counterexample a. But according to (13)

$$\langle c_0 c_1 \ldots c_{m+1} \rangle \equiv \sum_{1 \neq \sigma \in S_n} \beta_\sigma \langle c_0 c_{\sigma(1)} \cdots c_{\sigma(m)} c_{m+1} \rangle (\mathrm{mod}(J+V)).$$

The leading term of any product $\langle c_0 c_{\sigma(1)} \cdots c_{\sigma(m)} c_{m+1} \rangle$ strictly less than a as soon as $\sigma \neq 1$, from so that we have $a \in J + V$, contradicting the choice of a. Lemma 3 has been proven. □

The remaining portion of the proof of Theorem 2 completely repeats the proof of Theorem 1 with a reference to Lemma 3 in place of Lemma 2. □

4 SOME COROLLARIES

Now we list a number of consequences from the results obtained.

COROLLARY 1. *The growth of identities of any alternative PI - algebra is asymptotically less than $\frac{n!}{a^n}$ for of any constant a.*

This result follows from of Theorem 1 and of Lemma 1, since by definition the growth of identities of an algebra A is the growth of the sequence

$$c_n(A) = \dim \frac{P_n}{P_n \cap I},$$

where I is the ideal of identities of A in the free alternative algebra.

COROLLARY 2. *Let G be a finite solvable group of the automorphisms of an alternative algebra A over an arbitrary field such that* $\operatorname{char} F$ *is not a divisor of the order $|G|$. If the subalgebra of G-fixed points A^G is a PI-algebra with a non-trivial identity of degree d then A itself satisfies a non-trivial identity whose degree depends only on d and the order $|G|$.*

We omit the proof of this corollary since it is standard in all similar situations for various classes of algebras (see., for example, [4]). It should be remarked that A.P.Semenov [11] proved a similar result, with G arbitrary finite but without the connection between the degrees of identities in A^G and A.
The techniques developed above enable us also to consider the question about the existence of identities in matrices over alternative algebras. We prove first a more general result, of independent interest.

LEMMA 4. *Let A be a non-associative algebra, the growth of identities of which is asymptotically less than $\frac{n!}{a^n}$ for of any constants $a\rangle 0$. Let also B be a finite - dimensional algebra. Then the tensor product $A \otimes B$ satisfies a non-trivial identity of the form*

$$\sum_{\sigma \in S_n} \alpha_\sigma \langle z_{\sigma(1)} \dots z_{\sigma(n)} \rangle \equiv 0, \tag{17}$$

where $\langle z_1 \dots z_n \rangle$ stands for the left-normed product of the elements $z_1, \dots, z_n$ of a non-associative algebra.

Proof. Let $e_1, \dots, e_m$ be a basis of the algebra B, and $\mathcal{A}(X)$ a relatively free algebra of countable rank in the variety var A, where $X = \{x_i^j; i, j = 1, 2, \dots\}$. We notice first that if on the elements $x_1^1 \otimes e_1, \dots, x_1^m \otimes e_m, \dots, x_n^1 \otimes e_1, \dots, x_n^m \otimes e_m$ of the tensor product $\mathcal{A}(X) \otimes B$ a relation

$$g(x_1^1 \otimes e_1, \dots, x_1^m \otimes e_m, \dots, x_n^1 \otimes e_1, \dots, x_n^m \otimes e_m) = 0, \tag{18}$$

holds then also for any $a_i^j \in A$ we have a similar relation

$$g(a_1^1 \otimes e_1, \dots, a_1^m \otimes e_m, \dots, a_n^1 \otimes e_1, \dots, a_n^m \otimes e_m) = 0. \tag{19}$$

The reason is that for any algebras P, Q, R, S and any homomorphisms of algebras $\varphi : P \to R$, $\psi : Q \to S$ the natural linear map $\varphi \otimes \psi : P \otimes Q \to R \otimes S$ defined by

$(\varphi \otimes \psi)(p \otimes q) = \varphi(p) \otimes \psi(q)$, is a homomorphism of algebras :

$$(\varphi \otimes \psi)((p \otimes q)(p' \otimes q')) = (\varphi \otimes \psi)(pp' \otimes qq')) =$$
$$\varphi(pp') \otimes \psi(qq') = \varphi(p)\varphi(p') \otimes \psi(q)\psi(q') =$$
$$(\varphi(p) \otimes \psi(q))(\varphi(p') \otimes \psi(q')) = ((\varphi \otimes \psi)(p \otimes q))((\varphi \otimes \psi)(p' \otimes q')).$$

Since $\mathcal{A}(X)$ is a free algebra of the variety $\operatorname{var} A$, the map $x_j^i \mapsto a_j^i$ extends to a homomorphism $\varphi : \mathcal{A}(X) \to A$. Taking ψ equal to the identity map $B \to B$ we obtain a homomorphism of algebras $(\varphi \otimes \psi) : \mathcal{A}(X) \otimes B \to A \otimes B$.
We consider now in $\mathcal{A}(X)$ the elements

$$z_i = \sum_{j=1}^{m} x_i^j \otimes e_j, \quad j = 1, 2, \ldots .$$

If

$$f(z_1, \ldots, z_n) =$$
$$f(x_1^1 \otimes e_1 + \ldots + x_1^m \otimes e_m, \ldots, x_n^1 \otimes e_1 + \ldots + x_n^m \otimes e_m)$$
$$= 0,$$

then, as noticed above,

$$f(a_1^1 \otimes e_1 + \ldots + a_1^m \otimes e_m, \ldots, a_n^1 \otimes e_1 + \ldots + a_n^m \otimes e_m) = 0$$

for any $a_1^1, \ldots, a_n^m \in A$. Since any element from $A \otimes B$ is a sum of the form $c_1 \otimes e_1 + \cdots + c_m \otimes e_m$ then f is identically equal to zero on $A \otimes B$. In other words, it is sufficient to prove, that a relation of the form (17) holds for the elements selected by us: $z_1, z_2, \ldots$.
We consider a multilinear non-associative polynomial $f = f(y_1, \ldots, y_n)$. If f is a monomial then its value

$$f(x_1^1 \otimes e_1 + \ldots + x_1^m \otimes e_m, \ldots, x_n^1 \otimes e_1 + \ldots + x_n^m \otimes e_m)$$

can be expressed linearly through

$$f(x_1^{i_1}, \ldots, x_n^{i_n}) \otimes e_k$$

where $1 \leq i_1, \ldots, i_n \leq m$. Hence the value of any multilinear polynomial f in $z_1, \ldots, z_n$ can be linearly expressed through the tensors of the form

$$p(x_1^{i_1}, \ldots, x_n^{i_n}) \otimes e_k, \quad 1 \leq k, i_1, \ldots, i_n \leq m,$$

where $p(x_1^{i_1}, \ldots, x_n^{i_n})$ is a monomial in $x_1^{i_1}, \ldots, x_n^{i_n}$. We denote by $P_{n,q}$ the linear span of all monomials of $\mathcal{A}(X)$ in $x_1^{i_1}, \ldots, x_n^{i_n}$, if $q = (i_1, \ldots, i_n)$. By the hypothesis of Lemma, starting with some n the following inequality holds

$$\dim P_{n,q} < \frac{n!}{(2m)^n}.$$

But then

$$\dim \sum_{k=1}^{n} \sum_{q} P_{n,q} \otimes e_k \leq n! m^n \dim P_n < \frac{(n+1)!}{2^n} < n!.$$

This means that the left - normed products $\langle z_{\sigma(1)} \ldots z_{\sigma(n)} \rangle$ are linearly dependent, that is, satisfy the relation (17), and Lemma 4 has been proven.
An immediate consequence of Theorem 1 and Lemma 4 is the following.

THEOREM 3. *Let A be a alternative PI-algebra. Then the matrix algebra $M_t(A)$ satisfies an identity of the form*

$$\sum_{\sigma \in S_n} \alpha_\sigma \langle x_{\sigma(1)} \ldots x_{\sigma(n)} \rangle \equiv 0,$$

where $\langle x_1 \ldots x_n \rangle$ is a left - normed non - associative monomial in $x_1, \ldots, x_n$.

REFERENCES

[1] Bahturin, Yuri. Identities of algebras with actions of Hopf algebras. Lecture Notes in Pure Appl. Math. 198 (1998), 1 - 36.

[2] Bahturin, Yu.; Giambruno, A.; Riley, D., Group - graded algebras with polynomial identity, Israel J. Math., 104 (1998), 145 - 156.

[3] Bahturin, Yu.; Giambruno, A.; Zaicev, M., G - identities of associative algebras, Proc. Amer. Math. Soc., 127 (1998), pp. 63-69.

[4] Bahturin, Yu.; Zaicev, M., Identities of graded algebras, J. Algebra, 205 (1998), 1 - 12

[5] Bahturin, Yu.; Zaicev, M., Identities of graded special Jordan algebras, Vestnik MSU. Math., Mech., 1997, No. 6, 18 - 20..

[6] Bergen G., Cohen M., Actions of commutative Hopf algebras. Bull. London Math. Soc., 18(1986), 159 - 164.

[7] Grishkov, A.N., On the growth of varieties of Lie algebras , Mat. Zametki, 4 (1988), 51 - 54.

[8] Procesi, Claudio; Small, Lance, Endomorphism rings of modules over PI - algebras, Math. Z., 106 (1968), 178 - 180.

[9] Razmyslov, Yu. P., Identities of algebras and their representations, Nauka, Moscow, 1985.

[10] Regev A. Existence of identities in $A \otimes B$, Israel J. Math., 11(1972), 131-152.

[11] Semenov A.P., Subrings of invariants of a finite group of automorphisms of a Jordan ring, Siberian Math. J., 32 (1991), 203 - 211.

[12] Stanley, R.P., Enumerative Combinatorics, v.1, Wadsworth & Brooks/Cole, Adv. Books & Software, Monterey, CA, 1986.

[13] Zhevlakov, K.A.; Shestakov, I.P.; Shirshov, A.I.; Slin'ko, A.M., Rings that are nearly associative, Nauka, Moscow., 1978.

Invariant connections on symmetric spaces

PILAR BENITO* and CRISTINA DRAPER** Departamento de Matemáticas y Computación, Universidad de La Rioja, 26004 Logroño, Spain.

ALBERTO ELDUQUE*** Departamento de Matemáticas, Facultad de Ciencias, Universidad de Zaragoza, 50009 Zaragoza, Spain.

ABSTRACT: A classical problem in Differential Geometry, the determination of the invariant affine connections in the simply connected irreducible symmetric spaces, is equivalent to the algebraic problem of computing the set $\mathrm{Hom}_S(T \otimes_{\mathbb{R}} T, T)$ for any $\mathbb{Z}_2$-graded simple Lie algebra $L = S \oplus T$. The algebraic problem is solved using known information about the Lie triple system structure on T, because the simple $\mathbb{Z}_2$-graded Lie algebra $L = S \oplus T$ is just the embedding for the simple Lie triple system T. It turns out that the set of homomorphisms contains non trivial elements if and only if T is related to a simple Jordan algebra. Now it is possible to come back to the geometric context to describe the affine connections and express the holonomy and torsion and curvature tensors in algebraic terms.

1. INTRODUCTION.

In another paper of these proceedings [1], it is explained how the geometrical problem of describing the invariant affine connections on a reductive homogeneous space $M = G/H$ is equivalent to the algebraic problem of describing the multiplications $\alpha : \mathfrak{m} \times \mathfrak{m} \to \mathfrak{m}$ with $(\mathrm{Ad}\, h)\alpha(x, y) = \alpha\big((\mathrm{Ad}\, h)(x), (\mathrm{Ad}\, h)(y)\big)$ for any $h \in H$ and any $x, y \in \mathfrak{m}$, where $\mathfrak{m}$ is a vector subspace of the Lie algebra $\mathfrak{g}$ of G such that

$$\mathfrak{g} = \mathfrak{m} \oplus \mathfrak{h}$$

(where $\mathfrak{h}$ is the Lie subalgebra of $\mathfrak{g}$ which corresponds to the closed subgroup H

* Supported by the Spanish DGICYT (Pb 94-1311-C03-02) and by the Universidad de La Rioja (API-98/B15).

** Supported by a grant from the "Programa de Formación de Profesorado Universitario" (MEC), by the Spanish DGICYT (Pb 94-1311-C03-02) and by the Universidad de La Rioja (API-98/B15).

*** Supported by the Spanish DGICYT (Pb 94-1311-C03-03) and by the Universidad de La Rioja (API-98/B15).

of G) and $AdH(\mathfrak{m}) \subseteq \mathfrak{m}$, a condition equivalent to $[\mathfrak{h}, \mathfrak{m}] \subseteq \mathfrak{m}$ if H is connected. Such decomposition is called a *reductive decomposition*. This interplay between Geometry and Algebra was proved by Nomizu in [2].

The aim of this paper is to study the connection algebras $(\mathfrak{m}, \alpha)$ that appear in the irreducible symmetric spaces, a special type of reductive homogeneous spaces.

If M is a manifold and ∇ an affine connection, (M, ∇) is a *symmetric space* if for each point $p \in M$ there exists a central symmetry S_p with center at p. Any symmetric space is isomorphic to a space of the form G/H, where the closed subgroup $H \subseteq G$ is such that ${G_0}^\sigma \subseteq H \subseteq G^\sigma$, with σ an involutive automorphism of the Lie group G, G^σ the subgroup of fixed points under σ and ${G_0}^\sigma$ the connected component of the identity in G^σ.

For any symmetric space G/H, there exists a reductive decomposition $\mathfrak{g} = \mathfrak{h} \oplus \mathfrak{m}$ satisfying $[\mathfrak{m}, \mathfrak{m}] \subseteq \mathfrak{h}$, because $d\sigma$ provides a $\mathbb{Z}_2$-gradation ($\mathfrak{h}$ and $\mathfrak{g}$, as before, are the Lie algebras of H and G respectively). This kind of decomposition is called *a symmetric decomposition*. The converse is true.

For the simply connected irreducible symmetric spaces $M = G/H$, the Lie algebra $\mathfrak{g} = \mathfrak{h} \oplus \mathfrak{m}$ is graded-simple. Therefore the geometric problem of computing invariant affine connections in a simply connected irreducible symmetric space $M = G/H$ is equivalent, after applying Nomizu's result, to the algebraic problem of computing the set $\mathrm{Hom}_{\mathfrak{h}}(\mathfrak{m} \otimes_{\mathbb{R}} \mathfrak{m}, \mathfrak{m})$, where $\mathfrak{g} = \mathfrak{h} \oplus \mathfrak{m}$ is a graded-simple $\mathbb{Z}_2$-graded Lie algebra, because each homomorphism of $\mathfrak{h}$-modules $\tilde{\alpha} \in \mathrm{Hom}_{\mathfrak{h}}(\mathfrak{m} \otimes_{\mathbb{R}} \mathfrak{m}, \mathfrak{m})$ determines an $\mathbb{R}$-bilinear multiplication $\alpha : \mathfrak{m} \times \mathfrak{m} \to \mathfrak{m}$ such that $\mathrm{ad}\, \mathfrak{h}|_{\mathfrak{m}} \subseteq Der(\mathfrak{m}, \alpha)$ and, since H is connected, this is equivalent to $(\mathrm{Ad}\, h)\alpha(x, y) = \alpha((\mathrm{Ad}\, h)(x), (\mathrm{Ad}\, h)(y))$ for any $h \in H$ and any $x, y \in \mathfrak{m}$; that is, it determines a connection algebra and so an invariant affine connection.

In the next section, we will present briefly the solution of the general problem of determining the set $\mathrm{Hom}_S(T \otimes_F T, T)$ for any simple finite-dimensional $\mathbb{Z}_2$-graded Lie algebra $L = S \oplus T$ over an arbitrary field F of characteristic zero. In particular, in the real case we will have obtained an algebraic description of all invariant affine connections in irreducible symmetric spaces (this was done previously in the Riemannian case, using different methods, by Laquer [3,4]). This description will be used in section 3 to determine the holonomy algebra and give explicit formulas for the torsion and curvature tensors associated to each connection.

2. THE SOLUTION OF THE ALGEBRAIC PROBLEM.

The determination of the set $\mathrm{Hom}_S(T \otimes_F T, T)$ for any $\mathbb{Z}_2$-graded simple Lie algebra $L = S \oplus T$ has been done in [5], the purpose of this section is to present the main result in [5, Theorem 4.3] explaining briefly how the structure of a Lie triple system can be effectively used to compute those sets without decomposing $T \otimes_F T$.

From now on all the algebras and systems considered will be assumed to be finite-dimensional over a field F of characteristic 0.

2.1. Basic definitions.

A Lie triple system (L.t.s) is a vector space T endowed with a trilinear product $[x, y, z]$ satisfying:

i) $[x, x, y] = 0$,

ii) $[x, y, z] + [y, z, x] + [z, x, y] = 0$,

iii) $[x, y, -] \in \mathrm{Der}\, T$,

for any $x, y, z \in T$. The set $S = \mathrm{span}\,\langle [x, y, -] \mid x, y \in T\rangle$ is a Lie subalgebra of linear mappings on T and the $\mathbb{Z}_2$-graded Lie algebra $L(T) = S \oplus T$, whose bracket operation is given by:

$$[s_1, s_2] := s_1 s_2 \in S \text{ (the multiplication in } S)$$
$$[s, t] := s(t) \in T \text{ (the natural action)}$$
$$[t_1, t_2] := [t_1, t_2, -] \in S$$

for any $s, s_1, s_2 \in S$ and $t, t_1, t_2 \in T$, is called the *standard embedding* of T. Note that S and T are the even and odd part of the embedding and T is a faithful S-module. Thus any Lie triple system is nothing else but the odd part of a $\mathbb{Z}_2$-graded Lie algebra with product $[x, y, z] = [[x, y], z]$.

Among the Lie triple systems we are interested in the so called simple ones (without nontrivial ideals). They are just the odd part of simple $\mathbb{Z}_2$-graded Lie algebras (see [6]). Consequently the problem now is to compute the set $\mathrm{Hom}_S(T \otimes_F T, T)$ for any simple Lie triple system T with standard embedding $L(T) = S \oplus T$.

2.2. Examples.

It is not difficult to get elementary examples of simple L.t.s. We point out three of them which are specially relevant for our purposes. First we note that if T is a (simple) L.t.s. and $\mu \in F \setminus \{0\} = F^*$, the vector space T with the product $[x, y, z]^\mu = \mu[x, y, z]$ is a (simple) L.t.s. (in case $\mu \in F^2$ both systems are isomorphic).

Example 1. Any simple Lie algebra A is a simple L.t.s under the trilinear product $[x, y, z] = [[x, y], z]$ ($[\,,\,]$ the Lie bracket in A). In this case $S = \mathrm{ad}\, A = \mathrm{Der}\, A \simeq A$ and the standard embedding is $L(A) = \mathrm{Der}\, A \oplus A \simeq A \oplus A$. Moreover

A is the adjoint module for S and therefore the Lie bracket in A is a nonzero skew-symmetric element in $\mathrm{Hom}_S(A\otimes A, A)$. For any $\mu \in F^*$, we shall refer to the system $(A, [\ ,\ ,\ ]^\mu)$ as a L.t.s *of adjoint type*.

Over algebraically closed fields, it is known ([7,8]) that

$$\dim \mathrm{Hom}_A(A\otimes A, A) = \begin{cases} 2 & \text{if } A \text{ is of type } A_n\ (n\geq 2), \\ 1 & \text{otherwise.} \end{cases}$$

Consequently, $\mathrm{Hom}_S(A\otimes A, A)$ is spanned by the "Lie bracket in A" for types other than A_n $(n \geq 2)$. But, what happens in the remaining cases? The answer can be found in the following example.

Example 2. For any simple Jordan algebra J of degree $n \geq 2$ with generic trace t, the set J_0 of trace zero elements in J becomes a simple L.t.s via $[x, y, z] = (xz)y - x(zy)$ (the associator of x, z and y). Since $[x, y, -] = [R_y, R_x]$ $(R_x u = ux)$, it follows that $S = [R_J, R_J] = \mathrm{Der}\, J$ and the standard embedding for this system is $L(J_0) = \mathrm{Der}\, J \oplus J_0 \simeq \mathrm{Der}\, J \oplus R_{J_0}$. In this case we can obtain a symmetric element in $\mathrm{Hom}_S(J_0 \otimes J_0, J_0)$ by means of

$$\begin{aligned} J_0 \times J_0 &\longrightarrow J_0 \\ x\otimes y &\mapsto x\cdot y = xy - \frac{1}{n}t(xy)1 \end{aligned}$$

(that is, the projection of the element xy onto J_0) and this product is nontrivial if and only if $n \geq 3$. The systems $(J_0, [\ ,\ ,\ ]^\mu)$ will be said to be of *Jordan type*.

In case F is algebraically closed, if J is a simple Jordan algebra of degree $n \geq 3$ over F, then either:

i) J is the algebra of symmetric $n\times n$ matrices. In this case $\mathrm{Der}\, J$ is isomorphic to the set of skew-symmetric matrices ([9, Theorem VI.9]) and $L(J_0) \simeq sl(n, F)$.

ii) J is the algebra of symmetric $2n \times 2n$ matrices with respect to the standard symplectic involution. Again in this case $\mathrm{Der}\, J$ is isomorphic to the skew-symmetric matrices with respect to this involution [9, Theorem VI.9] and $L(J_0) \simeq sl(2n, F)$.

iii) J is the exceptional simple Jordan algebra, $\mathrm{Der}\, J$ is a Lie algebra of type F_4 and $L(J_0)$ is a Lie algebra of type E_6 [9, Section IV.11].

In all these cases, $\dim \mathrm{Hom}_S(J_0\otimes J_0, J_0) \geq 1$, and the main Theorem will assert that, in fact, it is 1.

iv) $J = Mat_n(F)^+$. In this case $J_0 = sl(n, F)$ and $\mathrm{Der}\, J = \mathrm{ad}\, J_0$. Then the triple system J_0 is both of adjoint type (A_{n-1}) and of Jordan type.
Therefore, $\mathrm{Hom}_S(T \otimes_F T, T)$ is spanned by the "Lie bracket in S" and the "Jordan product $\cdot$" for the adjoint type A_n $(n \geq 2)$. This is the answer to the question in Example 1.

The situation in iv) can be settled in a more general context over arbitrary fields of characteristic 0, as the following and final example ([9, Section V.7]) shows:

Example 3. Let (A, j) be a central simple associative involutorial algebra of degree $n \geq 3$ with j an involution of second kind and center $P = F[q]$, a quadratic extension of the base field or isomorphic to $F \times F$. Then the Jordan algebra $J = H(A, j)^+$ of j-symmetric elements of A is central simple of type A and degree n (that is, after extension by scalars it becomes a total $n \times n$ matrix algebra) and the derived algebra $S(A, j)_0 = [S(A, j), S(A, j)]$ of the Lie algebra $S(A, j)^-$ of j-skew elements of A is a central simple Lie algebra of type A_{n-1} [10, Theorem X.8] (Note that if $P = F \oplus F$, $A = B \oplus B^{op}$, with B a central simple associative algebra and j the exchange involution: $j(b_1, b_2) = (b_2, b_1)$. So we can identify $H(A, j)^+ \simeq B^+$ and $S(A, j)^- \simeq B^-$). The converse is also true: any central simple Jordan algebra of type A and degree $n \geq 3$ can be obtained in this manner.

If we consider the L.t.s. of Jordan type $T = J_0 = H(A, j)_0 = \{x \in H(A, j) \mid t(x) = 0\}$ (t the generic trace of $H(A, j)$), we have that $S = \mathrm{Der}\, J = \mathrm{ad}\, S(A, j)_0|_{H(A,j)_0}$ [9].

As the linear map $x \mapsto qx$ of J_0 onto $S(A, j)_0$ is an isomorphism of $S(A, j)_0$-modules, our L.t.s. is both of adjoint type and of Jordan type, and the bracket in $S(A, j)_0$ under this isomorphism provides the product

$$x * y = q[x, y],$$

which is a nonzero skew-symmetric element in $\mathrm{Hom}_{\mathrm{Der}\, J}(J_0 \otimes J_0, J_0)$. Consequently the dimension of this space is at least 2. We shall refer to the systems $(H(A, j)_0, [\, , \, , \,]^\mu)$ as L.t.s. of *adjoint-Jordan type*.

2.3. The centroid of a Lie triple system.

A natural and useful tool in the study of simple algebraic structures is the centroid. Given any L.t.s. T we define the centroid of T in the natural way as $\Gamma = \{\alpha \in End_F(T) \mid \alpha([x, y, z]) = [\alpha(x), y, z] = [x, \alpha(y), z] = [x, y, \alpha(z)] \quad \forall x, y, z \in T\}$. If T is simple, Γ is a field contained in the centroid of $L(T)$, and T over Γ remains simple by scalars extensions.

A technical result in [5, Lemma 4.1] shows that for any simple Lie triple system T with centroid Γ, in order to determine $\mathrm{Hom}_S(T \otimes_F T, T)$ it is enough to determine $\mathrm{Hom}_S(T \otimes_\Gamma T, T)$.

In this way, our problem is reduced to the special case of central simple L.t.s. for which we can extend scalars up to an algebraically closed field. Then, the problem becomes to compute the set of homomorphisms for simple L.t.s. over algebraically closed fields.

Now we shall sketch how to solve this situation.

2.4. Solution in the algebraically closed case.

In this paragraph let T be a simple L.t.s. over an algebraically closed field F and $L(T) = S \oplus T$ be its standard embedding.

The complete classification of such systems was obtained by Lister in 1952 [6]. In 1980 Faulkner [11] gave a method based on the use of the affine Dynkin diagrams (see [12, Ch.8]) that provides an useful and quicker classification. This one will be used here to deduce our results.

From [6,11] the algebra $L(T)$ is a simple ungraded algebra unless T is of adjoint type, and S is a reductive algebra with $\dim Z(S) \leq 1$, consequently the derived subalgebra $[S, S]$ of S is semisimple. In case $Z(S) = 0$, T is an irreducible selfdual module for S; otherwise $T = T_1 \oplus T_2$, T_i being irreducible dual S-modules.

The central idea in Faulkner's classification is to associate a diagram to each simple L.t.s. T, starting with the Dynkin diagram of the semisimple algebra $[S, S]$ and adding a node for each $[S, S]$-irreducible component of T (which represents its minimum weight). All the nodes in the diagrams are equipped by numerical labels representing the linear dependence of the roots of $[S, S]$ and the weights of T involved in the diagrams. These weighted diagrams contain the whole information on the Lie algebra S and the S-module T. They can be grouped into four types.

i) Diagrams representing simple L.t.s. of adjoint type. For these ones, the problem was already solved.

ii) Diagrams representing non irreducible simple L.t.s.. Given such a system T, the center of S is one-dimensional and $T = T_1 \oplus T_2$ with T_i S-irreducible. In this case, there exists a nonzero central element $z \in S$ with $\operatorname{ad} z|_{T_1} = 1$ and $\operatorname{ad} z|_{T_2} = -1$. Therefore $T \otimes T$ is the sum of the eigenspaces 2, 0, -2 and so $\operatorname{Hom}_S(T \otimes T, T) = 0$.

iii) Diagrams representing irreducible simple L.t.s with trivial 0-weight space, and

iv) Diagrams representing irreducible simple L.t.s. with nontrivial 0-weight space.

In the two latter cases the corresponding L.t.s. are selfdual modules and the determination of the set $\operatorname{Hom}_S(T \otimes T, T)$ can be reduced to the problem of bounding the dimension of a certain subspace in the 0-weight space of T (which is 0 for systems of type iii)) because of the following result [5, Lemma 3.3] and [8]:

LEMMA *Let S be a semisimple Lie algebra, H a Cartan subalgebra of S, $T = V(\lambda)$ an irreducible selfdual module with highest weight λ relative to H, v_λ and $v_{-\lambda}$ nonzero weight vectors for λ and $-\lambda$. For any root α relative to H let S_α be the corresponding root space. If T_0 is the 0-weight space in T, the linear map*

$$\begin{array}{rcl} \phi : \operatorname{Hom}_S(T \otimes T, T) & \longrightarrow & T_0^\lambda = \{v \in T_0 \mid S_\alpha \cdot v = 0 \quad \forall \alpha \perp \lambda\} \\ \varphi & \mapsto & \varphi(v_\lambda \otimes v_{-\lambda}) \end{array}$$

is well defined and one-to-one. That is, $\dim \mathrm{Hom}_S(T \otimes T, T) \leq \dim T_0^\lambda$.

So, given a simple L.t.s. T of type iii), from the previous Lemma we get that $\mathrm{Hom}_S(T \otimes T, T) = 0$; and given a T of type iv), by using the technical Lemma in [5, Lemma 3.4] and the information provided by the affine diagrams of types i), iii) and iv) it is possible to obtain that the space $\mathrm{Hom}_S(T \otimes T, T)$ is trivial or one-dimensional, turning out that $\mathrm{Hom}_S(T \otimes T, T) = 0$ unless T is a simple L.t.s. of Jordan type (the L.t.s. i), ii), iii) described in Example 2), in this case $\mathrm{Hom}_S(T \otimes T, T)$ is spanned by the product $\cdot$ (J is not of type A).

2.5. The main Theorem.

Now, taking into account the considerations above, we can establish the main result in [5, Theorem 4.3]:

THEOREM *Let T be a simple Lie triple system over a field F of characteristic zero with centroid Γ and let $L = S \oplus T$ be its standard embedding. Then $\mathrm{Hom}_S(T \otimes_F T, T) = 0$ unless either:*

a) *T is of adjoint type with S being a central simple Lie algebra over Γ of type different from A_n ($n \geq 2$). In this case $\mathrm{Hom}_S(T \otimes_F T, T) \simeq \mathrm{Hom}_S(S \otimes_F S, S)$, which is spanned over Γ by the Lie multiplication in S, or*

b) *there exists a central simple Jordan algebra J of degree $n \geq 3$ over Γ and a nonzero scalar $\mu \in \Gamma$ (which can be taken modulo Γ^2) such that T is the Lie triple system $(J_0, [\,,\,,\,]^\mu)$. In this case either:*

   i) *J is not of type A over Γ, then $\mathrm{Hom}_S(T \otimes_F T, T) = \mathrm{Hom}_{\mathrm{Der}\,J}(J_0 \otimes_F J_0, J_0)$ is spanned over Γ by the product $x \cdot y$,*

   ii) *$J = H(A, j)$ for some central simple associative involutorial algebra (A, j) of second kind over Γ. Let $P = \Gamma[q]$ ($0 \neq q^2 \in \Gamma$) be the center of A. Then $\mathrm{Hom}_S(T \otimes_F T, T) = \mathrm{Hom}_{\mathrm{Der}\,J}(J_0 \otimes_F J_0, J_0) = \mathrm{Hom}_{S(A,j)_0}(H(A, j)_0 \otimes_F H(A, j)_0, H(A, j)_0)$ is spanned over Γ by the products $\cdot$ and $*$.*

From the point of view of nonassociative algebras, this result points out the central role played by the multiplication $\cdot$, defined in the set J_0 of the trace zero elements in a Jordan algebra J, in the computation of the sets $\mathrm{Hom}_S(T \otimes T, T)$.

The restriction of the Theorem to the real field gives us the invariant affine connections on the irreducible symmetric spaces. So, in a sense, we can say that the Jordan algebras are responsible for the existence of noncanonical connections on the symmetric spaces.

3. APPLICATIONS TO DIFFERENTIAL GEOMETRY.

In this section we are going to compute some important geometric objects in a symmetric space, namely, the torsion and curvature tensors and the holonomy algebra associated to the invariant affine connections on it.

First of all, we will get expressions for these objects in any reductive homogeneous space in terms of the multiplication in the associated connection algebra. Afterwards we will specialize the previous computations to the symmetric spaces, because for these spaces we know the concrete connection algebras (thanks to the central Theorem in Section 2).

The torsion and curvature tensors are the most important tensors associated to an affine connection ∇, and they are given by the following formulas:

- Torsion tensor: $T(X,Y) = \nabla_X Y - \nabla_Y X - [X,Y]$
- Curvature tensor: $R(X,Y)Z = \nabla_X \nabla_Y Z - \nabla_Y \nabla_X Z - \nabla_{[X,Y]} Z$

Now, if $M = G/H$ is a reductive homogeneous space with reductive decomposition $\mathfrak{g} = \mathfrak{h} \oplus \mathfrak{m}$ and ∇ is a G-invariant affine connection, let $\alpha : \mathfrak{m} \times \mathfrak{m} \to \mathfrak{m}$ be the associated multiplication by Nomizu's result. In this case the torsion and curvature tensors can be expressed in terms of the multiplication α by means of:

$$T(X,Y) = \alpha(X,Y) - \alpha(Y,X) - [X,Y]_{\mathfrak{m}}$$
$$R(X,Y)Z = \alpha(X,\alpha(Y,Z)) - \alpha(Y,\alpha(X,Z)) - \alpha([X,Y]_{\mathfrak{m}}, Z) - [[X,Y]_{\mathfrak{h}}, Z]$$

for any $X, Y, Z \in \mathfrak{m}$, where $Z_{\mathfrak{m}}$ and $Z_{\mathfrak{h}}$ denote the projections of $Z \in \mathfrak{g}$ onto $\mathfrak{m}$ and $\mathfrak{h}$ respectively, and we are identifying $\mathfrak{m}$ with the tangent space $T_{\bar{e}}M$.

If M is symmetric, the projection over $\mathfrak{m}$ of the bracket of two elements in $\mathfrak{m}$ is zero, hence:

$$T(X,Y) = \alpha(X,Y) - \alpha(Y,X)$$
$$R(X,Y)Z = \alpha(X,\alpha(Y,Z)) - \alpha(Y,\alpha(X,Z)) - [[X,Y]_{\mathfrak{h}}, Z].$$

On the other hand, the holonomy algebra $\operatorname{Hol}\nabla$ is the smallest Lie subalgebra of $\operatorname{End}_{\mathbb{R}}(\mathfrak{m})$ containing $R(X,Y)$ for all $X, Y \in \mathfrak{m}$ and closed under commutation by ∇_X for any $X \in \mathfrak{m}$.

The previous facts lead us to introduce the following definitions:

Given a Lie triple system T with standard embedding $L(T) = S \oplus T$ and given $\alpha \in \operatorname{Hom}_S(T \otimes_F T, T)$, we define:

i) Torsion tensor: $T^\alpha(x,y) = \alpha(x,y) - \alpha(y,x)$

ii) Curvature tensor: $R^\alpha(x,y)z = \alpha(x,\alpha(y,z)) - \alpha(y,\alpha(x,z)) - [x,y,z]$

iii) Holonomy algebra: $\operatorname{Hol}^\alpha(T)$ = the smallest subalgebra of $\operatorname{End}_F T$ containing $R^\alpha(x,y)$ for all $x, y \in T$ and closed under commutation by the operators $\alpha(x,-)$.

We are going to compute all these concepts for any simple Lie triple system T over an arbitrary field F of characteristic zero. In the particular case $F = \mathbb{R}$, each $\alpha \in \mathrm{Hom}_S(T \otimes_{\mathbb{R}} T, T)$ corresponds to an invariant affine connection in the symmetric space so that T^α, R^α and Hol^α are the true torsion tensor, curvature tensor and holonomy algebra associated to the connection.

According to the main theorem in Section 2 we can assume that T is a central simple Lie triple system over F with standard embedding $L(T) = S \oplus T$ and the pair (T, α) has one of the following forms:

I) **Trivial case**: $\alpha = 0$

Clearly we have for any $x, y \in T$:

$$\begin{cases} T^\alpha(x,y) = 0 \\ R^\alpha(x,y) = -[x,y,-] \\ \mathrm{Hol}^\alpha(T) = \mathrm{alg}\langle [x,y,-] \mid x,y \in T\rangle = S \end{cases}$$

II) **Adjoint type**: there exists a central simple Lie algebra S and $\mu, \eta \in F^* = F \setminus \{0\}$ such that $T = S$ and for any $x, y \in T$:

$$\begin{cases} [x,y,-] = \mu \,\mathrm{ad}[x,y] \\ \alpha(x,y) = \eta[x,y] \end{cases}$$

where $[\,,\,]$ is the product in S.

Immediately we obtain in this case

$$\begin{cases} T^\alpha(x,y) = 2\eta[x,y] \\ R^\alpha(x,y) = (\eta^2 - \mu)\,\mathrm{ad}[x,y] \\ \mathrm{Hol}^\alpha(T) = \begin{cases} 0 & \text{if } \mu = \eta^2 \\ \mathrm{ad}\, S \simeq S & \text{if } \mu \neq \eta^2 \end{cases} \end{cases}$$

III) **Jordan type**: there exists a central simple Jordan algebra J of degree $n \geq 3$ with generic trace t and $\mu, \eta \in F^*$ such that $T = J_0 = \{x \in J \mid t(x) = 0\}$ is the set of trace zero elements and for any $x, y \in J_0$:

$$\begin{cases} [x,y,-] = \mu[R_x, R_y]|_{J_0} \\ \alpha(x,y) = \eta\, x \cdot y \end{cases}$$

where $x \cdot y = xy - \frac{1}{n}t(xy)1$ and xy is the product in J.

In order to compute the holonomy we need a previous result:

LEMMA *Let J be a central simple Jordan algebra of degree $n \geq 3$ with generic trace t and $J_0 = \{x \in J \mid t(x) = 0\}$ in which we consider the product $x \cdot y = xy - \frac{1}{n}t(xy)1$ and the trace form $t(x,y) = t(xy)$.*

Then the derivation algebra and the Lie multiplication algebra of $(J_0, \cdot)$ are the Lie algebras given by:

$$\mathrm{Der}(J_0, \cdot) = \{d|_{J_0} : d \in \mathrm{Der}\, J\} \quad (\simeq \mathrm{Der}\, J)$$

$$\mathrm{Lie}(J_0, \cdot) = sl(J_0)$$

Furthermore, $(J_0, \cdot)$ is simple.

Proof: If $(a, b, c) = (ab)c - a(bc) = [R_c, R_a](b)$ is the associator in J and $(a, b, c)^{\cdot} = (a \cdot b) \cdot c - a \cdot (b \cdot c)$ the associator in $(J_0, \cdot)$, it is straightforward to check that

$$(a, b, c)^{\cdot} = (a, b, c) - \frac{1}{n}(t(a, b)c - t(b, c)a) \qquad (*)$$

for any $a, b, c \in J_0$. Therefore for any $x, y \in J_0$

$$(x \cdot x, y, x)^{\cdot} = \frac{1}{n}(t(x, y)x \cdot x - t(x \cdot x, y)x).$$

For any $d \in \mathrm{Der}(J_0, \cdot)$, its action on this equality gives

$$(t(dx, y) + t(x, dy))\, x \cdot x - (t(d(x \cdot x), y) + t(x \cdot x, dy))\, x = 0$$

for any $x, y \in J_0$. Since the degree of J is ≥ 3, the set of elements $x \in J_0$ such that x and $x \cdot x$ are linearly independent is a dense subset in the Zariski topology, we obtain from the above that

$$t(dx, y) + t(x, dy) = 0$$

for any $x, y \in J_0$. That is, $\mathrm{Der}(J_0, \cdot)$ is contained in the orthogonal Lie algebra related to t: $so(J_0, t)$. Now, extending d to J by means of $d1 = 0$, we get immediately that $d \in \mathrm{Der}\, J$. Conversely, any derivation of J restricts to a derivation of $(J_0, \cdot)$, so that

$$\mathrm{Der}(J_0, \cdot) = \{d|_{J_0} : d \in \mathrm{Der}\, J\} \simeq \mathrm{Der}\, J\,.$$

It is known that J_0 is an irreducible $\mathrm{Der}\, J$-module, therefore J_0 is an irreducible $\mathrm{Der}(J_0, \cdot)$-module, which implies that $(J_0, \cdot)$ is simple and their derivations are inner ([13, Theorem 3.4] and [14, Lemma 1]), that is, $\mathrm{Der}(J_0, \cdot) \subseteq \mathrm{Lie}(J_0, \cdot)$.

We denote by $R^{\cdot}_x$ the multiplication operator in $(J_0, \cdot)$ given by $R^{\cdot}_x(y) = x \cdot y$. Since $trace(R_x)$ is a multiple of $t(x)$, we have $trace(R^{\cdot}_x) = trace(R_x) = 0$ if $x \in J_0$, hence $\mathrm{Lie}(J_0, \cdot) \subseteq sl(J_0)$.

In order to prove the converse, we note that (*) is equivalent to:

$$[R_x, R_y] = [R^{\cdot}_x, R^{\cdot}_y] - \frac{1}{n}(t(x, -)y - t(y, -)x)$$

for any $x, y \in J_0$. Hence

$$\begin{aligned}&\frac{1}{n}(t(x, -)y - t(y, -)x) = \\ &\qquad = [R^{\cdot}_x, R^{\cdot}_y] - [R_x, R_y] \in \mathrm{Lie}(J_0, \cdot) + \mathrm{Der}(J_0, \cdot) \subseteq \mathrm{Lie}(J_0, \cdot)\end{aligned}$$

since $[R_x, R_y]$ is always a derivation of J. But the linear maps on the left hand side above span the orthogonal Lie algebra $so(J_0, t)$, so that $so(J_0, t) \subseteq \mathrm{Lie}(J_0, \cdot)$. Now we consider the symmetric elements relative to the trace: $H = \{\varphi \in End_F(J_0) \mid trace(\varphi) = 0,\ t(\varphi x, y) = t(x, \varphi y) \quad \forall x, y \in J_0\}$. Then $sl(J_0) = H \oplus so(J_0, t)$, and H is an irreducible module for $so(J_0, t)$. As the operator $R_x^{\cdot}$ is symmetric with respect to the trace form since

$$t(x \cdot y, z) = t((xy)z) - \frac{1}{n} t(xy)t(z) = t((xy)z) = t(y(xz)) = t(y, x \cdot z),$$

it follows that $H \cap \mathrm{Lie}(J_0, \cdot) \neq 0$ and, by irreducibility, that $H \subseteq \mathrm{Lie}(J_0, \cdot)$ and we conclude that $\mathrm{Lie}(J_0, \cdot) = sl(J_0)$. ∎

We now proceed to the computations in this case. It is clear that $T^\alpha \equiv 0$. As regards curvature, we see at once that

$$\begin{aligned} R^\alpha(x, y) &= \eta^2 [R_x^{\cdot}, R_y^{\cdot}] - \mu [R_x, R_y] \\ &= (\eta^2 - \mu)[R_x, R_y] + \frac{\eta^2}{n}(t(x, -)y - t(y, -)x). \end{aligned}$$

As for the holonomy algebra, we notice that

$$R^\alpha(x, y) = (\eta^2 - \mu)[R_x, R_y] + \frac{\eta^2}{n}(t(x, -)y - t(y, -)x) \in \mathrm{Der}\, J + so(J_0, t),$$

which is contained in $\mathrm{Lie}(J_0, \cdot)$ and shows that $\mathrm{Hol}^\alpha(J_0) \subseteq \mathrm{Lie}(J_0, \cdot)$, because $\mathrm{Lie}(J_0, \cdot)$ is obviously closed under commutation by $\alpha(x, -) = \eta R_x^{\cdot}$. Besides, $\mathrm{Hol}^\alpha(J_0)$ is closed under commutation by $R_x^{\cdot}$ and these elements generate the Lie algebra $\mathrm{Lie}(J_0, \cdot)$, which yields that $\mathrm{Hol}^\alpha(J_0)$ is an ideal of $\mathrm{Lie}(J_0, \cdot) = sl(J_0)$; but this is a simple algebra, consequently $\mathrm{Hol}^\alpha(J_0)$ is 0 or $sl(J_0)$. As the degree of J is greater than 2, there exists $x \in J_0$ such that x and $x \cdot x = y$ are linearly independent, and for them we have $[R_x, R_y] = 0$ (Jordan identity) and so $R^\alpha(x, y) \neq 0$, therefore $\mathrm{Hol}^\alpha(J_0) = sl(J_0)$. Summarizing, we have obtained

$$\begin{cases} T^\alpha(x, y) = 0 \\ R^\alpha(x, y) = (\eta^2 - \mu)[R_x, R_y] + \frac{\eta^2}{n}(t(x, -)y - t(y, -)x) \\ \mathrm{Hol}^\alpha(T) = sl(J_0) \end{cases}$$

IV) **Adjoint-Jordan type**: there exists a simple associative algebra of degree $n \geq 3$ with an involution of second kind (A, j) and center $K = F[q]$ (quadratic extension of F or isomorphic to $F \times F$) where $q^2 \in F^*$, and $\mu, \eta, \nu \in F^*$ such that $T = H(A, j)_0 = \{x \in A \mid j(x) = x, t(x) = 0\}$ (t the generic trace of $H(A, j)$) and for any $x, y \in T$:

$$\begin{cases} [x, y, -] = \mu\, \mathrm{ad}[x, y]|_T \\ \alpha(x, y) = \eta\, x \cdot y + \nu q\, [x, y] \end{cases}$$

where $x \cdot y = \frac{1}{2}(xy + yx) - \frac{1}{n} t(xy)1$ with xy the product in A, and $[x, y] = xy - yx$ the bracket in A.

In this case first we will find out the derivation algebra and the Lie multiplication algebra, as in the Jordan case. As $\operatorname{ad} x$ is a derivation for both the Lie bracket and the Jordan product in A, it is clear that $\operatorname{ad} qx \in \operatorname{Der}(T,\alpha)$ for all $x \in T$. The converse is true, because

$$\operatorname{Der}(T,\alpha) \subseteq \operatorname{Der}((T,\alpha)^+) = \operatorname{Der}(H(A,J)_0,\cdot) = \operatorname{Der} H(A,j)$$

by the Lemma, where $(T,\alpha)^+$ denotes the algebra defined on T with multiplication given by $\alpha(x,y)+\alpha(y,x)$, and it is known [9, Theorem VI.9] that $\operatorname{Der} H(A,j) = \operatorname{ad} S(A,j)_0|_{H(A,j)_0}$, where $S(A,j) = \{x \in A \mid j(x) = -x\} = qH(A,j)$ and $S(A,j)_0$ its subset of trace zero elements . Therefore

$$\operatorname{Der}(T,\alpha) = \operatorname{ad} S(A,j)_0|_{H(A,j)_0} \simeq S(A,j)_0$$

since $S(A,j)_0$ is a simple Lie algebra. Hence $T = H(A,J)_0 = qS(A,j)_0$ is an irreducible $\operatorname{ad} S(A,j)_0 = \operatorname{Der} T$-module, consequently (T,α) is again a simple algebra and its derivations are inner. As for the Lie multiplication algebra, $\alpha(x,-)$ and $\alpha(-,x)$ are linear combinations of $R^{\cdot}_x$ (with zero trace, by the Lemma) and $\operatorname{ad} qx$, whence $\operatorname{Lie}(T,\alpha) \subseteq sl(T)$. But $\operatorname{Lie}(T,\alpha) \supseteq \operatorname{Lie}((T,\alpha)^+) = \operatorname{Lie}(H(A,j)_0,\cdot)$ and we saw in the Lemma above that $\operatorname{Lie}(H(A,j)_0,\cdot) = sl(H(A,J)_0)$, therefore

$$\operatorname{Lie}(T,\alpha) = sl(T).$$

We now turn to the computations of the tensors. We obtain that $T^\alpha(x,y) = 2\nu q[x,y]$, while

$$\begin{aligned} R^\alpha(x,y) &= [\alpha(x,-),\alpha(y,-)] - [x,y,-] \\ &= [\eta R^{\cdot}_x + \nu q \operatorname{ad} x, \eta R^{\cdot}_y + \nu q \operatorname{ad} y] - \mu \operatorname{ad}[x,y] \\ &= \eta^2[R^{\cdot}_x, R^{\cdot}_y] + \eta\nu q([R^{\cdot}_x, ad\, y] + [\operatorname{ad} x, R^{\cdot}_y]) + (\nu^2q^2 - \mu)\operatorname{ad}[x,y]. \end{aligned}$$

If we denote by l_x and r_x the right and left multiplications in the associative algebra A, it is easily checked using (*) that

$$\begin{aligned} [R^{\cdot}_x, R^{\cdot}_y] &= \frac{1}{4}[l_x + r_x, l_y + r_y] + \frac{1}{n}(t(x,-)y - t(y,-)x) \\ &= \frac{1}{4}ad[x,y] + \frac{1}{n}(t(x,-)y - t(y,-)x) \end{aligned}$$

and, since $\operatorname{ad} qx \in \operatorname{Der}(H(A,j)_0,\cdot)$ then $[\operatorname{ad} qx, R^{\cdot}_y] = R^{\cdot}_{q[x,y]}$.

Combining these equalities we get

$$R^\alpha(x,y) = \left(\frac{\eta^2}{4} + \nu^2q^2 - \mu\right)\operatorname{ad}[x,y] + 2\eta\nu R^{\cdot}_{q[x,y]} + \frac{\eta^2}{n}(t(x,-)y - t(y,-)x).$$

We note that

$$\begin{aligned} &R^\alpha(x,y) = \\ &= [\alpha(x,-),\alpha(y,-)] - \mu \operatorname{ad}[x,y] \in \operatorname{Lie}(T,\alpha) + \operatorname{Der}(T,\alpha) \subseteq \operatorname{Lie}(T,\alpha) \end{aligned}$$

and $\alpha(x,-) \in \mathrm{Lie}(T,\alpha)$, so $\mathrm{Hol}^\alpha(T) \subseteq \mathrm{Lie}(T,\alpha)$. Besides $\mathrm{Hol}^\alpha(T)$ is closed under commutation by $\alpha(x,-)$, but these operators generate $\mathrm{Lie}(T,\alpha)$, which forces $\mathrm{Hol}^\alpha(T)$ to be an ideal of $\mathrm{Lie}(T,\alpha)$. As we saw that $\mathrm{Lie}(T,\alpha) = sl(T)$, it is a simple algebra and as in the Jordan type we may conclude that

$$\begin{cases} T^\alpha(x,y) = 2\nu q[x,y] \\ R^\alpha(x,y) = \left(\frac{\eta^2}{4} + \nu^2 q^2 - \mu\right) \mathrm{ad}[x,y] + 2\eta\nu R_{q[x,y]} \\ \qquad\qquad + \frac{\eta^2}{n}(t(x,-)y - t(y,-)x) \\ \mathrm{Hol}^\alpha(T) = sl(T). \end{cases}$$

Notice that in case $K = F[q] = F \oplus F$, then $A = B \oplus B^{op}$, with B a central simple associative algebra, and j is the exchange involution. Thus $H(A,j)$ may be identified with B^+ and $S(A,j)$ with the Lie algebra B^- by means of the projection onto the first component. The formulae above remain valid for $T = B_0$ in this case, with $q = 1$ and with t the generic trace of B.

REFERENCES

1 A. Elduque and H.C. Myung. Octonions and affine connections on spheres. These proceedings.

2 K. Nomizu. Invariant affine connections on homogeneous spaces. Amer. J. Math. 76:33-65, 1954.

3 H.T. Laquer. Invariant affine connections on Lie groups. Trans. Amer. Math. Soc. 331:541-551, 1992.

4 H.T. Laquer. Invariant affine connections on symmetric spaces. Proc. Amer. Math. Soc. 115: 447-454, 1992.

5 P. Benito, C. Draper and A. Elduque. On some algebras related to simple Lie triple systems. Preprint.

6 W.G. Lister. A structure theory of Lie triple systems. Trans. Amer. Math. Soc. 72:217-242, 1952.

7 G.M. Benkart and J.M. Osborn. Flexible Lie-admissible algebras. J. Algebra 71:11-31, 1981.

8 J.R. Faulkner. Identity Classification in Triple Systems. J. Algebra 94:352–363, 1985.

9 N. Jacobson. Structure and representations of Jordan algebras. Amer. Math. Soc., Providence, Rhode Island 1968.

10 N.Jacobson. Lie algebras. Interscience, New York 1962.

11 J.R. Faulkner. Dynkin Diagrams for Lie Triple Systems. J. Algebra 62:384-392, 1980.

12 V.G. Kac. Infinite dimensional Lie algebras (Third ed.). Cambridge University Press 1990.

13 A. Elduque and F. Montaner. A Note on Derivations of Simple Algebras. J. Algebra 165:636-644, 1994.

14 A.A. Sagle and D.J. Winter. On homogeneous spaces and reductive subalgebras of simple Lie algebras. Trans. Amer. Math. Soc. 128:142-147, 1967.

E-ideals in Berstein algebras

A. CATALAN [2] Departamento de Matemática, Universidad de La Frontera, Casilla 54-D, Temuco, Chile, e-mail: acatalan@ufro.cl

ABSTRACT This paper is a natural continuation of [1] and [2]. In the first one we study E-ideals in arbitrary baric algebras and, in the second one, we limit ourselves to train algebras. In this paper we study E-ideals in Bernstein algebras

1 BERNSTEIN ALGEBRAS

In this section we present, without proofs, the main results concerning Bernstein algebras, necessary for understanding the results which follow.

Let F be a field of characteristic not 2, A a commutative but not necessarily associative algebra over F, $w : A \to F$ a homomorphism , $w \neq 0$. The ordered pair (A, w) is called a baric algebra and w is its weight function. The ideal $N = ker\ w$ has codimension 1 in A and if $d \in A$, $w(d) = 1$, then $A = Fd \oplus N$.

A commutative baric algebra (A, w) over the field F is a Bernstein algebra if for all $x \in A$, we have

$$(x^2)^2 = \omega(x)^2 x^2 \tag{1}$$

Every Bernstein algebra admits idempotent elements, namely, the elements $e = a^2 \in A$ with $\omega(a) = 1$. It is well known that the homomorphism w is unique. If $e \in A$ is a fixed idempotent in A, the Peirce decomposition of A relative to this idempotent e is $A = Fe \oplus U_e \oplus Z_e$ where $U_e = \{a \in kerw : ea = \frac{1}{2}a\}$ and

[2] Sponsored by FONDECYT N^0 1970890 and DIUFRO N^0 9806 Chile.

$Z_e = \{a \in kerw : ea = 0\}$. The following relations are satisfied

$$U_e^2 \subseteq Z_e, \quad U_e Z_e \subseteq U_e, \quad Z_e^2 \subseteq U_e \tag{2}$$

In spite of the fact that the subspaces U_e and Z_e depend on the choice of e, their dimensions are invariant. The ordered pair $(1+\dim U_e, \dim Z_e)$ is called the type of A.

The baric algebra (A, w) is a Bernstein algebra if and only if A admits a decomposition $A = Fe \oplus U_e \oplus Z_e$ such that

$$\begin{array}{llllll} i) & (n^2)^2 = 0 & ii) & (uz)^2 = 0 & iii) & u^3 = 0 \\ iv) & u(uz) = 0 & v) & uz^2 = 0 & & \end{array} \tag{3}$$

for every element $n \in N$, $u \in U_e$ and $z \in Z_e$, where e is an idempotent of weight 1 in A.

We include some linearizations of (3) that will we be used in this paper.

$$\begin{array}{ll} i) & (n_1n_2)(n_3n_4) + (n_1n_3)(n_2n_4) + (n_1n_4)(n_2n_3) = 0 \\ ii) & (u_1{}^2)u_2 + 2(u_2u_1)u_1 = 0 \\ iii) & (u_1u_2)u_3 + (u_2u_3)u_1 + (u_3u_1)u_2 = 0 \\ iv) & u_1(u_2z) + u_2(u_1z) = 0 \end{array} \tag{4}$$

for all elements $n_i \in N$, $u_i \in U_e$ and $z \in Z_e$.

Let $A = Fe \oplus U_e \oplus Z_e$ be a Bernstein algebra, not necessarily finite dimensional. Then the principal powers of $N = U_e \oplus Z_e$ are ideals of A. If $R_z : N \to N$ is the multiplication by z, $z \in Z_e$, then we have also the identities

$$(i) \quad uR_z^k(u^2) = 0, \quad (ii) \quad uR_z^k(u) = 0, \quad (iii) \quad uR_z^k(z) = 0 \tag{5}$$

for all $k \geq 1$, $u \in U_e$ and $z \in Z_e$.

We say that A is exceptional when $U_e^2 = 0$. Lyubich in [10] defines the concept of exceptional Bernstein algebra in the case of finite dimension. A Bernstein algebra A is Jordan when $Z_e^2 = 0$ and $(uz)z = 0$ for all $u \in U_e$ and $z \in Z_e$, or equivalently, $x^3 = \omega(x)x^2$ for all $x \in A$ (see [13], [14]). An important subclass of all Bernstein-Jordan algebras is formed by the normal ones. We say that a Bernstein algebra A is normal if we have $x^2y = \omega(x)xy$ in A, or equivalently, $U_eZ_e = Z_e^2 = 0$. A Bernstein algebra A is nuclear when $A^2 = A$, that is, $U_e^2 = Z_e$. It is easy to see that if A is any Bernstein algebra then A^2 is nuclear, showing the existence of enough nuclear algebras. In a general nuclear Bernstein algebra A, all proper ideals are in N. Also in the nuclear algebra the annihilator of N, denoted $ann(N)$, is the ideal $U_0 = \{u \in U_e : uU_e = 0\}$, see [7], this ideal always contains z^2 and $(uz)z$ for all $u \in U_e$ and $z \in Z_e$, see [12].

2 *E*-IDEALS IN TRAIN ALGEBRAS

In this section we present, without proof, the main results on E-ideals in train algebras, necessary for the development of this section.

If $p(x) = x^n + \gamma_1 w(x)x^{n-1} + \ldots + \gamma_{n-1} w(x)^{n-1} x$ is a train polynomial with coefficients $\gamma_i \in F$, we denote by $E_A(p)$ or $E_A(1, \gamma_1, \ldots, \gamma_{n-1})$ the ideal of A generated by all values $p(a)$ with $a \in A$. This ideal is contained in N as $w(p(a)) = 0$ for all $a \in A$. We let $\mathcal{E}_A$ be the set of all these ideals, called the E-ideals of A. Special cases of E-ideals were considered by Etherington in [5], namely, the ideals P and Q_λ. The set $\mathcal{E}_A$, once partially ordered by inclusion, has a greatest element, which is the E-ideal $E_A(1,-1)$ corresponding to the train polynomial $p(x) = x^2 - w(x)x$. The ideal $E_A(1,-1)$ is described, in terms of the Peirce decomposition, in some classes of baric algebras. For instance, if (A, w) is a Bernstein algebra with Peirce decomposition $A = Fe \oplus U_e \oplus Z_e$ then

$$E_A(1,-1) = (U_e Z_e + Z_e^2) \oplus Z_e \tag{6}$$

In [1, Prop.5,Cor.6] shows that the determination of $E_A(p)$ is reduced to the determination of the E-ideals of the indecomposable direct summands of A, when A satisfies ascending and descending conditions on ideals contained in N. In [1, Section 5] it is presented a set theoretical definition of train polynomial, in order to prove some properties of the set $\mathcal{E}_A$. Basically, it is proved that every train polynomial $p(x) = x^n + \gamma_1 w(x)^{n-1} + \ldots + \gamma_{n-1} w(x)^{n-1} x$ has a unique "canonical antecedent", which is $p(X) = X^n + \gamma_1 X^{n-1} + \ldots + \gamma_{n-1} X$, an ordinary polynomial in the associative algebra of polynomials $F[X]$, admiting the roots 0 and 1. As a consequence of this one-to-one correspondence, the concepts of divisibility, greatest common divisor and least common multiple can be defined for train polynomials, by means of their corresponding canonical antecedents. It was proved in [1] that if $p(x)$ and $q(x)$ are train polynomials such that $p(x)$ is a divisor of $q(x)$ then $E_A(q) \subseteq E_A(p)$. The train polynomial $x^2 - w(x)x$ is a divisor of all train polynomials. Moreover, if $r(x)$ is the greatest common divisor of $p(x)$ and $q(x)$ then $E_A(r) = E_A(p) + E_A(q)$. According to [1, Prop. 11], the E-ideal corresponding to the least common multiple of $p(x)$ and $q(x)$ is contained in $E_A(p) \cap E_A(q)$. The usual Euclidean algorithm is valid also for train polynomials, *via* their canonical antecedents. If $p(X)$ is the canonical antecedent of $p(x)$ and its roots are $0, 1, r_1, \ldots, r_{n-2}$ (which we assume to be in the field F), then we shall write a reduced and factorised form of $p(x)$, namely $p(x) = x(x-1)(x-r_1)\ldots(x-r_{n-2})$.
Let (A, w) be a commutative baric algebra over the field F and $p(x)$ a train polynomial, say

$$p(x) = x^n + \gamma_1 w(x) x^{n-1} + \ldots + \gamma_{n-1} w(x)^{n-1} x \tag{7}$$

so that $p(X) = X^n + \gamma_1 X^{n-1} + \ldots + \gamma_{n-1} X = X(X-1)\overline{p}(X)$ is its canonical antecedent. Let $p_1(X), \ldots, p_k(X)$ be the prime factors of $\overline{p}(X)$ in $F[X]$, so that $p(X) = X(X-1)\; p_1(X)^{n_1} \ldots p_k(X)^{n_k}$, for some integers $n_1, .., n_k$.

We present now the main results on E-ideals in train algebras.

PROPOSITION 1 *Let (A, w) be a train algebra with train equation (7) as above (i.e. which satisfy $p(x) = 0$ for all $x \in A$ and the equation is minimal). If $q(x)$ is a train polynomial (so that $q(x) = p(x)f(x) + r(x)$ by the Euclidean algorithm) then $E_A(q) = E_A(r)$.*

THEOREM 1 *Let (A, w) be a train algebra with train equation (7) and let $p(X) = X(X-1)p_1(X)^{n_1} \ldots p_k(X)^{n_k}$ be the prime decomposition of $p(X)$ in $F[X]$. The maximum number of E-ideals in A is $(1+n_1)(1+n_2)\ldots(1+n_k)$.*

The concept of (commutative) duplicate of a baric algebra was introduced by Etherington in [6] and treated recently by several authors. In particular, it is known that if (A, w) is any baric algebra such that A^2 is train with train equation $p(x) = 0$ then the duplicate A^D of A is train and satisfies the equation $p(x)x = 0$, see [11].

COROLLARY 1 *Suppose (A, w) is a baric algebra such that A^2 satisfies the train equation $p(x) = 0$ and let $p(X) = X(X-1)p_1(X)^{n_1}...p_k(X)^{n_k}$ be the prime fatorisation of $p(X)$. Then the number of E-ideals of A^D is, at most,*

(a) $2(1+n_1)...(1+n_k)$ *if no prime factor $p_i(X)$ is X*

(b) $(1+n_1)...(2+n_i)...(1+n_k)$ *if $p_i(X) = X$ for some i $(1 \leq i \leq k)$*

3 E-IDEALS IN BERNSTEIN ALGEBRAS.

In this section we present the main results concerning E-ideals in Bernstein algebras.

PROPOSITION 2 *Let $n = u + z$ be an arbitrary element in the kernel $N = U_e \oplus Z_e$ of a Bernstein algebra A. Then the principal powers of n are given recursively by $n^k = R_z^{k-2}(n^2) = R_z^{k-2}(u^2 + 2uz + z^2)$ for all $k \geq 2$.*

Proof: We prove by induction on k. Clearly the statement is true for $k = 2$. Supose by induction that the formula holds for k. Then $n^{k+1} = n^k n = R_z^{k-2}(u^2 + 2uz + z^2)(u+z) = uR_z^{k-2}(u^2 + 2uz + z^2) + zR_z^{k-2}(u^2 + 2uz + z^2) = uR_z^{k-2}(u^2 + 2uz + z^2) + R_z^{k-1}(u^2 + 2uz + z^2)$. But from (5) we have $uR_z^{k-2}(u^2 + 2uz + z^2) = uR_z^{k-2}u^2 + 2uR_z^{k-1}z + uR_z^{k-1}z = 0$. Then $n^{k+1} = R_z^{k-1}(u^2 + 2uz + z^2) = R_z^{k-1}(n^2)$ ■

PROPOSITION 3 *Let $a = e+n = e+u+z$ be an element of weight 1 in a Bernstein algebra $A = Fe \oplus U_e \oplus Z_e$. Then for all $k \geq 0$ we have*

$$(a^3 - a^2)(a - \frac{1}{2})^k = -\frac{1}{2}z^{k+2} + n^{k+3} \tag{8}$$

Proof: We use induction on k. In fact, we have $a^2 = e + (u + 2uz + z^2) + u^2$, $a^3 = e + (u + 2uz + \frac{1}{2}z^2 + (u^2 + 2uz + z^2)z) + u^2$ and so $a^3 - a^2 = -\frac{1}{2}z^2 + n^3$ so expression (8) is true for $k = 0$. Suppose, by induction hypothesis that (8) is true for $k = l - 1$, then

$$(a^3 - a^2)(a - \frac{1}{2})^{l-1} = -\frac{1}{2}z^{l+1} + n^{l+2} \tag{9}$$

If we multiply the above expression in (9) by the element a, we get

$$((a^3 - a^2)(a - \frac{1}{2})^{l-1})a = -\frac{1}{4}z^{l+1} + \frac{1}{2}n^{l+2} - \frac{1}{2}z^{l+2} + n^{l+3} \tag{10}$$

Now if from (10) subtract (9), multiplied by $\frac{1}{2}$, we have

$$(a^3 - a^2)(a - \frac{1}{2})^l = -\frac{1}{2}z^{l+2} + n^{l+3}$$

■

PROPOSITION 4 *Let $A = Fe \oplus U_e \oplus Z_e$ be a Bernstein algebra, not necessarily of finite dimension. Then each one of the principal powers N^k ($k \geq 1$) contains an E-ideal of A.*

Proof: Consider the train polynomials $q_k(x) = (x^3 - x^2)(x - \frac{1}{2})^k$, with $k \geq 0$. If $a = e + n = e + u + z \in A$ is any element of weight 1, then from (8) we have $q_k(a) = -\frac{1}{2}z^{k+2} + n^{k+3}$, which belongs to $N^{k+2} + N^{k+3} \subseteq N^{k+2}$ and as every principal power of N is an ideal of A, we have that the generators $q_k(a)$ of $E_A(q_k)$ belong to the ideal N^{k+2} and so $E_A(q_k) \subseteq N^{k+2}$ for all $k \geq 0$. Moreover, N contains $E_A(1,-1)$, as $a^2 - \omega(a)a \in N$. ■

PROPOSITION 5 *Let (A, ω) be a Bernstein algebra, not necessarily of finite dimension. Suppose that $N = U_e \oplus Z_e$ is nil of index r, that is, every element $n \in N$ satisfies $n^r = 0$. Then A satisfies the train equation $(x^3 - x^2)(x - \frac{1}{2})^{r-2} = 0$ and so A has at most $2r - 2$ E-ideals, which are determined by the train polynomials $f_k(x) = (x^2 - x)(x - \frac{1}{2})^k$ and $g_k(x) = (x^3 - x^2)(x - \frac{1}{2})^k$ for $k = 0, \ldots, r - 2$.*

Proof: Let $a = e+n = e+u+z \in A$. Then $q_k(a) = -\frac{1}{2}z^{k+2}+n^{k+3}$. As N is nil of index r we have for $k + 2 = r$, that is, $k = r - 2$, $q_{r-2}(a) = (a^3 - a^2)(a - \frac{1}{2})^{r-2} = 0$. Then (A, ω) is a train algebra with train equation $(x^3 - x^2)(x - \frac{1}{2})^{r-2} = 0$. Then A has at most $2r - 2$ E-ideals, corresponding to the train polynomials described above, according to our theorem 1. ■

Remark 1: The above proved fact (a Bernstein algebra not necessarily of finite dimension, with kernel nil of index r, is a train algebra of rank $\leq r - 2$) improves the similar result, obtained by Koulibaly and Ouattara [8], saying that under the same hypothesis, A is a train algebra of rank $\leq 2r - 3$.

Remark 2: In a Bernstein algebra $A = Fe \oplus U_e \oplus Z_e$ we have $E_A(1,-1) = (U_eZ_e + Z_e^2) \oplus Z_e$, from which we deduce that $\dim(U_eZ_e + Z_e^2)$ is an invariant of A. As a consequence, we have $E_A(1,-1) = N$ if and only if $U_eZ_e + Z_e^2 = U_e$. An important and not solved question in the theory of Bernstein algebras is to try to describe all E-ideals, generalizing the expression $E_A(1,-1) = (U_eZ_e + Z_e^2) \oplus Z_e$.

PROPOSITION 6 *Let A be an exceptional Bernstein algebra, not necessarily finite dimensional. Then*

a) The subspace U_e is an ideal of A

b) $N^k \subseteq U_e$ for all $k \geq 2$

c) If $a = e + u + z \in A$ then $a^k = e + u_k$ for all $k \geq 2$ where u_k is some element in U_e.

Proof: a) Follows from $U_e^2 = 0$, $U_eZ_e \subseteq U_e$ and the fact that U_e is invariant under multiplication by e.

b) Use induction on $k \geq 2$. In fact, $N^2 = (U_eZ_e + Z_e^2) \oplus U_e^2 = U_eZ_e + Z_e^2 \subseteq U_e$. Now if $N^k \subseteq U_e$ then $N^{k+1} = N^kN \subseteq U_eN \subseteq U_e$.

c) Follows immediatly from b) ■

PROPOSITION 7 *Suppose that A is an exceptional Bernstein algebra, not necessarily finite dimensional. Let the train polynomial $p(x) = x^n + \gamma_1\omega(x)x^{n-1} + \ldots + \gamma_{n-1}\omega(x)^{n-1}x$, where $\gamma_i \in F$. Then $E_A(p) \not\subseteq U_e$ if $\gamma_{n-1} \neq 0$ and $E_A(p) \subseteq U_e$ if $\gamma_{n-1} = 0$.*

Proof: We know from [1,Prop 1] that $E_A(p) \subseteq E_A(1,-1)$ for all $p(x)$. It rests to prove that if $\gamma_{n-1} \neq 0$ then $E_A(p) \not\subseteq U_e$ and if $\gamma_{n-1} = 0$ then $E_A(p) \subseteq U_e$. We consider $a = e + u + z \in A$ an arbitrary element of weight 1. Then $p(a) = (e+u_n)+\gamma_1(e+u_{n-1})+\ldots+\gamma_{n-1}(e+u+z) = u_n+\gamma_1 u_{n-1}+\ldots+\gamma_{n-1}u+\gamma_{n-1}z$. Then if $\gamma_{n-1} = 0$ we have $p(a) \in U_e$ and consequently $E_A(p) \subseteq U_e$. If $\gamma_{n-1} \neq 0$, $p(a) \notin U_e$. ■

PROPOSITION 8 *Let $A = Fe \oplus U_e \oplus Z_e$ be a Bernstein-Jordan algebra. Then its only E-ideals are $E_A(1,-1) = U_eZ_e \oplus Z_e$ and $E_A(1,-1,0) = 0$.*

Proof: Every Bernstein-Jordan algebra is train algebra satisfying $x^3 - x^2 = 0$ for all $x \in A$ of weight 1. This train equation $x^3 - x^2$ can be factored as $(x^2 - x)x$ and so by Theorem 1, there are in A two E-ideals $E_A(1,-1)$ and $E_A(1,-1,0)$. Clearly $E_A(1,-1,0) = 0$ and from (6) we have $E_A(1,-1) = U_eZ_e \oplus Z_e$. ■

COROLLARY 2 *The zigotic algebra $Z(n+1,2)$ has only 2 E-ideals.*

Proof: This is a Bernstein-Jordan algebra. ■

COROLLARY 3 *Every normal Bernstein algebra, not necessarily finite dimensional, has only 2 E-ideals, namely $E_A(1,-1) = Z_e$ and $E_A(1,-1,0) = 0$.*

PROPOSITION 9 *Every nuclear Bernstein algebra is a train, satisfying the train equation $x^4 - \frac{3}{2}\omega(x)x^3 + \frac{1}{2}\omega(x)^2x^2 = 0$*

Proof: In any Bernstein algebra $A = Fe \oplus U_e \oplus Z_e$, the principal powers of every element $a = e+u+z$ are given by $a^2 = e+u+u^2+2uz+z^2$, $a^3 = e+u+u^2+2uz+\frac{1}{2}z^2+u^2z+2(uz)z+z^3$ and $a^4 = e+u+u^2+2uz+\frac{1}{4}z^2+\frac{3}{2}u^2z+z^3+(u^2z)z+z^4$. But if A is nuclear, we have $(u^2z)z = 0$ and $z^3 = 0$ (consequently $z^4 = 0$) as $u^2z, z^2 \in Z_e^2 \subseteq Ann(N)$. From this it is easy to see that $a^4 - \frac{3}{2}a^3 + \frac{1}{2}a^2 = 0$ ■

Remark: This result was already proved by Ouattara [13].

PROPOSITION 10 *Every nuclear Bernstein algebra has at most 3 E-ideals.*

Proof: ¿¿From Prop. 9 it follows that every element of weight 1 satisfies the train polynomial $x^4 - \frac{3}{2}x^3 + \frac{1}{2}x^2 = 0$ which can be factored as $(x^2-x)x(x-\frac{1}{2}) = 0$. By [1, Theorem 1] there are at most 4 E-ideals in A, namely $E_A(1,-1) = (U_eZ_e + Z_e^2) \oplus Z_e = [U_e^3 + (U_e^2)^2] \oplus U_e^2$, $E_A(1,-1,0)$, $E_A(1,-\frac{3}{2},\frac{1}{2})$ and $E_A(1,-\frac{3}{2},\frac{1}{2},0)$. Let as determine $E_A(1,-1,0)$ and, for this, take $a = e+u+z \in A$ of weight 1. Then $a^2 = e+u+u^2+2uz+z^2$, $a^3 = e+u+u^2+2uz+\frac{1}{2}z^2+u^2z+2(uz)z$ so that $a^3 - a^2 = -\frac{1}{2}z^2 + 2u^2z + 2(uz)z$ is in U_e and we observe that the first two summands belong to $Z_e^2 \subseteq U_e$. An the other hand, let $b_1 = e+z$ and $b_2 = e-z$

where $z \in Z_e$. We have then $(b_2^3 - b_2^2) + (b_1^3 - b_1^2) = -z^2$ and so for every $z \in Z_e$, $z^2 \in E_A(1,-1,0)$ and as $z_1 z_2 = \frac{1}{2}((z_1+z_2)^2 - z_1^2 - z_2^2)$ it results that $Z_e^2 = (U_e^2)^2 \subseteq E_A(1,-1,0)$. Moreover $\frac{1}{2}((a^3 - a^2) + \frac{1}{2}z^2 + 2u^2 z) = (uz)z \in E_A(1,-1,0)$ for all $u \in U_e$ and $z \in Z_e$. We have that $E_A(1,-1,0)$ is the sum of Z_e^2, that in a nuclear algebra is an ideal contained in U_0, and the ideal generated by the elements $(uz)z$, also contained in U_0. Finally it is easy to prove that $E_A(1,-\frac{3}{2},\frac{1}{2}) = E_A(1,-1)$. ■

At this point, we can pose the following problem: Is it possible to describe the ideal generated by all elements of the form $(uz)z$, $u \in U_e$ and $z \in Z_e$, in terms of U_e and Z_e?

COROLLARY 4 *The maximum number of E-ideals of the duplicate of a Bernstein algebra is 6.*

Proof: For every Bernstein algebra A, A^2 is nuclear and satisfies the train equation $x^4 - \frac{3}{2}w(x)x^3 + \frac{1}{2}w(x)^2x^2 = 0$, whose canonical antecedent is $X(X-1)X(X-\frac{1}{2})$. By the above corollary, A^D satisfies the train equation $x^5 - \frac{3}{2}w(x)x^4 + \frac{1}{2}w(x)^2x^3 = 0$ whose canonical antecedent is $X(X-1)X^2(X-\frac{1}{2})$ and this implies that the number of E-ideals is at most 6. ■

4 EXAMPLES

We will show now, by means of examples, that there exist infinite dimensional Bernstein algebras with few E-ideals and also algebras with an infinite number of E-ideals.

Example 1 : Let A be infinite dimensional vector space over a field of characteristic not 2, with a basis $\{c_0, c_1, \ldots\}$ and let $w: A \to F$ by defined by $w(c_0) = 1$ and $w(c_i) = 0$ for $i \geq 1$. We introduce in A the following multiplication:

$$c_0^2 = c_0,\ c_0c_1 = 0,\ c_0c_i = \frac{1}{2}c_i\ (i \geq 2),\ c_1c_i = c_{i+1},\ \text{other products are zero}$$

It is easy to see that A is a Bernstein algebra with Peirce decomposition $A = Fe \oplus U \oplus Z$ where $U = \langle c_2, c_3 \ldots \rangle$, $Z = \langle c_1 \rangle$. It is an exceptional one as $U^2 = 0$. Moreover the principal powers of $N = U \oplus Z$, which are ideals of A, are given by $N^2 = U = \langle c_2, c_3, \ldots \rangle$, $N^3 = \langle c_3, c_4, \ldots \rangle$ and recursively $N^k = \langle c_k, c_{k+1}, \ldots \rangle$. So we have an infinite chain of ideals, all distinct: $N \supset N^2 \supset N^3 \supset \ldots$. Using Proposition 4, we conclude that A has an infinite number of E-ideals.

Example 2 : Let as consider the infinite dimensional Bernstein algebra $A = Fe \oplus U \oplus Z$ where $U = \langle u_1, \ldots, u_i, \ldots \rangle$ and $Z = \langle z_1, \ldots, z_j, \ldots \rangle$, with multiplication given by

$$e^2 = e,\ eu_i = \frac{1}{2}u_i,\ u_i^2 = z_i\ (i = 1, 2, \ldots),\ \text{other products are zero}$$

This algebra has only 2 E-ideals which are $E_A(1,-1) = Z$ and $E_A(1,-1,0) = 0$. This follows immediatly from Corollary 3.

We remark that the estimation of the number of elements of $\mathcal{E}_A$, if A is Bernstein but not train, is a more difficult question.

REFERENCES

[1] Catalán A.; Costa R.; Mallol C., E-ideals in baric algebras:basic properties. Proc. Edinburgh Math. Soc. 40: 491-503 (1997)

[2] Catalán A.; Costa R.; E-ideals in train algebras. International Journal of Math., Game Theory, and Algebra 8 (1): 59-71 (1998).

[3] Costa R.; Guzzo Jr. H., Indecomposable baric algebras . Linear Algebra Aplications. 183: 223-236 (1993).

[4] Etherington I.M.H., Genetic algebras, Proc. Roy. Soc. Edinb. 59: 242-258 (1939).

[5] Etherington I.M.H., Commutative train algebras of rank 2 and 3, J. London Math.Soc. 15: 136-149 (1940).

[6] Etherington I.M.H., Duplication of linear algebras, Proc. Edinb. Math. Soc. (2)6: 222-230 (1941).

[7] Gonzalez S. ; Martinez C., Idempotent elements in a Bernstein algebra, J.London Math. Soc. (2)42: 430-436 (1990).

[8] Koulibaly A.; Ouattara M., Structure et théorie de Frattini d'une algèbre de Bernstein, Algebras, Groups and Geometries 8: 145-158 (1991).

[9] Holgate P., Genetic algebras satisfying Bernstein's stationarity principle. J. London Math. Soc. 9: 613-623 (1975).

[10] Lyubich Ju. I., Mathematical Structures in Population Genetics. Springer-Verlag, (1993).

[11] Micali A.; Ouattara M., Dupliquée d'une algébre et le théorème d'Etherington. Linear Algebra Aplications 153: 193-207 (1991).

[12] Odoni R.W.K.; Straton A.E., Structure of Bernstein algebras in Algèbres Génétiques, ed. A. Micali. Hermann, Paris, (1987).

[13] Ouattara M., Sur les algèbres de Bernstein qui sont des T-algèbres. Linear Algebra Aplications 148: 171-178 (1991).

[14] Walcher S., Bernstein algebras which are Jordan algebras. Arch. Math. 50: 218-222 (1988).

[15] Worz A., Algebras in genetics, Lecture Notes in Biomath. 36, Springer-Verlag, Berlin. (1980).

Octonions and affine connections on spheres

ALBERTO ELDUQUE* Departamento de Matemáticas, Facultad de Ciencias, Universidad de Zaragoza, 50009 Zaragoza, Spain.

HYO CHUL MYUNG† Korea Institute for Advanced Study and KAIST, School of Mathematics, 207-43 Cheongryangri-dong Dongdaemun-gu, Seoul 130-012, Korea.

ABSTRACT: The interplay of the theory of octonions with affine connections on the spheres S^6, S^7 and S^{15} will be presented. Specifically, when these spheres are realized as the reductive homogeneous spaces $S^6 = G_2/SU(3)$, $S^7 = \mathrm{Spin}(7)/G_2$ and $S^{15} = \mathrm{Spin}(9)/\mathrm{Spin}(7)$, the structure of octonions can be effectively used to determine all left invariant affine connections on them. These connections on S^6 and S^7 are respectively obtained from a two parameters family of the compact vector color algebra and a one parameter family of the compact simple non-Lie Malcev algebra. If $\mathbb{O} = \mathbb{R} \oplus \mathbb{O}_0$ is the algebra of octonions (or Cayley algebra), then all Spin(9)-invariant affine connections on S^{15} are given by a three parameters family of $\mathbb{Z}_2$-graded products on $\mathbb{O}_0 \times \mathbb{O}$, with even part $\mathbb{O}_0$ and odd part $\mathbb{O}$.

1. INTRODUCTION.

Let G be a Lie group acting smoothly and transitively on a manifold M. Given any $p \in M$ and the isotropy subgroup at p: $H = \{g \in G : g \cdot p = p\}$, the homogeneous space $M \simeq G/H$ is said to be *reductive* in case there is a subspace $\mathfrak{m}$ of the Lie algebra $\mathfrak{g}$ of G such that

$$\mathfrak{g} = \mathfrak{h} \oplus \mathfrak{m} \tag{1}$$

(where $\mathfrak{h}$ is the Lie subalgebra of $\mathfrak{g}$ which corresponds to the closed subgroup H) and $(\mathrm{Ad}\, H)(\mathfrak{m}) \subseteq \mathfrak{m}$; so that $[\mathfrak{h}, \mathfrak{m}] \subseteq \mathfrak{m}$, and the converse is true if H is connected.

* Supported by the Spanish DGICYT (Pb 94-1311-C03-03) and by the Universidad de La Rioja (API-98/B15).

† Supported by KIAS Grant M98004.

Then the pairs (G, H) and $(\mathfrak{g}, \mathfrak{h})$ are called *reductive pairs*. Notice that for compact H, such a complement $\mathfrak{m}$ always exist (just take any complement of $\mathfrak{h}$ as a vector subspace of $\mathfrak{g}$ and "average" the corresponding projection operator).

An affine connection ∇ on a homogeneous space $M \simeq G/H$ is said to be *invariant* in case the left translation by any element of G is an affine transformation of ∇.

Nomizu ([1, Theorem 8.1]) proved in 1954 the following fundamental theorem for affine connections:

THEOREM 1. *Let $M \simeq G/H$ be a reductive homogeneous space with decomposition (1). Then the set of G-invariant affine connections on M is in bijection with the set of bilinear maps*

$$\alpha : \mathfrak{m} \times \mathfrak{m} \longrightarrow \mathfrak{m}$$

with $(\mathrm{Ad}\, h)(\alpha(x, y) = \alpha\big((\mathrm{Ad}\, h)(x), (\mathrm{Ad}\, h)(y)\big)$ for any $h \in H$ and any $x, y \in \mathfrak{m}$.

In other words, each invariant affine connection on M is determined by a bilinear product on $\mathfrak{m}$ which admits $\mathrm{Ad}\, H|_{\mathfrak{m}}$ as a subgroup of its group of automorphisms. This implies that $\mathrm{ad}\, \mathfrak{h}|_{\mathfrak{m}}$ is contained in the Lie algebra of derivations, and the converse is valid if H is connected.

Looking at $\mathfrak{m}$ as a module for the group H or the Lie algebra $\mathfrak{h}$ via the adjoint actions, Nomizu's Theorem asserts that the set of invariant affine connections on M is in bijection with the vector space $\mathrm{Hom}_H(\mathfrak{m} \otimes_{\mathbb{R}} \mathfrak{m}, \mathfrak{m})$, or with $\mathrm{Hom}_{\mathfrak{h}}(\mathfrak{m} \otimes_{\mathbb{R}} \mathfrak{m}, \mathfrak{m})$ if H is connected.

Nomizu proved this Theorem by identifying the tangent space T_pM of M at p with $\mathfrak{m}$ and by defining for each vector field $X \in \mathfrak{m} \subseteq \mathfrak{g}$ a vector field $\tilde{X}$ on a neighborhood of p in M, and then considered for each invariant affine connection ∇ on M the product given by

$$\alpha(X, Y) = \left(\nabla_{\tilde{X}} \tilde{Y}\right)_p .$$

Dealing in a more global way (see [2, Chapter IV]), one can consider for each left-invariant vector field $X \in \mathfrak{g}$ the global vector field X^+ defined on M by $X^+_m = \frac{d}{dt}(\exp tX) \cdot m$. Now, for any invariant affine connection ∇ on M we may identify $\mathfrak{m}$ with the tangent space T_pM by means of $X \mapsto X^+_p$ and associate with ∇ the product defined by

$$\alpha(X, Y) = \left(\nabla_{X^+} Y^+ - [X^+, Y^+]\right)_p . \tag{2}$$

This too gives a bijection as in Theorem 1.

The two basic tensors associated to any connection ∇, torsion and curvature, are then determined by their values at p (by invariance) and after the identification above of T_pM with $\mathfrak{m}$, they are given by the formulae:

$$\begin{aligned} T(X,Y) &= \alpha(X,Y) - \alpha(Y,X) - [X,Y]_{\mathfrak{m}} \\ R(X,Y)Z &= \alpha(X,\alpha(Y,Z)) - \alpha(Y,\alpha(X,Z)) \\ &\quad - \alpha([X,Y]_{\mathfrak{m}},Z) - [[X,Y]_{\mathfrak{h}},Z] \end{aligned} \tag{3}$$

where for any $X,Y \in \mathfrak{m}$, $[X,Y]_{\mathfrak{m}}$ and $[X,Y]_{\mathfrak{h}}$ denote the projections of the Lie bracket $[X,Y]$ onto $\mathfrak{m}$ and $\mathfrak{h}$, respectively, in the decomposition (1) and α is given by (2).

Two important invariant connections appear immediately associated to any reductive homogeneous space, the *canonical connection*, which corresponds to $\alpha(X,Y) = 0\ \forall X,Y \in \mathfrak{m}$, and the *natural connection*, which corresponds to $\alpha(X,Y) = \frac{1}{2}[X,Y]_{\mathfrak{m}}\ \forall X,Y \in \mathfrak{m}$. The natural connection is *symmetric*, that is, it has zero torsion.

According to Borel ([3, Theorem III] and [4, Théorème 3]) the only spheres that appear as homogeneous spaces in nonclassical ways are

$$S^6 = G_2/SU(3), \quad S^7 = \mathrm{Spin}(7)/G_2 \quad \text{and} \quad S^{15} = \mathrm{Spin}(9)/\mathrm{Spin}(7)\,. \tag{4}$$

The three of them are intimately related to the algebra of octonions, and this algebra can be used to determine (through Nomizu's Theorem) the invariant affine connections on them. This has been done in [5,6,7,8,9] and the aim of this work is to give a unified survey of these results.

Although much of the algebraic results on the algebras related to the octonions that will appear in what follows can be given in much more generality (for Cayley-Dickson algebras over arbitrary fields), we shall stick for simplicity to the real case. Thus the term algebra will always refer to a finite-dimensional algebra over $\mathbb{R}$.

We finish this introduction with a technical useful result. In all three homogeneous spaces in (4), $\mathfrak{m}$ in the decomposition (1) is unique by the following result (see [5, Proposition 2.3] or [7, Proposition 4]):

LEMMA 2. *Let $\mathfrak{h}$ be a simple subalgebra of a Lie algebra $\mathfrak{g}$ with* $\dim\mathfrak{g} < 2\dim\mathfrak{h}$. *Then there is a unique subspace $\mathfrak{m}$ of $\mathfrak{g}$ satisfying $\mathfrak{g} = \mathfrak{h}\oplus\mathfrak{m}$ and $[\mathfrak{h},\mathfrak{m}] \subseteq \mathfrak{m}$; namely, the orthogonal complement of $\mathfrak{h}$ relative to the Killing form of $\mathfrak{g}$.*

Proof: By Cartan's criterion, $\mathfrak{g} = \mathfrak{h} \oplus \mathfrak{m}$ for the orthogonal complement $\mathfrak{m}$ to $\mathfrak{h}$ relative to the Killing form. Let π be the projection of $\mathfrak{g}$ onto $\mathfrak{h}$ relative to this decomposition. In case $\mathfrak{g} = \mathfrak{h} \oplus \mathfrak{n}$ and $[\mathfrak{h},\mathfrak{n}] \subseteq \mathfrak{n}$ for some subspace $\mathfrak{n}$, $\pi(\mathfrak{n})$ is an $\mathfrak{h}$-submodule of $\mathfrak{h}$, which by simplicity and dimension count has to be trivial. Hence $\mathfrak{n} \subseteq \mathfrak{m}$ and $\mathfrak{n} = \mathfrak{m}$ since they both have the same dimension.

2. INVARIANT CONNECTIONS ON $S^6 = G_2/SU(3)$.

The first sphere to be considered is the six dimensional sphere S^6, realized as a quotient of the compact group of type G_2, which is the group of automorphisms of the algebra of octonions, by its subgroup $SU(3)$. To put this explicitly, some facts concerning the octonions will be recalled first.

The algebra of octonions can be defined as $\mathbb{O} = \mathbb{C} \oplus \mathbb{C}^3$ with the multiplication defined by

$$(\alpha + u)(\beta + v) = (\alpha\beta - \sigma(u,v)) + (\alpha v + \bar{\beta}u + u * v) \tag{5}$$

for any $\alpha, \beta \in \mathbb{C}$ and $u, v \in \mathbb{C}^3$, where $\sigma : \mathbb{C}^3 \times \mathbb{C}^3 \to \mathbb{C}^3$ is the usual hermitian form given by $\sigma((x_1, x_2, x_3), (y_1, y_2, y_3)) = x_1\bar{y}_1 + x_2\bar{y}_2 + x_3\bar{y}_3$, and the real product $*$ on $\mathbb{C}^3$ is defined by means of

$$\sigma(u, v * w) = \det(u, v, w)$$

for any $u, v, w \in \mathbb{C}^3$. The real algebra $(\mathbb{C}^3, *)$ is related to the so-called "color algebra" introduced in [10] (see [11]).

Given any $x = \alpha + u \in \mathbb{O}$, let $\bar{x} = \bar{\alpha} - u$. Then $t(x) = x + \bar{x}$ and $n(x) = x\bar{x}$ are real numbers and x satisfies the quadratic equation

$$x^2 - t(x)x + n(x)1 = 0\,. \tag{6}$$

Identifying $\mathbb{O}$ with $\mathbb{R}^8$, $n(x)$ turns out to be the euclidean norm in $\mathbb{R}^8$, hence $\mathbb{O}_0 = \{x \in \mathbb{O} : t(x) = 0\} = \mathbb{R}i \oplus \mathbb{C}^3$ can be identified with the euclidean space $\mathbb{R}^7$ and the six dimensional sphere S^6 lives inside $\mathbb{O}_0$ as $S^6 = \{x \in \mathbb{O}_0 : n(x) = 1\}$.

The group of automorphisms $\operatorname{Aut}\mathbb{O}$ is the compact Lie group of type G_2. Because of (6) it preserves the standard involution $x \mapsto \bar{x}$, the trace t and the norm n, so that it preserves $\mathbb{O}_0$ and $S^6 \subseteq \mathbb{O}_0$. Moreover, the action on S^6 is transitive (see [5, Theorem 6.5] or [12, exercise 6.9]).

Take the element $p = i \in S^6$. Any automorphism fixing i fixes $\mathbb{C}$ elementwise, so that it preserves $\mathbb{C}^3$, which is the orthogonal complement of $\mathbb{C}$ relative to n. By (5), it is an automorphism of the algebra $(\mathbb{C}^3, *)$, it is $\mathbb{C}$-linear on $\mathbb{C}^3$ and belongs to the unitary group. The definition of the product $*$ forces its determinant (over $\mathbb{C}$) to be 1. Hence it belongs to $SU(3)$, and conversely. Thus, the isotropy subgroup of the induced action $G_2 \times S^6 \to S^6$ at i is $SU(3)$ and $S^6 \simeq G_2/SU(3)$.

The Lie algebra of $G_2 = \operatorname{Aut}\mathbb{O}$ is $\mathfrak{g}_2 = \operatorname{Der}\mathbb{O}$, and the Lie algebra of the closed subgroup $SU(3)$ is $su(3) = \{d \in \operatorname{Der}\mathbb{O} : d(\mathbb{C}) = 0\}$. By Lemma 2, there is a unique complement of $su(3)$ invariant under the adjoint action of $su(3)$.

Now, by [13, III.8] we know that $\operatorname{Der}\mathbb{O}$ is the linear span of the maps

$$D_{x,y} = [L_x, L_y] + [L_x, R_y] + [R_x, R_y] = L_{[x,y]} - R_{[x,y]} - 3[L_x, R_y]\,,$$

where L_x and R_x denote the left and right multiplications by x. For $x, y \in \mathbb{O}_0$, the restriction $D_{x,y}|_{\mathbb{O}_0}$ is twice the map $d(x, y)$ considered in [5, (2.11)] and [6] (see [8, Lemma 6]).

For any $u \in \mathbb{C}^3$ $D_{i,u}(i) = i(ui) - ui^2 + i(iu) - i^2u + (iu)i - i^2u = 4u$. From this it follows that

$$\mathfrak{g}_2 = su(3) \oplus D_{i,\mathbb{C}^3},$$

and for any $D \in su(3)$ and $u \in \mathbb{C}^3$, $[D, D_{i,u}] = D_{D(i),u} + D_{i,D(u)} = D_{i,D(u)} \in D_{i,\mathbb{C}^3}$, so that $\mathfrak{m} = D_{i,\mathbb{C}^3} = su(3)^\perp$.

In $\mathbb{C}^3$ consider the new product given by $u \circ v = i(u * v)$. Actually, the map $(\mathbb{C}^3, \circ) \to (\mathbb{C}^3, *)$, $u \mapsto iu$ is an isomorphism of algebras. Now the identification of $\mathfrak{m}$ with the tangent space of S^6 at i, which is the orthogonal complement of $\mathbb{R}i$ in $\mathbb{O}_0$, that is $\mathbb{C}^3$, is given by

$$X = D_{i,u} \mapsto X_p^+ = \frac{d}{dt}|_{t=0}(\exp tD_{i,u})(i) = D_{i,u}(i) = 4u\,.$$

This is not just a vector space isomorphism (see [5, Proposition 4.2]):

PROPOSITION 3. *The linear map $(\mathfrak{m}, [\,,\,]_{\mathfrak{m}}) \to (\mathbb{C}^3, \circ)$, given by $D_{i,u} \mapsto 4u$, is an isomorphism of algebras and of $su(3)$-modules.*

Therefore, using that $SU(3)$ is connected and this Proposition we get

$$\mathrm{Hom}_{SU(3)}(\mathfrak{m} \otimes_{\mathbb{R}} \mathfrak{m}, \mathfrak{m}) \cong \mathrm{Hom}_{su(3)}(\mathbb{C}^3 \otimes_{\mathbb{R}} \mathbb{C}^3, \mathbb{C}^3)$$

and this is a real vector space of dimension 2, something that is checked by extending scalars up to $\mathbb{C}$ ([5, Section 5]). Since for any $\mu \in \mathbb{C}$, the map $\mathbb{C}^3 \times \mathbb{C}^3 \to \mathbb{C}^3$, $(u, v) \mapsto \mu(u \circ v)$ is $su(3)$-invariant, it follows that $\mathrm{Hom}_{su(3)}(\mathbb{C}^3 \otimes_{\mathbb{R}} \mathbb{C}^3, \mathbb{C}^3)$ is spanned over $\mathbb{C}$ by the product $\circ$. That is, it is spanned over $\mathbb{R}$ by the products $(u, v) \mapsto u \circ v$ and $(u, v) \mapsto i(u \circ v)$. It should be noticed here that the multiplication by i on $\mathbb{C}^3 \simeq T_iS^6$ corresponds to the classical invariant almost complex structure on S^6.

We have arrived at:

THEOREM 4. *There is a two parameters family of G_2-invariant affine connections on S^6.*

Through the isomorphism in Proposition 3, the formulae (3) for the torsion and curvature of the connection determined by the product $(u, v) \mapsto \mu(u \circ v)$ $(\mu \in \mathbb{C})$ are given by ([6, Theorem 9]):

$$\begin{aligned}
T(u, v) &= (2\mu - 1)(u \circ v),\\
R(u, v)w &= \left(|\mu|^2 - \frac{1}{4}\right)(u \circ (v \circ w) - v \circ (u \circ w))\\
&\quad - \left(\mu - \frac{1}{2}\right)(u \circ v) \circ w + n(v, w)u - n(u, w)v
\end{aligned}$$

where $n(u,v) = \frac{1}{2}(n(u+v) - n(u) - n(v))$ (the real part of $\sigma(u,v)$).

The only symmetric G_2-invariant affine connection is the natural one ($\mu = \frac{1}{2}$), and for it the curvature has a very simple expression: $R(u,v)w = n(v,w)u - n(u,w)v$ ([6, Corollary 10]). This natural connection is the usual Riemannian connection on S^6.

3. INVARIANT CONNECTIONS ON $S^7 = \mathrm{Spin}(7)/G_2$.

There is a triple product defined on the octonions with some nice properties, which is given by

$$\{x, y, z\} = (x\bar{y})z \tag{7}$$

for any $x, y, z \in \mathbb{O}$ (see [14] for the main features of this product). The group of automorphisms of this triple product turns out to be isomorphic to the compact group $B_3 = \mathrm{Spin}(7)$ under its spin representation. The action of this automorphism group is transitive on the seven dimensional sphere $S^7 = \{x \in \mathbb{O} : n(x) = 1\}$. There is a distinguished element in S^7, the unit element $1 \in \mathbb{O}$, and the isotropy subgroup at 1 is the subgroup of those automorphisms of the triple product that fix 1. From (7) with $y = 1$ it follows that this isotropy subgroup is exactly $G_2 = \mathrm{Aut}\,\mathbb{O}$.

The Lie algebra $\mathfrak{b}_3$ of B_3 is the Lie algebra of derivations of the triple product in (7), and the Lie subalgebra associated to the isotropy subgroup at 1 is the subalgebra of the derivations of $\mathbb{O}$ (these are the derivations of the triple product that annihilate 1).

¿From the alternative laws we have for any $u, x, y \in \mathbb{O}$:

$$\begin{aligned} L_u(xy) &= (L_u + R_u)(x)y - xL_u(y), \\ R_u(xy) &= -R_u(x)y + x(L_u + R_u)(y). \end{aligned} \tag{8}$$

Hence

$$\begin{aligned} (L_u + 2R_u)(xy) &= \mathrm{ad}_u(x)y + x(L_u + 2R_u)(y), \\ \mathrm{ad}_u(xy) &= (L_u + 2R_u)(x)y - x(2L_u + R_u)(y) \end{aligned} \tag{9}$$

where $\mathrm{ad}_u(x) = [u, x] = ux - xu$. For any $u \in \mathbb{O}_0$, let $E_u = L_u + 2R_u$. Also notice that $(2L_u + R_u)(\bar{x}) = -\overline{E_u(x)}$ for any $u \in \mathbb{O}_0$ and any $x \in \mathbb{O}$, since $\bar{u} = -u$. Hence, (9) implies that

$$\begin{aligned} E_u(\{x, y, z\}) = E_u((x\bar{y})z) &= \mathrm{ad}_u(x\bar{y})z + (x\bar{y})E_u(z) \\ &= (E_u(x)\bar{y})\,z - (x(2L_u + R_u)(\bar{y})) + (x\bar{y})E_u(z) \\ &= \{E_u(x), y, z\} + \{x, E_u(y), z\} + \{x, y, E_u(z)\} \end{aligned}$$

for any $u \in \mathbb{O}_0$ and any $x, y, z \in \mathbb{O}$. Thus, $E_u \in \mathfrak{b}_3$ for any $u \in \mathbb{O}_0$. Moreover, since $\{x, x, y\} = n(x)y$ for any $x, y \in \mathbb{O}$, it follows that $\mathfrak{b}_3 \subseteq o(\mathbb{O}, n)$, the orthogonal Lie algebra relative to the euclidean norm n; hence for any $D \in \mathfrak{b}_3$, $D(1) = u \in \mathbb{O}$ and

$D = (D - \frac{1}{3}E_u) + \frac{1}{3}E_u$, with $(D - \frac{1}{3}E_u)(1) = 0$, so that $D - \frac{1}{3}E_u \in \operatorname{Der}\mathbb{O} = \mathfrak{g}_2$. Therefore

$$\mathfrak{b}_3 = \mathfrak{g}_2 \oplus E_{\mathbb{O}_0}$$

and for any $D \in \mathfrak{g}_2$ and $u \in \mathbb{O}_0$, $[D, E_u] = [D, L_u + 2R_u] = L_{D(u)} + 2R_{D(u)} = E_{D(u)}$, so that $\mathfrak{m} = E_{\mathbb{O}_0}$ is the orthogonal complement to $\mathfrak{g}_2$ relative to the Killing form of $\mathfrak{b}_3$, because of Lemma 2.

Now, the identification of $\mathfrak{m} = E_{\mathbb{O}_0}$ with the tangent space to S^7 at 1: $T_1 S^7 = \mathbb{O}_0$, is given by

$$X = E_u \mapsto X_1^+ = \frac{d}{dt}|_{t=0}(\exp tE_u)(1) = E_u(1) = 3u\,.$$

For any $u, v \in \mathbb{O}_0$, [7, Lemma 2] shows that

$$\begin{aligned}[E_u, E_v] &= [L_u + 2R_u, L_v + 2R_v]\\ &= 2D_{u,v} - (L_{[u,v]} + 2R_{[u,v]}) = 2D_{u,v} - E_{[u,v]}\,.\end{aligned}$$

This immediately gives the following:

PROPOSITION 5. *The linear map* $(\mathfrak{m}, [\, ,\,]_{\mathfrak{m}}) \to (\mathbb{O}_0, -\frac{1}{3}[\, ,\,])$, *given by* $E_u \mapsto 3u$, *is an isomorphism of algebras and of* $\mathfrak{g}_2$*-modules.*

Notice that the algebra $(\mathbb{O}_0, -\frac{1}{3}[\, ,\,])$ is isomorphic to $(\mathbb{O}_0, [\, ,\,])$, which is a central simple non-Lie Malcev algebra. For any $x, y \in \mathbb{O}_0$, let us write $x \circ y = -\frac{1}{3}[x, y]$.

Using Proposition 5 and [7, Theorem 8] we get

$$\operatorname{Hom}_{G_2}(\mathfrak{m} \otimes_{\mathbb{R}} \mathfrak{m}, \mathfrak{m}) \cong \operatorname{Hom}_{\mathfrak{g}_2}(\mathbb{O}_0 \otimes_{\mathbb{R}} \mathbb{O}_0, \mathbb{O}_0)$$

and this last real vector space is spanned by the map given by $x \otimes y \mapsto [x, y]$ (or by $x \otimes y \mapsto x \circ y$). Then:

THEOREM 6. *There is a one parameter family of* Spin(7)*-invariant affine connections on* S^7.

Through the isomorphism in Proposition 5, the formulae for the torsion and curvature of the connection determined by the product given by $(u, v) \mapsto \mu(u \circ v)$ ($\mu \in \mathbb{R}$) are given, with the same computations as in [8, Theorem 7] but taking into account that $[u, v] = 2(u \times v)$ $\forall u, v \in \mathbb{O}_0$, by

$$\begin{aligned}T(u,v) &= (2\mu - 1)u \circ v,\\ R(u,v)w &= (1 - \mu - 2\mu^2)(u \circ v) \circ w + \frac{4}{3}(\mu^2 - 1)\big(n(u,w)v - n(v,w)u\big)\,.\end{aligned}$$

As for S^6, the only symmetric Spin(7)-invariant affine connection is the natural one ($\mu = \frac{1}{2}$) and for it the curvature tensor has the same very simple expression, since this is again the usual Riemannian connection on S^7.

4. INVARIANT CONNECTIONS ON $S^{15} = \mathrm{Spin}(9)/\mathrm{Spin}(7)$.

Let us consider now the nine dimensional euclidean space $\mathbb{R}^9 = \mathbb{R} \times \mathbb{O}$. Identify $\mathbb{O}$ with the subspace $0 \times \mathbb{O}$ and let $e = (1,0)$. Then $\mathbb{R}^9$ is the orthogonal direct sum of $\mathbb{R}e$ and $\mathbb{O}$. The linear map

$$\rho : \mathbb{R} \times \mathbb{O} \longrightarrow \mathrm{End}_{\mathbb{R}}(\mathbb{O} \times \mathbb{O}),$$
$$\alpha e + x \mapsto \begin{pmatrix} \alpha & L_x \\ L_{\bar{x}} & -\alpha \end{pmatrix}$$

where $\alpha \in \mathbb{R}$ and $x \in \mathbb{O}$ (that is $\rho(\alpha e + x)(u,v) = (\alpha u + xv, \bar{x}u - \alpha v)$), induces an isomorphism, also denoted by ρ, of the even Clifford algebra of $\mathbb{R}^9$ onto $\mathrm{End}_{\mathbb{R}}(\mathbb{O}\times\mathbb{O})$. The corresponding spin group Spin(9) will be identified in what follows with its image under ρ. It is not difficult to see that Spin(9) acts transitively on the sphere $S^{15} = \{(x,y) \in \mathbb{O} \times \mathbb{O} : n(x) + n(y) = 1\}$ (see [9] or [12, Theorem 14.79], although the way it is done in this last reference is slightly different).

By the well known triality principle (see [12, 14.19]) this allows to identify Spin(8) with the subgroup of Spin(9) consisting of the elements

$$\begin{pmatrix} \varphi_1 & 0 \\ 0 & \varphi_2 \end{pmatrix} \in \mathrm{End}_{\mathbb{R}}(\mathbb{O} \times \mathbb{O})$$

where φ_1, φ_2 are orthogonal transformations of $\mathbb{O}$ and there is a third orthogonal transformation φ_0 of $\mathbb{O}$ satisfying $\varphi_1(xy) = \varphi_0(x)\varphi_2(y)$ for any $x, y \in \mathbb{O}$.

Given an automorphism φ of the triple product in (7), we have for any $x, y \in \mathbb{O}$:

$$\varphi(xy) = \varphi(\{x,1,y\}) = \{\varphi(x), \varphi(1), \varphi(y)\} = \tilde{\varphi}(x)\varphi(y),$$

where $\tilde{\varphi}(x) = \varphi(x)\overline{\varphi(1)}$. Then $\varphi(1) = \tilde{\varphi}(1)\varphi(1)$, so that $\tilde{\varphi}(1) = 1$. Moreover, all this implies for any $x, y, z \in \mathbb{O}$ that

$$(\varphi(x)\overline{\varphi(y)})\varphi(z) = \varphi(\{x,y,z\}) = \varphi((x\bar{y})x) = \tilde{\varphi}(x\bar{y})\varphi(z),$$

so that $\tilde{\varphi}(xy) = \varphi(x)\overline{\varphi(\bar{y})}$ for any $x, y \in \mathbb{O}$. Let $\hat{\varphi}$ be defined by $\hat{\varphi}(x) = \overline{\varphi(\bar{x})}$. Then $\tilde{\varphi}(xy) = \varphi(x)\hat{\varphi}(y)$ for any $x, y \in \mathbb{O}$ and the group Spin(7), which is the group of automorphisms of the triple product, embeds in Spin(8) through

$$\varphi \mapsto \begin{pmatrix} \tilde{\varphi} & 0 \\ 0 & \hat{\varphi} \end{pmatrix}.$$

Let H be the isotropy subgroup of the action of Spin(9) on S^{15} at the point $(1,0) \in \mathbb{O} \times \mathbb{O}$. Then the arguments above show that, since $\tilde{\varphi}(1) = 1$, H contains a subgroup K isomorphic to Spin(7). Thus $\mathrm{Spin}(9)/K \to S^{15} : gK \mapsto g \cdot (1,0)$ is a covering map, and since S^{15} is simply connected it follows by dimension count that this is a bijection, so that $H = K \cong \mathrm{Spin}(7)$ (see [12, Theorem 14.79], or [9] for a proof in a wider context).

The Lie algebra $\mathfrak{b}_4$ of Spin(9) is graded over $\mathbb{Z}_2$ in a natural way:

$$\mathfrak{b}_4 = \mathfrak{g}_0 \oplus \mathfrak{g}_1 ,$$

where $\mathfrak{g}_0$ is the Lie algebra of Spin(8), spanned by the elements

$$\rho(x)\rho(y) - \rho(y)\rho(x) = \begin{pmatrix} L_x L_{\bar{y}} - L_y L_{\bar{x}} & 0 \\ 0 & L_{\bar{x}} L_y - L_{\bar{y}} L_x \end{pmatrix} \in \mathrm{End}_{\mathbb{R}}(\mathbb{O} \times \mathbb{O})$$

for $x, y \in \mathbb{O}$, and where $\mathfrak{g}_1$ is spanned by the elements

$$P_x = \rho(e)\rho(x) = \begin{pmatrix} 0 & L_x \\ -L_{\bar{x}} & 0 \end{pmatrix}$$

for $x \in \mathbb{O}$.

Notice that

$$\mathrm{span}\,\langle L_x L_{\bar{y}} - L_y L_{\bar{x}} : x, y \in \mathbb{O}\rangle = \mathrm{span}\,\langle L_{\bar{x}} L_y - L_{\bar{y}} L_x : x, y \in \mathbb{O}\rangle$$

is the orthogonal Lie algebra $o(\mathbb{O}, n)$. Now, by local triality ([13, Theorem 3.31]), for any $d \in o(\mathbb{O}, n)$, there are $d', d'' \in o(\mathbb{O}, n)$ such that for any $x, y \in \mathbb{O}$

$$d(xy) = d'(x)y + xd''(y) . \tag{10}$$

Using (8) it can be checked that $(L_x L_{\bar{y}} - L_y L_{\bar{x}})'' = L_{\bar{x}} L_y - L_{\bar{y}} L_x$, thus

$$\mathfrak{g}_0 = \left\{ M_d = \begin{pmatrix} d & 0 \\ 0 & d'' \end{pmatrix} : d \in o(\mathbb{O}, n) \right\} .$$

The (1,2)-entry of

$$\left[\begin{pmatrix} d & 0 \\ 0 & d'' \end{pmatrix}, \begin{pmatrix} 0 & L_x \\ L_{-\bar{x}} & 0 \end{pmatrix} \right]$$

is $dL_x - L_x d''$, which by (10) is exactly $L_{d'(x)}$. Hence

$$[M_d, P_x] = P_{d'(x)} \tag{11}$$

for any $d \in o(\mathbb{O}, n)$ and $x \in \mathbb{O}$.

The Lie subalgebra of the isotropy subgroup of the action of Spin(9) on S^{15} at the point $(1, 0)$ is

$$\mathfrak{h} = \{M_d : d \in o(\mathbb{O}, n) \text{ and } d(1) = 0\} = \{M_d : d \in o(\mathbb{O}_0, n)\} .$$

¿From [13, (3.79)] we have

$$o(\mathbb{O}, n) = \mathrm{Der}\,\mathbb{O} \oplus L_{\mathbb{O}_0} \oplus R_{\mathbb{O}_0} = \mathrm{Der}\,\mathbb{O} \oplus \mathrm{ad}_{\mathbb{O}_0} \oplus T_{\mathbb{O}_0} \tag{12}$$

where $T_x = \frac{1}{2}(L_x + R_x)$ for any $x \in \mathbb{O}$. Since for any $d \in \operatorname{Der}\mathbb{O} \oplus \operatorname{ad}_{\mathbb{O}_0}$, $d(1) = 0$, it follows from (12) that

$$o(\mathbb{O}_0, n) = \operatorname{Der}\mathbb{O} \oplus \operatorname{ad}_{\mathbb{O}_0} .$$

Now from the flexible law we get $[\operatorname{ad}_x, T_y] = T_{\operatorname{ad}_x(y)} = T_{[x,y]}$ for any $x, y \in \mathbb{O}$, so that

$$[d, T_x] = T_{d(x)} \tag{13}$$

for any $d \in o(\mathbb{O}_0, n)$ and $x \in \mathbb{O}_0$, and hence $T_{\mathbb{O}_0}$ is a subspace of $o(\mathbb{O}, n)$ invariant under the adjoint action of $o(\mathbb{O}_0, n) = \operatorname{Der}\mathbb{O} \oplus \operatorname{ad}_{\mathbb{O}_0}$. If we write

$$N_x = \begin{pmatrix} T_x & 0 \\ 0 & T_x'' \end{pmatrix} = \begin{pmatrix} T_x & 0 \\ 0 & \frac{1}{2}R_x \end{pmatrix} \quad \text{(because of (8),)}$$

we arrive at the decomposition

$$\mathfrak{b}_4 = \mathfrak{h} \oplus N_{\mathbb{O}_0} \oplus P_{\mathbb{O}} .$$

Besides, the subspace $\mathfrak{m} = N_{\mathbb{O}_0} \oplus P_{\mathbb{O}}$ is $\mathfrak{h}$-invariant, because of (11) and since $[M_d, N_x] = N_{d(x)}$ by (13) for any $d \in o(\mathbb{O}_0, n)$ and any $x \in \mathbb{O}_0$. From Lemma 2 we conclude that $\mathfrak{m} = N_{\mathbb{O}_0} \oplus P_{\mathbb{O}}$ is the orthogonal complement of $\mathfrak{h}$ relative to the Killing form of $\mathfrak{b}_4$.

The identification of $\mathfrak{m} = N_{\mathbb{O}_0} \oplus P_{\mathbb{O}}$ with the tangent space to S^{15} at $(1,0)$, $T_{(1,0)}S^{15} = \mathbb{O}_0 \times \mathbb{O}$ is given by

$$X = N_x + P_y \mapsto X^+_{(1,0)} = (N_x + P_y)(1,0) = \begin{pmatrix} T_x & L_y \\ -L_{\bar{y}} & \frac{1}{2}R_x \end{pmatrix} \begin{pmatrix} 1 \\ 0 \end{pmatrix} = \begin{pmatrix} x \\ -\bar{y} \end{pmatrix} .$$

As for the multiplication among the different elements involved, notice that

$$\begin{aligned} &[N_x, P_y] = P_{(T_x)'(y)} = \frac{1}{2}P_{xy} \quad \text{because of (11) and (8)}, \\ &[N_{x_1}, N_{x_2}] \in \mathfrak{h} \quad \text{since } [T_{x_1}, T_{x_2}](1) = 0, \\ &[P_{y_1}, P_{y_2}]_{\mathfrak{m}} = -N_{y_1\bar{y}_2 - y_2\bar{y}_1} \quad \text{(see [9])}, \end{aligned}$$

for any $x, x_1, x_2 \in \mathbb{O}_0$ and $y, y_1, y_2 \in \mathbb{O}$. From this we get immediately:

PROPOSITION 7. *The linear map*

$$\begin{aligned} (\mathfrak{m}, [\,,\,]_{\mathfrak{m}}) &\to (\mathbb{O}_0 \times \mathbb{O}, \diamond) : \\ N_x + P_y &\mapsto (x, -\bar{y}) , \end{aligned}$$

where

$$(x_1, y_1) \diamond (x_2, y_2) = (\bar{y}_2 y_1 - \bar{y}_1 y_2, \frac{1}{2}(y_1 x_2 - y_2 x_1)$$

for any $x_1, x_2 \in \mathbb{O}_0$, $y_1, y_2 \in \mathbb{O}$, *is an isomorphism of algebras and of* $\mathfrak{h} = o(\mathbb{O}_0, n)$-*modules, where* $\mathbb{O}_0$ *is given the natural structure of* $o(\mathbb{O}_0, n)$-*module, while* $\mathbb{O}$ *is the irreducible* $o(\mathbb{O}_0, n)$-*module given by the spin representation:* $d \cdot x = d''(x)$ *for any* $d \in o(\mathbb{O}_0, n)$ *and* $x \in \mathbb{O}$.

Using Proposition 7 one gets

$$\mathrm{Hom}_H(\mathfrak{m} \otimes_{\mathbb{R}} \mathfrak{m}, \mathfrak{m}) \cong \mathrm{Hom}_{\mathfrak{h}}\left((\mathbb{O}_0 \times \mathbb{O}) \otimes_{\mathbb{R}} (\mathbb{O}_0 \times \mathbb{O}), \mathbb{O}_0 \times \mathbb{O}\right),$$

and this last space is determined in [9]. It has dimension 3 and it is spanned by the products:

$$\begin{aligned}
(x_1, y_1) \diamond_0 (x_2, y_2) &= (\bar{y}_2 y_1 - \bar{y}_1 y_2, 0), \\
(x_1, y_1) \diamond_1 (x_2, y_2) &= (0, y_1 x_2), \\
(x_1, y_1) \diamond_2 (x_2, y_2) &= (0, y_2 x_1).
\end{aligned}$$

THEOREM 8. *There is a three parameters family of* Spin(9)-*invariant affine connections on* S^{15}.

Through the isomorphism given in Proposition 7 it is straightforward, although tedious, to compute the torsion and curvature tensors, as we did for S^6 and S^7, of the invariant affine connection determined by the product

$$\begin{aligned}
&(x_1, y_1) \star (x_2, y_2) \\
&\quad = \gamma_0 (x_1, y_1) \diamond_0 (x_2, y_2) + \gamma_1 (x_1, y_1) \diamond_1 (x_2, y_2) - \gamma_2 (x_1, y_1) \diamond_2 (x_2, y_2) \\
&\quad = (\gamma_0(\bar{y}_2 y_1 - \bar{y}_1 y_2), \gamma_1 y_1 x_2 - \gamma_2 y_2 x_1).
\end{aligned} \tag{14}$$

We will do it only for the torsion tensor:

$$\begin{aligned}
&T\left((x_1, y_1), (x_2, y_2)\right) \\
&\quad = (x_1, y_1) \star (x_2, y_2) - (x_2, y_2) \star (x_1, y_1) - (x_1, y_1) \diamond (x_2, y_2) \\
&\quad = \left((2\gamma_0 - 1)(\bar{y}_2 y_1 - \bar{y}_1 y_2), (\gamma_1 + \gamma_2 - \frac{1}{2})(y_1 x_2 - y_2 x_1)\right).
\end{aligned}$$

Hence there is a whole one parameter family of symmetric invariant affine connections: those determined by $\gamma_0 = \frac{1}{2} = \gamma_1 + \gamma_2$. The natural connection corresponds to $\gamma_0 = \frac{1}{2}$, $\gamma_1 = \gamma_2 = \frac{1}{4}$.

The usual Riemannian metric on S^{15} (which is invariant under the whole orthogonal group $O(16)$, so in particular it is Spin(9)-invariant) is given in $T_{(1,0)} S^{15} = \mathbb{O}_0 \times \mathbb{O}$ by $\langle (x_1, y_1), (x_2, y_2) \rangle = n(x_1, x_2) + n(y_1, y_2)$ (the usual scalar product on $\mathbb{R}^{15} = \mathbb{O}_0 \times \mathbb{O}$). According to [1, (13.1)] (where a minus sign is missing!) the corresponding Riemannian connection is determined by the product (through Proposition 7):

$$\frac{1}{2}(x_1, y_1) \diamond (x_2, y_2) + U\left((x_1, y_1), (x_2, y_2)\right)$$

where U is determined by

$$\begin{aligned}&2\langle U\left((x_1,y_1),(x_2,y_2)\right),(x_3,y_3)\rangle\\&\quad=\langle(x_1,y_1),(x_3,y_3)\diamond(x_2,y_2)\rangle+\langle(x_2,y_2),(x_3,y_3)\diamond(x_1,y_1)\rangle\,.\end{aligned}$$

Using that $n(ab,c)=n(b,\bar{a}c)=n(a,c\bar{b})$ and that $n(a,b)=n(\bar{a},\bar{b})$ for any $a,b,c\in\mathbb{O}$, it is easy to check that

$$U\left((x_1,y_1),(x_2,y_2)\right)=\left(0,\frac{3}{4}(y_1x_2+y_2x_1)\right)$$

whence the usual Riemannian connection on S^{15} corresponds to the product (14) with $\gamma_0=\frac{1}{2}$, $\gamma_1=1$ and $\gamma_2=\frac{1}{2}$ (see [9] for details).

REFERENCES

1 K. Nomizu. Invariant affine connections on homogeneous spaces. Amer. J. Math. 76:33-65, 1954.

2 D.V. Alekseevskij, V.V. Lychagin, A.M. Vinogradov. Geometry I: Basic Ideas and Concepts of Differential Geometry. In: R.V. Gamkrelidze (ed.), Encyclopaedia of Mathematical Sciences 28. Springer-Verlag, Berlin–Heidelberg 1991.

3 A. Borel. Some remarks about Lie groups transitive on spheres and tori. Bull. Amer. Math. Soc. 55:580-587, 1949.

4 A. Borel. Le plan projectif des octaves et les sphères comme espaces homogènes. Compt. Rendus Acad. Sci. Paris 230:1378-1380, 1950.

5 A. Elduque and H.C. Myung. Color algebras and affine connections on S^6. J. Algebra 149:234-261, 1992.

6 A. Elduque and H.C. Myung. Note on the Cayley algebra and connections on S^6. In: H.C. Myung, ed. Proc. Fifth International Conference on Hadronic Mechanics and Nonpotential Interactions (Part I), Cedar Falls 1990. Nova Science Pub., New York 1993, pp 139-147.

7 A. Elduque and H.C. Myung. The reductive pair (B_3,G_2) and affine connections on S^7. J. Pure Appl. Algebra 86:155-171, 1993.

8 A. Elduque and H.C. Myung. The compact Malcev algebra and torsion and curvature tensors of connections on S^7. In: S. González and H.C. Myung, eds. Nonassociative algebraic models, Zaragoza 1989. Nova Science Pub., New York 1992, pp 131-141.

9 A. Elduque and H.C. Myung. The reductive pair (B_4,B_3) and affine connections on S^{15}. In preparation.

10 G. Domokos and S. Kövesi-Domokos. The algebra of color. J. Math. Phys. 19:1477-1481, 1978.

11 A. Elduque and H.C. Myung. Colour algebras and Cayley–Dickson algebras. Proc. Royal Soc. Edinburgh. 125A:1287-1303, 1995.

12 F.R. Harvey. Spinors and Calibrations. Academic Press, San Diego 1990.

13 R.D. Schafer. An Introduction to Nonassociative Algebras. Academic Press, New York 1966.

14 A. Elduque. On a class of ternary composition algebras. J. Korean Math. Soc. 33:183-203, 1996.

Immediate applications of a new classification of finite dimensional simple Lie algebras

MOHAMMED EL HOUARI[1] Instituto de Matemática e Estatística – USP, Departamento de Matemática, Rua do Matão, 1010 – Cidade Universitaria, 05315-970 São Paulo–SP, Brasil, E–mail: elhouari@ime.usp.br

ABSTRACT We present three important applications of a new classification of simple Lie algebras. The first one consists in the determination of a scalar series $(C_q)_{q\geq 0}$ related to Casimir elements, associated with a simple Lie algebra. The second one consists in the determination of the spectral decomposition of the square tensor of a simple Lie algebra. Finally the third application consists in a simple characterization of Lie superalgebras of type $Q(n)$. We present also a brief presentation of the new classification mentioned above.

1 INTRODUCTION.

The classification of finite dimensional simple Lie algebras, inaugurated in [2] has been reconstitued using, on one hand, the calculus of C_3, the eigenvalue with which the cubic Casimir element acts on the Lie algebra, and on the other hand, the relationship of the scalar C_3 with the multiplicities of the elements of the spectrum of the quadratic Casimir element acting on the tensor square of the Lie algebra [4].

In section 2 we define generalized Casimir elements and give some of their properties. In particular, we introduce the corresponding scalar series $(C_q)_{q\geq 0}$.

In section 3 we introduce the classifying element of the new classification of simple Lie algebras, together with some of its properties. The aim of this section is to illustrate clearly how this classification is achieved. Some technical details have been omitted, nevertheless the interested reader should refer to the references [2, 3].

A natural curiosity leads us to determine the scalar series $(C_q)_{q\geq 0}$ of eigenvalues

[1]Research supported by *FAPESP, Brasil*, n^o : 97/0268 – 9.

along which the q-th order Casimir element of a simple Lie algebrs acts on itself (the action beeing induced from the adjoint action). This leads to a first application of the new classification. Section 4 is devoted to this.

As a second application we establish, in section 5, the spectral decomposition of the quadratic tensor of each simple Lie algebra relative to the quadratic Casimir element. In particular, we arrive at the well known decomposition into spectral components.

The third application of the new classification is given in section 6. It consists on the characterization of Lie superalgebras of type $Q(n)$.

Throughout this work, we consider a simple Lie algebra L over $\mathbf{C}$ with dimension N. We denote by $\mathcal{U}(L)$ its universal enveloping algebra. Let $(X_\alpha)_{\alpha \in I}$ be a fixed basis of L, I being a set of N indices. Finally Einstein's rule of summation over repeated indices is used.

2 SCALAR SERIES $(C_q)_{q\geq 0}$ ASSOCIATED WITH A SIMPLE LIE ALGEBRA.

As it is well known, the structure of the Lie algebra L is determined by its tensor of structure constants $(C^\gamma{}_{\alpha\,\beta})_{\alpha,\,\beta,\,\gamma\,\in\,I}$ given by:

$$[X_\alpha\ ,\ X_\beta]\ :=\ X_\gamma\ C^\gamma{}_{\alpha\,\beta} \tag{1}$$

satisfying the two relations :

$$C^\gamma{}_{\alpha\,\beta}\ =-\ C^\gamma{}_{\beta\,\alpha}\,\text{(skewsymmetry)} \tag{2}$$

$$C^s{}_{\alpha\,t}\,C^t{}_{\beta\,\gamma}\ +\ C^s{}_{\beta\,t}\,C^t{}_{\gamma\,\alpha}\ +\ C^s{}_{\gamma\,t}\,C^t{}_{\alpha\,\beta}\ =\ 0\quad\text{(Jacobi identity)} \tag{3}$$

The indices will be designated either by $x,\ y,\ z,\ \ldots,\ a,\ b,\ c,\ \ldots,\ \alpha,\ \beta,\ \gamma,$ $\ldots$ or $u,\ v,\ w, \ldots$

Let $g_{\cdot\,\cdot}$ be the metric tensor (Killing form):

$$g_{\alpha\beta} := \text{Trace}_L(adX_\alpha\ \circ adX_\beta) = C^v{}_{\alpha\,u}\,C^u{}_{\beta\,v} \tag{4}$$

$g_{\cdot\,\cdot}$ is symmetric and admits as a matrix the inverse $g^{\cdot\,\cdot}$ such that :

$$g_{\alpha\,\beta}\,g^{\beta\,\gamma}\ =\ \delta_\alpha{}^\gamma \tag{5}$$

By means of the metric tensor and its inverse, we can define the following tensors :

$$C^{\alpha\,\beta}{}_\gamma\ :=\ g^{\beta\,u}\,C^\alpha{}_{u\,\gamma},\quad C_{\alpha\,\beta\,\gamma} := g_{\alpha\,u}\,C^u{}_{\beta\,\gamma},\quad\text{and}\quad C^{\alpha\,\beta\,\gamma}\ :=\ g^{\gamma\,u}\,C^{\alpha\,\beta}{}_u \tag{6}$$

Finally, we have the duality relation :

$$C^\alpha{}_{\gamma\,\beta}\,C^{\lambda\,\gamma}{}_\mu\ =\ C^{\alpha\,\gamma}{}_\beta\,C^\lambda{}_{\gamma\,\mu} \tag{7}$$

Now in order to define the Casimir elements of higher orders (quadratic, cubic, quartic, ..., etc), we set:

$$\begin{aligned} g_{\alpha_1\,\alpha_2\,\cdots\,\alpha_q} &:= \text{Trace}_L\ (adX_{\alpha_1}\circ adX_{\alpha_2}\circ\ldots\circ adX_{\alpha_q}) \\ &= C^{u_q}{}_{\alpha_1\,u_1}\,C^{u_1}{}_{\alpha_2\,u_2}\ \cdots\ C^{u_{q-1}}{}_{\alpha_q\,u_q} \end{aligned} \tag{8}$$

For $q = 2$, we find again (4). We also define a tensor :

$$\begin{aligned} g^{\alpha_1\ \alpha_2\ \cdots\ \alpha_q} &:= \mathrm{Trace}_L\ (adX^{\alpha_1} \circ adX^{\alpha_2} \circ \cdots \circ adX^{\alpha_q}) \\ &= C^{\alpha_1\ u_1}{}_{u_q}\ C^{\alpha_2\ u_2}{}_{u_1}\ \cdots\ C^{\alpha_q\ u_q}{}_{u_{q-1}} \end{aligned} \tag{9}$$

DEFINITION 1 *The Casimir element of order q is the central element in the universal enveloping algebra given by :*

$$\widetilde{C_q} := g^{\alpha_1\ \alpha_2\ \cdots\ \alpha_q}\ X_{\alpha_1}\ X_{\alpha_2}\ \cdots\ X_{\alpha_q}\quad in\ \mathcal{U}(L) \tag{10}$$

$\widetilde{C_q}$ is independent of the choice of a basis in L. We associate with it the operator $\overline{C_q}$ given by:

$$\overline{C_q} := g^{\alpha_1\ \alpha_2\ \cdots\ \alpha_q}\ (adX_{\alpha_1} \circ adX_{\alpha_2} \circ \cdots \circ adX_{\alpha_q})\ \text{ in }\ \mathrm{End}_L(L) \tag{11}$$

which gives a $(q\ ,\ k)$–family of operators $\overline{C_q}^{\otimes k}$ lying in $\mathrm{End}_L(L^{\otimes k})$. Then, it follows from the Schur's lemma that the operators $(\overline{C_q})_{q\ \geq\ 0}$, given by (11), act as scalars on L. We denote by $(C_q)_{q\ \geq\ 0}$ the corresponding family of scalars given by:

$$\overline{C_q} = C_q\ Id_L\quad \text{for every } q \geq 0. \tag{12}$$

DEFINITION 2 *The scalars $(C_q)_{q\geq 0}$ in (12) constitute the scalar series of Casimir elements associated with the simple Lie algebra L.*

It will be shown in section 4 that the determination of the series $(C_k)_{k\geq 0}$ is completely obtained from the determination of the first four of such elements, that is to say for $q\ =\ 0,\ 1,\ 2,\ 3$.

DEFINITION 3 *The quadruple $(C_0\ ,\ C_1\ ,\ C_2\ ,\ C_3)$ is called the four–fold of L.*

In the following, we have to compute the *four-fold* defined above. The first three components of it are already known: $C_0 = N$ (= dimension of L), $C_1 = 0$ (by the semisimplicity of L) and $C_2 = 1$ (by Schur's lemma and normalization). For the computation of the fourth component C_3, we proceed as follows. First, we put $q = 3$ in (8) to (12), and we obtain respectively:

$$\begin{aligned} g_{\alpha\ \beta\ \gamma} &:= C^{w}{}_{\alpha\ u}\ C^{u}{}_{\beta\ v}\ C^{v}{}_{\gamma\ w} \\ &= g^{u\ u'}\ g^{v\ v'}\ g^{w\ w'}\ C_{\alpha\ u\ w'}\ C_{\beta\ v\ u'}\ C_{\gamma\ w\ v'} \end{aligned} \tag{13}$$

$$\begin{aligned} g^{\alpha\ \beta\ \gamma} &:= C^{\alpha\ u}{}_{w}\ C^{\beta\ v}{}_{u}\ C^{\gamma\ w}{}_{v} \\ &= g_{u\ u'}\ g_{v\ v'}\ g_{w\ w'}\ C^{\alpha\ u\ w'}\ C^{\beta\ v\ u'}\ C^{\gamma\ w\ v'} \end{aligned} \tag{14}$$

$$\widetilde{C_3}\ :=\ g^{\alpha\ \beta\ \gamma} X_\alpha\ X_\beta\ X_\gamma\ \text{ in } \mathcal{U}(L) \tag{15}$$

$$\overline{C_3}\ :=\ g^{\alpha\ \beta\ \gamma}\ (adX_\alpha \circ adX_\beta \circ adX_\gamma)\ \text{ in }\ \mathrm{End}_L(L) \tag{16}$$

The following theorem summarizes some of the properties and relations between these tensors:

THEOREM 1 ([4]) *We have the following relations :*

$$g_{\alpha\;\beta\;\gamma} = \frac{1}{2}\,C_{\alpha\;\beta\;\gamma},\quad g^{\alpha\;\beta\;\gamma} = \frac{1}{2}\,C^{\alpha\;\beta\;\gamma}\quad and\quad g^{\alpha\;\nu\;\beta}\,g_{\alpha\;\mu\;\beta} = -\,\frac{1}{4}\,\delta^{\nu}{}_{\mu} \quad (17)$$

$$\overline{C}_3 = -\frac{1}{4}\,Id_L,\quad that\ is\ to\ say\quad (\overline{C}_3)^{\mu}{}_{\lambda} := g^{\alpha\;\beta\;\gamma}\,C^{\mu}{}_{\alpha\;v}\,C^{v}{}_{\beta\;u}\,C^{u}{}_{\gamma\;\lambda} = -\frac{1}{4}\,\delta^{\mu}{}_{\lambda} \quad (18)$$

$$\widetilde{C}_3 = -\frac{1}{4}\,\widetilde{C}_2 \quad in\ \mathcal{U}(L) \quad (19)$$

REMARK 1 *An important application of (19) is that the cubic Casimir element acts as a scalar on any L-module* V *(not necessarly irreducible) whenever the quadratic Casimir element acts as a scalar on* V.

3 BRIEF PRESENTATION OF A NEW CLASSIFICATION OF FINITE DIMENSIONAL SIMPLE LIE ALGEBRAS.

3.1 System of eigenvalues associated to a simple Lie algebra.

To any simple Lie algebra L we shall associate a collection of scalars,which enables us to give a characterization of such Lie algebra.

Recall that gr$\mathcal{U}(L)$, by the Poincaré–Birkhoff–Witt theorem, is isomorphic to $\mathcal{S}(L)$, the symetric algebra associated with L. The corresponding isomorphism χ is given by the family $\chi_n : \mathcal{U}^n(L) \cong \mathcal{S}^n(L)$, $n \geq 0$, where $\mathcal{U}^n(L)$ is the set of homogeneous symetric elements of degree n, that is the set of canonical images of homogeneous symetric tensors of degree n in $\mathcal{U}(L)$. Moreover, $\chi(\mathcal{S}(L)^L) = \mathcal{Z}(L)$, $\mathcal{S}(L)^L$ denotes the elements of $\mathcal{S}(L)$ invariant under the adjoint action of L. In particular $\mathcal{Z}^2(L) = \chi(\mathcal{U}^2(L) \bigcap \mathcal{Z}(L))$, considered as an L–module, is isomorphic to $\mathcal{S}^2(L^*)$ (L being isomorphic to its dual L^*). Since this is in fact the space of non–degenerated L–invariant symetric bilinear forms, which is clearly one–dimensional according to Schur's lemma, it follows that the same holds for $\mathcal{Z}^2(L)$. Let w denote a generator of this space. w is called *the quadratic Casimir element associated with L*. We shall now study the action of w on $L^{\otimes 2}$, the tensor square of L. We denote by $\mathrm{w}^{\otimes 2}$ the corresponding operator. The operator $\mathrm{w}^{\otimes 2}$ stabilizes both the antisymetric and the symetric parts of $L^{\otimes 2}$ $(= \Lambda^2(L) \oplus \mathrm{S}^2(L))$, so that we have the corresponding decomposition $\mathrm{w}^{\otimes 2} = \mathrm{w}^{\wedge 2} \oplus \mathrm{w}^2$ (direct sum), where $\mathrm{w}^{\wedge 2}$ and w^2 are the corresponding operators to $\Lambda^2(L)$ respectively to $\mathrm{S}^2(L)$.

PROPOSITION 1 *1) The operator* $\mathrm{w}^{\otimes 2}$ *has the following linearization:*

$$\mathrm{w}^{\otimes 2} = 2\mathrm{Id}_{L^{\otimes 2}} - 2\xi \quad (20)$$

where $\xi \in \mathrm{End}_L(L^{\otimes 2})$ *is given by* $\xi(X \otimes Y) = -[X^\alpha, X] \otimes [X_\alpha, Y]$, *for* $X,\ Y \in L$, *that is:* $\xi^{\lambda\delta}_{\mu\nu} = -g^{\alpha\beta} C^{\lambda}{}_{\alpha\;\mu} C^{\delta}{}_{\beta\;\nu}$. *Here* ξ *is called the deployed Casimir element.*

2) The restriction of ξ *to* $\Lambda^2(L)$ *is given by:*

$$\overline{\xi} := \xi_{|\Lambda^2(L)} = \frac{\mathrm{u}}{2} \quad (21)$$

where u *is idempotent, whose image is isomorphic to* L.

3) The operator $\Lambda^2(L)$ decomposes as follows:

$$\mathrm{w}^{\wedge 2} = \mathrm{u} \oplus 2\mathrm{u}^{\perp} \tag{22}$$

where $\mathrm{u}^{\perp}$ is an element of $\mathrm{End}(\Lambda^2(L))$ *such that* $\mathrm{Id}_{|\Lambda^2(L)} = \mathrm{u} \oplus \mathrm{u}^{\perp}$.

Proof: 1) A straightforward computation gives:

$$\begin{aligned}
\mathrm{w}^{\otimes 2}(X_\mu \otimes X_\nu) &= g^{\alpha\,\beta} X_\alpha X_\beta (X_\mu \otimes X_\nu) \\
&= g_{\alpha\,\beta} X_\alpha([X_\beta\,,\,X_\mu] \otimes X_\nu \,+\, X_\mu \otimes [X_\beta\,,\,X_\nu]) \\
&= g_{\alpha\,\beta}[X_\alpha\,,\,[X_\beta\,,\,X_\mu]] \otimes X_\nu \\
&+ g_{\alpha\,\beta}[X_\beta\,,\,X_\mu] \otimes [X_\alpha\,,\,X_\nu] \\
&+ g_{\alpha\,\beta}[X_\alpha\,,\,X_\mu] \otimes [X_\beta\,,\,X_\nu] \\
&+ g_{\alpha\,\beta} X_\mu \otimes [X_\alpha\,,\,[X_\beta\,,\,X_\nu]]
\end{aligned}$$

Then the fact that w is normalized implies that the first and the fourth term of the last expression on the right side equal $X_\mu \otimes X_\nu$, while the other two terms equal $-\xi^{\lambda\delta}_{\mu\nu} X_\lambda \otimes X_\delta$.

2) By definition of the Lie algebra structure, L is injected in $\Lambda^2(L)$ via the L–homorphism $\iota \in \mathrm{Hom}(L, \Lambda^2(L))$. We also consider the L-homorphism $\jmath \in \mathrm{Hom}\,(\Lambda^2(L), L)$ given by: $\iota(X_\lambda) = (X_\mu \wedge X_\nu) C^{\nu\mu}_{\ \ \lambda}$ and $\jmath(X_\mu \wedge X_\nu) = X_\lambda C^{\lambda}_{\ \mu\nu}$. We set $\mathrm{u} = \jmath \circ \iota$, that is:

$$\mathrm{u}^{\lambda\delta}_{\mu\nu} = C^{\delta\lambda}_{\ \ \theta} C^{\theta}_{\ \mu\nu} = C^{\delta\lambda\theta} C_{\theta\mu\nu} \tag{23}$$

The operator u is an idempotent.

The restriction of ξ to $\Lambda^2(L)$ is given by $\overline{\xi} := \xi_{|\Lambda^2(L)} = \xi \circ \mathrm{P} = \mathrm{P} \circ \xi$, where P is the projector of $L^{\otimes 2}$ defined by $2\mathrm{P}(X \otimes Y) = X \otimes Y - Y \otimes X$. Thus we obtain:

$$\begin{aligned}
2\overline{\xi}^{\lambda\delta}_{\mu\nu} &= \overline{\xi}^{\lambda\delta}_{\mu\nu} - \overline{\xi}^{\lambda\delta}_{\nu\mu} \\
&= g^{\alpha\beta} C^{\lambda}_{\ \alpha\ \nu} C^{\delta}_{\ \beta\ \mu} - g^{uv} C^{\lambda}_{\ u\ \mu} C^{\delta}_{\ v\ \nu} \\
&= C^{\lambda\beta}_{\ \ \nu} C^{\delta}_{\ \beta\mu} - C^{\lambda v}_{\ \ \mu} C^{\delta}_{\ v\nu} \\
&= C^{\lambda\beta}_{\ \ \nu} C^{\delta}_{\ \beta\mu} - C^{\lambda\beta}_{\ \ \mu} C^{\delta}_{\beta\nu}\ .
\end{aligned}$$

Using the identity:

$$\begin{aligned}
-C^{\lambda\beta}_{\ \ \mu} C^{\delta}_{\beta\nu} &= -g_{\nu u} C^{\lambda\beta}_{\ \ \mu} C^{\ u\delta}_{\beta} \\
&= g_{\nu u}(C^{u\beta}_{\ \ \mu} C^{\ \delta\lambda}_{\beta} + C^{\delta\beta}_{\ \ \mu} C^{\ \lambda u}_{\beta}) \\
&= C^{\beta}_{\ \mu\nu} C^{\delta\lambda}_{\ \ \beta} + C^{\delta\beta}_{\ \ \mu} C^{\lambda}_{\ \nu\beta} \\
&= \mathrm{u}^{\lambda\delta}_{\mu\nu} + C^{\delta\beta}_{\ \ \mu} C^{\lambda}_{\ \nu\beta} \\
&= \mathrm{u}^{\lambda\delta}_{\mu\nu} - C^{\lambda}_{\ \beta\nu} C^{\delta\beta}_{\ \ \mu} \\
&= \mathrm{u}^{\lambda\delta}_{\mu\nu} - C^{\lambda\beta}_{\ \ \nu} C^{\delta}_{\ \beta\mu}
\end{aligned}$$

we obtain $2\overline{\xi}^{\lambda\delta}_{\mu\nu} = \mathrm{u}^{\lambda\delta}_{\mu\nu}$.

3) (22) follows immediately from the second part of the proposition. ♡

REMARK 2 *There is another proof of Proposition 1, based on graphic formalism (see [7]).*

Similarly, we introduce the operator $\varphi \in \mathrm{End}(\mathrm{S}^2(L))$ satisfying one of the following equivalent relations:

$$\mathrm{w}^2 = 2\mathrm{Id}_{\mathrm{S}^2} - 2\varphi\ , \quad \varphi = \xi_{|\mathrm{S}^2(L)} \text{ and } \xi = \frac{\mathrm{u}}{2} \oplus \varphi \tag{24}$$

REMARK 3 *The spectrum of φ, $\mathrm{Spec}(\varphi)$, consists of the scalars (of finite number) $\mathrm{T}_j := 1 - \frac{\mathrm{I}_j}{2}$, where the I_j are the elements of the spectrum of w^2. Clearly, there exists a unique T_j, say T_0, having the value 1, corresponding to $\mathrm{I}_0 = 0$, whose associated eigenspace is isomorphic to $\mathrm{Ker}(\mathrm{w}^2)$. This is just the trivial L-submodule of $\mathrm{S}^2(L)$, which is isomorphic to $\mathcal{Z}^2L$. But we know that the latter space is one-dimensonal, and is generated by the quadratic Casimir element* w. *Thus* $\mathrm{Spec}(\varphi) = \{1\} \bigcup (\{\mathrm{T}_j\}_{1 \le j \le \#\tau}$, *where* $\tau = \mathrm{Spec}(\varphi) \setminus \{1\}$.

Obviously, to any two isomorphic simple Lie algebras correspond the same collection τ. Conversely, two simple algebras with the same collection τ are necessarily isomorphic. This fact is not as obvious as the first statement, and further details may be found in chapters 4 and 5 of [2]. Thus we make the following defintion:

DEFINITION 4 *The colection τ, introduced above, is called the system of eigenvalues associated with a simple Lie algebra.*

3.2 The first basic formula associated with a simple Lie algebra.

Consider an endomorphism $\psi := -\xi$, where ξ is given by (20). Although the map ψ is never identically zero, its trace is null, moreover, $\mathrm{tr}_{L^{\otimes 2}}(\psi^2) = N\ (\neq 0)$. In fact $\psi^2(X_\lambda \otimes X_\delta) = \mathrm{g}^{\mu\ \nu}\mathrm{g}^{\alpha\ \beta}C^x{}_{\alpha\lambda}C^y{}_{\beta\delta}C^t{}_{\mu x}C^s{}_{\nu y}X_t \otimes X_s$, such that:

$$\mathrm{tr}_{\mathrm{L}^{\otimes 2}}(\psi^2) = \mathrm{g}^{\mu\ \nu}\mathrm{g}^{\alpha\ \beta}C^x{}_{\alpha\lambda}C^y{}_{\beta\delta}C^\lambda{}_{\mu x}C^\delta{}_{\nu y} = \mathrm{g}^{\mu\ \nu}\mathrm{g}^{\alpha\ \beta}\mathrm{g}_{\alpha\ \mu}\mathrm{g}_{\beta\ \nu} = \delta^\nu{}_\alpha\delta^\alpha{}_\nu = N \tag{25}$$

The formula (25) can even be generalized further if we write it in the form $\mathrm{tr}_{\mathrm{L}^{\otimes 2}}(\psi^2) = N \times 1$: we consider the sequence of successive compositions of ψ with itself, namely $\psi^{q+1} = \psi^q \circ \psi$, $q \ge 0$, and set $\psi^0 = \mathrm{Id}_{L^{\otimes 2}}$. Then we have the following proposition:

PROPOSITION 2 *For any integer $q \ge 0$, we have:*

$$\mathrm{Trace}_{\mathrm{L}^{\otimes 2}}(\psi^q) = N \times C_q \tag{26}$$

where the scalars C_q are given by (12).

<u>Proof</u>: The operator ψ may also be written as $\psi = \mathrm{ad}X^\lambda \otimes \mathrm{ad}X_\lambda$. Then the identity:

$$\begin{aligned}\mathrm{Trace}_{\mathrm{L}^{\otimes 2}}(\psi^q) = \mathrm{Trace}_{\mathrm{L}^{\otimes 2}}[(\mathrm{adX}^{\alpha_1} \circ \ldots \circ \mathrm{adX}^{\alpha_q}) \otimes (\mathrm{adX}_{\alpha_1} \circ \ldots \circ \mathrm{adX}_{\alpha_q})] \\ = g^{\alpha_1 \ldots \alpha_q}\ g_{\alpha_1 \ldots \alpha_q} \\ = \mathrm{Trace}_{\mathrm{L}}(g^{\alpha_1 \ldots \alpha_q}\ \mathrm{ad}X_{\alpha_1} \circ \ldots \circ \mathrm{ad}X_{\alpha_q}) \\ = \mathrm{Trace}_{\mathrm{L}}(C_q\ \mathrm{Id}_L) \\ = N \times C_q \quad \text{holds} \quad \heartsuit\end{aligned}$$

As the first application of (26), we obtain again $C_0 = N$, $C_1 = 0$, $C_2 = 1$ for $q = 0,\ 1,\ 2$ respectively.

Alternatively, we know that for any semisimple Lie algebra L and for any representation π (not necessarly irreducible) of L in a finite dimensional vector space V, the decomposition of V into its irreducible subspaces is associated with the diagonalizing the set of commuting operators $\pi(\mathrm{D})$ where D is in the center $\mathcal{Z}(L)$ of $\mathrm{U}(L)$. Morever, $\mathcal{Z}(L)$ is generated by a finite number of elements (equal to the rank of L). Here, we take $\mathrm{V} = L^{\otimes 2}$ and $\pi = \mathrm{ad}^{\otimes 2}$ to be the adjoint representation defined by the family of operators $(\Xi_\alpha)_{\alpha\in I}$, where $\Xi_\alpha := \mathrm{ad}X_\alpha \otimes \mathrm{Id}_L + \mathrm{Id}_L \otimes \mathrm{ad}X_\alpha$, which will be written simply in the form $\Xi_\alpha = X_\alpha \otimes \mathrm{Id}_L + \mathrm{Id}_L \otimes X_\alpha$ (in fact we will always denote the operator $\mathrm{ad}X$ by X, and $\mathrm{ad}X \circ \mathrm{ad}Y$ by XY, $X,\ Y \in L$). Then it easy to show that the operators ψ^q commute with the opertaors Ξ_α. Recall that for any simple Lie algebra L, $L^{\otimes 2}$ is not an irreducible L–module, but admits a decomposition into s irreducible L–submodules:

$$L^{\otimes 2} = L_1 \oplus L_2 \oplus \ldots \oplus L_s \tag{27}$$

Each component L_j appears as many times as its multiplicity.

Consider the partition of unity $(\mathrm{P}_j)_{1\le j\le s}$ in $\mathrm{End}(L^{\otimes 2})$, and $(\Xi^j_\alpha)_{\alpha\in I}$ the representations of L satisfying:

$$\begin{cases} \mathrm{Id}_{L^{\otimes 2}} = \sum_{j=1}^{j=s} \mathrm{P}_j \\ \mathrm{P}_j\mathrm{P}_i = \mathrm{P}_i\mathrm{P}_j = \delta_{i,j}\mathrm{P}_i, \quad 1 \le i,\ j \le s. \\ \Xi_\alpha \mathrm{P}_j = \mathrm{P}_j \Xi_\alpha = \Xi^j_\alpha, \quad \alpha \in I \text{ and } 1 \le\ j \le s. \\ \Xi_\alpha = \sum_{j=1}^{j=s} \mathrm{P}_j \Xi^j_\alpha \end{cases} \tag{28}$$

Since the Casimir elements of L act as scalars on any irreducible component of any L–module (in particular on those of $L^{\otimes 2}$), we obtain, in addition to $\mathrm{g}^{\alpha\,\beta}\mathrm{ad}X_\alpha \circ \mathrm{ad}X_\beta = \mathrm{Id}_L$:

$$\begin{cases} \mathrm{g}^{\alpha\,\beta}\Xi^j_\alpha\Xi^j_\beta = I_j \mathrm{Id}_{L_j} \ ,1 \le j \le s \ \text{ in } \mathrm{End}(L_j) \\ \mathrm{g}^{\alpha\,\beta}\Xi_\alpha\Xi_\beta = \sum_{j=s}^{j=1} I_j \mathrm{P}_j, \quad \alpha \in I \end{cases} \tag{29}$$

where I_j denotes the eigenvalue corresponding to the action of w on L_j.

As $\xi = -\psi$, (20) can also be written in the form:

$$\mathrm{g}^{\alpha\,\beta}\Xi_\alpha\Xi_\beta = 2\mathrm{g}^{\alpha\,\beta}X_\alpha \otimes X_\beta + (\mathrm{g}^{\alpha\,\beta}X_\alpha X_\beta) \otimes \mathrm{Id}_L + \mathrm{Id}_L \otimes (\mathrm{g}^{\alpha\,\beta}X_\alpha X_\beta) \tag{30}$$

This leads to an explicite expression of ψ:

$$2\psi = \mathrm{g}^{\alpha\,\beta}\Xi_\alpha\Xi_\beta - (\mathrm{g}^{\alpha\,\beta}X_\alpha X_\beta) \otimes \mathrm{Id}_L - \mathrm{Id}_L \otimes (\mathrm{g}^{\alpha\,\beta}X_\alpha X_\beta) \tag{31}$$

and by (28) we obtain:

$$\begin{cases} -\psi = \sum_{j=1}^{j=s} T_j \mathrm{P}_j \\ (-1)^q\psi^q = \sum_{j=1}^{j=s} (T_j)^q \mathrm{P}_j \end{cases} \tag{32}$$

for any integer $q \geq 0$, and where the $T_j = 1 - \frac{I_j}{2}$ are the eigenvalues of the operator $-\psi$ $(=\xi)$. The I_j are given by (29).

Then, by computing the traces in both sides of (32), and using (26), we get:

$$\begin{cases} (-1)^q C_q \times N = \sum_{j=1}^{j=s} (T_j)^q \mathrm{t}_j \ , \ q \geq 1 \\ N^2 = \sum_{j=1}^{j=s} \mathrm{t}_j \end{cases} \tag{33}$$

where C_q is given by (12) and $\mathrm{t}_j = \dim(L_j) = \mathrm{Trace}(\mathrm{P}_j)$.

More generally, for $V \otimes L$, where V is an irreducible L–module of dimension d, we similarly obtain the analogue of (33), that is:

THEOREM 2 *Given the data above, then, if $V \otimes L$ decomposes into s_V irreducible components: $V \otimes L = V_1 \oplus V_2 \oplus \ldots \oplus V_{s_V}$. Moreover, we have:*

$$\begin{cases} (-1)^q C_{V,\,q} \times d = \sum_{j=1}^{j=s_V} (T_{V,\,j})^q \mathrm{t}_{V,\,j} \ , \ q \geq 1 \\ qN \times d = \sum_{j=s_V}^{j=1} \mathrm{t}_{V,\,j} \end{cases} \tag{34}$$

where $C_{V,\,q}$ is the eigenvalue corresponding to the scalar action of the Casimir element of order q on V, $T_{V,\,j} = \frac{1}{2}(C_{V,\,2} + 1 - C_{V_j,\,2})$, and $\mathrm{t}_{V,\,j} = \dim V_j$. $\mathrm{t}_{V,j}$ is the multiplicity of $T_{V,\,j}$.

REMARK 4 *1) Formula (33) was obtained independently in [10].*

2) Fomula (33) is certainly true, for any integer $0 \leq q \leq 3$, for any L-module V (not necessarly irreducible) on which the quadratic Casimir element acts as a scalar.

3) Conjecture: Fomula (33) should be true for any L–module V (not necessarly irreducible) on which the quadratic Casimir element acts as a scalar, for any integer $q \geq 0$.

Formula (33) is of a great significance in the remainder of this paper. More precisely, it is its double reduction (see below) which plays the principle role.

First reduction of (33): We reduce (33) taking in account the skewsymetric subspace $\Lambda^2(L)$, using (21) and the normalization of the quadratic Casimir element. In fact, on one hand, the irreducibility of L as L–module implies that u is equal to one of the P_j in (28). The operator $\mathrm{u}^\perp$ is generally the sum of such operators, and on the other hand, the decomposition of $\psi_{|S^2(L)}$, relative to the partition of unity (28), always contains the factor $-$v, where v $\in \mathrm{End}(\psi_{|S^2(L)})$ is an idempotent equal to exactly one of the P_j, and whose image is the one–dimensional vector space generated by w, with the eigenvalue 0 (i.e 1 for ψ). Therefore, according to the previous discussion, we see that (33) first reduces to

$$\begin{cases} (-1)^q C_q \times N = \frac{N}{2^q} + 1 \sum_{j=1}^{j=s_1} (T_j)^q \mathrm{t}_j \ , \ q \geq 1 \\ \frac{(N-1)(N+2)}{2} = \sum_{j=s_1}^{j=1} \mathrm{t}_j \end{cases} \tag{35}$$

where s_1 is the number of non–irreducible components of $\mathrm{S}^2(L) \setminus \mathbf{C}.\mathrm{w}$, where $\mathbf{C}.\mathrm{w}$, is the one–dimensional vector space generated by w is obtained from the decomposition (27).

REMARK 5 *In order to find a relationship with the results of Section 2, it is necessary to reduce (33) further so as to take into account only the decomposition of $L^{\otimes 2}$ into its spectral components (which are not necessarly irreducible).*

Second reduction of (33): Let us define the equivalence relation T on the set of the irreducible components of $\mathrm{S}^2(L)$ other than the component Imv: $L_j \mathsf{T} L_k$ if and only if $I_j = I_k$. Thus, the L_j, $1 \leq j \leq s_1$, are regrouped modulo T into s_2 ($s_2 \leq s_1 \leq s$) classes $\overline{L}_j$, $1 \leq j \leq s_2$ given by:

$$\overline{L}_j = L_{j_1} \oplus \ldots \oplus L_{j_r}, \quad \text{with} \quad I_{j_1} = \ldots = I_{j_r} = I_j. \tag{36}$$

such that we have

$$\mathrm{S}^2(L) = \overline{L}_j \oplus \ldots \oplus \overline{L}_j \oplus \mathbf{C}.\mathrm{w} \tag{37}$$

Then, if $t_j = \dim \overline{L}_j = \sum_{k=1}^{k=r} n_{j_k}$, (35) and consequently (33) becomes:

$$\begin{cases} (-1)^q C_q \times N = \frac{N}{2^q} + 1 + \sum_{j=1}^{j=s_2} (T_j)^q \mathrm{t}_j \ , \ q \geq 1 \\ \frac{(N-1)(N+2)}{2} = \sum_{j=s_2}^{j=1} \mathrm{t}_j \end{cases} \tag{38}$$

Therfore we see that the eigenvalues T_j in (38) are the same as those introduced in Remark 3. So we can reformulate (38) in the following unified way:

THEOREM 3 *The elements of the collection τ and their multiplicities satisfy:*

$$\begin{cases} (-1)^q C_q = \frac{N}{2^q} \ + \ 1 \ + \ \sum_{\lambda \in \tau} \lambda^q \mathrm{t}_\lambda \ , \ q \geq 1 \\ \sum_{\lambda \in \tau} \mathrm{t}_\lambda = \frac{(N+2)(N-1)}{2} \end{cases} \tag{39}$$

where t_λ denotes the dimension of the eigenspace associated with λ.

REMARK 6 *Unlike (33), we cannot reduce (34) in order to obtain the analogue of (39), in the generalized case. Nevertheless, there exists a means of modifying (34), to overcome this difficulty (See [2, 3]). We will not need this generalization here.*

3.3 An overview of the characterization of a new classifying element and the classification theorem of simple Lie algebras.

Since s_2, the cardinal of τ, is less or equal to 3 [2], the system (39) yields, among other things, the multiplicities t_λ, $\lambda \in \tau$ as rational functions of the elements of τ. We recall that such dimensions were obtained independently in [2] , using different techniques. There, they are given by:

$$\mathrm{t}_\lambda = \frac{N(a\lambda - b)}{2\lambda^2} \prod_{\mu \in \mathrm{Spec}(\varphi) \setminus \{\lambda\}} \frac{1 - \lambda - \mu}{\lambda - \mu} \tag{40}$$

where $(a , b) = (-3 , -2)$ (respectively $= (3 , 1)$ if $\#\tau$ is even (respectively odd). Next we shall compare (40) with the expression (53), which will be established later using (39). We will see that (40) and (53), obviously, satisfy some complementarity which will be precised later, to enable us to understand certain phenomena that are analogous to singularities, for example encountered while using the methods of residues by the author of [2]. We shall examine the cases $s_2 = 1, 2, 3$, respectively. In each case, we will transfom (39) into a *Cramer system* after omitting the cases where singularities occur, that is when at least two or all the elements of τ coincide. The following concept will often be used:

DEFINITION 5 *A parasite eigenvalue is a scalar with the associated eigenspace identically null, and which, when added to the collection τ, does not contradict (38). We denote the completion of τ modulo the parasite eigenvalues by $\overline{\tau}$.*

Since we know that $(C_0 , C_1 , C_2 , C_3) = (N , 0 , 1 , -\frac{1}{4})$ (see Theorem 1), we obtain the following system from (38):

$$\begin{cases} \sum_{j=1}^{j=s_2} \mathrm{t}_j = \frac{(N-1)(N+2)}{2} \\ \sum_{j=1}^{j=s_2} T_j \mathrm{t}_j = -\frac{N+2}{2} \\ \sum_{j=1}^{j=s_2} (T_j)^2 \mathrm{t}_j = \frac{3N-4}{4} \\ \sum_{j=1}^{j=s_2} (T_j)^3 \mathrm{t}_j = \frac{N-8}{8} \end{cases} \tag{41}$$

1. **Case** $s_2 = 1$. Suppose $\tau = \{T_1\}$. In this case, (41) reduces to:

$$\begin{cases} \mathrm{t}_1 = \frac{(N-1)(N+2)}{2} \\ T_1 \mathrm{t}_1 = -\frac{N+2}{2}, \; (T_1)^2 \mathrm{t}_1 = \frac{3N-4}{4}, \; (T_1)^3 \mathrm{t}_1 = \frac{N-8}{8} \end{cases} \tag{42}$$

thus $N = 3$, and $(T_1 , \mathrm{t}_1) = (\frac{1}{2} , 5)$.

REMARK 7 *To study the cases $s_2 = 2$ and $s_2 = 3$, it is convenient to consider the system below, which is derived from (41):*

$$\begin{cases} \sum_{j=1}^{j=s_2} \mathrm{t}_j = \frac{(N-1)(N+2)}{2} & (43-1) \\ \sum_{j=1}^{j=s_2} (T_j - 1) \mathrm{t}_j = -\frac{N(N+2)}{2} & (43-2) \\ \sum_{j=1}^{j=s_2} (T_j - 1)^2 \mathrm{t}_j = \frac{5N}{4} & (43-3) \\ \sum_{j=1}^{j=s_2} (T_j - 1)^3 \mathrm{t}_j = -\frac{5N}{8} & (43-4) \end{cases} \tag{43}$$

2. **Case** $s_2 = 2$. Suppose $\tau = \{T_1 , T_2\}$. By subtitution in (43), we easily obtain:

PROPOSITION 3 *T_1 and T_2 are both different from 0 and $\frac{2}{3}$.*

Furthermore, the equations (43–2) and (43–3) yield:

$$\begin{cases} t_1 = \frac{2(N+2)T_2+5}{4(T_1-1)(T_1-T_2)} \times N \\ t_2 = \frac{2(N+2)T_1+5}{4(T_2-1)(T_2-T_1)} \times N \end{cases} \tag{44}$$

Then, by puting (44) into (43–4), we get:

$$4(N+2)\mathrm{P} + 10\mathrm{S} + 5 = 0 \tag{45}$$

where $\mathrm{P} = T_1 T_2$ and $\mathrm{S} = T_1 + T_2$.

Similarly, (43–1) gives:

$$(N-1)\mathrm{P} + \mathrm{S} + \frac{3N-4}{2(N+2)} = 0 \tag{46}$$

and by combining (45) and (46), we obtain:

$$\mathrm{S} = \frac{1}{6} \quad \text{and} \quad \mathrm{P} = -\frac{5}{3(N+2)} \tag{47}$$

Thus T_1 and T_2 satisfy the second degree equation $6T^2 - T - \frac{10}{N+2} = 0$. So we have:

$$T_1 = \frac{1-\sqrt{\Delta}}{12} \quad \text{and} \quad T_2 = \frac{1+\sqrt{\Delta}}{12} \tag{48}$$

$$\text{where} \quad \Delta = \frac{N+242}{N+2}$$

Whence, (44) becomes:

$$t_1 = \frac{3N(32+N+(N+2)\sqrt{\Delta})}{\Delta + 11\sqrt{\Delta}} \quad \text{and} \quad t_2 = \frac{3N(32+N-(N+2)\sqrt{\Delta})}{\Delta - 11\sqrt{\Delta}} \tag{49}$$

A necessary condition for t_1 and t_2 to be non negative integers (since they represent dimensions) is that $\Delta = \frac{N+242}{N+2}$ is a square in $\mathbf{Q}$. Let $\Delta = \frac{a^2}{b^2}$, where $\frac{a}{b}$ is an irreducible rational fraction. Then, there is a non negative integer ζ such that $N + 242 = \zeta \times a^2$ and $N + 2 = \zeta \times b^2$. Thus we obtain:

$$240 = 3 \times 5 \times 2^4 = \zeta \times (a+b) \times (a-b) \tag{50}$$

Without loss of generality, we may choose a and b such that $a + b > a - b$. Moreover, we see that $a+b$ and $a-b$ must have the same parity. Suppose, for example, that $a+b$ and $a-b$ are both even ; set $a = c+d$ and $b = c-d$, where c and d are some integers. Then (50) becomes: $60 = 3 \times 5 \times 2^2 = \zeta \times c \times d$. We note that c and d (like a and b) have no commun factor ; moreover, they are of different parities. Therefore, $c \times d$ is even, and we deduce that ζ is a divisor of 30. We proceed similarly when $a + b$ and $a - b$ are odd. All the obtained values are regrouped in the folowing table:

N	0	1	3	8	14	28	47	52	78	96	119
ζ	2	3	5	10	16	30	1	6	80	2	1
a	11	9	7	5	4	3	17	7	2	13	19
b	1	1	1	1	1	1	7	3	1	7	11

Table 1.1

N	133	190	248	287	336	484	603	782	1081	1680	3479
ζ	15	48	10	1	2	6	5	16	3	2	1
a	5	3	7	23	17	11	13	8	21	31	61
b	3	2	5	17	13	9	11	7	19	29	59

Table 1.2

If, in addition, we apply the constraints given by (49), that is the fact that t_1 and t_2 are non negative integers, table 1 yields a first refinement given by table 2:

Type of L [KCC]	A_1	A_2	G_2	D_4	F_4	E_6	E_7	E_8
N	3	8	14	28	52	78	133	248
t_1	5	27	77	300	1053	2430	7371	27000
t_2	0	8	27	105	324	650	1539	3875
T_1	$-\frac{1}{2}$	$-\frac{1}{2}$	$-\frac{1}{3}$	$-\frac{1}{4}$	$-\frac{1}{6}$	$-\frac{1}{9}$	$-\frac{1}{12}$	$-\frac{1}{30}$
T_2	$\frac{2}{3}$	$\frac{1}{2}$	$\frac{5}{12}$	$\frac{1}{3}$	$\frac{5}{18}$	$\frac{1}{4}$	$\frac{2}{9}$	$\frac{1}{5}$
m	2	3	4	6	9	12	18	30

Table 2.1

Type of L [KCC]	?	?	?
N	190	484	1081
t_1	15504	107892	559911
t_2	2640	9477	24909
T_1	$-\frac{1}{18}$	$-\frac{1}{54}$	$-\frac{1}{114}$
T_2	$\frac{5}{24}$	$\frac{5}{27}$	$\frac{10}{57}$
m	24	54	114

Table 2.2 (Virtual cases)

For a final refinement of table 2, consisting of eliminating all the values of N in table 2, whose the "corresponding" types are designed by the symbol "?", the reader should refer to [2, 3].

We can find again t_1 and t_2, already given by (49), in a more elegant form using (43–1) and (43–2), as follows:

$$\begin{cases} t_1 = \frac{(N-1)\sqrt{\Delta}-(N+11)}{4\sqrt{\Delta}} \times (N+2) \\ t_2 = \frac{(N-1)\sqrt{\Delta}+(N+11)}{4\sqrt{\Delta}} \times (N+2) \end{cases} \tag{51}$$

Then, applying (48), we arrive at all pairs $(T_1\ ,\ T_2)$ partially defining the system of eigenvalues with two elements (see table 2). Thereby, we introduce a parameter m, which is non negative, in such a way to present the system τ in the following unified form:

THEOREM 4 *Suppose that $\#\tau = 2$ for a simple Lie algebra of dimension N, then, there exists a parameter m, which is a non negative integer, such that:*

$$\begin{cases} T_1 = -\frac{1}{m} \text{ and } T_2 = -\frac{m+6}{6m} \\ \mathrm{t}_1 = \frac{5m^2(2m+3)(5m-6)}{(m+6)(m+12)} \text{ and } \mathrm{t}_2 = \frac{270m^2(m-2)(m+1)}{(m+6)^2(m+12)} \\ N = \frac{2(m+1)(5m-6)}{m+6} \end{cases} \tag{52}$$

REMARK 8 *1) We see from table 2 that the case " $\mathrm{s}_2 = 1$" is included in the case "$\mathrm{s}_2 = 2$". This is achieved by adjoining the parasite eigenvalue $\frac{2}{3}$ (compare with Proposition 3).*

2) The constraints imposed by (51) are not enough to limit the system of eigenvalues with two elements ; table 2 can be improved further. A final refinement would require the introduction of further concepts introduced in [2, 3].

3) Unlike (41), (40) can not be used to characterize even the case where $\#\tau = 1$!

4) Although (40) and (41) present a complementarity, they are not enough to give additional constraints to refine table 2 further. Nevertheless, their interaction cannot be avoided in the case $\mathrm{s}_2 = 3$.

3. **Case $\mathrm{s}_2 = 3$:** Suppose $\tau = \{T_1\ ,\ T_2\ ,\ T_3\}$. Resolving the system (43–1, 2, 3) gives:

$$\begin{cases} \mathrm{t}_1 = -\frac{4(N+2)T_2T_3+10(T_2+T_3)+5}{8(T_1-1)(T_2-T_1)(T_3-T_1)} \times N \\ \mathrm{t}_2 = -\frac{4(N+2)T_1T_3+10(T_1+T_3)+5}{8(T_2-1)(T_1-T_2)(T_3-T_2)} \times N \\ \mathrm{t}_3 = -\frac{4(N+2)T_1T_2+10(T_1+T_2)+5}{8(T_3-1)(T_1-T_3)(T_2-T_3)} \times N \end{cases} \tag{53}$$

As an immediate consequence of (53), one can see from (45) that the case "$\mathrm{s}_2 = 2$" is itself included in the case "$\mathrm{s}_2 = 3$". Actually, as soon as we assume that one of the components of the triple $(T_1\ ,\ T_2\ ,\ T_3)$ is a parasite eigenvalue, the two remaining components automatically satisfy (45). Indeed we will see in Proposition 4 that $\frac{1}{3}$ is the suitable parasite eigenvalue, which alows us to complete the system of eigenvalues of two elements.

Alternatively, (40) gives:

$$\begin{cases} t_0 = \frac{T_1T_2T_3}{(T_1-1)(T_2-1)(T_3-1)} \times N \\ t_1 = -\frac{(3T_1-1)(T_1+T_2-1)(T_1+T_3-1)}{2T_1(T_1-1)(T_1-T_2)(T_1-T_3)} \times N \\ t_2 = -\frac{(3T_2-1)(T_2+T_1-1)(T_2+T_3-1)}{2T_2(T_2-1)(T_2-T_1)(T_2-T_3)} \times N \\ t_3 = -\frac{(3T_3-1)(T_3+T_1-1)(T_3+T_2-1)}{2T_3(T_3-1)(T_3-T_1)(T_3-T_2)} \times N \end{cases} \tag{54}$$

But $t_0 = 1$; so we get:

$$(N-1)\sigma_3 - \sigma_2 - \sigma_1 - 1 = 0 \tag{55}$$

where $\sigma_k = \sigma_k(-T_1\ ,\ -T_2\ ,\ -T_3)$ is the k–th elementary symetric polynomial in $-T_1,\ \ -T_2$ and $-T_3$.

Similary, identifying the other equations in (53) with those in (54), and adding side–by–side the three resulting equations, we obtain:

$$12(N-1)\sigma_3 - 12\sigma_2 + 12\sigma_1 + \sigma_1^2 - 1 = 0 \tag{56}$$

which, together with (55), yields

$$4\sigma_1^2 + 24\sigma_1 + 11 = 0 \tag{57}$$

having roots $-\frac{11}{2}$ and $-\frac{1}{2}$. However, we reject the first root in accordance with a result in [2] ; hence we have:

THEOREM 5 *In the case* $s_2 = 3$, *where* $\tau = \{T_1\ ,\ T_2\ ,\ T_3\}$, *we get the following fundamental relation:*

$$T_1 + T_2 + T_3 = \frac{1}{2} \tag{58}$$

REMARK 9 *1) Relation (58) further describes the generic system of eigenvalues*
$\tau = \{T_1\ ,\ T_2\ ,\ T_3\} = \{T_1\ ,\ T_2\ ,\ \frac{1}{2} - T_1 - T_2\}$.
2) We see that $\frac{1}{2} \in \tau$ *if and only if* $N = 3$.
Moreover, if we omit this case, the identifications above, which gave (56), also yield the following unified equation:

$$\frac{T_j(T_j-1)(T_j+\frac{1}{2})}{(N-1)T_j+1} = c_\tau, \quad j = 1,\ 2,\ 3. \tag{59}$$

where c_τ *is a constant depending only on the system of eigenvalues* τ.

As a corollary to theorem 5, we have:

PROPOSITION 4 *1)* $T_1,\ T_2$ *and* T_3 *are each non zero.*
2) Besides the case of dimension 28, $\frac{1}{3}$ *is not an element of* τ.

Proof: 1) Suppose, on the contrary, that $T_3 = 0$ (say) is a permissible eigenvalue (i.e is not parasite); then (53) becomes:

$$\begin{cases} t_1 = -\frac{5(T_2+1)}{T_1(T_1-1)(T_1-T_2)} \times N \\ t_2 = -\frac{5(T_1+1)}{T_2(T_2-1)(T_2-T_1)} \times N \\ t_3 = \frac{4(N+2)T_1T_2+10(T_1+T_2)+5}{8T_1T_2} \times N \end{cases} \tag{60}$$

Thus (43–2) implies that $T_1 = -\frac{1}{2}$; whence $t_2 = 0$ (i.e T_2 is a parasite eigenvalue, $T_2 \neq 0$). Then, (53) may be reduced further to get

$$t_1 = \frac{5(T_2+1)}{3(2T_2-1)} \times N, \quad , \quad t_2 = 0 \text{ and } t_3 = \frac{N(N-3)}{2} \tag{61}$$

Since $T_1 + T_2 = -\frac{1}{2}$ ($\neq \frac{1}{6}$, compare with (47)), T_1 must be a parasite eigenvalue, which gives $T_2 = -\frac{1}{2}$ (recall that it also is a parasite eigenvalue). Then by (43–1), we have that $t_3 = \frac{N(N-3)}{2} = \frac{(N+2)(N-1)}{2}$, which gives $N = \frac{1}{2}$. This proves the first statement of the proposition.
2) Suppose, for example, that $T_3 = \frac{1}{3}$ is a permissible element of τ, and let $P = T_1T_2$ and $S = T_1 + T_2$. Then $S = \frac{1}{6}$ by (58). Moreover, by substituting $T_3 = \frac{1}{3}$ in the first relation of (55) ($t_0 = 1$), we get $P = -\frac{5}{3(N+2)}$. Thus S and P are exactly as given in (47), and this characterizes systems of eigenvalues having two elements. It then follows that one of the eigenvalues T_1, T_2 and T_3, must be parasite. A glance at table 2 shows that, with the exception of the case of dimension 28, $\frac{1}{3}$ is the parasite eigenvalue in question. When $N = 28$ (type D_4), $\frac{1}{3}$ is a permissible eigenvalue. If not, $\frac{1}{3}$ together with $\frac{2}{3}$ (the parasite eigenvalue in the case $\#\tau = 2$, see Proposition 3) forces the remaining eigenvalue to be equal to $-\frac{1}{2}$ and $N = 3$, a contradiction with $N = 28$. ♡

REMARK 10 *Using the fundamental relations (58), we see that the unified equation (59) guarantees that, the knowledge of two of the components of the quadruplet (T_1 , T_2 , T_3 , N) gives the two remaining components. In particular, if the dimension of a simple Lie algebra is fixed, the associated system of eigenvalues, including also the parasite ones, is completely determined by any one of its elements. It follows that the case $s_2 = 2$ is completely determined if the parameter N is given, since we already know in this case the partasite eigenvalue (i.e $\frac{1}{3}$).*

We recall that the quadratic Casimir element (respectively ξ) acts as a scalar on L ($\subset \Lambda^2(L)$) with the eigenvalue 1 (respectively $\frac{1}{2}$). Its associated eigenspace is a copy of L. So, it is legitimate to ask if $S^2(L)$ could also contain a copy of L, relative to the operator ξ. The answer to this question is no in general. More precisely, we have:

THEOREM 6 *If $\frac{1}{2} \in \tau$, then $\tau = \{\frac{1}{2}\ ,\ \frac{1}{m}\ ,\ -\frac{1}{m}\}$, where $|m| \in (\mathbb{N} + 4)$. In this case, $N = \dim(L) = m^2 - 1$*

Proof: Suppose, for example, $T_3 = \frac{1}{2}$ and let $P = T_1T_2$ and $S = T_1 + T_2$. Then, (53) and (54) become respectively:

$$\begin{cases} t_1 = \frac{(N+7)T_2+5}{2(T_1-1)(2T_1-1)(T_2-T_1)} \times N \\ t_2 = \frac{(N+7)T_1+T_3)+5}{2(T_2-1)(2T_2-1)(T_2-T_1)} \times N \\ t_3 = \frac{4(N+2)P+10S+5}{(2T_1-1)(T_2-1)} \times N \end{cases} \tag{62}$$

and

$$\begin{cases} t_0 = -\frac{P}{(T_1-1)(T_2-1)} \times N \quad , \quad t_1 = \frac{(3T_1-1)(S-1)}{2T_1(T_1-1)(T_2-T_1)} \times N \\ t_2 = \frac{(3T_2-1)(S-1)}{2T_2(T_2-1)(T_2-T_1)} \times N \quad , \quad t_3 = N \end{cases} \tag{63}$$

Comparing the relations in (62) to those in (63), we obtain:

$$\begin{cases} (N+1)P - S +_1 = 0 \\ (N+7)P - 6(S-1)T_1^2 + 5ST_1 + (1-S) = 0 \\ (N+7)P - 6(S-1)T_2^2 + 5ST_2 + (1-S) = 0 \\ (N+1)P + 3S + 1 = 0 \end{cases} \tag{64}$$

An immediate resolution of (64) implies that: $S = 0$, $P = -\frac{1}{N+1}$, $T_1^2 = T_2^2 = |P| = \frac{1}{N+1}$. This gives $T_1 = -T_2 = \frac{1}{m}$, with $N = m^2 - 1$, m being an integer satisfying the constraint of the theorem. ♡

Although the previously derived formulas give information about the nature of the systems of eigenvalues having two or three elements, they are incapable (except in the case where $\frac{1}{2} \in \tau$, treated in Theorem 6) to give an ultimate characterization. To do this, we would need to introduce further concepts. In order to keep things simple, we only consider here the partial classification of system of eigenvalues. Nevertheless, we recall the classification theorem stated below (as well as the summury of table 3, see below). The proof can be found in [2, 3].

THEOREM 7 (Classification Theorem, [2, 3, 7]) *Let L be a simple Lie algebra of dimension N over $\mathbf{C}$, and τ, the associated system of eigenvalues. Then one of the following cases holds:*

1. *Type $\mathcal{T}_1$) $\tau = \{-\frac{1}{2}\}$ and $N = 3$ $(m = 2)$.*

2. *Type $\mathcal{X}_\mathbf{m}$) $\tau = \{-\frac{1}{m}\ ,\ \frac{m+6}{6m}\}$, $m = 3,\ 4,\ 6,\ 9,\ 12,\ 18,\ 30$ and $N = \frac{2(m+1)(5m-6)}{m+6}$.*

3. *Type $\mathcal{Y}_\mathbf{m}$) $\tau = \{\frac{m+2}{2m}\ ,\ -\frac{2}{m}\ ,\ \frac{1}{m}\}$, $-m \in (\mathbb{N}+7) \cup \{5\}$ where $m \in (2\mathbb{N}+6)$ and $N = \frac{(m-1)(m-2)}{2}$.*

4. *Type $\mathcal{Z}_\mathbf{m}$) $\tau = \{-\frac{1}{m}\ ,\ \frac{1}{m}\ ,\ \frac{1}{2}\}$, $m \in (\mathbb{N}+4)$ and $N = m^2 - 1$.*

Moreover we have the following equivalences of types inside small dimensions:

$$\mathcal{X}_2 \cong \mathcal{T}_1 \;,\; \mathcal{Y}_{-3} \cong \mathcal{Y}_6 \;,\; \mathcal{Z}_4 \cong \mathcal{Y}_{-4} \;,\; \mathcal{Z}_2 \cong \mathcal{T}_1 \text{ and } \mathcal{Z}_3 \cong \mathcal{X}_3.$$

The following is an immediate consequence of Theorem 7:

PROPOSITION 5 *If a finite dimensional simple Lie algebra (over* $\mathbf{C}$*)* L *is such that* $\mathrm{S}^2(L)$ *admits an* L*-submodule isomorphic to* L*, then this* L*-submodule has multiplicity* 1 *and is of type* $\mathcal{A}_n$*, for some integer* $n \geq 2$.

The result in Proposition 5 was also proved independently in [9, 13]. In the following, there is the table[2]

EA–types	τ	**Domain of** m	**Dimension**
$\mathcal{T}_1$	$\{-\frac{1}{2}\}$		3
$\mathcal{X}_m$	$\{-\frac{1}{m}\,,\,\frac{m+6}{6m}\}$	3	8
		4	14
		6	28
		9	52
		12	78
		18	133
		30	248
$\mathcal{Y}_m$	$\{\frac{m+2}{2m}\,,\,-\frac{2}{m}\,,\,\frac{1}{m}\}$	$(\mathbb{N}+7)\cup\{5\}\cup(2\mathbb{N}+6)$	$\frac{(m-1)(m-2)}{2}$
$\mathcal{Z}_m$	$\{-\frac{1}{m}\,,\,\frac{1}{m}\,,\,\frac{1}{2}\}$	$\mathbb{N}+4$	m^2-1

Table 3

The correspondance between the two kinds of classification, that is between the **KCC** and the **EA**, is given by the following heuristic dictionary:

KCC–types	$\mathcal{A}_r$	$\mathcal{B}_r$	$\mathcal{C}_r$	$\mathcal{D}_r$
EA–types	$\mathcal{Z}_{r+1}$	$\mathcal{Y}_{-2r+1}$	$\mathcal{Y}_{2r+2}$	$\mathcal{Y}_{-2r+2}$
Domain of r	$r \geq 3$	$r \geq 3$	$r \geq 2$	$r \geq 3,\ r \neq 4$

Table 4.1 (Classical Lie algebras)

KCC–types	$\mathcal{G}_2$	$\mathcal{F}_4$	$\mathcal{E}_6$	$\mathcal{E}_7$	$\mathcal{E}_8$
EA–types	$\mathcal{X}_4$	$\mathcal{X}_9$	$\mathcal{X}_{12}$	$\mathcal{X}_{18}$	$\mathcal{X}_{30}$

Table 4.2 (Exceptional Lie algebras)

KCC–types	$\mathcal{A}_1$	$\mathcal{A}_2$	$\mathcal{D}_4$
EA–types	$\mathcal{T}_1$	$\mathcal{X}_3$	$\mathcal{X}_6$

Table 4.3 (Small dimensional Lie algebras)

4 APPLICATION I: Determination of the scalar series $(C_q)_{q\geq 0}$.

We begin by writing (32) in the form:

$$\prod_{k=1}^{s}(\psi + T_k) \tag{65}$$

[2]The *Killing–Cartan*'s classification (respectively the Angelopolous's one) is denoted along the paper by **KCC** (respectively by **EA**).

which gives, according to *Viète*'s formulas (relating the roots and the coefficients of a given polynomial):

$$\psi^s = \sum_{k=0}^{s-1} a_k^{(0)} \psi^k \tag{66}$$

where $a_0^{(0)} = 0$ and $a_k^{(0)} = (-1)^{s+1-k}\sigma_{s-k}(-T_1, \ldots, -T_s)$, with $\sigma_r(-T_1, \ldots, -T_s)$ denotes the r–th order elementary symmetric polynomial in $-T_1, \ldots -T_s$. Then, passing to the trace, the relation (26) in Proposition 2 gives:

$$C_s = \sum_{k=0}^{s-1} a_k^{(0)} C_k \tag{67}$$

The coefficients $a_k^{(0)}$ as in (66).

More generally, by multiplying (66) by the powers of ψ and after passage to the trace, we get:

$$C_{s+j} = \sum_{k=0}^{s-1} a_k^{(j)} C_k, \quad j \geq 1 \tag{68}$$

with the coefficients $a_k^{(j)}$ satisfy the following relations:

$$\begin{cases} a_0^{(j)} = a_{s-1}^{(j-1)} a_0^{(j-1)} \\ a_k^{(j)} = a_{k-1}^{(j-1)} + a_{s-1}^{(j-1)} a_k^{(j-1)} \\ j \geq 1 \text{ and } 1 \leq k \leq s-1 \end{cases} \tag{69}$$

Next, we use a double reduction of (67) and (68), as it was done for (33). To do this, we consider $\overline{\psi} := \psi + \frac{\mathrm{u}}{2} + \mathrm{v} = \sum_{j=0}^{s_1} T_j \mathrm{P}_j$, where u, v and P_j are the operators introduced in Subsection 3.2. Using the relations $\psi \circ \mathrm{u} = \mathrm{u} \circ \psi = -\frac{\mathrm{u}}{2}$ and $\psi \circ \mathrm{v} = \mathrm{v} \circ \psi = -\mathrm{v}$, we get:

$$\overline{\psi}^k = \psi^k + (-1)^{k+1}\left(\frac{\mathrm{u}}{2^k} + \mathrm{v}\right), \quad k \geq 1 \tag{70}$$

In particular we have $\overline{\psi}^{s_1+1} = \psi^{s_1+1} + (-1)^{s_1}\left(\frac{\mathrm{u}}{2^{s_1+1}} + \mathrm{v}\right)$. Moreover, the relation $\overline{\psi} = -\sum_{j=0}^{s_1} T_j \mathrm{P}_j$ gives:

$$\overline{\psi}^{s_1+1} = \sum_{k=0}^{s_1} b_k^{(0)} \overline{\psi}^k \tag{71}$$

where $b_0^{(0)} = 0$ and $b_k^{(0)} = (-1)^{s_1-k}\sigma_{s_1+1-k}(-T_1, \ldots, -T_{s_1})$ and $\sigma_r(-T_1, \ldots, -T_{s_1})$ denotes the r–th order elementary symmetric polynomial in $-T_1, \ldots -T_{s_1}$. We set by convention $b_{s_1+1}^{(0)} = -1$.

REMARK 11 *Unlike the coefficients $a_k^{(0)}$ in (66), $b_k^{(0)}$ in (71) take into consideration the eigenvalue 0, whose associated eigenspace is* $\mathrm{Im}(\mathrm{u}^{\perp})$ *(having dimension $\frac{N(N-3)}{2}$) and which is never zero except for the simple Lie algebra* sl(2).

Applying Remark 11, (66) becomes in terms of the coefficients $b_k^{(0)}$:

$$\psi^{s_1+1} = \sum_{k=0}^{s_1} b_k^{(0)} \psi^k + \sum_{k=0}^{s_1+1} (-1)^{k+1} b_k^{(0)} (\frac{u}{2^k} + v) \tag{72}$$

And by passing to the trace, we see that (67) reduces to

$$C_{s_1+1} = \sum_{k=0}^{s_1} b_k^{(0)} C_k + \sum_{k=0}^{s_1+1} (-1)^{k+1} b_k^{(0)} (\frac{1}{2^k} + \frac{1}{N}) \tag{73}$$

We now want to generalize (73) as was done with (69). First, observe that $\overline{\psi}$ satisfies the same formula as ψ, that is:

$$\overline{\psi}^{s_1+j+1} = \sum_{k=0}^{s_1} b_k^{(j)} \overline{\psi}^k, \quad j \geq 1 \tag{74}$$

where the $b_k^{(j)}$ satisfy relations similar to those given in (69), that is satisfied by the coefficients $a_k^{(j)}$. These are:

$$\begin{cases} b_0^{(j)} = b_{s_1}^{(j-1)} b_0^{(j-1)} \\ b_k^{(j)} = b_{k-1}^{(j-1)} + b_{s_1}^{(j-1)} b_k^{(j-1)} \\ j \geq 1 \text{ and } 1 \leq k \leq s_1 \end{cases} \tag{75}$$

By a repeated application of (70), (73) and (75) can be generalized, respectively, as follows. For $j \geq 1$:

$$\psi^{s_1+1+j} = \sum_{k=0}^{s_1} b_k^{(j)} \psi^k + \sum_{k=0}^{s_1} (-1)^{k+1} b_k^{(j)} (\frac{u}{2^k} + v) + (-1)^{s_1+j+1} (\frac{u}{2^{s_1+j+1}} + v) \tag{76}$$

$$C_{s_1+1+j} = \sum_{k=0}^{s_1} b_k^{(j)} C_k + \sum_{k=0}^{s_1} (-1)^{k+1} b_k^{(j)} (\frac{1}{2^k} + \frac{1}{N}) + (-1)^{s_1+j+1} (\frac{1}{2^{s_1+j+1}} + \frac{1}{N}) \tag{77}$$

Finally, applying the second reduction introduced in Section 3, (71) and (77) become as stated in the following theorem:

THEOREM 8 *For any simple Lie algebra over* $\mathbf{C}$ *of dimension* N*, with associated system of eigenvalues* τ *(containing* s_2 *elements), the associated scalar series* $(C_q)_{q\geq 0}$ *is given by:*

$$\begin{cases} C_{s_2+1} = \sum_{k=0}^{s_2} d_k^{(0)} C_k \; + \; \sum_{k=0}^{s_2+1} (-1)^{k+1} d_k^{(0)} (\frac{1}{2^k} + \frac{1}{N}) \\ C_{s_2+j+1} = \sum_{k=0}^{s_2} d_k^{(j)} C_k \; + \; \sum_{k=0}^{s_2} (-1)^{k+1} d_k^{(j)} (\frac{1}{2^k} + \frac{1}{N}) \; + \\ \qquad (-1)^{s_2+j+1} (\frac{1}{2^{s_2+j+1}} + \frac{1}{N}), \qquad j \geq 1 \end{cases} \tag{78}$$

where the $d_k^{(j)}$ *satisfy the following relations:*

$$\begin{cases} d_0^{(0)} = 0 \text{ and } d_k^{(0)} = (-1)^{s_2-k} \sigma_{s_2+1-k}(-T_1 \, , \, \ldots \, , \, -T_{s_2}) \\ d_0^{(k)} = d_{s_2}^{(k-1)} d_0^{(k-1)} \text{ and } d_j^{(k)} = d_{j-1}^{(k-1)} + d_{s_2}^{(k-1)} d_j^{(k-1)}. \end{cases} \tag{79}$$

$\sigma_r(-T_1\ , \ \dots\ , \ -T_{s_2})$ denotes the r-th order elementary symmetric polynomial in $-T_1\ , \dots\ , \ -T_{s_2}$.

Proof: To obtain formulas (78)-(79), we begin by writing $\overline{\psi} = -\sum_{j=1}^{s_2} T_j \overline{\mathrm{P}}_j$, where $\overline{\mathrm{P}}_j = \sum_{k,\ T_k = T_j} T_j \mathrm{P}_j$, and we proceed as it was done with (33) to produce (39) in Theorem 3. ♡

REMARK 12 *Since $s_2 = \#\tau$ does not exceed 3, for any finite dimensional simple Lie algebra [2], formulas (78)-(79) are sufficient to determine the series $(C_q)_{q\geq 0}$, from the quadruple $(C_0\ , \ C_1\ , \ C_2\ , \ C_3)$ (given in Theorem 1). An illustration of the determination of this series is given below.*

Examples of calculations: To do explicite calculations, it is sufficient to know the scalars $d_k^{(0)}$, $0 \leq k \leq s_2$; the other $d_k^{(j)}$, $0 \leq k \leq s_2$, $j \geq 1$ being given by the recurrence relations in (79). We shall put the results of these calculations in tables 5 to 7:

1. **Case $s_2 = 1$.** In this case, we have $d_0^{(0)} = 0$, $d_1^{(0)} = \frac{1}{2}$ and $d_2^{(0)} = -1$.

2. **Case $s_2 = 2$.** In this case, we have $d_0^{(0)} = 0$, $d_1^{(0)} = -\mathrm{P}$, $d_2^{(0)} = -\mathrm{S}$ and $d_3^{(0)} = -1$, S and P being, respectively, the sum and the product of the two elements of τ, and are given by (47). Using table 2, we get table 5:

Domain of m in $\mathcal{X}_m$	$d_0^{(0)}$	$d_1^{(0)}$	$d_2^{(0)}$	$d_3^{(0)}$
3	0	$\frac{1}{6}$	$-\frac{1}{6}$	-1
4	0	$\frac{5}{48}$	$-\frac{1}{6}$	-1
6	0	$\frac{1}{18}$	$-\frac{1}{6}$	-1
9	0	$\frac{5}{162}$	$-\frac{1}{6}$	-1
12	0	$\frac{1}{48}$	$-\frac{1}{6}$	-1
18	0	$\frac{1}{81}$	$-\frac{1}{6}$	-1
30	0	$\frac{1}{150}$	$-\frac{1}{6}$	-1

Table 5

3. **Case $s_2 = 3$.** Suppose $\tau = \{T_1\ , \ T_2\ , \ T_3\}$. In this case, we have $d_0^{(0)} = 0$, $d_1^{(0)} = -T_1T_2T_3$, $d_2^{(0)} = -(T_1T_2 + T_2T_3 + T_1T_3)$ and $d_3^{(0)} = -\frac{1}{2}$. Using table 3, we get table 6:

EA–types	$d_0^{(0)}$	$d_1^{(0)}$	$d_2^{(0)}$	$d_3^{(0)}$	$d_4^{(0)}$
$\mathcal{Z}_m$	0	$\frac{1}{2m^2}$	$\frac{1}{m^2}$	$-\frac{1}{2}$	-1
$\mathcal{Y}_m$	0	$\frac{m+2}{m^3}$	$-\frac{m+6}{2m^2}$	$\frac{1}{2}$	-1

Table 6

We leave it to the reader use tables 5 and 6 to do the necessary calculations. Nevertheless, for the sake of illustration, we give details here of such calculations for sl(2) and sl(3).

Case of sl(2). In this case, we have $d_0^{(0)} = 0$, $d_1^{(0)} = \frac{1}{2}$, $d_2^{(0)} = -1$ and the $d_1^{(j)}$

satisfying $d_0^{(j)} = 0$ and $d_1^{(j)} = (d_1^{(j)})^2$, such that $d_0^{(j)} = 0$ and $d_1^{(j)} = \frac{1}{2^{2^j}}, j \geq 0$. So (79) gives:

$$C_j = \frac{5}{3 \times 2^j} + (-1)^j(\frac{1}{2^j} + \frac{1}{3}), \quad j \geq 0 \tag{80}$$

Case of sl(3). In this case, we have $d_0^{(0)} = 0$, $d_1^{(0)} = -d_2^{(0)} = \frac{1}{6}$, $d_3^{(0)} = -1$ and the $d_j^{(j)}$, $j = 1, 2$ satisfying $d_0^{(j)} = 0$, $d_1^{(j)} = d_2^{(j-1)}d_1^{(j-1)}$ and $d_2^{(j)} = d_1^{(j-1)} + (d_2^{(j-1)})^2$. So (78) gives:

$$C_j = \frac{1}{8 \times 3^{j-3}} + (-1)^j(\frac{1}{2^{j-1}} + \frac{1}{8}), \quad j \geq 1 \tag{81}$$

REMARK 13 *It can be easily shown, as it was done above (i.e for the cases* sl(2) *and* sl(3)*), that for exceptional Lie algebras (i.e when* $s_2 = 2$*), equations (78)-(79) enable us to calculate the last component of the quadruplet* $(C_0\ ,\ C_1\ ,\ C_2\ ,\ C_3)$ *from the other three. This means, in some sense, that the type of such Lie algebras is not intrinsic, that is, it depends - in a sense to be specified - on another kind of Lie algebras (which are nothing but the classical ones). In fact, classically, we realize the exceptional Lie algebras as derivations of the so called classical Lie algebras; in which case, the knowledge of the above quadruplet is both necessary and sufficient for deducing the* $(C_q)_{q\geq 4}$.

5 APPLICATION II: Spectral decomposition of the tensor square of a simple Lie algebra.

Recall that the factors of this decomposition are the eigenspaces associated with the the spectrum of the operator $\mathrm{w}^{\otimes 2} = \mathrm{w}^{\wedge 2} \oplus \mathrm{w}^2)$. According to what follows (see Section 3), the general form of this decomposition of $L^{\otimes 2}$, deduced from (27), is:

$$L^{\otimes 2} = \mathrm{Imu} \oplus \mathrm{Imu}^{\perp} \oplus \overline{L}_1 \oplus \ldots \oplus \overline{L}_{s_2} \oplus \mathrm{Imv} \tag{82}$$

where the $\overline{L}_j$ are given by (36).

REMARK 14 *The L-module structure of the factors on the right-hand side of (82) being deduced from the one of* $L^{\otimes 2}$ *, implies that the above decomposition is equivalent to that obtained in terms of the underlying vector spaces. Therefore (82) can be rewritten in the form:*

$$L^{\otimes 2} = N^{\otimes 2} = N \oplus \frac{N(N-3)}{2} \oplus \mathrm{t}_1 \oplus \ldots \oplus \mathrm{t}_{s_2} \oplus 1 \tag{83}$$

with $\mathrm{t}_j = \dim \overline{L}_j$, $j = 1, \ldots, s_2$.

1. **Case of** $s_2 = 1$**.** Let, for example, $\mathrm{t}_2 = \mathrm{t}_3 = 0$. In this case (83) becomes:

$$\mathrm{sl}(2)^{\otimes 2} = 3^{\otimes 2} = 3 \oplus 5 \oplus 1 \tag{84}$$

2. **Case of** $s_2 = 2$**.** Let $\mathrm{t}_3 = 0$, for example, and (83) can then be easily interpreted using table 2. As an illustration, we have the following example:

$$\mathcal{G}_2{}^{\otimes 2} = 14^{\otimes 2} = 14 \oplus 77 \oplus 77 \oplus 27 \oplus 1 \tag{85}$$

3. **Case of $s_2 = 3$.** In this case, using (53) as well as table 3, the multiplicities can be summarized in the following table:

EA–types	t_1	t_2	t_3
$\mathcal{Z}_m$	$\frac{(m+3)(m-1)m^2}{4}$	$\frac{(m-3)(m+1)m^2}{4}$	m^2-1
$\mathcal{Y}_m$	$\frac{m(m-2)(m^2-1)}{24}$	$\frac{(m+1)(m-2)(m-3)(m-4)}{12}$	$\frac{(m-1)(m-4)}{2}$

Table 7

This allows us, with the help of table 4, to interprete (83). For the sake of illustration, we consider only simple Lie algebras of type A_r, $r \geq 3$, for which we have:

$$\begin{cases} m = r+1 \quad \text{and} \quad N = r(r+2) \\ t_1 = \frac{r(r+4)(r+1)^2}{4}, \ t_2 = \frac{(r^2-4)(r+1)^2}{4} \ \text{and} \ t_3 = r(r+2) \end{cases} \tag{86}$$

such that we achieve the following decomposition:

$$\begin{aligned} A_r^{\otimes 2} = (r(r+2))^{\otimes 2} = r(r+2) \oplus \frac{r(r-1)(r+2)(r+3)}{2} \oplus \\ \frac{r(r+4)(r+1)^2}{4} \oplus \frac{(r^2-4)(r+1)^2}{4} \oplus r(r+2) \oplus 1 \end{aligned} \tag{87}$$

REMARK 15 *The determination of the scalar series $(C_q)_{g\geq 0}$ can be done using the basic formula (39) (See Theorem 3) and with the help of tables 1 to 7. The manner of calculation presented in Section 4 is advantageous, in the sence that we only make use of the classifying element but the use of the manner mentioned above requires, in addition, the spectral decomposition described in Section 5.*

6 APPLICATION III: A new realization of Lie superagebras of type $\mathcal{Q}(n)$.

The definition of a Lie superalgebra, which is most suitable for our needs, is as follows:

DEFINITION 6 *A Lie superalgebra (over $\mathbf{C}$) consists of a $\mathbf{Z}_2$-graded vector space $L = \mathcal{G} \oplus V$, where $\mathcal{G}$ is a Lie algebra and V a $\mathcal{G}$-module; $\mathcal{G}$ and V are related by a homomorphism of $\mathcal{G}$-modules $\Theta : S^2(V) \longrightarrow \mathcal{G}$ satisfying the following relations:*

$$\begin{aligned} \Theta(v_1\,,\, v_2)(v_3) + \Theta(v_2\,,\, v_3)(v_1) + \Theta(v_3\,,\, v_1)(v_2) = 0 \\ \textit{where } (v_i\,,\, v_j) = \frac{1}{2}(v_i \otimes v_j \oplus v_j \otimes v_i) \textit{ in } V^{\otimes 2}. \end{aligned} \tag{88}$$

Recall that the Killing form Υ_L on the Lie superalgebra L is such that:

$$\Upsilon_L(X\,,\,Y) = \Upsilon_{\mathcal{G}}(X\,,\,Y) - \Upsilon_V(X\,,\,Y) \tag{89}$$

for all $X,\ Y \in \mathcal{G}, \Upsilon_{\mathcal{G}} = \Upsilon_{L|_{\mathcal{G}}}$ is symetric and Υ_V is skew–symetric.

In the case of a simple Lie algebra (i.e one which does not admit any proper graded ideal), the Killing form Υ_L is either non–degenerate or zero [11]. We recall that the classification of finite dimensional simple Lie superalgebras has been finished in [8] (see also [11]).

In this final section, we prove, differently and easier, a well–known result, as a direct consequence of Proposition 5. This result shows the importance of the new classification presented in Section 3 ; it allows us to find again the characterization of a class of simple Lie superalgebras denoted in the literature (see for example [8]) by $\mathcal{Q}(n)$. More precisely, we have:

THEOREM 9 *Let $L = \mathcal{G} \oplus \mathrm{V}$ be a finite dimensional simple Lie superalgebra over* $\mathbf{C}$*, satisfying the folowing conditions:*

1. *$\mathcal{G}$ is simple Lie algebra.*
2. *V is an irreducible $\mathcal{G}$-module.*
3. *The Killing form Υ_L is zero.*

Then, L is isomorphic to the Lie superalgebra:

$$\mathcal{Q}(n) = \mathrm{sl}(n+1) \oplus \mathrm{ad}_{\mathrm{sl}(n+1)}, \quad \textit{for some integer } n \geq 2. \tag{90}$$

Proof: By 1) (respectively 3)) Υ_{V} is a multiple of $\Upsilon_{\mathcal{G}}$ (respectively of multiplicity 1). Also, 2) implies that V is equivalent to the adjoint representation of g [1]. Thus, it reduces to the case where $\mathcal{G}$ is simple and $\mathrm{V} \cong \mathrm{ad}_{\mathcal{G}} \cong \mathcal{G}$. So, Defintion 7 shows that there exists homomorphism of $\mathcal{G}$–modules $\mathrm{S}^2(\mathcal{G}) \longrightarrow \mathrm{g}$ (which is actually an epimorphism [8]). Therefore, $\mathrm{S}^2(\mathcal{G})$ must contain a sub- $\mathcal{G}$–module isomorphic to $\mathcal{G}$, that is to say the decompostion in (37) contains a copy of $\mathcal{G}$. This, together with Proposition 5, proves the desired result. ♡

REMARK 16 *Theorem 9 makes us consider the possibility of applying the new classification, presented in Section 3, to Lie superalgebras. The first step in that direction has been made in [6], where the methods introduced in [5] for Lie algebras have been generalized to the super case. In particular the case of Lie superalgebras* $\mathrm{sl}(m|n)$, $m-n \neq 0$ *and* $\mathrm{osp}(m|n)$, $m-n \neq 2$ *are studied there in detail. This gives a hope that a transcription of the new classification presented in Section 3 to Lie superalgebras is highly possible.*

REFERENCES

[1] E–M. Andreev, E–B. Vinberg & A–G. Elashivili. Orbits of greatest dimension in semisimple linear Lie groups. Func. Anal. Appl. 1, pp. 257–261, 1967.

[2] E. Angelopoulos. Classification des algèbres de Lie simples. Private communication, 1991.

[3] M. El Houari. Nouvelle classification des (super)algèbres de Lie simples par le biais de leurs invariants tensoriels. Thesis of Doctorate, University Paris XI, Dec. 1994.

[4] M. El Houari. On Casimir elements of simple Lie algebras. Journal "Algebras, Groups & Geometries", pp 309–326, 1999. .

[5] M. El Houari. Tensor invariants associated with classical Lie algebras: Illustration of a new classification of simple Lie algebras. Journal of "Algebras, Groups & Geometries", Vol. 14, pp 423–446, 1997.

[6] M. El Houari. Tensor invariants associated with "Good" Lie superalgebras : Towards a new classification of simple Lie superalgebras. To appear in Journal "Algebras, Groups & Geometries".

[7] M. El Houari. On a new classification of simple Lie algebras. To appear in Journal "Algebras, Groups & Geometries".

[8] V. Kac. Lie superalgebras. Advances in Math. 26, pp. 8–96 p. 2115, 1977.

[9] M. Krämer. Eine Klassifikation Bestimmter Untergruppen kompakter Zusammen–hängender Lie Gruppen. Comm. Alg., Vol. 3, pp. 691–737 1975.

[10] S. Okubo. Casimir invariants and vector operators in simple and classical Lie algebras. J. Math. Phys., Vol. 18, pp. 2382–2394, 1977.

[11] M. Scheunert. An introduction to the theory of Lie superalgebras. Lecture notes in mathematics, Vol. 716, 1979.

[12] M. Scheunert, W. Nahm & V. Rittenberg. Classification of simple graded Lie algebras whose Lie algebra is reductive. J. Math. Phys, Vol. 17, 1976.

[13] E–B. Vinberg & A–L. Onishik. Seminar on algebraic groups and Lie groups. Moscow, 1969.

An approach to integrable theories in any dimension: the role of non-semisimple Lie algebras

LUIZ A. FERREIRA Instituto de Física Teórica - IFT/UNESP, Rua Pamplona 145, 01405-900 São Paulo-SP, Brazil

ABSTRACT It is presented an approach to construct a zero curvature representation for integrable theories in spacetimes of dimension $d+1$ by the introduction of a d-form connection. It generalizes in a quite natural way the well known Lax and Zakharov-Shabat equations for two dimensional integrable models. We show that some sufficient conditions to make the zero curvature equation local involve, in an interesting way, non-semisimple Lie algebras and their representations. The new generalized zero curvature conditions can be used to represent the equations of motion of some relativistic invariant field theories of physical interest. Our approach leads to new methods of constructing conserved currents and solutions. We discuss in detail the example of the $2+1$ dimensional CP^1 model. For one of its submodels we explicitly construct an infinite number of nontrivial local conserved currents.

1 INTRODUCTION

The equations of motion for many models in two dimensions such as the sine-Gordon model can be formulated in terms of the zero curvature condition

$$[\partial_0 + A_0, \partial_1 + A_1] = 0 \; . \tag{1.1}$$

where the potentials A_μ, $\mu = 0, 1$, take value on some (finite or infinite) Lie algebra, and where ∂_μ are derivatives w.r.t. the space-time coordinates. This formulation first appeared in the work on the nonlinear Schrödinger equation (NLS) by Zakharov and Shabat [1] in 1971 and opened the way to obtaining many results on the NLS system including solutions and conserved charges.

Soliton theory is now a mature area of research and we have a very good understanding of classical soliton solutions. A common feature of multi-soliton solutions in most of soliton theories is that they are associated to vertex operator representations of affine Lie algebras and they correspond to some special points on the orbit, under the dressing transformation group, of some vacuum solutions [2, 3, 4, 5]. Most of these developments have used in an essential way the zero curvature condition (1.1), but unfortunately these remarkable developments have been only achieved for theories on one or two dimensional space-time. This formulation appears not to generalize directly to dimensions higher than two, although there are some exceptional cases, like self-dual Yang-Mills [6] and a few nonrelativistic systems [7].

In this talk I present a work developed in collaboration with O. Alvarez from Univ. of Miami (USA) and J. Sánchez Guillén of Univ. de Santiago de Compostela (Spain) [8], where we show that some of the methods and results used in two dimensional integrable models, like the existence of a infinite number of conserved charges, can indeed be systematically generalized to theories in higher dimensions. Our approach is to find the counterpart of the zero curvature condition (1.1) in $1+1$ dimensional models. It is true that the equation (1.1) is the compatibility or integrability condition for covariant derivatives. However, eq.(1.1) is also a conservation law in the sense that the path ordered exponential of the connection A_μ along a closed curve is a constant independent of the curve. That leads to a Gauss type law stating that there is a conserved flux through the curve. We generalize the zero curvature (1.1) to a $d+1$ dimensional space-time by looking for a conserved flux through a d dimensional surface, associated to a rank d antisymmetric tensor. Consequently, we are generalizing the concept of integrability to higher dimensions, from the point of view of conservation laws. However, the geometrical and algebraic structures of our approach confirms the deep connection between conservation laws and integration of the equations of motion by leading to interesting methods for constructing large classes of solutions.

The zero curvature equation we get in higher dimensions is in general non-local. However, we present some sufficient conditions to make that zero curvature local. Among those conditions is the requirement that the connections should involve a non-semisimple Lie algebra and a representation (finite or infinite) of it. The number of conserved currents turns out to be equal to the dimension of that representation.

Our approach has been applied to some relativistic invariant field theories of physical interest in $2+1$ dimensions like the BF theories, Chern-Simons, $2+1$ gravity and the CP^1 model, in $3+1$ dimensions like self-dual Yang-Mills theory and the Bogomolny equations [8], and also to non-linear sigma models on homogeneous spaces, defined on space-time of any dimension [9]. Here we shall discuss in detail the case of the $2+1$ dimensional CP^1 model.

2 THE APPROACH TO INTEGRABLE THEORIES IN ANY DIMENSION

The central point of the approach of [8] is to generalize the zero curvature condition in two dimensions guided by the fact that it embodies conservation laws. Indeed, consider a connection A_μ and a curve Γ on a two dimensional space time, and define the quantity W through the equation

$$\frac{dW}{d\sigma} + A_\mu \frac{dx^\mu}{d\sigma} W = 0 \tag{2.1}$$

with σ parametrizing Γ. Then the zero curvature condition (1.1) is the sufficient condition for the quantity W to be path independent as long as its end points are kept fixed. Therefore, if suitable boundary conditions are imposed on the fields, like periodic ones where space-time can be taken as $\mathbf{R} \times S^1$ for instance, then any power N of the path ordered exponential $\mathrm{Tr}\left(P\exp(\int_{S^1} A_x(x,t)dx)\right)^N$ is conserved in time.

The basic idea in [8] to bring such concepts to higher dimensions, is to introduce quantities integrated over hypersurfaces and to find the conditions for them to be independent of deformations of the hypersurfaces which keep their boundaries fixed. Such an approach will certainly lead to conservation laws in a manner very similar to the two dimensional case. However, the main problem of that it is how to introduce non-linear zero curvatures keeping things as local as possible. The way out is to introduce auxiliary connections to allow for parallel transport. The number of possibilities of implementing those ideas increase with the dimensionality of space-time. However, the simplest scenario is that where, in a space-time of dimension $d+1$, one introduces a rank d antisymmetric tensor $B_{\mu_1\mu_2\ldots\mu_d}$ and a vector A_μ. The idea can perhaps be best stated using a formulation in "loop space". On a $d+1$ dimensional space-time M one considers the space $\Omega^{d-1}(M, x_0)$ of $d-1$ dimensional closed hypersurfaces based at a fixed point $x_0 \in M$. One then introduces on such "higher loop space" a 1-form $\mathcal{A}$ which is basically the quantity $W^{-1}B_{\mu_1\mu_2\ldots\mu_d}W$ integrated over the closed hypersurfaces (see [8] for details). The quantity W is defined in terms of the vector A_μ through (2.1). However, for W to be independent of the way one integrates it from x_0 to a given point on the hypersurface, one has to assume that A_μ is flat, i.e.

$$F_{\mu\nu} = [\partial_\mu + A_\mu, \partial_\nu + A_\nu] = 0 \; ; \qquad \mu, \nu = 0, 1, 2 \ldots d \tag{2.2}$$

Roughly speaking a d dimensional closed hypersurface in M, based at x_0, corresponds to a (one dimensional) loop in $\Omega^{d-1}(M, x_0)$. Therefore, the condition to have things independent of deformation of hypersurfaces translates in such "higher loop space" to the zero curvature condition for $\mathcal{A}$, namely

$$\mathcal{F} = \delta\mathcal{A} + \mathcal{A} \wedge \mathcal{A} = 0 \tag{2.3}$$

The relation (2.3) (together with (2.2)) is the generalization of the zero curvature (1.1) to higher dimensions proposed in [8]. Although (2.3) is local in $\Omega^{d-1}(M, x_0)$, it is highly non-local in the space-time M. Again in [8] it is presented some basic manners of introducing *local* conditions which are sufficient for the vanishing of $\mathcal{F}$. The relevant local conditions for the applications in this paper are the following.

Let $\mathcal{G}$ be a Lie algebra and R be a representation of it. We introduce the nonsemisimple Lie algebra $\mathcal{G}_R$ as

$$\begin{aligned} [T_a \, , \, T_b] &= f_{ab}^c T_c \\ [T_a \, , \, P_i] &= P_j R_{ji}(T_a) \\ [P_i \, , \, P_j] &= 0 \end{aligned} \tag{2.4}$$

where T_a constitute a basis of $\mathcal{G}$ and P_i a basis for the abelian ideal P (representation space). The fact that R is a matrix representation, i.e.

$$[R(T_a)\ ,\ R(T_b)] = R([T_a\ ,\ T_b]) \tag{2.5}$$

follows from the Jacobi identities.

We take the connection A_μ to be in $\mathcal{G}$ and the rank d antisymmetric tensor $B_{\mu_1\mu_2\ldots\mu_d}$ to be in P, i.e.

$$A_\mu = A_\mu^a T_a\ , \qquad B_{\mu_1\mu_2\ldots\mu_d} = B^i_{\mu_1\mu_2\ldots\mu_d} P_i \tag{2.6}$$

Then a set of sufficient *local* conditions for the vanishing of the curvature $\mathcal{F}$ in (2.3) is given by

$$D_\mu \tilde{B}^\mu = 0\ ; \qquad F_{\mu\nu} = 0 \tag{2.7}$$

where we have introduced the covariant derivative

$$D_\mu \cdot \equiv \partial_\mu\ \cdot + [A_\mu\ ,\ \cdot\,] \tag{2.8}$$

and the dual of $B_{\mu_1\mu_2\ldots\mu_d}$ as

$$\tilde{B}^\mu \equiv \frac{1}{d!}\,\varepsilon^{\mu\mu_1\mu_2\ldots\mu_d}\, B_{\mu_1\mu_2\ldots\mu_d} \tag{2.9}$$

The relations (2.7) constitute the *local* generalization to higher dimensions of the zero curvature condition (1.1). They lead to local conservation laws. Indeed, since the connection A_μ is flat it can be written as

$$A_\mu = -\partial_\mu W\, W^{-1} \tag{2.10}$$

and consequently (2.7) imply that the currents

$$J_\mu \equiv W^{-1}\, \tilde{B}^\mu\, W \tag{2.11}$$

are conserved

$$\partial_\mu\, J^\mu = 0 \tag{2.12}$$

The zero curvature conditions (2.7) are invariant under the gauge transformations

$$\begin{aligned} A_\mu &\rightarrow g\, A_\mu\, g^{-1} - \partial_\mu g\, g^{-1} \\ \tilde{B}_\mu &\rightarrow g\, \tilde{B}_\mu\, g^{-1} \end{aligned} \tag{2.13}$$

and

$$\begin{aligned} A_\mu &\rightarrow A_\mu \\ \tilde{B}_\mu &\rightarrow \tilde{B}_\mu + \varepsilon_{\mu\mu_1\ldots\mu_d} D^{\mu_1}\alpha^{\mu_2\ldots\mu_d} \equiv \tilde{B}_\mu + D^\nu \tilde{\alpha}_{\mu\nu} \end{aligned} \tag{2.14}$$

where we have introduced the dual $\tilde{\alpha}_{\mu\nu} \equiv \varepsilon_{\mu\nu\mu_2\ldots\mu_d}\alpha^{\mu_2\ldots\mu_d}$. In (2.13) g is an element of the group obtained by exponentiating the Lie algebra $\mathcal{G}$. The transformations

(2.14) are symmetries of (2.7) as a consequence of the fact that the connection A_μ is flat, i.e. $[D_\mu\,,\,D_\nu]=0$. In addition, the parameters $\alpha^{\mu_1\cdots\mu_{d-1}}$ take values in the abelian ideal P.

The currents (2.11) are invariant under the transformations (2.13), but under (2.14) they transform as

$$J_\mu \to J_\mu + \varepsilon_{\mu\mu_1\ldots\mu_d}\partial^{\mu_1}\left(W^{-1}\,\alpha^{\mu_2\ldots\mu_d}\,W\right) = J_\mu + \partial^\nu\left(W^{-1}\,\tilde{\alpha}_{\mu\nu}\,W\right) \qquad (2.15)$$

The transformations (2.13) and (2.14) do not commute and their algebra is isomorphic to the non-semisimple algebra $\mathcal{G}_R$ introduced in (2.4).

3 THE EXAMPLE OF THE CP^1 MODEL

To illustrate the ideas presented in the previous section we now apply those methods to the CP^1 model in $2+1$ dimensions. The main point is to show how its classical equations of motion can be written in the form of the local zero curvature condition (2.7). For more examples see [8, 10, 11, 9].

The CP^1 model has one complex scalar field u and its equation of motion is (see for instance [12])

$$(1+\mid u\mid^2)\,\partial^2 u = 2u^*\,(\partial_\mu u)\,(\partial^\mu u) \qquad (3.1)$$

where $\partial^2=\partial_\mu\partial^\mu=\partial_0^2-\partial_1^2-\partial_2^2$.

It can be obtained from the Lagrangian

$$\mathcal{L}=\frac{\partial_\mu u\partial^\mu u^*}{(1+\mid u\mid^2)^2} \qquad (3.2)$$

The CP^1 model is invariant under the action of $SO(3)$. Associated with the isometry

$$u\to e^{i\theta}\,u \qquad (3.3)$$

one has Noether current

$$J_\mu^{\text{Noet.}}=\frac{1}{(1+\mid u\mid^2)^2}\,(u^*\partial_\mu u - u\partial_\mu u^*) \qquad (3.4)$$

There are two more conserved Noether currents coming from the remaining isometries. They are the real and imaginary parts of the current

$$j_\mu=\frac{\partial_\mu u + u^2\partial_\mu u^*}{(1+\mid u\mid^2)^2} \qquad (3.5)$$

This model also has a topological charge given by[1]

$$J_\mu^{\text{Top.}}\equiv -\frac{i}{2\pi}\epsilon_{\mu\nu\lambda}\frac{\partial^\nu u^*\partial^\lambda u}{(1+\mid u\mid^2)^2} \qquad (3.6)$$

[1] In fact, any current of the form $\epsilon_{\mu\nu\lambda}f(u,u^*)\partial^\nu u^*\partial^\lambda u$, with $f(u,u^*)$ an arbitrary function of u and u^*, is trivially conserved. Any such current arises as the pullback of a 2-form on CP^1 and is thus manifestly closed. The cohomology class $H^2(CP^1,\mathbf{R})$ is one dimensional so there is only one independent conserved charge of this type.

Notice that the Lagrangian and equations of motion of the CP^1 model are invariant under the discrete symmetry

$$u \rightarrow \frac{1}{u} \tag{3.7}$$

The CP^1 model is the same as the $O(3)$ sigma model which is formulated in terms of three real scalar fields $\phi = (\phi_1, \phi_2, \phi_3)$ [13, 14]. It is defined by the Lagrangian and constraint

$$\mathcal{L} = \frac{1}{2g} (\partial_\mu \phi)^2 \qquad\qquad \phi^2 = 1 \tag{3.8}$$

By adding to the Lagrangian a Lagrange multiplier term $\lambda(\phi^2 - 1)$, one gets the equations of motion to be

$$\partial^2 \phi + (\partial_\mu \phi \cdot \partial^\mu \phi)\phi = 0 \tag{3.9}$$

The $O(3)$ sigma model is invariant under the global $SO(3)$ transformations $\phi \rightarrow M \cdot \phi$, with M being orthogonal 3×3 matrices. The corresponding Noether charges are (write $M = \exp(i\omega_j T_j)$, with $[T_i\,, T_j] = i\epsilon_{ijk} T_k$)

$$j^i_\mu \equiv \epsilon^{ijk} \phi^j \partial_\mu \phi^k \qquad\qquad i, j, k = 1, 2, 3 \tag{3.10}$$

Notice, however, that such currents are constrained by

$$\sum_{i=1}^{3} \phi^i j^i_\mu = 0 \tag{3.11}$$

The relation between the two models is achieved by the stereographic projection of the ϕ field on the sphere to the u field on the plane. One has

$$u \equiv u_1 + i u_2 = \frac{\phi_1 + i\phi_2}{1 - \phi_3} \tag{3.12}$$

or

$$\phi = \frac{1}{1 + \mid u \mid^2} \left(u + u^*, -i(u - u^*), \mid u \mid^2 - 1\right) \tag{3.13}$$

Substituting (3.12) into (3.1) one can verify that CP^1 is indeed equivalent to (3.9) after both are σ-models of the two sphere in the round metric.

These models have Bogomolny type bounds which leads to first order differential equations. Belavin and Polyakov [13] have shown that the energy functional

$$E = \int d^2x \left((\partial_0 \phi)^2 + (\partial_m \phi)^2 \right) \qquad m = 1, 2 \tag{3.14}$$

satisfies the bound $E \geq 8\pi Q$, where Q is the topological charge

$$Q = \frac{1}{8\pi} \int d^2x \, \epsilon_{ijk} \epsilon_{mn} \phi_i \partial_m \phi_j \partial_n \phi_k \tag{3.15}$$

with $i, j, k = 1, 2, 3$, and $m, n = 1, 2$. The conditions for saturating the bound are

$$\begin{aligned} \partial_0 \phi_i &= 0 \\ \partial_m \phi_i &= \epsilon_{ijk} \epsilon_{mn} \phi_j \partial_n \phi_k \end{aligned} \tag{3.16}$$

Any solution of these first order differential equations are solutions of the $O(3)$ sigma model. In terms of the (static) u fields, eqs. (3.16) are the Cauchy-Riemann equations

$$\partial_{x_1} u_1 = \partial_{x_2} u_2 \qquad \partial_{x_2} u_1 = -\partial_{x_1} u_2 \tag{3.17}$$

and therefore any meromorphic function leads to a static solution of the CP^1 model.

Of particular interest is the baby-Skyrmion solution which has one unit of topological charge and corresponds to

$$u = x_1 + i x_2 \tag{3.18}$$

or

$$\phi = \left(\frac{x_1}{r} \sin f(r), \, \frac{x_2}{r} \sin f(r), \, \cos f(r) \right) \; ; \qquad r^2 \equiv x_1^2 + x_2^2 \tag{3.19}$$

with

$$f = 2 \text{ arctg} \frac{1}{r} \tag{3.20}$$

We now show that the CP^1 model constitutes an example of the local integrability conditions (2.7). We introduce the potentials

$$A_\mu = \frac{1}{(1+ \mid u \mid^2)} \left(-i \partial_\mu u \, T_+ - i \partial_\mu u^* \, T_- + (u \partial_\mu u^* - u^* \partial_\mu u) \, T_3 \right) \tag{3.21}$$

$$\tilde{B}_\mu = \frac{1}{(1+ \mid u \mid^2)} \left(\partial_\mu u \, P_1^{(1)} - \partial_\mu u^* \, P_{-1}^{(1)} \right) \tag{3.22}$$

where, following (2.4), T_3, $T_\pm$ are the generators of the $sl(2)$ algebra

$$[T_3 \, , T_\pm] = \pm T_\pm \, , \qquad [T_+ \, , T_-] = 2 \, T_3 \tag{3.23}$$

and $P_1^{(1)}$ and $P_{-1}^{(1)}$ (together with $P_0^{(1)}$) transform under the triplet representation of it. For present and for future use, the commutation relations associated to a generic spin-j representation of $sl(2)$ ($m = -j, -j+1, \ldots, j-1, j$) are given by

$$[T_3 \, , P_m^{(j)}] = m \, P_m^{(j)} \tag{3.24}$$

$$[T_\pm \, , P_m^{(j)}] = \sqrt{j(j+1) - m(m \pm 1)} \; P_{m \pm 1}^{(j)} \tag{3.25}$$

$$[P_m^{(j)} \, , P_{m'}^{(j')}] = 0 \tag{3.26}$$

One can easily check that the gauge potential (3.21) is flat independent of the equations of motion of the CP^1 model

$$F_{\mu\nu} = 0 \, ; \qquad \text{off shell.} \tag{3.27}$$

Indeed, (3.21) can be written as in (2.10) with many choices of W's. Two particularly useful choice are[2]

$$W = W_1 \equiv e^{iuT_+}\, e^{\varphi T_3}\, e^{iu^* T_-}\,; \qquad \text{and} \qquad W = W_2 \equiv e^{iu^* T_-}\, e^{-\varphi T_3}\, e^{iuT_+} \tag{3.28}$$

where $\varphi = \ln(1+ \mid u \mid^2)$.

On the other hand one can check that, as a consequence of the CP^1 equations of motion,

$$D_\mu \tilde{B}^\mu = 0\,, \qquad \text{on shell.} \tag{3.29}$$

In fact, (3.29) is equivalent to the CP^1 equations of motion.

The conserved currents (2.12) in this case are (using $W = W_1$ or $W = W_2$)

$$J_\mu = W^{-1}\, \tilde{B}_\mu\, W = j_\mu\, P_1^{(1)} - i\sqrt{2}\, J_\mu^{\text{Noet.}}\, P_0^{(1)} - j_\mu^*\, P_{-1}^{(1)} \tag{3.30}$$

where $J_\mu^{\text{Noet.}}$ and j_μ are respectively given by (3.4) and (3.5).

3.0.1 A simpler submodel of CP^1

In obtaining (3.29) it was important that $P_1^{(1)}$ and $P_{-1}^{(1)}$ were respectively the highest and the lowest spin states of the representation. If we now release that restriction by introducing

$$\tilde{B}_\mu^{(j)} = \frac{1}{(1+\mid u \mid^2)}\left(\partial_\mu u\, P_1^{(j)} - \partial_\mu u^*\, P_{-1}^{(j)}\right)\,; \qquad \text{with } j \text{ integer and } j \geq 1 \tag{3.31}$$

one can easily verify that $D_\mu \tilde{B}_\mu^{(j)} = 0$, with A_μ being the same as in (3.21)):

$$\begin{aligned} 0 \;=\; & \frac{1}{(1+|u|^2)^2}\left\{\sqrt{j(j+1)-2}\left(-i\partial_\mu u \partial^\mu u P_{+2}^{(j)} + i\partial_\mu \bar{u}\partial^\mu \bar{u} P_{-2}^{(j)}\right)\right. \\ & + \left.\left((1+|u|^2)\partial^2 u - 2\bar{u}\partial_\mu u \partial^\mu u\right) P_{+1}^{(j)} + \left((1+|u|^2)\partial^2 \bar{u} - 2u\partial_\mu \bar{u}\partial^\mu \bar{u}\right) P_{-1}^{(j)}\right\} \end{aligned}$$

As previously discussed, the $j = 1$ case leads to the CP^1 equations of motion. On the other hand, for all $j > 1$, we see that $D^\mu \tilde{B}_\mu^{(j)} = 0$ *are* the equations of motion of the submodel of CP^1 [12]

$$\partial^2\, u = 0\,; \qquad \partial_\mu u\, \partial^\mu u = 0 \tag{3.32}$$

Any solution of the above is a solution of the CP^1 model. Conversely, one can easily verify that any (static) solution of the Cauchy-Riemann equations or the Belavin-Polyakov equations (3.17) is a static solution of (3.32). Indeed, the baby-Skyrmion (3.18) solves (3.32).

[2]Notice that the commutation relations (3.23)-(3.26) are compatible with the hermiticity conditions, $T_3^\dagger = T_3$, $T_\pm^\dagger = T_\mp$ and $P_m^{(j)\,\dagger} = (-1)^m P_{-m}^{(j)}$. Therefore, $A_\mu^\dagger = -A_\mu$, $\tilde{B}_\mu^\dagger = \tilde{B}_\mu$, and $W_1^\dagger = W_2^{-1}$. This guarantees that J_μ and $J_\mu^{(j)}$, defined in (3.30) and (3.33) respectively, are hermitian since $J_\mu^{(j)} = W_i^{-1}\, \tilde{B}_\mu^{(j)}\, W_i$, $i = 1, 2$, and $J_\mu^{(j)\,\dagger} = W_1^\dagger\, \tilde{B}_\mu^{(j)}\, W_1^{-1\,\dagger} = W_2^{-1}\, \tilde{B}_\mu^{(j)}\, W_2 = J_\mu^{(j)}$. Notice that, W_1 and W_2 are elements of the group $SL(2,\mathbb{C})$, and not of $SU(2)$, but iA_μ does belong to the algebra of $SU(2)$. In addition, the mapping $\sigma(T_3) = -T_3$, $\sigma(T_\pm) = -T_\mp$, and $\sigma(P_m^{(j)}) = (-1)^m P_{-m}^{(j)}$, is an automorphism of order 2 of the algebra (3.23)-(3.26).

So, from (2.12) we expect $2j+1$ conserved currents. Since we are free to choose j we get in fact an infinite number of conserved currents. Denote

$$J_\mu^{(j)} = W^{-1}\,\tilde{B}_\mu^{(j)}\,W \equiv \sum_{m=-j}^{j} J_\mu^{(j,m)}\,P_m^{(j)} \tag{3.33}$$

The explicit new infinite currents are therefore given just by the elements $\mathcal{A}_{m,\pm 1}$ of the adjoint action $WP_n^{(j)}W^{-1} = \sum_{m=-j}^{j} \mathcal{A}_{mn} P_m^{(j)}$. The construction of the currents is in fact related to the existence of a singlet state which corresponds to the generator T_3 of the $SU(2)$ algebra [9]. We do not give the general construction here but just present the explicit form of the first currents. For the case $j=1$, we obviously get the same as in (3.30). For the case $j=2$ we get

$$\begin{aligned} J_\mu^{(2,2)} &= -\frac{2\,i\,u\,(\partial_\mu u + u^2\,\partial_\mu u^*)}{(1+\mid u\mid^2)^3} \\ J_\mu^{(2,1)} &= \frac{(1-3\mid u\mid^2)\,\partial_\mu u + (3-\mid u\mid^2)u^2\,\partial_\mu u^*}{(1+\mid u\mid^2)^3} \\ J_\mu^{(2,0)} &= -\frac{i\sqrt{6}\,(1-\mid u\mid^2)\,(u^*\,\partial_\mu u - u\,\partial_\mu u^*)}{(1+\mid u\mid^2)^3} \end{aligned} \tag{3.34}$$

with $J_\mu^{(2,-m)} = (-1)^m J_\mu^{(2,m)\dagger}$, $m = 1,2$. For the case $j=3$ we have

$$\begin{aligned} J_\mu^{(3,3)} &= -\frac{\sqrt{15}\,u^2\,(\partial_\mu u + u^2\,\partial_\mu u^*)}{(1+\mid u\mid^2)^4} \\ J_\mu^{(3,2)} &= \frac{i\sqrt{10}\,u\,((-1+2\mid u\mid^2)\,\partial_\mu u + (-2+\mid u\mid^2)\,u^2\,\partial_\mu u^*)}{(1+\mid u\mid^2)^4} \\ J_\mu^{(3,1)} &= \frac{(1-8\mid u\mid^2+6\mid u\mid^4)\,\partial_\mu u + (6-8\mid u\mid^2+\mid u\mid^4)u^2\,\partial_\mu u^*}{(1+\mid u\mid^2)^4} \\ J_\mu^{(3,0)} &= -\frac{2\,i\sqrt{3}\,(1-3\mid u\mid^2+\mid u\mid^4)\,(u^*\,\partial_\mu u - u\,\partial_\mu u^*)}{(1+\mid u\mid^2)^4} \end{aligned} \tag{3.35}$$

with $J_\mu^{(3,-m)} = (-1)^m J_\mu^{(3,m)\dagger}$, $m = 1,2,3$.

The submodel (3.32) has another representation for the integrability conditions (2.7). Let P, Q and $\mathbb{1}$ be the generators of the Heisenberg algebra

$$[P\,,\,Q] = \mathbb{1} \tag{3.36}$$

Notice that, within the scheme of (2.4) one can take $\mathcal{G} = \{Q\}$ and $R = \{P\,,\,\mathbb{1}\}$. Define

$$A_\mu = g(u)\partial_\mu u\,Q\,, \qquad \tilde{B}_\mu = f(u)\partial_\mu u\,P \tag{3.37}$$

where f and g are arbitrary nonvanishing functions of u. One can easily verify that the connection A_μ is flat, i.e. $F_{\mu\nu} = 0$, and that

$$D_\mu \tilde{B}^\mu = \left(\frac{df}{du}\partial_\mu u \partial^\mu u + f\partial^2 u\right) P - fg\partial_\mu u\partial^\mu u\,\mathbb{1} \tag{3.38}$$

Therefore, the solutions of (3.32) solve the equations (2.7).

The conserved currents obtained from such representation are of the form

$$J_\mu = G(u)\partial_\mu u \tag{3.39}$$

where $G(u)$ is any function of u (and not of u^*). One has $\partial_\mu J^\mu = 0$ as a consequence of the equations of motion (3.32). Clearly, the complex conjugate of (3.39) provide another set of conserved currents.

REFERENCES

[1] V.E. Zakharov and A.B. Shabat, Zh. Exp. Teor. Fiz. 61 (1971) 118-134; english transl. Soviet Phys. JETP 34 (1972) 62-69.

[2] O.Babelon and D.Bernard Int. J. Mod. Phys. A8 (1993) 507-543, hep-th/9206002.

[3] V.G. Kac, Infinite dimensional Lie algebras (3rd ed.), Cambridge University Press (1990); V.G. Kac and M. Wakimoto, Exceptional hierarchies of soliton equations, in "Proceedings of Symposia in Pure Mathematics", Vol. 49 (1989) 191.

[4] M. Jimbo and T. Miwa, Publ. RIMS, Kyoto Univ. 19 (1983) 943;
E. Date, M. Jimbo, M. Kashiwara, and T. Miwa, Publ. RIMS, Kyoto Univ. 18 (1982) 1077.

[5] L.A.Ferreira, J.L.Miramontes, J. Sánchez Guillén, *J. Math. Phys.* 38 (1997) 882.

[6] A.A. Belavin and V.E. Zakharov, Phys. Rev. Lett. 35 (1977).

[7] S. Novikov et al. Theory of Solitons, Cont. Sov. Maths. N.Y. (1983).

[8] O. Alvarez, L. A. Ferreira and J. Sánchez Guillén, Nuclear Physics B529 (1998) 689; hep-th/9710147.

[9] L. A. Ferreira and E. E. Leite; Integrable theories in any dimension and homogeneous spaces; IFT-P/067/98; hep-th/9810067.

[10] D. Gianzo, J.O. Madsen, and J. Sánchez Guillén, Integrable chiral theories in (2+1)-dimensions; US-FT-3-98; hep-th/9805094; to appear in Nuclear Physics B.

[11] K. Fujii and T. Suzuki, Non-linear sigma models in (1+2)-dimensions and an infinite number of conserved currents, hep-th/9802105; Some useful formulas in non-linear sigma models in (1+2)-dimensions, hep-th/9804004; K. Fujii, Y. Homma and T. Suzuki; Non-linear grassmann sigma models in any dimension and an infinite number of conserved currents; hep-th/9806084; Submodels of non-linear grassmann sigma models in any dimension and conserved currents, exact solutions; hep-th/9809149.

[12] R.S. Ward, *Phys. Lett.* 158B (1985) 424-428.

[13] A.A. Belavin and A.M. Polyakov, JETP Lett. 22 (1975) 245-247.

[14] F. Wilczek and A. Zee, Phys. Rev. Lett.51 (1983) 2250-2252.

Versal deformations of Lie algebras

ALICE FIALOWSKI* Department of Applied Analysis, Eötvös Loránd University, Múzeum krt. 6-8, H-1088 Budapest, Hungary

INTRODUCTION

In this talk we consider deformations of finite or infinite dimensional Lie algebras over a field of characteristic 0. By deformations of such an object we mean the affine algebraic manifold of all Lie brackets. One wants to consider the quotient of this variety by the action of the group GL. But it is well-known that in the category of algebraic varieties the quotient by a group action does not always exist. Specifically, there is in general no universal deformation of a Lie algebra L with a commutative algebra base A with the property that for any other deformation of L with base B there exists a unique homomorphism $f: B \to A$ that induces an equivalent deformation. If such a homomorphism exists (but it usually not unique) we call the deformation of L with base A a *versal* deformation.

Deformations of various mathematical objects are at the center of current research. The notion of local and infinitesimal deformations of a complex analytic manifold first appeared in the works of Kodaira and Spencer (1958, [KS]). In particular, they proved that infinitesimal deformations can be parameterized by a related cohomology group. The deformation theory of compact complex manifolds was developed by Kuranishi (1965, [Ku]) and Palamodov (1976, [P]). The basic ideas of deformation theory of analytic structures motivated the deformation theory of algebraic manifolds, which was developed by Artin (1969, [A]) and Schlessinger (1968, [Sch]). Deformations of arbitrary rings and associative algebras and connected to

* This work was partially supported by the Hungarian grants OTKA T030823 and FKFP 0170/1999.

them cohomology questions were first studied by Gerstenhaber (1964-68, [Ge]). The most actively studied field was cohomology and deformations of commutative local rings. This was then applied to problems in analytic and algebraic geometry. See Grauert-Kerner 1964 [GK], Illusie 1971 [I], Palamodov 1976 [P], Laudal 1979 [La], etc). Nowadays there is a well-developed general theory (Kontsevich 1994, [Ko].

The deformation theory of Lie algebras is studied much less. Some general questions of the theory were considered by Nijenhuis and Richardson (1966-68, [NR]) but not much is known about deformations of a given Lie algebra. The versal property of deformations does not follow from the classical theory.

A more general deformation theory for Lie algebras follows from Schlessinger's work [Sch]. If we consider deformations with base *specA*, where A is a local algebra, this set-up is adequate to study the problem of "universality" among formal deformations. This was worked out for Lie algebras in [Fi1], [Fi2]; in [Fi2] I showed that in this case under some minor restrictions there exists a *versal* element. The problem is to construct this element.

There is confusion – and errors – in the literature when one tries to describe *all* the nonequivalent deformations of a given Lie algebra. What is usually done, is to compute infinitesimal deformations and try to extend them to real deformation. But the question remains: how many nonequivalent ways one can extend an infinitesimal deformation to a real deformation. The lack of this investigation led to a number of incorrect results. Finding a versal element in different situations is crucial in mathematical physics, especially in string theory.

For a proper procedure, one needs an appropriate theory of Massey operations and use Harrison cohomology. In this talk we are going to outline a construction which gives a method to compute a versal deformation of a given Lie algebra by describing the base of this deformation.

The work was done in collaboration with Dmitry Fuchs. The detailed procedure with proofs appeared in [FiFu].

1. DEFINITIONS AND FUNDAMENTAL EXAMPLE FOR INFINITESIMAL DEFORMATION

Let L be a Lie algebra over a characteristic 0 field $\mathbb{K}$, and let A be a commutative algebra with identity over $\mathbb{K}$ with a fixed augmentation $\varepsilon: A \to \mathbb{K}$, $\varepsilon(1) = 1$; we set $\operatorname{Ker}\varepsilon = \mathfrak{m}$. To avoid transfinite induction, we will assume that $\dim(\mathfrak{m}^k/\mathfrak{m}^{k+1}) < \infty$ for all k.

DEFINITION 1.1. A *deformation* λ of L with base $(A, \mathfrak{m})$, or simply with base A, is a Lie A-algebra structure on the tensor product $A \otimes_{\mathbb{K}} L$ with the bracket $[\,,\,]_\lambda$, such that

$$\varepsilon \otimes \mathrm{id}: A \otimes L \to \mathbb{K} \otimes L = L$$

is a Lie algebra homomorphism. (We usually abbreviate $\otimes_{\mathbb{K}}$ to $\otimes$.) See [Fi1], [Fi2].

EXAMPLE. If $A = \mathbb{K}[t]$, then a deformation of L with base A is the same as an algebraic 1-parameter deformation of L.

DEFINITION 1.2.Two deformations of a Lie algebra L with the same base A are called *equivalent* if there exists a Lie algebra isomorphism between the two copies of $A \otimes L$ with the two Lie algebra structures, compatible with $\varepsilon \otimes \mathrm{id}$. A deformation with base A is called *local* if the algebra A is local, and it is called *infinitesimal* if, in addition to this, $\mathfrak{m}^2 = 0$.

DEFINITION 1.3. Let A be a complete local algebra (completeness means that $A = \varprojlim_{n\to\infty}(A/\mathfrak{m}^n)$, where $\mathfrak{m}$ is the maximal ideal in A). A *formal deformation* of L with base A is a Lie A-algebra structure on the completed tensor product $A\hat{\otimes} L = \varprojlim_{n\to\infty}((A/\mathfrak{m}^n)\otimes L)$ such that

$$\varepsilon\hat{\otimes}\,\mathrm{id}\colon A\hat{\otimes} L \to \mathbb{K}\otimes L = L$$

is a Lie algebra homomorphism (see [Fi2]) .

The above notion of equivalence is extended to formal deformations in an obvious way.

EXAMPLE. If $A = \mathbb{K}[[t]]$ then a formal deformation of L with base A is the same as a formal 1-parameter deformation of L. See [Ge], [NR].

Let A' be another commutative local algebra with identity over $\mathbb{K}$ with a fixed augmentation $\varepsilon'\colon A' \to \mathbb{K}$, and let $\varphi\colon A \to A'$ be an algebra homomorphism with $\varphi(1) = 1$ and $\varepsilon'\circ\varphi = \varepsilon$.

DEFINITION 1.4. If a deformation λ of L with base $(A,\mathfrak{m})$ is given, then the *push-out* $\varphi_*\lambda$ is the deformation of L with base $(A', \mathfrak{m}' = \operatorname{Ker}\varepsilon')$, which is the Lie algebra structure

$$[a_1'\otimes_A(a_1\otimes l_1), a_2'\otimes_A(a_2\otimes l_2)]' = a_1'a_2'\otimes_A[a_1\otimes l_1, a_2\otimes l_2],$$

$$a_1', a_2'\in A',\ a_1, a_2\in A,\ l_1, l_2\in L,$$

on $A'\otimes L = (A'\otimes_A A)\otimes L = A'\otimes_A(A\otimes L)$. Here A' is regarded as an A-module with the structure $a'a = a'\varphi(a)$, and the operation $[\,,\,]$ in the right hand side of the formula refers to the Lie algebra structure λ on $A\otimes L$.

The push-out of a formal deformation is defined in a similar way.

FUNDAMENTAL EXAMPLE: Assume that the Lie algebra L satisfies the condition

$$\dim H^2(L;L) < \infty.$$

This is true, for example, if $\dim L < \infty$.

Consider the algebra

$$\mathcal{A} = \mathbb{K}\oplus H^2(L;L)'$$

with the second summand being an ideal with zero multiplication ($'$ means the dual). Fix some homomorphism

$$\mu\colon H^2(L;L)\to C^2(L;L) = \operatorname{Hom}(\Lambda^2 L, L)$$

which takes a cohomology class into a cocycle representing this class. Define a Lie algebra structure on

$$\mathcal{A}\otimes L = (\mathbb{K}\otimes L)\oplus(H^2(L;L)'\otimes L) = L\oplus\operatorname{Hom}(H^2(L;L),L)$$

by the formula

$$[(l_1,\varphi_1),(l_2,\varphi_2)] = ([l_1,l_2],\psi),$$

where

$$\psi(\alpha) = \mu(\alpha)(l_1,l_2) + [\varphi_1(\alpha),l_2] + [l_1,\varphi_2(\alpha)],$$

$$l_1,l_2 \in L,\ \varphi_1,\varphi_2 \in \operatorname{Hom}(H^2(L;L),L)),\ \alpha \in H^2(L;L).$$

(The Jacobi identity for this operation is implied by $\delta\mu(\alpha) = 0$.) This determines a deformation of L with base $\mathcal{A}$ which is clearly infinitesimal.

PROPOSITION 1.5. *Up to equivalence, this deformation does not depend on the choice of μ.*

The proof follows from direct computation.

Denote the infinitesimal deformation of L constructed above by η_L.

The main property of η_L is its (co-)universality in the class of infinitesimal deformations.

PROPOSITION 1.6. *For any infinitesimal deformation λ of the Lie algebra L with a finite-dimensional base A there exists a unique homomorphism $\varphi: \mathbb{K}\oplus H^2(L;L)' \to A$ such that λ is equivalent to the push-out $\varphi_*\eta_L$.*

SKETCH OF THE PROOF. Take $\xi \in \mathfrak{m}'$, or, equivalently, $\xi \in A'$ and $\xi(1) = 0$. For $l_1,l_2 \in L$ set

$$\alpha_{\lambda,\xi}(l_1,l_2) = (\xi \otimes \mathrm{id})[1\otimes l_1, 1\otimes l_2]_\lambda \in \mathbb{K}\otimes L = L.$$

LEMMA. *The cochain $\alpha_{\lambda,\xi} \in C^2(L;L)$ is a cocycle.*

Let $a_{\lambda,\xi} \in H^2(L;L)$ be the cohomology class of the cocycle $\alpha_{\lambda,\xi}$. The correspondences $\xi \mapsto \alpha_{\lambda,\xi}, \xi \mapsto a_{\lambda,\xi}$ define homomorphisms

$$\alpha_\lambda: \mathfrak{m}' \to C^2(L;L), \qquad \delta\circ\alpha_\lambda = 0,$$

$$a_\lambda: \mathfrak{m}' \to H^2(L;L).$$

We claim that

(i) the deformation λ is fully determined by α_λ;

(ii) the deformations λ, λ' are equivalent if and only if $a_\lambda = a_{\lambda'}$;

(iii) if $\varphi = \mathrm{id}\oplus a'_\lambda: \mathbb{K}\oplus H^2(L;L)' \to \mathbb{K}\oplus\mathfrak{m}$, then $\varphi_*\eta_L$ is equivalent to λ.

For the details see [FiFu].

Let A be a local algebra with $\dim(A/\mathfrak{m}^2) < \infty$. Obviously, $A/\mathfrak{m}^2$ is local with the maximal ideal $\mathfrak{m}/\mathfrak{m}^2$, and $(\mathfrak{m}/\mathfrak{m}^2)^2 = 0$. Recall that the dual space $(\mathfrak{m}/\mathfrak{m}^2)'$ is called the *tangent space* of A; we denote it by TA.

DEFINITION 1.7. Let λ be a deformation of L with base A. Then the mapping

$$a_{\pi_*\lambda}: TA = (\mathfrak{m}/\mathfrak{m}^2)' \to H^2(L;L),$$

where π is the projection $A \to A/\mathfrak{m}^2$, is called the *differential* of λ. Denote it by $d\lambda$.

The differential of a formal deformation is defined in a similar way.

It is clear from the construction that equivalent deformations or formal deformations have equal differentials.

It is not possible to construct a local or formal deformation of a Lie algebra with a similar universality property in the class of local or formal deformations. But it becomes possible for an appropriate weakening of this property.

DEFINITION 1.8. A formal deformation η of a Lie algebra L with base B is called *versal* if

(i) for any formal deformation λ of L with any (local) base A there exists a homomorphism $f: B \to A$ such that the deformation λ is equivalent to $f_*\eta$;

(ii) in the notations of (i), if A satisfies the condition $\mathfrak{m}^2 = 0$, then f is unique (see [Fil]).

2. OBSTRUCTIONS TO EXTENDING DEFORMATIONS

Let λ be a deformation of a Lie algebra L with a finite-dimensional local base A, and let $0 \to \mathbb{K} \xrightarrow{i} B \xrightarrow{p} A \to 0$ be a 1-dimensional extension of A. It is known [Harr] that such an extension corresponds to a Harrison 2-cohomology class $f \in H^2_{\text{Harr}}(A;\mathbb{K})$.

Let $I = i \otimes \text{id}: L = \mathbb{K} \otimes L \to B \otimes L$ and $P = p \otimes \text{id}: B \otimes L \to A \otimes L$. Let also $E = \hat{\varepsilon} \otimes \text{id}: B \otimes L \to \mathbb{K} \otimes L = L$, where $\hat{\varepsilon}$ is the augmentation of B. The Lie algebra structure $[\,,\,]_\lambda$ in $A \otimes L$ can be lifted to a B-bilinear operation $\{\,,\,\}: \Lambda^2 B \to B$ such that

(i) $P\{l_1, l_2\} = [P(l_1), P(l_2)]_\lambda$ for any $l_1, l_2 \in B \otimes L$,
(ii) $\{I(l), l_1\} = I[l, E(l_1)]$ for any $l \in L$, $l_1 \in B \otimes L$.

This operation $\{\,,\,\}$ partially satisfies the Jacobi identity:

$$\varphi(l_1, l_2, l_3) := \{l_1, \{l_2, l_3\}\} + \{l_2, \{l_3, l_1\}\} + \{l_3, \{l_1, l_2\}\} \in \operatorname{Ker} P.$$

Remark that φ is multilinear and skew-symmetric, and $\varphi(l_1, l_2, l_3) = 0$ if $l_1 \in \operatorname{Ker} E$. Hence φ determines a multilinear form

$$\bar{\varphi}: \Lambda^3 L = \Lambda^3((B \otimes L)/\operatorname{Ker} E) \to \operatorname{Ker} P = L,$$

that is an element $\bar{\varphi}$ of $C^3(L; L)$. It is easy to check that $\delta\bar{\varphi} = 0$.

Let $\{\,,\,\}'$ be another B-bilinear operation $\Lambda^2 B \to B$ satisfying the conditions (i), (ii).

It is easy to check that if $\bar{\varphi}, \bar{\varphi}' \in C^3(L; L)$ are the cochains corresponding to $\{\,,\,\}, \{\,,\,\}'$, then

$$\bar{\varphi}' - \bar{\varphi} = \delta\psi.$$

with some $\psi \in C^2(L; L)$.

Let $\mathcal{O}_\lambda(f) \in H^3(L; L)$ be the cohomology class of the cochain $\bar{\varphi}$. It is obvious that

$$\mathcal{O}_\lambda: H^2_{\text{Harr}}(A; \mathbb{K}) \to H^3(L; L),\ f \mapsto \mathcal{O}_\lambda(f)$$

is a linear map.

We get the following

PROPOSITION 2.1. *The deformation λ with base A can be extended to a deformation of L with base B if and only if $\mathcal{O}_\lambda(f) = 0$.*

DEFINITION 2.2. The cohomology class $\mathcal{O}_\lambda(f)$ is called the *obstruction* to the extension of the deformation λ from A to B.

Suppose now that $\mathcal{O}_\lambda(f) = 0$, that is the deformation λ is extendible to a deformation with base B. We are going to study the set of all possible extensions.

Let μ, μ' be deformations of L with base B such that $p_*\mu = p_*\mu' = \lambda$. Then we know that the difference $[\,,\,]_{\mu'} - [\,,\,]_\mu$ determines and is determined by a certain cochain $\psi \in C^2(L; L)$. Since $[\,,\,]_{\mu'}$ and $[\,,\,]_\mu$ both satisfy the Jacobi identity, $\delta\psi = 0$. Moreover, it is easy to check that if we replace any of the structures $[\,,\,]_\mu, [\,,\,]_{\mu'}$ with an equivalent one, then the cocycle ψ will be replaced by a cohomologous one. That means that $H^2(L; L)$ operates transitively on the set of these equivalence classes.

It is known [Harr] that $H^1_{\mathrm{Harr}}(A; M)$ is the space of derivations $A \to M$. From this it follows that if A is local algebra with maximal ideal $\mathfrak{m}$, then $H^1_{\mathrm{Harr}}(A; \mathbb{K}) = (\mathfrak{m}/\mathfrak{m}^2)' = TA$.

On the other hand, $H^1_{\mathrm{Harr}}(A; M)$ may also be interpreted as the group of automorphisms of any given extension $0 \to M \xrightarrow{i} B \xrightarrow{p} A \to 0$ of A. (An automorphism of this extension is an algebra automorphism $f: B \to B$ such that $f \circ i = i$, $p \circ f = p$.) [Harr].

The group of automorphisms of the extension also operates on the set of equivalence classes of deformations.

PROPOSITION 2.3. If $r: B \to B$ determines an automorphism of the extension $0 \to \mathbb{K} \xrightarrow{i} B \xrightarrow{p} A \to 0$ which corresponds to an element $h \in H^1_{\mathrm{Harr}}(A; \mathbb{K}) = TA$, then for any deformation μ of L with base B such that $p_*\mu = \lambda$, the difference between $[\,,\,]_{r_*\mu}$ and $[\,,\,]_\mu$ is a cocycle of the cohomology class $d\lambda(h)$.

In other words, these two operations are related to each other by the differential $d\lambda: TA \to H^2(L; L)$.

The proof is obvious.

COROLLARY. *Suppose that the differential $d\lambda: TA \to H^2(L; L)$ is onto. Then the group of automorphisms of the extensions $0 \to \mathbb{K} \xrightarrow{i} B \xrightarrow{p} A \to 0$ operates transitively on the set of equivalence classes of deformations μ of L with base B such that $p_*\mu = \lambda$. In other words, μ is unique up to an isomorphism and an automorphism of the extension $0 \to \mathbb{K} \to B \to A \to 0$.*

REMARK 1. The relation between the second Harrison cohomology group of a finite-dimensional local commutative algebra A and extensions of A may be described in terms of one remarkable extension. This is the extension

$$0 \to H^2_{\mathrm{Harr}}(A; \mathbb{K})' \to C \to A \to 0, \tag{1}$$

where the operation of A on $H^2_{\mathrm{Harr}}(A; \mathbb{K})'$ is induced by the operation of A on $\mathbb{K}$, and the cocycle

$$f_A : S^2A \to H^2_{\mathrm{Harr}}(A; \mathbb{K})'$$

is defined as the dual of a homomorphism

$$\mu : H^2_{\mathrm{Harr}}(A; \mathbb{K}) \to (S^2A)',$$

which takes a cohomology class to a cocycle from this class. This extension does not depend, up to an isomorphism, on the choice of μ and possesses the following partial (co-)universality property.

PROPOSITION 2.4. *Let M be an A-module with $\mathfrak{m}M = 0$. Then the extension* (1) *admits a unique homomorphism into an arbitrary extension $0 \to M \to B \to A \to 0$ of A.*

REMARK 2. We recall another result of Harrison.

PROPOSITION 2.5. ([Harr], Theorems 11 and 18). *Let $A = \mathbb{K}[x_1, \dots, x_n]$ be a polynomial algebra, and let $\mathfrak{m}$ be the ideal of polynomials without constant terms. If an ideal I of A is contained in $\mathfrak{m}^2$, then*

$$H^2_{\text{Harr}}(A/I; \mathbb{K}) \cong (I/(\mathfrak{m} \cdot I))'.$$

Harrison's work contains also an explicit construction of the above homomorphism, which implies the following description of the canonical extension

$$0 \to H^2_{\text{Harr}}(B; \mathbb{K})' \to C \to B \to 0$$

of $B = A/I$ (see Remark 1).

PROPOSITION 2.6. *If A, $\mathfrak{m}$, and I are as in Proposition* 2.5, *then the preceding extension for $B = A/I$ is*

$$0 \to I/(\mathfrak{m} \cdot I) \xrightarrow{i} A/(\mathfrak{m} \cdot I) \xrightarrow{p} A/I \to 0,$$

where i and p are induced by the inclusions $I \to A$ and $\mathfrak{m} \cdot I \to I$.

3. CONSTRUCTION

Suppose that $\dim H^2(L; L) < \infty$.

Let $C_0 = \mathbb{K}$, $C_1 = \mathbb{K} \oplus H^2(L; L)'$, and let

$$0 \to H^2(L; L)' \xrightarrow{i_1} C_1 \xrightarrow{p_1'} \mathbb{K} \to 0$$

be the canonical splitting extension. Denote the universal infinitesimal deformation η_L of L with base C_1 constructed in Section 1 by η_1. Suppose that for some $k \geq 1$ we have already constructed a finite-dimensional commutative algebra C_k and a deformation η_k of L with base C_k. Using the cocycle f_{C_k} consider the extension

$$0 \to H^2_{\text{Harr}}(C_k; \mathbb{K})' \xrightarrow{\bar{i}_{k+1}} \bar{C}_{k+1} \xrightarrow{\bar{p}_{k+1}'} C_k \to 0 \tag{2}$$

introduced in Section 2 (see (1)).

The obstruction to the extension of η_k is

$$\mathcal{O}_{\eta_k}(f_{C_k}) \in H^2_{\text{Harr}}(C_k, \mathbb{K})' \otimes H^3(L; L).$$

This gives a map

$$\omega_k : H^2_{\text{Harr}}(C_k, \mathbb{K}) \to H^3(L; L).$$

Set

$$C_{k+1} = \bar{C}_{k+1}/\bar{i}_{k+1} \circ \omega'_k(H^3(L;L)').$$

Obviously, the extension (2) factors to an extension

$$0 \to (\operatorname{Ker}\omega_k)' \xrightarrow{i_{k+1}} C_{k+1} \xrightarrow{p'_{k+1}} C_k \to 0. \tag{3}$$

Here all algebras C_k are local. Since the algebra C_k is finite-dimensional, the cohomology $H^2_{\mathrm{Harr}}(C_k;\mathbb{K})$ is also finite-dimensional, and hence C_{k+1} is finite-dimensional.

PROPOSITION 3.1. *The deformation η_k admits an extension to a deformation with base C_{k+1}, and this extension is unique up to an isomorphism and an automorphism of an extension* (3).

SKETCH OF THE PROOF. We know that the obstruction to the extension of the deformation η_k of L from C_k to C_{k+1} is a homomorphism $\operatorname{Ker}\omega_k \to H^3(L;L)$, and it is easy to show that it is precisely the restriction of ω_k. Hence it is equal to 0, and the proposition follows.

Let us choose an extended deformation and denote it by η_{k+1}.

The induction yields a sequence of finite-dimensional algebras

$$\mathbb{K} \xleftarrow{p'_1} C_1 \xleftarrow{p'_2} \dots \xleftarrow{p'_k} C_k \xleftarrow{p'_{k+1}} C_{k+1} \xleftarrow{p'_{k+2}} \dots,$$

and a sequence of deformations η_k of L such that $(p'_{k+1})_*\eta_{k+1} = \eta_k$. (Here p'_k are algebra projections.)

Taking the projective limit, we obtain a formal deformation η of L with base $C = \varprojlim_{k\to\infty} C_k$.

Denote the space $H^2(L;L)$ briefly by $\boldsymbol{H}$. Assume that $\dim \boldsymbol{H} < \infty$. Let $\mathfrak{m}$ be the maximal ideal in $\mathbb{K}[[\boldsymbol{H}']]$.

PROPOSITION 3.2. $C_k = \mathbb{K}[[\boldsymbol{H}']]/I_k$ *where*

$$\mathfrak{m}^2 = I_1 \supset I_2 \supset \dots,\ I_k \supset \mathfrak{m}^{k+1}.$$

PROOF. By construction,

$$C_1 = \mathbb{K} \oplus \boldsymbol{H}' = \mathbb{K}[[\boldsymbol{H}']]/\mathfrak{m}^2.$$

Suppose that we already know that

$$C_k = \mathbb{K}[[\boldsymbol{H}']]/I_k,\ \mathfrak{m}^2 \supset I_k \supset \mathfrak{m}^{k+1}.$$

Then, according to Proposition 2.6,

$$\bar{C}_{k+1} = \mathbb{K}[\boldsymbol{H}']/(\mathfrak{m}\cdot I_k),$$

and by construction C_{k+1} is the quotient of $\bar{C}_{k+1}$ over an ideal contained in $I_k/(\mathfrak{m}\cdot I_k) \subset \mathfrak{m}^2/(\mathfrak{m}\cdot I_k)$. Hence

$$C_{k+1} = \mathbb{K}[\boldsymbol{H}']/I_{k+1}, \text{ where } \mathfrak{m}^2 \supset I_{k+1} \supset \mathfrak{m}\cdot I_k \supset \mathfrak{m}^{k+2}.$$

This completes the proof.

COROLLARY. *For $k \geq 2$ the projection $p'_k: C_k \to C_{k-1}$ implies an isomorphism $TC_k \to TC_{k-1}$. In particular, the space TC_k does not depend on k; $TC_k = TC_1 = H^2(L; L)$. More precisely, for any $k \geq 1$ the differential $d\eta_k: TC_k \to H^2(L; L)$ is an isomorphism.*

PROPOSITION 3.3. *$C = \mathbb{K}[[\boldsymbol{H}']]/I$, where I is an ideal contained in $\mathfrak{m}^2$. Since $\mathbb{K}[[\boldsymbol{H}']]$ is Noetherian, then I is finitely generated.*

PROOF. By construction, $C = \varprojlim_{k\to\infty} C_k$. From Proposition 3.2 we get an epimorphism

$$\varprojlim_{k\to\infty} \mathbb{K}[[\boldsymbol{H}']]/\mathfrak{m}^{k+1} \to \varprojlim_{k\to\infty} C_k,$$

that is

$$\mathbb{K}[[\boldsymbol{H}']] \to C,$$

and

$$C = \mathbb{K}[[\boldsymbol{H}']]/I, \text{ where } I = \cap I_k = \varprojlim I_k.$$

THEOREM 3.4. *If $\dim H^2(L; L) < \infty$, then the formal deformation η is a versal deformation of L.*

For the proof see [FiFu].

THEOREM 3.5. *If $\dim H^2(L; L) < \infty$, then the base of the versal formal deformation of L is formally embedded in $H^2(L; L)$, that is, it may be described in $H^2(L; L)$ by a finite system of formal equations.*

The proof follows directly from Proposition 3.3.

4. PROCEDURE TO COMPUTE THE BASE OF THE DEFORMATION

Consider the sequence

$$\mathbb{K} \xrightarrow{p_1} C_1 \xrightarrow{p_2} \dots \xrightarrow{p_k} C_k \xrightarrow{\bar{p}_{k+1}} \bar{C}_{k+1}$$

where p_i are embeddings.

Recall that all $C_i, \bar{C}_i$ are finite-dimensional algebras,

$$C_1 = \mathbb{K} \oplus H^2(L; L)';$$

there is an extension

$$0 \to H^2_{\text{Harr}}(C_k; \mathbb{K})' \xrightarrow{\bar{i}_{k+1}} \bar{C}_{k+1} \xrightarrow{\bar{p}'_{k+1}} C_k \to 0$$

(see equation (2)) and an obstruction homomorphism

$$\omega_k: H^2_{\text{Harr}}(C_k; \mathbb{K}) \to H^3(L; L).$$

Recall also that

$$C_{k+1} = \bar{C}_{k+1}/\text{Im}\,(\bar{i}_{k+1} \circ \omega'_k).$$

Let $\mathfrak{m}_i, \bar{\mathfrak{m}}_i$ be the maximal ideals in $C_i, \bar{C}_i$. Then we also have the sequence

$$\mathfrak{m}_1 \xrightarrow{p_2} \mathfrak{m}_2 \xrightarrow{p_3} \dots \xrightarrow{p_k} \mathfrak{m}_k \xrightarrow{\bar{p}_{k+1}} \bar{\mathfrak{m}}_{k+1}.$$

Consider the dual sequence

$$\mathfrak{m}_1' \xleftarrow{p_2'} \mathfrak{m}_2' \xleftarrow{p_3'} \dots \xleftarrow{p_k'} \mathfrak{m}_k' \xleftarrow{\bar{p}_{k+1}'} \bar{\mathfrak{m}}_{k+1}'.$$

This is a sequence of successively embedded cocommutative coassociative coalgebras. Put $\bar{\mathfrak{m}}_{k+1}' = F$, $\mathfrak{m}_1' = F_0$, $\mathfrak{m}_k' = F_1$. Then

$$F_0 = H^2(L;L),\ F/F_1 = H^2_{\mathrm{Harr}}(C_k;\mathbb{K}).$$

We choose the grading in F to be trivial ($\deg\varphi = 0$ for any $\varphi \in F$).

THEOREM 4.1. $2\omega_k \in \langle \mathrm{id}\rangle_F$.

Here ω_k is contained in the Massey F-product of id if there exists a map $\alpha: \mathfrak{m}_k' \to C^2(L;L)$ with

$$\delta\alpha = \mu \circ (\alpha \otimes \alpha) \circ \Delta$$

where $\Delta: F \to F \otimes F$ is the comultiplication and the diagrams

$$\begin{array}{ccc} H^2(L,L) & \xrightarrow{\alpha|_{H^2(L,L)}} & \mathrm{Ker}\,\delta \\ \downarrow{\scriptstyle \mathrm{id}} & & \downarrow{\scriptstyle \pi} \\ H^2(L,L) & \xrightarrow{\mathrm{id}} & H(L,L) \end{array} \qquad \begin{array}{ccc} \bar{\mathfrak{m}}_{k+1}' & \xrightarrow{\mu\circ(\alpha\otimes\alpha)\circ\Delta} & \mathrm{Ker}\,\delta \\ \downarrow{\scriptstyle \pi} & & \downarrow{\scriptstyle \pi} \\ H^2_{\mathrm{Harr}} & \xrightarrow{2\omega_k} & H(L,L) \end{array}$$

are commutative (the vertical maps labeled by π denote the projections of each space onto the quotient space), see [FuL] and [FiFu].

SKETCH OF THE PROOF. One can show that $\omega_k = \frac{1}{2}b$, where $b: H^2_{\mathrm{Harr}} \to H^3(L,L)$ from $\langle\mathrm{id}\rangle_F$ is represented by $\mu \circ (\alpha\otimes\alpha)\circ\Delta: F \to C^3(L;L)$. Moreover, an arbitrary element of $\langle\mathrm{id}\rangle_F$ is equal to $2\omega_k$ for an appropriate extension of the deformation $\eta_1 = \eta_L$ of L with base C_1 to a deformation η_k of L with base C_k. From this the theorem follows.

Begin now with the *first two steps* of this inductive computation. Use the notation $H^2(L;L) = \boldsymbol{H}$, and also denote by $\mathfrak{m}$ the maximal ideal of the polynomial algebra $\mathbb{K}[\boldsymbol{H}']$. As before, we assume that $\dim \boldsymbol{H} < \infty$. We denote the map $\alpha: \mathfrak{m}_k' \to C^2(L;L)$ by α_k.

We have

$$C_1 = \mathbb{K} \oplus \boldsymbol{H}' = \mathbb{K}[\boldsymbol{H}']/\mathfrak{m}^2,$$

and hence

$$\mathfrak{m}_1 = \mathfrak{m}/\mathfrak{m}^2 = \boldsymbol{H}',\ \mathfrak{m}_1' = \boldsymbol{H}.$$

By Remark 2 in Section 2 we also have

$$\bar{C}_2 = \mathbb{K}[\boldsymbol{H}']/\mathfrak{m}^3,$$

so

$$\bar{\mathfrak{m}}_2 = \mathfrak{m}/\mathfrak{m}^3, \ \bar{\mathfrak{m}}_2' = \boldsymbol{H} \oplus S^2\boldsymbol{H}.$$

The map

$$\alpha_1 \colon \mathfrak{m}_1' = \boldsymbol{H} \to C^2(L; L)$$

takes a cohomology class into a representing cocycle. Hence the map

$$\mu \circ (\alpha_1 \otimes \alpha_1) \circ \Delta \colon \bar{\mathfrak{m}}_2' \to C^3(L; L), \tag{4}$$

where $\Delta \colon \bar{\mathfrak{m}}_2' \to \mathfrak{m}_1' \otimes \mathfrak{m}_1'$ is the comultiplication, acts as zero on $\boldsymbol{H}$ (because $\Delta|\boldsymbol{H} = 0$) and takes $\xi\eta \in S^2\boldsymbol{H}$ (where $\xi, \eta \in \boldsymbol{H}$) into the product of the chosen cocycles $\alpha_1(\xi), \alpha_1(\eta)$ representing ξ, η. Easy to see that the image of the map (4) belongs to $\operatorname{Ker}\delta$, and the composition of this map with the projection $\pi \colon \operatorname{Ker}\delta \to H^3(L; L)$ acts as zero on $\boldsymbol{H}$ and coincides with the multiplication $[\,,\,] \colon S^2\boldsymbol{H} \to H^3(L; L)$ on $S^2\boldsymbol{H}$. Hence

$$\mathfrak{m}_2' = \boldsymbol{H} \oplus \operatorname{Ker}([\,,\,] \colon S^2\boldsymbol{H} \to H^3(L; L)),$$

$$\mathfrak{m}_2 = \frac{\mathfrak{m}}{\mathfrak{m}^3 + J_2}, \text{ where } J_2 = \operatorname{Im}([\,,\,]')$$

$$C_2 = \frac{\mathbb{K}[\boldsymbol{H}']}{\mathfrak{m}^3 + J_2}.$$

Note that if $\dim H^3(L; L) = q$, then J_2 is an ideal in $\mathbb{K}[\boldsymbol{H}']$ generated by (at most) q quadratic polynomials.

Furthermore,

$$\bar{C}_3 = \frac{\mathbb{K}[\boldsymbol{H}']}{\mathfrak{m}^4 + (\mathfrak{m} \cdot J_2)},$$

and hence

$$\bar{\mathfrak{m}}_3 = \frac{\mathfrak{m}}{\mathfrak{m}^4 + (\mathfrak{m} \cdot J_2)}, \ \bar{\mathfrak{m}}_3' = \boldsymbol{H} \oplus S^2\boldsymbol{H} \oplus K.$$

Here $K \subset S^3\boldsymbol{H}$ is the intersection of kernels of the maps

$$f_\varphi \colon S^3\boldsymbol{H} \to H^3(L; L), \ \varphi \in \boldsymbol{H}',$$

$$f_\varphi(\xi\eta\zeta) = \varphi(\xi)[\eta, \zeta] + \varphi(\eta)[\zeta, \xi] + \varphi(\zeta)[\xi, \eta].$$

The map

$$\alpha_2 \colon \mathfrak{m}_2' = \boldsymbol{H} \oplus \operatorname{Ker}[\,,\,] \to C^2(L; L)$$

coincides with α_1 on $\boldsymbol{H}$ and takes $\sum \xi_i\eta_i \in \operatorname{Ker}[\,,\,]$ ($\xi_i, \eta_i \in \boldsymbol{H}$) into a two-dimensional cochain whose coboundary is $\sum[\alpha_1(\xi_i), \alpha_1(\eta_i)]$. Hence the composition

$$\mu \circ (\alpha_2 \otimes \alpha_2) \circ \Delta \colon \bar{\mathfrak{m}}_3' \to C^3(L; L), \tag{5}$$

where $\Delta \colon \bar{\mathfrak{m}}_3' \to \mathfrak{m}_2' \otimes \mathfrak{m}_2'$ is the comultiplication, coincides with the map (4) on $\boldsymbol{H} \oplus S^2\boldsymbol{H}$ and takes $\sum \xi_i\eta_i\zeta_i$ into

$$[\alpha_1(\xi_i), \alpha_2(\eta_i, \zeta_i)] + [\alpha_1(\eta_i), \alpha_2(\zeta_i, \xi_i)] + [\alpha_1(\zeta_i), \alpha_2(\xi_i, \eta_i)].$$

One can show that the latter is a cocycle, and the composition of the map (5) and the projection $\pi \colon \operatorname{Ker}\delta \to H^3(L; L)$ acts as zero on $\boldsymbol{H}$, as $[\,,\,]$ on $S^2\boldsymbol{H}$, and as the "triple Massey product" on K. The kernel of this composition is $\mathfrak{m}_3'$, and $\mathfrak{m}_3$ is the dual of this kernel. Thus, by construction,

$$\mathfrak{m}_3 = \frac{\mathfrak{m}}{\mathfrak{m}^4 + J_3}, \quad C_3 = \frac{\mathbb{K}[\boldsymbol{H}']}{\mathfrak{m}^4 + J_3},$$

where $J_3 \cap S^2\boldsymbol{H}' = J_2 \cap S^2\boldsymbol{H}'$.

Describe now the *k-th induction step*. Suppose that we have already constructed

$$C_k = \frac{\mathbb{K}[\boldsymbol{H}']}{\mathfrak{m}^{k+1} + J_k}, \quad \mathfrak{m}_k = \frac{\mathfrak{m}}{\mathfrak{m}^{k+1} + J_k}, \quad \alpha_k : \mathfrak{m}_k' \to C^2(L; L).$$

Then we have

$$\bar{C}_{k+1} = \frac{\mathbb{K}[\boldsymbol{H}']}{\mathfrak{m}^{k+2} + (\mathfrak{m} \cdot J_k)}, \quad \bar{\mathfrak{m}}_{k+1} = \frac{\mathfrak{m}}{\mathfrak{m}^{k+2} + (\mathfrak{m} \cdot J_k)},$$

$$\bar{\mathfrak{m}}_{k+1}' \subset \boldsymbol{H} \oplus S^2\boldsymbol{H} \oplus \ldots \oplus S^{k+1}\boldsymbol{H}.$$

The image of the composition

$$\mu \circ (\alpha_k \otimes \alpha_k) \circ \Delta : \bar{\mathfrak{m}}_{k+1}' \to C^3(L; L), \tag{6}$$

where $\Delta : \bar{\mathfrak{m}}_{k+1}' \to \mathfrak{m}_k' \otimes \mathfrak{m}_k'$ is the comultiplication, is contained in $\operatorname{Ker} \delta$ and the composition

$$\pi \circ \mu \circ (\alpha_k \otimes \alpha_k) \circ \Delta : \bar{\mathfrak{m}}_{k+1}' \to H^3(L; L)$$

acts as zero on $\mathfrak{m}_k'$. We put

$$\mathfrak{m}_{k+1}' = \operatorname{Ker}(\pi \circ \mu \circ (\alpha_k \otimes \alpha_k) \circ \Delta) \supset \mathfrak{m}_k'.$$

The map $\alpha_k : \mathfrak{m}_k' \to C^2(L; L)$ is extended to the map $\alpha_{k+1} : \mathfrak{m}_{k+1}' \to C^2(L; L)$ such that $\delta \circ \alpha_{k+1}$ is the restriction of the map (6). The dual to $\mathfrak{m}_{k+1}'$ is

$$\mathfrak{m}_{k+1} = \frac{\mathfrak{m}}{\mathfrak{m}^{k+2} + J_{k+1}},$$

and we put

$$C_{k+1} = \mathbb{K} \oplus \mathfrak{m}_{k+1} = \frac{\mathbb{K}[\boldsymbol{H}']}{\mathfrak{m}^{k+2} + J_{k+1}}.$$

This completes the construction.

PROPOSITION 4.2. *If* $\dim H^3(L; L) = q$, *then the ideal* $I = \overleftarrow{\lim} I_k = \overleftarrow{\lim} J_k$ *from Proposition 3.3 has at most q generators. Less formally, the base of the versal deformation of L is the zero locus of a formal map* $H^2(L; L) \to H^3(L; L)$.

REFERENCES

[A] Artin, M., "Algebraization of formal moduli I," Global Analysis, (Papers in Honor of K. Kodaira), edited by D. C. Spencer and S. Iyanaga, University of Tokyo Press, Tokyo; Princeton University Press, Princeton, N.J. 1969, 21–71.

[Fi1] Fialowski, A., "Deformations of Lie Algebras," Math. USSR- Sbornik, 55:2 (1986), 467–473.

[Fi2] Fialowski, A., "An example of formal deformations of Lie algebras," NATO Advanced Study Institute on Deformation Theory of Algebras and Structures and Applications (1986 : Castelvecchio Pascoli, Italy), edited by Michiel Hazewinkel and Murray Gerstenhaber (Il Ciocco, 1986). Proceedings, NATO Adv. Sci. Inst. Ser. C: Math. Phys. Sci., 247, Kluwer Acad. Publ., Dordrecht, 1988, 375–401.

[FiFu] Fialowski, A., Fuchs, D., "Construction of Miniversal Deformations of Lie Algebras," Journal of Functional Analysis 161:1 (1999), 76–110.

[FuL] Fuchs, D., Lang, L., "Massey products and Deformations," To appear in Journal of Pure and Applied Algebra.

[Ge] Gerstenhaber, M., "On the deformation of rings and algebras, I, III," Ann. Math. 79 (1964), 59–103; 88 (1968), 1–34.

[GK] Grauert, H., Kerner, H., "Deformationen von Singularitäten komplexer Raume," Math. Annalen 153 (1964), 236-260.

[Harr] Harrison, D. K., "Commutative algebras and cohomology," Trans. Amer. Math. Soc. 104 (1962), 191–204.

[I] Illusie, L., "Complexe cotangent et déformations I," Lect. Notes in Math. 239, Springer (1971).

[KS] Kodaira K., Spencer, D.C., "On deformations of complex analytic structures I-II," Ann. of Math., 67 (1958), 326–466.

[Ko] Kontsevich, M., Topics in Algebra: Deformation Theory, Lecture Notes, Univ. Calif. Berkeley (1994).

[Ku] Kuranishi, M., "New proof for the existence of locally complex families of complex structures," in Proceedings of the Conference on Complex Analysis, Minneapolis, Minn., 1964, edited by A. Aeppli, E. Calabi, H. Röhrl, Springer-Verlag, Berlin-Heidelberg-New York 1965, 142–154.

[La] Laudal, O. A., "Formal Moduli of Algebraic Structures," Lect. Notes 754, Springer (1979).

[NR] Nijenhuis, A., Richardson, R. W., "Cohomology and deformations in graded Lie algebras," Bull. Amer. Math. Soc., 72 (1966), 1–29.

[P] Palamodov, V. P., "Deformations of complex spaces," Russian Math. Surveys, 31 (1976).

[Sch] Schlessinger, M., "Functors of Artin rings," Trans. Amer. Math. Soc. 130 (1968), 208–222.

EMAIL: Alice Fialowski `fialowsk@cs.elte.hu`

Derivations in second order Bernstein algebras

M.A. GARCIA-MUÑIZ, C. MARTÍNEZ*. Departamento de Matemáticas. Universidad de Oviedo. 33007 Oviedo, España.

0 INTRODUCTION

We shall consider commutative, finite dimensional algebras over an infinite field K, char $K \neq 2$. A baric algebra (A, ω) is a K-algebra having a nonzero homomorphism of algebras $\omega : A \to K$ (weight homomorphism).

We will denote by x^i the principal powers of an element $x \in A$, defined inductively by $x^1 = x$, $x^{i+1} = xx^i$ for $i \geq 1$. And $x^{[i]}$ denotes the plenary powers of x, defined by $x^{[1]} = x$, $x^{[i+1]} = (x^{[i]})^2$ for $i \geq 1$.

A Bernstein algebra is a baric algebra such that the identity $(x^2)^2 = \omega(x)^2 x^2$ holds for every $x \in A$.

A second order Bernstein algebra is a baric algebra that satisfies the identity $x^{[4]} = \omega(x)^4 x^{[3]}$. We list some known properties of second order Bernstein algebras.

If (A, ω) is a second order Bernstein algebra, the weight homomorphism is unique. So $Ker\,\omega$ is an ideal of codimension one uniquely determined.

There are always nonzero idempotent elements. We will denote $E(A)$ the set of nonzero idempotent elements in A. If e is one of them then $\omega(e) = 1$ and A has a decomposition $A = Ke + Ker\,\omega$ with $Ker\,\omega = U_e + Z_e$ where $U_e = \{x \in Ker\,\omega \mid ex = \frac{1}{2}x\}$ and $Z_e = \{x \in Ker\,\omega \mid e(ex) = 0\}$. We denote $C_{1e} = \{z \in Z_e \mid ez = 0\}$.

The following identities hold: $U_e^2 \subset Z_e$, $(Ker\,\omega)C_{1e} \subset U_e + C_{1e}$, $eu^3 + u(eu^2) = 0$, $e(uz) + u(ez) \in U_e$, $ez^2 + 2(ez)^2 \in U_e$, $z^2(ez) \in U_e$, $uz^2 + 4(uz)(ez) = 2e[uz^2 + 4(uz)(ez)] + 2u[ez^2 + 2(ez)^2]$, $u^3 + 2u(eu^2) \in Z_e$, for every $u \in U_e$, $z \in Z_e$.

The behaviour of products in second order Bernstein algebras is not so well known as in the Bernstein case and identities involving elements are quite more

* Partially supported by DGICYT PB94-1311-C03-01 and FC-97-PBPGI97-03

complicated.

The set $\{(e+u+u^2)^2 \mid u \in U_e\}$ is contained in $E(A)$ and the equality of both sets holds if $dim\, A = 3$ or 4 but not in general.

The dimensions of the Peirce components U_e and Z_e do not depend on the particular idempotent considered in A. So the couple $(1+r, s)$, $r = dim\, U_e$, $s = dim\, Z_e$, is an invariant called type of A $(1+r+s = dim\, A)$.

Let $A = Ke + U_e + Z_e$ be a power-associative second order Bernstein algebra. Then we have:

1) For every $u \in U_e$, $z \in Z_e$ the following identities hold:
$$eZ_e = 0, \tag{1}$$
$$U_eZ_e \subset U_e, \tag{2}$$
$$Z_e^2 \subset Z_e, \tag{3}$$
$$u^2z = 2u(uz), \tag{4}$$
$$uz^2 = 2z(uz), \tag{5}$$
$$z^4 = 0, \tag{6}$$
$$u^3 = 0. \tag{7}$$
2) In particular, if A is a Jordan algebra then $z(z'z^2) = 0$ for all $z, z' \in Z_e$.
3) $E(A) = \{(e+u+u^2)^2 \mid u \in U_e\}$. If $f = (e+\bar{u}+\bar{u}^2)^2 \in E(A)$ with $\bar{u} \in U_e$ then $U_f = \{u + 2\bar{u}u \mid u \in U_e\}$, $Z_f = \{z - 2\bar{u}z \mid z \in Z_e\}$.

A derivation of an algebra A is a linear map $D : A \to A$ that verifies $D(xy) = D(x)y + xD(y)$ for every $x, y \in A$. We denote $Der(A)$ the Lie algebra of derivations of A.

There are several results about the derivation algebra of a Bernstein algebra (A, ω). Worz-Busekros has proved that if $(1+r, d)$ is the type of such an algebra then $\dim_K Der(A) \leq r + r^2 + d^2$ and the equality is satisfied only if $(Ker\, \omega)^2 = 0$. S. González and C. Martínez (see [4]) find Bernstein algebras A with $Der(A) = 0$ and determine their properties.

In the present paper we will study the algebra of derivations $Der(A)$ in a second order Bernstein algebra A. A bound for $dim_K Der(A)$ will be found. This bound is reached in the case of a second order Bernstein algebra that verifies: a) $eZ \neq 0$ for some idempotent element or b) $(U+Z)^2 \subset Z$ or c) A is power-associative. Some particular cases of power-associative second order Bernstein algebras with $Der(A) \neq 0$ are studied. We will prove that in a Jordan second order Bernstein algebra $Der(A) \neq 0$, and we will give a description of the set of inner derivations of A.

1 DERIVATIONS

Let (A, ω) be a second order Bernstein algebra and $D \in Der(A)$. We can characterize D by a triple $(\tilde{u}, f, g)$ in a similar way to Bernstein algebras (see [4]).

THEOREM 1.1. (see [2]) Let $A = Ke + U + Z$ be a second order Bernstein algebra. Let $D : A \to A$ be a derivation. Then D defines and is defined uniquely by a triple $(\tilde{u}, f, g)$ with $\tilde{u} = D(e) \in U$; $f : U \to U$, $g : Z \to Z$ linear maps which verify

$$D(u) = f(u) + [2D(e)u + 4e(D(e)u)] \in U + Z,$$

$$D(z) = -4[D(e)(ez) + e(D(e)z)] + g(z) \in U + Z,$$

1) $g(uu') = f(u)u' + uf(u') + 2[D(e)u + 2e(D(e)u)]u' + 2[D(e)u' + 2e(D(e)u')]u + 4[D(e)(e(uu')) + e(D(e)(uu'))]$,
2) $4f(e(e(uz))) + g(uz - 4e(e(uz))) = f(u)z + ug(z) + 2[D(e)u + 2e(D(e)u)]z - 4u[D(e)(ez) + e(D(e)z)] - 8[D(e)(e(e(uz))) + 2e(D(e)(e(e(uz))))] + 4[D(e)(e(uz - 4e(e(uz)))) + e(D(e)(uz - 4e(e(uz))))]$,
3) $4f(e(e(zz'))) + g(zz' - 4e(e(zz'))) = g(z)z' + zg(z') - 4z'[D(e)(ez) + e(D(e)z)] - 4z[D(e)(ez') + e(D(e)z')] - 8[D(e)(e(e(zz'))) + 2e(D(e)(e(e(zz'))))] + 4[D(e)(e(zz' - 4e(e(zz')))) + e(D(e)(zz' - 4e(e(zz'))))]$,
4) $g(ez) = eg(z) + D(e)z - 4e(e(D(e)z))$, for every $u, u' \in U$; $z, z' \in Z$.

Also the linear map $Der(A) \to U \times End_K(U) \times End_K(Z)$, $D \to (\tilde{u}, f, g)$, is injective.

Let (A, ω) be a Bernstein algebra of type $(1 + r, d)$. It is known that $\dim_K Der(A) \leq r + r^2 + d^2$ and we have the equality if and only if A is trivial ($(Ker\,\omega)^2 = 0$).

Now let (A, ω) be a second order Bernstein algebra with type $(1 + r, d)$, $r = dim_K U$, $d = dim_K Z$; it is known that type of A is an invariant of A. By Theorem 1.1 we have $\dim_K Der(A) \leq r + r^2 + d^2$. Let us see that the equality is imposible.

PROPOSITION 1.1. Let $A = Ke + U + Z$ be a Bernstein algebra of orden less than or equal to two. Then

$$dim_K Der(A) = r + r^2 + d^2 \text{ if and only if } (Ker\,\omega)^2 = 0 \text{ and } eZ = 0.$$

In this case A is a Bernstein algebra.

Proof:

Notice that, when $(Ker\,\omega)^2 = 0$, condition $eZ = 0$ does not depend on the chosen idempotent.

Let $\dim_K Der(A) = r + r^2 + d^2$. Then the linear map of Theorem 1.1

$$\theta : Der(A) \to U \times End_K(U) \times End_K(Z), \; D \to (\tilde{u}, f, g),$$

is a bijection and every triple $(\tilde{u}, f, g)$ defines a derivation.

Considering $D \equiv (0, 0_U, id_Z)$, we get $U^2 = 0$, $Z^2 = 0$ and $UZ \subset Z$.

With $D \equiv (0, id_U, 0_Z)$ we get $UZ = 0$.

Let us suppose $eZ \neq 0$. Then there is $z_1 \in Z$ such that $ez_1 \neq 0$; since z_1, ez_1 are linearly independent, we can consider a basis of Z that contains these elements, $\{z_1, z_2 = ez_1, \ldots, z_d\}$, and define a linear map $g : Z \to Z$ by $g(z_i) = z_1$ $\forall i = 1, \ldots, d$. Then the triple $(0, f, g)$, with any f, does not verify condition 4 in Theorem 1.1, which is a contradiction.

Conversely, if $eZ = 0$ and $(Ker\,\omega)^2 = 0$, every triple $(\tilde{u}, f, g)$ satisfies conditions 1, 2, 3 and 4 in Theorem 1.1, and defines a derivation. ◇

THEOREM 1.2. Let (A, ω) be a second order Bernstein algebra of type $(1 + r, d)$, such that for some idempotent element e, $eZ \neq 0$; $A = Ke + U + Z$. We denote $\dim_K(eZ) = d_2$, $\dim_K C_1 = d_1$. Then

$$dim_K Der(A) \leq r + r^2 + (d_2^2 + d_1^2)$$

and this bound is reached if and only if $(Ker\,\omega)^2 = 0$.

Proof:

Since $eZ \subset C_1 \subset Z$, we have $d_2 \leq d_1 \leq d$ and $d_2 + d_1 = d$.

If a derivation D is associated to a triple $(\tilde{u}, f, g)$, we can consider the linear map $h : Z \to Z$ **(uniquely defined by $\tilde{u}$)** given by $h(z) = \tilde{u}z - 4e(e(\tilde{u}z))$. Notice that

$$g(ez) = eg(z) + h(z) \quad \text{for all } z \in Z. \tag{8}$$

We will consider a basis of $Z = \underbrace{eZ + C_2'}_{=C_1} + \underbrace{C_1'}_{dim\ d_2}$ of the form:

$$B = \{ez_1, \ldots, ez_{d_2}, z_{d_2+1}, \ldots, z_{d_1}, z_1, \ldots, z_{d_2}\},$$

such that $\{z_1, \ldots, z_{d_2}\}$ is a basis of C_1' a complementary of C_1 in Z, $\{ez_1, \ldots, ez_{d_2}\}$ is a basis of eZ and $\{z_{d_2+1}, \ldots, z_{d_1}\}$ is a basis of C_2' a complementary of eZ in C_1. In order to know g we need to know the images by g of any of the elements in the basis B.

1) If $ez_i \in eZ$, $g(ez_i) = eg(z_i) + h(z_i) = [eg(z_i) + a_1(ez_1) + \ldots + a_{d_2}(ez_{d_2})] + a_{d_2+1}z_{d_2+1} + \ldots + a_{d_1}z_{d_1} + b_1z_1 + \ldots + b_{d_2}z_{d_2}$. So the coordenates of $g(ez_i)$ in $z_{d_2+1}, \ldots, z_{d_1}, z_1, \ldots, z_{d_2}$ are the same as the corresponding coordenates of $h(z_i)$. That is, in order to know $g(ez_1), \ldots, g(ez_{d_2})$ we need to know the first d_2 coordenates of these vectors. In total we have to determine no more than $d_2d_2 = d_2^2$ parameters.
2) If $z_i \in \{z_{d_2+1}, \ldots, z_{d_1}\}$ then $z_i \in C_1$ and consequently $h(z_i) \in eZ$. By (8), $0 = g(ez_i) = eg(z_i) + h(z_i)$. This implies that the last d_2 coordenates of $g(z_i)$ are given by $h(z_i)$. In other words, the first d_1 coordenates of $g(z_i)$ determine (together with h) all coordenates of $g(z_i)$. In total we need at most $(d_1 - d_2)d_1 = d_1^2 - d_2d_1$ parameters.
3) Finally if $z_i \in \{z_1, \ldots, z_{d_2}\}$ and $h(z_i) = a_1(ez_1) + \ldots + a_{d_2}(ez_{d_2}) + a_{d_2+1}z_{d_2+1} + \ldots + a_{d_1}z_{d_1} + b_1z_1 + \ldots + b_{d_2}z_{d_2}$ as before, using results in the part 1 and (8) we have

$$g(ez_i) = ez' + a_{d_2+1}z_{d_2+1} + \ldots + a_{d_1}z_{d_1} + b_1z_1 + \ldots + b_{d_2}z_{d_2} = eg(z_i) + h(z_i)$$

where $ez' \in eZ$ is known because $g(ez_i)$ in known by part 1. So $eg(z_i) = e(z' - a_1z_1 - \ldots - a_{d_2}z_{d_2}) = ez''$, and $z'' \in Z$, that is $eg(z_i)$, is known in terms of h and $g(ez_i)$. This implies that the last d_2 coordenates of $g(z_i)$ are known, and only the first d_1 coordenates have to be given. In other words, we have not more than d_1d_2 parameters to be determined in order to know $g(z_1), \ldots g(z_{d_2})$.

We have proved that given the triple $(\tilde{u}, f, g)$, associated to $D \in Der(A)$, once we know $\tilde{u}$ and f, we need not more than $d_2^2 + (d_1^2 - d_1d_2) + d_1d_2 = d_2^2 + d_1^2$ parameters to fix g. So $dim_K Der(A) \leq r + r^2 + (d_2^2 + d_1^2)$.

Let us see that this bound is reached if and only if $(Ker\,\omega)^2 = 0$.

If $(Ker\,\omega)^2 = 0$, a derivation D is determined by $(\tilde{u}, f, g)$ if and only if it verifies condition (8) with $h(z) = 0$ for all $z \in Z$, that is, if $g(ez) = eg(z)$ for every $z \in Z$. This means that g can be defined taking in any arbitrary way the first d_2 coordenates of vectors $g(ez_i)$, $i = 1, \ldots, d_2$, and the first d_1 coordenates with respect to the basis B of vectors $g(z_i)$, $i = 1, \ldots, d_1$. Therefore we have $\dim_K Der(A) = r + r^2 + (d_2^2 + d_1^2)$.

If $\dim_K Der(A) = r + r^2 + (d_2^2 + d_1^2)$, then every $\tilde{u} \in U$ and every $f \in End_K(U)$ are part of a triple $(\tilde{u}, f, g)$ defining a derivation. In particular if $\tilde{u} = 0$ then $h(z) = 0$ and condition (8) reduces to $eg(z_i) = g(ez_i)$, for $i = 1, \ldots, d_1$. This assures that $(0, id_U, 0_Z)$ and $(0, 0_U, id_Z)$ are derivations what implies, in the way used in Proposition 1.1, that $(Ker\,\omega)^2 = 0$. ◇

Notice that for another idempotent f satisfying also $fZ_f \neq 0$ we get (in general) a different bound since d_1 and d_2 are not invariant (only $d = d_1 + d_2$ is invariant).

This result can be improved in the case of a second order Bernstein algebra such that $(U + Z)^2 \subset Z$. This condition does not depend on the idempotent e (see [16]).

THEOREM 1.3. Let (A, ω) be a second order Bernstein algebra with type $(1+r, d)$ such that $(U + Z)^2 \subset Z$ and for some idempotent element e, $eZ = 0$, $Z^2 \neq 0$. Then $\dim_K Der(A) \leq r + r^2 + (d^2 - 2d + 2)$.

Proof:

If $A = Ke + U + Z$, since $(U + Z)^2 \subset Z$, we know that $z^{[3]} = 0$ for every $z \in Z$ (see [6]). If $z \in Z$ satisfies $z^2 \neq 0$, then z, z^2 are linearly independent. So we can consider a basis of Z of the form $\{z_1 = z, z_2 = z^2, z_3, \ldots, z_d\}$.

Let $D \equiv (\tilde{u}, f, g) \in Der(A)$; then, by Theorem 1.1 and the identity $u(ez) + e(uz) = 0$ for all $u \in U$, $z \in Z$, that holds in this type of algebras, $g \in Der(Z)$. Let us see that $Der(Z)$ has dimension less or equal than $d^2 - 2d + 2$.

We consider the vector subspace of $End_K(Z)$, $G = K(g_{12}, g_{22}, \ldots, g_{d2}, g_{13}, g_{14}, \ldots, g_{1d})$, where

$$g_{i2}(z_2) = z_i,\ g_{i2}(z_h) = 0,\ h \neq 2,\ i = 1, \ldots, d,$$

$$g_{1j}(z_j) = z_1,\ g_{1j}(z_h) = 0,\ h \neq j,\ j = 3, \ldots, d.$$

It can be easily proved that $\dim_K G = 2d - 2$.

Let us see that $G \cap Der(Z) = 0$; Let $g \in G \cap Der(Z)$, $g = \alpha_{12} g_{12} + \alpha_{22} g_{22} + \cdots + \alpha_{d2} g_{d2} + \beta_{13} g_{13} + \cdots + \beta_{1d} g_{1d}$. Then $g(z_2) = 2z_1 g(z_1) = 0$ and so $\alpha_{i2} = 0$ for $i = 1, \ldots, d$; we obtain also that $g(Z) \subset K(z_1)$. For $j = 3, \ldots, d$ we have $g(z_1 z_j) = g(z_1) z_j + z_1 g(z_j) = \beta_{1j} z_2 \in K(z_1)$; so $\beta_{1j} = 0$ and $g = 0$.

Therefore $\dim_K Der(Z) \leq d^2 - 2d + 2$ and $\dim_K Der(A) \leq r + r^2 + (d^2 - 2d + 2)$. ◇

Corollary 1.1. Let (A, ω) be a second order Bernstein algebra of type $(1+r, d)$ such that $(U + Z)^2 \subset Z$. Then $\dim_K Der(A) \leq r + r^2 + d^2 - 2$.

Proof:

If there is an idempotent e such that $eZ \neq 0$ then we can apply Theorem 1.2.

If $eZ = 0$ for every $e \in E(A)$ and Peirce decomposition $A = Ke + U + Z$, then $Z^2 \neq 0$, since, otherwise, we could find an idempotent element $f = e + u + u^2 + 2u^3$

such that $fZ_f \neq 0$. Now we can apply Theorem 1.3. ◊

2 LOWER BOUNDS

In a second order Bernstein algebra the lower bound for $\dim_K Der(A)$ is zero. There are algebras with $eZ \neq 0$ for some idempotent element e verifying $Der(A) = 0$ (see Example 2.2). And there are algebras with $eZ = 0$ for every idempotent e verifying $Der(A) = 0$ as in the following example.

EXAMPLE 2.1. Let $A = K(e, z_1, \ldots, z_5)$ be the second order Bernstein algebra with products $ez_i = 0$ for $i = 1, \ldots, 5$, $z_1^2 = z_1z_5 = z_2z_3 = z_4$, $z_1z_2 = z_2z_4 = z_5$, $z_1z_3 = z_1z_4 = z_2^2 = z_2z_5 = z_3$ and the other products equal to zero. Then $Der(A) = 0$.

In some particular cases it is known that $Der(A) \neq 0$.

Corollary 2.1. Let $A = Ke + U + Z$ be a second order Bernstein algebra.
1) If $UZ = 0 = Z^2$ then $Der(A) \neq 0$.
2) If $UZ = 0 = U^2$, $eZ \neq 0$, $U \neq 0$, then $Der(A) \neq 0$.
Proof:

In case 1 we have $eZ \neq 0$ because otherwise A is a Bernstein algebra. If $U = 0$ then $(Ker\, w)^2 = 0$ and $\dim_K Der(A) = d_1^2 + d_2^2 \neq 0$ by Theorem 1.2. If $U \neq 0$, every triple $(\tilde{u}, 0_U, 0_Z)$ defines a derivation and then $0 \neq r \leq dim_K Der(A)$.

In case 2 every triple $(\tilde{u}, 0_U, 0_Z)$ defines a derivation. ◊

If $U = 0$, there are algebras verifying $Der(A) = 0$ and algebras such that $Der(A) \neq 0$, as we can see in the following examples. Notice that if $U = 0$ then there is only one idempotent element in A.

EXAMPLE 2.2. (char $K \neq 3$) Let $A = K(e, z_1, \ldots, z_5)$ be the second order Bernstein algebra with products $e^2 = e$, $ez_1 = ez_2 = z_5$, $z_1z_2 = z_2z_4 = z_3$, $z_1z_3 = z_4$, $z_1z_4 = z_2z_3 = z_5$, $z_1z_5 = -z_3$ and the other products equal to zero. Then $Der(A) = 0$.

EXAMPLE 2.3. Let $A = K(e, z_1, \ldots, z_5)$ be the second order Bernstein algebra with products $e^2 = e$, $ez_1 = z_5$, $z_1z_2 = z_2z_4 = z_3$, $z_1z_3 = z_4$, $z_2z_3 = z_5$, $z_1z_5 = -z_3$ and the other products equal to zero. Then $Der(A) \neq 0$.

PROPOSITION 2.1. Let $A = Ke + Z$ be a second order Bernstein algebra of type $(1, d)$ such that $eZ = 0$ and $Z^k = 0$ for some k. Then $Der(A) \neq 0$.
Proof:

We have $Z^{k-1} \neq 0$. Then $k > 2$ because in other case A is a Bernstein algebra. In particular $Z^2 \neq Z$. If W is a suplementary of Z^2 in Z, $Z = Z^2 + W$, every linear map $g \in End_K(Z)$ such that $g(Z^2) = 0$, $g(W) \subset Z^{k-1}$ is a derivation of Z.

Proposition is proved. ◊

3 RIGHT MULTIPLICATIONS

A derivation D of an algebra A is called inner if $D \in \mathcal{L}(A)$, being $\mathcal{L}(A)$ the Lie algebra generated by right and left multiplications R_x, L_x, $x \in A$.

Let us see necessary and sufficient conditions for R_u, R_z, with $u \in U$, $z \in Z$, to be derivations in a second order Bernstein algebra. We can easily prove the following two propositions.

PROPOSITION 3.1. Let $A = Ke + U + Z$ be a second order Bernstein algebra. If $\bar{u} \in U$ then $R_{\bar{u}}$ is a derivation if and only if the following relations hold: $\bar{u}U \subset C_1$, $\bar{u}Z = 0$, $(\bar{u}u)u = 0$, $(\bar{u}u)z = \bar{u}(uz)$, $\bar{u}Z^2 = 0$, for every $u \in U$, $z \in Z$; in this case the triple associated to $R_{\bar{u}}$ is $(\frac{1}{2}\bar{u}, 0_U, 0_Z)$.

PROPOSITION 3.2. Let $A = Ke + U + Z$ be a second order Bernstein algebra. If $\bar{z} \in Z$ then $R_{\bar{z}}$ is a derivation if and only if the following relations hold: $e\bar{z} = 0$, $\bar{z}U \subset U$, $(ez)\bar{z} = 0 = e(z\bar{z})$, $\bar{z}u^2 = 2u(\bar{z}u)$, $\bar{z}(uz) = (\bar{z}u)z + (\bar{z}z)u$, $\bar{z}z^2 = 2z(\bar{z}z)$, for every $u \in U$, $z \in Z$.

The following examples give second order Bernstein algebras in which there are non zero derivations of type R_u or R_z.

EXAMPLE 3.1. Let $A = K(e, u, z_1, z_2, z_3)$ be the second order Bernstein algebra with products $e^2 = e$, $eu = \frac{1}{2}u$, $u^2 = z_1z_2 = z_3$ and the other products equal to zero. Then R_u is a derivation of A.

EXAMPLE 3.2. Let $A = K(e, z_1, \ldots, z_5)$ be the second order Bernstein algebra with products $e^2 = e$, $eZ = 0$, $z_1z_2 = z_2z_4 = z_3$, $z_1z_3 = z_4$, $z_2z_3 = z_5$, $z_1z_5 = -z_3$ and the other products equal to zero. Then $R_{\bar{z}}$ is a derivation of A if and only if $\bar{z} \in K(z_3)$.

In a Bernstein algebra $A = Ke + U + Z$, it is known that if $x = \bar{u} + \bar{z}$ and R_x is a derivation, then $R_{\bar{u}}$ and $R_{\bar{z}}$ are derivations too.

This is not true in general for a second order Bernstein algebra, even if $eZ = 0$ for some idempotent element. We can see that in the following example.

EXAMPLE 3.3. Let $A = K(e, u_1, u_2, z_1, z_2, z_3)$ be the second order Bernstein algebra with products $e^2 = e$, $eu_i = \frac{1}{2}u_i$, $u_1^2 = z_2^2 = z_1$, $u_1u_2 = z_3$, $u_2^2 = z_2$, $u_2z_3 = \frac{1}{2}z_1$ and the other products equal to zero. Then if $x = u_1 + z_2$, R_x is a derivation of A but R_{u_1} is not.

4 DERIVATIONS IN POWER-ASSOCIATIVE SECOND ORDER BERNSTEIN ALGEBRAS

Let $char\, K \neq 2, 3$. We will consider power-associative second order Bernstein algebras. In these case Theorem 1.1 can be reformulated in an easier way.

PROPOSITION 4.1. Let $A = Ke + U + Z$ be a power-associative second order Bernstein algebra, $D : A \to A$ a derivation. Then D defines and is defined by a triple $(\tilde{u}, f, g)$ with $\tilde{u} \in U$; $f : U \to U$, $g : Z \to Z$ linear maps such that $D(e) = \tilde{u}$, $D(u) = f(u) + 2D(e)u$, $D(z) = -2D(e)z + g(z)$, and
1) $g(uu') = f(u)u' + uf(u')$,
2) $f(uz) = f(u)z + ug(z)$,
3) $g(zz') = g(z)z' + zg(z')$, for every $u \in U$, $z \in Z$.

Let A be a power-associative second order Bernstein algebra. We will find an upper bound of $\dim_K Der(A)$. Notice that $Z^2 \neq 0$ is needed because, in other case, identities (4), (5) and (7) would imply that A is a Bernstein algebra.

THEOREM 4.1. Let (A, ω) a second order Bernstein algebra of type $(1 + r, d)$ that is Jordan and such that $z^3 = 0$ for every $z \in Z$. Then

$$dim_K Der(A) \leq r + r^2 + (d^2 - 2d + 2).$$

Proof:
Let $z \in Z$ such that $z^2 \neq 0$. Then z, z^2 are linearly independent and we have a basis of Z, $\{z_1 = z, z_2 = z^2, z_3, \ldots, z_d\}$. By Proposition 4.1, if $(\tilde{u}, f, g)$ defines a derivation then $g \in Der(Z)$. The result can be proved in the same way that Theorem 1.3 was proved. ◇

Notice that a power-associative second order Bernstein algebra such that $z^3 = 0$ for every $z \in Z$ is a Jordan algebra (see [2]).

The following Example shows that bounds in Theorems 1.3 and 4.1 can be reached.

EXAMPLE 4.1. Let U and Z be vector spaces of dimensions r and d respectively, $\{z_1, z_2, \ldots, z_d\}$ a basis of Z. We define a product in $A = Ke + U + Z$ by $e^2 = e$, $eu = \frac{1}{2}u$ for all $u \in U$, $z_1^2 = z_2$ and the other products equal to zero. Then A satisfies the hypotheses of Theorem 4.1 and $\dim_K Der(A) = r + r^2 + (d^2 - 2d + 2)$.

THEOREM 4.2. Let (A, ω) be a power-associative second order Bernstein algebra of type $(1 + r, d)$. If there is an element $z \in Z$ with $z^3 \neq 0$, then

$$dim_K Der(A) \leq r + r^2 + (d^2 - 4d + 6).$$

Proof:
Let $z \in Z$ such that $z^3 \neq 0$. Then z, z^2, z^3 are linearly independent. We can consider $\{z_1 = z, z_2 = z^2, z_3 = z^3, z_4, \ldots, z_d\}$ a basis of Z. By Proposition 4.1, if

$(\tilde{u}, f, g)$ defines a derivation then $g \in Der(Z)$. Let us see that dimension of $Der(Z)$ is bounded by $d^2 - 4d + 6$.

Let us consider $G=K(g_{i2}, g_{i3}, g_{1j}, g_{2j})$ the vector subspace of $End_K(Z)$ spanned by g_{i2}, g_{i3}, $i = 1, \ldots, d$ and g_{1j}, g_{2j}, $j = 4, \ldots, d$, where

$$g_{i2}(z_2) = z_i,\ g_{i2}(z_h) = 0,\ h \neq 2;$$

$$g_{i3}(z_3) = z_i,\ g_{i3}(z_h) = 0,\ h \neq 3;$$

$$g_{1j}(z_j) = z_1,\ g_{1j}(z_h) = 0,\ h \neq j;$$

$$g_{2j}(z_j) = z_2,\ g_{2j}(z_h) = 0,\ h \neq j.$$

It can be easily proved that $\dim_K G = 4d - 6$.

If $g \in G \cap Der(Z)$, then $g(z_2) = 2z_1 g(z_1) = 0$, $g(z_3) = g(z_1)z_2 + z_1 g(z_2) = 0$. So coefficients of g in g_{i2}, g_{i3}, $i = 1, \ldots, d$, are zero and $g(Z) \subset K(z_1, z_2)$. Similarly from $g(z_1 z_j) = z_1 g(z_j)$ and $g(z_2 z_j) = z_2 g(z_j)$ it follows that coefficients of g with respect to g_{1j}, g_{2j}, $j = 4, \ldots, d$, are zero. Then $g = 0$.

Hence $\dim_K Der(Z) \leq d^2 - 4d + 6$ and $\dim_K Der(A) \leq r + r^2 + (d^2 - 4d + 6)$.
◇

Again the bound in Theorem 4.2 can be reached as the following example shows.

EXAMPLE 4.2. Let U and Z be vector spaces of dimensions r and d respectively, $d \geq 3$, $\{z_1, z_2, \ldots, z_d\}$ a basis of Z. We define a product in $A = Ke + U + Z$ by $e^2 = e$, $eu = \frac{1}{2}u$ for all $u \in U$, $z_1^2 = z_2$, $z_1 z_2 = z_3$ and the other products equal to zero. Then A is a Jordan second order Bernstein algebra and $\dim_K Der(A) = r + r^2 + (d^2 - 4d + 6)$.

PROPOSITION 4.2. Let (A, ω) be a power-associative second order Bernstein algebra. Then $\dim_K Der(A) \leq r + [t_1^2 + r(r - t_1)] + [t_2^2 + d(d - t_2)]$, where $t_1 = dim_K UZ$ and $t_2 = dim_K(U^2 + Z^2)$.

Proof:

Notice that in a power-associative second order Bernstein algebra, $\dim_K UZ$ and $\dim_K(U^2 + Z^2)$ do not depend on the particular idempotent element (see [2]).

Let $D \equiv (\tilde{u}, f, g) \in Der(A)$. It follows from Proposition 4.1 that $f(UZ) \subset UZ$ and $g(U^2 + Z^2) \subset U^2 + Z^2$. ◇

We will study now the existence of non zero derivations in a power-associative second order Bernstein algebra. We will see that such derivations exist if $U \neq 0$.

Corollary 4.1. Let (A, ω) be a power-associative second order Bernstein algebra of type $(1 + r, d)$, $A = Ke + U + Z$. Then

1) If $U \neq 0$ then $Der(A) \neq 0$.
2) If $U = 0$ and $Z^k = 0$ for some integer k then $Der(A) \neq 0$.

Proof:

If $U \neq 0$ then the triple $(\tilde{u}, 0_U, 0_Z)$, $\tilde{u} \neq 0$, defines a non zero derivation. Case 2 is a consequence of Proposition 2.1. ◇

In particular, if $Der(A) = 0$ then $Ker\, w$ is not nilpotent. So we have the following corollary.

Corollary 4.2. Let (A, ω) be a power-associative second order Bernstein algebra. If $Der(A) = 0$ then A is not a genetic algebra.

Let $A = Ke + Z$ be a power-associative second order Bernstein algebra with $U = 0$. The existence of non zero derivations in A is equivalent to the existence of non zero derivations in a power-associative nilalgebra of nilindex 4 as we can see in the following theorem.

THEOREM 4.3. Let $A = Ke + Z$ be a power-associative second order Bernstein algebra. Then A is determined (up to isomorphism) by a power-associative nilalgebra of nilindex four, Z, and $Der(A) \cong Der(Z)$.

Proof:

Since $U = 0$, there exists a unique idempotent e in A, $A = Ke + Z$, Z is a power-associative algebra and, by identity (6), $x^4 = 0$ for every $x \in Z$.

Conversely, if Z is a power-associative nilalgebra of nilindex four we can extend the product in Z to a product in $A = Ke + Z$ by $e^2 = e$, $ez = 0$ for every $z \in Z$. If w is the weight homomorphism given by $w(e) = 1$, $w(z) = 0$, then (A, w) is a second order Bernstein algebra. Since identities (1) to (6) are satisfied, A is power-associative (see [7]). If Z and Z' are isomorphic algebras, also are A and A'.

We have an isomorphism $Der(A) \to Der(Z)$ given by $D \to D|_Z$. ◇

Now we have to worry about the existence of derivations in a power associative nilalgebra B of nilindex 4. Although we have some partial results (in the case $B^2B^2 = 0$), this is still an open problem. We will suppose that B is not a Jordan algebra because this case will be studied in Section 6.

Let B be a commutative, 5-dimensional K-algebra, char$K \neq 2, 3, 5$, that is not Jordan and satisfies $B^2B^2 = 0$ and $x^4 = 0$ for all $x \in B$. Then B it is a power-associative algebra (see [12]).

The example given by Suttles satisfies these conditions. We will prove that, up to isomorphism, it is the only one. Futhermore $\dim_K Der(B) = 5$.

Lemma 4.1. Let B be a commutative, 5-dimensional $K-$algebra such that $B^2B^2 = 0$ and $x^4 = 0$ for every $x \in B$. If B is not Jordan then $\dim_K B^2 = 3$.

Proof:

If $x^3 = 0$ for every $x \in B$, then B is a Jordan algebra. So we can consider an element $z \in B$ such that $z^3 \neq 0$. Then z, z^2, z^3 are linearly independent and consequently $\dim_K B^2 \geq 2$.

There exist linearly independent elements ω, $\omega' \in B$ such that $\omega(\omega^2\omega') \neq 0$, $\omega, \omega' \notin B^2$. So $\dim_K B^2 \leq 3$.

We will prove that if $\dim_K B^2 = 2$ then B is a Jordan algebra. Indeed, let $\dim_K B^2 = 2$. Then $\{z^2, z^3\}$ is a basis of B^2 and can be extended to a basis of B, $\{z, z^2, z^3, \omega_1, \omega_2\}$. If we apply identities

$$x^3x_1 + x(x^2x_1) + 2x(x(xx_1)) = 0, \tag{9}$$

$$x(xx_1^2) + x_1(x^2x_1) + 2x(x_1(xx_1)) + 2x_1(x(xx_1)) = 0, \tag{10}$$

with $x = z$, $x_1 = \omega_1$ and with $x = z$, $x_1 = \omega_2$, to the products $z\omega_1 = az^2 + bz^3$, $z\omega_2 = cz^2 + dz^3$, $z^2\omega_1 = ez^2 + fz^3$, $z^2\omega_2 = gz^2 + hz^3$, $z^3\omega_1 = iz^2 + jz^3$, $z^3\omega_2 = kz^2 + lz^3$, we get $i = j = k = e = g = l = 0$. Then $z^3 \in$ An B and B is a Jordan algebra. ◇

PROPOSITION 4.3. Let B be a commutative, 5-dimensional $K-$algebra such that $B^2B^2 = 0$ and $x^4 = 0$ for every $x \in B$. Supose that B is not Jordan. Then there exists a basis of B, $\{z, z^2, z^3, w, w^2\}$, with products $zw = a_1z^2 + a_2z^3 + a_3w^2$, $zw^2 = b_1z^2 + b_2z^3$, $z^2w = c_1z^2 + c_2z^3 + c_3w^2$, $z^3w = d_1z^2 + d_2z^3$, $w^3 = e_1z^2 + e_2z^3 + e_3w^2$ satisfying $c_3, b_1 \neq 0$ and

$$a_2 = \frac{1 - 4a_1a_3}{2c_3},\ c_1 = -c_3b_2 - 4a_3b_1,\ c_2 = \frac{2a_3(c_3b_2 + 4a_3b_1)}{c_3},$$

$$d_1 = -c_3b_1,\ d_2 = 2a_3b_1,\ e_1 = \frac{-(c_3b_2 + 2a_3b_1)^2}{c_3},$$

$$e_2 = \frac{2a_3(c_3b_2 + 2a_3b_1)^2}{c_3^2},\ e_3 = c_3b_2 + 2a_3b_1.$$

Proof:

Linearizing the identity $x^4 = 0$ we get
$x_1(x^2x_2) + x_2(x^2x_1) + 2x_1(x(xx_2)) + 2x_2(x(xx_1)) + 2x(x_1(xx_2)) + 2x(x_2(xx_1)) + 2x(x(x_1x_2)) = 0.$ (11)

By Lemma 4.1 $\dim_K B^2 = 3$ and there exists an element $z \in B$ such that $z^3 \neq 0$, $z \notin B^2$ and z, z^2, z^3 are linearly independent.

We can extend $\{z^2, z^3\}$ to a basis of B^2, $\{z^2, z^3, w^2\}$. Then $\{z, w, z^2, z^3, w^2\}$ is a basis of B. Indeed, if $w = \alpha z + \beta z^2 + \gamma z^3 + \delta w^2$ then $w^2 = \alpha^2z^2 + 2\alpha\beta z^3 + 2\alpha\delta zw^2$, $zw^2 = \alpha^2z^3 + 2\alpha\delta z(zw^2)$ and $z(zw^2) = 2\alpha\delta z[z(zw^2)] = 0$ by (9); it follows that $zw^2 = \alpha^2z^3$ and $w^2 = \alpha^2z^2 + (2\alpha\beta + 2\alpha^3\delta)z^3$ which is a contradiction.

Let us denote $zw = a_1z^2 + a_2z^3 + a_3w^2$, $zw^2 = b_1z^2 + b_2z^3 + b_3w^2$, $z^2w = c_1z^2 + c_2z^3 + c_3w^2$, $z^3w = d_1z^2 + d_2z^3 + d_3w^2$, $w^3 = e_1z^2 + e_2z^3 + e_3w^2$.

1) Applying identity (9) with $x = z$, $x_1 = w^2$ we get $b_3 = 0$.

2) Using identity (9) with $x = z$, $x_1 = w$ we have $d_3 = 0$, and
$d_1 + c_3b_1 = 0,$ (12)
$d_2 + c_1 + c_3b_2 + 2a_3b_1 = 0.$ (13)

3) From $w^4 = 0$ we get
$e_1c_1 + e_2d_1 + e_3e_1 = 0,$ (14)
$e_1c_2 + e_2d_2 + e_3e_2 = 0,$ (15)
$e_1c_3 + e_3^2 = 0.$ (16)

4) Applying identity (9) to $x = w$, $x_1 = z^3$ we get
$d_1c_1(c_1 + d_2) + d_1(d_1c_2 + d_2^2) + d_1c_3e_1 = 0,$ (17)
$d_1c_2(c_1 + d_2) + d_2(d_1c_2 + d_2^2) + d_1c_3e_2 = 0,$ (18)
$d_1c_3(c_1 + d_2) + d_1c_3e_3 = 0.$ (19)

5) Applying identity (9) to $x = w$, $x_1 = z^2$ we get
$c_1(c_1^2 + c_2d_1) + c_2d_1(c_1 + d_2) + c_1c_3e_1 = 0,$ (20)
$c_2(c_1^2 + c_2d_1) + c_2d_2(c_1 + d_2) + c_1c_3e_2 = 0,$ (21)
$c_3(c_1^2 + c_2d_1) + c_1c_3e_3 = 0.$ (22)

6) Applying identity (10) to $x = z$, $x_1 = w$ we get

$c_1^2 + c_2d_1 + c_3e_1 + 2a_1b_1c_3 + 2a_3b_1e_3 + 2a_1d_1 + 2a_3b_1c_1 + 2a_3b_2d_1 = 0,$ (23)
$b_1 + c_1c_2 + c_2d_2 + c_3e_2 + 2a_1c_1 + 2a_1b_2c_3 + 2a_2d_1 + 2a_3e_1 + 2a_3b_2e_3 + 2a_1d_2 +$
$2a_3b_1c_2 + 2a_3b_2d_2 = 0,$ (24)
$c_1c_3 + c_3e_3 + 2a_3b_1c_3 = 0.$ (25)

7) Applying identity (9) to $x = w$, $x_1 = z$ we have
$e_3b_1 + b_1c_1 + b_2d_1 + 2a_1c_1^2 + 2a_2d_1c_1 + 2a_1c_2d_1 + 2a_2d_2d_1 + 2a_1c_3e_1 = 0,$ (26)
$e_1 + e_3b_2 + b_1c_2 + b_2d_2 + 2a_1c_1c_2 + 2a_2d_1c_2 + +2a_1c_2d_2 + 2a_2d_2^2 + 2a_1c_3e_2 = 0,$ (27)
$b_1c_3 + 2a_1c_1c_3 + 2a_2d_1c_3 + 2a_1c_3e_3 = 0.$ (28)

8) Applying identity (11) to $x = w$, $x_1 = z$, $x_2 = z^3$ we obtain
$c_1 + 2d_2 + c_3b_2 = 0.$ (29)
Since A is not Jordan it follows that $b_1 \neq 0$ and $c_3 \neq 0$. By (12) $d_1 \neq 0$.
The rest of relations can be obtained from identities (12), (13), (29), (19), (22), (16), (14) and (28). Coefficients a_1, a_3, b_1, b_2, c_3 can be arbitrarily taken. ◇

The example given by Suttles is a particular case of this proposition.

Corollary 4.3. Let B be the $K-$algebra in Proposition 4.3 with $a_1 = \frac{1}{4}$, $a_3 = \frac{1}{2}$, $b_1 = 1$, $b_2 = 0$, $c_3 = 1$. Then there is another basis of B, $\{z_1, z_2, z_3, z_4, z_5\}$, such that the only non zero products are $z_1z_2 = z_3 = -z_1z_5 = z_2z_4$, $z_1z_3 = z_4$, $z_2z_3 = z_5$.
Proof:
Let $\{z, z^2, z^3, w, w^2\}$ be the basis used in Proposition 4.3. We can define the new basis by $z_1 = -z + \frac{1}{2}w^2$, $z_2 = -\frac{1}{2}z^2 + z^3 + w$, $z_3 = -\frac{3}{4}z^2 + \frac{3}{4}z^3$, $z_4 = \frac{3}{4}z^3$, $z_5 = \frac{3}{4}z^2 - \frac{3}{4}z^3 - \frac{3}{4}w^2$.◇

PROPOSITION 4.4. Let B be a commutative, 5-dimensional $K-$algebra such that $B^2B^2 = 0$ and $x^4 = 0$ for every $x \in B$. Supose that B is not Jordan. Then there exists a basis of B, $\{y_1, y_2, y_3, y_4, y_5\}$, such that the only non zero products are $y_1y_2 = y_3 = -y_1y_5 = y_2y_4$, $y_1y_3 = y_4$, $y_2y_3 = y_5$. Hence two arbitrary algebras satisfying the above conditions are isomorphic.
Proof:
Let $\{z, z^2, z^3, w, w^2\}$ be the basis used in Proposition 4.3. We define the basis of B, $\{y_1, y_2, y_3, y_4, y_5\}$ as follows:
If $a_1c_3 + 2a_3^2b_1 \neq 0$ then

$$y_1 = -\frac{1}{b_1}z - \frac{b_2}{2b_1^2}z^2 + \frac{1}{2b_1^2}w^2,$$

$$y_2 = -\frac{1}{2c_3^2}z^2 + \frac{c_1}{2c_3^2d_1}z^3 + \frac{1}{c_3}w,$$

$$y_3 = h\left(-z^2 + \frac{2a_3}{c_3}z^3\right),$$

$$y_4 = \frac{h}{b_1}z^3,$$

$$y_5 = \frac{h}{c_3}\left(\alpha z^2 - \frac{2a_3}{c_3}\alpha z^3 - c_3w^2\right),$$

where $h = \frac{1}{b_1 c_3}\left(a_1 + \frac{2a_3^2 b_1}{c_3}\right)$ and $\alpha = b_2 c_3 + 2a_3 b_1$.

If $a_1 c_3 + 2a_3^2 b_1 = 0$ then

$$y_1 = \frac{1}{c_3}z + \frac{1-2a_3}{2c_3^2}z^2 + \frac{\gamma}{c_3}z^3 - \frac{1}{c_3}w,$$

$$y_2 = -\frac{1}{b_1}z - \frac{b_2}{2b_1^2}z^2 + \frac{1}{2b_1^2}w^2,$$

$$y_3 = -\frac{1}{2b_1 c_3}z^2 + \frac{a_3}{b_1 c_3^2}z^3,$$

$$y_4 = \frac{1}{2b_1 c_3^2}\left(-\alpha z^2 + \left(-1 + \frac{2a_3}{c_3}\alpha\right)z^3 + c_3 w^2\right),$$

$$y_5 = \frac{1}{2b_1^2 c_3}z^3,$$

with γ defined by $2d_1\gamma = 1 - 2a_1 - \frac{c_1(1-2a_3)}{c_3}$. ◇

PROPOSITION 4.5. Let B be a $K-$algebra satisfying the hypothesis of Proposition 4.3. Then $\dim_K Der(B) = 5$.

Proof:

We will find the derivation algebra in the case of algebras with $b_1 = c_3 = 1$.

Let $\{z, z^2, z^3, w, w^2\}$ be a basis of B with the multiplication table given by Proposition 4.3. A derivation $D \in Der(B)$ is determined by

$$D(z) = \alpha_1 z + \alpha_2 z^2 + \alpha_3 z^3 + \beta_1 w + \beta_2 w^2,$$

$$D(w) = \lambda_1 z + \lambda_2 z^2 + \lambda_3 z^3 + \mu_1 w + \mu_2 w^2,$$

and the following five relations hold:

1) $\alpha_1 + \mu_1 = 0,$
2) $\alpha_2 + \left(a_2 - 2a_3^2(b_2 + 2a_3)\right)\beta_1 + b_2\beta_2 + a_3\mu_1 = 0,$
3) $\alpha_1 + 2\left(a_1 + a_3(b_2 + 2a_3)\right)\beta_1 + 2\beta_2 - 2a_3\lambda_1 - 2\lambda_2 - 2\mu_1 - 2(b_2 + 2a_3)\mu_2 = 0,$
4) $2a_1\lambda_1 - 2(b_2 + 4a_3)\lambda_2 - 2\lambda_3 + (b_2 + 2a_3)\mu_1 - 2(b_2 + 2a_3)^2\mu_2 = 0,$
5) $a_1\alpha_1 + (b_2 + 4a_3)\alpha_2 + \alpha_3 + \left(2a_1^2 - a_2(b_2 + 2a_3)\right)\beta_1 + \left(2a_1 + (b_2 + 2a_3)^2\right)\beta_2 +$ $(2a_1a_3 - 1)\lambda_1 - 2a_3(b_2 + 4a_3)\lambda_2 - 2a_3\lambda_3 - a_1\mu_1 - \left(2a_3(b_2 + 2a_3)^2 + 1\right)\mu_2 = 0.$

These equations are linearly independent and proposition is proved. ◇

PROPOSITION 4.6. Let B be an algebra as in Proposition 4.3. Then

1) $\mathrm{An}\, B = 0$.
2) If $x \in B$, R_x is a derivation if and only if $x \in K\left(z^2 - \frac{2a_3}{c_3}z^3\right)$.
   In particular, if $x \neq 0$ then $R_x \neq 0$.

Proof:

1) Let $x = \alpha z + \beta z^2 + \gamma z^3 + \delta \omega + \eta w^2 \in An\, B$. ¿From $z^3x = 0$, $z^2x = 0$, $zx = 0$ and $wx = 0$ we obtain $x = 0$.
2) Let $x = \alpha z + \beta z^2 + \gamma z^3 + \delta \omega + \eta w^2 \in B$. If $D \equiv R_x$ is a derivation of B then: $D(z^2) = 2zD(z)$ implies $\delta = 0$ and $\alpha + 2b_1\eta = 0$.

$D(zw) = D(z)w + zD(w)$ implies $\eta = 0$ and $\alpha = 0$.
$D(w^2) = 2wD(w)$ implies $\gamma = -\frac{2a_3}{c_3}\beta$. So $x = \beta\left(z^2 - \frac{2a_3}{c_3}z^3\right)$.
Conversely, for such elements x, R_x is a derivation. ◇

Now we will drop the condition $\dim_K B = 5$, that is, B will denote a commutative, finite dimensional K-algebra such that $B^2B^2 = 0$, $x^4 = 0$ for all $x \in B$, and is not Jordan. Let $B = W + B^2$ be a direct sum of vector spaces. We will prove that $\dim_K B^2 \geq 3$, $\dim_K W \geq 2$ and that $Der(B) \neq 0$ when $\dim_K B^2 = 3$ or $\dim_K W = 2$.

PROPOSITION 4.7. Let B be a commutative, finite dimensional K-algebra such that $B^2B^2 = 0$, $x^4 = 0$ for all $x \in B$, and is not Jordan. Then,
1) $\dim_K B^2 \geq 3$,
2) $\dim_K B/B^2 \geq 2$,
3) there exist linearly independent elements z, z^2, z^3, w, w^2.

Proof:
We will use the following identities (linearizations of $x^4 = 0$):

$$x^3x' + x(x^2x') + 2x(x(xx')) = 0, \quad (30)$$

$$x(xx_1^2) + x_1(x^2x_1) + 2x(x_1(xx_1)) + 2x_1(x(xx_1)) = 0. \quad (31)$$

1) If $x^3 = 0$ for every $x \in B$ then B is Jordan. So there exists an element $z \in B$ such that $z^3 \neq 0$.

Let $z \in B$ be such that $z^3 \neq 0$. Then z, z^2, z^3 are linearly independent and hence $\dim_K B^2 \geq 2$.

Let us suppose that $\dim_K B^2 = 2$. Then $\{z^2, z^3\}$ is basis of B^2. Let $B = W + B^2$ be a direct sum of vector spaces. Applying (30) with $x = z, x' = w$, $w \in W$, we obtain $z^3w \in K(z^3)$; from (31) with $x = z, x_1 = w$ we get $w(wz^2) \in K(z^3)$ and then $z^2w \in K(z^3)$ and $z^3w = 0$. Hence $z^3 \in An\, B$ and B is a Jordan algebra. This contradiction proves that $\dim_K B^2 \geq 3$.

2) There exist linearly independent elements ω, $\omega' \in B$ such that $\omega(\omega^2\omega') \neq 0$, $\omega, \omega' \notin B^2$. So $\dim_K W \geq 2$.

3) Let us consider $z \in B$ such that $z^3 \neq 0$. Then z, z^2, z^3 are linearly independent elements.

Since $\dim_K B^2 \geq 3$ we can find w^2 such that $\{z^2, z^3, w^2\}$ are linearly independent. Then z, z^2, z^3, w^2 are also linearly independent and so are z, z^2, z^3, w, w^2. Suppose $w = \alpha z + \beta z^2 + \gamma z^3 + \delta w^2$. Then $w^2 = \alpha^2 z^2 + 2\alpha\beta z^3 + 2\alpha\delta z w^2$, $zw^2 = \alpha^2 z^3 + 2\alpha\delta z(zw^2)$ and $z(zw^2) = 2\alpha\delta z[z(zw^2)] = 0$ by (30). This implies $zw^2 = \alpha^2 z^3$ and $w^2 = \alpha^2 z^2 + (2\alpha\beta + 2\alpha^3\delta)z^3$ which contradicts linear independence of $\{z^2, z^3, w^2\}$. ◇

PROPOSITION 4.8. Let B be a commutative, finite dimensional K-algebra such that $B^2B^2 = 0$, $x^4 = 0$ for all $x \in B$, and that is not Jordan. If $\dim_K B/B^2 = 2$ then $Der(B) \neq 0$.

Proof:
We will use the identity

$$x_1(x^2x_2) + x_2(x^2x_1) + 2x_1(x(xx_2)) + 2x_2(x(xx_1)) + 2x(x_1(xx_2)) + 2x(x_2(xx_1)) + 2x(x(x_1x_2)) = 0. \quad (32)$$

By the previous proposition there exist linearly independent elements z, z^2, z^3, w, w^2 and $\{z, w\}$ is a basis of W, where $B = B^2 + W$ is a direct sum of vector spaces.

Let us see that the right multiplication $D \equiv R_{z^3w}$ is a derivation. Since $D(z) = z(z^3w)$, $D(w) = w(z^3w)$, $D(B^2) = 0$, D is a derivation of B if and only if $zD(z) = 0$, $wD(w) = 0$, and $zD(w) + D(z)w = 0$.

$zD(z) = 0$ and $wD(w) = 0$ follow from (30) and we can deduce $zD(w) + D(z)w= 0$ applying (32) with $x = w$, $x_1 = z$, $x_2 = z^3$.

If $R_{z^3w} \neq 0$, then $\operatorname{Der} B \neq 0$.

If $z^3x \in An\,B$ for all $x \in B$ then R_{z^3} is a derivation. If $R_{z^3} \neq 0$ then $\operatorname{Der} B \neq 0$.

If $R_{z^3} = 0$ then $z^3 \in An\,B$ and the linear map g, defined by $g(z) = z^3$ and $g(u) = 0$ for any other basic element of B, is a non zero derivation. ◇

PROPOSITION 4.9. Let B be a commutative, finite dimensional K-algebra such that $B^2B^2 = 0$, $x^4 = 0$ for all $x \in B$, and that is not Jordan. If $\dim_K B^2 = 3$ then $Der(B) \neq 0$.

Proof:

Let $B = W + B^2$ be a direct sum of vector spaces. We will find a basis of W, $\{z, w, \sigma_1, \cdots, \sigma_s\}$, such that $z^3\sigma_i = 0$ for all $i = 1, \ldots, s$.

By Proposition 4.7 there exist linearly independent elements z, z^2, z^3, w, w^2 in B. Then $B^2 = K(z^2, z^3, w^2)$ and elements z, z^2, z^3, w, w^2 generate a subalgebra isomorphic to the algebra in Proposition 4.3; we can assume $b_1 = c_3 = 1$, $d_1 = -1$, $a_1 = a_3 = b_2 = 0$, and so the multiplication table is: $zw = \frac{1}{2}z^3$, $z^2w = w^2$, $z^3w = -z^2$, $zw^2 = z^2$ and the other products equal to zero.

Let $\sigma \in W$. We denote $z\sigma = a_1z^2 + a_2z^3 + a_3w^2$, $z^2\sigma = b_1z^2 + b_2z^3 + b_3w^2$, $z^3\sigma = c_1z^2 + c_2z^3 + c_3w^2$, $w\sigma = d_1z^2 + d_2z^3 + d_3w^2$, $w^2\sigma = e_1z^2 + e_2z^3 + e_3w^2$, $\sigma^2 = f_1z^2 + f_2z^3 + f_3w^2$.

Applying identity (30) with $x = z, x' = \sigma$; $x = w, x' = \sigma$; $x = \sigma, x' = z^3$; $x = \sigma, x' = w^2$; $x = \sigma, x' = z^2$; identity (31) with $x = z, x_1 = \sigma$; identity (32) with $x = z, x_1 = w, x_2 = \sigma$ and identity (31) with $x = w, x_1 = \sigma$, we obtain $a_3 = b_1 = c_2 = c_3 = e_2 = e_3 = 0$, $c_1 = -b_3$, $e_1 = b_2 = 2d_2$, $a_2 = \frac{1}{2}(2d_3 - b_3)$, $f_3 = b_3(2d_3 - b_3)$, $f_2 = 2d_2(2d_3 - b_3)$.

Let $\{z, w, \sigma_1', \cdots, \sigma_s'\}$ be a basis of W. For every element σ_i' we have the same relations. In particular, $z^3\sigma_i' = -b_3^{(i)}z^2$. If $\sigma_i = -b_3^{(i)}w + \sigma_i'$, then $z^3\sigma_i = 0$ for $i = 1, \ldots, s$. We have found a basis of W, $\{z, w, \sigma_1, \cdots, \sigma_s\}$ such that $z^3\sigma_i = 0$ for $i = 1, \ldots, s$.

We can see that the right multiplication $D \equiv R_{z^3w}$ is a derivation.

If $R_{z^3w} \neq 0$, then $\operatorname{Der} B \neq 0$.

If $R_{z^3w} = 0$, then $z^3w \in An\,B$ and if $z^3w \neq 0$ then $\operatorname{Der} B \neq 0$.

If $z^3w = 0$ then $z^3 \in An\,B$ and, as we have seen in the proof of Proposition 4.8, there are non zero derivations in this case. ◇

5 INNER DERIVATIONS IN POWER-ASSOCIATIVE SECOND ORDER BERNSTEIN ALGEBRAS

The following Corollary follows from Propositions 3.1 and 3.2 applying identities (1) to (7).

Corollary 5.1. Let $A = Ke + U + Z$ be a power-associative second order Bernstein algebra; $\bar{u} \in U$, $\bar{z} \in Z$. Then

1) $R_{\bar{u}}$ is a derivation if and only if $\bar{u}Z = 0$.
2) $R_{\bar{z}}$ is a derivation if and only if $(U\bar{z})Z = 0$ and $R_{\bar{z}}|_Z$ is a derivation.

PROPOSITION 5.1. Let $A = Ke + U + Z$ be a power-associative second order Bernstein algebra. If $\bar{u} \in U$ is such that $\bar{u}Z = 0$ then $R_{\bar{u}u_1}$ is a derivation of A for every $u_1 \in U$.

Proof:

Using identities (4) to (7) we will deduce:

1) Let $u \in U, z \in Z$. Since $\bar{u}Z = 0$ we have $[(\bar{u}u_1)u]z = u[z(\bar{u}u_1)] - (uz)(\bar{u}u_1) = u[z(\bar{u}u_1)] + u_1[\bar{u}(uz)] = u[\bar{u}(u_1z)] + u_1[\bar{u}(uz)] = -(u_1z)(u\bar{u}) + u_1[\bar{u}(uz)] = -(u_1z)(u\bar{u}) + u_1[z(u\bar{u})] = z[u_1(u\bar{u})] = -z[u(u_1\bar{u})]$. Hence $((\bar{u}u_1)U)Z = 0$.
2) Let $z_1, z_2 \in Z$; then $(\bar{u}u_1)(z_1z_2) = [(z_1z_2)u_1]\bar{u} = [(u_1z_1)z_2]\bar{u} + [(u_1z_2)z_1]\bar{u} = z_2[(u_1z_1)\bar{u}] + z_1[\bar{u}(u_1z_2)] = z_2[z_1(u_1\bar{u})] + z_1[z_2(u_1\bar{u})]$. Hence $R_{\bar{u}u_1}|_Z$ is a derivation.

Proposition follows now from Corollary 5.1. ◇

In a power-associative second order Bernstein algebra, as in a Bernstein algebra, $R_{\bar{u}+\bar{z}}$ is a derivation if and only if $R_{\bar{u}}$ and $R_{\bar{z}}$ are derivations. This result is not true in general.

THEOREM 5.1. Let (A, ω) be a power-associative second order Bernstein algebra, $A = Ke + U + Z$. Let $x \in Ker\,\omega$, $x = \bar{u} + \bar{z}$ with $\bar{u} \in U$, $\bar{z} \in Z$. Then R_x is a derivation if and only if $R_{\bar{u}}$ and $R_{\bar{z}}$ are derivations.

Proof:

If $R_x \in Der(A)$, then $(ez)x = (ex)z + e(zx)$ for all $z \in Z$ implies $\bar{u}z = 0$; that is, $\bar{u}Z = 0$.

Similarly $(uz)x = (ux)z + u(zx)$ for all $u \in U$, $z \in Z$, and identities (4) and (5) imply $(u\bar{z})z = 0$; so $(U\bar{z})Z = 0$.

Finally $(zz')x = (zx)z' + z(z'x)$ for all $z, z' \in Z$ implies $(zz')\bar{z} = (z\bar{z})z' + z(z'\bar{z})$. Hence $R_{\bar{u}}$ and $R_{\bar{z}}$ are derivations. ◇

It is known that if an algebra A is Z_2- graded, $A = A_0 + A_1$, then also is $Der(A) = D_0(A) + D_1(A)$, with

$$D_0(A) = \{D \in Der(A) \mid D(A_0) \subset A_0,\ D(A_1) \subset A_1\},$$

$$D_1(A) = \{D \in Der(A) \mid D(A_0) \subset A_1,\ D(A_1) \subset A_0\}.$$

If a Bernstein algebra A is Jordan (that is, power-associative) then A and $Der(A)$ are Z_2- graded. Let us see that if A is a power-associative second order Bernstein algebra then A and $Der(A)$ are Z_2- graded.

PROPOSITION 5.2. Let $A = Ke + U + Z$ be a power-associative second order Bernstein algebra. Then A and $Der(A)$ are Z_2- graded, and $Der(A) = D_0(A) +$

$D_1(A)$, where

$$D_0(A) = \{D \in Der(A) \mid D(e) = 0\},$$

$$D_1(A) = \{D \in Der(A) \mid D(U) \subset Z, D(Z) \subset U\}.$$

Proof:

If $A_0 = Ke + Z$ and $A_1 = U$, $A = A_0 + A_1$ defines a Z_2- gradation in A. Then $Der(A) = D_0(A) + D_1(A)$. If $D \in D_0(A)$ then $D(e) \in U \cap A_0$, that is, $D(e) = 0$. Conversely, if $D \in Der(A)$ satisfies $D(e) = 0$, then $D(u) = f(u) \in U$ and $D(z) = g(z) \in Z$, by Proposition 4.1. That is, $D \in D_0(A)$. $\diamond$

PROPOSITION 5.3. Let (A, ω) be a power-associative second order Bernstein algebra of type $(1 + r, d)$, $A = Ke + U + Z$. Then there exists a bijective linear map $\varphi : D_1(A) \to U$ defined by $D \to D(e)$. In particular $dim_K D_1(A) = r$ and $D_1(A) = \{D_{\tilde{u}} := (\tilde{u}, 0_U, 0_Z) \mid \tilde{u} \in U\}$.

Proof:

By Proposition 4.1, every $D \in D_1(A)$ has an associated triple $(\tilde{u}, 0_U, 0_Z)$ with $\tilde{u} = D(e)$. Conversely, every triple $(\tilde{u}, 0_U, 0_Z)$ gives a derivation in $D_1(A)$. So $D_1(A) = \{D_{\tilde{u}} := (\tilde{u}, 0_U, 0_Z) \mid \tilde{u} \in U\}$. $\diamond$

Let us denote $Inn(A) = Der(A) \cap \mathcal{L}(A)$ the set of inner derivations of A.

Corollary 5.2. Let $A = Ke + U + Z$ be a power-associative second order Bernstein algebra. Then $D_1(A) \subset Inn(A)$.

Proof:

Let $D \in D_1(A)$. Then $D \equiv (\tilde{u}, 0_U, 0_Z)$ with $D(e) = \tilde{u}$. It can be easily proved that $D = [R_{4e}, R_{\tilde{u}}]$; so $D \in Inn(A)$. $\diamond$

Therefore $Inn(A) = D_1(A) + (D_0(A) \cap Inn(A))$ and A has some derivation that is not inner if and only if there exists some derivation that is not inner in $D_0(A)$.

6 JORDAN ALGEBRAS

Now we will consider a Jordan second order Bernstein algebra A. Since a Jordan algebra is power-associative we can apply to these algebras the results already obtained. In particular we know an upper bound of $\dim_K Der(A)$.

Corollary 6.1. Let (A, ω) be a Jordan second order Bernstein algebra. Then

1) $Der(A) \neq 0$.
2) If $U \neq 0$, there are non zero inner derivations R_u.

Proof:

The assertion 1 follows from Corollary 4.1 together with the nilpotency of $Ker\,\omega$ in a Jordan second order Bernstein algebra.

Since $Ker\,\omega$ is nilpotent and $UZ \subset U$, we can always find $0 \neq u \in U$ with $uZ = 0$. Then R_u is a non zero derivation by Corollary 5.1. $\diamond$

It is known that in Jordan algebras A, the Lie algebra generated by right multiplications is $\mathcal{L}(A) = R(A) + [R(A), R(A)] = \{R_x + \sum[R_y, R_z] \mid x, y, z \in A\}$ and that for every $y, z \in A$, $[R_y, R_z]$ is a derivation. Then $Inn(A) = \{R_x + \sum[R_y, R_z] \mid x, y, z \in A,\ R_x\, derivation\}$.

Corollary 6.2. Let $A = Ke + U + Z$ be a Jordan second order Bernstein algebra. Then $Inn(A) = \{R_{\bar{z}} + \sum_{i=1}^{n}[R_{x_i}, R_{y_i}] \mid n \in N, x_i, y_i \in A, \bar{z} \in Z, (U\bar{z})Z = 0, R_{\bar{z}}|_Z$ is a derivation $\}$.

Proof:

It is enough to prove that a derivation $R_{\bar{u}}$ is a commutator. Let $\bar{u} \in U$ such that $\bar{u}Z = 0$. Then $R_{\bar{u}} = [R_{2e}, R_{\bar{u}}]$. ◇

PROPOSITION 6.1. Let $A = Ke + U + Z$ be a Jordan second order Bernstein algebra, $x = \alpha e + u + z$, $y = \alpha' e + u' + z' \in A$, $\bar{u} \in U$, $\bar{z} \in Z$. If $D = [R_x, R_y]$ then

1) $eD = \frac{1}{4}(\alpha u' - \alpha' u) + \frac{1}{2}(uz' - u'z) \in U$.
2) $\bar{u}D = [(\bar{u}u)u' - (\bar{u}u')u + (\bar{u}z)z' - (\bar{u}z')z] + [\frac{1}{2}(\alpha u' - \alpha' u)\bar{u} + (uz' - u'z)\bar{u}]$, where the first bracket belongs to U and the second one belongs to Z.
3) $\bar{z}D = [\frac{1}{2}(\alpha' u - \alpha u')\bar{z} + (u'z - uz')\bar{z}] + [(\bar{z}u)u' - (\bar{z}u')u + (\bar{z}z)z' - (\bar{z}z')z]$, where the first bracket belongs to U and the second one belongs to Z.
4) $D = D_{u^*} + ([R_u, R_{u'}] + [R_z, R_{z'}])$, where $u^* = \frac{1}{4}(\alpha u' - \alpha' u) + \frac{1}{2}(uz' - u'z)$ and $D_{u^*} \equiv (u^*, 0_U, 0_Z)$. Notice that $D_{u^*} \in D_1(A)$ and $[R_u, R_{u'}] + [R_z, R_{z'}] \in D_0(A)$.

Proof:

1) to 3) can be obtained directly, applying identities (4) and (5).

Let us denote $u^* = \frac{1}{4}(\alpha u' - \alpha' u) + \frac{1}{2}(uz' - u'z) \in U$ and $D_{u^*} \equiv (u^*, 0_U, 0_Z)$. If $\bar{u} \in U$, $\bar{z} \in Z$, then $eD_{u^*} = u^* = \frac{1}{4}(\alpha u' - \alpha' u) + \frac{1}{2}(uz' - u'z)$, $\bar{u}D_{u^*} = 2u^*\bar{u} = \frac{1}{2}(\alpha u' - \alpha' u)\bar{u} + (uz' - u'z)\bar{u}$, $\bar{z}D_{u^*} = -2u^*\bar{z} = \frac{1}{2}(\alpha' u - \alpha u')\bar{z} + (u'z - uz')\bar{z}$. Therefore, $[R_x, R_y] = D_{u^*} + ([R_u, R_{u'}] + [R_z, R_{z'}])$ follows from 1) to 3) in the proposition.

$D_{u^*} \in D_1(A)$ by Proposition 5.3 and the derivations $[R_u, R_{u'}]$, $[R_z, R_{z'}]$ satisfy $e[R_u, R_{u'}] = e[R_z, R_{z'}] = 0$, so they and their sum belong to $D_0(A)$. ◇

Corollary 6.3. Let $A = Ke + U + Z$ be a Jordan second order Bernstein algebra. Then $Inn(A) = D_U + \{R_{\bar{z}} + \sum[R_u, R_{u'}] + \sum[R_z, R_{z'}], u, u' \in U; z, z', \bar{z} \in Z, (\bar{z}U)Z = 0, R_{\bar{z}}|_Z$ is a derivation $\}$, where $D_U = \{(\bar{u}, 0_U, 0_Z) \mid \bar{u} \in U\} = D_1(A)$.

Corollary 6.4. Let $A = Ke + U + Z$ be a Jordan second order Bernstein algebra such that $UZ = 0$. Then $Inn(A) = R_U + \{R_{\bar{z}} + \sum[R_z, R_{z'}]; z, z', \bar{z} \in Z, R_{\bar{z}}|_Z$ is a derivation $\}$, where $R_U = \{R_u \mid u \in U\} = D_1(A)$.

Proof:

If $UZ = 0$ then $[R_u, R_{u'}] = 0$ for every $u, u' \in U$. Futhermore $D_{\bar{u}} \equiv (\bar{u}, 0_U, 0_Z)$ is equal to $2R_{\bar{u}}$. ◇

THEOREM 6.1. Let $A = Ke + U + Z$ be a Jordan second order Bernstein algebra of type $(1 + r, d)$ such that $UZ = 0 = Z^3$. Then

1) $Inn(A) = \{R_x \mid x \in Ker\,\omega\} = R_U + R_Z$, with $R_U = D_1(A)$ and $R_Z = \{R_z \mid z \in Z\} \subset D_0(A)$.

2) $dim_K Inn(A) = r + d - dim_K(An\, A)$.

3) There are derivations of A that are not inner.

Proof:

1) Now $[R_z, R_{z'}] = 0$ for every $z, z' \in Z$ and $R_{\bar{z}}|_Z$ is a derivation for every $\bar{z} \in Z$. Then $Inn(A) = R_U + R_Z$.

2) We have $dim_K Inn(A) = dim_K R_U + dim_K R_Z$. By Proposition 5.3 $dim_K R_U = dim_K D_1(A) = r$. The set $\{R_z / z \in Z\}$ is a vector subspace of $Der(A)$. We can define a linear map φ from Z onto $\{R_z / z \in Z\}$ by $\varphi(z) = R_z$. Then $Ker\,\varphi = \{z \in Z / R_z = 0\} = An\, A$. So $dim_K R_Z = d - dim_K(An\, A)$ and $dim_K Inn(A) = r + d - dim_K(An\, A)$.

3) $UZ = 0 = Z^3$ implies that $Z^2 \subset An\, A$ and $U^2 \subset An\, A$ by (4).

Let $Z = An\, A + W$ be a direct sum of vector spaces. If $D \equiv (0, f = id_U, g)$, where $g(a+w) = 2a + w$ for every $a \in AnA$ and $w \in W$, then D is a derivation in A by Proposition 4.1. Indeed, g is linear and satisfies $g(uu') = 2uu' = f(u)u' + uf(u')$, $g(zz') = 2zz' = g(z)z' + zg(z')$. Notice that D belongs to $D_0(A)$ because $D(e) = 0$.

Let us see that $D \neq 0$ is not inner. Since $Z^2 \neq 0$ (otherwise A is a Jordan Bernstein algebra), $W \neq 0$ and so there exists $0 \neq w \in W$. Then $D(w) = w$ and $D \neq 0$. Since any inner derivation $R_u + R_z$ satisfies $(R_u + R_z)(w) = R_z(w) \in Z^2 \subset An\, A$ and $D(w) = w \notin An\, A$, we can conclude that D is not inner derivation. ◇

References

[1] R. Andrade, Ivo Basso, Alicia Labra, Derivations in some Bernstein algebras of order 2, Algebras, Groups and Geometries, 14, 31-39, 1997.

[2] M.A. García Muñiz, Derivaciones en Algebras de Bernstein de orden dos, Tesis doctoral, Universidad de Oviedo, 1997.

[3] S. González, Inner derivations of Bernstein algebras, Linear Algebra and its Applications, 170, 206-210, 1992.

[4] S. González and C. Martínez, Bernstein algebras with zero derivation algebra, Linear Algebra and its Applications, 191, 235-244, 1993.

[5] S. González, J. C. Gutierrez and C. Martínez, On Bernstein algebras of n-th order, Ed. Kluwer, 158-164, 1992.

[6] S. González, C. Martínez and P. Vicente, Some special classes of 2nd-order Bernstein algebras, J. Alg. 167(3), 855-868, 1994.

[7] S. González, C. Martínez and P. Vicente, Power-Associative and Jordan 2nd-order Bernstein algebras. Nova J. Alg. Geom. 2(4), 367-382, 1993.

[8] N. Jacobson, Abstract derivation and Lie algebras, Trans. Amer. Math. Soc., 42, 206-224, 1937.

[9] N. Jacobson, Derivation algebras and multiplication algebras of semisimple Jordan-algebras, Ann. of Math., (2) 50, 866-874, 1949.

[10] I. Kaplansky, Algebras with many derivations, Aspects of Math. Appl., 431-438, 1986.

[11] C. Martínez, Inner derivations in Jordan-Bernstein algebras, Hadronic Mechanics and Nonpotencial Interaction, Nova Science Publ. New York, 217-228, 1992.

[12] M. Osborn, Varieties of algebras, Advances in Math. 8, 163-369, 1972.

[13] R.D. Schafer, An introduction to nonassociative algebras, Academic Press, New York, 1966.

[14] R.D. Schafer, Inner derivations of non-associative algebras, Bull. Amer. Math. Soc., 55, 769-776, 1949.

[15] D. Suttles, A counterexample to a conjeture of Albert, Notices Amer. Math. Soc. 19 (1972), A-566. Abstract 72T-A169.

[16] P. Vicente, Idempotentes en álgebras de Bernstein de orden dos, Tesis doctoral, Universidad de Oviedo, 1992.

[17] K.A. Zhevlakov, A.M.Slinko, I.P.Shestakov, A.I.Shirshov, Rings that are nearly associative, Academic Press, New York, 1982.

Algebraic structure of the non-abelian Toda models[1]

J. F. GOMES, F. E. MENDONÇA DA SILVEIRA, A. H. ZIMERMAN *Instituto de Física Teórica - UNESP, Rua Pamplona, 145 , 01405-900, Sao Paulo, SP, Brazil*

G. M. SOTKOV *Departamento de Campos e Partículas - CBPF Rua Dr. Xavier Sigaud, 150 -Urca 22290-180, Rio de Janeiro - RJ, Brazil*

ABSTRACT The algebraic conditions that specific gauged $WZNW$ model have to satisfy in order to give rise to Non-Abelian Toda models with *non singular* metric with or without *torsion* are found. The classical algebras of symmetries corresponding to grade one rank 2 and 3 singular NA-Toda models are derived.

The object of this talk is to discuss the structure of two dimensional G_0-Toda models in terms of a $G_0 \subset G$ embedding and a G-invariant $WZNW$-model. The simplest class of models is the Abelian Toda, which is known to be completely integrable and conformal invariant. These models correspond to a Abelian subgroup $G_0 \subset G$ and their symmetries generate the W_n-algebra (n=rank G).

The non-Abelian Toda models, in turn are connected to non-abelian embeddings $G_0 \subset G$ and describe a string theory propagating on a specific background, containing tachyons, dilatons $\Phi(X)$ and, possibly axions. Its general action is of the form $(i,j = 1,\cdots,D;\ \mu,\nu = 0,1)$:

$$S=\int d^2z\{(G_{ij}(X)\eta_{\mu\nu}+\epsilon_{\mu\nu}B_{ij}(X))\,\partial_\mu X^i\partial_\nu X^j-\alpha' R^{(2)}\Phi(X)+\text{Tach. potential}\}.$$

The structures of the background metric G_{ij} and of the anti-symmetric term B_{ij}(torsion) depend upon the embedding $G_0 \subset G$, where G_0 is now non-Abelian and classify the models according to *singular* or *non-singular* metrics and the presence

[1]Talk given by J.F. Gomes

of *axionic* or *torsionless* terms. Its symmetries are described by a non-local algebra, which corresponds to the semi-classical limit of a mixed parafermionic and W-algebra structure, denoted by V-algebra (see [1],[2], [3] and [4]).

Consider the G-invariant $WZNW$-model, describing a 2-D conformal invariant field theory on a group manifold, with fields parametrizing the group element

$$g(z,\bar{z}) = \exp(\alpha^a(z,\bar{z})T^a), \tag{1}$$

where $[T^a, T^b] = f^{abc}T^c$ is the Lie bracket satisfied by the generators of the Lie algebra of G.

The $WZNW$-action is given by

$$S_{WZNW} = \frac{k}{4\pi}\int d^2z Tr(g^{-1}\partial g g^{-1}\bar{\partial} g) + \frac{k}{12\pi}\int_B \epsilon_{ijk} Tr(g^{-1}\partial_i g g^{-1}\partial_j g g^{-1}\partial_k g), \tag{2}$$

where the topological term denotes a surface integral over a ball B identified as space-time.

The equations of motion are

$$\bar{\partial} J = \partial \bar{J} = 0, \tag{3}$$

where the currents $J = g^{-1}\partial g$ and $\bar{J} = -\bar{\partial} g g^{-1}$ satisfy the Kac-Moody algebra

$$\{J^a(x), J^b(y)\} = f^{abc}J^c(x)\delta(x-y) + k\delta^{ab}\partial_x(x-y).$$

We now seek for a field theory in the submanifold $G_0 \subset G$ preserving conformal invariance. In order to eliminate unwanted degrees of freedom, corresponding to the tangent space of G/G_0, we may either implement constraints upon specific components of J and $\bar{J}$ [5] or, equivalently, propose a gauge invariant action and eliminate degrees of freedom by choice of gauge [7]. In this note we shall follow the second approach.

In order to construct a gauge invariant action, we first discuss a systematic procedure in defining the G_0 subgroup, in terms of a grading operator.

Let Q be a grading operator, decomposing the Lie algebra $\mathcal{G}$ into graded subspaces, i. e.,

$$\mathcal{G} = \oplus_i \mathcal{G}_i, \qquad [Q, \mathcal{G}_i] = i\mathcal{G}_i, \qquad [\mathcal{G}_i, \mathcal{G}_j] \in \mathcal{G}_{i+j}. \tag{4}$$

It is clear that $\mathcal{G}_0$ is the subalgebra of $\mathcal{G}$.

Consider now the Gauss decomposition of the group element $g \in G$, i. e.,

$$g = N_- g_0 M_+, \tag{5}$$

where $N_- = \exp \mathcal{G}_{<0}$, $g_0 = \exp \mathcal{G}_0$ and $M_+ = \exp \mathcal{G}_{>0}$. The Toda models correspond to the co-set $N_-\backslash G/M_+$, and we seek for an action invariant under

$$g \longrightarrow g' = \alpha_- g \alpha_+, \tag{6}$$

so that N_- and M_+ may be eliminated by choice of gauge.

Introduce gauge fields $A = A_- \in \mathcal{G}_<$ and $\bar{A} = \bar{A}_+ \in \mathcal{G}_>$, transforming as

$$A \longrightarrow A' = \alpha_- A \alpha_-^{-1} + \alpha_- \partial \alpha_-^{-1}, \qquad \bar{A} \longrightarrow \bar{A}' = \alpha_+^{-1}\bar{A}\alpha_+ + \bar{\partial}\alpha_+^{-1}\alpha_+. \tag{7}$$

An invariant action under transformations (6) and (7) is

$$\begin{aligned} S^{gauged}_{WZNW}(g,A,\bar{A}) &= S_{WZNW}(g) \\ &- \frac{k}{2\pi}\int dz^2 Tr\left(A(\bar{\partial}gg^{-1}-\epsilon_+)+\bar{A}(g^{-1}\partial g-\epsilon_-)+Ag\bar{A}g^{-1}\right), \end{aligned}$$

where $\epsilon_\pm$ are constant generators of grade ± 1, respectively. Since the action S^{gauged}_{WZNW} is invariant, we may choose $\alpha_- = N_-^{-1}$ and $\alpha_+ = M_+^{-1}$, so that

$$\begin{aligned} S^{gauged}_{WZNW}(g,A,\bar{A}) &= S^{gauged}_{WZNW}(g_0,A',\bar{A}') \\ &= S_{wznw}(g_0) - \frac{k}{2\pi}\int dz^2 Tr[A'\epsilon_+ + \bar{A}'\epsilon_- + A'g_0\bar{A}'g_0^{-1}], \quad (8) \end{aligned}$$

since from the graded structure $Tr(A\bar{\partial}g_0g_0^{-1}) = Tr(\bar{A}g_0^{-1}\partial g_0) = 0$.

After integrating the quadratic terms in the functional integral

$$Z = \int DAD\bar{A}\exp(-F) \quad \sim \quad \exp(-S_{eff}), \tag{9}$$

where

$$F = -\frac{k}{2\pi}\int Tr(A'\epsilon_+ + \bar{A}'\epsilon_- + A'g_0\bar{A}'g_0^{-1}), \tag{10}$$

we find the effective action

$$S_{eff} = S_{WZNW}(g_0) - \frac{k}{2\pi}\int Tr(\epsilon_+ g_0 \epsilon_- g_0^{-1}). \tag{11}$$

1. ABELIAN TODA MODELS.

Consider the grading operator

$$Q = \sum_{i=1}^{r}\frac{2}{\alpha_i^2}H_{\lambda_i}, \tag{12}$$

where $r = rank\mathcal{G}$, H is the Cartan subalgebra of $\mathcal{G}$, α_i is the ith simple root of $\mathcal{G}$ and λ_i is the ith fundamental weight of $\mathcal{G}$, which satisfies

$$\frac{2\lambda_i\cdot\alpha_j}{\alpha_i^2} = \delta_{ij}, \qquad TrH_{\lambda_i}H_{\alpha_j} = \frac{2}{\alpha_i^2}\delta_{ij}$$

In this case, the Abelian subalgebra of $\mathcal{G}$, of grade zero, is

$$\mathcal{G}_0 = u(1)^r = \{h_1, h_2, \ldots, h_r\}, \tag{13}$$

where h_i, defined as $h_i = \frac{2}{\alpha_i^2}H_{\alpha_i}$, and satisfy $[h_i, E_{\alpha_j}] = k_{ji}E_{\alpha_j}$; k_{ji} denote the Cartan matrix of $\mathcal{G}$. Taking all this into account, the group element in G_0 may be parameterized as $g_0 = \exp(\phi_i h_i)$.

The most general constant generators of grade ± 1 are

$$\epsilon_\pm = \sum_{i=1}^{r} c_{\pm i}E_{\pm\alpha_i} \tag{14}$$

and the potential term may be evaluated as

$$Tr(\epsilon_+ g_0 \epsilon_- g_0^{-1}) = \sum_{i=1}^{r} \frac{2}{\alpha_i^2} c_i c_{-i} \exp(k_{ij}\phi_j). \tag{15}$$

Therefore, the action is given by

$$S_{WZNW}^{gauged} = \int d^2z \left(\partial\phi_i \bar{\partial}\phi_j Tr(h_i h_j) - \sum_{i=1}^{r} \frac{2}{\alpha_i^2} c_i c_{-i} \exp(k_{ij}\phi_j) \right). \tag{16}$$

At this point, let us remark that if $c_a = c_{-a} = 0$, then

$$\epsilon_\pm = \sum_{i \neq a} c_{\pm i} E_{\pm\alpha_i}.$$

The structure of this Abelian Toda model is such that, by a change of variables

$$g_0 = \exp(\sum_{i=1}^{a-1} \varphi_i h_i + \varphi \lambda_a \cdot H + \sum_{i=a+1}^{r} \varphi_i h_i), \tag{17}$$

the action decomposes into two Toda models, together with a free field, associated to the generator $g_0^0 = \lambda_a \cdot H$, such that $[g_0^0, \epsilon_\pm] = 0$. Hence, the action can be written as

$$S_{WZNW}^{gauged} = S^{(1)}(\varphi_1, \varphi_2, ..., \varphi_{a-1}) + S^{(2)}(\varphi_{a+1}, \varphi_{a+2}, ..., \varphi_r) + \frac{1}{2}\int d^2z \partial\varphi\bar{\partial}\varphi,$$

where $S^{(1)}$ and $S^{(2)}$ denote the abelian Toda models associated to the decomposed Dynkin diagram of G, by deleting the ath point.

2. NON-ABELIAN TODA MODELS.

Consider now the grading operator (see for instance ref. [1])

$$Q_a = \sum_{i \neq a} \frac{2}{\alpha_i^2} H_{\lambda_i}, \tag{18}$$

which determines the non-Abelian subalgebra of grade zero

$$\mathcal{G}_0 = u(1)^{r-1} \otimes sl(2). \tag{19}$$

The group element of grade zero may be then, parameterized as

$$g_0 = \exp(\chi E_{-\alpha_a}) \exp(\sum_{i=1}^{r} \phi_i h_i) \exp(\psi E_{\alpha_a}). \tag{20}$$

Now, take

$$\epsilon_\pm = \sum_{i \neq a} c_{\pm i} E_{\pm\alpha_i}, \qquad [\epsilon_\pm, \lambda_a \cdot H] = 0 \tag{21}$$

such that

$$Tr(\epsilon_+ g_0 \epsilon_- g_0^{-1}) = \sum_{i \neq a} \frac{2}{\alpha_i^2} c_i c_{-i} \exp(\sum_{j \neq a} k_{ij}\phi_j),$$

as before. The pure $WZNW$-action for g_0 acquires a contribution from the non-abelian structure,

$$S_{WZNW} = \int d^2z \left(\sum_{i,j=1}^{r} \partial\phi_i \bar{\partial}\phi_j Tr(h_i h_j) + 2\partial\chi\bar{\partial}\psi \exp(k_{ai}\phi_i) \right). \tag{22}$$

Notice that an attempt to decouple a free field from the action by changing the variables to

$$g_0 = \exp(\chi E_{-\alpha}) \exp(\sum_{i=1}^{a-1} \varphi_i h_i + \varphi\lambda_a \cdot H + \sum_{i=a+1}^{r} \varphi_i h_i) \exp(\psi E_{\alpha_a}) \tag{23}$$

decouples only part of the kinetic term, leading to

$$\mathcal{L}_{WZNW}(g_0) = \sum_{i,j \neq a} \partial\phi_i \bar{\partial}\phi_j Tr(h_i h_j) + \partial\varphi\bar{\partial}\varphi + 2\partial\chi\bar{\partial}\psi \exp(\sum_{i \neq a} k_{ai}\varphi_i + \varphi).$$

It will become clear that the elimination of field φ is responsible for the non-trivial singular metric (of black hole-type) in the string action. In order to gauge away the extra degree of freedom, we need to modify the action to become invariant under transformations within the subspace $\mathcal{G}_0^0 = \lambda_a \cdot H$ of grade zero, i. e.,

$$g_0 \longrightarrow g_0' = \alpha_0 g_0 \alpha_0, \tag{24}$$

where $\alpha_0 = \exp(\varphi\lambda_a \cdot H)$. Following the same line of reasoning as before, we now introduce the new gauge fields $A = A_- + A_0$, $\bar{A} = \bar{A}_+ + \bar{A}_0$, where $A_0, \bar{A}_0 \in \mathcal{G}_0^0$ transform as

$$A \longrightarrow A' = \alpha_0^{-1} A \alpha_0 + \alpha_0^{-1} \partial\alpha_0, \qquad \bar{A} \longrightarrow \bar{A}' = \alpha_0 \bar{A} \alpha_0^{-1} + \bar{\partial}\alpha_0 \alpha_0^{-1},$$

$$A_0 \longrightarrow A_0' = A_0 + \alpha_0^{-1}\partial\alpha_0, \qquad \bar{A}_0 \longrightarrow \bar{A}_0' = \bar{A}_0 + \bar{\partial}\alpha_0 \alpha_0^{-1}.$$

The invariant action, under transformations generated by $H_{(-,0)}$ (in left) and $H_{(+,0)}$ (in right) is given by

$$\begin{aligned} S_{WZNW}^{gauged}(g, A, \bar{A}) &= S_{WZNW}(g) - \frac{k}{2\pi} \int d^2z Tr \left(A(\bar{\partial} g g^{-1} - \epsilon_+) \right. \\ &+ \left. \bar{A}(g^{-1}\partial g - \epsilon_-) + A g \bar{A} g^{-1} + A_0 \bar{A}_0 \right). \end{aligned} \tag{25}$$

The redundant fields, corresponding to the nilpotent subalgebras $\mathcal{G}_{<0}$, $\mathcal{G}_{>0}$ and to $\mathcal{G}_0^0$, may be eliminated by subsequent gauge transformations, generated by α_-, α_+ and α_0, respectively. Therefore, we arrive at

$$S_{WZNW}^{gauged}(g, A, \bar{A}) = S_{WZNW}^{gauged}(g_0^f, A', \bar{A}'), \tag{26}$$

where

$$g_0^f = \exp(\chi E_{-\alpha_a})\exp(\sum_{i\neq a}\phi_i h_i)\exp(\psi E_{\alpha_a}). \tag{27}$$

Due to the trace properties

$$\mathrm{Tr}A\bar{\partial}g_0^f(g_0^f)^{-1} = \mathrm{Tr}A_0\bar{\partial}g_0^f(g_0^f)^{-1}, \qquad \mathrm{Tr}\bar{A}(g_0^f)^{-1}\partial g_0^f = \mathrm{Tr}\bar{A}_0(g_0^f)^{-1}\partial g_0^f$$
$$\mathrm{Tr}Ag_0^f\bar{A}'(g_0^f)^{-1} = \mathrm{Tr}A_0 g_0^f\bar{A}_0(g_0^f)^{-1} + \mathrm{Tr}A_- g_0^f\bar{A}_+(g_0^f)^{-1}.$$

the action decomposes into three parts, i. e.,

$$S_{WZNW}^{gauged} = S_{WZNW}(g_0^f) + F_0 + F_\pm, \tag{28}$$

where

$$F_0 = -\frac{k}{2\pi}\int d^2z Tr\left(A_0\bar{\partial}g_0^f(g_0^f)^{-1} + \bar{A}_0(g_0^f)^{-1}\partial g_0^f + A_0 g_0^f\bar{A}_0(g_0^f)^{-1} + A_0\bar{A}_0\right), \tag{29}$$

$$F_\pm = -\frac{k}{2\pi}\int dz^2 Tr\left(A_-\epsilon_+ + \bar{A}_+\epsilon_- + A_- g_0^f\bar{A}_+(g_0^f)^{-1}\right),$$

and the functional integral now factorizes into

$$Z = \int DA_0 D\bar{A}_0\exp(-F_0)\int DA_- D\bar{A}_+\exp(-F_\pm). \tag{30}$$

Explicitly making use of the given parameterization for g_0^f and writing the gauge fields as $A_0 = a_0(z,\bar{z})H_{\lambda_a}$, $\bar{A}_0 = \bar{a}_0(z,\bar{z})H_{\lambda_a}$, we find

$$F_0 = -\frac{k}{2\pi}\int dz^2\left[\frac{a_0\bar{a}_0}{2\lambda_a^2}\Delta_a - (a_0\chi\bar{\partial}\psi + \bar{a}_0\psi\partial\chi)\exp(k_{ai}\phi_i)\right], \tag{31}$$

where $\Delta_a = 1 + \frac{\chi\psi\exp(k_{ai}\phi_i)}{2\lambda_a^2}$.

The total effective action, then, becomes

$$\begin{aligned} S_{eff} &= -\frac{k}{2\pi}\int dz^2\left(\frac{1}{2}\partial\phi_i\bar{\partial}\phi_j Tr(h_i h_j) + \frac{\partial\chi\bar{\partial}\psi}{\Delta_a}\exp(k_{ai}\phi_i)\right. \\ &- \left.\sum_{i,j\neq a}^{r}\frac{2}{\alpha_i^2}c_i c_{-i}\exp(-k_{ij}\phi_j)\right). \end{aligned} \tag{32}$$

The non-symmetric term may be written as

$$\partial\chi\bar{\partial}\psi = g^{\mu\nu}\partial_\mu\chi\partial_\nu\psi - \epsilon_{\mu\nu}\partial_\mu\chi\partial_\nu\psi.$$

and we can rewrite the effective action (discarding the total derivative term) in the form

$$\begin{aligned} S_{eff} = \int d^2z &\left(\frac{1}{2}g^{\mu\nu}\partial_\mu\phi_i\partial_\nu\phi_j Tr(h_i h_j) + g^{\mu\nu}\partial_\mu\chi\partial_\nu\psi\frac{\exp(k_{ai}\phi_i)}{\Delta_a}\right. \\ &- \left.\frac{1}{2}\epsilon_{\mu\nu}k_{ai}\partial_\mu\phi_i(\chi\partial_\nu\psi - \psi\partial_\nu\chi)\frac{\exp(k_{aj}\phi_j)}{\Delta_a} - \sum_{i,j\neq a}^{r}\frac{2}{\alpha_i^2}c_i c_{-i}\exp(-k_{ij}\phi_j)\right) \end{aligned} \tag{33}$$

As we shall demonstrate below (see [3], [4]) the symmetries of the non-Abelian Toda model (33), are generated by non-local algebraic structures due to the gauge invariance within the zero grade subspace.

Let us remark that for $\mathcal{G} = sl(2)$ and for $\phi_i = 0$ the action coincides with Witten's black hole action [9], i. e.,

$$S = \int dz^2 g^{\mu\nu} \frac{\partial_\mu \chi \partial_\nu \psi}{1 + \chi\psi}.$$

3. NO-TORSION THEOREM.

We now discuss the *generalized* non-Abelian Toda model, as well as the "no-torsion theorem". To do so, first we take the very same grading operator

$$Q_a = \sum_{i \neq a}^{r} \frac{2}{\alpha_i^2} H_{\lambda_i}$$

and consider *the most general constant generators* of grade ± 1, i. e.,

$$\epsilon_\pm = \sum_{i \neq a}^{r} c_{\pm i} E_{\pm\alpha_i} + b_\pm E_{\pm(\alpha_a + \alpha_{a+1})} + d_\pm E_{\pm(\alpha_a + \alpha_{a-1})}. \tag{34}$$

It is clear that if $c_{\pm i}, b_\pm, d_\pm \neq 0$, there shall be *no* g_0^0 *commuting* with $\epsilon_\pm$, since that require an orthogonal direction to all roots appearing in $\epsilon_\pm$. These are the generalized *non-singular* NA-Toda models of ref. [12]. The NA-Toda models of singular metric $G_{ij}(X)$ correspond to the cases when $g_0^0 = U(1)$ and we *impose* it as a subsidiary constraint [2]. Depending upon the choice of the constants $c_{\pm i}, b_\pm$ and $d_\pm$ we distinguish four families of *singular* NA-Toda models:

(i)$b_\pm = d_\pm = 0,\ g_0^0 = \frac{2}{\alpha_a^2} H_{\lambda_a}$;
(ii)$c_{\pm(a-1)} = c_{\pm(a+1)} = 0,\ g_0^0 = \frac{2}{\alpha_a^2} H_{\lambda_a} - \frac{2}{\alpha_{a-1}^2} H_{\lambda_{a-1}} - \frac{2}{\alpha_{a+1}^2} H_{\lambda_{a+1}}$;
(iii)$c_{\pm(a+1)} = d_\pm = 0,\ g_0^0 = \frac{2}{\alpha_a^2} H_{\lambda_a} - \frac{2}{\alpha_{a+1}^2} H_{\lambda_{a+1}}$;
(iv)$b_\pm = c_{\pm(a-1)} = 0,\ g_0^0 = \frac{2}{\alpha_a^2} H_{\lambda_a} - \frac{2}{\alpha_{a-1}^2} H_{\lambda_{a-1}}$.

Of course, if $c_{\pm j} = 0,\ j \neq a, a \pm 1$, we find $g_0^0 = H_{\lambda_j}$. However, since $[H_{\lambda_j}, E_{\pm\alpha_a}] = 0$, there will be no singular metric present and this case shall be negleted. Cases (i) and (ii) are equivalent, since they are related by the Weyl reflection $\sigma_{\alpha_a}(\alpha_{a\pm1}) = \alpha_a + \alpha_{a\pm1}$ and the corresponding fields are related by non-linear change of the variables. This case has already been discussed in refs. [3] and [4], and shown to present always the *antisymmetric term*, originated by the presence of $e^{k_{ai}\varphi_i}$ in Δ_a and in the kinetic trem as well. Since we are removing all dependence in g_0^0, when parameterizing g_0^f, cases (iii) and (iv) may be studied together with

$$g_0^f = \exp(\chi E_{-\alpha_a}) \exp(\Phi(H)) \exp(\psi E_{\alpha_a}) \tag{35}$$

where $\Phi(H) = \sum_{i=1}^{a-2} \varphi_i h_i + \chi_- \varphi_- H + \chi_+ \varphi_+ H + \sum_{i=a+2}^{r} \varphi_i h_i$,

[2] If we leave g_0^0 unconstrained the resulting model belongs again to the non singular NA-Toda class of models [12].

$$\chi_-^{iii} = \alpha_{a-1} + \alpha_a, \qquad \chi_+^{iii} = \alpha_{a+1}, \tag{36}$$

$$\chi_-^{iv} = \alpha_{a-1}, \qquad \chi_+^{iv} = \alpha_a + \alpha_{a+1} \tag{37}$$

for cases *iii* and *iv* respectively, and $g_0^0 = H_y$, such that $Tr(H_{\chi\pm} g_0^0) = 0$. It is straightforward to evaluate the integral

$$F_0 = -\frac{k}{2\pi}\int [2y^2 a_0 \bar{a}_0 \Delta_a - \frac{2}{\alpha_a^2}(a_0 \chi \bar{\partial}\psi + \bar{a}_0 \psi \partial \chi)\exp(\Phi(\alpha_a))],$$

where

$$\Phi(\alpha_a) = \sum_{i=1}^{a-2} k_{ai}\varphi_i + \alpha_a \cdot \chi_- \varphi_- + \alpha_a \cdot \chi_+ \varphi_+ + \sum_{i=a+2}^{r} k_{ai}\varphi_i$$

and $\Delta_a = 1 + \frac{\chi\psi \exp(\Phi(\alpha_a))}{2y^2}$. The corresponding effective action takes the form

$$\begin{aligned} S &= \int d^2 z \left(g^{\mu\nu}\partial_\mu \varphi_i \partial_\nu \varphi_j Tr(h_i h_j) + \frac{exp(\Phi(\alpha_a))}{y^2 \Delta_a} g^{\mu\nu}\partial_\mu \psi \partial_\nu \chi \right. \\ &- \left. \frac{1}{2}\epsilon_{\mu\nu}\partial_\mu \Phi(\alpha_a)(\psi\partial_\nu \chi - \chi \partial_\nu \psi)\frac{exp(\Phi(\alpha_a))}{\Delta_a} - V \right) \end{aligned} \tag{38}$$

where the potential is given in terms of the constant elements $\epsilon_\pm$ and $\mathcal{G}_0^f$ as $V = Tr[\epsilon_+ g_0^f \epsilon_- g_0^{f^{-1}}]$. Now, if we consider Lie algebras whose Dynkin diagrams connect only nearest neighbours, i. e.,

$$\Phi(\alpha_a) = \alpha_a \cdot \chi_- \varphi_- + \alpha_a \cdot \chi_+ \varphi_+, \tag{39}$$

then the "no-torsion condition" implies $\Phi(\alpha_a) = 0$.

Considering case (iii), we have

$$\alpha_a \cdot \chi_- = \alpha_a \cdot (\alpha_{a-1} + \alpha_a) = 0, \tag{40}$$

$$\alpha_a \cdot \chi_+ = \alpha_a \cdot (\alpha_{a+1}) = 0. \tag{41}$$

In this case, the only solution for both equations above is to take $a = r$ (in such a way that $\alpha_{r+1} = 0$) and $G = B_r$ (so that $\alpha_{r-1} \cdot \alpha_r = -\alpha_r^2 = -1$). This is precisely the case proposed by Leznov and Saveliev [10] and subsequently discussed by Gervais and Saveliev [1] and also by Bilal [2], for the particular case of B_2.

For case (iv), the "*no-torsion condition*" requires that

$$\alpha_{a-1} \cdot \alpha_a = 0, \qquad \alpha_a \cdot (\alpha_a + \alpha_{a+1}) = 0,$$

which are satisfied by $a = 1$ and also by $G = C_2$, since $\alpha_{a-1} = 0$ and also $\alpha_1 \cdot \alpha_2 = -\alpha_1^2 = -1$, respectively.

In general, the "*no-torsion condition*", i. e., $\Phi(\alpha_a) = 0$, may be expressed in terms of the structure of the co-set $G_0/G_0^0 = \frac{u(1)^{r-1}\otimes sl(2)}{u(1)}$. The crucial ingredient for the appearence of $\Phi(\alpha_a)$ arises from the conjugation

$$Tr(A_0 g_0^f \bar{A}_0 (g_0^f)^{-1} + A_0\bar{A}_0) = 2\lambda_a^2\left(1 + \frac{2}{\alpha_a^2}\frac{\chi\psi\exp(\Phi(\alpha_a))}{2\lambda_a^2}\right).$$

Henceforth, if all generators belonging to the Cartan subalgebra parameterizing g_0^f commute with $E_{\pm\alpha_a}$, then $\Phi(\alpha_a) = 0$, and therefore the structure of the co-set

$$\frac{G_0}{G_0^0} = \frac{u(1)^{r-1}\otimes sl(2)}{u(1)} = u(1)^{r-1}\otimes\frac{sl(2)}{u(1)} \tag{42}$$

is the general condition for the *absence* of the *antisymmetric* term in the action.

4. ALGEBRA OF SYMMETRIES.

The symmetries of the action (38) is expressed in terms of its algebra of conserved currents. Those are obtained from the Hamiltonian reduction of the WZW model when particular set of constraints and gauge fixing conditions are implement. The algebra of the remaining currents can be derived using Dirac brackets or following the method employed in refs. [3] and [4], where the explicit form of the algebra of symmetries $V_{n+1}^{(1,1)}$ for $a = 1$ and $\mathcal{G} = A_n$ (cases i and ii) were derived . For B_2 we consider $Q_1 = \frac{2\lambda_1 H}{\alpha_1^2}$ and the two Non-Abelian Toda models can be defined by choosing $\epsilon_\pm^i = E_{\pm\alpha_1}$ and $\epsilon_\pm^{iii} = E_{\pm(\alpha_1+\alpha_2)}$. The remaining currents are $T_i = J_{\alpha_1}$, $V_i^+ = \sqrt{2}J_{\alpha_1+\alpha_2}$ and $V_i^- = -\sqrt{2}J_{-\alpha_2}$ and $T_{iii} = J_{\alpha_1+\alpha_2}$, $V_{iii}^+ = \frac{1}{\sqrt{2}}J_{\alpha_1+2\alpha_2}$ and $V_{iii}^- = -\frac{1}{\sqrt{2}}J_{\alpha_1}$ respectively. Their algebra is given by

$$\{T(\sigma), T(\sigma')\} = 2T(\sigma')\delta'(\sigma-\sigma') - T'(\sigma')\delta(\sigma-\sigma') - \frac{1}{2}\delta'''(\sigma-\sigma') \tag{43}$$

$$\{T(\sigma), V^\pm(\sigma')\} = sV^\pm(\sigma')\delta'(\sigma-\sigma') - V^{\pm\prime}(\sigma')\delta(\sigma-\sigma') \tag{44}$$

$$\{V^\pm(\sigma), V^\pm(\sigma')\} = tV^\pm(\sigma)V^\pm(\sigma')\epsilon(\sigma-\sigma') \tag{45}$$

for $s = \frac{3}{2}$ and 2, $t = \frac{1}{4}$ and 1 corresponding to cases (i) and (iii) respectively. We also have

$$\begin{aligned}\{V_i^\pm(\sigma), V_i^\mp(\sigma')\} &= -\frac{1}{4}(V_i^\pm(\sigma)V_i^\mp(\sigma') - 2V_i^\pm(\sigma)'V_i^\mp(\sigma))\epsilon(\sigma-\sigma')\\ &\pm T(\sigma')\delta(\sigma-\sigma') \mp \delta''(\sigma-\sigma')\end{aligned} \tag{46}$$

and

$$\begin{aligned}\{V_{iii}^\pm(\sigma), V_{iii}^\mp(\sigma')\} &= -(V_{iii}^\pm(\sigma)V_{iii}^\mp(\sigma') - 2V_{iii}^\pm(\sigma)'V_{iii}^\mp(\sigma))\epsilon(\sigma-\sigma')\\ &- T'(\sigma')\delta(\sigma-\sigma') + 2T(\sigma')\delta'(\sigma-\sigma') - \frac{1}{2}\delta'''(\sigma-\sigma')\end{aligned} \tag{47}$$

for cases (i) and (iii) respectively. The nonlocal V-algebra of the B_2 (case iii) was derived in [2] by explicit construction of the conserved currents. The algebra of symmetries of the B_3 (case *iii*) *torsionless* NA-Toda model is an example of a new type of *nonlocal quadratic* algebra of VW-type which mixes the features of V- and W-algebras. It is generated by two nonlocal currents $V^{\pm}$ of spin 3 and two local, T and W of spin 2 and 4 respectively (all $V^{\pm}$ and W are primary fields):

$$\{V^{\pm}(\sigma), V^{\pm}(\sigma')\} = \frac{1}{2}V^{\pm}(\sigma)V^{\pm}(\sigma')\epsilon(\sigma-\sigma') \tag{48}$$

$$\begin{aligned}
\{V^{\pm}(\sigma), V^{\mp}(\sigma')\} &= -\frac{1}{2}V^{\pm}(\sigma)V^{\mp}(\sigma')\epsilon(\sigma-\sigma') \\
&+ (\frac{1}{5}T'''(\sigma) - \frac{4}{25}(T^2(\sigma))' - W'(\sigma))\delta(\sigma-\sigma') \\
&+ (\frac{9}{10}T''(\sigma) - \frac{8}{25}T^2(\sigma) - 2W(\sigma))\delta'(\sigma-\sigma') \\
&+ \frac{3}{2}T'(\sigma)\delta''(\sigma-\sigma') + T(\sigma)\delta'''(\sigma-\sigma') - \frac{1}{2}\delta^{(v)}(\sigma-\sigma')
\end{aligned}$$

$$\begin{aligned}
\{W(\sigma), V^{\mp}(\sigma')\} &= -\frac{7}{5}V^{\pm}(\sigma)\delta'''(\sigma-\sigma') - \frac{7}{5}(V^{\pm}(\sigma))'\delta''(\sigma-\sigma') \\
&- \left(\frac{3}{5}(V^{\pm}(\sigma))'' - \frac{26}{25}T(\sigma)V^{\pm}(\sigma)\right)\delta'(\sigma-\sigma') \\
&- \left(\frac{1}{10}(V^{\pm}(\sigma))''' - \frac{1}{2}T'(\sigma)V^{\pm}(\sigma) + \frac{9}{25}T(\sigma)(V^{\pm}(\sigma))'\right)\delta(\sigma-\sigma')
\end{aligned}$$

$$\begin{aligned}
\{W(\sigma), W(\sigma')\} &= \frac{1}{20}\delta^{(vii)}(\sigma-\sigma') + \frac{7}{25}T(\sigma)\delta^{(v)}(\sigma-\sigma') + \frac{7}{25}T'(\sigma)\delta^{(iv)}(\sigma-\sigma') \\
&+ \sum_{i=4}^{7} A_i(\sigma)\delta^{(7-i)}(\sigma-\sigma')
\end{aligned}$$

where

$$\begin{aligned}
A_4(\sigma) &= \frac{21}{25}T''(\sigma) - \frac{49}{125}T^2(\sigma) + \frac{3}{5}W(\sigma) \\
A_5(\sigma) &= \frac{14}{25}T'''(\sigma) - \frac{147}{250}(T^2(\sigma))' + \frac{9}{10}W'(\sigma) \\
A_6(\sigma) &= \frac{1}{5}T^{(iv)}(\sigma) - \frac{113}{200}(T^2(\sigma))'' + \frac{213}{500}T(\sigma)T''(\sigma) + \frac{27}{50}(T'(\sigma))^2 + \frac{72}{625}T^3(\sigma) \\
&+ \frac{1}{2}W''(\sigma) - \frac{14}{25}T(\sigma)W(\sigma) - 3V^+(\sigma)V^-(\sigma) \\
A_7(\sigma) &= \frac{3}{100}T^{(v)}(\sigma) - \frac{9}{100}(T^2(\sigma))''' - \frac{3}{10}(T(\sigma)T''(\sigma))' + \frac{243}{500}T'(\sigma)T''(\sigma) \\
&+ \frac{81}{250}T(\sigma)T'''(\sigma) + \frac{9}{100}(T^3(\sigma))' - \frac{243}{5000}T(\sigma)(T^2(\sigma))' + \frac{1}{10}W'''(\sigma) \\
&- \frac{7}{25}(T(\sigma)W(\sigma))' - \frac{3}{2}(V^+(\sigma)V^-(\sigma))'
\end{aligned}$$

The *axionic* A_3-NA-Toda model corresponding to $Q = (\lambda_1 + \lambda_3)\cdot H$, $\epsilon_{\pm} = E_{\alpha_1} + E_{\alpha_3}$, and $g_0^0 = \lambda_2 \cdot H$ provides an example of a *new algebraic structure* of

UV-type. The $V_4^{(2,1)}$-algebra is generated by three nonlocal currents $V^{\pm}$ and V^0 of spin 2 and a local one (stress tensor) T of spin 2:

$$\begin{aligned}\{V^{\pm}(\sigma),V^{\pm}(\sigma')\} &= \frac{1}{8}\epsilon(\sigma-\sigma')V^{\pm}(\sigma)V^{\pm}(\sigma'),\\ \{V^{0}(\sigma),V^{\pm}(\sigma')\} &= \frac{1}{8}\epsilon(\sigma-\sigma')V^{0}(\sigma)V^{\pm}(\sigma')\end{aligned}$$

$$\begin{aligned}\{V^{0}(\sigma),V^{0}(\sigma')\} &= -\frac{1}{4}\epsilon(\sigma-\sigma')[V^{+}(\sigma)V^{-}(\sigma')+V^{-}(\sigma)V^{+}(\sigma')]\\ &+ 2T(\sigma')\partial_{\sigma'}\delta(\sigma-\sigma')+\delta(\sigma-\sigma')\partial_{\sigma'}T(\sigma')-4\partial^3_{\sigma'}\delta(\sigma-\sigma')\end{aligned}$$

$$\begin{aligned}\{V^{-}(\sigma),V^{+}(\sigma')\} &= -\frac{1}{8}\epsilon(\sigma-\sigma')[V^{0}(\sigma)V^{0}(\sigma')+V^{-}(\sigma)V^{+}(\sigma')]\\ &+ T(\sigma')\partial_{\sigma'}\delta(\sigma-\sigma')+\frac{1}{2}\delta(\sigma-\sigma')\partial_{\sigma'}T(\sigma')-2\partial^3_{\sigma'}\delta(\sigma-\sigma')\end{aligned} \tag{49}$$

By another choice of gauge fixing conditions [8] (by nonlocal gauge transformations that relates two different gauge fixings) one can transform this algebra into nonlocal *rational* algebra $VU_4^{(2,1)}$ generated by the nonlocal $V^{\pm}$ currents and now V_0 is local but appears in the denominator of the r.h.s. of (49). The origin of this novelty is that together with the $\epsilon_{\pm}$ invariant subalgebra g_0^0 there exists one extra $\epsilon_{\pm}$ *invariant subalgebra* K^- i.e.

$$[\epsilon_{\pm},E_{-\alpha_1-\alpha_2}-E_{-\alpha_2-\alpha_3}]=0 \tag{50}$$

It is important to note that relaxing the g_0^0 constraint, the algebra of symmetries of the corresponding *nonsingular* NA-Toda is generated by three local currents $V^{\pm},T$ of spin 2, one nonlocal V_0 $(s=2)$ and one local spin 1 current J:

$$\{V^0(\sigma),J(\sigma')\}=\{V^{\pm}(\sigma),V^{\pm}(\sigma')\}=0,\quad \{V^0(\sigma),V^{\pm}(\sigma')\}=\frac{1}{2}V^{\pm}(\sigma)V^0(\sigma')\epsilon(\sigma-\sigma'),$$

$$\{J(\sigma),V^{\pm}(\sigma')\}=\mp\frac{1}{4}V^{\pm}(\sigma)\delta(\sigma-\sigma'),\qquad \{J(\sigma),J(\sigma')\}=\frac{1}{8}\partial_\sigma\delta(\sigma-\sigma'),$$

$$\begin{aligned}\{V^{\pm}(\sigma),V^{\mp}(\sigma')\} &= -\frac{1}{2}\partial^3_\sigma\delta(\sigma-\sigma')+\mathcal{T}(\sigma)\partial_\sigma\delta(\sigma-\sigma')\\ &+ \frac{1}{2}\partial_\sigma\mathcal{T}(\sigma)\delta(\sigma-\sigma')-\frac{1}{4}V^0(\sigma)V^0(\sigma')\epsilon(\sigma-\sigma'),\end{aligned}$$

$$\begin{aligned}\{V^{0}(\sigma),V^{0}(\sigma')\} &= -\partial^3_\sigma\delta(\sigma-\sigma')+2\mathcal{T}(\sigma)\partial_\sigma\delta(\sigma-\sigma')\\ &+ \partial_\sigma\mathcal{T}(\sigma)\delta(\sigma-\sigma')\\ &- [V^+(\sigma)V^-(\sigma')+V^+(\sigma')V^-(\sigma)]\epsilon(\sigma-\sigma').\end{aligned} \tag{51}$$

where $\mathcal{T}(\sigma)=T-4J^2$. Again by suitable change of the gauge fixing [8], this algebra takes the form of rational $U_4^{(1,2)}$-algebra (the current V^0 appears in the denominators in the V^+V^--Poisson brackets).

A complete discussion of the symmetries of all grade $l = 1$ (i.e. $\epsilon_\pm \in \mathcal{G}_1$), $Q = \sum_{i \neq a} \lambda_i \cdot H$ NA-Toda models (including the nonsingular cases $g_0^0 = 0$) is given in our forthcoming paper [8] (see also ref. [11])

Acknowledgments F.E. Mendonça da Silveira thanks FAPESP for financial support. G.M.Sotkov thanks FAPESP, IFT-UNESP and DCP-CBPF for financial support and hospitality. This work was supported in part by CNPq.

REFERENCES

[1] J.-L. Gervais and M. V. Saveliev, Phys. Lett. 286B (1992) 271

[2] A. Bilal, Nucl. Phys. B422 (1994) 258

[3] J. F. Gomes, G. M. Sotkov and A. H. Zimerman, Parafermionic reductions of WZW model, hepth 9803234, to appear in Ann. of Phys. (1999)

[4] J. F. Gomes, G. M. Sotkov and A. H. Zimerman, $SL(2,R)_q$ Symmetries of Non-Abelian Toda theories, hepth 9803122, Phys. Lett. 435B (1998) 49

[5] L. Feher, L. O'Raifeartaigh, P. Ruelle, I. Tsutsui and A. Wipf, Phys. Rep. 222, (1992),1

[6] J. Balog, L. Feher, L. O'Raifeartaigh, P. Forgas and A. Wipf, Ann. of Phys. 203 (1990) 76

[7] K. Gawedzki, A. Kupiainen, Nucl. Phys. B320 (1989) 625

[8] J. F. Gomes, F.E. M. da Silveira, G. M. Sotkov and A. H. Zimerman, "Symmetries of Non-Abelian Toda Models ", IFT preprint in preparation

[9] E. Witten,Phys. Rev. D44 (1991) 314

[10] A. N. Leznov and M. V. Saveliev, Commun. Math. Phys. 89 (1983) 59

[11] J. F. Gomes, F.E. M. da Silveira, G. M. Sotkov and A. H. Zimerman, " Quantum V-algebras", G.M. Sotkov talk at this conference

[12] J. Balog, L. Feher, L. O'Raifeartaigh, P. Forgas and A. Wipf, Ann. of Phys. 203 (1990) 76

Alternative rings of small order and the hunt for Moufang circle loops

EDGAR G. GOODAIRE Memorial University of Newfoundland, St. John's, Newfoundland, Canada A1C 5S7, edgar@math.mun.ca

ABSTRACT Let $(R, +, \cdot)$ be a nilpotent alternative ring and, for $x, y \in R$, define $x \circ y = x + y + xy$. Then $(R, \circ)$ is a Moufang loop called a "circle loop". Circle groups, which are not hard to exhibit, have sparked the interest of a number of people for more than forty years. Relatively little attention has been directed at the not associative situation, however. Indeed, the scarcity of nilpotent alternative rings (which are not associative) makes even the discovery of Moufang circle loops difficult. The search for Moufang circle loops of small order has led to a recent enumeration of small order alternative rings and alternative algebras of small finite dimension which is described here.

1 INTRODUCTION AND HISTORICAL CONTEXT

A ring is *alternative* if $x(xy) = x^2y$ and $(yx)x = yx^2$ for all x and y. A loop is *Moufang* if $((xy)z)y = x(y(zy))$ for all x, y and z. The units (invertible elements) of an alternative ring with 1 form a Moufang loop [8, Section II.5.3], but this paper is concerned with another connection between alternative rings and Moufang loops.

The *circle operation* on an alternative ring A is defined by setting

$$x \circ y = x + y + xy$$

for $x, y \in A$. An element $x \in A$ is *quasi-regular* if $x \circ y = y \circ x = 0$ for some $y \in A$. The set of all quasi-regular elements in A is denoted Quasi(A).

If A has a unity, then x is quasi-regular if and only if $1+x$ is invertible since $(1+x)(1+y)=1$ is equivalent to $x \circ y = 0$. It is thus easy to see that nilpotent elements are quasi-regular. In fact, if $x^n = 0$, then

$$-x + x^2 - x^3 + \cdots + (-1)^{n-1}x^{n-1}$$

is a quasi-inverse for x.

THEOREM 1 (GOODAIRE, 1987) *For any alternative ring A, $(\text{Quasi}(A), \circ)$ is a Moufang loop.*

In the particular case that the algebra A is *Jacobson radical*—$\text{Quasi}(A) = A$—we call $(A, \circ)$ a *circle loop* and ask

QUESTION 2 *When is a Moufang loop a circle loop?*

In other words, given a Moufang loop $(L, \cdot)$, does there exist an alternative ring $(A, +, \cdot)$ such that $(A, \circ) \cong (L, \cdot)$? This question has a substantial history in the associative case.

Any abelian group $(G, +)$ is a circle group. In fact, let $(A, +) = (G, +)$ and define $xy = 0$ for $x, y \in G$. Then $x \circ y = x + y + xy = x + y$, so $(A, \circ) = (G, +)$. One of the early nontrivial results on this subject is due to L. Kaloujnine [15] (see also [17, Theorem 1.6.7]).

THEOREM 3 (KALOUJNINE, 1954) *Let G be a p-group of class 2, p an odd prime. Then G is the circle group of some nilpotent ring.*

Here is an idea of the proof. Since p is odd, every $x \in G$ has a unique square root. Denote this element $\sqrt{x}$. Kaloujnine succeeds in showing that

$$x \cdot y = xyx^{-1}y^{-1}$$
$$x + y = xy\sqrt{y \cdot x}$$

provide definitions of multiplication and addition in G so that $(G, +, \cdot)$ is an associative ring with circle group isomorphic to G.

A result of R. L. Kruse shows that not every p-group is a circle group [16].

THEOREM 4 (KRUSE, 1970) *Let G be a p-group. If G is the circle group of a nilpotent ring, then G has a central series*

$$\{1\} = Z_0 \subseteq Z_1 \subseteq Z_2 \subseteq \cdots \subseteq Z_c = G$$

with $[Z_i : Z_{i-1}] \geq p^2$ for $i = 2, \ldots, c$.

In particular, a group of order p^4 is a circle group if and only if it is abelian or nilpotent of class 2. It is easy to see that a nilpotent ring of nilpotency index $c+1$ has a circle group which is nilpotent of class at most c. Thus it is natural to look for nilpotent groups of class c which are circle groups of nilpotent rings of class $c+1$. In 1973, J. C. Ault and J. F. Watters found necessary and sufficient conditions for a nilpotent class 2 group to be the circle group of a nilpotent ring of index 3 and showed that any finite nilpotent class 2 group is a circle group [1]. (See also [18] for a slight correction in the Ault and Watters proof.)

THEOREM 5 *A nilpotent group G of class 2 is the circle group of a nilpotent ring of index 3 if and only if there is a map $m: G \times G \rightarrow \mathcal{Z}(G)$ (the centre of G) such that*

(i) $m(gh, k) = m(g, k)m(h, k)$,

(ii) $m(g, hk) = m(g, h)m(g, k)$,

(iii) $m(m(g, h), k) = m(g, m(h, k)) = 1$ *and*

(iv) $m(g, h)[m(h, g)]^{-1} = (g, h)$

for all $g, h, k \in G$, where (g, h) denotes the commutator $g^{-1}h^{-1}gh$.

Ten years later, K. Tahara and A. Hosomi found necessary and sufficient conditions for a finite group which is nilpotent of class 3 to be the circle group of a nilpotent ring of index 4 [14]. Recently, A. A. Bovdi used Theorem 5 to establish another result of this sort [3].

THEOREM 6 (BOVDI, 1996) *If G is a nilpotent group of class 2 and exponent 4, then G is the circle group of a nilpotent ring of index 3 and characteristic 2.*

2 MOUFANG CIRCLE LOOPS WHICH ARE NOT ASSOCIATIVE

What about loops which are not groups? The situation is very different. It is impossible, for instance, for a Moufang loop which is nilpotent of class 2 (and not associative) to be the circle loop of an alternative ring of index 3 because rings of nilpotency index 3 are associative.

Is a commutative Moufang loop a circle loop? Since trivial multiplication implies associativity, there is no immediate answer as with the case of abelian groups. The real answer is sometimes, but not always. There are nilpotent commutative alternative rings which are not associative (see Examples 16 below, nos. 3 and 4) so there exist commutative Moufang circle loops. On the other hand, there exist commutative Moufang loops of order 3^4 which are not associative [19, Section IV.5], but none of these is a circle loop because of the following result [9].

THEOREM 7 (GOODAIRE, 1987) *Any Moufang circle loop of order p^4, p prime, is a group.*

Thus, it is of interest to ask

QUESTION 8 *When is a commutative Moufang loop a circle loop?*

Perhaps the only result which guarantees that a certain kind of Moufang loop is a circle loop is this [9].

THEOREM 9 *Let L be a Moufang loop in which $x \mapsto x^2$ is a permutation. Suppose*

i. every commutator (x, y) is in the nucleus of L;

ii. every associator (x, y, z) is in the centre of L;

iii. $((x,y),z) = (x,y,z)^2$ *for every* $x,y,z \in L$.

Then L *is a circle loop.*

(The associator (x,y,z) of three elements x,y,z in a loop is defined by $(xy)z = [x(yz)](x,y,z)$.)

One class of Moufang loops which may be easier to handle than those which are commutative are the so-called RA loops of Chein and Goodaire [7]. An *RA loop* is a (necessarily Moufang) loop which has an alternative (but not associative) loop ring in characteristic different from 2. The best source of information about these loops and their loop rings is the book by E. G. Goodaire, E. Jespers and C. Polcino Milies [8]. Yongxin Zhou has recently established an interesting relationship between RA loops and Moufang loops arising from the circle operation [21]. See also [13].

THEOREM 10 (ZHOU, 1998) *Let* S *be a commutative associative ring with unity and characteristic different from 2. Choose any* $\alpha,\beta,\gamma \in S$ *such that* $\alpha-1,\beta-1,\gamma-1 \in \mathrm{Quasi}(S)$. *Construct the Cayley-Dickson algebra*

$$A = S(\alpha,\beta,\gamma) = S + Si + Sj + Sk + S(ij) + S(ik) + S(jk) + S[(ij)k],$$

where $i^2 = \alpha$, $j^2 = \beta$, $k^2 = \gamma$. *Then* $i-1, j-1, k-1 \in \mathrm{Quasi}(A)$ *and the subloop of* $(\mathrm{Quasi}(A), \circ)$ *generated by* $\mathrm{Quasi}(S)$ *and* $i-1, j-1, k-1$ *is an RA loop. Any RA loop is the homomorphic image of such a loop.*

Notice, however, that in this theorem neither L, nor $(\mathrm{Quasi}(A), \circ)$ is a circle loop. So we ask

QUESTION 11 *Can an RA loop* L *be a circle loop?*

The answer is no if $|L| \leq 64$. This fact is a corollary of a more general theorem which relies on the following lemma of R. L. Kruse and David T. Price [17] which assumed, but does not require, associativity.

LEMMA 12 (KRUSE AND PRICE, 1969) *If* S *is a subring of a nilpotent ring* R *such that* $R = S + R^2$, *then* $R = S$.

Proof. There is a minimal positive integer t such that $R^t = S^t$. Induction shows that $R^n \subseteq S^n + R^{n+1}$ for all $n \geq 1$. If $t > 1$,

$$R^{t-1} \subseteq S^{t-1} + R^t \subseteq S^{t-1} + S^t \subseteq S^{t-1}$$

and so $R^{t-1} = S^{t-1}$, a contradiction. □

COROLLARY 13 *If* R *is a nilpotent alternative ring and the additive group* R/R^2 *can be generated by at most two elements, then* R *is associative.*

Proof. Assume $R/R^2 = \langle R^2 + a, R^2 + b\rangle$ and apply Lemma 12 with S the subring of R generated by a and b. A theorem of Artin states that in an alternative ring, the subring generated by any two elements is associative. (See, for example, the book by R. D. Schafer [20, Theorem 3.1].) It follows that S is associative. □

In what follows, we use $[x, y, z]$ to denote the associator of elements x, y, z in an alternative ring and $[A, B, C]$ to denote the linear span of all associators of the form $[a, b, c]$, $a \in A$, $b \in B$, $c \in C$. We denote the ring commutator of x and y by $[x, y]$. The nucleus, centre and commutator-associator subloops of a loop L are denoted by $\mathcal{N}(L)$, $\mathcal{Z}(L)$ and L', respectively.

THEOREM 14 *Let L be a Moufang loop with L' of order 2 and $\mathcal{N}(L) = \mathcal{Z}(L)$. Suppose L is the circle loop of an alternative ring R with $R^5 = \{0\}$. Then L is a group.*

Proof. We begin with the elementary remark that the nucleus of L and of R consist of the same elements and the same is true for the centre of each structure. In the present situation then, the nucleus and centre of R coincide. Since $R^5 = \{0\}$, $[R^3, R, R] = 0$, so R^3 is in the nucleus of R and thus central. Let $L' = \{0, s\}$. (Note that 0 is the identity of L.) Then s is central of order 2 [8, Corollary III.1.5] so $s \circ s = 0$.

Assume that L is not a group. Then s is a loop associator and, as such, is in R^3 (see equation (5) in [9]). Since $R^5 = \{0\}$, $s^2 = 0$. Thus $0 = s \circ s = s + s + s^2 = 2s$. Moreover, for any $x, y \in R$ which do not commute,

$$\begin{aligned}[x, y] &= xy - yx = x \circ y - y \circ x = x \circ y - s \circ x \circ y \\ &= x \circ y - \{s + x \circ y + s(x \circ y)\} = -s - s(x \circ y)\end{aligned} \tag{1}$$

which is in R^3 and hence central.

Now let

$$f(x, y, z, w) = [wx, y, z] - x[w, y, z] - [x, y, z]w \tag{2}$$

denote the Kleinfeld function. Since

$$f(x, y, z, w) = [[x, y], z, w] + [[z, w], x, y]$$

[5, Lemma 2.1] and, here, commutators are central, f is identically 0.

We claim that R^2 is also central. If not, choose a noncentral (and hence nonnuclear) $x \in R^2$ and also elements y and z so that $[x, y, z] \neq 0$. Let w be any element of R. Then $w[x, y, z]$ is in $R^5 = \{0\}$. However,

$$\begin{aligned}[x, y, z] &= (xy)z - x(yz) = (x \circ y) \circ z - x \circ (y \circ z) \\ &= (x \circ y) \circ z - s \circ (x \circ y) \circ z \\ &= (x \circ y) \circ z - \{s + (x \circ y) \circ z + s[(x \circ y) \circ z]\} \\ &= -s - s[(x \circ y) \circ z]\end{aligned} \tag{3}$$

so

$$0 = w[x, y, z] = -ws - (ws)[(x \circ y) \circ z] = -ws$$

because $(ws)[(x \circ y) \circ z] \in R^5$. Thus $Rs = sR = 0$. ¿From (2) and the fact that $f \equiv 0$, we obtain

$$[wx, y, z] = x[w, y, z] + [x, y, z]w = 0$$

since $x[w, y, z]$ and $[x, y, z]w$ are each in sR, because of (3). It follows that wx is nuclear (and hence central) for all $w, x \in R$. As claimed, R^2 is indeed central.

Finally, we employ the identity

$$[xy, z] = x[y, z] + [x, z]y + 3[x, y, z] \tag{4}$$

which holds in any alternative ring (use the fact that the associator is an alternating function of its arguments and direct computation). Let x, y, z be any three elements of R. Since R^2 is central, $[xy, z] = 0$. Since $x[y, z]$ and $[x, z]y$ are each in Rs, by (1), each of these elements is also 0. Use (3) to write $[x, y, z] = -s - st$ for some $t \in R$ and then note that $[x, y, z] = -s$ because $sR = 0$. By (4), we have $0 = 3[x, y, z] = -3s = s$, since $2s = 0$. This implies that $s = 0$, which is not the case. □

COROLLARY 15 *No circle loop of order $n \leq 64$ is an RA loop.*

Proof. Let L be an RA loop. Then L is Moufang but not a group, L' has order 2 and $\mathcal{N}(L) = \mathcal{Z}(L)$ [8, Theorem IV.1.8]. Suppose that L is a circle loop. Then there exist operations $+$ and $\cdot$ on L such that $(L, +, \cdot)$ is a nilpotent alternative ring which is not associative. For convenience, call this ring R and write the abelian group $(R, +)$ as the direct sum of cyclic groups R_p of prime power order. Each R_p is an ideal and R is the direct product of these ideals. At least one of these R_p is not associative, so its circle loop—call it L_p—is a subloop of L which is not associative; hence L_p is RA. By this reasoning, it follows that we may assume that L has order a power of p and, since RA loops of odd order are abelian groups [8, Corollary V.1.2], it must be that $p = 2$.

Assume $|R| \leq 2^6$. By Corollary 13, the additive group R/R^2 has order at least 8, so $|R^2| \leq 8$. Since $R \supseteq R^2 \supseteq R^3 \cdots \supseteq \{0\}$ must be strictly decreasing, $R^5 = \{0\}$. The result now follows from Theorem 14. □

The results of this section suggest that to find classes of Moufang loops which are not groups and which can be identified *a priori* as circle loops is likely very difficult. One approach would be to find some nilpotent alternative rings and to seek properties shared by the corresponding circle loops. We conclude this section by citing some examples of nilpotent alternative rings which are not associative.

EXAMPLES 16

1. Let F be a field of characteristic $\neq 2$. Construct the Cayley-Dickson algebra $F(0,0,0)$. The orthogonal complement of 1, namely, $Fi + Fj + Fk + F(ij) + F(ik) + F(jk) + F[(ij)k]$, is a nilpotent alternative algebra. (A multiplication table for this algebra appears in a 1944 article by R. H. Bruck who, coincidentally, used this and similar algebras of higher dimension to construct inverse property quasigroups by means of the circle operation [4, Section 5].)

2. Let F be a field of characteristic 2 and L be a (Moufang) loop of order 2^n such that FL is an alternative algebra. Then the *augmentation ideal* of FL, that is, the ideal $\Delta(L) = \{\sum_{g \in L} \alpha_g g \mid \sum \alpha_g = 0\}$, is nilpotent [10].

3. Let A be the six-dimensional algebra over a field of characteristic 3 with basis $\{x_1, x_2, \ldots, x_6\}$ and multiplication consistent with distributivity and

$x_1x_2 = x_2x_1 = x_4$, $x_2x_3 = x_3x_2 = x_5$, $x_1x_5 = x_5x_1 = x_6$, $x_3x_4 = x_4x_3 = -x_6$, all remaining $x_ix_j = 0$. Then A is a commutative nilpotent alternative algebra which is not associative. This example came originally to the author's attention in a private communication from Harry Smith who credited Kevin McCrimmon. It also appears in work of M. I. Badalov, who found all nilpotent, not associative, alternative algebras of dimension $n \leq 7$ over a field [2].

4. The author and D. A. Robinson have recently found a way to construct commutative alternative rings (which are not associative) [11]. Some of these rings are also nilpotent.

3 ALTERNATIVE RINGS OF SMALL ORDER

One difficulty with the nilpotent rings just mentioned is that they are too big for our purposes. There are 158 Moufang loops of order less than 64, for instance [6], and we would like to know if any of these is a circle loop. To find an answer, we have embarked upon a study of alternative rings of small finite order and alternative algebras of small finite dimension. As noted in the proof of Corollary 15, the search for such rings immediately reduces to a search for rings of prime power order. We consider the nilpotent case first.

For a prime p, any alternative ring of order p or p^2 is associative by Artin's theorem because its additive group can be generated by at most two elements. If R is nilpotent of order p^3, then $R^2 \subsetneqq R$, so as a subgroup of $(R, +)$, R^2 has order 1, p or p^2. In each of these cases, the additive group R/R^2 can be generated by at most two elements, so R is associative by Corollary 13. If R is nilpotent of order p^4, then R^2 has order 1, p, p^2 or p^3. By Corollary 13, the cases p^2 and p^3 imply associativity. If $|R^2| = 1$ or p, then $R^3 = \{0\}$ and, again, R is associative.

THEOREM 17 *A nilpotent alternative ring of order p^n, $n \leq 5$, is associative.*

Proof. The remarks preceding this theorem show that we need consider only the case $n = 5$. Because of Corollary 13, we may assume that $|R^2| = p^2$ and that R/R^2 is a vector space of dimension 3 over the field F_p of p elements. We may also assume that $R^3 \neq \{0\}$; thus $|R^3| = p$ and $R^4 = \{0\}$. Let $\{R^2 + a, R^2 + b, R^2 + c\}$ be a basis for R/R^2 and take any $x, y, z \in R$. Since $R^4 = \{0\}$ and the associator is an alternating function of its arguments, we have $[x, y, z] = \gamma[a, b, c]$ for some integer γ. Thus, it is sufficient to show that a, b, c may be chosen such that $[a, b, c] = 0$. Let $R^3 + u$ generate R^2/R^3. As suggested by [2, Proposition 1], we note that the map $R^2 + x \mapsto ux$ is well defined since $uR^2 = \{0\}$, so the kernel of this map, as a subspace of R/R^2, has dimension at least 2. Similarly, the map $R^2 + x \mapsto xu$ has a kernel of dimension at least 2. It follows that we may select linearly independent elements $R^2 + c$ and $R^2 + a$ such that $uc = au = 0$. Extend to a basis $\{R^2 + a, R^2 + b, R^2 + c\}$ of R/R^2. Since $ab \in R^2$, we have $ab = \alpha u$ modulo R^3 for some integer α and so $(ab)c = \alpha uc = 0$. Similarly, $a(bc) = 0$, so $[a, b, c] = 0$. □

COROLLARY 18 *Any alternative nil ring of order less than 64 is associative.*

COROLLARY 19 *Any Moufang circle loop of order less than 64 is a group.*

We finish by describing the situation with respect to (small) alternative rings which are not nilpotent. Here our investigations were more successful. The reader is referred to [12] for proofs and further details.

THEOREM 20 *Over any field, there are precisely two alternative but not associative algebras of dimension 4. These algebras, which we label 4/1 and 4/2, are anti-isomorphic, neither has a unity and each has a nonzero idempotent. With respect to a basis* $\{e, x_1, x_2, x_3\}$, *multiplication consistent with the distributive laws is defined by*

4/1: $e^2 = e$, $ex_1 = x_1$, $x_2e = x_2$, $x_2x_3 = x_1$, $x_3e = x_3$, $x_3x_2 = -x_1$

4/2: $e^2 = e$, $ex_2 = x_2$, $ex_3 = x_3$, $x_1e = x_1$, $x_2x_3 = x_1$, $x_3x_2 = -x_1$

with all unspecified products 0. For any prime p, the only alternative rings of order p^4 *which are not associative are algebras over the field of p elements.*

THEOREM 21 *There are eleven alternative algebras of dimension 5 over a field which are not associative. All possess an idempotent* $e \neq 0, 1$. *Precisely one has a unity and, with respect to a basis* $\{1, e, x_1, x_2, x_3\}$, *this algebra, 5/0, is defined by*

5/0: $e^2 = e$, $ex_1 = x_1$, $x_2e = x_2$, $x_2x_3 = x_1$, $x_3e = x_3$, $x_3x_2 = -x_1$.

With respect to a basis $\{e, x_1, x_2, x_3, x_4\}$, *the other ten algebras are defined as follows.*

5/1: $e^2 = e$, $ex_1 = x_1$, $ex_3 = x_3$, $ex_4 = x_4$, $x_1e = x_1$, $x_2e = x_2$, $x_3x_4 = x_2$, $x_4x_3 = -x_2$

5/2: $e^2 = e$, $ex_3 = x_3$, $ex_4 = x_4$, $x_1{}^2 = x_1$, $x_2e = x_2$, $x_3x_4 = x_2$, $x_4x_3 = -x_2$

5/3: $e^2 = e$, $ex_1 = x_1$, $ex_2 = x_2$, $x_1e = x_1$, $x_3e = x_3$, $x_3x_4 = x_2$, $x_4e = x_4$, $x_4x_3 = -x_2$

5/4: $e^2 = e$, $ex_2 = x_2$, $x_1{}^2 = x_1$, $x_3e = x_3$, $x_3x_4 = x_2$, $x_4e = x_4$, $x_4x_3 = -x_2$

5/5: $e^2 = e$, $ex_2 = x_2$, $ex_3 = x_3$, $ex_4 = x_4$, $x_1e = x_1$, $x_2x_3 = x_1$, $x_3x_2 = -x_1$

5/6: $e^2 = e$, $ex_3 = x_3$, $ex_4 = x_4$, $x_1e = x_1$, $x_1x_2 = x_3$, $x_2e = x_2$, $x_2x_1 = -x_3$

5/7: $e^2 = e$, $ex_3 = x_3$, $ex_4 = x_4$, $x_1e = x_1$, $x_2e = x_2$, $x_3x_4 = x_1$, $x_4x_3 = -x_1$

5/8: $e^2 = e$, $ex_1 = x_1$, $x_2e = x_2$, $x_2x_3 = x_1$, $x_3e = x_3$, $x_3x_2 = -x_1$, $x_4e = x_4$

5/9: $e^2 = e$, $ex_3 = x_3$, $ex_4 = x_4$, $x_2e = x_2$, $x_3x_4 = x_2$, $x_4x_3 = -x_2$

5/10: $e^2 = e$, $ex_2 = x_2$, $x_3e = x_3$, $x_3x_4 = x_2$, $x_4e = x_4$, $x_4x_3 = -x_2$

unspecified products always 0. For any prime p, there are thirteen alternative rings of order p^5 *which are not associative. In addition to eleven algebras over* F_p *defined as above, there are two rings with additive structure* $Z_{p^2} \oplus Z_p \oplus Z_p \oplus Z_p$ *and, if corresponding generators are* e, x_1, x_2 *and* x_3, *respectively, these rings are defined by*

5/11: $e^2 = e$, $ex_2 = x_2$, $x_1e = x_1$, $x_2x_3 = x_1$, $x_3x_2 = -x_1$

5/12: $e^2 = e$, $ex_1 = x_1$, $x_2e = x_2$, $x_2x_3 = x_1$, $x_3e = x_3$, $x_3x_2 = -x_1$.

REFERENCES

[1] J. C. Ault and J. F. Watters, Circle groups of nilpotent rings, Amer. Math. Monthly 80 (1973), 48–52.

[2] M. I. Badalov, Nilpotent alternative algebras, Algebra i Logika 23 (1984), no. 3, 167–181, English translation.

[3] A. A. Bovdi, On circle groups of nilpotent rings of characteristic 2, Period. Math. Hungar. 32 (1996), no. 1-2, 31–34.

[4] R. H. Bruck, Some results in the theory of quasigroups, Trans. Amer. Math. Soc. 55 (1944), 19–52.

[5] R. H. Bruck and E. Kleinfeld, The structure of alternative division rings, Proc. Amer. Math. Soc. 2 (1951), 878–890.

[6] Orin Chein, Moufang loops of small order, Mem. Amer. Math. Soc. 13 (1978), no. 197, 1–131.

[7] Orin Chein and Edgar G. Goodaire, Loops whose loop rings are alternative, Comm. Algebra 14 (1986), no. 2, 293–310.

[8] E. G. Goodaire, E. Jespers, and C. Polcino Milies, Alternative loop rings, North-Holland Math. Studies, vol. 184, Elsevier, Amsterdam, 1996.

[9] Edgar G. Goodaire, Circle loops of radical alternative rings, Algebras Groups Geom. 4 (1987), 461–474.

[10] ———, The radical of a modular alternative loop algebra, Proc. Amer. Math. Soc. 123 (1995), no. 11, 3289–3299.

[11] Edgar G. Goodaire and D. A. Robinson, Commutative alternative rings: A construction, preprint.

[12] Edgar G. Goodaire and Yongxin Zhou, Alternative rings of small order, preprint.

[13] ———, Cayley-Dickson algebras and RA loops, Comm. Algebra, to appear.

[14] Ken Ichi Tahara and Akinori Hosomi, On the circle groups of finite nilpotent rings, Groups—Korea 1983 (Kyoungju, 1983), Lecture Notes in Math., vol. 1098, Springer, Berlin–New York, 1981, pp. 161–179.

[15] L. Kaloujnine, Zum Problem der Klassifikation der endlichen metabelschen p-Gruppen, Wiss. Z. Humboldt-Univ. Berlin Math.-Natur. Reihe 4 (1954/55), 1–7.

[16] R. L. Kruse, On the circle group of a nilpotent ring, Amer. Math. Monthly 77 (1970), 168–170.

[17] Robert L. Kruse and David T. Price, Nilpotent rings, Gordon and Breach, New York, 1969.

[18] Yogesh Kumar, On a theorem of Ault and Watters, Math. Today XII (1994), 51–52.

[19] H. O. Pflugfelder, Quasigroups and loops: Introduction, Heldermann Verlag, Berlin, 1990.

[20] R. D. Schafer, An introduction to nonassociative algebras, Academic Press, New York, 1966.

[21] Yongxin Zhou, Alternative algebras and RA loops, 1998-99, PhD thesis, Memorial University of Newfoundland.

A note on linearization of some identities[1]

H. GUZZO JR. [1] Universidade de São Paulo, Instituto de Matemática e Estatística, Caixa Postal 66281, 05315-970, São Paulo, Brazil, E-mail: guzzo@ime.usp.br

P. VICENTE [2] Universidade de León, Campus de Vegazana, Departamento de Matemáticas, 24071, León, Spain, E-mail: demmvm@isidoro.unileon.es

ABSTRACT In this paper we present a method to linearize an identity. This consists of differentiating with regard to the variable we want to linearize and then replacing the derivative of this variable with the new variable that will appear in the linearization.

1 INTRODUCTION

The process of linearization of identities is very useful to the theory of non-associative algebra. In several classes of algebras, for instance Jordan and alternative, satisfying some identities, their linearizations can bring us important results (for more details see [1], [2], [3], [4] and [5]).

The linearization of identities applied to the Genetic Algebras is a very important tool. For example, the Peirce decomposition and the product of Peirce subspaces, for Bernstein and train algebras, are obtained from them. It consists on "partially deriving" the identity with respect to a variable, say x, to be linearized - minding the non-associativity and, sometimes, non-commutativity of the algebra - and consider the partial derivative of x as a new variable.

This method has been already used in [1] and [5] for algebras in general. In this note we aim to apply the process to Genetic Algebras, but the new proof holds true for algebras in general.

[1] This paper was written while this author was visiting the University of León, Spain. Sponsored by CNPq, proc. 453477/97-6 (NV).

[2] Partially supported by DGICYT, PB94-1311-C03-01.

2 THE LINEARIZATION

Let F be a field with char$(F) = 0$. Given A an algebra over F, $f_i: A \longrightarrow K_i$ where $K_i = A$ or $K_i = F$, $1 \le i \le m$, are arbitrary functions, we define recursively $(f_1, \ldots, f_m): A \oplus \ldots \oplus A \longrightarrow K$, where $K = A$ if some f_i takes values in A and $K = F$ otherwise, by $(f_1) = f_1$ and

$$(f_1, \ldots, f_m)(a_1, \ldots, a_m) = \frac{1}{m}\sum_{i=0}^{m-1}(f_1, \ldots, f_{m-1})(a_{\sigma^i(1)}, \ldots, a_{\sigma^i(m-1)})f_m(a_{\sigma^i(m)})$$

where $(a_1, \ldots, a_m) \in A \oplus \ldots \oplus A$ and σ is the cycle $(1\ 2\ \ldots\ m)$ in the symmetric group S_m. If all f_i are linear functions then $(f_1, \ldots, f_m)$ is a m-linear symmetric function, see [6].

Let $f, g: A \longrightarrow K$ as above, $a, b \in A$. We shall use the following notations, for $m, p \ge 1$: $f^{(1)} = f$; $f^{(m)} = (f, \ldots, f)$; $(f^{(m)}, g^{(p)}) = (f, \ldots, f, g, \ldots, g)$; $(f^{(m)}, g^{(0)}) = f^{(m)}$; $(f^{(0)}, g^{(p)}) = g^{(p)}$; $a^{(1)} = a$; $a^{(m)} = (a, \ldots, a)$; $(a^{(m)}, b^{(p)}) = (a, \ldots, a, b, \ldots, b)$; $(a^{(m)}, b^{(0)}) = a^{(m)}$ and $(a^{(0)}, b^{(p)}) = b^{(p)}$.

By [6, Lemma 2.2] for all $a, b \in A$ and $m \ge 1$ we have,

$$(a+b)^m = \sum_{i=0}^{m}\binom{m}{i} id_A^{(m)}\left(a^{(m-i)}, b^{(i)}\right), \tag{1}$$

where id_A is the identity function on A.

DEFINITION 2.1 *Given* $a, a_1, \ldots, a_i, \in A$, $f(x) = \sum_{j=0}^{m} b_j x^j$ *and* $g(x) = \sum_{k=0}^{r} c_k x^k$, *where* $b, b_j, c_k \in A \cup F$ $(j = 1, \ldots, m;\ k = 1, \ldots, r)$ *and* $b_0, c_0 \in A$. *We define,*

1. $\partial_{a,x}(x) = a,\ \partial_{a,x}(b) = 0;$
2. $\partial_{a,x}(x^n) = \partial_{a,x}(x^{n-1})\,x + x^{n-1}\partial_{a,x}(x);$
3. $\partial_{a,x}(f(x)) = \sum_{j=0}^{m} b_j \partial_{a,x}(x^j);$
4. $\partial_{a,x}(f(x)g(x)) = \partial_{a,x}(f(x))\,g(x) + f(x)\partial_{a,x}(g(x));$
5. $\partial^0_{a,x}(f(x)) = f(x),\ \partial^i_{a_1 \ldots a_i, x}(f(x)) = \partial^{i-1}_{a_1 \ldots a_{i-1}, x}\left(\partial_{a_i, x}(f(x))\right),\ i \ge 1;$ $\partial^i_{a,x}(f(x)) = \partial^i_{\underbrace{a \ldots a}_{i\ \text{times}}, x}(f(x)).$

LEMMA 2.1 *For all* $n \ge 1$ *and* $0 \le k \le i \le n+1$, *we have*

$$\partial^i_{y,x}\left(x^{n+1}\right) = k\partial^{i-1}_{y,x}(x^n)\,y + \partial^{i-k}_{y,x}\left(\partial^k_{y,x}(x^n)\,x\right).$$

Proof: By induction on n. ■

For $k = i$ in Lemma 2.1, we have

$$\partial^i_{y,x}\left(x^{n+1}\right) = i\partial^{i-1}_{y,x}(x^n)\,y + \partial^i_{y,x}(x^n)\,x. \tag{2}$$

LEMMA 2.2 *For all $n \geq 1$, we have*

1. $id_A^{(n)}\left(y^{(n)}\right) = y^n$;
2. $\binom{n}{1} id_A^{(n)}\left(x^{(n-1)}, y\right) = \partial_{y,x}\left(x^n\right)$;
3. $\partial_{y,x}^n\left(x^n\right) = n!\,y^n$.

Proof: By induction on n. ■

LEMMA 2.3 *For all $n \geq 1$ and $0 \leq i \leq n$, we have*

$$i!\binom{n}{i} id_A^{(n)}\left(x^{(n-i)}, y^{(i)}\right) = \partial_{y,x}^i\left(x^n\right).$$

Proof: We will prove by induction on n. The Lemma 2.2 shows that the assumption is true for $i = 0, 1, n$ and for $n = 1, 2$. Suppose that is true for $n \geq 3$. Given i, $2 \leq i < n$, then

$$i!\binom{n+1}{i} id_A^{(n+1)}\left(x^{(n+1-i)}, y^{(i)}\right) =$$

$$\frac{i!}{n+1}\binom{n+1}{i}\left[i\, id_A^{(n)}\left(x^{(n+1-i)}, y^{(i-1)}\right) y + (n+1-i) id_A^{(n)}\left(x^{(n-i)}, y^{(i)}\right)\right] =$$

$$\frac{i!\,i}{n+1}\binom{n+1}{i} id_A^{(n)}\left(x^{(n-(i-1))}, y^{(i-1)}\right) y +$$

$$\frac{i!\,(n+1-i)}{n+1}\binom{n+1}{i} id_A^{(n)}\left(x^{(n-i)}, y^{(i)}\right) x =$$

$$\frac{i!\,i}{n+1}\binom{n+1}{i}\frac{1}{(i-1)!}\binom{n}{i-1}^{-1} \partial_{y,x}^{i-1}\left(x^n\right) y +$$

$$\frac{i!(n+1-i)}{n+1}\binom{n+1}{i}\frac{1}{i!}\binom{n}{i}^{-1} \partial_{y,x}^{i}\left(x^n\right) =$$

$$i\partial_{y,x}^{i-1}\left(x^n\right) y + \partial_{y,x}^i\left(x^n\right) x = \partial_{y,x}^i\left(x^{n+1}\right).$$ ■

COROLLARY 2.1 *For all $n \geq 1$ and $0 \leq i \leq n$, we have $(n-i)!\partial_{y,x}^i\left(x^n\right) = i!\partial_{x,y}^{n-i}\left(y^n\right)$.*

LEMMA 2.4 *For $f(x) = \sum_{n=0}^{m} a_n x^n$, where $a_n \in F \cup A, (n = 1, \ldots, m)$, $a_0 \in A$ and $\alpha \in F$, we have*

$$\sum_{i=0}^{m}\frac{\alpha^i}{i!}\partial_{y,x}^i\left(f(x)\right) = \sum_{n=0}^{m}\sum_{i=0}^{n}\frac{\alpha^i}{i!}\partial_{y,x}^i\left(a_n x^n\right).$$

Proof: By induction on m. ■

THEOREM 2.1 *If $f(x) = \sum_{n=0}^{m} a_n x^n$, then $f(x+\alpha y) = \sum_{i=0}^{m}\frac{\alpha^i}{i!}\partial_{y,x}^i\left(f(x)\right)$, where $\alpha \in F$, $a_n \in A \cup F$, $(n = 1, \ldots, m)$ and $x, y, a_0 \in A$.*

Proof: By (1), we have

$$f(x+\alpha y)=\sum_{n=0}^{m}a_n(x+\alpha y)^n=\sum_{n=0}^{m}\sum_{i=0}^{n}a_n\alpha^i\binom{n}{i}id_A^{(n)}\left(x^{(n-i)},y^{(i)}\right),$$

By Lemmas (2.3) and (2.4)

$$f(x+\alpha y)=\sum_{n=0}^{m}\sum_{i=0}^{n}\frac{\alpha^i}{i!}\partial^i_{y,x}(a_nx^n)=\sum_{i=0}^{m}\frac{\alpha^i}{i!}\partial^i_{y,x}(f(x)).$$ ■

COROLLARY 2.2 *If $f(x)$ is an identity in A of degree m, then the first linearization of $f(x)$ is $\partial^i_{y,x}(f(x))$ $i=0,\ldots,m$.*

COROLLARY 2.3 *If $f(x)=x^m$ is an identity in A, then the first linearization of $f(x)$ is $\partial^i_{y,x}(f(x))$ $i=0,\ldots,\left[\frac{m}{2}\right]$.*

LEMMA 2.5 *If $f(x)$ has degree m and $g(x)$ has degree n, then*

$$\sum_{i+j=k}\frac{k!}{i!j!}\partial^i_{y,x}(f(x))\,\partial^j_{y,x}(g(x))=\partial^k_{y,x}(f(x)g(x))\,,\ 0\le k\le m+n.$$

Proof: By induction on k. ■

LEMMA 2.6 *For all $n\ge 1$, $m\ge 1$ and $0\le i\le n+m$, we have*

$$(n+m-i)!\,\partial^i_{y,x}(x^nx^m)=i!\,\partial^{n+m-i}_{x,y}(y^ny^m).$$

Proof: By Lemma 2.5 and Corollary 2.1, we have

$$\partial^i_{y,x}(x^nx^m)=\sum_{r+s=i}\frac{i!}{r!s!}\partial^r_{y,x}(x^n)\,\partial^s_{y,x}(x^m)=$$

$$\frac{i!}{(n+m-i)!}\sum_{r+s=i}\frac{(n+m-i)!}{(n-r)!(m-s)!}\partial^{n-r}_{x,y}(y^n)\,\partial^{n-s}_{x,y}(y^m)=\frac{i!}{(n+m-i)!}\partial^{n+m-i}_{x,y}(y^ny^m).$$ ■

THEOREM 2.2 *If $f(x)$ has degree m and $g(x)$ has degree n, then*

$$f(x+\alpha y)g(x+\alpha y)=\sum_{k=0}^{n+m}\frac{\alpha^k}{k!}\partial^k_{y,x}(f(x)g(x))\,,$$

where $\alpha\in F$ and $y\in A$.

Proof: By Theorem 2.1 and Lemma 2.5, we have

$$f(x+\alpha y)g(x+\alpha y)=\left(\sum_{i=0}^{m}\frac{\alpha^i}{i!}\partial^i_{y,x}(f(x))\right)\left(\sum_{j=0}^{n}\frac{\alpha^j}{j!}\partial^j_{y,x}(g(x))\right)=$$

$$\sum_{i=0}^{m}\sum_{j=0}^{n}\frac{\alpha^{i+j}}{i!j!}\partial^i_{y,x}(f(x))\,\partial^j_{y,x}(g(x))=\sum_{k=0}^{n+m}\frac{\alpha^k}{k!}\left(\sum_{i+j=k}\frac{k!}{i!j!}\partial^i_{y,x}(f(x))\,\partial^j_{y,x}(g(x))\right)$$

$$= \sum_{k=0}^{n+m} \frac{\alpha^k}{k!} \partial^k_{y,x} (f(x)g(x)). \qquad \blacksquare$$

COROLLARY 2.4 *If $f(x)g(x)$ is an identity in A, then the first linearization of $f(x)g(x)$ is $\partial^i_{y,x} (f(x)g(x))$ $i = 0, \ldots, m+n$.*

COROLLARY 2.5 *If $x^n x^m$ is an identity in A, then the first linearization is $\partial^i_{y,x} (x^n x^m)$ $i = 0, \ldots, [\frac{m+n}{2}]$.*

DEFINITION 2.2 *Let $\lambda: A \longrightarrow K$ be an homomorphism of algebras where $K = A$ or $K = F$. We define:*

1. $\partial^i_{a_1 \ldots a_i, x} (\lambda(f(x))) = \lambda(\partial^i_{a_1 \ldots a_i, x} (f(x)))$;
2. $\partial_{a,x} (\lambda(f(x))g(x)) = \partial_{a,x} (\lambda(f(x)))\, g(x) + \lambda(f(x))\partial_{a,x} (g(x))$.

With a similar proof of the Theorem 2.2, we have:

PROPOSITION 2.1 *Let $\lambda: A \longrightarrow K$ be an homomorphism of algebras where $K = A$ or $K = F$, $f(x)$ of degree m and $g(x)$ of degree n. If $\lambda(f(x))g(x)$ is an identity in A, then the first linearization of $\lambda(f(x))g(x)$ is $\partial^i_{y,x} (\lambda(f(x))g(x))$ $i = 0, 1, \ldots, m+n$.*

COROLLARY 2.6 *Let $\lambda_i: A \longrightarrow K_i$ where $K_i = A$ or $K_i = F$, $1 \leq i \leq n$ be homomorphisms of algebras. If $f(x) = a_1\lambda_1(f_1(x))g_1(x) + \ldots + a_n\lambda_n(f_n(x))g_n(x)$ has degree m and is an identity in A, then the first linearization of f is $\partial^i_{y,x} (f(x)) = a_1\partial^i_{y,x} (\lambda_1(f_1(x))g_1(x)) + \ldots + a_n\partial^i_{y,x} (\lambda_n(f_n(x))g_n(x))$, for $i = 0, 1, \ldots, m$.*

OBSERVATION: If $f_1(x, y)$ is a first linearization we will obtain a second linearization having $\partial^i_{z,x} (f_1(x, y))$ and so on.

EXAMPLE 2.1 *Let A a Jordan algebra. We have $x^2(xy) = x(x^2y)$ all $x, y \in A$. We have all the linearizations of $f = x^2(xy) - x(x^2y)$, (see [7])*

1. The first linearization:
   (a) $2(xz)(xy) + x^2(zy) = z(x^2y) + 2x((xz)y)$;
   (b) $2z^2(xy) + 2(xz)(zy) + 2(xz)(zy) = 2z((xz)y) + 2z((xz)y) + 2x(z^2y)$;

   We obtain only an identity

$$2(xz)(xy) + x^2(zy) = z(x^2y) + 2x((xz)y).$$

2. The second linearization: $2(tz)(xy) + 2(xz)(ty) + 2(xt)(zy) = 2z((xt)y) + 2t((xz)y) + 2x((tz)y)$ or

$$(xy)(tz) + (xz)(ty) + (xt)(zy) = x((tz)y) + z((xt)y) + t((xz)y).$$

EXAMPLE 2.2 *Let (A, ω) a Bernstein algebra. We have $(x^2)^2 = \omega(x)^2x^2$. We have all the linearizations of $f = (x^2)^2 - \omega(x)^2x^2$, (see [8]).*

1. *The first linearization:*

(a) $2(xy)x^2 + 2x^2(xy) = 2\omega(xy)x^2 + 2\omega(x)^2xy;$

$$2x^2(xy) = \omega(xy)x^2 + \omega(x)^2xy;$$

(b) $2x^2y^2 + 4(xy)(xy) = 2\omega(xy)xy + 2\omega(xy)xy + \omega(x)^2y^2 + \omega(y)^2x^2;$

$$2x^2y^2 + 4(xy)^2 = \omega(x)^2y^2 + 4\omega(xy)xy + \omega(y)^2x^2.$$

2. *The second linearization:*

   $2(zy)x^2 + 4(xy)(xz) = \omega(zy)x^2 + 2\omega(xy)xz + 2\omega(xz)xy + \omega(x)^2zy.$

3. *The third linearization:*

   $4(zy)(xt)+4(ty)(xz)+4(xy)(tz)=2\omega(zy)xt+2\omega(ty)xz+2\omega(tz)xy+2\omega(xz)ty+ 2\omega(xy)zy + 2\omega(xt)(zy).$

REFERENCES

[1] N Jacobson. Structure and representations of Jordan algebras, American Mathematical Society, Coloquium Publications, U.S.A., pp 27-31, 1968.

[2] YI Lyubich. Mathematical Structures in Population Genetic, Biomathematics 22, Springer, Berlin-Heidelberg-New York,pp 86-106, 1992.

[3] RD Schafer. Introduction to nonassociative algebras, Academic Press, New York, pp 27, 91, 129, 1966.

[4] A Wörz. Algebras in Genetic (Lecture Notes in Biomathematics 36, Springer, Berlin-Heidelberg-New York, 1980.

[5] KA Zhevlakov, AM Slin'ko, IP Shestakov, AI Shirshov. Rings that are nearly associative, Inst.Math.Acad.Sci. of the U.S.S.R., Siberian Branch Novosibirsk, U.S.S.R., Academic Press, New York, pp 9-17, 1982.

[6] H Guzzo Jr. The Peirce decomposition for commutative train algebras, Comm. in Algebra, 22 (14), 5745-5757, 1994.

[7] S Walcher. Bernstein algebras which are Jordan algebras, Arch. Math. 50, 218-222, 1988.

[8] A Micali, MT Alcalde. Sur les algèbres de Bernstein Proc. London Math.Soc. 58 (3), 51-68, 1989.

The commuting nucleus of right alternative rings

IRVIN ROY HENTZEL Department of Mathematics, Iowa State University, Ames, Iowa 50011-2064 USA, hentzel@iastate.edu

ABSTRACT. Let R be a prime right alternative ring with commuting nucleus U. We show that if U is not contained in the center, then $R- := (R, +, [\ ,\ ])$ is nilpotent of index at most 11. Thus R is a generalization of strongly $(-1, 1)$ rings (i.e. right alternative and $[R, [R, R]] = 0$). We assume characteristic different from 2 and 3.

1 INTRODUCTION

The study of the commuting nucleus of right alternative rings has been done by Hentzel, Kleinfeld, Pchelintsev, Smith, and Thedy. The difficulty occurs because the commuting nucleus U of a right alternative ring R does not absorb associators. We define:

The associator: $(x, y, z) := (xy)z - x(yz)$.
The commutator: $[x, y] := xy - yx$.
The commuting nucleus of $R := \{u \text{ in } R | [u, R] = 0\}$.
The center of $R := \{u \text{ in } U | (u, R, R) = (R, u, R) = (R, R, u) = 0\}$.
We generalize the concept of commuting to the weaker conditions on v of:

$$(x, x, v) = 0,$$
$$(x, y, v) + (y, v, x) + (v, x, y) = 0,$$
$$[x, [x, v]] = 0$$

for all x, y, in R. We study additive subgroups V of R which satisfy the above three conditions and which absorb both associators and commutators. From the existence of such a nonzero set V, we obtain an ideal I and an ideal J such that $IJ = JI = 0$, where $(R, R, V) \subset J$ and $R- := (R, +, [\ ,\ ])$ is nilpotent of index at most 11 modulo I. If $(R, R, V) \neq 0$ and R is prime, then we conclude that $I = 0$. Thus $R-$ is nilpotent.

The Pchelintsev example [5] of a prime strongly $(-1, 1)$ ring where the commuting center is not the center proves that such rings exist. In that example, the associated bracket algebra is nilpotent of index 3.

Throughout this paper R will be a right alternative ring. U will be the commuting nucleus. V will be an additive subgroup under discussion. We use A, I, and J for subsets of R which will be defined in terms of V.
We define:

$$S(a,b,c) := (a,b,c) + (b,c,a) + (c,a,b).$$
$$M(a,b,c) := (a,b,c) + (b,a,c).$$
$$B(a,b,c) := [a,[b,c]] + [b,[a,c]].$$

2 THE CONSEQUENCES OF THE DEFINITION OF V

We assume that R contains an additive subgroup V satisfying:

a) $M(R,R,V) = 0$.
b) $S(R,R,V) = 0$.
c) $B(R,R,V) = 0$.
d) $[R,V] \subset V$.
e) $(R,R,V) \subset V$.

This says that V absorbs associators and commutators and that V has some other properties as well. These properties are:

a) $(x,x,v) = 0$.
b) $(a,b,v) + (b,v,a) + (v,a,b) = 0$.
c) $[x,[x,v]] = 0$.

for all v in V and a, b, x in R. In [7] Thedy discussed elements v satisfying a) which he called the N_β nucleus. Elements satisfying a) and b) are said to be in the $(-1,1)$ nucleus and are discussed in [2].

We give some well-known identities which hold in any non-associative ring. They can be checked by expanding them.

LEMMA 0: In any nonassociative ring the following identities hold:

a) $[ab,c] = a[b,c] + [a,c]b + (a,b,c) - (a,c,b) + (c,a,b)$.
b) $[a,[b,c]] + [b,[c,a]] + [c,[a,b]] = -S(a,b,c) + S(b,a,c)$.

We give some well-known consequences of the right alternative law.

LEMMA 1: The following identities hold in any right alternative ring.

a) $(a,b,b) = 0$ (right alternative law itself).
b) $(a,bc,d) + (a,dc,b) = (a,c,d)b + (a,c,b)d$.
c) $(ab,c,d) + (a,b,[c,d]) = a(b,c,d) + (a,c,d)b$.
d) $([a,b],c,d) + (a,b,[c,d]) - (b,a,[c,d]) = [a,(b,c,d)] - [b,(a,c,d)]$.
e) $((a,b,c),d,e) = ((a,d,e),b,c) + (a,(b,d,e),c) + (a,b,(c,d,e))$
$\qquad -(a,b,c)[d,e] + (a,b,c[d,e]) - (a,b,[d,e])c$.
f) $2(a,b,cd) - 2(a,b,d)c - 2(a,b,c)d = (a,[b,c],d) + (a,b,[c,d]) - (a,c,[b,d])$.
g) $2((a,b,c),d,e) = 2((a,d,e),b,c) + 2(a,(b,d,e),c) + 2(a,b,(c,d,e))$
$\qquad +(a,[b,c],[d,e]) + (a,b,[c,[d,e]]) - (a,c,[b,[d,e]])$.
h) $((a,b,c),c,b) = (a,b,c)[b,c]$.

PROOF: Part a) is from the definition. Part b) is [7, Eq. (2)]. Part c) [7, Eq.(4)]. Part d) is [7, Eq.(5)]. Part e) is [7, Eq.(6)].
We prove part f).

$$
\begin{aligned}
&(a,[b,c],d)+(a,b,[c,d])-(a,c,[b,d])\\
&\quad=(a,bc-cb,d)+(a,b,cd-dc)-(a,c,bd-db)\\
&\quad=2(a,b,cd)-\{(a,c,bd)+(a,b,cd)\}-\{(a,d,bc)+(a,b,dc)\}\\
&\qquad+\{(a,d,cb)+(a,c,db)\}\\
&\quad=2(a,b,cd)-\{(a,c,d)b+(a,b,d)c\}-\{(a,d,c)b+(a,b,c)d\}\\
&\qquad+\{(a,d,b)c+(a,c,b)d\}\text{ (using Lemma 1(b))}\\
&\quad=2(a,b,cd)-2(a,b,d)c-2(a,b,c)d.
\end{aligned}
$$

Part g) is a consequence of e) and f). Part h) is [3, Lemma 4, p.941].

We now give some useful consequences of the definitions of V.

LEMMA 2: The following results hold for v in V and a, b, c, x in R.

a) $(a,b,v)=-(b,a,v)$.
b) $(v,a,b)=-2(a,b,v)=2(b,a,v)$.
c) $[a,[b,v]]=-[b,[a,v]]$.
d) $[ab,v]=a[b,v]+[a,v]b$.
e) $[[a,b],v]=2[a,[b,v]]$.
f) $(x,x,a)v=(x,x,va)$.
g) $[(x,x,b),v]=3(b,x,[x,v])$.
h) $[x,(x,b,v)]=-(x,[x,b],v)=-(x,b,[x,v])$.
i) $[S(a,b,c),v]=0$.
j) $[a,[b,[c,v]]]=0$.
k) $[[a,[b,c]],v]=0$.
l) $2[a,(b,c,v)]=([a,b],c,v)-([b,c],a,v)+([c,a],b,v)$ $+2(a,b,[c,v])-2(b,c,[a,v])+2(c,a,[b,v])$.
m) $((a,a,b),b,v)=0$.

PROOF: Part a): $0=M(a,b,v)=(a,b,v)+(b,a,v)$.

Part b):
$$
\begin{aligned}
0=S(a,b,v)&=(a,b,v)+(b,v,a)+(v,a,b)\\
&=(a,b,v)-(b,a,v)+(v,a,b)\quad\text{(by the right alternative law)}\\
&=2(a,b,v)+(v,a,b)\quad\text{(by part a)).}
\end{aligned}
$$

So $(v,a,b)=-2(a,b,v)=2(b,a,v)$ by part a).

Part c): $0=B(a,b,v)=[a,[b,v]]+[b,[a,v]]$.

Part d):
$$
\begin{aligned}
[ab,v]&=a[b,v]+[a,v]b+(a,b,v)-(a,v,b)+(v,a,b)\quad\text{(by Lemma 0 a))}\\
&=a[b,v]+[a,v]b+(a,b,v)+(a,b,v)-2(a,b,v)\\
&\qquad\text{(by right alternative law and part b))}\\
&=a[b,v]+[a,v]b.
\end{aligned}
$$

Part e):
$$
\begin{aligned}
[[a,b],v]&=[[a,v],b]+[a,[b,v]]\quad\text{(by part d))}\\
&=-[b,[a,v]]+[a,[b,v]]\quad\text{(since commutators anti-commute)}\\
&=2[a,[b,v]]\quad\text{(by part c)).}
\end{aligned}
$$

Part f) is [7, Lemma 1]. It requires only that $M(R,R,v)=0$.

Part g): $[(x,x,b),v] = ([x,v],x,b) + (x,[x,v],b) + (x,x,[b,v])$ (by part d))
$= ([x,v],x,b) + (x,[x,v],b)$
(since $[b,v] \subset V$ and $(x,x,V) = 0$)
$= 2(b,x,[x,v]) + (b,x,[x,v])$
(by part b), the right alternative law, part a), and $[R,V] \subset V$)

$= 3(b,x,[x,v])$.

Part h): Using Lemma 1 d):
$([x,b],x,v) + (x,b,[x,v]) - (b,x,[x,v]) = [x,(b,x,v)] - [b,(x,x,v)]$. Using part a) and part b) this becomes:

$$(*) \qquad -(x,[x,b],v) + 2(x,b,[x,v]) = -[x,(x,b,v)].$$

Starting again with Lemma 1 d):
$([x,v],x,b) + (x,v,[x,b]) - (v,x,[x,b]) = [x,(v,x,b)] - [v,(x,x,b)]$. From part a), part b), part g) and the right alternative law we get
$-2(x,b,[x,v]) - (x,[x,b],v) + 2(x,[x,b],v) = -2[x,(x,b,v)] + 3(b,x,[x,v])$.
Using part a) and $[R,V] \subset V$ gives:

$$(**) \qquad +(x,[x,b],v) + (x,b,[x,v]) = -2[x,(x,b,v)].$$

Adding (*) and (**) gives: $3(x,b,[x,v]) = -3[x,(x,b,v)]$. Substituting this into equation (**) gives

$$(x,[x,b],v) = -[x,(x,b,v)].$$

Part i): $[S(a,b,c),v] = S([a,v],b,c) + S(a,[b,v],c) + S(a,b,[c,v])$ (by part d))
$= S(b,c,[a,v]) + S(c,a,[b,v]) + S(a,b,[c,v])$
(by the shift symmetry property of S)
$= 0$ (since $[R,V] \subset V$ and $S(R,R,V) = 0$).
Part j) and part k) will be now proved simultaneously. $[[a,[b,c]],v] = 4[a,[b,[c,v]]]$ by part e). The right hand side alternates on a,b,c by part c) and $[R,V] \subset V$. We conclude that the left hand side also alternates on a,b,c. By Lemma 0 b)

$$[a,[b,c]] + [b,[c,a]] + [c,[a,b]] = -S(a,b,c) + S(b,a,c).$$

If we commute the above with v, the right hand side is zero by part i). By the alternating property, the equation becomes $3[[a,[b,c]],v] = 0$.
Part l) is [2. eq.(20)] using part a) and $[R,V] \subset V$.
We now prove part m):

$$2((a,b,v),a,b) + 2((a,b,a),v,b) = 2(a,b,v)[b,a] + 2(a,b,a)[b,v]$$
(by Lemma 1 h)).

$-((v,a,b),b,a) = -(v,a,b)[a,b]$ (by Lemma 1 h)).
$((v,a,b),b,a) = -2((a,b,v),b,a)$ (by Lemma 2 b)).
$-(v,a,b)[a,b] = +2(a,b,v)[a,b]$ (by Lemma 2 b)).

Adding gives: $2((a,a,b),b,v) = -2(a,a,b)[b,v]$.
$= -2(a,a,[b,v]b)$ (by Lemma 2 f))
$= -(a,a,b[b,v]) - (a,a,[b,v]b) + (a,a,[b,[b,v]])$.
$= -(a,a,[bb,v])$ (by Lemma 2 d))

$= 0.$

This finishes the proof of Lemma 2.

We now prove some more involved properties involving the elements of V. We define a new product on the right alternative right R by defining $R- := (R, +, [\,,\,])$. We define H^n inductively in $R-$ by

$$H^1 = R, \text{ and } H^n = \sum_{i+j=n} [H^i, H^j].$$

LEMMA 3: The following results hold for v in V and a, b, c, d, e, f in R.

a) $[a, (b, c, [d, [e, v]])]$ alternates on its non-v arguments.
b) $[a, [b, (c, d, [e, v])]]$ alternates on its non-v arguments.
c) $[a, [b, (c, d, [e, [f, v]])]] = 0$.
d) $(H^i, H^j, v) = 0$ when $i + j \geq 6$.
e) $([[a, b], [c, d]], e, v) = 0$.
f) $([a, b], [c, d], [e, v]) = 0$.
g) $[a, (b, [c, d], [e, v])] = 0$.
h) $([a, [b, [c, d]]], e, v)$ alternates on the non-v arguments.

PROOF: Part a):

$$\begin{aligned}
[x, (x, c, [d, [e, v]])] &= -(x, c, [x, [d, [e, v]]]) && \text{(from Lemma 2 h))}\\
&= 0 && \text{(from Lemma 2 j)).}\\
[a, (x, x, [d, [e, v]])] &= 0 && \text{(from } M(R, R, V) = 0 \text{ and } [R, V] \subset V).\\
3[a, (b, x, [x, [e, v]])] &= [a, [(x, x, b), [e, v]]] && \text{(from Lemma 2 g))}\\
&= 0 && \text{(from Lemma 2 j)).}\\
[a, (b, c, [x, [x, v]])] &= 0 && \text{(from } B(R, R, V) = 0).
\end{aligned}$$

Part b):

$$\begin{aligned}
[x, [x, (c, d, [e, v])]] &\subset B(R, R, V)\\
& && \text{(since } [R, V] \subset V \text{ and } (R, R, V) \subset V)\\
&= 0 && \text{(since } B(R, R, V) = 0).\\
[a, [x, (x, d, [e, v])]] &= -[a, (x, d, [x, [e, v]])] && \text{(by Lemma 2 h))}\\
&= 0 && \text{(by Lemma 3 a)).}\\
[a, [b, (x, x, [e, v])]] &\subset [R, [R, M(R, R, V)]] && \text{(since } [R, V] \subset V)\\
&= 0 && \text{(since } M(R, R, V) = 0).\\
3[a, [b, (c, x, [x, v])]] &= [a, [b, [(x, x, c), v]]] && \text{(by Lemma 2 g))}\\
&= 0 && \text{(from Lemma 2 j)).}
\end{aligned}$$

Part c):

$$\begin{aligned}
2[a, [b, (c, d, [e, [f, v]])]] &= [[a, b], (c, d, [e, [f, v]])]\\
& \qquad \text{(by Lemma 2 e) and the fact that } (R, R, [R, [R, V]]) \subset V)\\
&= -[f, (c, d, [e, [[a, b], v]])] && \text{(by part a))}\\
&= -2[f, (c, d, [e, [a, [b, v]]])] && \text{(by Lemma 2 e))}\\
&= 0 && \text{(by Lemma 2 j)).}
\end{aligned}$$

Part d): Using Lemma 1 d) and Lemma 2 a) we can take any (H^i, H^j, v) and rewrite it into a sum of expressions where the commutator products are moved onto the element v or pushed outside of the associator.

Then using Lemma 2 e), expressions of the form $[H^i, v]$ can be written as expressions of the form $\ldots[R,[R,[R,v]]]\ldots$ where the number of R's is i. Since $[R,[R,[R,v]]]=0$ by Lemma 2 j), all products which have 3 or more R's bracketed around the v, or three or more R's bracketed around the associator (R,R,V) will be zero. This and part c) show that $(H^i, H^j, v) = 0$ if $i+j \geq 6$.

An important special case of this is $(H^5, R, V) = 0$.

Now parts e), f), g), h) can all be expanded in this same way to produce sums of expressions of the form $[R,[R,(R,R,[R,v])]]$ and $[R,(R,R,[R,[R,v]])]$. By part a) and part b) these expressions alternate on the non-v positions. We conclude that the expressions in parts e), f), g), h) alternate on the non-v positions. But part e) and part f) cannot alternate since there is an even permutation $(ac)(bd)$ of the arguments which changes them to their negative. We conclude that the expressions in part e) and part f) are already zero. We prove part g):

$$\begin{aligned} &([a,b],[c,d],[e,v]) && [a,(b,[c,d],[e,v])] \\ &+(a,b,[[c,d],[e,v]]) = && -[b,(a,[c,d],[e,v])] \\ &-(b,a,[[c,d],[e,v]]) && \text{(by Lemma 1 d)).} \end{aligned}$$

Now $([a,b],[c,d],[e,v]) = 0$ by part f) and $[[c,d],[e,v]] = 2[c,[d,[e,v]]] = 0$ by Lemma 2 e), Lemma 2 j), and $[R,V] \subset V$. Thus the left hand side is zero. The right hand sides alternate on their non-v arguments. We have shown that $0 = 2[a,(b,[c,d],[e,v])]$. This completes the proof of Lemma 3.

3 THE IDEALS GENERATED BY V

Having postulated the existence of V, we define A, I and J by:
$A := \{a \text{ in } R | [a,(R,R,V)] = (a,R,V) = 0\}$.
$I := \{a \text{ in } A | a(R,R,V) = 0\}$.
$J :=$ the ideal of R generated by (R,R,V).

REMARK: It will turn out that for any V, the set A will contain:

$$[x,(y,y,x)],\ (x,x,[x,y]),\ [[x,y],[z,w]].$$

A will also contain elements which are constructed using elements of V and so depend on the set V. I is an ideal, and A takes on all the properties of a V-set modulo the ideal I. These properties are $[R,A]+(R,R,A) \subset A$ and $M(R,R,A)+S(R,R,A)+B(R,R,A) \subset I$.

It also happens that $IJ = JI = 0$ which gives the result that in a prime ring, either $I = 0$ or $J = 0$. The existence of a V for which $(R,R,V) \neq 0$ means that J is not zero. We will show that $R-$ is nilpotent modulo I. If I is zero, we get that R is nilpotent under the bracket product. The Pchelintcev example is nilpotent under the bracket product of index 3.

We first note that from Lemma 2 a) and Lemma 2 b), for any permutation of its three arguments, the associator (A,R,V) is zero. Also $I(R,R,V) = (R,R,V)I = 0$ since $I(R,R,V) = 0$ and $[I,(R,R,V)] \subset [A,(R,R,V)] = 0$ by definition.

THEOREM 1: I is an ideal of R. The proof will be in 3 parts.

a) $[IR+RI,(R,R,V)] = 0$.
b) $(IR+RI,R,V) = 0$.
c) $(IR+RI)(R,R,V) = 0$.

PROOF:

Part a): $[IR+RI,(R,R,V)] \subset [I,(R,R,V)]R + I[R,(R,R,V)]$
$+[R,(R,R,V)]I + R[I,(R,R,V)]$ (by Lemma 2 d))
$\subset I[R,(R,R,V)] + [R,(R,R,V)]I$
(since $I(R,R,V) = (R,R,V)I = 0$)
$= 0$ (since $[R,(R,R,V)] \subset (R,R,V)$ by Lemma 2 l))
$[R,V] \subset V$, and $I(R,R,V) = (R,R,V)I = 0$).

Part b):

$$(IR+RI,R,V) \subset (I,R,[R,V]) + (R,I,[R,V]) + I(R,R,V) + R(I,R,V) + (I,R,V)R + (R,R,V)I$$

(by Lemma 1 c))

$$\subset (I,R,V) + (R,I,V) + I(R,R,V) + R(I,R,V) + (I,R,V)R + (R,R,V)I$$

(since $[R,V] \subset V$)

$$\subset (I,R,V) + (R,I,V) + R(I,R,V) + (I,R,V)R$$

(since $I(R,R,V) = (R,R,V)I = 0$)

$$= 0$$

(since $I \subset A$ implies $(I,R,V) = (R,I,V) = 0$).

Part c): We have already shown in part a) and part b) of this proof that $IR+RI \subset A$. We will now show that $(IR+RI)(R,R,V) = 0$.

$$(IR+RI)(R,R,V) \subset (R,R,V)(IR) + (RI)(R,R,V)$$

(since $[RI+IR,(R,R,V)] = 0$ by part a))

$$\subset ((R,R,V)I)R + ((R,R,V),I,R) + R(I(R,R,V)) + (R,I,(R,R,V))$$

$$\subset (V,I,R) + (R,I,V)$$

(since $(R,R,V)I = I(R,R,V) = 0$ and $(R,R,V) \subset V$)

$$= 0$$

(since $I \subset A$ and $(V,A,R) = (R,A,V) = 0$).

This finishes the proof of Theorem 1.

THEOREM 2: A satisfies the following for any V.

(a) $M(R,R,A) \subset I$.
(b) $S(R,R,A) \subset I$.
(c) $B(R,R,A) \subset I$.
(d) $[R,A] \subset A$.
(e) $(R,R,A) \subset A$.

PROOF: In this proof we let "v" be a generic element of V, "a" be a generic element of A, and "b", "c", "d", "x" be elements of R. We let w be a generic element of (R,R,V). The parts of Theorem 2 are proved in the order (d), (e), (a), (b), (c).

Part d): We have to show that $[[R,A],(R,R,V)] = 0$ and $([R,A],R,V) = 0$.
First: $[[R,A],(R,R,V)] \subset [R,[A,(R,R,V)]]$ (by Lemma 2 e))
$= 0$ (by the definition of A).

Second: $([R,A],R,V) \subset (R,A,[R,V]) + (A,R,[R,V]) + [R,(A,R,V)] + [A,(R,R,V)]$
(by Lemma 1 d))
$\subset (R,A,V) + (A,R,V) + [R,(A,R,V)] + [A,(R,R,V)]$
(since $[R,V] \subset V$)
$= 0$ (since $(R,A,V) = (A,R,V) = [A,(R,R,V)] = 0$).

To prove part e), we have to show that $[(R,R,A),(R,R,V)] = 0$ and $((R,R,A),R,V) = 0$.

First: $[(R,R,A),(R,R,V)]$
$\subset ([R,(R,R,V)],R,A) + (R,[R,(R,R,V)],A) + (R,R,[A,(R,R,V)])$
(by Lemma 2 d))
$\subset (V,R,A) + (R,V,A) + (R,R,[A,(R,R,V)])$
(since $[R,(R,R,V)] \subset V$ because $(R,R,V) \subset V$ and $[R,V] \subset V$)
$= 0$ (since $(V,R,A) = (R,V,A) = 0$ and $[A,(R,R,V)] = 0$).

Second: We shall first show that $((A,R,R,),R,V) = 0$. From Lemma 1 g) we have:

$$
\begin{aligned}
2((a,b,c),d,v) &= 2((a,d,v),b,c) + 2(a,(b,d,v),c) + 2(a,b,(c,d,v)) \\
&\quad + (a,[b,c],[d,v]) + (a,b,[c,[d,v]]) - (a,c,[b,[d,v]]) \\
&\subset ((A,R,V),R,R) + (A,V,R) + (A,R,V) \\
&\quad + (A,R,V) + (A,R,V) + (A,R,V) \\
&\qquad \text{(since } (R,R,V) \subset V \text{ and } [R,V] \subset V) \\
&= 0 \ \text{(since } (A,R,V) = (A,V,R) = 0).
\end{aligned}
$$

We next notice that:

$$
\begin{aligned}
((x,x,a),R,V) &\subset ((x,x,R),a,V) \ \text{(by Lemma 2 m))} \\
&= 0 \ \text{(since } (R,A,V) = 0).
\end{aligned}
$$

From the right alternative law, we have that $S(a,b,c)$ is an alternating function of its three arguments. Therefore $-S(a,b,c) + S(b,a,c) = -2S(a,b,c)$. Now from part d) and Lemma 0 b), we have $S(a,b,c) = (a,b,c) + (b,c,a) + (c,a,b)$ is in A. Therefore

$$
\begin{aligned}
0 &= ((a,b,c) + (b,c,a) + (c,a,b),d,v) && \text{(since } (A,R,V) = 0) \\
&= (\ (b,c,a) - (c,b,a),d,v) && \text{(since } ((A,R,R),R,V) = 0) \\
&= 2((b,c,a),d,v) && \text{(since } ((x,x,a),R,V) = 0).
\end{aligned}
$$

This shows that $((R,R,A),R,V) = 0$ and finishes the proof that $(R,R,A) \subset A$.

Part a): $(x,x,a)(b,c,v) = (x,x,(b,c,v)a)$ (by Lemma 2 f))
$= (x,x,(ba,c,v) + (b,a,[c,v]) - b(a,c,v))$ (by (Lemma 1 c))
$\subset (x,x,V)$
(since $(R,R,V) \subset V, [R,V] \subset V$ and $(A,R,V) = 0$)
$= 0$ (since $M(R,R,V) = 0$).

Part b): We first notice that $A(R,R,V) \subset V$, since

$a(b,c,v) = (ab,c,v) + (a,b,[c,v]) - (a,c,v)b$ (by Lemma 1 c))

$\subset (R,R,V)$ (since $[R,V] \subset V$ and $(A,R,V) = 0$)

$\subset V$ (since $(R,R,V) \subset V$).

We now let w be a generic element of (R,R,V).

$(a,b,c)w = (aw,b,c) - a(w,b,c) + (a,w,[b,c])$ (by Lemma 1 c))

$= -2(b,c,aw) + 2a(b,c,w)$

(by Lemma 2 b), using the previous result that $aw \subset V$, and the fact that $(A,w,R) = 0$ since $w \in V$)

$= -2(b,c,w)a - 2(b,c,a)w$
$-(b,[c,a],w) - (b,c,[a,w]) + (b,a,[c,w])$
$+2a(b,c,w)$ (from Lemma 1 f))

$= -2(b,c,a)w + (R,A,V)$

(since $[A,(R,R,w)] = 0$, $[R,A] \subset A$, $([A,w] \subset [a,(R,R,V)] = 0$, and $[R,w] \subset V$)

$= -2(b,c,a)w$ (since $(R,A,V) = 0$).

We conclude that $\{(a,b,c) + 2(b,c,a)\}w = 0$. Since

$$S(a,x,y) = (a,x,y) + (x,y,a) + (y,a,x) = \{(a,x,y) + 2(x,y,a)\} - M(x,y,a)$$

we have $S(a,x,y)w = 0$ using part a).

Part c): Letting w be a generic element in (R,R,V)

$0 = [x,[x,aw]]$ (since aw is in V from the proof of part b))

$= [[aw,x],x]$ (since the commutator is anti-commutative)

$= [a[w,x] + [a,x]w, x]$ (by Lemma 0 a) since the associator terms are all zero)

$= a[[w,x],x] + [a,x][w,x] + [a,x][w,x] + [[a,x],x]w$
(by Lemma 0 a) since the associator terms are all zero)

$= 0 + 2[x,a][x,w] + [x,[x,a]]w$
(since $[[w,x],x] = [x,[x,w]] \subset [x,[x,V]] = 0$ and $[[a,x],x] = [x,[x,a]]$).

Also: $0 = [[xx,a],w]$ (since $[R,A] \subset A$ and $[A,w] = 0$)

$= [x[x,a] + [x,a]x, w]$

(by Lemma 0 a) since the associator terms are in A by part e) and commute with $w = (R,R,V)$)

$= x[[x,a],w] + [x,w][x,a] + [x,a][x,w] + [[x,a],w]x$ (by Lemma 2 d))

$= [x,w][x,a] + [x,a][x,w]$ (since $[[R,A],w] \subset [A,w] = 0$)

$= 2[x,a][x,w]$

(since $[[R,R],[R,V]] \subset [R,[R,[R,V]]] = 0$ by Lemma 2 e) and Lemma 2 j)).

Putting the above parts together we get $[x,[x,a]](R,R,V) = 0$ which is part c).

THEOREM 3: These are elements of A.

a) $V \subset A$.
b) H^n for $n \geq 5$.
c) $[[R,R],[R,R]]$.
d) $[x,[x,[R,R]]], [x,[R,[x,R]]], [R,[x,[x,R]]]$ for all x in R.
e) $[x,(y,y,x)]$ for all x,y in R.
f) $(x,x,[x,y])$ for all x,y in R.

PROOF: Part a) Let w and v be elements of V. We have to show that $(w,x,v) = 0$ and $[w,(b,c,v)] = 0$. By Lemma 2 a) and Lemma 2 b) any associator with 2 arguments from V is zero. Thus $(w, x, v) = 0$. Next, from Lemma 1 d)

$$([w,b],c,v) + (w,b,[c,v]) - (b,w,[c,v]) = [w,(b,c,v)] - [b,(w,c,v)]$$

since $[V, R] \subset V$, every associator but one contains two elements of V. Thus $[w,(b,c,v)] = 0$.

Part b) Use Lemma 3 d) and Lemma 2 k).
Part c) Use Lemma 3 e) and Lemma 2 k).
Part d) Use Lemma 3 h) and Lemma 2 k).
The proof of part e) requires five steps.

e1) $[a,(b,b,a)] = -S(ab,a,b)$.
e2) $[S(ab,a,b),v] = 0$.
e3) $(a,[b,ab],[c,v]) + (b,[ab,a],[c,v]) + (ab,[a,b],[c,v]) = 0$.
e4) $(a,b,[ab,[c,v]]) + (b,ab,[a,[c,v]]) + (ab,a,[b,[c,v]]) = 0$.
e5) $(S(ab,a,b),c,v) = 0$.

Part e1):

$$\begin{aligned} S(ab,a,b) &= (ab,a,b) + (a,b,ab) + (b,ab,a) \\ &= a(b,a,b) + (a,a,b)b - (a,b,[a,b]) + (a,b,ab) + (b,b,a)a \\ &\qquad \text{(by Lemma 1 c) and Lemma 1 b))} \\ &= a(b,a,b) + (a,ba,b) - (a,b,ab) + (a,b,ba) + (a,b,ab) + (b,b,a)a \\ &= [a,(b,a,b)]. \end{aligned}$$

Part e2) is a consequence of Lemma 2 i).
For the proof of e3) we expand each term separately.

$(a,[b,ab],[c,v]) = (a,[b,a]b,[c,v])$
(by Lemma 0 a) where all the associator terms cancel by Lemma 2 m))

$$\begin{aligned} &= -([b,a]b,a,[c,v]) \quad (\text{since } M(R,R,V)=0) \\ &= -[b,a](b,a,[c,v]) - ([b,a],a,[c,v])b + ([b,a],b,[a,[c,v]]) \quad \text{(by Lemma 1 c))} \\ &= -[b,a](b,a,[c,v]) + (b,a,[a,[c,v]])b \quad \text{(by Lemma 2 h) and Lemma 2 j)).} \end{aligned}$$

$(b,[ab,a],[c,v]) = (b,a[b,a],[c,v])$
(by Lemma 0 a) where all the associator terms cancel by Lemma 2 m))

$$\begin{aligned} &= -(a[b,a],b,[c,v]) \quad (\text{by } M(R,R,V)=0) \\ &= -a([b,a],b,[c,v]) - (a,b,[c,v])[b,a] + (a,[b,a],[b,[c,v]]) \end{aligned}$$

(by Lemma 1 c))

$$= -a(a,b,[b,[c,v]]) - (a,b,[c,v])[b,a] \quad \text{(by Lemma 2 h) and Lemma 2 j)).}$$

$(ab,[a,b],[c,v]) = a(b,[a,b],[c,v]) + (a,[a,b],[c,v])b - (a,b,[[a,b],[c,v]])$
(by Lemma 1 c))

$$= -a(b,a,[b,[c,v]]) + (a,b,[a,[c,v]])b$$

(by Lemma 2 h), Lemma 2 e), and Lemma 2 j)).
Now adding these three terms gives: $-[\,[a,b],(a,b,[c,v])]$

$$\begin{aligned} &= -2[a,[b,(a,b,[c,v])]] \quad \text{(by Lemma 2 e))} \\ &= -2(a,b,[a,[b,[c,v]]]) \quad \text{(by Lemma 2 h))} \\ &= 0 \quad \text{(by Lemma 2 j)).} \end{aligned}$$

For the proof of e4), we expand each term separately.

$$\begin{aligned} 2(a,b,[ab,[c,v]]) &= 2(a,b,a[b,[c,v]]) + (a,b,[a,[c,v]]b) \quad \text{(by Lemma 2 d))} \\ &= 2(a,b,a)[b,[c,v]] + 2(a,b,[b,[c,v]])a \\ &\quad +(a,[b,a],[b,[c,v]]) + (a,b,[a,[b,[c,v]]]) - (a,a,[b,[b,[c,v]]]) \\ &\quad +2(a,b,b[a,[c,v]]) \quad \text{(by Lemma 1 f) and Lemma 2 j))} \\ &= -2(a,a,b)[b,[c,v]] + 2(a,b,[b,[c,v]])a + 2(a,b,[a,[c,v]])b \end{aligned}$$

(by Lemma 2 h), Lemma 2 j) and Lemma 1 b))

$= -2(a,a,[b,[c,v]]b) + 2(a,b,[b,[c,v]])a + 2(a,b,[a,[c,v]])b$ (by Lemma 2 f))

$= -(a,a,[bb,[c,v]]) + 2(a,b,[b,[c,v]])a + 2(a,b,[a,[c,v]])b$

(by Lemma 2 d) and $[b,[b,V]] = 0$)

$= 2(a,b,[b,[c,v]])a + 2(a,b,[a,[c,v]])b$ (since $(a,a,V) = 0$).

$2(b,ab,[a,[c,v]]) = 2(b,a,[a,[c,v]])b$ (by Lemma 1 c), Lemma 2 a), Lemma 2 j)).

$2(ab,a,[b,[c,v]]) = 2a(b,a,[b,[c,v]])$ (by Lemma 1 c), Lemma 2 j)).

Now the sum of these three is

$-2[a,(a,b,[b,[c,v]])] = 0$ (by Lemma 2 h) and Lemma 2 j)).

We now prove part e5).

$$\begin{aligned}2(S(ab,a,b),c,v) &= 2S((ab,c,v),a,b) + 2S((ab,(a,c,v),b) + 2S(ab,a,(b,c,v))\\ &\quad +(ab,[a,b],[c,v]) + (ab,a,[b,[c,v]]) - (ab,b,[a,[c,v]])\\ &\quad +(a,[b,ab],[c,v]) + (a,b,[ab,[c,v]]) - (a,ab,[b,[c,v]])\\ &\quad +(b,[ab,a],[c,v]) + (b,ab,[a,[c,v]]) - (b,a,[ab,[c,v]]) \quad \text{(by Lemma 1 g))}\\ &= \quad +(a,[b,ab],[c,v]) + 2(a,b,[ab,[c,v]])\\ &\quad +(b,[ab,a],[c,v]) + 2(b,ab,[a,[c,v]])\\ &\quad +(ab,[a,b],[c,v]) + 2(ab,a,[b,[c,v]])\end{aligned}$$

because $(R,R,V) \subset V$ and $S(R,R,V) = 0$, and because $[R,[R,V]] \subset V$ and $(x,x,v) = 0$. Now the first column sums to zero by e3) and the second column sums to zero by e4).

The proof of part f) takes these six steps.

f1) $[(x,x,[x,y]),v] = 0$.

f2) $(x,y,(x,y,v)) = 0$.

f3) $(x,y,(x,[x,z],v)) = +(x,z,(x,[x,y],v))$.

f4) $(x,y,(x,[x,z],v)) = -(x,z,(x,[x,y],v))$.

f5) $(x,y,(x,[x,z],v)) = 0$.

f6) $((x,x,[x,y]),z,v) = 0$.

Part f1): $[(x,x,[x,y]),v] = 3([x,y],x,[x,v])$ (by Lemma 2 g))

$= 3(y,x,[x,[x,v]])$ (by Lemma 2 h))

$= 0$.

Part f2):

$0 = 2((x,x,y),y,v)$ (from Lemma 2 m))

$= 2((x,y,v),x,y) + 2(x,(x,y,v),y) + 2(x,x,(y,y,v))$

$+(x,[x,y],[y,v]) + (x,x,[y,[y,v]]) - (x,y,[x,[y,v]])$ (from Lemma 1 g))

$= -4(x,y,(x,y,v)) - 2(x,y,(x,y,v))$ (by Lemma 2 b) and Lemma 2 h))

$= -6(x,y,(x,y,v))$.

Part f3):

$(x,y,(x,[x,z],v)) = -(x,y,[x,(x,z,v)])$ (by Lemma 2 h))

$= -(x,[x,y],(x,z,v))$ (by Lemma 2 h))

$= +(x,z,(x,[x,y],v))$ (by f2)).

Part f4):

$(x,y,(x,[x,z],v)) = -(x,y,[x,(x,z,v)])$ (by Lemma 2 h))

$= +[x,(x,y,(x,z,v))]$ (by Lemma 2 h))

$= -[x,(x,z,(x,y,v))]$ (by f2))

$= +(x,z,[x,(x,y,v)])$ (by Lemma 2 h))

$= -(x,z,(x,[x,y],v))$ (by Lemma 2 h)).

Part f5): is a consequence of f3) and f4).

Part f6):

$$2((x,x,[x,y]),z,v) =$$
$$2((x,z,v),x,[x,y]) + 2(x,(x,z,v),[x,y]) + 2(x,x,([x,y],z,v))$$
$$+(x,[x,[x,y]],[z,v]) + (x,x,[[x,y],[z,v]]) - (x,[x,y],[x,[z,v]])$$

(from Lemma 1 g))

$$= -4(x,[x,y],(x,z,v)) - 2(x,[x,y],(x,z,v)) \quad \text{(from Lemma 2 b), h))}$$
$$= -6(x,[x,y],(x,z,v)).$$
$$= -6(x,y,[x,(x,z,v)]) \quad \text{(by Lemma 2h))}$$
$$= +6(x,y,(x,[x,z],v)) \quad \text{(by Lemma 2 h))}$$
$$= 0 \text{ by f5)}.$$

THEOREM 4: These are some elements of I.

a) $[A,A] \subset I$.
b) $[R,V][R,V] \subset I$.
c) $(x,x,[x,[x,y]]) \in I$ for all x,y in R.

PROOF: Part a). $[A,A] \subset A$ by Theorem 2 d). We have to show $[A,A](R,R,V) = 0$.

$$[A,A](R,R,V) \subset ([A,A]R,R,V) \quad \text{(by Lemma 1 c))}$$
$$\subset (A[R,A],R,V)$$

(by Lemma 0 a), Theorem 2 d), and Theorem 2 e))

$$= 0 \quad \text{(by Lemma 1 c))}.$$

The proof of part b) takes six steps.
We set v and w be elements of V, a, b, c, x elements of R.

b1) $w(a,b,v) = (wa,b,v) = (aw,b,v) = (a,wb,v) = (a,bw,v) = (a,b,v)w$.
b2) $[a,v](b,c,w)]$ alternates on arguments a,b,c.
b3) $[va,(b,c,w)] = v[a,(b,c,w)]$.
b4) $[a,v](b,c,w) = [a,w](b,c,v)$.
b5) $[a,v][b,w]$ alternates on v and w.
b6) $[a,v][b,w](R,R,V) = 0$.

Part b1) follows from Lemma 1 c), Lemma 2 a) and Lemma 2 b) using $[R,V] \subset V$ and $(V,V,R) = 0$.
For the proof of b2) we only need show $[x,v](x,c,w) = 0$.

$$0 = ([xx,v],c,w) = (x[x,v] + [x,v]x,c,w) \quad \text{(by Lemma 2 d))}$$
$$= 2[x,v](x,c,w) \text{ by b1)}.$$

Part b3) follows from Lemma 0 a) and $[V,(R,R,V)] = (R,V,V) = 0$.

Part b4) starts by using Lemma 2 1) applied to $2[va,(b,c,w)]$. Then using b1) on each of the terms, the v is brought out to the left.

$$2[va,(b,c,w)] = ([va,b],c,w) - ([b,c],va,w) + ([c,va],b,w)$$
$$+2(va,b,[c,w]) - 2(b,c,[va,w]) + 2(c,va,[b,w]) \text{ (by Lemma 2 1))}.$$

Term by term, we can apply Lemma 0 a), Lemma 1 c) and b1) to bring the v to the left leaving at most one correction factor.

$$([va,b],c,w) = \quad v([a,b],c,w) + [v,b](a,c,w).$$
$$-([b,c],va,w) = \quad -v([b,c],a,w).$$

$$\begin{aligned}
([c,va],b,w) &= v([c,a],b,w)+[c,v](a,b,w).\\
2(va,b,[c,w]) &= 2v(a,b,[c,w]).\\
-2(b,c,[va,w]) &= -2v(b,c,[a,w])-2(b,c,v)[a,w].\\
2(c,va,[b,w]) &= 2v(c,a,[b,w]).
\end{aligned}$$

So we get using Lemma 2 1), b1) and b2) that

$$2[va,(b,c,w)] = 2v[a,(b,c,w)] + 2[a,v](b,c,w) - 2[a,w](b,c,v).$$

Now from b3) we get $[a,v](b,c,w) = [a,w](b,c,v)$ or that $[a,v](b,c,w)$ is symmetric on v and w.

We prove b5) by showing that $[a,v][b,w]$ alternates on a and b. From Lemma 2 d) we have $[[a,v],w] = -[[v,a],w] = -2[v,[a,w]] = 2[[a,w],v] = -2[[w,a],v] = -4[w,[a,v]] = 4[[a,v],w]$. Therefore $3[[a,v],w] = 0$. Now using $[[R,V],V] = 0$ several times we calculate:

$$\begin{aligned}
0 &= [[xx,v],w].\\
&= [x[x,v]+[x,v]x,w] \quad \text{(by Lemma 2 d))}\\
&= [x,w][x,v]+[x,v][x,w] \quad \text{(by Lemma 2 d))}\\
&= 2[x,w][x,v] \quad \text{(by Lemma 2 e) and Lemma 2 j))}
\end{aligned}$$

Now since $[[a,v],[b,w]] = [a,[v,[b,w]]] = 0$ by Lemma 2 j) we have $[a,v][b,w] = [b,w][a,v] = -[a,w][b,v]$ since it alternates on a and b.

Now the proof of b6) looks at $[R,V][R,V](R,R,V)$. This product alternates on the first two V positions and is symmetric on the last two V positions. This is impossible. We conclude that the product is zero.

For part c) we first prove that $(a,a,b)(b,c,v) = 0$.

$$\begin{aligned}
0 &= (a,a,(bb,c,v)).\\
&= (a,a,b(b,c,v)+(b,c,v)b-(b,b,[c,v])) \qquad \text{(by Lemma 1 c))}\\
&= 2(a,a,b)(b,c,v) \qquad \text{(by Lemma 2 f)).}
\end{aligned}$$

$$\begin{aligned}
\text{Now } (x,x,[x,[x,b]])(c,d,v) &= -(x,x,c)([x,[x,b]],d,v).\\
&= -\{(x,x,c)+(c,x,x)\}([x,[x,b]],d,v).\\
&= +\{(x,c,x)+(c,x,x)\}([x,[x,b]],d,v).\\
&= -\{(x,c,d)+(c,x,d)\}([x,[x,b]],x,v).\\
&= 0 \text{ since } ([x,[x,b]],x,v) = (b,x,[x,[x,v]]) = 0\\
&\qquad \text{(by Lemma 2 h)).}
\end{aligned}$$

There are many other elements of I which are consequences of the fact that A/I satisfies all the V-hypotheses of R/I. If an expression involving V is zero, that same expression will be in I if the V's are replaced by A's.

THEOREM 5: The following elements are in I.

a) $[R,[R,[R,A]]] \subset I$.
b) $[H^n,A] \subset I$, if $n \geq 3$.
c) $[R,A][R,A](R,R,A) \subset I$.
d) $(x,x,H^5) \subset I$.
e) $[H^i,H^j] \subset I$, if $i \geq 3$ and $j \geq 5$.
f) $[R,[R,[R,H^i]]] \subset I$, if $i \geq 5$.
g) $[[H^2,H^2],[H^2,H^2]] \subset I$.

PROOF: Part a) is Lemma 2 j). Part b) is Lemma 2 k). Part c) is Theorem 4 b). Part d) follows from Theorem 3 b). Part e) is from part b) and Theorem 3 b). Part f) is from part a) and Theorem 3 b). Part g) is Theorem 3 c) and Theorem 4 a).

THEOREM 6: If R is a prime right alternative ring with a V-set V, then either $(R,R,V)=0$ or $I=0$.

PROOF: Let $J=(V,R,R)+(V,R,R)R$. J is an ideal. Since $IJ=JI=0$ we conclude using the fact that R is prime, that $I=0$ or $J=0$. Since $(R,R,V)\subset J$, we get $I=0$ or $(R,R,V)=0$.

THEOREM 7: If R is a prime right alternative ring with a V-set V such that $(R,R,V)\neq 0$. Then $R-=(R,+,[\ \ ])$ is nilpotent of index at most 11.

PROOF: By Theorem 6, we must have that $I=0$. From Theorem 5 e) and Theorem 5 f) we have:

$$[H^i,H^j]=0 \text{ if } i\geq 3 \text{ and } j\geq 5.$$

$$[R,[R,[R,H^i]]]=0 \text{ if } i\geq 5.$$

$$
\begin{aligned}
H^9 &\subset [H^1,H^8]+[H^2,H^7]+[H^3,H^6]+[H^4,H^5]\\
&\subset [R,A] && \text{(by Theorem 3 b)).}\\
H^{10} &\subset [H^1,H^9]+[H^2,H^8]+[H^3,H^7]+[H^4,H^6]+[H^5,H^5]\\
&\subset [H^1,H^9]+[H^2,A] && \text{(by Theorem 3 b))}\\
&\subset [R,[R,A]]+[H^2,A] && \text{(by the previous statement)}\\
&\subset [R,[R,A]] && \text{(by Lemma 2 e)).}\\
H^{11} &\subset [H^1,H^{10}]+[H^2,H^9]+[H^3,H^8]+[H^4,H^7]+[H^5,H^6]\\
&\subset [H^1,H^{10}]+[H^2,H^9] && \text{(by Theorem 5 e))}\\
&\subset [R,[R,[R,A]]]+[H^2,[R,A]] && \text{(by previous statements)}\\
&=0 && \text{(by Lemma 2 j) and Lemma 2 e)).}
\end{aligned}
$$

4 IDENTITIES IN A RIGHT ALTERNATIVE RING

We now give four identities which hold in any right alternative ring. In these identities, the elements x,a,b,c,d,v represent any elements in R. In particular, the element v is not special. We give them without proof because the known proofs are too long. They may be checked with a computer algebra system like Albert. Up till now, the proofs were rather short and were unsuitable for Albert because they dealt with non-generic elements from V which Albert is not able to handle. These next four identities can be proved by Albert. I also wish to display the actual identities so that in later proofs, one can see their interrelationships.

THEOREM 8: The following identities hold in any right alternative ring:

(a)
$$
\begin{aligned}
&-2(x,x,(v,a,b))+2((x,x,v),a,b)-2M(x,(x,a,b),v)\\
&+M(x,[a,b],v).x+x.M(x,[a,b],v)+[(x,x,v),[a,b]]\\
&-M(xx,[a,b],v)+M(x,[[a,b],x],v).
\end{aligned}
$$

(b)
$$
\begin{aligned}
&2S(a,b,(v,c,d))-2(S(a,b,v),c,d)-2S(a,b,v)[c,d]\\
&+2S((a,c,d),b,v)-2S((b,c,d),a,v)\\
&-S(v,b,a[c,d])-M(b,a[c,d],v)\\
&+S(v,a,b[c,d])+M(a,b[c,d],v)\\
&+S(v,b,[c,d])a+M(b,[c,d],v)a\\
&-S(v,a,[c,d])b-M(a,[c,d],v)b.
\end{aligned}
$$

$$\begin{aligned}
&- 2[a,(b,c,v)] + ([a,b],c,v) + 2(a,b,[c,v]) \\
\text{(c)}\qquad &+ (c,b,[a,v]) + (c,a,[b,v]) \\
&+ M(ab,c,v) + M(ba,c,v) - 2M(b,c,v)a - 2M(a,c,v)b.
\end{aligned}$$

$$\begin{aligned}
&[a,[a,(c,d,v)]] - [a,[a,(d,c,v)]] + ([a,[a,v]],c,d) \\
&- B(a,(a,c,d),v) + 2a.S(v,c,d)a - aa.S(v,c,d) \\
\text{(d)}\qquad &- S(v,c,d)a.a + S([v,a],a,[c,d]) + [S(v,a,[c,d]),a] \\
&+ (a,a,[[c,d],v]) + M(aa,[c,d],v) - M(a,[c,d],v)a \\
&- a.M(a,[c,d],v).
\end{aligned}$$

5 CREATING A V SET FROM THE FIRST FOUR CONDITIONS

If we have an additive subgroup V_0 such that $M(R,R,V_0) = 0$, $S(R,R,V_0) = 0$, $B(R,R,V_0) = 0$, and $[R,V_0] \subset V_0$. Then V_0 is not necessarily a V-set because we do not have V_0 closed under taking associators. However, if we define inductively

$$V_1 = V_0 + (R,R,V_0),$$
$$V_{n+1} = V_n + (R,R,V_n)$$

then the next theorem shows that $V = \bigcup_{i\geq 0} V_i$ will be a V set.

THEOREM 9: If V_0 satisfies $M(R,R,V_0) = 0$, $S(R,R,V_0) = 0$, $B(R,R,V_0) = 0$ and $[R,V_0] \subset V_0$, then $V = \bigcup_{i\geq 0} V_i$ where $V_{n+1} = V_n + (R,R,V_n)$ is a V-set.

PROOF: We will show by induction that V satisfied

a) $M(R,R,V) = 0$.
b) $S(R,R,V) = 0$.
c) $[R,V] \subset V$.
d) $B(R,R,V) = 0$.

While part e), $(R,R,V) \subset V$ is a consequence of the union. It is given that V_0 satisfies a), b), c), d). We will use $\langle S\rangle$ to represent the ideal generated by the set S. The particular use will be to focus the readers attention on the reason why all the terms are zero. If all the terms are in the ideal generated by S, and S is zero, then all the terms are zero.

Assume that $S(a,b,V_n) = 0$ and $M(a,b,V_n) = 0$. Then one gets that $(R,R,V_n) \subset (V_n,R,R)$. This shows that

$$V_{n+1} \subset V_n + (V_n,R,R).$$

Theorem 8 a) says that $(x,x,V_{n+1}) \subset \langle M(R,R,V_n)\rangle = 0$ by induction.

Theorem 8 b) says that $S(R,R,V_{n+1}) \subset \langle S(R,R,V_n)\rangle + \langle M(R,R,V_n)\rangle = 0$ by induction.

So far we have proved that $S(R,R,V)=0$ and $M(R,R,V)=0$. This proof is also given in [2, Theorem 1 page 3148].

Theorem 8 c) says that $[R,V_{n+1}]\subset V_n+\langle M(R,R,V_n)\rangle\subset V_n$ by induction. This proves that $[R,V]\subset V$.

Theorem 8 d) says if v in V_n, then $[a,[a,(c,d,v)-(d,c,v)]]=0$.
But since $(c,d,v)+(d,v,c)+(v,c,d)=0$ we conclude that $(v,c,d)=-(c,d,v)+(d,c,v)$. Therefore $[a,[a,(v,c,d)]]=0$ and so $[a,[a,v_{n+1}]]=0$. This proves that $B(R,R,V)=0$.

6 THE COMMUTING NUCLEUS

We now apply the above analysis to the commuting nucleus

$$U=\{u\in R|[u,R]=0\}.$$

THEOREM 10: Let $U=\{u \text{ in } R|[u,R]=0\}$, then U is contained in a V set.

PROOF: We have to show four results before we quote Theorem 9.

a) $M(R,R,U)=0$.
b) $S(R,R,U)=0$.
c) $[R,U]\subset U$.
d) $B(R,R,U)=0$.

If u is in U,

$$0=[xx,u]=x[x,u]+[x,u]x+(x,x,u)-(x,u,x)+(u,x,x) \quad \text{(from Lemma 0 a))}$$
$$=2(x,x,u).$$

So $M(R,R,U)=0$.

$$\begin{aligned}0&=[ab,u]\\&=a[b,u]+[a,u]b+(a,b,u)-(a,u,b)+(u,a,b) && \text{(from Lemma 0 a))}\\&=(a,b,u)+(b,u,a)+(u,a,b) && \text{(using part a))}\\&=S(a,b,u).\end{aligned}$$

So $S(R,R,U)=0$.

It is immediate that $[R,U]=0\subset U$ and $B(R,R,U)=0$. From Theorem 9, there is a V set containing U.

REMARK: Notice that the definition of $I=\{a \text{ in } A|a(R,R,V)=0\}$. It is possible for R to have an identity element which could be in V. But I is not the annihilator of V, but only the annihilator of (R,R,V). Thus the existence of an identity element in V does not imply that I is zero. Notice also that $J=(R,R,V)+(R,R,V)R$ and only $(R,R,V)\subset J$, not necessarily V itself. Thus V can have an identity element and J need not be all of R.

THEOREM 11: If R is a prime right alternative ring and the commuting nucleus of R is not contained in the center of R, then $R- := (R,+,[\ ,\ ])$ is nilpotent of index at most 11.

PROOF: Apply Theorem 7 and Theorem 10.

THEOREM 12: If R is a prime right alternative ring and the commuting nucleus of R is not contained in the center of R, then $R- := (R,+,[\ ,\ ])$ is solvable.

PROOF: $I = 0$ and $[[H^2, H^2], [H^2, H^2]] \subset I$ by Theorem 10 and Theorem 5 g).

ACKNOWLEDGEMENTS

The author is indebted to Erwin Kleinfeld and Harry F. Smith for ideas and stimulation in this paper. The approach to the problem was through intensive computation using Iowa State University's VINCENT Computer Network.

REFERENCES

1. D.P. Jacobs, S.V. Muddana and A.J. Offutt. A computer algebra system for nonassociative identities. Hadronic Mechanics and Nonpotential Interactions. Proceedings of the Fifth International Conference, Cedar Falls, Myung, H.C. (Ed.), Nova Science Publishers, Inc., New York, 1993, pp. 185-195.
2. I.R. Hentzel, E. Kleinfeld and H.F. Smith. Commutative center = center in a prime finitely generated right alternative ring. Comm. Algebra 25(10):3147-3153, 1997.
3. E. Kleinfeld. Right alternative rings. Proc. Am. Math. Soc. 4:939-944, 1953.
4. E. Kleinfeld and H.F. Smith. On centers and nuclei in prime right alternative rings. Comm. Algebra 22(3):829-855, 1994.
5. S.V. Pchelintsev. Prime algebras and absolute divisors of zero (Russian). Izv. Akad. Nauk SSSR Ser. Mat. 50:79-100, 1986.
6. R.E. Roomeldi. On simple right alternative rings (Russian). Tartu Riikl, U1. Toimetised 700:60-70, 1985.
7. A. Thedy. Right alternative rings. J. Algebra 37:1-43, 1975.

A construction of Jordan superalgebras from Jordan–Lie triple systems

NORIAKI KAMIYA Department of Mathematics, Shimane University,Japan

SUSUMU OKUBO Department of Physics, University of Rochester, New York

0. INTRODUCTION

The concept of the standard embedding plays an important role in the theory of a construction of Lie algebras from triple systems, in particular, Frudenthal-Kantor triple systems and Lie triple systems are intimately related to Lie algebras as well as Lie superalgebras (for example, [3], [5], [7], etc.).

The notation of a Jordan-Lie triple system is introduced by authors ([6]) and is investigated for the quasi-classicality.

It is interesting for us to study how the standard embedding behaves for Jordan-Lie triple systems with familiar connection to Jordan superalgebras.

Our aim is to give an unified explicit frame work for the construction of simple Jordan superalgebras and is to characterize it by concept of our triple systems.

In this paper, we are concerned with algebra and triple systems which are finite dimensional over a field Φ of characteristic$\neq$ 2.

1.

Let V be a superspace. Suppose that a triple product $[\ .,\ .,\ .] : V \otimes V \otimes V \to V$ satisfies conditions

$$\sigma([x, y, z]) = \{\sigma(x) + \sigma(y) + \sigma(z)\}\text{mod}(2) \tag{1.0}$$

$$[xyz] = -\delta(-1)^{xy}[yxz] \tag{1.1}$$

$$(-1)^{xz}[xyz] + (-1)^{yx}[yzx] + (-1)^{zx}[zxy] = 0 \tag{1.2}$$

$$[uv[xyz]] = [[uvx]yz] + (-1)^{(u+v)x}[x[uvy]z] + \delta(-1)^{(u+v)(x+y)}[xy[uvz]] \tag{1.3}$$

for $\delta = \pm 1$.

REMARK. For $\delta = 1$, this equations mean the Lie triple system. For $\delta = -1$, the triple system is intimately related to Jordan-superalgebra as studied in ([6]), so that we called it a Jordan–Lie supertriple system (abbreviated as JLTS hereafter).

It is often convenint to stydy both the case of $\delta = 1$ and $\delta = -1$, and we call the triple system satisfying above conditions (1.0) $\sim$ (1.3) a δ Jordan-Lie supertriple system. If a triple product satisfies (1.1) and (1.3), but not necessarily (1.2), call the system a generalized δ-Jordan Lie triple system.

DEFINITION. A generalized δJLTS is called quasi-classical, if there exists a bilinear non-degenerate form $< .| . >$ satisfying conditions:

$$< x|y >= 0 \ unless \ \sigma(x) = \sigma(y) \tag{i}$$

$$< y|x >= (-1)^{xy} < x|y > \tag{ii}$$

$$< y|[xuv] >= (-1)^{xy+uv} < x|[yvu] > . \tag{iii}$$

DEFINITION. Let V be a generalized δJLTS. We introduce the left multiplication operations: $V \otimes V \to \text{End } V$ by

$$L(x, y)z = [xyz]$$

with product given by

$$L(x, y) \bullet L(u, v) := L(x, y)L(u, v) - \delta(-1)^{(u+v)(x+y)} L(u, v)L(x, y) \tag{1.4}$$

DEFINITION. Let $L(x, y)$ be as in the above. We introduce a vector space M which a linear span of all $L(x, y)'s$, i,e,

$$M = \{X|X = \sum_{j,k} C_{j,k}L(x_j, x_k), x_j, x_k \in V\} \tag{1.5}$$

for constants $C_{j,k}$ and write $M = L(V, V)$.

PROPOSITION 1([6]). Let V be a quasi-clasical generalized $\delta JLTS$. Then the vector space M defined by (1.5) is a quasi- clasical Jordan superalgebra for $\delta = -1$.

When V is a JLTS,we can construct a large Jordan-superalgebra as follows. That is, we consider a direct sum

$$L_0 = V \oplus M = V \oplus L(V, V).$$

We introduce a bilinear product $X \bullet Y$ $(X, Y \in L_0)$ in L_0 by

$$\begin{aligned} x \bullet y &:= L(x, y) \in M \\ L(x, y) \bullet z &:= -\delta(-1)^{(x+y)z} z \bullet L(x, y) \\ &= [xyz] \end{aligned}$$

for $x, y, z \in V$ in addition to $L(x,y) \bullet L(u,v)$ given by (1.4).

THEOREM 2. Let L_0 be given as in the above, then L_0 is a Jordan superalgebra for $\delta = -1$. Moreover, L_0 is quasi-classical, if V is quasi-classical.

REMARK ([3][4][5]) For $\delta = 1, L_0$ is the standard embedding Lie superalgebra associated with Lie supertriple system V.

EXAMPLE 1.1. Let $x * y$ be an associative and commutative superalgebra, i.e.,

$$\begin{aligned} (x * y) * z &= x * (y * z), \\ y * x &= (-1)^{xy}(x * y). \end{aligned}$$

Then $xyz := (x * y) * z$ defines a generalized JLTS for $\delta = -1$.

EXAMPLE 1.2. Let V be a Z_2-graded vector space equipped with $< .| . >$ and let $< .| . >$ be a nondegenerate bilinear form satisfying

$$\begin{aligned} < x|y > &= 0 \quad \text{unless } \sigma(x) = \sigma(y) && (1) \\ < x|y > &= (-1)^{xy} < y|x > . && (2) \end{aligned}$$

Then

$$xyz := < y|z > x - \delta(-1)^{xy} < x|z > y$$

defines a quasi-clasical Lie supertriple system for $\delta = 1$, and this triple product defines a quasi-classical generalized Jordan-Lie supertriple system for $\delta = -1$.

EXAMPLE 1.3. Let $W = \sum_i \oplus W_i$ be a supervector space equipped with a super-symmetric bilinear form

$$< x_i, y_j >= (-1)^{ij} < y_j, x_i > \delta_{i,j}.$$

Then

$$\begin{aligned} < x_i y_j z_k >: &= < x_i, y_j > z_k + < y_j, z_k > x_i \\ &- (-1)^{(i+j)k} < z_k, x_i > y_j \end{aligned}$$

defines a $(-1,1)$-Jordan supertriple system on W, (for the details, to see [4]), that is,

$$< x_i y_j z_k >= (-1)^{(i+j)k+ij} < z_k y_j x_i > \qquad (1.6a)$$

$$< x_i y_j < z_k v_l w_m >>=<< x_i y_j z_k > v_l w_m > \qquad (1.6b)$$

$$-(-1)^{(i+j)k+ij} < z_k < y_j x_i v_l > w_m > +(-1)^{(k+l)(i+j)} < z_k v_l < x_i y_j w_m >> .$$

2

In this section, we shall give several examples of simple Jordan superalgebras.

BC. Let a superspace $W = W_{\bar{0}} \oplus W_{\bar{1}}$ is equipped with a nondegenerate bilinear form $< , >$ whose restriction to $V_{\bar{0}}$ is symmetric, whose restriction to $V_{\bar{1}}$ is skew symmetric, and for which $< W_{\bar{0}}, W_{\bar{1}} >= 0$. We set

$$J_\alpha = \{a \in \mathrm{End}_\alpha(W) | < ax, y >= (-1)^{\alpha x} < x, ay >\}.$$

Then $J = J_{\bar{0}} \oplus J_{\bar{1}}$ is a subalgebra of the full linear Jordan superalgebra.

This example can be constructed from generalized Jordan–Lie triple systems for $\delta = -1$ as follows.

¿From Example 1.2,it follows that

$$< xyz >=< y|z > x + (-1)^{xy} < x|z > y$$

is a quasi-classical generalized Jordan-Lie supertriple system for $\delta = -1$ satisfying

$$< L(x,y)u|v >= (-1)^{(x+y)u} < u|L(x,y)v > .$$

Thus from Proposition 1, we obtain,

$$M = \{X|X = \sum C_{jk}L(x_j, x_k), x_j, x_k \in W\}$$

is a Jordan superalgebra.

D. A superspace $V = V_{\bar{0}} \oplus V_{\bar{1}}$ equipped with a bilinear form

$$< x_i, y_j >= (-1)^{ij} < y_j, x_i > \delta_{i,j}, x_i, x_j \in V_{\bar{0}} \text{or } V_{\bar{1}}$$

becomes a Jordan superalgebra if we let e denote some vector of unit length in $V_{\bar{0}}$ and set:

$$a \circ b =< e, a > b+ < e, b > a- < a, b > e.$$

This example can be constructed from Example 1.3 as follows.

In Example 1.3, if we put $y_j = e \in V_0$, then we have

$$< x_i e z_k >:=< x_i, e > z_k+ < e, z_k > x_i - (-1)^{ik} < z_k, x_i > e.$$

Thus this identity means that the identity

$$x_i \circ z_k =< e, x_i > z_k+ < e, z_k > x_i- < x_i.z_k > e.$$

E. The 27-dimensional Albert Jordan algebra of Hermite matrices of order three over the Cayley numbers.

For this construction, many mathematicians have studied, for example, A,A, Albert, N.Jacobson, M.Koecher and their students (for example, to see [3]). We note that this Jordan algebra has a structure of a Jordan triple system with respect to the triple product

$$[xyz] := x(yz) - y(xz) + (xy)z.$$

K. The 3-dimensional simple Jordan superalgebra $< a > \oplus < \xi, \eta >$ with multiplication table

$$a^2 = a, a \circ \xi = \frac{1}{2}\xi, a \circ \eta = \frac{1}{2}\eta, \xi \circ \eta = a.$$

This example can be contructed by the concept of standard embedding from Theorem 2. as follows.

We define the triple product in $V_{\bar{1}} =< \xi, \eta >$ as follows;

$$\begin{aligned} L(\xi,\eta): &= \xi \circ \eta = a \\ L(\xi,\eta): &= -L(\eta,\xi), \\ L(\xi,\xi) &= L(\eta,\eta) := 0 \\ L(\xi,\eta)\xi &:= \frac{1}{2}\xi, L(\xi,\eta)\eta = \frac{1}{2}\eta. \end{aligned}$$

Then the triple system $V_{\bar{1}} =< \xi, \eta >$ is a Jordan-Lie triple system for $\delta = -1$ with degree one.

In fact, we have

$$L(\xi,\eta)\xi + L(\eta,\xi)\xi + L(\xi,\xi)\eta = 0,$$

$$L(\xi,\eta) \bullet L(\xi,\eta)\xi = L(L(\xi,\eta)\xi,\eta)\xi + L(\xi, L(\xi,\eta)\eta)\xi.$$

Thus $L_0 := L(V_{\bar{1}}, V_{\bar{1}}) \oplus V_{\bar{1}} = V_{\bar{0}} \oplus V_{\bar{1}}$ is a Jordan superalgebra.

For the detail argument about the construction of other examples of simple Jordan superalgebras by means of our triple systems, we will discuss in force coming paper.

In the end of this section, we describe another characterization of the definition of Jordan superalgebras due to V.G.Kac.

THEOREM 3. Let V be a superalgebra equipped with $xy = (-1)^{xy}yx$. Then the following relations are equivalent;

$$L(a,b,c) := \tag{1}$$
$$(-1)^{ac}[L_{ab}, L_c] + (-1)^{ba}[L_{bc}, L_a] + (-1)^{cb}[L_{ca}, L_b] = 0,$$
$$F(x,y,z,w) := \tag{2}$$
$$(-1)^{(x+w)z}(xy,w,z) + (-1)^{(y+w)x}(yz,w,x) + (-1)^{(z+w)y}(zx,w,y) = 0,$$

for $a, b, c, x, y, z, w \in V$, where $L_a(b) := ab$ and $(x,y,z) := (xy)z - x(yz)$.

REFERENCES

1. L.Hogben and V.G.Kac, The correct multiplication table for the exceptional Jordan superalgebra F,Comm.Alg.11(10)(1983),1155–1156,

2. V.G.Kac,Classification of simple Z-graded Lie superalgebra and simple Jordan superalgebras, Comm.Alg.5(1977),1375–1400,

3. N.Kamiya,A structure theory of Freudentha–Kantor triple systems III, Mem. Fac.Sci., Shimane Univ.23(1989)33–51,

4. N.Kamiya,On (ε, δ)– Freudenthal–Kantor triple systems,International Symposium on a Nonassociative Algebras and related topics, Hiroshima,Japan(1991) 65–75,World Scientific Pub.,

5. S.Okubo and N.Kamiya, Quasi–calssical Lie superalgebra and Lie supertriple systems,Univ.of Rochester Report UR–1462(1996),

6. S.Okubo and N.Kamiya, Jordan Lie superalgebra and Jordan–Lie triple systems, J. Alg. 198(1997) 388-411,

7. S.Okubo, Para–statistics as Lie super triple systems, J.Math.Phys.35(1994) 2785–2803,

8. R.D.Schafer, An introduction to Nonassociative algebras, Academic Press, 1996,

Current adress of first author;
Center for Mathematical Sciences
Aizu University
Aizuwakamatsu city, JAPAN

Rings with (x,y,z) = (x,z,y)

ERWIN KLEINFELD and MARGARET KLEINFELD Mathematics Department, University of Iowa , Iowa City, IA 52242

1 INTRODUCTION.

The earliest work on algebras satisfying the identity in the title belongs to A.A. Albert in the context of rings of (γ, δ) type, in particular rings of type (1,1). Albert wasn't much interested in algebras unless they were power associative and so added that hypothesis whenever it was needed. Power associativity is not implied by $(x, y, z) = (x, z, y)$ [3]. However the additional assumptions $(x, x, x) = 0$ and characteristic $\neq 2, 3$ are sufficient to give power associativity. With these assumptions Albert's proved that nil, finite dimensional algebras are nilpotent. Thus if an algebra of this kind is not nilpotent it must contain an idempotent e. Kokoris then showed if the algebra is simple and not associative then $e = 1$ [6]. The classification of the simple algebras and simple rings with idempotent $e \neq 0$ as associative followed soon after [2]. This was extended to all simple rings by Ng [7].

Novikov rings satisfy $(x, y, z) = (x, z, y)$ and additionally $x \cdot yz = y \cdot xz$, but not $(x, x, x) = 0$. This additional identity can be weakened to $(x, y, yz) = y(x, y, z)$ without much loss [4]. Zelmanov [11] and especially Osborn [8], [9], have done interesting work on these rings, but the simple ones need not be associative.

In a different direction Y. Paul [10] and Hentzel, Jacobs, and myself [1], have studied rings satisfying $(x, y, z) = (x, z, y)$ and $(w, [x, y], z) = 0$. The latest generalization in this direction involves rings satisfying $(x, y, z) = (x, z, y)$ and $(x, (y, y, y), z) = 0$ [5]. The simple rings are still associative if one adds the assumption characteristic $\neq 5$.

In the following we revisit rings satisfying $(x, y, z) = (x, z, y)$ and $(x, x, x) = 0$, and present a much streamlined argument which shows that simple and the existence of a left identity implies associative.

2 MAIN SECTION.

THEOREM 1 *Let R be a ring satisfying $(x,y,z) = (x,z,y)$ and $(x,x,x) = 0$, and having characteristic $\neq 2,3$. If the associator ideal of R has a left identity element e, then R must be associative.*

The proof is based on the following lemma proved in [2].

LEMMA 1 *In a ring satisfying $(x,y,z) = (x,z,y)$ and $(x,x,x) = 0$, and having characteristic $\neq 2,3$, let J be the right ideal generated by (a,b,a). Then $J(a,b,a) = 0$.*

Let A be the associator ideal of R. Then A consists of all finite sums of associators and right multiples of associators. Define B as the sum of right ideals J generated by (x,y,x) as x and y range over R. Clearly $B \subseteq A$. For all x,y,z in R, we have $-(x+z,y,x+z)+(x,y,x)+(z,y,z) = -(x,y,z)-(z,y,x) = -(x,y,z)-(z,x,y)$. Linearization of $(x,x,x) = 0$ yields $-(x,y,z)-(z,x,y) = (y,z,x)$, so that $(y,z,x) \in B$. Thus $A \subseteq B$ and so $A = B$. By hypothesis A has a left identity element e, so e is also a left identity element of B. We write

$$e = a_1 + a_2 + \cdots + a_n,$$

where each a_i belongs to some right ideal J_i generated by (x_i, y_i, x_i). Multiplying on the right by (x_n, y_n, x_n), we obtain

$$(x_n, y_n, x_n) = a_1(x_n, y_n, x_n) + \cdots + a_{n-1}(x_n, y_n, x_n) + a_n(x_n, y_n, x_n).$$

Now using the lemma, $a_n(x_n, y_n, x_n) = 0$ and we can let $b_i = a_i(x_n, y_n, x_n) \in J_i$, so that

$$(x_n, y_n, x_n) = (b_1 + b_2 + \cdots + b_{n-1}) \in J_1 + J_2 + \cdots + J_{n-1}.$$

It follows that $J_n \subseteq J_1 + J_2 + \cdots + J_{n-1}$, and thus $e \in J_1 + J_2 + \cdots + J_{n-1}$. Repetition of this process leads to $e \in J_1$. Multiplying both sides on the right by (x_1, y_1, x_1) and using the lemma gives $e(x_1, y_1, x_1) \in J_1(x_1, y_1, x_1) = 0$. Since e is a left identity on B, this gives $(x_1, y_1, x_1) = 0$. But now (x_1, y_1, x_1) is the generator of J_1, so $J_1 = 0$, whence $e = 0$. For 0 to be a left identity of an ideal, the ideal must be zero, so we have shown that $A = B = 0$, and R is associative. This shows that simple rings of type (1,1) cannot have a left identity element unless they are associative. Kokoris' result lets one extend this result to any nonzero idempotent.

Ng has obtained a more general result [7]. His proof (also based on the above lemma from [2]) is quite complicated and requires some modifications suggested by Armin Thedy in Math. Reviews 86a:17001.

REFERENCES

[1] I. R. Hentzel, D.P. Jacobs, and E. Kleinfeld, Rings with $(a,b,c) = (a,c,b)$ and $(a,[b,c],d) = 0$: A case study using Albert, International J. Computer Math. 49(1993), 19-27.

[2] Erwin Kleinfeld, Simple algebras of type (1,1) are associative, Canad. J. Math. 13(1961), 129-148.

[3] Erwin Kleinfeld, On rings satisfying $(x, y, z) = (x, z, y)$, Algebras, Groups and Geometries 4(1987), 129-138.

[4] Erwin Kleinfeld and Harry F. Smith, A generalization of Novikov rings, Nonassociative algebra and its applications, 1994 Kluwer academic publishers, S. Gonzalez (ed.), 219-222.

[5] Erwin Kleinfeld and Margaret Kleinfeld, A Generalization of Rings of Type (1,1), to appear.

[6] L. A. Kokoris, On rings of (γ, δ) type, Proc. Amer. Math. Soc. 9(1958), 897-904.

[7] Seong Nam Ng, Local nilpotence of the associator ideal of a (1,1) algebra, Comm. Algebra 12(1984), 1125-1137.

[8] J. M. Osborn, Novikov algebras, Nova J. Algebra Geom. 1(1992), 1-14.

[9] J. M. Osborn, Simple Novikov algebras with an idempotent, Comm. Algebra 20(1992), 2729-2753.

[10] Y. Paul, Prime rings satisfying $(x, y, z) = (x, z, y)$, Proceedings of the Symposium on Algebra and Number Theory, Kochi, Kerala, India, July 1990, 91-95.

[11] E. I. Zelmanov, On a class of local translation invariant Lie Algebras, English translation, Soviet Math. Dokl.35(1987), 216-218.

On the semisimplicity of special Jordan superalgebras

CARLOS GOMEZ-AMBROSI Departamento de Matemáticas, Universidad de Zaragoza, 50009 Zaragoza, Spain

JESUS LALIENA Departamento de Matemáticas y Computación, Universidad de La Rioja, 26004 Logroño, Spain

ABSTRACT: We provide a definition of the Jacobson radical of a Jordan super- algebra, and study some relationships between the semisimplicity of an associative superalgebra (with superinvolution) and that of its related Jordan superalgebras.

INTRODUCTION

McCrimmon proved in [4] that the Jacobson radical of a Jordan algebra of the form A^+, where A is an associative algebra, coincides with the Jacobson radical of the associative algebra A, and that the Jacobson radical of the Jordan algebra $H(A,*)$ of $*$-symmetric elements of an associative algebra A with involution $*$, coincides with the symmetric part of the Jacobson radical of the associative algebra A. In fact, McCrimmon reduces these theorems to the following ones: A is semisimple as

The authors have been supported by the DGICYT, PB 97-1291-C03. The second author has also been supported by the Universidad de La Rioja API-97/B05.

an associative algebra if and only if A^+ is semisimple as a Jordan algebra, and if A is a semisimple associative algebra with involution $*$ then $H(A,*)$ is a semisimple Jordan algebra. The purpose of this note is to show that, under some restrictions, analogous theorems in the setting of superalgebras are true.

In his paper, McCrimmon deduces the theorems cited above from what he calls the First Herstein Construction, which associates with any Jordan ideal in a Jordan algebra of the form A^+ an associative ideal in the associative algebra A, and the Second Herstein Construction, which associates with any Jordan ideal in the Jordan algebra $H(A,*)$ of $*$-symmetric elements an associative $*$-ideal in the associative algebra A with involution $*$. This constructions had been recently "translated" into the language of superalgebras by Montaner and the first author in [2]. So, in some sense, this note should be considered as an appendix to that paper. The reader is also refered to [2] for the basic definitions and examples introductory to superalgebras and superinvolutions needed here.

Specifically, we will prove that if A is a prime noncommutative associative superalgebra then A is semisimple if and only if A^+ is semisimple (section 3), and that if A is a $*$-prime associative superalgebra with superinvolution $*$ such that A_0 is not commutative, then A is semisimple if and only if $H(A,*)$ is semisimple (section 4). Note that the restrictions imposed to the superalgebras are the same as the ones needed in the Herstein Constructions for superalgebras given in [2]. We provide counterexamples to show that these restrictions are in fact necessary. Stronger results concerning deeper relationships between the Jacobson radicals had not yet been proved, and are merely stated as conjectures.

Throughout this note Φ will denote a commutative unital ring of scalars with $\frac{1}{2} \in \Phi$ and all algebras will be Φ-algebras.

Let $A = A_0 \oplus A_1$ be an associative superalgebra. The map $p: A \to A$ given by $p(a_0 + a_1) = a_0 - a_1$ is an automorphism of order 2 of the superalgebra A, called the *parity automorphism* of A. The set of even elements of A is $A_0 = \{a \in A \mid p(a) = a\}$, and the set of odd elements of A is $A_1 = \{a \in A \mid p(a) = -a\}$. The group $G = \{1, p\}$ is a group of order 2 acting on the algebra A, whose algebra of fixed points is $A^G = A_0$. Notice that a graded ideal of A is the same thing as an ideal of A stable under G (or, equivalently, under p). This point of view will be important in order to apply the results obtained by Montgomery in [6].

In all that follows, $J(A)$ will denote the Jacobson radical of the algebra A. Notice that $J(A)$ is stable under p for it is stable under all automorphisms of the algebra A, hence $J(A)$ is a graded ideal of A.

1 THE JACOBSON RADICAL OF AN ASSOCIATIVE SUPERALGEBRA

In this section we provide a definition of the Jacobson radical of an associative superalgebra via modular graded left ideals, and characterize it in terms of quasi-invertible elements, following the discussion given by Rowen in [7]. It is not difficult to check that our definition agrees with that given by Draper in [1] via annihilators of irreducible graded modules or primitive graded ideals, and in fact our major goal will be to prove that all those definitions agree with the classical Jacobson radical of the superalgebra viewed as an algebra. The notion of semisimplicity which arises from this definition is also the same that the one used by Józefiak in [3] for finite-dimensional associative superalgebras (cf. [1]).

A graded left ideal L of A is *modular* if there exists $e \in A$ such that $a - ae \in L$ for all $a \in A$; e is called a *right unit (modulo L)*.

LEMMA 1. *If $e = e_0 + e_1$ is a right unit modulo L, then e_0 is a right unit modulo L.*

Proof. Let $a = a_0 + a_1 \in A$. Since L is modular, $a_0 - a_0(e_0 + e_1) \in L$, hence $a_0 - a_0e_0 \in L_0$. On the other hand, $a_1 - a_1(e_0 + e_1) \in L$, hence $a_1 - a_1e_0 \in L_1$. Now $a - ae_0 = (a_0 - a_0e_0) + (a_1 - a_1e_0) \in L$ and so e_0 is a right unit modulo L. ∎

In all that follows, a right unit modulo L will be assumed to be even without further reference.

A modular graded left ideal L with right unit e_0 is proper if and only if $e_0 \notin L$, for, otherwise, $a \in ae_0 + L \subseteq L$ for all $a \in A$, proving $L = A$. Consequently, Zorn's lemma shows that any modular graded left ideal $L \neq A$ is contained in a maximal modular graded left ideal ($\neq A$), with the same right unit. Thus one can define the *Jacobson radical* of the superalgebra A as $J_s(A) = \bigcap$ {maximal modular graded left ideals of A}, where the subscript "s" stands for "super".

We can characterize $J_s(A)$ in terms of left quasi-invertible elements, where we say that an element $a \in A$ is *left quasi-invertible* (or left quasi-regular) if there exists an element $b \in A$ such that $a + b - ba = 0$. We call b a *left quasi-inverse* of a. If $A^1 = \Phi 1 + A$ is the associative superalgebra obtained by the formal adjoining of an identity element to A, then a is left quasi-invertible with left quasi-inverse b if and only if $1 - a$ is left invertible with left inverse $1 - b$. We will use "q.i." as an abbreviation of "quasi-invertible". A graded left ideal is *left quasi-invertible* if all its HOMOGENEOUS elements are left q.i. The definitions of right and two-sided quasi-invertibility, for elements as well as for ideals, are the corresponding ones.

LEMMA 2. *Let $a_0 \in A_0$, $a_1 \in A_1$. Then*
(i) a_0 is left q.i. in A if and only if a_0 is left q.i. in A_0.
(ii) a_1 is left q.i. in A if and only if a_1^2 is left q.i. in A_0.

Proof. (i) If a_0 is left q.i. in A, there exist $b_0 \in A_0$, $b_1 \in A_1$ such that $a_0 + (b_0 + b_1) - (b_0 + b_1)a_0 = 0$, hence $a_0 + b_0 - b_0a_0 = 0$ and a_0 is left q.i. in A_0. The converse is clear.

(ii) If a_1 is left q.i. in A, there exist $b_0 \in A_0$, $b_1 \in A_1$ such that $a_1 + (b_0 + b_1) - (b_0 + b_1)a_1 = 0$, hence $b_0 = b_1a_1$ and $a_1 + b_1 - b_0a_1 = 0$. Multiplying the last equation by a_1 on the right and using the first one we get $0 = a_1^2 + b_1a_1 - b_0a_1^2 = a_1^2 + b_0 - b_0a_1^2$, thus a_1^2 is left q.i. in A_0. Conversely, if a_1^2 is left q.i. in A_0, there exists $b_0 \in A_0$ such that $a_1^2 + b_0 - b_0a_1^2 = 0$. If we put $b_1 = b_0a_1 - a_1$ then it is easy to see that $a_1 + (b_0 + b_1) - (b_0 + b_1)a_1 = 0$, hence a_1 is left q.i. in A. ∎

Lemma 2 gives us the following useful characterization.

LEMMA 3. *A graded left ideal L is left q.i. in A if and only if L_0 is left q.i. in A_0.*

PROPOSITION 1. *$J_s(A)$ is a left q.i. graded left ideal which contains every q.i. graded left ideal.*

Proof. If $a_0 \in J_s(A)_0$ is not left q.i. then $L = \{b - ba_0 \mid b \in A\}$ is a proper modular graded left ideal, for a_0 is a right unit modulo L; but L is contained in a maximal modular graded left ideal not containing a_0, contradiction. Thus $J_s(A)$ is left q.i. by lemma 3.

Now let B be any q.i. graded left ideal. We claim that B is contained in every maximal modular graded left ideal L. Indeed, otherwise $e_0 \in B + L$, where e_0 is a right unit modulo L. Writing $e_0 = b_0 + a_0$ for $b_0 \in B_0$, $a_0 \in L_0$, we see that $a_0 = e_0 - b_0 \in L_0$. But $b_0 - b_0 e_0 \in L_0$ by hypothesis, so $e_0 - b_0 e_0 \in L_0$. Take c_0 such that $b_0 + c_0 - c_0 b_0 = 0$. Then $e_0 = e_0 - (b_0 + c_0 - c_0 b_0) e_0 = (e_0 - b_0 e_0) - c_0(e_0 - b_0 e_0) \in L_0$, contrary to $L \neq A$. Thus $B \subseteq \bigcap$ {maximal modular graded left ideals of $A\} = J_s(A)$. ■

LEMMA 4. *Let L be a graded left ideal. Then all homogeneous elements in L are left q.i. if and only if all elements in L are left q.i.*

Proof. Assume that all homogeneous elements in L are left q.i., and let $a = a_0 + a_1 \in L$. Since a_0 is left q.i. in A, a_0 is left q.i. in A_0, hence there exists $b_0 \in A_0$ such that $a_0 + b_0 - b_0 a_0 = 0$. $a_1 - b_0 a_1 \in L_1$ for L is a left ideal, hence $a_1 - b_0 a_1$ is left q.i. in A and so there exist $c_0 \in A_0$, $c_1 \in A_1$ such that $(a_1 - b_0 a_1) + (c_0 + c_1) - (c_0 + c_1)(a_1 - b_0 a_1) = 0$. Finally, straightforward calculations show that $(a_0 + a_1) + (b_0 + c_0 + c_1 - (c_0 + c_1) b_0) - (b_0 + c_0 + c_1 - (c_0 + c_1) b_0)(a_0 + a_1) = 0$, therefore $a = a_0 + a_1$ is left q.i. ■

THEOREM 1. $J_s(A) = J(A)$.

Proof. By lemma 4, $J_s(A)$ is a left q.i left ideal of the algebra A, hence $J_s(A) \subseteq J(A)$. Conversely, $J(A)$ is a left q.i. graded ideal of the superalgebra A, hence $J(A) \subseteq J_s(A)$. ■

Notice that $J(A)$ is then the maximal q.i. graded ideal of the superalgebra A.

We call an associative superalgebra *semisimple* if its Jacobson radical is zero.

2 THE JACOBSON RADICAL OF A JORDAN SUPERALGEBRA

In this section we provide a definition of the Jacobson radical of a Jordan superalgebra in terms of quasi-invertible elements, following the ideas posed by McCrimmon in [5] (see also [8]).

Let B be a Jordan superalgebra. An element $a \in B_0$ is called *quasi-invertible* if there exists an element $b \in B_0$ such that, in the Jordan superalgebra $B^\sharp = \Phi 1 + B$ obtained by the formal adjoining of an identity element to B, the element $1 - a$ is invertible with inverse element $1 - b$. The element b is called a *quasi-inverse* for the element a. If B is a special Jordan superalgebra which is embedded as a subsuperalgebra of A^+, where A is an associative superalgebra, then an element $a \in B_0$ is quasi-invertible in B with quasi-inverse $b \in B_0$ if and only if a is quasi-invertible in A with quasi-inverse b (in the associative sense of the previous section, see [8]). A graded ideal K of B is called *quasi-invertible* if all elements in K_0 are q.i. in B.

Notice that $a \in B_0$ is q.i. in the Jordan superalgebra B if and only if a is q.i. in the Jordan algebra B_0, and also that a graded ideal K of B is q.i. in B if and only if K_0 is q.i. in B_0. Now the proof of the following facts carries over almost verbatim from the ungraded case (see [5] or [8]).

LEMMA 5. *The sum of two q.i. ideals is a q.i. ideal.*

LEMMA 6. *If a is q.i. in B, it is q.i. in any quotient B/K (equivalently, $\varphi(a)$ is q.i. in any homomorphic image $\varphi(B)$). If $a+K$ is q.i. in B/K, where K is a q.i. ideal, then a is q.i. in B.*

Since the condition that an ideal be q.i. is defined elementwise, by Zorn's lemma we can find a maximal q.i. ideal J. By lemma 5, J contains all q.i. ideals. In particular, J is unique and is the sum of all q.i. ideals. Lemma 6 shows that B/J has no nonzero q.i. ideals since the inverse image of such an ideal would be a q.i. ideal in B properly containing J. J will be called the *Jacobson radical* $J(B)$ of the Jordan superalgebra B. We have established the following theorem.

THEOREM 2. *For any Jordan superalgebra B there exists a unique maximal q.i. ideal $J(B)$, the Jacobson radical of B, which contains all q.i. ideals. The quotient $B/J(B)$ contains no nonzero q.i. ideals: $J(B/J(B)) = 0$.*

We call a Jordan superalgebra *semisimple* if its Jacobson radical is zero.

3 SEMISIMPLICITY OF A^+

Let A be an associative superalgebra and let us consider its Jacobson radical $J(A)$. Since $J(A)$ is a q.i. graded ideal of A, it is also a q.i. graded ideal of A^+, hence $J(A) \subseteq J(A^+)$, and in particular $J(A)_0 \subseteq J(A^+)_0$.

PROPOSITION 2. $J(A)_0 = J(A^+)_0$.

Proof. By theorem 7 of [6], $J(A)_0 = J(A_0)$. On the other hand, by theorem 1 of [4], $J(A_0) = J(A_0^+)$. Now $J(A^+)_0$ is a q.i. ideal of A_0^+, hence $J(A^+)_0 \subseteq J(A_0^+) = J(A)_0$. ∎

THEOREM 3. *If A is a prime noncommutative associative superalgebra, then A is semisimple if and only if A^+ is semisimple.*

Proof. If A is semisimple then $J(A) = 0$ and $J(A^+)_0 = J(A)_0 = 0$ by proposition 2. Thus $J(A^+) = J(A^+)_1$ is a graded ideal of A^+ such that $J(A^+)_1 \circ J(A^+)_1 = 0$. But A^+ is prime by the corollary to the First Herstein Construction for superalgebras that can be found in [2], hence $J(A^+) = J(A^+)_1 = 0$ and A^+ is semisimple.

The converse is obvious. ∎

We will now provide a counterexample of a commutative prime associative superalgebra for which the previous theorem is not true.

COUNTEREXAMPLE. Let Φ be a field and consider the superalgebra $Q(1)$ over Φ (cf. example 1.2 of [2] or examples 2.5 of [3]). This superalgebra is simple, hence prime, and commutative. It is clear that $J(Q(1)) = 0$, and so $Q(1)$ is semisimple. However, $Q(1)_1$ being a nonzero ideal of $Q(1)^+$, $J(Q(1)^+) = Q(1)_1 \neq 0$, therefore $Q(1)^+$ is not semisimple.

Although we have not yet proved that always $J(A) = J(A^+)$, we believe that this is the case at least under the conditions stated in the previous theorem. So we make the following conjecture.

CONJECTURE 1. *If A is a prime noncommutative associative superalgebra, then $J(A) = J(A^+)$.*

4 SEMISIMPLICITY OF H(A,*)

Now let A be an associative superalgebra with superinvolution $*$, and let $H = H(A,*) = \{a \in A \mid a^* = a\}$ be the set of symmetric elements of A, which is a subsuperalgebra of the Jordan superalgebra A^+. In this section, J will denote the Jacobson radical $J(A)$ of the associative superalgebra A, and $J(H)$ will denote the Jacobson radical of the Jordan superalgebra H. Moreover, $H(J) = J \cap H$ is the set of symmetric elements in J.

Since J is a q.i. graded ideal of A, $H(J)$ is a q.i. graded ideal of H, hence $H(J) \subseteq J(H)$, and in particular $H(J)_0 \subseteq J(H)_0$.

PROPOSITION 3. $H(J)_0 = J(H)_0$.

Proof. By theorem 7 of [6], $J_0 = J(A_0)$, thus $H(J)_0 = H(J_0) = H(J(A_0))$. Moreover, by theorem 3 of [4], $J(H_0) = H(J(A_0))$, hence $H(J)_0 = J(H_0)$. Finally, $J(H)_0$ is a q.i. ideal of H_0, so $J(H)_0 \subseteq J(H_0) = H(J)_0$. ∎

THEOREM 4. *If A is a $*$-prime associative superalgebra with superinvolution $*$ such that A_0 is not commutative, then A is semisimple if and only if $H = H(A,*)$ is semisimple.*

Proof. If A is semisimple then $J = 0$ and $J(H)_0 = H(J)_0 = 0$ by proposition 3. Thus $J(H) = J(H)_1$ is a graded ideal of H such that $J(H)_1 \circ J(H)_1 = 0$. But H is prime by the corollary to the Second Herstein Construction for superalgebras that can be found in [2], hence $J(H) = J(H)_1 = 0$ and H is semisimple.

Conversely, if H is semisimple then $J(H) = 0$, and so $H(J) = 0$. Moreover, it is easy to check that J is a $*$-ideal of A, hence $J = 0$ by lemma 6.3 of [2], and A is semisimple. ∎

We will now provide a counterexample of a $*$-prime associative superalgebra with superinvolution, and with commutative even part, for which the previous theorem is not true.

COUNTEREXAMPLE. Let Φ be a field and consider the superalgebra $M(1)$ over Φ (cf. example 1.1 of [2] or examples 2.5 of [3]), provided with the transpose superinvolution *trp* (cf. example 2.2 of [2]). This superalgebra is simple, hence prime and $*$-prime, and its even part is commutative. It is clear that $J(M(1)) = 0$, and so $M(1)$ is semisimple. However, H_1 being a nonzero ideal of $H = H(M(1), trp)$, $J(H) = H_1 \neq 0$, therefore H is not semisimple.

Although we have not yet proved that always $H(J) = J(H)$, we believe that this is the case at least under the conditions stated in the previous theorem. So we make the following conjecture.

CONJECTURE 2. *If A is a $*$-prime associative superalgebra with superinvolution $*$ such that A_0 is not commutative, then $H(J) = J(H)$.*

REFERENCES.

[1] C. Draper, Superálgebras Asociativas, Tesina de Licenciatura, Universidad de Málaga, 1997.

[2] C. Gómez-Ambrosi, F. Montaner, On Herstein's constructions relating Jordan and associative superalgebras, preprint.

[3] T. Józefiak, Semisimple superalgebras, Lecture Notes in Math. No. 1352, Springer -Verlag, Berlin, 1987, 96-113.

[4] K. McCrimmon, On Herstein's theorems relating Jordan and associative algebras, J. Algebra 13 (1969), 382-392.

[5] K. McCrimmon, The radical of a Jordan algebra, Proc. Nat. Acad. Sci. USA 62 (1969), 671-678.

[6] S. Montgomery, The Jacobson radical and fixed rings of automorphisms, Comm. Algebra 4 (5) (1976), 459-465.

[7] L.H. Rowen, Ring Theory, Vol. I, Academic Press, Boston, 1988.

[8] K.A. Žhevlakov, A.M. Slin'ko, I.P. Shestakov, A.I. Shirshov, Rings that are Nearly Associative, Academic Press, New York, 1982.

Bernstein superalgebras of dimension four

S. GONZÁLEZ, C. LÓPEZ-DIAZ, C. MARTÍNEZ †. Departamento de Matemáticas. Universidad de Oviedo. 33007 Oviedo, España.

0 INTRODUCTION

Bernstein algebras appear when we try to translate the description of all stationary operators in biological heritage into algebraic language (see [5]).

The origin of superalgebras lies in physics and today they are an important part of mathematics. They are 2-graded algebras and describe supersymmetry, what explains the name superalgebras (see [6]).

In this paper we classify Bernstein superalgebras of dimension 4 in a similar way to the followed in dimensions 2 and 3 (see [2]). We will assume that the odd part is different of zero since the case in which the odd part is zero reduces to the classification of Bernstein algebras of dimension 5, already known (see [4]). By simplicity we will consider only the case of nonzero product. There are 30 4-dimensional Bernstein superalgebras (with odd part different of zero) nonisomorphic as Bernstein superalgebras, one for each of the 30 possible gradings. In order to work in a natural way for superalgebras, we have considered the natural definition of Bernstein superalgebra using the variety of Bernstein algebras without weight. (see [3]).

In this classification of Bernstein superalgebras of dimension 4, there are cases in which Bernstein algebras without weight appear. Each of them corresponds to a Bernstein algebra of dimension 5. In those cases that the superalgebra is not

† Partially supported by DGICYT PB94-1311-CO3-01 and FC-97-PBPGI97-03

a Bernstein algebra without weight, we have studied which of them are Jordan algebras.

In the paper, F denotes a field, char $F \neq 2$.

1 BASIC CONCEPTS

We will give in this section the most important definitions and results on Bernstein superalgebras that appear in [3].

DEFINITION 1.1. Let A be a commutative algebra over a field F and let $\sigma : A \longrightarrow F$ be a nonzero homomorphism of algebras. The algebra A is called Bernstein algebra if for all $x \in A$, $(x^2)^2 = \sigma(x)^2 x^2$. Then, σ is called weight homomorphism.

PROPOSITION 1.1. (See [7]) Let A be a Bernstein algebra over a field F and $e \neq 0$ an idempotent element in A (whose existence is known). Then, the Peirce decomposition is:

$$A = Fe \oplus U_e \oplus W_e$$

where $N = U_e \oplus W_e$ is the kernel of σ, $U_e = \{ x \in A \,/\, ex = \frac{1}{2}x \}$ and $W_e = \{ x \in A \,/\, ex = 0 \}$.

THEOREM 1.2. (See [7]) Let $A = Fe \oplus U \oplus W$ be a Bernstein algebra over a field F, where $U = U_e$ and $W = W_e$. Then:

a) $U^2 \subseteq W$, $W^2 \subseteq U$, $UW \subseteq U$.

b) $u^3 = 0$, $u(uw) = 0$, $uw^2 = 0$, $(u^2)^2 = 0$, $(uw)^2 = 0$.
$\forall\, u \in U, \forall\, w \in W$.

Moreover, dimensions of U and W do not depend on e.

DEFINITION 1.2. Let $A = U + W$ be a commutative algebra over a field F which is a direct sum of vector subspaces U and W. Then, A is called Bernstein algebra without weight or (U, W)-Bernstein algebra, if A verifies conditions $a)$ and $b)$ in Theorem 1.2. We will refer to them as B.a.w.w.

PROPOSITION 1.3. (See [3]) Let A' be a Bernstein algebra with weight σ. Let A be the kernel of σ. Then, for every nonzero idempotent element e in A', A is (U_e, W_e)-Bernstein algebra.

PROPOSITION 1.4. (See [3]) Let $A = U + W$ be a commutative algebra over a field F which is a direct sum of subspaces U and W. Let us extend the multiplication in A to $A(e) = Fe \oplus A$ by:

$$e^2 = e,\ eu = \frac{1}{2}u,\ ew = 0\ \forall\, u \in U,\ \forall\, w \in W.$$

Let $\sigma : A(e) \longrightarrow F$ be the homomorphism of algebras, such that $\sigma(\alpha e+u+w) = \alpha$. Then, A is a B.a.w.w. if and only if $A(e)$ is a Bernstein algebra.

DEFINITION 1.3. Let $A = U+W$ be a B.a.w.w. with $dim\,U = n$ and $dim\,W = m$. The pair (n,m) is called type of A.

DEFINITION 1.4. (See [3]) Let $A = U + W$ be a (U,W)-Bernstein algebra and $M = X+Y$ a commutative module over A. M is called (X,Y)-Bernstein module over A or graded Bernstein module over A if the algebra $A+M$, under the product $(a+m)(b+n) = ab+mb+an, \quad \forall\, a+m, b+n \in A+M$, is a $((U+X),(W+Y))$-Bernstein algebra.

DEFINITION 1.5. Let $A = A_0 + A_1$ be a superalgebra over a field F. Let $G = alg\langle 1, e_i \,/\, e_i^2 = 0,\ e_ie_j = -e_je_i,\ i \in N\rangle$ be the Grassman associative algebra over F which is a superalgebra with $G = G_0 + G_1$, where $G_0 = vect_F\langle e_{i_1}e_{i_2}...e_{i_k} \,/\, k \text{ even}\rangle$ and $G_1 = vect_F\langle e_{i_1}e_{i_2}...e_{i_k} \,/\, k \text{ odd}\rangle$. Then, $A = A_0 + A_1$ is called Bernstein superalgebra if $G(A) = G_0 \otimes A_0 + G_1 \otimes A_1$, the Grassman envelope of A, is a B.a.w.w.

Remark. If $A = A_0 + A_1$ is a Bernstein superalgebra, then $A_0 = U + W$ is a B.a.w.w., $A_1 = \overline{U} + \overline{W}$ is a graded Bernstein module over A_0 and A verifies the superidentities that are obtained applying Kaplansky principle to the complete linearizations of those identities defining B.a.w.w. (see [3]).

PROPOSITION 1.5. (See [3]) Let A be a Bernstein superalgebra, $A = A_0 + A_1 = (U+W) + (\overline{U}+\overline{W}) = (U+\overline{U}) + (W+\overline{W})$. Consider $L(A) = L = \{\, u+\overline{u} \in U+\overline{U} \mid (u+\overline{u})(U+\overline{U}) = 0 \,\}$. Then:

1) L is a graded ideal of A and $L^2 = 0$.
2) A/L verifies the superidentity:

$$(x_1x_2)x_3 + (-1)^{dx_2dx_3}(x_1x_3)x_2 + (-1)^{dx_1(dx_2+dx_3)}(x_2x_3)x_1 = 0.$$

Hence A/L is a Jordan superalgebra.

2 BERNSTEIN SUPERALGEBRAS OF DIMENSION FOUR

In this section we classify Bernstein superalgebras of dimension four up to isomorphism of Bernstein superalgebras. To do this, we give the classification of B.a.w.w. of dimensions two and three, in a convenient way. In all cases that will follow, products not explicitly given are zero.

The following lemma can be easily proved.

LEMMA 2.1. Every B.a.w.w. A of dimension two and with nonzero product is isomorphic, as B.a.w.w., to one of the following types:

[1] $A = U + W = Fu + Fw$ with $u^2 = w$.
[2] $A = U + W = Fu + Fw$ with $uw = u$.
[3] $A = U + W = Fu + Fw$ with $w^2 = u$.
[4] $A = U + W = Fu + Fw$ with $w^2 = u = uw$.

LEMMA 2.2. Every B.a.w.w. A of dimension three and with nonzero product is isomorphic, as B.a.w.w., to one of the following types:

[1] $A = U + W = U + Fw$, where $\{u_1, u_2\}$ is a basis of U, with $w^2 = u_1$.
[2] $A = U + W = U + Fw$, where $\{u_1, u_2\}$ is a basis of U, with $u_1 w = u_1$.
[3] $A = U + W = U + Fw$, where $\{u_1, u_2\}$ is a basis of U, with $u_1 w = u_2$.
[4] $A = U + W = U + Fw$, where $\{u_1, u_2\}$ is a basis of U, with $u_1 w = u_1$ and $u_2 w = \lambda u_1 + \beta u_2$.
[5] $A = U + W = U + Fw$, where $\{u_1, u_2\}$ is a basis of U, with $u_1 w = u_2$ and $u_2 w = \beta u_1 + \lambda u_2$ with $x^2 - \lambda x - \beta$ irreducible in $F[x]$.
[6] $A = U + W = U + Fw$, where $\{u_1, u_2\}$ is a basis of U, with $u_1 w = u_1$ and $w^2 = u_1$.
[7] $A = U + W = U + Fw$, where $\{u_1, u_2\}$ is a basis of U, with $u_1 w = u_2$ and $w^2 = u_2$.
[8] $A = U + W = U + Fw$, where $\{u_1, u_2\}$ is a basis of U, with $u_1 w = u_1$ and $w^2 = u_2$.
[9] $A = U + W = U + Fw$, where $\{u_1, u_2\}$ is a basis of U, with $u_1 w = u_2$ and $w^2 = u_1$.
[10] $A = U + W = U + Fw$, where $\{u_1, u_2\}$ is a basis of U, with $u_1 w = u_1$, $u_2 w = \lambda u_1 + \beta u_2$ and $w^2 = u_1$.
[11] $A = U + W = U + Fw$, where $\{u_1, u_2\}$ is a basis of U, with $u_1 w = u_2$, $u_2 w = \beta u_1 + \lambda u_2$ and $w^2 = u_1$.
[12] $A = U + W = U + Fw$, where $\{u_1, u_2\}$ is a basis of U, with $u_1^2 = w$, $u_2^2 = \delta w$.
[13] $A = U + W = Fu + W$, where $\{w_1, w_2\}$ is a basis of W, with $uw_1 = u$.
[14] $A = U + W = Fu + W$, where $\{w_1, w_2\}$ is a basis of W, with $w_1^2 = u$.
[15] $A = U + W = Fu + W$, where $\{w_1, w_2\}$ is a basis of W, with $w_1^2 = u$ and $w_2^2 = \beta u$.
[16] $A = U + W = Fu + W$, where $\{w_1, w_2\}$ is a basis of W, with $uw_1 = u$, $w_1^2 = \lambda u$ and $w_2^2 = \delta u$.
[17] $A = U + W = Fu + W$, where $\{w_1, w_2\}$ is a basis of W, with $u^2 = w_1$.

$\delta \in F$, $\beta \in F^*$ and $\lambda \in \{0, 1\}$.

Proof:

Let A be a B.a.w.w. of dimension three and with nonzero product. We have the following possibilities:

I) $A = U + W$ where $\dim U = 2$, $\{u_1, u_2\}$ is a basis of U and $W = Fw$. The product in A is:

$$\begin{array}{cccc} & u_1 & u_2 & w \\ u_1 & \alpha w & \beta w & \kappa_1 u_1 + \kappa_2 u_2 \\ u_2 & \beta w & \delta w & \tau_1 u_1 + \tau_2 u_2 \\ w & \kappa_1 u_1 + \kappa_2 u_2 & \tau_1 u_1 + \tau_2 u_2 & \sigma_1 u_1 + \sigma_2 u_2 \end{array}$$

with $\alpha\kappa_i = \alpha\tau_i = \alpha\sigma_i = 0$, $\beta\kappa_i = \beta\tau_i = \beta\sigma_i = 0$ and $\delta\kappa_i = \delta\tau_i = \delta\sigma_i = 0$ $\forall i \in \{1,2\}$.

We will distinguish two cases:

Case $U^2 = 0$.

1) $UW = 0$. Then A is isomorphic as B.a.w.w. to an algebra with the only nonzero product $w^2 = u_1$.

2) $W^2 = 0$. There are two possibilities:

2.1 dim $UW = 1$. In this case, if we consider $p(x)$ the minimum polynomial of $R_w : U \longrightarrow U$, we have that A is isomorphic as B.a.w.w. to an algebra with the only nonzero product $u_1 w = u_1$ if $p(x) = x(x-1)$, and A is isomorphic as B.a.w.w. to an algebra with the only nonzero product $u_1 w = u_2$ if $p(x) = x^2$.

2.2 dim $UW = 2$. Then A is isomorphic as B.a.w.w. to one of the following algebras:

i) $u_1 w = u_1$ and $u_2 w = \lambda u_1 + \beta u_2$.

ii) $u_1 w = u_2$ and $u_2 w = \beta u_1 + \lambda u_2$.

In both cases $\lambda \in \{0,1\}$ and $\beta \in F^*$.

3) dim UW = dim W^2 = dim $(UW + W^2) = 1$. Here we have that A is isomorphic as B.a.w.w. to one of the following algebras:

i) $u_1 w = u_1$ and $w^2 = u_1$.

ii) $u_1 w = u_2$ and $w^2 = u_2$.

4) dim UW = dim $W^2 = 1$ and dim $(UW + W^2) = 2$. In this case A is isomorphic as B.a.w.w. to one of the following algebras:

i) $u_1 w = u_1$ and $w^2 = u_2$.

ii) $u_1 w = u_2$ and $w^2 = u_1$.

5) dim $W^2 = 1$ and dim $UW = 2$. In this case A is isomorphic as B.a.w.w. to one of the following algebras:

i) $u_1 w = u_1$, $u_2 w = \beta u_2$ and $w^2 = u_1$.

ii) $u_1 w = u_2$, $u_2 w = \beta u_1 + \lambda u_2$ and $w^2 = u_1$.

In both cases $\lambda \in \{0,1\}$ and $\beta \in F^*$.

Case $U^2 \neq 0$.

Hence $UW = W^2 = 0$. Then, A is isomorphic as B.a.w.w. to the algebra with product: $u_1^2 = w$ and $u_2^2 = \delta w$ with $\delta \in F$.

II) $A = U + W$ where dim $W = 2$, $\{w_1, w_2\}$ is a basis of W and $U = Fw$. The product in A is:

$\cdot$	u	w_1	w_2
u	$\alpha_1 w_1 + \alpha_2 w_2$	κu	τu
w_1	κu	σu	ρu
w_2	τu	ρu	ηu

with $\alpha_i \kappa = \alpha_i \tau = \alpha_i \sigma = \alpha_i \rho = \alpha_i \eta = 0$, $\forall\, i \in \{1,2\}$.

Again we will distinguish two cases:

Case $U^2 = 0$.

1) $UW = 0$. Then A is isomorphic as B.a.w.w. to one of the following algebras with the only nonzero products:

i) $w_1^2 = u$.

ii) $w_1^2 = u$ and $w_2^2 = \beta u$ where $\beta \in F^*$.

2) $W^2 = 0$.Then A is isomorphic to an algebra with nonzero product $uw_1 = u$.

3) dim $W^2 =$ dim $UW = 1$. In this case A is isomorphic as Bernstein algebra without weight to the algebra with product: $uw_1 = u$, $w_1^2 = \lambda u$ and $w_2^2 = \delta u$, where $\delta \in F$ and $\lambda \in \{0,1\}$.

Case $U^2 \neq 0$.

Now we have $W^2 = 0 = UW$. Hence A is isomorphic as Bernstein algebra without weight to the algebra with nonzero product $u^2 = w_1$.

Similarly to the classification of Bernstein superalgebras of dimension two and three (see [2]), the following theorems can be proved. Products given are not zero.

THEOREM 2.3. Every Bernstein superalgebra $A = A_0 + A_1$ of dimension four, $A_1 \neq \{0\}$ and $A^2 \neq \{0\}$, is isomorphic as Bernstein superalgebra to a superalgebra of the following list:

[I] $dimA_0 = 3$ and $dimA_1 = 1$.

1) $A_0 = U + W$ where $\dim U = 2$ with $\{u_1, u_2\}$ basis of U, $W = Fw$, $A_1 = F\overline{u} = \overline{U}$ with $A_0^2 = \{0\}$ and $\overline{u}w = \overline{u}$.
2) A_0 is [1] in Lemma 2.2, $A_1 = F\overline{u} = \overline{U}$ and $\overline{u}w = \overline{u}$.
3) A_0 is [2] in Lemma 2.2, $A_1 = F\overline{u} = \overline{U}$ and $\overline{u}w = \lambda\overline{u}$.
4) A_0 is [3] in Lemma 2.2, $A_1 = F\overline{u} = \overline{U}$ and $\overline{u}w = \overline{u}$.
5) A_0 is [4] in Lemma 2.2, $A_1 = F\overline{u} = \overline{U}$ and $\overline{u}w = \lambda\overline{u}$.
6) A_0 is [5] in Lemma 2.2, $A_1 = F\overline{u} = \overline{U}$ and $\overline{u}w = \overline{u}$.
7) A_0 is [6] in Lemma 2.2, $A_1 = F\overline{u} = \overline{U}$ and $\overline{u}w = \lambda\overline{u}$.
8) A_0 is [7] in Lemma 2.2, $A_1 = F\overline{u} = \overline{U}$ and $\overline{u}w = \overline{u}$.
9) A_0 is [8] in Lemma 2.2, $A_1 = F\overline{u} = \overline{U}$ and $\overline{u}w = \lambda\overline{u}$.
10) A_0 is [9] in Lemma 2.2, $A_1 = F\overline{u} = \overline{U}$ and $\overline{u}w = \overline{u}$.
11) A_0 is [10] in Lemma 2.2, $A_1 = F\overline{u} = \overline{U}$ and $\overline{u}w = \lambda\overline{u}$.
12) A_0 is [11] in Lemma 2.2, $A_1 = F\overline{u} = \overline{U}$ and $\overline{u}w = \overline{u}$.
13) A_0 is [13] in Lemma 2.2, $A_1 = F\overline{u} = \overline{U}$,

   [13.1] $\overline{u}w_1 = \alpha\overline{u}$.

[13.2] $\overline{u}w_2 = \overline{u}$.

14) A_0 is [14] in Lemma 2.2, $A_1 = F\overline{u} = \overline{U}$,

[14.1] $\overline{u}w_1 = \overline{u}$.

[14.2] $\overline{u}w_2 = \overline{u}$.

15) A_0 is [15] in Lemma 2.2, $A_1 = F\overline{u} = \overline{U}$ and $\overline{u}w_1 = \overline{u}$.

16) A_0 is [16] in Lemma 2.2, $A_1 = F\overline{u} = \overline{U}$,

[16.1] $\overline{u}w_1 = \alpha\overline{u}$.

[16.2] $\overline{u}w_2 = \overline{u}$.

17) A_0 is [17] in Lemma 2.2, $A_1 = F\overline{u} = \overline{U}$ and $w_2\overline{u} = \overline{u}$.

18) A_0 any of the algebras in Lemma 2.2, $A_1 = F\overline{u} = \overline{U}$ and $A_1^2 = A_0A_1 = 0$.

19) A_0 any of the algebras in Lemma 2.2, $A_1 = F\overline{w} = \overline{W}$ and $A_1^2 = A_0A_1 = 0$.

20) $A_0 = U + W$ where dim $W = 2$ with $\{w_1, w_2\}$ a basis of W, $U = Fu$, $A_1 = F\overline{u} = \overline{U}$ with $A_0^2 = \{0\}$ and $w_1\overline{u} = \overline{u}$.

21) $A_0 = W$ with basis $\{w_1, w_2, w_3\}$, $A_1 = F\overline{u} = \overline{U}$ and $w_1\overline{u} = \overline{u}$.

[II] $dimA_0 = 2$ and $dimA_1 = 2$.

1) $A_0 = U$ with basis $\{u_1, u_2\}$, $A_1 = F\overline{u} + F\overline{w} = \overline{U} + \overline{W}$ with $A_0^2 = \{0\}$,

[1.1] $u_1\overline{u} = \overline{w}$.

[1.2] $u_1\overline{w} = \overline{u}$.

[1.3] $\overline{u}\,\overline{w} = u_1$.

[1.4] $u_1\overline{u} = \overline{w}$, $\overline{u}\,\overline{w} = \kappa u_1$.

[1.5] $u_1\overline{w} = \overline{u}$, $\overline{u}\,\overline{w} = \kappa u_1$.

[1.6] $u_2\overline{u} = \overline{w}$, $\overline{u}\,\overline{w} = u_1$.

[1.7] $u_2\overline{w} = \overline{u}$, $\overline{u}\,\overline{w} = u_1$.

2) $A_0 = U$ with basis $\{u_1, u_2\}$, $A_1 = \overline{W}$ with basis $\{\overline{w}_1, \overline{w}_2\}$, $A_0^2 = \{0\}$ and $\overline{w}_1\overline{w}_2 = u_1$.

3) $A_0 = Fu + Fw = U + W$, $A_1 = \overline{U}$ with basis $\{\overline{u}_1, \overline{u}_2\}$, $A_0^2 = \{0\}$,

[3.1] $w\overline{u}_1 = \overline{u}_1$.

[3.2] $w\overline{u}_1 = \overline{u}_2$.

[3.3] $w\overline{u}_1 = \overline{u}_1$, $w\overline{u}_2 = \alpha\overline{u}_2$.

[3.4] $w\overline{u}_1 = \overline{u}_2$, $w\overline{u}_2 = \alpha\overline{u}_1 + \beta\overline{u}_2$ where $\beta = 1$ or 0.

[3.5] $\overline{u}_1\overline{u}_2 = w$.

[3.6] $w\overline{u}_1 = \overline{u}_2$, $\overline{u}_1\overline{u}_2 = kw$.

4) $A_0 = Fu + Fw = U + W$, $A_1 = F\overline{u} + F\overline{w} = \overline{U} + \overline{W}$, $A_0^2 = \{0\}$,

[4.1] $u\overline{u} = \overline{w}$.

[4.2] $\overline{u}\,\overline{w} = u$.

[4.3] $u\overline{w} = \overline{u}$.

[4.4] $w\overline{u} = \overline{u}$.

[4.5] $w\overline{w} = \overline{u}$.

[4.6] $u\overline{u} = \overline{w}$, $\overline{u}\,\overline{w} = \sigma u$.

[4.7] $u\overline{w} = \overline{u} = w\overline{u}$.

[4.8] $u\overline{w} = \overline{u} = w\overline{w}$.

[4.9] $u\overline{w} = \overline{u}$, $\overline{u}\,\overline{w} = \sigma u$.

[4.10] $w\overline{u} = \overline{u} = w\overline{w}$.

[4.11] $w\overline{u} = \overline{u}$, $\overline{u}\,\overline{w} = u$.

[4.12] $w\overline{w} = \overline{u}$, $\overline{u}\,\overline{w} = u$.

[4.13] $u\overline{w} = \overline{u} = w\overline{u} = w\overline{w}$.

[4.14] $u\overline{w} = \overline{u} = w\overline{u}$, $\overline{u}\,\overline{w} = \sigma u$.

[4.15] $u\overline{w} = \overline{u} = w\overline{w}$, $\overline{u}\,\overline{w} = \sigma u$.

[4.16] $w\overline{u} = \overline{u} = w\overline{w}$, $\overline{u}\,\overline{w} = \sigma u$.

[4.17] $u\overline{w} = \overline{u} = w\overline{u}$, $w\overline{w} = \kappa\overline{u}$, $\overline{u}\,\overline{w} = \sigma u$.

5) $A_0 = Fu + Fw = U + W$, $A_1 = \overline{W}$ with basis $\{\overline{w}_1, \overline{w}_2\}$, $A_0^2 = \{0\}$ and $\overline{w}_1\,\overline{w}_2 = u$.

6) A_0 is [1] in Lemma 2.1, $A_1 = \overline{U}$ with basis $\{\overline{u}_1, \overline{u}_2\}$ and $\overline{u}_1\overline{u}_2 = w = u^2$.

7) A_0 is [1] in Lemma 2.1, $A_1 = \overline{U} + \overline{W} = F\overline{u} + F\overline{w}$,

[7.1] $u\overline{w} = \overline{u}$.

[7.2] $w\overline{w} = \overline{u}$.

[7.3] $u\overline{w} = \overline{u}$, $w\overline{w} = \overline{u}$.

8) A_0 is [2] in Lemma 2.1, $A_1 = \overline{U}$ with basis $\{\overline{u}_1, \overline{u}_2\}$

[8.1] $w\overline{u}_1 = \alpha\overline{u}_1$.

[8.2] $w\overline{u}_1 = \overline{u}_2$.

[8.3] $w\overline{u}_1 = \alpha\overline{u}_1$, $w\overline{u}_2 = \overline{u}_2$.

[8.4] $w\overline{u}_1 = \overline{u}_2$, $w\overline{u}_2 = \alpha\overline{u}_1$.

9) A_0 is [2] in Lemma 2.1, $A_1 = \overline{U} + \overline{W} = F\overline{u} + F\overline{w}$,

[9.1] $u\overline{w} = \overline{u}$.

[9.2] $w\overline{u} = \delta\overline{u}$.

[9.3] $w\overline{w} = \overline{u}$.

[9.4] $\overline{u}\,\overline{w} = u$.

[9.5] $u\overline{w} = \overline{u}$, $w\overline{u} = \delta\overline{u}$.

[9.6] $u\overline{w} = \overline{u}$, $w\overline{w} = \overline{u}$.

[9.7] $u\overline{w} = \overline{u}$, $\overline{u}\,\overline{w} = \sigma u$.

[9.8] $w\overline{u} = \delta\overline{u}$, $w\overline{w} = \overline{u}$.

[9.9] $w\overline{u} = \delta\overline{u}$, $\overline{u}\,\overline{w} = u$.

[9.10] $w\overline{w} = \overline{u}$, $\overline{u}\,\overline{w} = \sigma u$.

[9.11] $u\overline{w} = \overline{u}$, $w\overline{u} = \delta\overline{u}$, $w\overline{w} = \overline{u}$.

[9.12] $u\overline{w} = \overline{u}$, $w\overline{u} = \delta\overline{u}$, $\overline{u}\,\overline{w} = \sigma u$.

[9.13] $u\overline{w} = \overline{u}$, $w\overline{w} = \overline{u}$, $\overline{u}\,\overline{w} = \sigma u$.

[9.14] $u\overline{w} = \overline{u} = w\overline{w}$, $w\overline{u} = \delta\overline{u}$, $\overline{u}\,\overline{w} = \sigma u$.

10) A_0 is [2] in Lemma 2.1, $A_1 = \overline{W}$ with basis $\{\overline{w}_1, \overline{w}_2\}$ and $\overline{w}_1\,\overline{w}_2 = u$.

11) A_0 is [3] in Lemma 2.1, $A_1 = \overline{U}$ with basis $\{\overline{u}_1, \overline{u}_2\}$,

[11.1] $w\overline{u}_1 = \overline{u}_1$.

[11.2] $w\overline{u}_1 = \overline{u}_2$.

[11.3] $\overline{u}_1\overline{u}_2 = w$.

[11.4] $w\overline{u}_1 = \overline{u}_1,\ w\overline{u}_2 = \alpha\overline{u}_2$.

[11.5] $w\overline{u}_1 = \overline{u}_1,\ w\overline{u}_2 = \overline{u}_1 + \alpha\overline{u}_2$.

[11.6] $w\overline{u}_1 = \overline{u}_1,\ \overline{u}_1\overline{u}_2 = w$.

[11.7] $w\overline{u}_1 = \overline{u}_2,\ w\overline{u}_2 = \overline{u}_1$.

[11.8] $w\overline{u}_1 = \overline{u}_2,\ \overline{u}_1\overline{u}_2 = w$.

[11.9] $w\overline{u}_1 = \overline{u}_1,\ w\overline{u}_2 = \alpha\overline{u}_2,\ \overline{u}_1\overline{u}_2 = w$.

[11.10] $w\overline{u}_1 = \overline{u}_1,\ w\overline{u}_2 = \overline{u}_1 + \alpha\overline{u}_2,\ \overline{u}_1\overline{u}_2 = \beta w$.

[11.11] $w\overline{u}_1 = \overline{u}_2,\ w\overline{u}_2 = \overline{u}_1,\ \overline{u}_1\overline{u}_2 = \beta w$.

12) A_0 is [3] in Lemma 2.1, $A_1 = \overline{U} + \overline{W} = F\overline{u} + F\overline{w}$,

[12.1] $u\overline{w} = \overline{u}$.

[12.2] $w\overline{u} = \overline{u}$.

[12.3] $w\overline{w} = \overline{u}$.

[12.4] $\overline{u}\,\overline{w} = u$.

[12.5] $u\overline{w} = \overline{u} = w\overline{u}$.

[12.6] $u\overline{w} = \overline{u},\ w\overline{w} = \kappa\overline{u}$.

[12.7] $u\overline{w} = \overline{u},\ \overline{u}\,\overline{w} = \sigma u$.

[12.8] $w\overline{u} = \overline{u},\ w\overline{w} = \overline{u}$.

[12.9] $w\overline{u} = \overline{u},\ \overline{u}\,\overline{w} = u$.

[12.10] $w\overline{w} = \overline{u},\ \overline{u}\,\overline{w} = u$.

[12.11] $u\overline{w} = \overline{u} = w\overline{u},\ w\overline{w} = \kappa\overline{u}$.

[12.12] $u\overline{w} = \overline{u} = w\overline{u},\ \overline{u}\,\overline{w} = \sigma u$.

[12.13] $u\overline{w} = \overline{u} = w\overline{w},\ \overline{u}\,\overline{w} = \sigma u$.

[12.14] $w\overline{u} = \overline{u} = w\overline{w},\ \overline{u}\,\overline{w} = \sigma u$.

[12.15] $u\overline{w} = w\overline{u} = \overline{u},\ w\overline{w} = \kappa\overline{u},\ \overline{u}\,\overline{w} = \sigma u$.

13) A_0 is [3] in Lemma 2.1, $A_1 = \overline{W}$ with basis $\{\overline{w}_1, \overline{w}_2\}$ and $\overline{w}_1\overline{w}_2 = u$.

14) A_0 is [4] in Lemma 2.1, $A_1 = \overline{U}$ with basis $\{\overline{u}_1, \overline{u}_2\}$,

[14.1] $w\overline{u}_1 = \alpha\overline{u}_1$.

[14.2] $w\overline{u}_1 = \overline{u}_2$.

[14.3] $w\overline{u}_1 = \alpha\overline{u}_1,\ w\overline{u}_2 = \beta\overline{u}_2$.

[14.4] $w\overline{u}_1 = \alpha\overline{u}_1,\ w\overline{u}_2 = u_1 + \alpha\overline{u}_2$.

[14.5] $w\overline{u}_1 = \overline{u}_2,\ w\overline{u}_2 = \beta\overline{u}_1$.

15) A_0 is [4] in Lemma 2.1, $A_1 = \overline{U} + \overline{W} = F\overline{u} + F\overline{w}$,

[15.1] $u\overline{w} = \overline{u}$.

[15.2] $w\overline{u} = \delta\overline{u}$.

[15.3] $w\overline{w} = \overline{u}$.

[15.4] $\overline{u}\,\overline{w} = u$.

[15.5] $u\overline{w} = \overline{u},\ w\overline{u} = \delta\overline{u}$.

[15.6] $u\overline{w} = \overline{u},\ w\overline{w} = \kappa\overline{u}$.

[15.7] $u\overline{w} = \overline{u},\ \overline{u}\,\overline{w} = \sigma u$.

[15.8] $w\overline{u} = \delta\overline{u},\ w\overline{w} = \overline{u}$.

[15.9] $w\overline{u} = \delta\overline{u},\ \overline{u}\,\overline{w} = u$.

[15.10] $w\overline{w} = \overline{u}$, $\overline{u}\,\overline{w} = \sigma u$.

[15.11] $u\overline{w} = \overline{u}$, $w\overline{u} = \delta\overline{u}$, $w\overline{w} = \kappa\overline{u}$.

[15.12] $u\overline{w} = \overline{u}$, $w\overline{u} = \delta\overline{u}$, $\overline{u}\,\overline{w} = \sigma u$.

[15.13] $u\overline{w} = \overline{u}$, $w\overline{w} = \kappa\overline{u}$, $\overline{u}\,\overline{w} = \sigma u$.

[15.14] $w\overline{u} = \delta\overline{u}$, $w\overline{w} = \overline{u}$, $\overline{u}\,\overline{w} = \sigma u$.

[15.15] $u\overline{w} = \overline{u}$, $w\overline{u} = \delta\overline{u}$, $w\overline{w} = \kappa\overline{u}$, $\overline{u}\,\overline{w} = \sigma u$.

16) A_0 is [4] in Lemma 2.1, $A_1 = \overline{W}$ with basis $\{\overline{w}_1, \overline{w}_2\}$ and $\overline{w}_1\overline{w}_2 = u$.

17) A_0 any of the algebras in Lemma 2.1, $A_1 = \overline{U}$ with basis $\{\overline{u}_1, \overline{u}_2\}$ and $A_1^2 = A_0A_1 = 0$.

18) A_0 any of the algebras in Lemma 2.1, $A_1 = \overline{U} + \overline{W} = F\overline{u} + F\overline{w}$ and $A_1^2 = A_0A_1 = 0$.

19) A_0 any of the algebras in Lemma 2.1, $A_1 = \overline{W}$ with basis $\{\overline{w}_1, \overline{w}_2\}$ and $A_1^2 = A_0A_1 = 0$.

20) $A_0 = W$ with basis $\{w_1, w_2\}$, $A_1 = \overline{U}$ with basis $\{\overline{u}_1, \overline{u}_2\}$, $A_0^2 = \{0\}$,

[20.1] $w_1\overline{u}_1 = \overline{u}_1$.

[20.2] $w_1\overline{u}_1 = \overline{u}_2$.

[20.3] $w_1\overline{u}_1 = \overline{u}_1$ and $w_1\overline{u}_2 = \alpha\overline{u}_2$.

[20.4] $w_1\overline{u}_1 = \overline{u}_1 = w_2\overline{u}_1$.

[20.5] $w_1\overline{u}_1 = \overline{u}_1$ and $w_2\overline{u}_1 = \overline{u}_2$.

[20.6] $w_1\overline{u}_1 = \overline{u}_1 = w_2\overline{u}_2$.

[20.7] $w_1\overline{u}_1 = \overline{u}_1$, $w_2\overline{u}_2 = \overline{u}_2$.

[20.8] $w_1\overline{u}_1 = \overline{u}_2$, $w_1\overline{u}_2 = k_1\overline{u}_1 + k_2\overline{u}_2$.

[20.9] $w_1\overline{u}_1 = \overline{u}_2$, $w_2\overline{u}_1 = (1-k)\overline{u}_1 + k\overline{u}_2$.

[20.10] $w_1\overline{u}_1 = \overline{u}_2$, $w_2\overline{u}_1 = -\overline{u}_1 + \overline{u}_2$.

[20.11] $w_1\overline{u}_1 = \overline{u}_2$, $w_2\overline{u}_2 = (1-k)\overline{u}_1 + k\overline{u}_2$.

[20.12] $w_1\overline{u}_1 = \overline{u}_2$, $w_2\overline{u}_2 = -\overline{u}_1 + \overline{u}_2$.

[20.13] $w_1\overline{u}_1 = \overline{u}_1$, $w_1\overline{u}_2 = \alpha\overline{u}_2$, $w_2\overline{u}_1 = A\overline{u}_1 + B\overline{u}_2$ with $A, B \in \{0,1\}$.

[20.14] $w_1\overline{u}_1 = \overline{u}_2$, $w_1\overline{u}_2 = k_1\overline{u}_1 + k_2\overline{u}_2$, $w_2\overline{u}_1 = A\overline{u}_1 + B\overline{u}_2$ with $A, B \in \{0,1\}$.

[20.15] $w_1\overline{u}_1 = \overline{u}_1$, $w_1\overline{u}_2 = \alpha\overline{u}_2$, $w_2\overline{u}_2 = A\overline{u}_1 + \overline{u}_2$ with $A \in \{0,1\}$.

[20.16] $w_1\overline{u}_1 = \overline{u}_2$, $w_1\overline{u}_2 = k_1\overline{u}_1 + k_2\overline{u}_2$, $w_2\overline{u}_2 = \overline{u}_1 + B\overline{u}_2$ with $B \in \{0,1\}$.

[20.17] $w_1\overline{u}_1 = \overline{u}_1$, $w_1\overline{u}_2 = \alpha\overline{u}_2$, $w_2\overline{u}_1 = A\overline{u}_1 + B\overline{u}_2$ with $A, B \in \{0,1\}$, $w_2\overline{u}_2 = \kappa_1\overline{u}_1 + \kappa_2\overline{u}_2$ where $\kappa_2 \neq 0$.

[20.18] $w_1\overline{u}_1 = \overline{u}_2$, $w_1\overline{u}_2 = k_1\overline{u}_1 + k_2\overline{u}_2$, $w_2\overline{u}_1 = A\overline{u}_1 + B\overline{u}_2$ with $A, B \in \{0,1\}$, $w_2\overline{u}_2 = \kappa_1\overline{u}_1 + \kappa_2\overline{u}_2$ where $\kappa_1 \neq 0$.

[20.19] $w_1\overline{u}_1 = \overline{u}_1$, $w_1\overline{u}_2 = \alpha\overline{u}_2$, $w_2\overline{u}_1 = \delta_1\overline{u}_1 + \delta_2\overline{u}_2$ where $\delta_2 \neq 0$.

[20.20] $w_1\overline{u}_1 = \overline{u}_2$, $w_1\overline{u}_2 = k_1\overline{u}_1 + k_2\overline{u}_2$, $w_2\overline{u}_2 = \overline{u}_1 + B\overline{u}_2$ with $B \in \{0,1\}$, $w_2\overline{u}_1 = \delta_1\overline{u}_1 + \delta_2\overline{u}_2$ where $\delta_1 \neq 0$.

[20.21] $\overline{u}_1\,\overline{u}_2 = w_1$.

[20.22] $w_1\overline{u}_1 = \overline{u}_2$, $\overline{u}_1\,\overline{u}_2 = w_2$.

[20.23] $w_2\overline{u}_1 = \alpha\overline{u}_2$, $\overline{u}_1\,\overline{u}_2 = w_2$.

[20.24] $w_1\overline{u}_1 = \overline{u}_1$, $w_2\overline{u}_1 = \alpha\overline{u}_2$, $\overline{u}_1\,\overline{u}_2 = w_2$.

[20.25] $w_1\overline{u}_1 = \overline{u}_1 + \alpha\overline{u}_2$, $w_2\overline{u}_1 = \overline{u}_1 + \beta\overline{u}_2$, $\overline{u}_1\,\overline{u}_2 = w_2$.

[20.26] $w_1\overline{u}_1 = \overline{u}_2$, $w_1\overline{u}_2 = \overline{u}_2$, $\overline{u}_1\,\overline{u}_2 = w_2$.

[20.27] $w_1\overline{u}_1 = \overline{u}_2$, $w_2\overline{u}_2 = \overline{u}_1$, $\overline{u}_1\,\overline{u}_2 = w_2$.

[20.28] $w_1\overline{u}_1 = \overline{u}_1$, $w_1\overline{u}_2 = \alpha\overline{u}_1$, $w_2\overline{u}_1 = \frac{1}{\alpha}\overline{u}_2$, $w_2\overline{u}_2 = \overline{u}_2$, $\overline{u}_1\,\overline{u}_2 = w_2$.

[20.29] $w_1\overline{u}_1 = \overline{u}_1 + \alpha\overline{u}_2$, $w_1\overline{u}_2 = \lambda\overline{u}_1 + \sigma\overline{u}_2$, $w_2\overline{u}_1 = \overline{u}_1 + \beta\overline{u}_2$, $w_2\overline{u}_2 = \chi\overline{u}_1 + \delta\overline{u}_2$, $\overline{u}_1\,\overline{u}_2 = w_2$.

[20.30] $w_1\overline{u}_1 = \alpha\overline{u}_2$, $w_1\overline{u}_2 = \overline{u}_2$, $w_2\overline{u}_1 = \overline{u}_2$, $w_2\overline{u}_2 = \frac{1}{\alpha}\overline{u}_1$, $\overline{u}_1\,\overline{u}_2 = w_2$.

21) $A_0 = W$ with basis $\{w_1, w_2\}$, $A_1 = F\overline{u} + F\overline{w} = \overline{W} + \overline{U}$, $A_0^2 = \{0\}$,

[21.1] $w_1\overline{u} = \overline{u}$.

[21.2] $w_1\overline{w} = \overline{u}$.

[21.3] $w_1\overline{w} = \overline{u} = w_1\overline{u}$.

[21.4] $w_1\overline{u} = \overline{u} = w_2\overline{u}$.

[21.5] $w_1\overline{u} = \overline{u} = w_2\overline{w}$.

[21.6] $w_1\overline{w} = \overline{u} = w_2\overline{w}$.

[21.7] $w_1\overline{u} = \overline{u} = w_2\overline{u} = w_1\overline{w}$.

[21.8] $w_1\overline{u} = \overline{u} = w_2\overline{w} = w_1\overline{w}$.

[21.9] $w_1\overline{u} = \overline{u} = w_2\overline{u} = w_1\overline{w}$, $w_2\overline{w} = \overline{u}$.

[III] $dimA_0 = 1$, $A_0 = U = Fu$ and $dimA_1 = 3$.

1) $A_1 = \overline{U} + F\overline{w} = \overline{U} + \overline{W}$ with $\{\overline{u}_1, \overline{u}_2\}$ basis of $\overline{U}$,

[1.1] $u\overline{w} = \overline{u}_1$.

[1.2] $\overline{u}_1\overline{w} = u$.

[1.3] $u\overline{w} = \overline{u}_1$, $\overline{u}_1\overline{w} = \alpha u$.

[1.4] $u\overline{w} = \overline{u}_1$, $\overline{u}_2\overline{w} = u$.

[1.5] $\overline{u}_1\overline{w} = u = \overline{u}_2\overline{w}$.

[1.6] $\overline{u}_1\overline{w} = \alpha u$, $u\overline{w} = \overline{u}_1$ and $\overline{u}_2\overline{w} = u$.

[1.7] $u\overline{u}_2 = \overline{w}$.

[1.8] $\overline{u}_2\overline{w} = u$.

[1.9] $u\overline{u}_2 = \overline{w}$, $\overline{u}_2\overline{w} = \alpha u$.

[1.10] $u\overline{u}_1 = \overline{w} = u\overline{u}_2$, $\overline{u}_1\overline{w} = \alpha u = \overline{u}_2\overline{w}$.

2) $A_1 = F\overline{u} + \overline{W} = \overline{U} + \overline{W}$ with $\{\overline{w}_1, \overline{w}_2\}$ basis of $\overline{W}$,

[2.1] $u\overline{w}_1 = \overline{u}$.

[2.2] $\overline{u}\,\overline{w}_1 = u$.

[2.3] $\overline{w}_1\overline{w}_2 = u$.

[2.4] $u\overline{w}_1 = \overline{u} = u\overline{w}_2$.

[2.5] $u\overline{w}_1 = \overline{u}$, $\overline{u}\,\overline{w}_1 = \alpha u$.

[2.6] $u\overline{w}_1 = \overline{u}$, $\overline{u}\,\overline{w}_2 = u$.

[2.7] $u\overline{w}_1 = \overline{u}$, $\overline{w}_1\overline{w}_2 = u$.

[2.8] $\overline{u}\,\overline{w}_1 = u = \overline{u}\,\overline{w}_2$.

[2.9] $\overline{u}\,\overline{w}_1 = u = \overline{w}_1\overline{w}_2$.

[2.10] $u\overline{w}_1 = \overline{u} = u\overline{w}_2$, $\overline{u}\,\overline{w}_1 = \alpha u$.

[2.11] $u\overline{w}_1 = \overline{u} = u\overline{w}_2$, $\overline{w}_1\overline{w}_2 = u$.

[2.12] $u\overline{w}_2 = \overline{u}$, $\overline{u}\,\overline{w}_1 = u$, $\overline{u}\,\overline{w}_2 = \alpha u$.

[2.13] $u\overline{w}_2 = \overline{u}$, $\overline{u}\,\overline{w}_1 = u$, $\overline{w}_1\overline{w}_2 = u$.

[2.14] $u\overline{w}_2 = \overline{u}$, $\overline{u}\,\overline{w}_2 = \alpha u$, $\overline{w}_1\overline{w}_2 = u$.

[2.15] $\overline{u}\,\overline{w}_1 = u$, $\overline{u}\,\overline{w}_2 = u$, $\overline{w}_1\overline{w}_2 = u$.

[2.16] $u\overline{w}_1 = \overline{u} = u\overline{w}_2$, $\overline{u}\,\overline{w}_1 = \alpha u$, $\overline{u}\,\overline{w}_2 = \beta u$.

[2.17] $u\overline{w}_2 = \overline{u}$, $\overline{u}\,\overline{w}_1 = u$, $\overline{u}\,\overline{w}_2 = \alpha u$, $\overline{w}_1\overline{w}_2 = u$.

[2.18] $u\overline{w}_1 = \overline{u} = u\overline{w}_2$, $\overline{u}\,\overline{w}_2 = \alpha u$, $\overline{w}_1\overline{w}_2 = u$.

[2.19] $u\overline{w}_1 = \overline{u} = u\overline{w}_2$, $\overline{u}\,\overline{w}_1 = \beta u$, $\overline{u}\,\overline{w}_2 = \alpha u$, $\overline{w}_1\overline{w}_2 = u$.

[2.20] $u\overline{u} = \overline{w}_2$.

[2.21] $u\overline{u} = \overline{w}_2$, $\overline{u}\,\overline{w}_1 = u$.

[2.22] $u\overline{u} = \overline{w}_2$, $\overline{u}\,\overline{w}_2 = \alpha u$.

[2.23] $u\overline{u} = \alpha\overline{w}_2$, $\overline{u}\,\overline{w}_1 = u = \overline{u}\,\overline{w}_2$.

3) $A_1 = \overline{W}$ with basis $\{\overline{w}_1, \overline{w}_2, \overline{w}_3\}$,

[3.1] $\overline{w}_1\overline{w}_2 = u$.

[3.2] $\overline{w}_1\overline{w}_2 = u = \overline{w}_1\overline{w}_3$.

[3.3] $\overline{w}_1\overline{w}_2 = u = \overline{w}_1\overline{w}_3 = \overline{w}_2\overline{w}_3$.

[IV] $dimA_0 = 1$, $A_0 = W = Fw$ and $dimA_1 = 3$.

1) $A_1 = \overline{U}$ with basis $\{\overline{u}_1, \overline{u}_2, \overline{u}_3\}$,

[1.1] $w\overline{u}_1 = \overline{u}_1$.

[1.2] $w\overline{u}_1 = \overline{u}_2$.

[1.3] $w\overline{u}_1 = \overline{u}_1 = w\overline{u}_2$.

[1.4] $w\overline{u}_1 = \overline{u}_3 = w\overline{u}_2$.

[1.5] $w\overline{u}_1 = \overline{u}_1$, $w\overline{u}_2 = \overline{u}_3$.

[1.6] $w\overline{u}_1 = \overline{u}_1 = w\overline{u}_2 = w\overline{u}_3$.

[1.7] $w\overline{u}_1 = \overline{u}_2 = w\overline{u}_3$, $w\overline{u}_2 = \overline{u}_3$.

[1.8] $w\overline{u}_1 = \overline{u}_2$, $w\overline{u}_2 = \overline{u} \in \overline{U}$, $w\overline{u}_3 = \overline{u}' \in \overline{U}$.

[1.9] $\overline{u}_1\overline{u}_2 = w$.

[1.10] $\overline{u}_1\overline{u}_2 = w = \overline{u}_1\overline{u}_3$.

[1.11] $\overline{u}_1\overline{u}_2 = w = \overline{u}_1\overline{u}_3 = \overline{u}_2\overline{u}_3$.

2) $A_1 = \overline{U} + F\overline{w} = \overline{U} + \overline{W}$ with $\{\overline{u}_1, \overline{u}_2\}$ basis of $\overline{U}$,

[2.1] $w\overline{u}_1 = \overline{u}_1$.

[2.2] $w\overline{u}_1 = \overline{u}_2$.

[2.3] $w\overline{w} = \overline{u}_1$.

[2.4] $w\overline{u}_1 = \overline{u}_1 = w\overline{u}_2$.

[2.5] $w\overline{u}_1 = \overline{u}_1$, $w\overline{w} = \overline{u}_2$.

[2.6] $w\overline{u}_1 = \overline{u}_1$, $w\overline{w} = \overline{u}_1$.

[2.7] $w\overline{u}_1 = \overline{u}_2$, $w\overline{u}_2 = \alpha\overline{u}_2$, $w\overline{w} = \overline{u}_2$.

[2.8] $w\overline{u}_1 = \overline{u}_1$, $w\overline{u}_2 = \overline{u}_1$, $w\overline{w} = \overline{u}_2$.

[2.9] $w\overline{u}_1 = \overline{u}_2$, $w\overline{u}_2 = \alpha\overline{u}_2$, $w\overline{w} = \beta\overline{u}_1 + \delta\overline{u}_2$.

[2.10] $w\overline{u}_1 = \overline{u}_2$, $w\overline{u}_2 = \alpha\overline{u}_1 + \chi\overline{u}_2$, $w\overline{w} = \beta\overline{u}_1 + \delta\overline{u}_2$.

[2.11] $\overline{u}_1\,\overline{u}_2 = w$.

[2.12] $w\overline{u}_1 = \overline{u}_2$, $w\overline{u}_2 = \alpha\overline{u}_1 + \chi\overline{u}_2$, $\overline{u}_1\,\overline{u}_2 = w$.

3) $A_1 = F\overline{u} + \overline{W} = \overline{U} + \overline{W}$ with $\{\overline{w}_1, \overline{w}_2\}$ basis of $\overline{W}$,

[3.1] $w\overline{u} = \overline{u}$.

[3.2] $w\overline{w}_1 = \overline{u}$.

[3.3] $w\overline{u} = \overline{u} = w\overline{w}_1$.

[3.4] $w\overline{w}_1 = \overline{u} = w\overline{w}_2$.

[3.5] $w\overline{u} = w\overline{w}_1 = \overline{u} = w\overline{w}_2$.

Proof:

To obtain the list above we consider all the possible dimensions and gradings in A_0 and A_1 being $A_1 \neq \{0\}$ and $A^2 \neq \{0\}$. We use the superidentities which hold in a Bernstein superalgebra. They are obtained applying Kaplansky principle to the complete linearizations of the identities which hold in a B.a.w.w. (see [3]). We use these superidentities getting the conditions for nonisomorphic Bernstein superalgebras (see [2]) and ordering the possibilities according to cases in Lemma 2.1 and Lemma 2.2.

THEOREM 2.4. Every Bernstein superalgebra $A = A_0 + A_1$ of dimension four satisfies some of the following conditions:

[I] A is a B.a.w.w.

[II] A is a Jordan superalgebra.

[III] A is isomorphic as Bernstein superalgebra to one of the following list:

i) $A_0 = Fu + Fw = U + W$ and $A_1 = F\overline{u} + F\overline{w} = \overline{U} + \overline{W}$,with product:

[1] $u\overline{u} = \overline{w}$, $\overline{u}\,\overline{w} = \sigma u$.

[2] $u\overline{w} = \overline{u}$, $\overline{u}\,\overline{w} = \sigma u$.

[3] $w\overline{u} = \overline{u}$, $\overline{u}\,\overline{w} = u$.

[4] $w\overline{w} = \overline{u}$, $\overline{u}\,\overline{w} = u$.

[5] $u\overline{w} = \overline{u} = w\overline{u}$, $\overline{u}\,\overline{w} = \sigma u$.

[6] $w\overline{u} = \overline{u} = w\overline{w}$, $\overline{u}\,\overline{w} = \sigma u$.

[7] $u\overline{w} = \overline{u} = w\overline{u}$, $w\overline{w} = \kappa\overline{u}$, $\overline{u}\,\overline{w} = \sigma u$.

[8] $uw = u$, $\overline{u}\,\overline{w} = u$.

[9] $uw = u$, $u\overline{w} = \overline{u}$, $\overline{u}\,\overline{w} = \sigma u$.

[10] $uw = u$, $w\overline{u} = \delta\overline{u}$, $\overline{u}\,\overline{w} = u$.

[11] $uw = u$, $w\overline{w} = \overline{u}$, $\overline{u}\,\overline{w} = \sigma u$.

[12] $uw = u$, $u\overline{w} = \overline{u}$, $w\overline{u} = \delta\overline{u}$, $\overline{u}\,\overline{w} = \sigma u$.

[13] $uw = u$, $u\overline{w} = \overline{u}$, $w\overline{w} = \overline{u}$, $\overline{u}\,\overline{w} = \sigma u$.

[14] $uw = u$, $u\overline{w} = \overline{u} = w\overline{w}$, $w\overline{u} = \delta\overline{u}$, $\overline{u}\,\overline{w} = \sigma u$.

[15] $w^2 = u$, $\overline{u}\,\overline{w} = u$.

[16] $w^2 = u$, $u\overline{w} = \overline{u}$, $\overline{u}\,\overline{w} = \sigma u$.

[17] $w^2 = u$, $w\overline{u} = \overline{u}$, $\overline{u}\,\overline{w} = u$.

[18] $w^2 = u$, $w\overline{w} = \overline{u}$, $\overline{u}\,\overline{w} = u$.

[19] $w^2 = u$, $u\overline{w} = \overline{u} = w\overline{u}$, $\overline{u}\,\overline{w} = \sigma u$.

[20] $w^2 = u$, $u\overline{w} = \overline{u} = w\overline{w}$, $\overline{u}\,\overline{w} = \sigma u$.
[21] $w^2 = u$, $w\overline{u} = \overline{u} = w\overline{w}$, $\overline{u}\,\overline{w} = \sigma u$.
[22] $w^2 = u = uw$, $\overline{u}\,\overline{w} = u$.
[23] $w^2 = u = uw$, $u\overline{w} = \overline{u}$, $\overline{u}\,\overline{w} = \sigma u$.
[24] $w^2 = u = uw$, $w\overline{u} = \delta\overline{u}$, $\overline{u}\,\overline{w} = u$.
[25] $w^2 = u = uw$, $w\overline{w} = \overline{u}$, $\overline{u}\,\overline{w} = \sigma u$.
[26] $w^2 = u = uw$, $u\overline{w} = \overline{u}$, $w\overline{u} = \delta\overline{u}$, $\overline{u}\,\overline{w} = \sigma u$.
[27] $w^2 = u = uw$, $u\overline{w} = \overline{u}$, $w\overline{w} = \kappa\overline{u}$, $\overline{u}\,\overline{w} = \sigma u$.
[28] $w^2 = u = uw$, $w\overline{u} = \delta\overline{u}$, $w\overline{w} = \overline{u}$, $\overline{u}\,\overline{w} = \sigma u$.
[29] $w^2 = u = uw$, $u\overline{w} = \overline{u}$, $w\overline{u} = \delta\overline{u}$, $w\overline{w} = \kappa\overline{u}$, $\overline{u}\,\overline{w} = \sigma u$.

ii) $A_0 = Fu + Fw = U + W$ and $A_1 = \overline{W}$ with basis $\{\overline{w}_1, \overline{w}_2\}$, and product $w^2 = u$, $uw = u$ and $\overline{w}_1\,\overline{w}_2 = u$

iii) $A_0 = Fu = U$ and $A_1 = \overline{U} + \overline{W}$, $dim\,\overline{W} = 2$ with basis $\{\overline{w}_1, \overline{w}_2\}$, and $\overline{U} = F\overline{u}$, where the product is $\overline{w}_1\,\overline{w}_2 = u$, $\overline{u}\,\overline{w}_1 = \beta u$, $\overline{u}\,\overline{w}_2 = \alpha u$, $u\overline{w}_1 = \overline{u}$ and $u\overline{w}_2 = \overline{u}$

Proof:

We have checked in what cases the identities which define B.a.w.w. hold in A. In the other cases we have checked if the superidentity

$$(x_1x_2)x_3 + (-1)^{dx_2dx_3}(x_1x_3)x_2 + (-1)^{dx_1(dx_2+dx_3)}(x_2x_3)x_1 = 0$$

holds in A or we have seen if the superideal L in Proposition 1.5 is zero.

There are, up to isomorphism of Bernstein superalgebras, twenty-nine superalgebras in case [III] i) of the previous theorem that are not Jordan superalgebras. So, considering the Grassman envelopes, there exist at least twenty-nine nonisomorphic classes of B.a.w.w. of the same type which are not Jordan.

REFERENCES

[1] S Bernstein. Solution of a mathematical problem connected with the theory of heredity. Ann. Sci. Ukraine 1, pp. 84-114 (1923); Ann. Math. Statis. 13, pp. 53-61 (1942).

[2] S González, C López-Díaz, C Martínez. Bernstein superalgebras of low dimension. Comm. in Algebra (1998). To appear.

[3] S González, C López-Díaz, C Martínez, IP Shestakov. Bernstein superalgebras and their supermodules. Journal of algebra (1998). To appear.

[4] JC Gutiérrez. Clasificación de álgebras de Bernstein en función del tipo y de la dimensión. Tesis doctoral. Universidad de Oviedo (1994).

[5] P Holgate. Genetic algebras satisfying Bernstein's stationary principle. J. London Math. Soc. (2) 9, pp. 613-623 (1975).

[6] IP Shestakov, EI Zelmanov. Prime alternative superalgebras and nilpotency of the radical of a free alternative algebra. Izv. Acad. Nauk SSSR, Ser. mat., 54, n. 4 (1990). English translation: Math. USSR Izvestiya, 37, pp. 16-36 (1991).

[7] A Wörz-Busekros. Algebras in Genetics. Lecture Notes in Biomathematics 36, Springer, New York (1980).

[8] KA Zhevlakov, AM Slinko, IP Shestakov, AI Shirshov. Rings that are nearly associative. Acad. Press, New York (1982).

Banach–Bernstein algebras and their applications

YU. LYUBICH Department of Mathematics, Technion, Haifa 32000, Israel

In this paper we introduce a new structure in the topological algebra combining the Banach algebra structure (which belongs itself to the topological algebra) and the Bernstein algebra structure (which is purely algebraic itself).

By a standard definition, a Banach algebra is a real or complex algebra A provided with a norm $\|.\|$ such that the multiplication is continuous and the linear normed space $(A, \|.\|)$ is complete. No special algebraic properties such as commutativity, associativity or unitality are supposed. The continuity of the multiplication is equivalent to the inequality

$$\|xy\| \leq C\,\|x\|\,\|y\| \tag{1}$$

with a constant $C > 0$. Without loss of generality one can assume that $C = 1$, i.e.

$$\|xy\| \leq \|x\|\,\|y\| \tag{2}$$

since $C\,\|x\|$ is an equivalent norm satisfying (2).

Now let A be an algebra. Any nonzero algebra homomorphism from A into the basis field is called a weight of A. In the context of topological algebras one has to require for any weight to be continuous. Thus, one can define a *Banach baric algebra* as a baric algebra with continuous weight.

DEFINITION 1. *A Banach-Bernstein algebra is a commutative Banach algebra A with a continuous weight ω such that the identity*

$$(x^2)^2 = \omega^2(x)x^2 \tag{3}$$

holds.

Thus, this is a Banach baric algebra which is also a Bernstein algebra with respect to the same weight.

Let us give some examples.

EXAMPLE 1. *Let A be a finite dimensional real or complex Bernstein algebra. This is a Banach-Bernstein algebra with respect to any norm in A.*

EXAMPLE 2. *Let A be a Banach space. For any linear continuous functional $\omega \neq 0$ and any linear continuous operator P such that*

$$P^2 = P, \quad P^*\omega = \omega \tag{4}$$

the formula

$$xy = \frac{\omega(y)Px + \omega(x)Py}{2} \tag{5}$$

provides A with a structure of a Banach-Bernstein algebra.

In particular, for a given ω one can take a vector e such that $\omega(e) = 1$ and let $P = \omega \otimes e$, i.e. $Px = \omega(x)e$. Following (5) we get

$$xy = \omega(x)\omega(y)e. \tag{6}$$

This Banach-Bernstein algebra is called *constant* since $xy = e$ for all x and y such that $\omega(x) = \omega(y) = 1$.

Another particular case is the identity operator E as P in (5), so that

$$xy = \frac{\omega(y)x + \omega(x)y}{2}. \tag{7}$$

This is the so-called *unity* algebra with given ω.

From now on we consider a Banach-Bernstein algebra A. We denote by I the set of all nonzero idempotents. This is just the image of the *unit hyperplane* $H = \{x : \omega(x) = 1\}$ under the quadratic mapping $Qx = x^2$. The *barideal* $\operatorname{Ker}\omega$ we denote by B. For any $e \in I$ we have

$$A = \operatorname{Lin}\{e\} \oplus B \tag{8}$$

and then

$$B = U_e \oplus W_e \tag{9}$$

where $U_e = \operatorname{Im} L_e$, $W_e = \operatorname{Ker} L_e$, the linear operator $L_e : B \to B$ is a continuous projection defined as $L_e y = 2ey$ $(y \in B)$.

LEMMA 1. *The sets H, I, B, U_e, W_e are closed.*

Proof. Each of them is the zero set of a continuous mapping, for example, $U_e = \operatorname{Ker}(E - L_e)$ and so on. □

The direct decomposition (9) is *topological* in the sense that the corresponding projection L_e is continuous. The same is true for (8), since the projection $x \mapsto \omega(x)e$ is continuous.

LEMMA 2. *The set I is connected.*

Proof. $I = \operatorname{Im}(Q|H)$, H is connected, Q is continuous. □

In the finite-dimensional case the numbers $m = \dim U_e + 1$, $\delta = \dim W_e$ are independent of e because of Lemma 2 and continuity of L_e with respect to e. The pair (m, δ) is the *type* of the Bernstein algebra. The previous argument can also be applied to the infinite-dimensional Banach-Bernstein algebra with a relevant definition of dimension.

For any Banach space X its *topological dimension* dimt$X$ can be defined as the minimal cardinality of the subsets whose linear spans are dense. If X is finite-dimensional then, obviously, its topological dimension coincides with its algebraic dimension: dimt$X$ = dim$X$. In general, dimt$X$ $\leq$dim$X$ since the last one is the minimal cardinality of the linearly independent subsets.

If P is a continuous projection then its *rank* , by definition, is the topological dimension of its image. It is known that the rank is stable in the sense that there exists $\epsilon = \epsilon(P) > 0$ such that $\operatorname{rank} P' = \operatorname{rank} P$ for all continuous projections P' with $\|P' - P\| < \epsilon$.

LEMMA 3. *The topological dimensions of subspaces U_e and W_e are independent of e.*

Proof. The projection $L_e : B \to B$ is a continuous function of e, namely,

$$\|L_{e'} - L_e\| \leq 2 \left\|e' - e\right\| \tag{10}$$

by (2). Therefore $\operatorname{rank} L_{e'} = \operatorname{rank} L_e$, i.e. $\operatorname{dimt} U_{e'} = \operatorname{dimt} U_e$, if $\left\|e' - e\right\| < \frac{1}{2}\epsilon(L_e)$.We see that the funciton $\operatorname{dimt} U_e$ is locally constant on the connected set I. Hence, $\operatorname{dimt} U_e$ is globally constant.

Changing L_e for $(E - L_e)$ we conclude that $\operatorname{dimt} W_e = \operatorname{rank}(E - L_e)$ is also constant. □

Lemma 3 allows us to define the *type* of a Banach-Bernstein algebra as the pair (m, δ) where $m = \operatorname{dimt} U_e + 1$, $\delta = \operatorname{dimt} W_e$ for any $e \in I$. Let us stress again that m and δ are, in general, some infinite cardinal numbers.

Now we introduce a subclass of Banach-Bernstein algebras which is in a sense the closest to the finite-dimensional ones. Recall that a continuous mapping of a Banach space into itself is called *compact* if the images of all bounded sets are relatively compact.

DEFINITION 2. *A Banach algebra is called compact if the mapping $x \longmapsto x^2$ is compact.*

Obviously, every finite-dimensional algebra is compact. Moreover, if the square of a Banach algebra is finite-dimensional then the algebra is compact.

THEOREM 1. *A Banach-Bernstein algebra A is compact if and only if the subalgebra A^2 is finite-dimensional.*

Proof. We only need to prove the "only if" part of the statement, since the "if" part is obvious.

Since

$$L_e y = 2ey = (e + y)^2 - e - y^2, \tag{11}$$

the projection $L_e : B \to B$ is compact. Therefore the subspace $U_e = \operatorname{Im} L_e$ is finite-dimensional. Indeed, $L_e|U_e = E|U_e$ hence, the unit ball in U_e is compact.

Since the algebra A is Bernstein, we have

$$A^2 = \mathrm{Lin}\{e\} \oplus U_e \oplus U_e^2 \tag{12}$$

and we see that A^2 is finite-dimensional. □

As an application we consider the algebra generated by the quadratic integral operator

$$(Qf)(z) = \int_0^1 \int_0^1 K(s,t,z) f(s) f(t) ds dt \tag{13}$$

in the Banach space $C[0,1]$ of scalar (real or complex) continuous functions $f(z)$, $0 \le z \le 1$, provided with the standard norm

$$\|f\| = \max_{0 \le z \le 1} |f(z)| . \tag{14}$$

The kernel $K(s,t,z)$ is supposed to be continuous in s,t,z and symmetric in s,t : $K(t,s,z) = K(s,t,z)$.

LEMMA 4. *The operator Q in $C[0,1]$ is compact.*

Proof. Obviously, Q is continuous. Now suppose that f runs over a ball $\|f\| \le M$. Then Qf runs over a bounded set in $C[0,1]$,

$$\|Qf\| \le \mu M^2 \tag{15}$$

where $\mu = \max_{s,t,z} |K(s,t,z)|$. Moreover, this set is uniformly continuous since

$$|(Qf)(z_1) - (Qf)(z_2)| \le M^2 \max_{s,t} |K(s,t,z_1) - K(s,t,z_2)| \tag{16}$$

and $K(s,t,z)$ is uniformly continuous. By the classical Arzela-Ascoli Theorem the set $\{Qf : \|f\| \le M\}$ is relatively compact in $C[0,1]$. □

If we provide the Banach space $C[0,1]$ with the multiplication

$$(f \circ g)(z) = \int_0^1 \int_0^1 K(s,t,z) f(s) g(t) ds dt \tag{17}$$

we get a commutative Banach algebra A_K since

$$\|f \circ g\| \le C \|f\| \, \|g\| . \tag{18}$$

Obviously, $f \circ f = Qf$, so Q is the quadratic mapping in the algebra A_K. In these terms Lemma 4 says that the algebra A_K is compact.

Now we suppose that the kernel $K(s,t,z)$ is *Bernstein* in the following sense: first of all,

$$\int_0^1 K(s,t,z) dz = 1 \tag{19}$$

and, secondly,

$$Q^2 f = Qf \tag{20}$$

for all f such that

$$\int_0^1 f(x)dx = 1. \tag{21}$$

THEOREM 2. *For any Bernstein kernel* $K(s,t,z)$ *the subspace* $\mathrm{Lin}(\mathrm{Im}\, Q)$ *is finite-dimensional.*

Proof. The linear functional

$$\omega[f] = \int_0^1 f(x)dx \tag{22}$$

is continuous. This is a weight in the algebra A_K by (17) and (19). Thus, A_K can be considered as a Banach baric algebra. Moreover, A_K is Bernstein. Indeed, if $\omega[f] \neq 0$ then $\omega[\hat{f}] = 1$ for the normalized function $\hat{f} = f/\omega[f]$. Then (20) is true for $\hat{f}$ and then we get

$$Q^2 f = \omega^2[f] Qf. \tag{23}$$

since Q is homogenuous of degree 2.

The last identity can be extended to the functions f with $\omega[f] = 0$ since (23) is true for $f_\tau = f + \tau$, $\tau \neq 0$ and one can pass to the limit as $\tau \to 0$.

In algebraic terms (23) means that the algebra A_K is Bernstein. Thus, A_K is a Banach-Bernstein algebra. By Lemma 4 the algebra is compact. By Theorem 1 the subalgebra A_K^2 is finite-dimensional. It remains to note that $A_K^2 = \mathrm{Lin}(\mathrm{Im}\, Q)$. □

COROLLARY 1. *Any Bernstein kernel is of the form*

$$K(s,t,z) = e_0(z) + \sum_{j=1}^{r-1} K_j(s,t) e_j(z) \tag{24}$$

where K_j *are linearly independent continuous functions,* $K_j(t,s) = K_j(s,t)$, *and* e_j *are linearly independent continuous functions such that*

$$\int_0^1 e_0(z)dz = 1, \int_0^1 e_j(z)dz = 0 \quad (1 \leq j \leq r-1). \tag{25}$$

Here $r = \dim(\mathrm{Lin}(\mathrm{Im} Q))$.

Actually, in (24) e_0 is an idempotent in A_K, $\omega[e_0] = 1$ and $\{e_j\}_0^{r-1}$ is a basis in A_K^2 chosen under condition $\omega[e_j] = 0 \quad (1 \leq j \leq r-1)$. Formula (24) follows from the expansion

$$Qf = \sum_{j=0}^{r-1} C_j[f] e_j \tag{26}$$

where

$$C_j[f] = \int_0^1 (Qf)(z)e_j^*(z)dz \quad (0 \le j \le r-1), \tag{27}$$

$\{e_j^*\}_0^{r-1}$ is a system in duality to $\{e_j\}_0^{r-1}$, i.e.

$$\int_0^1 e_j(z)e_k^*(z)dz = \delta_{jk}, \tag{28}$$

and

$$K_j(s,t) = \int_0^1 K(s,t,z)e_j^*(z)dz. \tag{29}$$

Since one can take $e_0^*(z) \equiv 1$ from the very beginning, we have

$$K_0(s,t) = \int_0^1 K(s,t,z)dz = 1 \tag{30}$$

COROLLARY 2. $r = 1$ *if and only if* $K(s,t,z)$ *only depends on* z.

This case is especially interesting because of the following

THEOREM 3. *If a Bernstein kernel* $K(s,t,z)$ *is pointwise positive then* K *only depends on* z.

This theorem was formulated by S.N. Bernstein [1] with no proof. Now we can prove it on base of the above developed general theory combininig with geometrical arguments from [2] (see also [3], Secion 5.2) we gave for the similar finite-dimensional statement. We omit the proof of Theorem 3 here. Let us only remark that $r = 1$ means that the corresponding Banach-Bernstein algebra is constant.

REFERENCES.

1. S.N. Bernstein, Solution of a mathematical problem related to the theory of inheritance, Uchenye Zapiski Nauchno-issledovatel'skich Kafedr Ukrainy, 1 (1924), 83-115 (in Russian).

2. Yu.I. Lyubich, On a theorem of S.N. Bernstein, Siberian Math. J., 14 (1973), 474.

3. Yu.I. Lyubich, Mathematical Structures in Population Genetics, Springer-Verlag, 1992.

Jordan superalgebras

C. MARTÍNEZ*. Departamento de Matemáticas. Universidad de Oviedo. 33007 Oviedo, España.

0 INTRODUCTION

Jordan superalgebras were introduced by V. G. Kac (see [6]) and I. Kaplansky (see [11]). Finite dimensional simple Jordan superalgebras over algebraically closed fields of zero characteristic have been classified by V. G. Kac ([6], [3]) (see also I. Kantor [9]). In his classification theorem Kac used the corresponding result for Lie superalgebras and the relations between Jordan and Lie structures.

In [15] M. Racine and E. Zelmanov obtained a classification of simple Jordan superalgebras over fields of characteristic $p \neq 2$ whose even part is semisimple. The infinite dimensional case has been considered by V. Kac, C. Martinez and E. Zelmanov in [7] where Z-graded simple Jordan superalgebras over algebraically closed fields of zero characteristic with the dimensions of the homogeneous components uniformely bounded are classified.

Here we will pay special attention to finite dimensional modular Jordan superalgebras whose even part is not semisimple.

1 DEFINITIONS AND EXAMPLES

Let K be a field of characteristic not 2, $\Gamma =< 1, g_i | i = 1, 2, ... >$ the Grassmann (or exterior) algebra over K on a countable set of generators g_i, with $g_i^2 = 0$, $g_i g_j = -g_j g_i$ if $i \neq j$. Then $\Gamma = \Gamma_0 + \Gamma_1$ is a $Z/2Z$-graded algebra, that is, a superalgebra,

* Partially supported by DGICYT PB94-1311-C03-01 and FC-97-PBPGI97-03

where the even part Γ_0 is spanned by 1 and the products of even length in the generators and Γ_1 is spanned by all products in generators of odd length, that is:

$$\Gamma_0 = K(1, g_{i_1}g_{i_2}\cdots g_{i_{2r}}, r \geq 1, i_1 < i_2 < \cdots < i_{2r}),$$

$$\Gamma_1 = K(g_{i_1}g_{i_2}\cdots g_{i_{2r-1}}, r \geq 1, i_1 < i_2 < \cdots < i_{2r-1}).$$

If $\mathcal{V}$ is a homogeneous variety of algebras, that is, a class of K-algebras satisfying a certain set of homogeneous identities (see [19]), $A = A_0 + A_1$ is a $\mathcal{V}$-superalgebra if its Grassmann envelope $\Gamma(A) = A_0 \otimes \Gamma_0 + A_1 \otimes \Gamma_1$ belongs to $\mathcal{V}$. Notice that the Grassmann envelope is a subalgebra of the tensor product $A \otimes \Gamma$.

In this way one can define Lie superalgebras, Jordan superalgebras, etc. Let us notice that associative superalgebras are just $Z/2Z$-graded associative algebras. However a Jordan (resp. Lie) superalgebra is not a Jordan (resp. Lie) algebra. A commutative superalgebra is often called supercommutative algebra.

Let us look now at some important examples of superalgebras

EXAMPLES (see [15])

1) The set of $(n+m) \times (n+m)$ matrices over K, $\mathcal{M}_{n+m}(K)$ is an associative superalgebra with the diagonal components $\mathcal{M}_n(K)$ and $\mathcal{M}_m(K)$ as the even part and the off-diagonal components as the odd part.

2) If $A = A_0 + A_1$ is an associative superalgebra, A^+ with the product $a_\alpha \cdot b_\beta = 1/2(a_\alpha b_\beta + (-1)^{\alpha\beta} b_\beta a_\alpha)$ is a Jordan superalgebra, where $a_\alpha \in A_\alpha$, $b_\beta \in A_\beta$, $\alpha, \beta = 0, 1$.

If A is a simple associative not commutative superalgebra, then A^+ is a simple Jordan superalgebra.

If

$$A = \{\begin{pmatrix} a & b \\ b & a \end{pmatrix} | a, b \in \mathcal{M}_n(K)\},$$

then the superalgebra A^+ is denoted $Q_n(K)$ and is simple if $n > 1$.

3) If $V = V_0 \oplus V_1$ is a $Z/2Z$-graded K-vector space, a superform $(,)$ on V is a bilinear form on $V = V_0 \perp V_1$ such that its extension $\Gamma((,))$ to $\Gamma(V)$ is a symmetric bilinear form over Γ_0. Let $J = J(V, q) = J_0 + J_1$, $J_0 = Ke + V_0$ and $J_1 = V_1$ with the product given by $ex = x$, $vw = (v, w)e$ for $x \in J, v, w \in V$. J is called the Jordan superalgebra of the superform $(,)$ and it is simple iff $(,)$ is nondegenerate.

4) If $A = A_0 + A_1$ is an associative superalgebra, a superinvolution * of A is a graded linear map $^* : A \to A$ such that $a^{**} = a$, $(a_\alpha b_\beta)^* = (-1)^{\alpha\beta} b_\beta^* a_\alpha^*$.

If $J = \mathcal{H}(A, *)$ denotes the set of symmetric elements of A, J is a Jordan subsuperalgebra of A^+.

For $A = \mathcal{M}_{n+2m}(K)$ as in example 1 and * the superinvolution induced by the superform

$$\begin{pmatrix} I_n & 0 \\ 0 & S_m \end{pmatrix},$$

where

$$S = \begin{pmatrix} 0 & 1 \\ -1 & 0 \end{pmatrix},$$

the Jordan superalgebra $\mathcal{H}(A,*)$ is called "orthosymplectic superalgebra" and it is denoted $osp_{n,2m}(K)$. If $m \neq 0$ then $osp_{n,2m}(K)$ is simple.

5) For

$$A = \mathcal{M}_2(B) = \begin{pmatrix} a & b \\ c & d \end{pmatrix} | a,b,c,d \in B\},$$

B an associative algebra with involution * we can get a superinvolution of A (that we will denote also *) by

$$\begin{pmatrix} a & b \\ c & d \end{pmatrix}^* = \begin{pmatrix} d^* & -b^* \\ c^* & a^* \end{pmatrix}.$$

Then $\mathcal{H}(A,*)$ is a subsuperalgebra of A^+. If $B = \mathcal{M}_n(K)$ and $*$ is the transpose involution

$$\mathcal{H}(A,*) = \{\begin{pmatrix} a & b \\ c & d \end{pmatrix} \in \mathcal{M}_{2n}(K) | d = a^*,\ b^* = -b,\ c^* = c\}$$

will be denoted $P_n(K)$. If $n > 1$ then $P_n(K)$ is a simple Jordan superalgebra.

6) The Kaplansky superalgebra $K_3 = J_0 + J_1$, where $J_0 = Ke$ and $J_1 = Kx + Ky$ with $e^2 = e$, $ex = 1/2x$, $ey = 1/2y$ and $[x,y] = e$ is also a simple Jordan superalgebra.

7) For $t \in K$, let us denote $D_t = J_0 + J_1$ where $J_0 = Ke_1 + Ke_2$ and $J_1 = Kx + Ky$ with the product $e_i x = 1/2x$, $e_i y = 1/2y$ and $[x,y] = e_1 + te_2$. Then D_t is a Jordan superalgebra that is simple if and only if $t \neq 0$.

8) The 10-dimensional Kac superalgebra $K_{10} = J_0 + J_1$ has even part $J_0 = (Ke_1 + \sum_{i=1}^4 Kv_i) + Ke_2$ and odd part $J_1 = \sum_{i=1,2}(Kx_i + Ky_i)$ with:
$e_i^2 = e_i$, $e_1 v_i = v_i$, $v_1 v_2 = 2e_1 = v_3 v_4$, $e_i x_j = 1/2x_j$,
$e_i y_j = 1/2y_j$, $y_1 v_1 = x_2$, $y_2 v_1 = -x_1$, $x_1 v_2 = -y_2$, $x_2 v_2 = y_1$,
$x_2 v_3 = x_1$, $y_1 v_3 = y_2$, $x_1 v_4 = x_2$, $y_2 v_4 = y_1$, $[x_i, y_i] = e_1 - 3e_2$,
$[x_1, x_2] = v_1$, $[x_1, y_2] = v_3$, $[x_2, y_1] = v_4$, $[y_1, y_2] = v_2$ and the other products are zero or obtained by symmetry or skew-symmetry of one of the above products.
K_{10} is a Jordan superalgebra and it is simple if ch$K \neq 3$.

The following classification Theorem was proved in [15]
THEOREM 1.1. If $J = J_0 + J_1$ is a finite dimensional central simple Jordan superalgebra over an algebraically closed field K of characteristic not 2,3, J_0 semisimple and $J_1 \neq (0)$, then J is isomorphic to one of the following superalgebras:

1) K_3 (Kaplansky superalgebra) or K_{10} (Kac superalgebra),
2) $J(V,q)$ the superalgebra of a nondegenerate superform,
3) $\mathcal{H}(A,*) = \{a \in A | a^* = a\}$, $(A,*)$ simple as a superalgebra with superinvolution ($\mathcal{M}_{n,m}(K)$, osp(n,2m), $P_n(K)$, $Q_n(K)$),

4) $D_t, 0 \neq t \in K$.

2 Z-GRADED SIMPLE JORDAN SUPERALGEBRAS

A superconformal algebra is a Z-graded simple Lie superalgebra that contains the Virasoro algebra in the even part and such that dimensions of all the homogeneous components, $\dim L_i$, are uniformely bounded. (see [7], [8] and [13])

The Tits-Kantor-Koecher construction (see [10], [16]) establish connections between Lie algebras (resp. Lie superalgebras) and Jordan algebras (resp. Jordan superalgebras). So, in the process of classification of superconformal algebras, it seems natural to study Z-graded simple Jordan superalgebras with the dimensions of homogeneous components $\dim J_i$ uniformely bounded. This was done in [7]. Before formulating the main result here, let us have a look at some examples and constructions.

EXAMPLES

1) Let $\mathcal{G}$ denote now a Z/nZ-graded finite dimensional superalgebra, $\mathcal{G} = \mathcal{G}_0 + \mathcal{G}_1 + \cdots + \mathcal{G}_{n-1}$. The loop superalgebra corresponding to $\mathcal{G}$ is $\mathcal{L}(\mathcal{G}) = \sum_{i \in Z} \mathcal{G}_{\bar{i}} \otimes t^i$. If $\mathcal{G}$ is a simple Jordan (Lie) superalgebra then $\mathcal{L}(\mathcal{G})$ is a graded simple Jordan (Lie) superalgebra and dimensions of the graded components, $dim\mathcal{L}(\mathcal{G})_i$, $i \in Z$, are uniformely bounded.

2) Let J be a graded simple Jordan superalgebra and $K(J) = J^- + [J^-, J^+] + J^+$ the Lie superalgebra associated to J via the Tits-Kantor-Koecher construction (see [10], [16]). J is said to be of Cartan type if J contains a graded subsuperalgebra B of finite codimension such that the corresponding subspace in the K-K-T construction $B^- + [B^-, J^+] + [J^-, B^+] + B^+$ is a subsuperalgebra of $K(J)$ of finite codimension. Thus $K(J)$ is a Lie superalgebra of Cartan type (see [7]).

3) Let $V = V_{\bar{0}} + V_{\bar{1}}$ be a $Z/2Z$-graded vector space in which both $V_{\bar{0}} = \oplus_{i \in Z} V_{\bar{0}i}$ and $V_{\bar{1}} = \oplus_{i \in Z} V_{\bar{1}i}$ are represented as direct sums of finite dimensional vectors spaces such that the dimensions of the subspaces $V_i = V_{\bar{0}i} + V_{\bar{1}i}$, $i \in Z$ are uniformly bounded. If $<,>$ denotes a nondegenerate supersymmetric superform on V, the Jordan superalgebra of the superform $J = F1 + V$ is a graded simple Jordan of the desired type.

4) Let $A = A_{\bar{0}} + A_{\bar{1}}$ be an associative commutative superalgebra. If $a \in A_{\bar{i}}$ then we denote $|a| = i$. By a bracket on A we mean a bilinear mapping $\{,\} : A \times A \to A$. Let us consider a direct sum of vector spaces $J = J(A, \{,\}) = A + Ax$ and define a multiplication on J in the following way. For arbitrary elements $a, b \in A$ their product in J is the product ab in A, $a(bx) = (ab)x$, $(bx)a = (-1)^{|a|}(ba)x$, $(ax)(bx) = (-1)^{|b|}\{a, b\}$.

The $Z/2Z$-gradation on A can be extended in a natural way to a $Z/2Z$-gradation on $J = J_0 + J_1$ by $J_{\bar{0}} = A_{\bar{0}} + A_{\bar{1}}x$, $J_{\bar{1}} = A_{\bar{1}} + A_{\bar{0}}x$. Then J is a superalgebra (see [10] and [11]) that is called Kantor Double of $(A, \{,\})$.

A bracket $\{,\}$ on A is called Jordan bracket if the Kantor double $J(A, \{,\})$ is a Jordan superalgebra (see [12]).

A bracket $\{,\} : A \times A \to A$ is called Poisson bracket if $(A, \{,\})$ is a Lie superalgebra and $\{ab, c\} = a\{b, c\} + (-1)^{|b||c|}\{a, c\}b$ for every $a, b, c \in A$.

An arbitrary Poisson bracket is a Jordan bracket. From the Jordan identity it follows that if $\{,\} : A \times A \to A$ is a Jordan bracket then:

(i) $D : A \to A, \quad a \to \{a, 1\}$ is a derivation of A,

(ii) $\{a, bc\} = \{a, b\}c + (-1)^{|a||b|}b\{a, c\} - D(a)(bc)$.

If the superalgebra A is generated by elements $\{a_i\}_i$ then a Jordan bracket $\{,\}$ is determined by the derivation D and by values $\{a_i, a_j\}$.

For example, if $A = K[t^{-1}, t, \xi_1, \ldots, \xi_n]$, where t is a Laurent variable and $\xi_1, \ldots, \xi_n$ are Grassmann variables, we have two brackets defined by:

a) $D = \frac{\partial}{\partial t}t$, $[\xi_i, \xi_j] = -\delta_{ij}$, $[t, \xi_i] = 0$. Neveu-Schwarz bracket,

b) $D = \frac{\partial}{\partial t}$, $[\xi_i, \xi_j] = -\delta_{ij}$, $[t, \xi_i] = 0$. Ramond bracket

Let $n \geq 1$, $V = V_0 + \ldots + V_{n-1}$ a finite dimensional Z/nZ-graded vector space over F. The gradation on V can be uniquely extended to a Z/nZ-gradation on the Grassmann algebra, $G(V) = \sum_{i=0}^{n-1} G(V)_i$.

Let $A = \sum_{i \in Z} G(V)_{\bar{i}} \otimes t^i = \mathcal{L}(G(V))$. If $\{,\} : A \times A \to A$ is a Jordan bracket, the element x in the Kantor Double construction is given degree i such that $2i \in nZ$ and $[A_j, A_k] \subseteq A_{j+k+2i}$ for arbitrary $j, k \in Z$, then the Kantor Double $J = A + Ax$ is a Z-graded Jordan superalgebra having all dimensions $dim(J_i)$, $i \in Z$, uniformly bounded.

It has been proved in [7] that the above given examples "nearly" cover the collection of graded simple Jordan superalgebras having dimensions of graded component uniformly bounded. In a concrete way, the following theorem was proved:

THEOREM 2.1. Let $J = \sum_{i \in Z} J_i$ be an infinite dimensional Z-graded simple Jordan superalgebra with 1 over an algebraically closed field K of characteristic zero such that dimensions of J_i are uniformly bounded. Then J is isomorphic to one of the following superalgebras:

1) a loop superalgebra $\mathcal{L}(\mathcal{G})$, where $\mathcal{G} = \mathcal{G}_0 + \mathcal{G}_1 + \cdots + \mathcal{G}_{n-1}$ is a finite dimensional simple Z/nZ- graded Jordan superalgebra,
2) a Jordan superalgebra $F1 + V$ of a nondegenerate supersymmetric superform in a Z-graded vector space $V = V_{\bar{0}} + V_{\bar{1}}$,
3) a Kantor Double $J = A + Ax$ of an associative commutative superalgebra $A = \sum_{i \in Z} G(V)_{\bar{i}} \otimes t^i$ with a Jordan bracket, where $V = V_0 + \cdots + V_{n-1}$ is a Z/nZ- graded finite dimensional vector space. If n is odd then there is one (up to isomorphism) Jordan superalgebra of this type with x being of degree 0. If n is even then there are two Jordan brackets on A, of Ramond and Neveu-Schwarz types respectfully (see above) leading to two nonisomorphic Jordan superalgebras, one with x having degree 0 and one with x having degree $\frac{-n}{2}$,
4) a Jordan superalgebra of Cartan type,

5) two exceptional Jordan superalgebra whose Tits-Kantor Koecher construction are isomorphic to the exceptional superalgebras $CK(6)$ of Neveu-Schwarz and Ramond types (see [2]).

3 MODULAR SUPERALGEBRAS

In what follows we will consider $J = J_0 + J_1$ a simple Jordan superalgebra over an algebraically closed field of characteristic $p > 2$ whose even part is not semisimple, that is, the remaining case in [15]. These superalgebras were classified in [14]. Arguments in [14] were inspired by those used in the previous paper [7], but some notable differences appear between the papers.

It is known that for every element $x \in J_1$ the operator $R(x)^2 : J \to J,\ y \to (yx)x$ is a derivation.

Let $\mathcal{D}$ be the linear span of $\{R(x)^2 | x \in J_1\}$, I be a maximal $\mathcal{D}$-invariant nilpotent ideal of J_1 and $\bar{A} = J_0/I$.

Then $\bar{A}$ is $\mathcal{D}$- semisimple, that is, $\bar{A}$ has no nonzero $\mathcal{D}$-invariant nilpotent ideals. It can be proved that in this case $\bar{A}$ is a direct sum of $\mathcal{D}$-simple ideals: $\bar{A} = J_0/I = \bar{A^{(1)}} \oplus \cdots \oplus \bar{A^{(r)}}$, where each algebra $\bar{A^{(i)}}$ is $\mathcal{D}$-simple.

The following theorem of Block can be applied:

THEOREM 3.1. (see [1]). If A is a differentially simple ring (or K-algebra) with a minimal ideal, then either A is simple or there is a simple ring (or K-algebra) S of prime characteristic and a positive integer n such that $A = S \otimes B_{n,p}(K)$, where $B_{n,p}(K)$ denotes the truncated polynomial algebra (commutative, associative with unit) in n variables, that is, $B(K) = K[X_1, \ldots, X_n]/(X_1^p, \ldots, X_n^p)$.

Comming back to our situation we know that the even part J_0 of the simple Jordan superalgebra J has an ideal I such that the quotient algebra $\bar{A} = J_0/I$ is a direct sum of s Jordan algebras whose structure is given by the previous theorem of Block.

It can be proved that the number s of direct summands in the expression of $\bar{A}$ is at most 2 and it is equal to 1 if the ideal I is not zero.

If we denote by N the nilpotent radical of J_0 (under our assumption $N \neq (0)$) we have the following possibilities for J_0:

1. If the ideal I is zero then J_0 is one of the following algebras:

a) $J_0 = K[a_1, \ldots, a_n]$ the algebra of truncated polynomials in n variables (so it is associative and commutative),

b) $J_0 = (K1 + V) \otimes K[a_1, \ldots, a_n]$, that is, J_0 is the tensor product of a simple Jordan algebra of a nondegenerate bilinear form over a finite dimensional vector space V and the algebra of truncated polynomials in n variables,

c) $J_0 = K[a_1, \ldots, a_n] \oplus K[b_1, \ldots, b_m]$, the direct sum of two algebras of truncated polynomials in n and m variables respectively, where $n \leq m$, $n \geq 0$ and

$m > 0$. (That is, we exclude the case $n = 0 = m$ because $J_0 = K \oplus K$ is semisimple, but the case $n = 0 < m$, $J_0 = Ke \oplus K[b_1, \ldots, b_m]$ is not excluded.

2. If $I \neq (0)$ then the following possibilities appear:

a) $\bar{A} = J_0/I = F1 + V$ a simple Jordan algebra of a bilinear form,

b) $\bar{A} = J_0/I = B \otimes F[a_1, \ldots, a_n]$ is the tensor product of a Jordan algebra of a bilinear form $B = F1 + V$ and $F[a_1, \ldots, a_n]$, the algebra of truncated polynomials in n variables.

It can be proved that in case 1 a) J_0 acts in an associative way on J_1, that is, $(ya)b = y(ab)$ for every $a, b \in J_0$, $y \in J_1$. This implies that J is a Kantor Double of $A = J_0$ with respect to some bracket $\{,\}$.

In case 1 b) the only possibility is obtained for $dimV = 3$. A Jordan superalgebra J of Cheng-Kac type is obtained for any number of truncated variables and a derivation d of $C = F[a_1, \ldots, a_n]$. The Jordan superalgebra obtained in this way is simple if and only if C does not contain proper d-invariant ideals.

In case 1 c) the simplicity of J gives $n = m$. So, in particular, the case $J_0 = Ke \oplus K[b_1, \ldots, b_m]$ is shown to be impossible. Again the Jordan superalgebra J is a Kantor double of an associative and commutative superalgebra with respect to some bracket.

In both cases 2 the corresponding Jordan superalgebra J is a Kantor double.

So we can formulate the main result in [14] as

THEOREM 3.2. Let $J = J_0 + J_1$ be a finite dimensional simple Jordan superalgebra over an algebraically closed field K of characteristic $p > 2$ whose even part J_0 is not semisimple. Then there exist two integers m, n and a Jordan bracket on the algebra $S = K[a_1, \ldots, a_m, \xi_1, \ldots, \xi_n]$ such that $J = S + Sx$ is a Kantor Double or J is an exceptional Jordan superalgebra that corresponds to the Cheng-Kac superalgebra.

If we accept the general principle that the study of finite dimensional algebras (or superalgebras) in prime characteristic helps to understand infinite dimensional algebras (resp. superalgebras) in characteristic zero, it is natural to formulate the following

CONJECTURE

Let $J = \sum_{i \in Z} J_i$ be a Z-graded simple Jordan superalgebra over an algebraically closed field K of characteristic zero. Let us assume that J is of finite growth, that is, $\dim J_i < |i|^c + d$, where c, d are constants. Then either J can be described by Theorem 2.1 or it is a Kantor double of $K[t_1, \ldots, t_m, \xi_1, \ldots, \xi_n]$, where $t_1, \ldots, t_m$ are even Laurent variables and $\xi_1, \ldots, \xi_n$ are odd Grassmann variables.

REFERENCES

[1] R. Block, Determination of the differentially simple rings with a minimal ideal, Ann. of Math. 90, (1969), 433-459.

[2] S. Cheng and V. Kac, A New N = 6 Superconformal Algebra, Preprint.

[3] L. Hogben and V. Kac, The correct multiplication table for the exceptional Jordan superalgebra F, Comm. in Algebra 11, (1983), 1155-1156.

[4] N. Jacobson, Structure and Representation of Jordan Algebras, A.M.S. Colloq.Publ. vol 39, Providence (1969).

[5] V.G. Kac, Lie superalgebras, Adv. Math. 28,(1977), 8-96.

[6] V.G. Kac, Classification of simple Z-graded Lie superalgebras and Jordan superalgebras, Comm. in Algebra 5, (1977), 1375-1400.

[7] V.G. Kac, C. Martinez and E. Zelmanov,Graded Simple Jordan superalgebras of growth one, Submitted.

[8] V.G. Kac and J.W. van de Leur, On classification of superconformal algebras, Strings 88, World Sci., (1988), 77-106.

[9] I.L. Kantor, Connection between Poisson brackets and Jordan and Lie superalgebras, in Lie Theory, Differential Equations and Representation Theory, Publications CRM, Montreal, (1990), 213-225.

[10] I.L. Kantor, Jordan and Lie superalgebras determined by a Poisson algebra, Amer. Math. Soc. Transl. (2), 151, (1992), 55-79.

[11] I. Kaplansky, Graded Jordan algebras I. Preprint.

[12] D. King and K. McCrimmon, The Kantor Construction of Jordan Superalgebras, Comm. in Algebra 20 (1), (1992), 109-126.

[13] C. Martinez, Graded simple Jordan Algebras and Superalgebras, To appear in Contemporary Mathematics.

[14] C. Martinez and E. Zelmanov, Modular simple Jordan superalgebras. Preprint.

[15] M. Racine and E. Zelmanov, Finite dimensional simple Jordan superalgebras with semisimple even part. Preprint.

[16] J. Tits, Une classe d'algèbres de Lie en relation avec les algèbres de Jordan, Indag. Math. 24 (1962), 530-535.

[17] E. Zelmanov, On prime Jordan Algebras II, Siberian Math. J. 24, (1983), 73-85.

[18] E. Zelmanov, Lie algebras with a finite grading, Math. USSR Sbornik 52, no 2 (1985), 347-385.

[19] K.A. Zhevlakov, A.M. Slin'ko, I.P. Shestakov, A.I. Shirshov, Rings that are nearly associative, Academic Press, New York (1982).

Derivations of structurable H*-algebras

J. MARTÍNEZ MORENO Departamento de Análisis Matemático, Facultad de Ciencias, Universidad de Granada, 18071 Granada, Spain. E-mail: jmmoreno@goliat.ugr.es

1 INTRODUCTION

Structurable algebras were introduced by Allison in [1], where the foundations for a fine structure theory of such algebras, later completed by Schafer [18] and Smirnov [19], were laid. In fact, the results by Allison-Schafer-Smirnov give a complete structure theory for semisimple finite-dimensional structurable algebras. Structurable algebras were defined as unital nonassociative algebras with linear involution τ satisfying a concrete identity of degree 4 involing the given involution. One of the main attractions of structurable algebras is the fact that, in the finite-dimensional case, they give by means of a modified Kantor-Koecher-Tits construction, all the isotropic simple Lie algebras [2]. Structurable algebras are close to the Jordan algebras. In fact, structurable algebras contain all Jordan algebras with the identity operator as involution, alternative algebras with any involution, algebras constructed from an [involutive associative algebra]-valued hermitian form in such a way that generalize the quadratic Jordan algebras of a bilinear form, and some other more sophisticate algebras which we will see later.

A long time before the appearance of structurable algebras, Steen introduced in [20] particular types of complex Banach algebras called associative H*-algebras and Ambrose [3], under the assumption of zero annihilator obtained for these algebras a complete structure theory. Ambrose's theory has been extended to the more familiar classes of nonassociative algebras, such as

[1]Talk in "IV International Conference on Non Associative Algebra and its Applications", Sao Paulo, 19 to 24 of July, 1998

Lie, Jordan, alternative, noncommutative Jordan, Malcev and structurable H*-algebras. Paying attention to the class of complex structurable H*-algebras, such structure theory can be regarded as an infinite-dimensional extension by the above cited one by Allison Schafer and Smirnov. (See [5] [7])

The determination of the Lie algebra, $Der\,(A,\tau)$, of all derivations which commute with the linear involution on a structurable H*-algebra (A,τ), with zero annihilator, is the final goal of this talk. From the structure theory [7] the problem reduces to the case where A is topologically simple and τ is isometric. Again the structure theory gives that such algebras are associative or Jordan or finite-dimensionals or arising from involutive Hilbert modules. In first, second and third cases $Der\,(A,\tau)$ is essentially known, so the problem reduces to the last case.

In the study of derivations of structurable H*-algebras constructed over involutive Hilbert modules over associative H*-algebras, appear in a natural way "differential operators" as they appear in the study of derivations of prime associative algebras with non zero socle. As is well known, if B is such an algebra over a field K, there exists a division algebra Λ over K, left and right (respectively) vector spaces X and Y over Λ, and a non degenerated Λ-bilinear mapping $<.,.>: X \times Y \to \Lambda$ such that B is a subalgebra of $L_Y(X)$ containing $FL_Y(X)$, where $L_Y(X)$ denotes the algebra of all operators on X which have adjoint with respect to $<.,.>$ and $FL_Y(X)$ is the set of those elements in $L_Y(X)$ which have finite rank over Λ. Then every derivation D on B is of the form $D(b) = bF - Fb$ where F is a suitable "continuous differential operator" on X [11, Theorem IV.14.3]. That F is a differential operator on X means that F is a K-linear mapping from X to X satisfying

$$F(\alpha x) = \widehat{F}(\alpha)\, x + \alpha F(x)$$

for all α in Λ and x in X and a suitable single-valued mapping $\widehat{F} : \Lambda \to \Lambda$, while the continuity of F refers to the natural topology on X given by the duality $(X, Y, <.,.>)$ and, in fact, it can be characterized "algebraically " by the existence of a mapping $F' : Y \to Y$ satisfying

$$< F(x), y > = < x, F'(y) > + \widehat{F}(< x, y >)$$

for all x in X and y in Y [11, Proposition IV.14.1].

The concept of differential operator can be extended verbatim to the context of modules over associative algebras. We characterize the continuity of a differential operator on a Hilbert module in a way very similar to the above mentioned "algebraic" characterization of the continuity of differential operators on a left vector space over a division algebra paired with a right vector space. We also prove that the set, $BDiff\,(W)$, of all continuous differential operators on a Hilbert module W is weak-operator closed and it is generated in the weak-operator topology by "elementary" bounded differential operators. Finally if (A,τ) is a structurable H*-algebra constructed from an involutive Hilbert module $(W,\square)$, then the set $Der\,(A,\tau)$ is in a one-to-one correspondence with the set of those elements in $BDiff\,(W)$ which are skew with respect to a natural involution on the Banach algebra $\mathcal{BL}(W)$.

2 STRUCTURE THEORY

Throughout this talk the word algebra will mean nonassociative complex algebra. A involution on an algebra A will be a linear mapping τ from A to A such that $\tau^2 = Id$ and $\tau(xy) = \tau(y)\tau(x)$ for all x, y in A. We denote by $H := \{x \in A : \tau(x) = x\}$ the set of τ-hermitian elements and by $S := \{x \in A : \tau(x) = -x\}$ the set of τ-skew-hermitian elements. H and S are linear subspaces of A with $A = H \oplus S$.

DEFINITION 2.1 *A structurable algebra is an algebra with linear involution* (A, τ) *satisfying*

1. $(s, x, y) = -(x, s, y) = (x, y, s)$
2. $(a, b, c) - (c, a, b) = (b, a, c) - (c, b, a)$
3. $\frac{2}{3}[[a^2, a], b] = (b, a^2, a) - (b, a, a^2)$

for all x, y *in* A, s *in* S, *and* a, b, c *in* H.

Here $(.,.,.)$ (resp. $[.,.]$) denotes the associator (resp. the commutator) on A.

The easiest examples of structurable algebras are alternative algebras with an involution and Jordan algebras with identity operator as involution.

DEFINITION 2.2 *An H^*-algebra is an algebra $\mathcal{E}$ with a conjugate-linear algebra involution * and a complete inner product* $(. \mid .)$ *satisfying* $(ef \mid g) = (e \mid gf^*) = (f \mid e^*g)$ *for all* e, f, g *in* $\mathcal{E}$.

The fundamental papers on general nonasssociative H*-algebras are [8] and [9]. We recall that the product of any H*-algebra $\mathcal{E}$, is continuous, so (by multiplying the inner product by a suitable positive number if necessary) $\mathcal{E}$ is a complete normed algebra in the usual sense (fact that will be assumed in what follows), and that if $\mathcal{E}$ has zero annihilator ($x\mathcal{E} = \mathcal{E}x = 0 \Rightarrow x = 0$), then the topology of the Hilbert norm is the only complete normable topology on $\mathcal{E}$ which makes continuous the product of $\mathcal{E}$ [13, Remark 2.8 (i)]. As a consequence, isomorphisms and antiisomorphisms between H*-algebras with zero annihilator are automatically continuous, in particular involutions on an H*-algebra with zero annihilator are continuous.

A first structure theorem for H*-algebras with a linear involution ($\mathcal{E}$,τ) (in particular for structurable H*-algebras) reduces the study of such algebras to the case which $\mathcal{E}$ is topologically τ-simple (with nonzero product and without nonzero proper τ-invariant closed ideals) and τ is a *-involution ($\tau(x^*) = \tau(x)^*$) [7, Section 1]. Concretely we have the following theorems

THEOREM 2.3 (Cabrera-Martínez-Rodríguez) *Let* $(\mathcal{E}, \tau)$ *be an H^*-algebra with involution, and assume that $\mathcal{E}$ has zero annihilator. Then* $(\mathcal{E}, \tau)$ *is the l_2-sum of a suitable family of H^*-algebras with involution* $\{(\mathcal{E}_\lambda, \tau_\lambda)\}_{\lambda \in \Lambda}$ *such that* $(\mathcal{E}_\lambda, \tau_\lambda)$ *is topologically τ_λ-simple for all* $\lambda \in \Lambda$.

THEOREM 2.4 (Cabrera-Martínez-Rodríguez) *Let $(\mathcal{E},\tau)$ be an H^*-algebra with involution, and assume that $\mathcal{E}$ has zero annihilator. Then there are a *-involution $\widehat{\tau}$ on $\mathcal{E}$ and an automorphism ϕ of $\mathcal{E}$ such that $\tau = \phi\widehat{\tau}\phi^{-1}$ [Clearly $(\mathcal{E},\tau)$ is structurable if and only if $(\mathcal{E},\widehat{\tau})$ is also].*

Every isometric linear involution on an H*-algebra with zero annihilator is a *-involution, and the converse is true if the H*-algebra is topologically simple (with nonzero product and without nonzero proper closed ideals) [5, 1.2.2].

If $\mathcal{E}$ is any topologically simple H*-algebra and λ is a positive number, then $\mathcal{E} \times \mathcal{E}^o$ (where $\mathcal{E}^o$ denotes the opposite algebra of $\mathcal{E}$) with inner product $((x,y) \mid (z,t)) = (x \mid z) + \lambda(y \mid t)$, H*-algebra involution $(x,y)^* = (x^*,y^*)$, and linear involution $\tau(x,y) = (y,x)$, is a topologically τ-simple H*-algebra with *-involution τ (denoted $B(\lambda)$). If one forgets the involution τ, clearly $B(\lambda)$ is not topologically simple. There are no other examples of nontopologically simple topologically τ-simple H*-algebras with *-involution. Moreover $B(\lambda)$ is structurable if and only if $\mathcal{E}$ is alternative [7, Theorem 1.5]. Topologically simple alternative H*-algebras are well known, they are the complex octonions $\mathbb{O}$ or the H*-algebra of all Hilbert-Schmidt operators on a suitable complex Hilbert space H.

If $\mathbb{D}$ denotes one of the four complex composition algebras ($\mathbb{C}$, $\mathbb{C}^2$, $M_2(\mathbb{C})$ and $\mathbb{O}$) endowed with their natural structures of H*-algebras and isometric involution [7, Theorem 4.6], then the tensor algebraic product $\mathbb{D}\otimes\mathbb{O}$ with linear involution τ as the tensor product of the Cayley involutions, can be seen as a topologically τ-simple structurable H*-algebra under the inner product and involution given by:

$$
\begin{aligned}
(x_1 \otimes y_1 \mid x_2 \otimes y_2) &= (x_1 \mid x_2)(y_1 \mid y_2) \\
(x \otimes y)^* &= x^* \otimes y^*.
\end{aligned}
$$

Moreover τ is clearly isometric. Another important examples of topologically simple finite-dimensional structurable H*-algebras are the structurable H*-algebras asociated to Springer H*-algebras $\mathbb{C}$, $H_3(\mathbb{D})$ (where $H_3(\mathbb{D})$ is the algebra of all hermitian 3 by 3 matrix over $\mathbb{D}$) and the Smirnov H*-algebra $\mathbb{T}(\mathbb{O})$ ([7] and [5]). Finally a relevant collection of structurable H*-algebras can be constructed from particular types of Hilbert modules called involutive Hilbert modules and defined as below.

In 1968 Saworotnow [14] introduced the concept of Hilbert module as an extension of the concept of complex Hilbert space, where the basic field $\mathbb{C}$ is replaced by a complex associative H*-algebra $\mathcal{E}$ with zero annihilator. The set $TC(\mathcal{E}) = \{ef : e,f \in \mathcal{E}\}$ is a *-invariant ideal of $\mathcal{E}$ (called the trace class of $\mathcal{E}$) which is a Banach *-algebra under suitable norm and there exists a canonical commutative linear form on $TC(\mathcal{E})$ denoted by tr (called trace), related to the inner product of $\mathcal{E}$ by $tr(ef) = (e \mid f^*)$ (see [15],[16]).

A **Hilbert module** over $\mathcal{E}$ means a faithful left module W over $\mathcal{E}$ together with a mapping $[.\mid.] : W \times W \to TC(\mathcal{E})$ satisfying, for $w_1, w_1', w_2 \in W, e \in \mathcal{E}$ and $\lambda \in \mathbb{C}$ the following properties:

1. $[\lambda w_1 \mid w_2] = \lambda[w_1 \mid w_2]$
2. $[w_1 + w_1' \mid w_2] = [w_1 \mid w_2] + [w_1' \mid w_2]$

3. $[e \circ w_1 \mid w_2] = e[w_1 \mid w_2]$
4. $[w_1 \mid w_2]^* = [w_2 \mid w_1]$
5. For each nonzero w in W there is a nonzero f in $\mathcal{E}$ such that

$$[w \mid w] = f^* f$$

6. W is a Hilbert space under its "$\mathbb{C}$-valued inner product" given by

$$(w_1 \mid w_2) = tr([w_1 \mid w_2]).$$

Let W be a Hilbert module over an associative H*-algebra $\mathcal{E}$ with zero annihilator. Assume that τ is an isometric linear involution on $\mathcal{E}$ and that $\Box$ is a conjugate-linear vector space involution on W satisfying:

1. $\tau([w_1 \mid w_2]) = [w_2^{\Box} \mid w_1^{\Box}]$
2. $(e \circ w)^{\Box} = \tau(e^*) \circ w^{\Box}$

for all w_1, w_2, w in W and e in $\mathcal{E}$. Then we say that $(W, \Box)$ is an **involutive Hilbert module** over $(\mathcal{E}, \tau)$.

Involutive Hilbert modules over topologically τ-simple associative H*-algebras with isometric linear involution τ well known [7, Theorem 3.10].

If $(W, \Box)$ is an involutive Hilbert module over $(\mathcal{E}, \tau)$, then the Hilbert space $A = \mathcal{E} \oplus^{l_2} W$, with product

$$(e_1 + w_1)(e_2 + w_2) = (e_1 e_2 + [w_2 \mid w_1^{\Box}]) + (\tau(e_1) \circ w_2 + e_2 \circ w_1)$$

H*-algebra involution

$$(e + w)^* = e^* + w^{\Box},$$

and linear involution

$$\tau(e + w) = \tau(e) + w,$$

is a structurable H*-algebra with zero annihilator and isometric linear involution. Moreover, A is topologically simple if and only if $(\mathcal{E}, \tau)$ is topologically τ-simple and either $\mathcal{E} \neq \mathbb{C}$ or $W \neq \mathbb{C}$ [7].

Now we can state the following structure theorem for topologically τ-simple structurable H*-algebras with τ *-involution.

THEOREM 2.5 (Cabrera-Martínez-Rodríguez) *The only topologically τ-simple structurable H^*-algebras with τ *-involution are the following:*

1. *H^*-algebras of the form $B(\lambda)$*
2. *The H^*-algebras, $HS(H)$, of all Hilbert-Schmidt operators on a suitable complex Hilbert space H endowed with an isometric linear involution.*

3. *The topologically simple Jordan H^*-algebras $H_3(\mathbb{O})$ and $H(\mathcal{E},\tau) = \{x \in \mathcal{E} : \tau(x) = x\}$, where $(\mathcal{E},\tau)$ is a topologically τ-simple associative H^*-algebra with isometric linear involution τ.*

4. *The H^*-algebra $\mathbb{D} \otimes \mathbb{O}$.*

5. *The 35-dimensional simple Smirnov H^*-algebra $\mathbb{T}(\mathbb{O})$.*

6. *The structurable H^*-algebras asociated to Springer H^*-algebras $\mathbb{C}$ and $H_3(\mathbb{D})$.*

7. *The structurable H^*-algebras constructed from an involutive Hilbert module $(W,\square)$, over a topologically τ-simple associative H^*-algebra $(\mathcal{E},\tau)$ with τ isometric and either $\mathcal{E} \neq \mathbb{C}$ or $W \neq \mathbb{C}$.*

See [7] and [5].

REMARK 2.6 *If H is a complex Hilbert space with a conjugation $\square$, then $(H,\square)$ is an involutive Hilbert module over $\mathbb{C}$, and its associated structurable H^*-algebra agrees with the quadratic Jordan H^*-algebra associated to the bilinear form given by $< x, y >:= (x \mid y^\square)$. This algebra is simple unless the dimension of H equals one.*

3 DERIVATIONS

From the structure theory seen in the above section, we have that the problem of the determination of the Lie algebra, $Der(A,\tau)$, of all derivations which commute with the linear involution, on a structurable H*-algebra (A,τ) with zero annihilator, reduces to the case where A is topologically simple and τ is isometric. Such algebras are associative or Jordan or finite dimensional or associated to involutive Hilbert modules. In the three first cases $Der(A,\tau)$ are essentially known. In fact:

It is well known that derivations of the algebra $HS(H)$ are the operators of the form $[.,T] : F \mapsto TF - FT$, where T is a bounded linear operator on the Hilbert space H. On the other hand isometric linear involutions on $HS(H)$ are the mappings of the form $\tau_J : F \mapsto JF^\bullet J^{-1}$ where $F^\bullet$ denotes the operator adjoint of F and J is a suitable conjugation or anticonjugation on H [6]. Hence

$$Der(HS(H),\tau_J) = \{[.,T] : \tau_J(T) = -T, T \in \mathcal{BL}(H)\}$$

where we have used that $T + \tau_J(T)$ is in the center of $HS(H)$ which is equals zero if H is infinite-dimensional or $\mathbb{C}1$ if H is finite-dimensional [8, Theorem 1.2].

Given an algebra A, we will denote by $Der(A)$ the Lie algebra of all derivations on A.

If the algebra is an infinite-dimensional Jordan algebra of the form $H(\mathcal{E},\tau)$, then by [10, Corollary on page 180] $Der(H(\mathcal{E},\tau))$ is the set of all restrictions to $H(\mathcal{E},\tau)$ of the elements in $Der(\mathcal{E},\tau)$.

The derivations of the finite-dimensional exceptional central simple Jordan algebra $H_3(\mathbb{O})$ are inner and $Der(H_3(\mathbb{O}))$ is a central simple Lie algebra of 52 dimensions [12, Theorem IX.11.17].

If $\mathbb{D}$ is $\mathbb{C}$ or $\mathbb{C}^2$, then $Der(\mathbb{D},\tau) = Der(\mathbb{D}) = 0$ and if $\mathbb{D} = \mathbb{O}$ by Cartan-Jacobson [17, Theorem 3.28 and Corollary 3.29] $Der(\mathbb{O})$ is a 14 dimensional central simple Lie algebra of type G, and every derivation is inner. Therefore, since in every simple structurable algebra (A,τ) over a field of characteristic zero $Der(A,\tau) = Inder(A,\tau)$, where $Inder(A,\tau)$ is the linear span of the set of all inner derivations (operators of the form $D_{x,y}(z) = \frac{1}{3}[[x,y] + [\tau(x),\tau(y)], z] + (z,y,x) - (z,\tau(x),\tau(y)))$ [2, Corollary 7], we have $Der(\mathbb{O},\tau) = Inder(\mathbb{O},\tau) = Der(\mathbb{O})$.

For the H*-algebras $\mathbb{D}\otimes\mathbb{O}$ we have that the mapping $(D_1, D_2) \mapsto D_1 \otimes Id_{\mathbb{O}} + Id_{\mathbb{D}} \otimes D_2$ is a Lie algebra isomorphism from $Der(\mathbb{D}\otimes\mathbb{O},\tau)$ to $Der(\mathbb{D}) \oplus Der(\mathbb{O})$ [1, Example 8, iv)].

Again by [2, Corollary 7], all derivations on $\mathbb{T}(\mathbb{O})$ are inner and it can be proved [5, Lemma 2.6] that $Der(\mathbb{T}(\mathbb{O}),\tau)$ is isomorphic to $Der(\mathbb{O})$.

Finally if (A,τ) is the H*-algebra asociated to Springer H*-algebras $\mathbb{C}$, $H_3(\mathbb{D})$, every derivation is inner, and $Der(A,\tau)$ is also well known ([1, Page 149] and [2, Page 1870]).

Thus the problem of the determination of the Lie algebra $Der(A,\tau)$, reduces to the algebras associated to involutive Hilbert modules.

DEFINITION 3.1 *Let W be a faithful left module over an associative algebra $\mathcal{E}$. A differential operator on W is a linear operator F from W into W such that for each e in $\mathcal{E}$, there exists some element $\widehat{F}(e)$ in $\mathcal{E}$ such that*

$$F(e \circ x) = \widehat{F}(e) \circ x + e \circ F(x)$$

for all x in W.

Note that for each e in $\mathcal{E}$, the element $\widehat{F}(e)$ in $\mathcal{E}$ is uniquely determined because of the faithfulness of W, so $\widehat{F}$ becomes a well defined operator on $\mathcal{E}$. It is easy to check that $\widehat{F}$ is a derivation on $\mathcal{E}$. Examples of differential operators on W are the operators S_e ($e \in \mathcal{E}$) defined by $S_e(x) = e \circ x$ for all x in W, and also all $\mathcal{E}$-linear operators on W (linear operators on W commuting with S_e for all e in $\mathcal{E}$).

If we denote by $Diff(W)$ the set of all differential operators on W, it is easy to see that $Diff(W)$ is a Lie subalgebra of the associative algebra of all linear operators on W and the mapping $F \mapsto \widehat{F}$ from $Diff(W)$ into $Der(\mathcal{E})$ is an homomorphism. Moreover, the set $L_{\mathcal{E}}(W)$ of all $\mathcal{E}$-linear operators on W and $S_{\mathcal{E}} := \{S_e : e \in \mathcal{E}\}$ are ideals of $Diff(W)$.

The following Theorem gives an "algebraic" characterization of the continuity of differential operators.

THEOREM 3.2 (Cabrera-El Marrakchi-Martínez-Rodríguez) *Let W be a Hilbert module over an H^*-algebra with zero annihilator. A differential operator F on W is continuous if and only if there exists a mapping F' from W into W such that*

$$[F(x) \mid y] = [x \mid F'(y)] + \widehat{F}[x \mid y]$$

for all x, y in W. Moreover, in this case $F' = F^\bullet$, and $F^\bullet$ is a continuous differential operator on W.

See [4, Theorem 1.6]

Our next result gives a structure theorem for bounded differential operators on Hilbert modules. $BDiff\,(W)$ will denote the set of all bounded diferential operators on the Hilbert module W.

THEOREM 3.3 (Cabrera-El Marrakchi-Martínez-Rodríguez) *If W is a Hilbert module over an H^*-algebra $\mathcal{E}$ with zero annihilator, then $BDiff\,(W)$ is a weak-operator closed Lie subalgebra of $\mathcal{BL}(W)$, and we have*

$$BDiff\,(W) = BL_{\mathcal{E}}\,(W) + \overline{S_{\mathcal{E}}}^{wot},$$

where $\overline{}^{wot}$ denotes closure for the weak-operator topology on $\mathcal{BL}\,(W)$.

See [4, Theorem 3.2].

Let $(W, \square)$ be an involutive Hilbert module over $(\mathcal{E}, \tau)$, let F be a differential operator on W, and consider the linear operator $F^{\square}$ on W defined by $F^{\square}(x) = F(x^{\square})^{\square}$ for all x in W. $F^{\square}$ is a differential operator on W and the mapping $\eta(F) = F^{\bullet^{\square}}$ defines an isometric linear involution on the Banach-Lie algebra $BDiff\,(W)$ [4, Page 2936]. We will denote by $BDiff\,(W, \square)$ the closed Lie subalgebra of those elements in $BDiff\,(W)$ which are skew with respect to η. For $F \in BDiff\,(W, \square)$, we denote by D_F the mapping from A to A defined by $D_F(e + x) = \widehat{F}(e) + F(x)$.

Our final Theorem determines the derivations of structurable H*-algebras (A, τ) constructed from involutive Hilbert modules. Except for two cases, $Der\,(A)$ is in a one-to-one correspondence with the set of those bounded differential operators on the parameterizing involutive Hilbert module W which are skew with respect to η.

THEOREM 3.4 (Cabrera-El Marrakchi-Martínez-Rodríguez) *Let $(W, \square)$ be an involutive Hilbert module over a topologically τ-simple H^*-algebra $(\mathcal{E}, \tau)$ with isometric linear involution τ, an let (A, τ) denote its associated structurable H^*-algebra. Then:*

1. *The mapping $F \mapsto D_F$ is a bicontinuous Lie algebra isomorphism from $BDiff\,(W, \square)$ onto $Der\,(A, \tau)$.*

2. *$Der\,(A, \tau) = Der\,(A)$ unless A is either $M_2(\mathbb{C})$ or $\mathbb{O}$ and τ is the non Cayley involution on A.*

See [4, Theorem 4.4].

REFERENCES

[1] B. N. Allison, A class of nonassociative algebras with involution containing the class of Jordan Algebras, Math. Ann. 237, (1979), 133-156.

[2] B. N. Allison, Models of isotropic simple Lie algebras, Comm. Algebra 7 (1979), 1835-1875.

[3] W. Ambrose, Structure Theorems for a special class of Banach algebras, Trans. Amer. Math. Soc. 57 (1945), 364-386.

[4] M. Cabrera, A. El Marrakchi, J. Martínez, and A. Rodríguez, Bounded differential operators on Hilbert modules and derivations of structurable H*-algebras, Comm. Algebra 21 (1993), 2905-2945.

[5] M. Cabrera, A. El Marrakchi, J. Martínez, and A. Rodríguez, An Allison-Kantor-Koecher-Tits construction for Lie H*-algebras, J. Algebra 164 (1994), 361-408.

[6] M. Cabrera, J. Martínez, and A. Rodríguez, A note on real H*-algebras, Math. Proc. Cambridge Philos. Soc. 105 (1989), 131-132.

[7] M. Cabrera, J. Martínez, and A. Rodríguez, Structurable H*-algebras, J. Algebra 147 (1992), 19-62.

[8] J. A. Cuenca and A. Rodríguez, Isomorphisms of H*-algebras, Math. Proc. Cambridge Philos. Soc. 97 (1985), 93-99.

[9] J. A. Cuenca and A. Rodríguez, Structure theory for noncommutative Jordan H*-algebras, J. Algebra 106 (1987), 1-14.

[10] I. N. Herstein, Rings with involution, Chicago Lectures in Mathematics (ed. I. Kaplansky), The University of Chicago Press, Chicago and London, 1976.

[11] N. Jacobson, Structure of rings, AMS Coll. Public. vol. 37, Providence, R.I., 1968.

[12] N. Jacobson, Structure and representations of Jordan algebras, Amer. Math. Soc. Coll. Publ. vol 39, Providence, R. I., 1968.

[13] A. Rodríguez, The uniqueness of the complete norm topology in complete normed nonassociative algebras, J. Funt. Anal. 60 (1985), 1-15.

[14] P. P. Saworotnow, A generalized Hilbert space, Duke Math. J. 35 (1968), 191-197.

[15] P. P. Saworotnow, Trace-class and centralizers of an H*-algebra, Proc. Amer. Math. Soc. 26 (1970), 101-104.

[16] P.P. Saworotnow and J. C. Friedell, Trace-class for arbitrary H*-algebras. Proc. Amer. Math. Soc. 26 (1970), 95-100.

[17] R. D. Schafer, An introduction to nonassociative algebras, Academic Press, New York, 1966.

[18] R. D. Schafer, On structurable algebras, J. Algebra 92 (1985), 400-412.

[19] O. M. Smirnov, Simple and Semisimple structurable algebras, Algebra and Logic 29 (1990), 337-395.

[20] S. W. P. Steen, Introduction to the theory of operators. V. Metric rings, Proc. Cambridge Philos. Soc. 36 (1940), 139-149.

Lie mappings of C^*-algebras

MARTIN MATHIEU Department of Pure Mathematics, School of Mathematics and Physics, The Queen's University of Belfast, Belfast BT7 1NN, Northern Ireland. E-mail: m.m@qub.ac.uk

ABSTRACT We give an outline of the arguments showing that a C^*-algebra essentially allows only trivial Lie derivations and Lie isomorphisms.

Let A be a C^*-algebra. By a Lie mapping of A, we understand a linear mapping from A into itself that is compatible with the natural Lie structure on A, i.e., the commutator bracket $[x, y] = xy - yx$, x, $y \in A$. Specifically, θ is called a *Lie homomorphism* if $\theta([x, y]) = [\theta x, \theta y]$ for all x, $y \in A$, and d is called a *Lie derivation* if $d([x, y]) = [dx, y] + [x, dy]$ for all x, y. In this paper, we survey some recent results describing the structure of such mappings in terms of linear mappings compatible with the associative structure of A. These results are obtained via new techniques in operator theory of C^*-algebras that emerged to be useful in various other instances too, cf. [2], [3], [16], [20]. Let us first of all summarise these ideas.

The traditional approach to study linear operators on C^*-algebras often starts by considering mappings on von Neumann algebras first, and then extending the results using the embedding of a C^*-algebra A into its enveloping von Neumann algebra A''. This route has a number of advantages, notably, the rich structure of the projection lattice of a von Neumann algebra can be exploited and the extension to A'', which is nothing but the second dual space of A, is natural. The main drawbacks of this approach are, firstly, the operators studied have to be bounded necessarily and, secondly, it is not automatic – and sometimes will be wrong – that the extension T'' of the operator T on A will have the same properties as T has.

Under these constraints, however, the traditional method works pretty well; e.g., results for Lie and Jordan derivations comparable to ours were obtained in [12].

The approach we propose uses the concept of *local multipliers*. This notion goes back to Elliott [10] and Pedersen [18] in the 1970s but has not been developed further until recently. It can, however, been realised as an analytic analogue of the concept of the *symmetric ring of quotients* of a semiprime ring which is by now fairly well understood [7], [13]. It is this point of view that turned out to be most fruitful in applications to operators on C^*-algebras. The basic idea is to replace A'' by a C^*-algebra which is, up to the norm closure, constructed purely in algebraic terms; this type of enlargement is called *the bounded central closure of* A, denoted by ${}^cA$. A C^*-algebra A is said to be *boundedly centrally closed* if ${}^cA = A$; $A \mapsto {}^cA$ turns out to be in fact a closure operation, i.e., ${}^{cc}A = {}^cA$. Like von Neumann algebras, boundedly centrally closed C^*-algebras behave very well, in particular with respect to central projections, and the class is wide enough, specifically, it contains all von Neumann algebras. We now first study operators on this class of C^*-algebras where the smoothest results can be obtained and using another extension procedure obtain the general results. Developing this method and, of course, providing a good understanding of the bounded central closure and more generally, the *local multiplier algebra* of a C^*-algebra is the contents of [4], where the details of the proofs of the subsequent results can be found.

1 THE RESULTS

In order to avoid too many preliminary definitions and notation we shall state the results describing the structure of Lie homomorphisms and derivations only for boundedly centrally closed C^*-algebras; the general formulations can be found in [4]. The close-at-hand examples of Lie mappings are of course their associative counterparts: Every derivation of A is a Lie derivation, and every homomorphism of A is a Lie homomorphism. It is readily seen that also the negative of each anti-homomorphism provides an example of a Lie homomorphism. Since we start from an associative context, it is natural to ask whether there can be any other examples. In fact, this question was popularised in the setting of (simple) rings by Herstein in his seminal paper [11]. Herstein's programme asking for a full understanding of additive mappings on rings that are compatible with the derived non-associative structures, notably the Jordan and the Lie structure, is by now almost completed, see e.g. [6]. The description in the setting of C^*-algebras is as follows.

THEOREM 1. *Let A be a unital, boundedly centrally closed C^*-algebra such that K_2 is essential. For each Lie isomorphism $\theta\colon A \to A$ there exist uniquely determined linear mappings $\psi\colon A \to A$ and $\tau\colon A \to Z(A)$, the centre of A, with the property that $\tau(xy) = \tau(yx)$ for all $x, y \in A$ such that*

$$\theta = \psi + \tau$$

and $\psi_{|pA}$ is an isomorphism whereas $\psi_{|(1-p)A}$ is the negative of an anti-isomorphism for a unique central projection p in A.

THEOREM 2. *Let A be a unital, boundedly centrally closed C^*-algebra such that K_2 is essential. For each Lie derivation $d\colon A \to A$ there exist a uniquely determined* *derivation δ on A and a unique linear mapping $\tau: A \to Z(A)$ with the property that $\tau(xy) = \tau(yx)$ for all $x, y \in A$ such that*

$$d = \delta + \tau.$$

We shall explain the definition of the ideal K_2 as well as its significance in the next section. A linear mapping τ from A into its centre $Z(A)$ with the property that $\tau(xy) = \tau(yx)$ for all $x, y \in A$ will be called a *centre-valued trace* of A. (Note that we do not require τ to be positive as is customary in C^*-algebra theory.) The two results above state that, apart from the additive perturbation by a centre-valued trace, every Lie mapping of A is made up by the corresponding associative mappings, or, put more briefly, there are no proper Lie mappings of C^*-algebras. This was also observed in [12] for bounded Lie derivations. Since every centre-valued trace is both a Lie homomorphism and a Lie derivation, its appearance is inevitable to take care of the commutative part of A.

Theorem 1 was recently obtained by Brešar and Miers in the case of von Neumann algebras [9] whereas Theorem 2 in this case is much older [17].

2 OUTLINE OF THE PROOFS

The starting point in both the proofs of Theorem 1 and 2 is the definition of a bilinear mapping that captures the deviation from being multiplicative, in the case of a Lie homomorphism, or being derivative, in the case of a Lie derivation. We put $B(x,y) = \theta(\theta^{-1}x\,\theta^{-1}y)$ respectively, $B(x,y) = d(xy) - x(dy) - (dx)y$ for all $x, y \in A$. Letting $q(x) = B(x,x)$ we obtain a quadratic mapping on A which is *commuting*, that is, $[x, q(x)] = 0$ for all $x \in A$. This is readily checked in either case.

In [2] we obtained a representation theorem for *linear* commuting mappings; it turns out that an analogous result can be proven for quadratic commuting mappings though it is somewhat harder to achieve.

THEOREM 3. *Let q be a quadratic commuting mapping on a unital, boundedly centrally closed C^*-algebra A such that K_2 is essential. Then there exist a unique element $\lambda \in Z(A)$, a unique linear mapping $\mu: A \to Z(A)$, as well as a unique mapping $\nu: A \to Z(A)$ such that*

$$q(x) = \lambda\, x^2 + \mu(x)\, x + \nu(x) \qquad (x \in A).$$

A result of this type seems to have first appeared in [8] where the situation of prime rings with certain additional properties is treated. In fact, this paper contains a similar study of Lie mappings in the algebraic setting. Theorem 3 is obtained for semiprime rings in [6], and some additional work is needed to derive the C^*-case from the purely algebraic result (cf. [4] and [5]). The algebraic manipulations are fairly involved, but we want to use this opportunity to indicate how the ideal K_2 comes into play.

Suppose that a representation of q as in Theorem 3 above has been obtained, with λ, μ and ν lying in, respectively mapping into, the centre of some algebra B containing A (B need not be a C^*-algebra). We shall see that λ necessarily maps a certain ideal of A into A; this will be the ideal called K_2. We denote by $[u,v,w] = uvw - wvu$ the *3-commutator* of $u,v,w \in B$. Starting from the representation in Theorem 3,

$$q(x) = \lambda\, x^2 + \mu(x)\, x + \nu(x) \qquad (x \in A)$$

we take the commutator with $y \in A$ and obtain, as $\nu(x)$ is central for every x,

$$[q(x), y] = \lambda\, [x^2, y] + \mu(x)\, [x, y] \qquad (x, y \in A).$$

We now take the 3-commutator $[\,\cdot\,, z, [x,y]]$ for arbitrary $z \in A$ and arrive at

$$[[q(x), y], z, [x, y]] = \lambda\, [[x^2, y], z, [x, y]] \qquad (x, y, z \in A) \tag{1}$$

as $\mu(x)$ is central as well and $[[x,y], z, [x,y]] = 0$. Since λ is in the centre this entails that

$$\lambda\, A\, [[x^2, y], z, [x, y]]\, A \subseteq A \qquad (x, y, z \in A).$$

Let K_2 be the ideal generated by all elements of the form $[[x^2,y], z, [x,y]]$ in A. For historical reasons, we call this ideal *the second Kaplansky ideal*; the corresponding *first* Kaplansky ideal is nothing but the usual commutator ideal K_1 of A. By the above, we can define an A-bimodule mapping f from K_2 into A via $f(u) = \lambda\, u$ for all $u \in K_2$. (In the algebraic terminology, λ is the element in the *extended centroid* $C(A)$ corresponding to the A-bimodule mapping f.) The main point distinguishing the analytic from the algebraic setting now is that we want f to be bounded. If it is, then we can extend it to an A-bimodule mapping on the closure $\overline{K_2}$ of K_2. Since every closed ideal in a C^*-algebra is idempotent, this extension will actually map into $\overline{K_2}$, in other words, λ is nothing but a central multiplier of the closed ideal $\overline{K_2}$. In order to be able to talk about continuity we, of course, need B to be a normed algebra; it is the *local multiplier algebra* $M_{\mathrm{loc}}(A)$ *of* A playing the role of B. Indeed, it is here also where we need that the ideal K_2 is *essential* which by definition means that K_2 has non-zero intersection with every other non-zero ideal of A. This requirement is equivalent to the condition that there is no open set of primitive ideals of A containing K_2. If A is a von Neumann algebra, this in turn is equivalent to the condition that A has no direct summands isomorphic to the complex numbers or the 2×2-matrices. (This is the condition appearing in [9].)

At this stage, the theory of algebras satisfying polynomial identities comes in. This shouldn't be a surprise as we want to control the *degree of non-commutativity* of A somehow, and *PI*-theory is one of the standard devices. The following lemma is also connected with Kaplansky's name; he observed the case $n = 1$ more than 40 years ago.

LEMMA. *The following conditions on a C^*-algebra A are equivalent.*

(a) *A does not satisfy the standard polynomial identity S_{2n};*

(b) *A has an irreducible representation of dimension greater than n;*

(c) *There is an element $x \in A$ such that $x^n \neq 0$ and $x^{n+1} = 0$.*

In this case, x can be chosen such that $\|x^n\| = 1$ and $\|x\| \leq \frac{n^2}{n-1}$ $(n > 1)$.

The standard polynomial identity S_m, $m \in \mathbf{N}$ is defined by

$$S_m = \sum_{\sigma \in S_m} \operatorname{sign}(\sigma)\, t_{\sigma(1)} \cdots t_{\sigma(m)}$$

where S_m is the symmetric group on $\{1, \ldots, m\}$; in some sense, more general identities can be reduced to the standard ones (cf. [19, II.4.1]). An algebra A is said *to satisfy the identity* S_m if $S_m(x_1, \ldots, x_m) = 0$ for all choices of $x_1, \ldots, x_m \in A$ in place of the non-commuting free variables $t_1, \ldots, t_m$.

This lemma is a nice combination of some standard arguments in *PI*-algebra and C^*-algebra theory with a result in [15]. For $n = 2$, the assumption (a) of the lemma is fulfilled if K_2 is non-zero. Hence, choosing $x \in A$ such that $x^3 = 0$ while $\|x^2\| = 1$ and $\|x\| \leq 4$ we put $y = xx^{*2}$ and $z = x$. In this case

$$\left\|[[x^2, y], z, [x, y]]\right\| = \left\|x^2x^{*2}x^2x^{*2}x^2\right\| = \left\|x^2\right\|^5 = 1$$

which is the key in proving that the above A-bimodule mapping f is

bounded. (If it was unbounded, the right hand side of (1) would have to become unbounded on the unit ball of A whereas the left hand side would remain bounded, roughly spoken.)

A similar reasoning applied to $\mu(x)$ shows that, for each $x \in A$, $\mu(x)$ necessarily has to be a central multiplier of the closed commutator ideal $\overline{K_1}$, which is automatically essential if K_2 is.

We now arrived at the point where we need to say more about boundedly centrally closed C^*-algebras. A C^*-algebra A is said to be *boundedly centrally closed* if every (bounded) A-bimodule mapping defined on an (essential) closed ideal of A with values in A can be extended to an A-bimodule mapping defined on all of A. If A is unital, every such mapping is then given by the multiplication with a central element in A. Hence, A is boundedly centrally closed if and only if it is self-injective in the category of Banach A-bimodules. There is a number of valuable equivalent conditions, cf. [2], all making essential use of the local Dauns-Hofmann theorem [1]. Examples of boundedly centrally closed C^*-algebras include all von Neumann algebras, more generally all AW^*-algebras, every prime C^*-algebra [14], and every algebra of continuous functions on a Stonean space. The class of boundedly centrally closed C^*-algebras behaves well under some natural constructions; e.g., every hereditary C^*-subalgebra of a boundedly centrally closed C^*-algebra is boundedly centrally closed itself [4]. The *bounded central closure* ${}^cA$ of an arbitrary C^*-algebra adds to A all the A-bimodule mappings defined on essential closed ideals of A, suitably identified via an inductive limit, into a larger C^*-algebra; ${}^cA$ itself then is boundedly centrally closed.

It is now clear that all the constituents of the quadratic mapping q in Theorem 3 belong to $Z(A)$ if A is boundedly centrally closed. We thus make the following Ansatz. In the case of a Lie derivation d we put $\delta = d + \frac{1}{2}\mu$ and prove that δ is in fact a derivation on A. Necessarily, $\tau = -\frac{1}{2}\mu$ will then be a centre-valued trace of A. The case of a Lie isomorphism of A is more involved. We first put $\varphi = \lambda\,\theta + \frac{1}{2}\mu\,\theta$. We then show that φ is a Jordan isomorphism of A using the results in [2]. Since it turns out that λ is invertible, we can put $\psi = \lambda^{-1}\varphi$ and $\tau = -\frac{\lambda^{-1}}{2}\mu\,\theta$. Again, τ automatically is a centre-valued trace. The decomposition of ψ via the central projection p in A is finally achieved by means of the following result, which appears to be new in this generality.

THEOREM 4. *Let $\varphi: A \to B$ be a surjective Jordan homomorphism between the C^*-algebras A and B. Then there exists a central projection q in the multiplier* *algebra of the closed commutator ideal of B such that*

$$\varphi(xy) = q\,\varphi(x)\varphi(y) + (1-q)\,\varphi(y)\varphi(x) \qquad (x, y \in A).$$

This result will be applied in Theorem 1 by putting $p = \varphi^{-1}(q)$.

The approach outlined above is typical for working with local multipliers. Quite often, one first of all has to leave the realm of C^*-algebras to find new elements as solutions to certain operator equations in a ring-theoretic enlargement which is not normable. In a second step one then comes back to the framework of C^*-algebras by showing the boundedness of the elements previously found as above; cf. also [16]. Arguments along the above lines can also be used to extend the representation theorem for bi-commutativity preserving linear mappings from the case of von Neumann algebras considered in [9] to the much wider class of boundedly centrally closed C^*-algebras [5].

REFERENCES

[1] P. Ara and M. Mathieu, A local version of the Dauns-Hofmann theorem, Math. Z. 208 (1991), 349–353.

[2] P. Ara and M. Mathieu, An application of local multipliers to centralizing mappings of C^*-algebras, Quart. J. Math. Oxford (2) 44 (1993), 129–138.

[3] P. Ara and M. Mathieu, On the central Haagerup tensor product, Proc. Edinburgh Math. Soc. 37 (1994), 161–174.

[4] P. Ara and M. Mathieu, Local multipliers of C^*-algebras, book in preparation.

[5] R. Banning, Kommutativitätserhaltende Abbildungen auf C^*-Algebren, Doctoral Thesis, Tübingen 1998.

[6] R. Banning and M. Mathieu, Commutativity preserving mappings on semiprime rings, Commun. Algebra 25 (1997), 247–265.

[7] K. I. Beidar, W. S. Martindale and A. V. Mikhalev, Rings with generalized identities, Marcel Dekker, New York Basel Hong Kong, 1996.

[8] M. Brešar, Commuting traces of biadditive mappings, commutativity preserving mappings and Lie mappings, Trans. Amer. Math. Soc. 335 (1993), 525–546.

[9] M. Brešar and C. R. Miers, Commutativity preserving mappings of von Neumann algebras, Canad. J. Math. 45 (1993), 695–708.

[10] G. A. Elliott, Automorphisms determined by multipliers on ideals of a C^*-algebra, J. Funct. Anal. 23 (1976), 1–10.

[11] I. N. Herstein, Lie and Jordan structures in simple, associative rings, Bull. Amer. Math. Soc. 67 (1961), 517–531.

[12] B. E. Johnson, Symmetric amenability and the nonexistence of Lie and Jordan derivations, Math. Proc. Cambridge Phil. Soc. 120 (1996), 455–473.

[13] V. K. Kharchenko, Automorphisms and derivations of associative rings, Kluwer Acad. Publ., Dordrecht, 1991.

[14] M. Mathieu, Elementary operators on prime C^*-algebras, I, Math. Ann. 284 (1989), 223–244.

[15] M. Mathieu, More properties of the product of two derivations of a C^*-algebra, Bull. Austral. Math. Soc. 42 (1990), 115-120.

[16] M. Mathieu, Derivations implemented by local multipliers, Proc. Amer. Math. Soc. 126 (1998), 1133–1138.

[17] C. R. Miers, Lie derivations of von Neumann algebras, Duke Math. J. 40 (1973), 403–409.

[18] G. K. Pedersen, Approximating derivations on ideals of C^*-algebras, Invent. math. 45 (1978), 299–305.

[19] C. Procesi, Rings with polynomial identities, Marcel Dekker, New York, 1973.

[20] D. W. B. Somerset, The local multiplier algebra of a C^*-algebra, II, preprint 1998.

Sur des anneaux commutatifs à puissances associatives de caractéristique première p > 0

JOÃO CARLOS DA MOTTA FERREIRA Universidade Federal de Mato Grosso do Sul, Departamento de Matemática, Caixa Postal 549, 79070-900 Campo Grande, MS, Brasil

ARTIBANO MICALI Université Montpellier II, Département des Sciences Mathématiques, Place Eugène Bataillon, 34095 Montpellier, France

Le but de cet article est l'étude des anneaux commutatifs U, non nécessairement associatifs, à puissances associatives, de caractéristique première $p > 0$, tels que la p-ième puissance soit un morphisme d'anneaux $U \to U$, $x \mapsto x^p$, dont l'image est contenue dans le noyau associatif $N(U)$. Par la suite, on notera $\mathcal{C}$ cette classe d'anneaux.

En ce qui concerne les questions étudiées (nilradical, anneaux semi-simples et décomposition primaire), on obtient des résultats analogues à ceux obtenus dans [5] et [11] pour les anneaux commutatifs et associatifs.

Dans [10], en suivant les idées classiques de McCoy, l'auteur donne une théorie générale d'idéaux premiers pour un anneau non associatif arbitraire. Mais son théorème de décomposition de Lasker-Noether assure seulement l'unicité de la décomposition et non l'existence. Pour de telles raisons, nous avons été conduits à choisir des généralisations plus naturelles de l'Algèbre Commutative classique.

ABSTRACT. The aim of this paper is essentially to show that some notions of Commutative Algebra, as noetherianity or primary decomposition, can be studied in some classes of commutative nonassociative rings. In concret, the class $\mathcal{C}$ that consists of commutative power-associative rings U of prime characteristic $p > 0$ such that the map $U \to U$ defined by $x \mapsto x^p$ is a ring homomorphism and the image is contained in the associative nucleus $N(U)$. We apply our results in the study of alternative rings of characteristic 3 and Jordan and Moufang rings of characteristic 2. We will follow Zhevlakov' ideas in studying alternative rings with finiteness conditions (see [12], chapter 12).

1. LE NILRADICAL.

Dans ce paragraphe on étend, aux anneaux de la classe $\mathcal{C}$, des résultats connus pour les anneaux commutatifs et associatifs. Par la suite, on caractérise le nilradical (nilidéal maximal) de tels anneaux comme étant l'ensemble de ses éléments nilpotents et les anneaux semi-simples comme des anneaux associatifs, résultats analogues à ceux des anneaux commutatifs et associatifs (cf. [5], [11]).

LEMME 1.1. *Soit U un anneau commutatif et S un sous-ensemble non vide de U. Le plus petit idéal de U contenant S est l'ensemble I des sommes finies d'éléments de U de la forme $\pm u_1(\ldots(u_{n-1}u_n)\ldots)$ où il existe un indice i, $1 \leq i \leq n$, tel que u_i soit dans S.*

En effet, si $\{I_\alpha\}_{\alpha\in A}$ est la famille non vide des idéaux de U contenant S, $\cap_{\alpha\in A} I_\alpha$ est le plus petit idéal de U contenant S. D'autre part, tout élément de I de la forme $\pm u_1(\ldots(u_{n-1}u_n)\ldots)$ est contenu dans I_α pour tout α dans A ou encore, $I \subset \cap_{\alpha\in A} I_\alpha$. Montrons que I est un idéal de U contenant S donc que $I \supset \cap_{\alpha\in A} I_\alpha$ ou encore, $I = \cap_{\alpha\in A} I_\alpha$. En effet, comme $S \neq \emptyset$, il existe un élément u dans S donc $0 = u - u$ est dans I. De plus, étant donnés deux éléments de I de la forme $\pm u_1(\ldots(u_{m-1}u_m)\ldots)$ et $\pm v_1(\ldots(v_{n-1}v_n)\ldots)$ alors $(\pm u_1(\ldots(u_{m-1}u_m)\ldots))+(\pm v_1(\ldots(v_{n-1}v_n)\ldots))$ est dans I ce qui entraîne que I est un sous-groupe additif de U. Comme, pour tout u dans U et tout élément de I de la forme $\pm u_1(\ldots(u_{m-1}u_m)\ldots)$, $u(\pm u_1(\ldots(u_{m-1}u_m)\ldots)) = \pm u(u_1(\ldots(u_{m-1}u_m)\ldots))$ est dans I, on conclut que I est un idéal de U.

THÉORÈME 1.2. *Soit U un anneau de la classe $\mathcal{C}$. L'ensemble R des éléments nilpotents de U est un nilidéal de U et l'anneau quotient U/R ne possède aucun élément nilpotent non nul.*

En effet, d'après les hypothèses du théorème et par récurrence sur $n \geq 1$ on a, dans U, les identités

(1) $(x+y)^{p^n} = x^{p^n} + y^{p^n}$, pour tout entier $n \geq 1$ et

(2) $(xy)^{p^n} = x^{p^n}y^{p^n}$, pour tout entier $n \geq 1$.

D'après les identités (1) et (2), on conclut que l'ensemble des éléments nilpotents dans U est un nilidéal de U. De plus, considérons un élément x dans U tel que l'élément $[x]$ dans U/R soit nilpotent. Il existe alors un entier positif m tel que $[x]^m = [0]$. Il s'ensuit que $x^m \in R$ et donc pour un entier positif r convenable on a $(x^m)^r = x^{mr} = 0$, c'est à dire, $x \in R$. Par conséquent, $[x] = [0]$.

2. CARACTÉRISATION DU NILRADICAL.

Soit U un anneau commutatif. On dit qu'un idéal P de U est un *idéal premier* si quels que soient les éléments x et y dans U tels que $xy \in P$, alors x ou y est dans P.

THÉORÈME 2.1. *Soit U un anneau de la classe $\mathcal{C}$. Le nilradical R de U est alors l'intersection des idéaux premiers de U.*

Considérons l'intersection des idéaux premiers de U, $\cap_{\mu \in M} P_\mu$. Si x est un élément nilpotent de U et si P_μ, $\mu \in M$, est un idéal premier, il existe un entier $n \geq 1$ tel que $x^n = 0 \in P_\mu$, soit $x \in P_\mu$. Donc, $R \subset \cap_{\mu \in M} P_\mu$. Réciproquement, soit g un élément de U non nilpotent et considérons l'ensemble Ω des idéaux I de U tels que $g^n \notin I$, pour tout entier $n \geq 1$; Ω est non vide car $\{0\} \in \Omega$ et d'après le lemme de Zorn, l'ensemble Ω, ordonné par inclusion, possède un élément maximal P. Montrons que P est un idéal premier. En effet, considérons des éléments x, y non dans P. Alors, les idéaux $P + (x)$ et $P + (y)$ engendrés par l'ensemble P et les éléments x et y, respectivement, contiennent proprement l'idéal P, donc n'appartiennent pas à Ω. Ainsi, il existe des entiers m, $n \geq 1$ tels que $g^n \in P + (x)$ et $g^m \in P+(y)$. Il s'ensuit que $g^{p(m+n)} \in P+(xy)$ donc l'idéal $P+(xy)$ n'appartient pas à Ω, c'est à dire, xy n'appartient pas à P. Par conséquent, on a un idéal premier P tel que $g \notin P$ de U, donc $g \notin \cap_{\mu \in M} P_\mu$.

3. ANNEAUX SEMI-SIMPLES.

Tout d'abord un résultat concernant les anneaux semi-simples :

THÉORÈME 3.1. *Soit U un anneau de la classe $\mathcal{C}$. Si U est semi-simple, alors U est un anneau associatif.*

En effet, si p est un entier premier impair, on a

$$\begin{aligned}(x, y, z)^p &= ((xy)z - x(yz))^p = ((xy)z)^p - (x(yz))^p = \\ &= (x^p y^p) z^p - x^p (y^p z^p) = 0\end{aligned}$$

quels que soient les éléments x, y et z dans U. Ainsi, $(x, y, z) \in R = \{0\}$, donc $(x, y, z) = 0$. Démonstration analogue pour $p = 2$.

EXEMPLE 3.2. Soient F un corps commutatif, et U une F-algèbre de dimension 4 dont la table de multiplication, relative à une base $\{e_1, \ldots, e_4\}$, s'écrit $e_1 e_1 = e_2$, $e_1 e_2 = e_2 e_1 = e_3$, $e_1 e_3 = e_3 e_1 = e_4$, tous les autres produits étant nuls. Il s'agit d'une algèbre commutative où la 5-ième puissance x^5 est un morphisme dont l'image est dans le noyau associatif $N(U)$. A partir d'un calcul direct, on vérifie que U n'est pas une algèbre à puissances associatives.

4. DÉCOMPOSITION PRIMAIRE.

Dans ce paragraphe on généralise aux anneaux de classe $\mathcal{C}$, la théorie connue concernant la décomposition primaire pour les anneaux commutatifs et associatifs (cf. [5], [11]).

LEMME 4.1. *Soit U un anneau commutatif.*

(i) Considérons des idéaux premiers $P_1, \ldots, P_n$ et I un idéal contenu dans $\bigcup_{i=1}^n P_i$. Il existe alors un indice i, $1 \leq i \leq n$, tel que $I \subset P_i$.

(ii) Considérons des idéaux $I_1, \ldots, I_n$ et P un idéal premier qui contient $\bigcap_{i=1}^n I_i$. Il existe alors un indice i, $1 \leq i \leq n$, tel que $I_i \subset P$. Si $P = \bigcap_{i=1}^n I_i$, il existe un indice i, $1 \leq i \leq n$, tel que $P = I_i$.

(i) On démontrera cette assertion par récurrence sur n dans la proposition

$$P(n): \text{si pour tout indice } i, 1 \leq i \leq n,\ I \not\subset P_i, \text{ alors } I \not\subset \cup_{i=1}^{n} P_i.$$

En effet, $P(1)$ est évidente. Supposons que pour un entier positif $n \geq 1$ la proposition $P(n)$ soit vérifiée ; pour des idéaux premiers $P_1, \ldots, P_n, P_{n+1}$ et I un idéal tel que $I \not\subset P_i$, pour $i = 1, \ldots, n, n+1$, il existe un élément $x_i \in I$ avec $x_i \notin P_j$, pour tout $j \neq i$. S'il existait un indice i, $1 \leq i \leq n+1$, tel que $x_i \notin P_i$, alors $x_i \notin \cup_{j=1}^{n+1} P_j$. Dans le cas contraire, pour tout i, $1 \leq i \leq n+1$, $x_i \in P_i$ et considérons l'élément

$$x = \sum_{i=1}^{n+1} x_1(x_2(\ldots(x_{i-1}(x_{i+1}(\ldots(x_n x_{n+1})\ldots)))\ldots))$$

où, dans cette somme, le terme correspondant à $i = 1$ est $x_2(\ldots(x_n x_{n+1})\ldots)$ et celui pour $i = n+1$ est $x_1(x_2(\ldots(x_{n-1}x_n)\ldots)$. Il s'ensuit que $x \in I$ et que $x \notin P_i$ $(i = 1, \ldots, n+1)$, car les idéaux P_i $(i = 1, \ldots, n+1)$ sont premiers. On conclut ainsi que $I \not\subset \cup_{i=1}^{n} P_i$.

(ii) Supposons que $I_i \not\subset P$ pour tout i, $1 \leq i \leq n$; il existe alors un $x_i \in I_i$, tel que $x_i \notin P$, donc $x = x_1(x_2(\ldots(x_{n-1}x_n)\ldots)) \in \cap_{i=1}^{n} I_i$. D'après l'hypothèse sur l'idéal P, on a $x \notin P$, donc $\cap_{i=1}^{n} I_i \not\subset P$. Si $P = \cap_{i=1}^{n} I_i$, , il existe un indice i, $1 \leq i \leq n$, tel que $I_i \subset P$. Comme $P \subset I_i$, alors $P = I_i$. □

Soit U un anneau commutatif et à puissances associatives. On dit qu'un idéal propre Q de U est un *idéal primaire* de U si quels que soient les élément x et y dans U vérifiant $xy \in Q$, alors $x \in Q$ ou il existe un entier $n = n(y) \geq 1$ tel que $y^n \in Q$.

PROPOSITION 4.2. *Soit U un anneau de la classe $\mathcal{C}$. Si Q est un idéal primaire de U, l'ensemble*

$$P = \{x | x \in U, \text{ il existe un entier } n = n(x) \geq 1 \text{ tel que } x^n \in Q\}$$

est le plus petit idéal premier de U contenant Q et noté $\sqrt{Q}$.

En effet, il est évident que $0 \in P$. Considérons des éléments x, y dans P. Comme $(x+y)^{p^n} = x^{p^n} + y^{p^n}$, pour tout entier $n \geq 1$, il s'ensuit que P est un sous-groupe additif de U. D'autre part, quels que soient les éléments $u \in U$ et $x \in P$, on a $(ux)^{p^n} = u^{p^n} x^{p^n}$, pour tout entier $n \geq 1$, donc P est un idéal de U. Montrons que P est un idéal premier. En effet, considérons des éléments x, $y \in U$ tels que $xy \in P$; il existe alors un entier $n \geq 1$ tel que $(xy)^{p^n} \in Q$. Donc $x^{p^n} \in Q$ ou il existe un entier $m \geq 1$ tel que $(y^{p^n})^m \in Q$. Ainsi, $x \in P$ ou $y \in P$, donc P est un idéal premier et il contient naturellement l'idéal Q. Finalement, pour tout idéal premier I de U tel que $Q \subset I$, soit $x \in P$; il existe alors un entier $n \geq 1$ tel que $x^n \in Q \subset I$ et comme I est un idéal premier de U, on a $x \in I$, donc $P \subset I$. Par conséquent, P est le plus petit idéal premier de U contenant Q. □

Si Q est un idéal primaire de U et $P = \sqrt{Q}$, on dira aussi que Q est un *idéal P-primaire*.

PROPOSITION 4.3. *Soit U un anneau de la classe $\mathcal{C}$. Si $Q_1, \ldots, Q_r$ sont des idéaux P-primaires, alors l'idéal $Q = \cap_{i=1}^r Q_i$ est aussi P-primaire.*

En effet, $Q \subset Q_1 \subsetneq U$ et donc $Q \subsetneq U$. Ensuite, considérons des éléments x, y dans U tels que $xy \in Q$. Si $x \notin Q$, il existe un indice $1 \leq i_0 \leq r$ tel que $x \notin Q_{i_0}$ donc il existe un entier $n_0 = n_0(i_0, x) \geq 1$ tel que $y^{n_0} \in Q_{i_0}$. Ainsi, $y \in P$ et donc il existe des entiers $n_1 = n_1(y) \geq 1, \ldots, n_r = n_r(y) \geq 1$ tels que $y^{n_i} \in Q_i$ $(i = 1, \ldots, r)$. En prenant $n = \max\{n_1, \ldots, n_r\}$, on a $y^n \in Q_i$ $(i = 1, \ldots, r)$, c'est-à-dire, $y^n \in Q$. Cela nous dit que Q est un idéal primaire. Montrons qu'il est P-primaire. En effet, si $x \in \sqrt{Q}$, il existe un entier $n = n(x) \geq 1$ tel que $x^n \in Q$, donc $x^n \in Q_i$ $(i = 1, \ldots, r)$. Ainsi, $x \in P$ et $\sqrt{Q} \subset P$. D'autre part, si x est dans P, il existe un entier $n \geq 1$ tel que $x^n \in Q_i$ $(i = 1, \ldots, r)$, donc $x^n \in Q$, c'est-à-dire, $x \in \sqrt{Q}$. Par conséquent, $\sqrt{Q} = P$. □

Soit U un anneau commutatif et à puissances associatives. On dit que un idéal propre Q de U possède une *décomposition primaire* s'il existe un nombre fini d'idéaux P_i-primaires Q_i de U, $1 \leq i \leq r$, tels que $Q = \cap_{i=1}^r Q_i$. Une telle décomposition est appelée *décomposition primaire minimale* de Q si : (i) $P_1, \ldots, P_r$ sont des idéaux premiers distincts ; (ii) pour tout $j, 1 \leq j \leq r$, on a $\cap_{i=1, i \neq j}^r Q_i \not\subset Q_j$. On dit qu'un idéal Q est un *idéal décomposable* de U s'il possède une décomposition primaire.

THÉORÈME 4.4. *Soit U un anneau de la classe $\mathcal{C}$. Soient Q un idéal décomposable de U et $Q = \cap_{i=1}^r Q_i = \cap_{j=1}^s Q_j^*$, deux décompositions primaires minimales de Q, avec $\sqrt{Q_i} = P_i$ $(i = 1, \ldots, r)$ et $\sqrt{Q_j^*} = P_j^*$ $(j = 1, \ldots, s)$. Alors, $r = s$ et $\{P_1, \ldots, P_r\} = \{P_1^*, \ldots, P_s^*\}$.*

Supposons que P_1 soit maximal dans l'ensemble des premiers appartenant aux idéaux primaires Q_i $(i = 1, \ldots, r)$ et Q_j^* $(j = 1, \ldots, s)$ et que $P_1 \neq P_j^*$ pour $j = 1, \ldots, s$. D'après le lemme 4.1-(i), il existe un élément $a \in P_1$ tel que $a \notin P_i$ $(i = 2, \ldots, r)$ et $a \notin P_j^*$ $(j = 1, \ldots, s)$. Ainsi, il existe un entier $n_0 \geq 1$ tel que $a^{n_0} \in Q_1$ et, pour tout entier $n \geq 1$, $a^n \notin Q_i$ $(i = 2, \ldots, r)$ et $a^n \notin Q_j^*$ $(j = 1, \ldots, s)$. Montrons que $\cap_{i=2}^r Q_i = \cap_{j=1}^s Q_j^*$. En effet, c'est évident que $\cap_{j=1}^s Q_j^* \subset \cap_{i=2}^r Q_i$. D'autre part, pour tout élément $x \in \cap_{i=2}^r Q_i$, on a $a^{n_0}x \in \cap_{i=1}^r Q_i$, donc $a^{n_0}x \in Q_j^*$ $(j = 1, \ldots, s)$. Cela nous dit que $x \in Q_j^*$ $(j = 1, \ldots, s)$, c'est-à-dire, $\cap_{i=2}^r Q_i = \cap_{j=1}^s Q_j^*$, contradisant l'hypothèse de minimalité sur l'expression de Q. On a ainsi montré que P_1 figure dans l'ensemble $\{P_1^*, \ldots, P_s^*\}$, disons $P_1 = P_1^*$. Montrons que $\cap_{i=2}^r Q_i = \cap_{i=2}^s Q_j^*$. En effet, comme $P_1 (= P_1^*) \neq P_j^*$ $(j = 2, \ldots, s)$, d'après le lemme 4.1-(i), on a

$$P_1 \not\subset \cup_{i=2}^r P_i \ \cup \ \cup_{j=2}^s P_j^* \ ,$$

donc il existe un élément $a \in P_1$ tel que $a \notin P_i$ $(i = 2, \ldots, r)$ et $a \notin P_j^*$ $(j = 2, \ldots, s)$. Ainsi, il existe un entier $n_1 \geq 1$ tel que $a^{n_1} \in Q_1 \cap Q_1^*$ et, pour tout entier $n \geq 1$, $a^n \notin Q_i$ $(i = 2, \ldots, r)$ et $a^n \notin Q_j^*$ $(j = 2, \ldots, s)$. Par conséquent, pour tout élément $x \in \cap_{i=2}^r Q_i$, on a $a^{n_1}x \in \cap_{i=1}^r Q_i = \cap_{j=1}^s Q_j^*$. Il s'ensuit que, $x \in \cap_{j=2}^s Q_j^*$ donc $\cap_{i=2}^r Q_i \subset \cap_{j=2}^s Q_j^*$. De même, on montre que $\cap_{j=2}^s Q_j^* \subset \cap_{i=2}^r Q_i$. Ce processus peut se répéter un nombre fini de fois et il nous conduit à l'assertion du théorème. □

Soient U un anneau commutatif et à puissances associatives et Q un idéal décomposable de U avec une décomposition primaire minimale $Q = \cap_{i=1}^{r} Q_i$, où $\sqrt{Q_i} = P_i$ $(i = 1, \ldots, r)$. L'ensemble $\{P_1, \ldots, P_r\}$, qui est indépendant du choix de la décomposition primaire minimale de Q, d'après le théorème 4.4, est appelé *ensemble des idéaux premiers associés à Q*.

PROPOSITION 4.5. *Soient U un anneau de la classe $\mathcal{C}$, Q un idéal décomposable de U et $Q = \cap_{i=1}^{r} Q_i$ avec $\sqrt{Q_i} = P_i$ $(i = 1, \ldots, r)$ une décomposition primaire minimale de Q. Pour tout idéal premier P de U avec $Q \subset P$, il existe un idéal premier P_i associé de Q, $1 \leq i \leq r$, tel que $Q \subset P_i \subset P$.*

Si P est un idéal premier de U avec $Q \subset P$, alors $\cap_{i=1}^{r} P_i \subset P$. En effet, on a $\cap_{i=1}^{r} Q_i = Q \subset P$, donc pour tout élément $x \in \cap_{i=1}^{r} P_i$, il existe un entier $n \geq 1$ tel que $x^n \in \cap_{i=1}^{r} Q_i$. Ainsi, $x^n \in P$ et $x \in P$. D'après la proposition 4.1-(ii), il existe un indice i, $1 \leq i \leq r$, tel que $P_i \subset P$. Par conséquent, P contient un idéal premier associé à Q. □

Ainsi, les idéaux premiers minimaux de U qui contiennent l'idéal Q sont précisément les éléments minimaux de l'ensemble des idéaux premiers associés à Q.

THÉORÈME 4.6. *Soient U un anneau de la classe $\mathcal{C}$, Q un idéal décomposable de U et $Q = \cap_{i=1}^{r} Q_i = \cap_{j=1}^{r} Q_j^*$ avec $\sqrt{Q_i} = \sqrt{Q_j^*} = P_i$ $(i = 1, \ldots, r)$ deux décompositions primaires minimales de Q. Alors, pour tout indice i, $1 \leq i \leq r$, dans lequel P_i est un idéal premier minimal associé à Q on a $Q_i = Q_i^*$, c'est-à-dire, les idéaux primaires correspondants aux idéaux premiers minimaux associés à Q sont déterminés de la façon unique.*

On suppose que P_i soit minimal dans l'ensemble $\{P_1, \ldots, P_r\}$. Pour démontrer l'unicité des idéaux primaires dans la décomposition, observons que pour chaque $j \neq i$, il existe un élément $a_j \in P_j$ et $a_j \notin P_i$. Considérons l'élément $a = (\ldots(((\ldots(a_1a_2)\ldots)a_{i-1})a_{i+1})\ldots)a_r$. Alors, $a \in P_j$ $(j = 1, \ldots, i-1, i+1, \ldots, r)$ et $a \notin P_i$. Ainsi, il existe un entier $m \geq 1$ tel que $a^m \in Q_j$ $(j \neq i)$ et pour tout entier $n \geq 1$, $a^n \notin Q_i$ et $a^n \notin Q_i^*$. Si $x \in Q_i$, alors $a^m x \in \cap_{i=1}^{r} Q_i = \cap_{j=1}^{r} Q_j^*$. En particulier, on a $a^m x \in Q_i^*$, donc $x \in Q_i^*$, c'est-à-dire, $Q_i \subset Q_i^*$. De même, on montre que $Q_i^* \subset Q_i$. □

Soit U un anneau commutatif. Un idéal Q de U est dit *irréductible* si pour des idéaux R et S de U vérifiant la condition $Q = R \cap S$, on a $Q = R$ ou $Q = S$. De plus, on dira que U est un *anneau noethérien* si toute chaîne croissante $I_1 \subset I_2 \subset \ldots \subset I_n \subset \ldots$ d'idéaux de U est stationnaire, c'est-à-dire, il existe un entier $n_0 \geq 1$ tel que pourtout $n \geq n_0$ on ait $I_n = I_{n_0}$.

LEMME 4.7. *Soit U un anneau commutatif. Si U est noethérien, tout idéal propre de U est une intersection finie d'idéaux irréductibles.*

La démonstration est analogue à celle du lemme 7.11 de [5].

LEMME 4.8. *Soit U un anneau de la classe $\mathcal{C}$. Si U est un anneau noethérien, alors tout idéal irréductible Q de U est primaire.*

En considérant l'anneau quotient U/Q, il suffit de démontrer que si l'idéal zéro est irréductible, alors celui-ci est primaire. En effet, supposons que $\{0\}$ ne soit pas un idéal primaire ; il existe alors des éléments a et b dans U tels que $ab = 0$, vérifiant les deux assertions suivantes : i) $a \neq 0$ et $b^n \neq 0$, pour tout entier $n \geq 1$; ii) $b \neq 0$ et $a^n \neq 0$, pour tout entier $n \geq 1$. Considérons la suite d'ensembles $N_n = \{x | x \in U, xa^{pn} = 0\}$ $(n = 1, 2, \ldots)$ et montrons que les N_n $(n = 1, 2, \ldots)$ sont des idéaux de U. En effet, il est évident que $0 \in N_n$ pour tout entier $n \geq 1$. D'autre part, quels que soient les éléments x et y dans N_n, on a $(x+y)a^{pn} = xa^{pn} + ya^{pn} = 0$, donc $x + y \in N_n$. De plus, si $x \in U$ et y est dans N_n, on a $(xy)a^{pn} = x(ya^{pn})$ car $a^p \in N(U)$ donc $(xy)a^{pn} = 0$ et xy appartient à N_n $(n = 1, 2, \ldots)$. Il s'ensuit aussi que $N_n \subset N_{n+1}$ $(n = 1, 2, \ldots)$. En effet, pour tout $x \in N_n$, on a $xa^{pn} = 0$ donc $xa^{p(n+1)} = x(a^{pn}a^p) = (xa^{pn})a^p = 0$, c'est-à-dire, $x \in N_{n+1}$. Comme U est un anneau noethérien, il existe un entier $n_0 \geq 1$ tel que $N_{n_0} = N_m$ pour tout entier $m \geq n_0$. Observons que $(b^p) \cap (a^{pn_0}) = \{0\}$. En effet, si $u \in (b^p)$, alors $u = \Sigma \pm c_1(c_2(\ldots(c_m b^p)\ldots))$, somme finie où les c_k sont dans U, d'après le lemme 1.1, donc $ua^p = 0$. De même, si $u \in (a^{pn_0})$, alors $u = \Sigma \pm d_1(d_2(\ldots(d_m a^{pn_0})\ldots))$, somme finie où les d_k sont dans U. Ainsi,

$$\begin{aligned} 0 = ua^p &= (\Sigma \pm d_1(d_2(\ldots(d_m a^{pn_0})\ldots)))a^p = ((\Sigma \pm d_1(d_2(\ldots(d_{m-1}d_m)\ldots)))a^{pn_0})a^p \\ &= (\Sigma \pm d_1(d_2(\ldots(d_{m-1}d_m)\ldots)))a^{pn_0+p}, \end{aligned}$$

c'est-à-dire, $\Sigma \pm d_1(d_2(\ldots(d_{m-1}d_m)\ldots)) \in N_{n_0+1} = N_{n_0}$, donc $u = 0$. Cela nous dit que $\{0\}$ n'est pas un idéal irréductible car $(b^p) \neq \{0\}$ et $(a^{pn_0}) \neq \{0\}$.

THÉORÈME 4.9. *Soit U un anneau de la classe $\mathbf{C}$. Si U est un anneau noethérien, tout idéal Q de U possède une décomposition primaire.*

Il s'agit d'une conséquence immédiate des lemmes 4.7 et 4.8.

5. APPLICATIONS AUX ANNEAUX ALTERNATIFS ET COMMUTATIFS EN CARACTÉRISTIQUE 3.

On sait que tout anneau alternatif et commutatif de caractéristique différente de 3 est associatif mais cela n'est plus vrai en caractéristique 3. En effet, soit U la nilalgèbre commutative de dimension 6 sur un corps F de caractéristique 3 dont la table de multiplication relative à une base $\{e_1, \ldots, e_6\}$ s'écrit $e_1e_3 = e_5$, $e_1e_4 = e_6$, $e_2e_3 = 2e_4$, $e_2e_5 = e_6$, les autres produits étant nuls. Il est immédiat de vérifier que U est une algèbre alternative (et commutative) mais non associative (cf. [1], exemple 1.2.3). La raison d'avoir choisi une nilalgèbre commutative de dimension 6 vient du fait que toute nilalgèbre alternative de dimension ≤ 5 est associative (cf. [6], proposition 4.2), indépendemment de la caractéristique du corps de base. Pour cette classe d'anneaux, on a les résultats suivants:

PROPOSITION 5.1. *Tout anneau U de caractéristique 3 commutatif et alternatif est un anneau à puissances associatives où la 3-ième puissance x^3 est un morphisme d'anneaux dont l'image est dans le noyau associatif $N(U)$.*

On sait que tout sous-anneau d'un anneau alternatif engendré par deux éléments est associatif. Ainsi, U est un anneau à puissances associatives et vérifie les identités

$$(i) \qquad (x+y)^3 = x^3 + y^3 \quad \text{et} \quad (xy)^3 = x^3y^3,$$

quels que soient les éléments x et y dans U. De plus, d'après les identités de Moufang, on a

$$(x^3y)z = (x(x(xy)))z = (x(xy)x)z = x((xy)(xz)) = x(x(yz)x) = x^3(yz),$$

quels que soient les éléments x, y et z dans U, c'est-à-dire,

$$(ii) \qquad (x^3, y, z) = 0.$$

D'après les identités (i) et (ii), on conclut que la 3-ième puissance x^3 est un morphisme d'anneaux dont l'image est dans le noyau associatif $N(U)$.

PROPOSITION 5.2. *Soit U un anneau de caractéristique 3 commutatif et alternatif. L'ensemble R des éléments nilpotents de U est un nilidéal de U et l'anneau quotient U/R ne possède aucun élément nilpotent non nul.*

C'est une conséquence directe du théorème 1.2.

PROPOSITION 5.3. *Soit U un anneau de caractéristique 3 commutatif et alternatif. Le nilradical R de U est l'intersection des idéaux premiers de U.*

Il s'agit d'une conséquence directe du théorème 2.1.

PROPOSITION 5.4. *Soit U un anneau de caractéristique 3 commutatif et alternatif semi-simple. Alors, U est un anneau associatif.*

Il s'agit d'une conséquence directe du théorème 3.1.

PROPOSITION 5.5. *Soit U un anneau de caractéristique 3 commutatif et alternative. Si U est noethérien, tout idéal de U possède une décomposition primaire.*

Il s'agit d'une conséquence directe du théorème 4.9.

6. APPLICATIONS AUX ANNEAUX DE JORDAN EN CARACTÉRISTIQUE 2.

D'après [9], on sait que si F est un corps de caractéristique 2 avec au moins quatre éléments distincts, toute F-algèbre de Jordan (commutative) U est une algèbre à puissances associatives où la deuxième puissance x^2 est dans le noyau associatif $N(U)$. Ainsi, en considérant la sous-classe des F-algèbres de Jordan vérifiant l'identité $(xy)^2 = x^2y^2$, on obtient les résultats suivants :

PROPOSITION 6.1. *Soient F un corps de caractéristique 2 avec au moins quatre éléments distincts et U une F-algèbre de Jordan vérifiant l'identité $(xy)^2 = x^2y^2$, quels que soient les éléments x et y dans U. L'ensemble R des éléments nilpotents de U est alors un nilidéal de U et l'algèbre quotient U/R ne possède aucun élément nilpotent non nul.*

C'est une conséquence directe du théorème 1.2.

PROPOSITION 6.2. *Soient F un corps de caractéristique 2 avec au moins quatre éléments distincts et U une F-algèbre de Jordan vérifiant l'identité $(xy)^2 = x^2y^2$, quels que soient les éléments x et y dans U. Le nilradical R de U est alors l'intersection des idéaux premiers de U.*

C'est une conséquence directe du théorème 2.1.

PROPOSITION 6.3. *Soient F un corps de caractéristique 2 avec au moins quatre éléments distincts et U une F-algèbre de Jordan semi-simple vérifiant l'identité $(xy)^2 = x^2y^2$, quels que soient les éléments x et y dans U. Alors U est une algèbre associative.*

Il s'agit d'une conséquence directe du théorème 3.1.

PROPOSITION 6.4.. *Soient F un corps de caractéristique 2 avec au moins quatre éléments distincts et U une F-algèbre de Jordan vérifiant l'identité $(xy)^2 = x^2y^2$, quels que soient les éléments x et y dans U. Si U est une algèbre noethérienne, tout idéal de U possède une décomposition primaire.*

Il s'agit d'une conséquence directe du théorème 4.9.

Un anneau U est dit une *généralisation d'anneaux alternatifs* $I(GAA\,I)$, s'il vérifie les identités ci-dessous (cf. [8]) :

(i) $(wx, y, z) + (w, x, [y, z]) = w(x, y, z) + (w, y, z)x$;
(ii) $([w, x], y, z) + (w, x, yz) = y(w, x, z) + (w, x, y)z$;
(iii) $(x, x, x) = 0$.

En considérant la sous-classe des anneaux de caractéristique 2, commutatifs et vérifiant l'identité $(xy)^2 = x^2y^2$, on obtient les résultats suivants :

PROPOSITION 6.5. *Soit U un anneau de caractéristique 2 commutatif et GAA I vérifiant l'identité $(xy)^2 = x^2y^2$, quels que soient les éléments x et y dans U. Alors, U est un anneau à puissances associatives où la 2-ième puissance x^2 est un morphisme dont l'image est dans le noyau associatif $N(U)$.*

D'après la commutativité de U et l'identité (i), on a $x^2x = xx^2$ et $x^2x^2 = x^3x = xx^3$. Comme $x^{n+1} = x^nx$, pour $n = 1, 2, \ldots$, alors $x^ix^j = x^{i+j}$ pour $i + j = 3$, $0 < i, j < 3$, et $i + j = 4$, $0 < i, j < 4$. On procède par la récurrence sur $i + j$. Supposons que $x^ix^j = x^{i+j}$, quels que soient $0 < i, j$ avec $i + j < n$ et soit $i + j = n$ pour $n \geq 5$. D'après (i), $(x^2, x^{n-2-i}, x^i) = 0$ ce qui nous amène à $x^{n-i}x^i = x^2x^{n-2}$ car $0 < i < n - 2$. D'autre part, $(x^2, x, x^{n-3}) = 0$ et $(x, x^2, x^{n-3}) = 0$, d'après (i). Donc, $x^3x^{n-3} = x^2x^{n-2}$ et $x^3x^{n-3} = xx^{n-1} = x^n$. Ainsi, $x^n = x^2x^{n-2}$ et donc on a $x^{n-i}x^i = x^n$ pour $0 < i < n - 2$. Si $i = n - 2$, on a aussi $x^2x^{n-2} = x^n$ et si $i = n - 1$, alors $xx^{n-1} = x^n$, d'après la définition. Donc, $x^ix^j = x^{i+j}$ pour $i + j = n$, $0 < i, j < n$, et U est un anneau à puissances associatives.

Ensuite, en prenant $w = x$ dans l'identité (i), il s'ensuit que $(x^2, y, z) = 0$, quels que soient les éléments x, y et z dans U. A partir des identités $(x + y)^2 = x^2 + y^2$ et $(xy)^2 = x^2y^2$, on conclut que la 2-ième puissance x^2 est un morphisme dont l'image est dans le noyau associatif $N(U)$.

PROPOSITION 6.6. *Soit U un anneau de caractéristique 2 commutatif et GAA I, vérifiant l'identité $(xy)^2 = x^2y^2$, quels que soient les éléments x et y dans U. L'ensemble R des éléments nilpotents de U est alors un nilidéal de U et l'anneau quotient U/R ne possède aucun élément nilpotent non nul.*

C'est une conséquence direct du théorème 1.2.

PROPOSITION 6.7. *Soit U un anneau de caractéristique 2 commutatif et GAA I, vérifiant l'identité $(xy)^2 = x^2y^2$, quels que soient les éléments x et y dans U. Le nilradical R de U est l'intersection des idéaux premiers de U.*

C'est une conséquence directe du théorème 2.1.

PROPOSITION 6.8. *Soit U un anneau de caractéristique 2 commutatif, GAA I et semi-simple, vérifiant l'identité $(xy)^2 = x^2y^2$, quels que soient les éléments x et y dans U. Alors, U est un anneau associatif.*

C'est une conséquence directe du théorème 3.1.

PROPOSITION 6.9. *Soit U un anneau de caractéristique 2 commutatif et GAA I, vérifiant l'identité $(xy)^2 = x^2y^2$, quels que soient les éléments x et y dans U. Si U est noethérien, tout idéal de U possède une décomposition primaire.*

C'est une conséquence directe du théorème 4.9.

EXEMPLE 6.10. Soient F un corps de caractéristique 2 avec au moins quatre éléments distincts et U une F-algèbre de dimension 4 dont la table de multiplication, relative à une base $\{e_1, \ldots, e_4\}$, s'écrit $e_1e_2 = e_2e_1 = e_3$, $e_3e_3 = e_4$, tous les autres produits étant nuls. Il s'agit d'une algèbre de Jordan. A partir d'un calcul direct, on constate que l'identité $(xy)^2 = x^2y^2$ n'est pas vérifiée.

Cet exemple nous montre que pour les algèbres de Jordan en caractéristique 2, l'application $U \to N(U)$ définie par $x \mapsto x^2$ *n'est pas toujours un morphisme d'algèbres* alors que si U est un anneau alternatif et commutatif de caractéristique 3, l'application $U \to N(U)$ définie par $x \mapsto x^3$ *est toujours un morphisme d'anneaux* (cf. proposition 5.1). On verra, par la suite, que ce type d'assertion est encore vraie pour les algèbres de Moufang commutatives en caractéristique 2.

7. NOTE SUR LES ALGÈBRES DE MOUFANG COMMUTATIVES.

Soient F un corps et U une F-algèbre flexible, c'est-à-dire, $x(yx) = (xy)x$, quels que soient x, y dans U. On dira que U est une *algèbre de Moufang* (cf. [1], [2]) si U est une algèbre qui vérifie les *identités de Moufang*, à savoir, $(xzx)y = x(z(xy))$, $y(xzx) = ((yx)z)x$ et $(xy)(zx) = x(yz)x$, quels que soient x, y et z dans U.

C'est un fait général que les algèbres de Moufang forment une classe d'algèbres "intermédiaires" entre les algèbres alternatives et celles de Jordan n.c., c'est-à-dire, non nécessairement commutatives, et ces trois classes d'algèbres (alternatives $\Longrightarrow$ Moufang $\Longrightarrow$ Jordan n.c.) sont distinctes. On renvoie à la bibliographie pour des exemples.

Si U est une algèbre de Moufang commutative, on sait (cf. [3], proposition 1.4) que $(xy)^3 = x^3y^3$ quels que soient x, y dans U et que pour tout x dans U, x^3 est dans $N(U)$. Ces résultats sont indépendants de la caractéristique du corps de base. Si l'on suppose que F soit de caractéristique 3, on n'a pas, en général, $(x+y)^3 = x^3 + y^3$ pour x et y dans U.

EXEMPLE 7.1. Considérons l'algèbre de Moufang commutative et non alternative U sur un corps F dont la table de multiplication relative à une base $\{e_1, e_2, e_3, e_4\}$ s'écrit $e_1^2 = e_3$, $e_1e_2 = e_3$, $e_1e_3 = e_4$, $e_2^2 = e_4$, tous les autres produits étant nuls (cf. [3], exemple 1.2). Si l'on suppose, de plus, que la caractéristique de F soit 3, on a $(e_1+e_2)^3 = 0$ alors que $e_1^3 + e_2^3 = e_4$.

Ainsi, si F est un corps de caractéristique 3 et U une F-algèbre de Moufang commutative, l'application $U \to U$ définie par $x \mapsto x^3$ n'est pas, en général,un morphisme d'algèbres bien que son image soit dans $N(U)$. Par contre, une assertion analogue est vraie si la caractéristique de F est 2, c'est-à-dire, on a le résultat suivant

PROPOSITION 7.2. *Soient F un corps de caractéristique 2 avec au moins quatre éléments distincts et U une F-algèbre de Moufang commutative. Alors U est une algèbre à puissances associatives où la 2-ième puissance x^2 est un morphisme dont l'image est dans le noyau associatif $N(U)$.*

En effet, on sait que toute algèbre de Moufang commutative est une algèbre à puissances associatives et de Jordan et d'après [9], on sait que si F est un corps de caractéristique 2 avec au moins quatre éléments distincts, toute F-algèbre de Jordan possède la deuxième puissance x^2 dans le noyau associatif $N(U)$. Ensuite, à partir de l'identité de Moufang au milieu, on a $(xy)^2 = (yx)(xy) = yx^2y = (x^2y)y = x^2y^2$ quels que soient x, y dans U et comme l'on a aussi $(x+y)^2 = x^2 + y^2$, il s'ensuit que la deuxième puissance x^2 est un morphisme.

PROPOSITION 7.3. *Soient F un corps de caractéristique 2 avec au moins quatre éléments distincts et U une F-algèbre de Moufang commutative. L'ensemble R des éléments nilpotents de U est alors un nilidéal de U et l'algèbre quotient U/R ne possède aucun élément nilpotent non nul.*

Cette proposition est une conséquence de la proposition 6.1 car toute algèbre de Moufang commutative est une algèbre de Jordan.

PROPOSITION 7.4. *Soient F un corps de caractéristique 2 avec au moins quatre éléments distincts et U une F-algèbre de Moufang commutative. Le nilradical R de U est alors l'intersection des idéaux premiers de U.*

Cette proposition est une conséquence de la proposition 6.2.

PROPOSITION 7.5. *Soient F un corps de caractéristique 2 avec au moins quatre éléments distincts et U une F-algèbre de Moufang commutative. Si U est une algèbre semi-simple alors U est une algèbre associative.*

Il s'agit ici d'une conséquence de la proposition 6.3. Dans [3], proposition 3.1, l'auteur démontre cette proposition pour un corps quelconque mais en supposant que l'algèbre soit de dimension finie.

PROPOSITION 7.6. *Soient F un corps de caractéristique 2 avec au moins quatre éléments distincts et U une F-algèbre de Moufang commutative. Si U est une algèbre noethérienne, tout idéal de U possède une décomposition primaire.*

C'est une conséquence de la proposition 6.4.

BIBLIOGRAPHIE

[1] L. Acosta, Sur les algèbres de Moufang, Thèse de Doctorat (1992), Université Montpellier II.

[2] L. Acosta, J.C. da Motta Ferreira, L.A. de Capua et A. Micali, Sur les algèbres de Moufang, Journal de Mathématiques du Maroc 1 (1993), 23–34.

[3] L. Acosta, Sur les algèbres de Moufang commutatives, Revista Colombiana de Matemáticas 29 (1995), 43-48.

[4] A.A. Albert, Power associative rings, Trans. Amer. Math. Soc. 64 (1948), 552–593.

[5] M.F. Atiyah and I.G. Macdonald, Introduction to Commutative Algebra, Addison-Wesley Publishing Company, Inc., 1969.

[6] L.A. de Capua et A. Micali, Sur les algèbres alternatives, Cahiers Mathématiques, Montpelleir 39 (1992), 348–370.

[7] I.N. Herstein, Noncommmutative rings, The Carus Mathematical Monographs, Number 15, John Wiley and Sons, Inc., 1968.

[8] E. Kleinfeld, Generalization of alternative rings I, J. Algebra 18 (1971), 304–325.

[9] L.A. Kokoris, Power associative rings of characteristic two, Proc. Amer. Math. Soc. 6 (1955), 705–710.

[10] H.C. Myung, On prime ideals and primary decompositions in a nonassociative ring, Osaka J. Math. 9 (1972), 41–47.

[11] R.Y. Sharp, Steps in Commutative Algebra, London Mathematical Society, Student Texts 19, Cambridge University Press, 1990.

[12] K.A. Zhevlakov, A.M. Slin'ko, I.P. Shestakov and A.I. Shirshov, Rings that are nearly associative, Academic Press, New York 1982.

Principal Q-bundles

ALEXANDER I. NESTEROV Departamento de Física, CUCEI, Universidad de Guadalajara, Blvd. M. García Barragán y Calz. Olímpica, Guadalajara, Jalisco, C.P. 44460, México. E-mail: nesterov@udgserv.cencar.udg.mx.

1 INTRODUCTION

The concept of fibre bundles, which has been introduced into differential geometry in the fifties, is applied not only to various fields of mathematics, such as: transformation group theory; Lie group theory; differential geometry; differential topology; characteristic classes, etc.; but to physics (gauge field theory in particular) as well. The definition of fibre bundles is given in terms of transformation groups and may consider the theory of fibre bundles as an application of the theory of transformation group, Lie group in a smooth case.

Lie group theory has been extraordinarily successfull in the development of modern physics, in particular, in the description of the strong, weak, and electromagnetic interactions. However, available difficulties to unify all interactions prompt to look for a mathematical structure beyond groups. For instance, the quantum group approach is a construction of this type where the Lie group symmetry is replaced by a quantum group symmetry and the latter reduced to the standard one in the some limit. An other possibility is a non-associative generalization of the Lie group, such as quasigroups and smooth loops.

Nowadays the non-associative structures begin to play an important role in the physics, for instance, non-associative objects such as 3-cocycles are linked with the chiral anomalies in field theory, there is a violation of the Jacobi identity in the quantum mechanics with the Dirac monopole, etc. Keeping in mind more general approaches to field theory, it would be reasonable to consider the generalized theory of fibre bundles based on the non-associative extension of Lie groups. This will allow to construct non-associative gauge theories which involve a higher "non-associative gauge symmetry".

In our paper we present the nonassociative generalization of the theory of principal fibre bundles said to be principal Q-bundles. The main results were obtained in [7].

2 PRELIMINARIES

The development of the theory of smooth loops started in 1955 with the appearence of the remarkable paper by A.I. Mal'cev on analytic loops [1]. Recently the infinitesimal theory of smooth local loops, similar to the Lie group – Lie algebras theory, has been developed by L.V. Sabinin [2, 3, 4, 5]. (Historical review the reader can find in [5, 6]. See also the references therein.)

Let $\langle Q, \cdot\rangle$ be a groupoid with a binary operation $(a, b) \mapsto a \cdot b$ and Q a smooth manifold. A groupoid $\langle Q, \cdot\rangle$ is called a *quasigroup* if equations $a{\cdot}x = b$, $y{\cdot}a = b$ have a unique solutions: $x = a\backslash b$, $y = b/a$. A *loop* is a quasigroup with a two-sided identity $a{\cdot}e = e{\cdot}a = a, \forall a \in Q$. A loop $\langle Q, \cdot, e\rangle$ with a smooth operation $\phi(a, b) := a{\cdot}b$ is called a *smooth loop*.

Let $\langle Q, \cdot, e\rangle$ be a smooth local loop with a neutral element e. We define

$$L_a b = R_b a = a{\cdot}b, \quad l_{(a,b)} = L^{-1}_{a\cdot b} \circ L_a \circ L_b, \quad \hat{l}_{(a,b)} = L_{a\cdot b} \circ l_{(a,b)} \circ L^{-1}_{a\cdot b} \qquad (1)$$

where L_a is a *left translation*, R_b is a *right translation*, $l_{(a,b)}$ is an *associator* and $\hat{l}_{(a,b)}$ is an *adjoint associator*.

Let $T_e(Q)$ be the tangent space of Q at the neutral element e. Then for each $X_e \in T_e(Q)$, we construct a smooth vector field on Q

$$X_b = L_{b*}X_e, \quad b \in Q,$$

where $L_{b*} : T(Q) \to T(Q)$ denotes the differential of the left translation. Note that X_b depends smoothly on both variables $b \in Q$ and $X_e \in T_e(Q)$ and satisfies the relation

$$\begin{aligned} L_{a*}X_b = L_{a*}L_{b*}X_e = (L_a \circ L_b)_* X_e &= (\hat{l}_{(a,b)} \circ L_{a\cdot b})_* X_e \\ &= \hat{l}_{(a,b)*} L_{a\cdot b*} X_e = \hat{l}_{(a,b)*} X_{a\cdot b} \end{aligned} \qquad (2)$$

DEFINITION 2.1: A vector field X on Q which satisfies the relation $L_{a*}X_b = \hat{l}_{(a,b)*}X_{a\cdot b}$ for any $a, b \in Q$ is called *left quasi-invariant* or *left fundamental* vector field.

Let $\mathcal{G}$ denote the set of left quasi-invariant vector fields on Q, which becomes a vector space under operations

$$(X + Y)_a = X_a + Y_a, \quad X, Y \in \mathcal{G},$$
$$(\alpha X)_a = \alpha X_a, \quad \alpha \in \mathbb{R}.$$

The vector space $\mathcal{G}$ can be equipped with a Lie commutator, which leads to the notion of quasialgebra.

DEFINITION 2.2: We define a *quasialgebra* $\mathfrak{q}$ on Q as the vector space of left quasi-invariant vector fields with the Lie commutator as the operation.

Let Γ_i, $i = 1, 2, \ldots, r$ be a basis of the left quasi-invariant fields. Then we have

$$[\Gamma_i, \Gamma_j] = C^p_{ij}(a)\Gamma_p, \tag{3}$$

where $C^p_{ij}(a)$ are the *structure functions*.

DEFINITION 2.3: A map $Q \longrightarrow Q$,

$$Ad_b(a) = L_a^{-1} \circ R_b^{-1} \circ L_{a \cdot b} \tag{4}$$

will be called *Ad-map.*

Evidently the map (4) leaves invariant the neutral element e and generates the map $T_e(Q) \longrightarrow T_e(Q)$,

$$\mathsf{Ad}_b(a) := \big(Ad_b(a)\big)_* = L^{-1}_{a*} R^{-1}_{b*} L_{a \cdot b*}. \tag{5}$$

DEFINITION 2.4: A vector-valued 1-form

$$\omega(V_a) := L^{-1}_{a*} V_a = V_e, \ V_e \in T_e(Q) \tag{6}$$

is called *canonical* Ad-form.

THEOREM 2.1: The 1-form ω is left quasi-invariant form and transforms with respect to left (right) translations as

$$(L^*_b \omega) V_a = (l_{(b,a)})_* \omega(V_a), \tag{7}$$

$$(R^*_b \omega) V_a = \mathsf{Ad}^{-1}_b(a) \omega(V_a). \tag{8}$$

Proof. For left translations one holds

$$(L^*_b \omega) V_a = \omega(L_{b*} V_a) = L^{-1}_{b \cdot a*} L_{b*} L_{a*} V_e = l_{(b,a)*} \omega(V_a),$$

thus (7) takes place. For right translations we get

$$\begin{aligned}(R^*_b \omega) V_a &= \omega(R_{b*} V_a) = L^{-1}_{a \cdot b*} R_{b*} L_{a*} L^{-1}_{a*} V_a \\ &= L^{-1}_{a \cdot b*} R_{b*} L_{a*} \omega(V_a) = \mathsf{Ad}^{-1}_b(a) \omega(V_a),\end{aligned}$$

and the proof is completed. □

3 PRINCIPAL Q-BUNDLES

3.1 Basic definitions

Let $\mathfrak{M}$ be a manifold and $\langle Q, \cdot, e \rangle$ a smooth two-sided loop. A *principal (loop) Q-bundle* (a principal quasigroup bundle with the structure loop Q) is a triple $(P, \pi, \mathfrak{M})$ such that the following conditions hold:

(1) $\mathfrak{M}$ is a coset space, $\mathfrak{M} = P/Q$, where P/Q is the orbit space of the Q-action on P, and the *canonical projection* $\pi : P \to \mathfrak{M}$ is a C^∞-surjection of P onto $\mathfrak{M}$;

(2) P is *locally trivial*: for each $x \in \mathfrak{M}$ there exists an open neighborhood U and a diffeomorphism $\Phi : \pi^{-1}(U) \to U \times Q$, such that for any point $u \in \pi^{-1}(U)$ and any element $q \in Q$ it has the form $\Phi(u) = (\pi(u), \varphi(u))$, where φ is the map from $\pi^{-1}(U)$ to Q.

(3) Q acts on P freely on the right: $(p, a) \in P \times Q \to \tilde{R}_a p \in P$; by definition $\varphi(\tilde{R}_a p) = R_a \varphi(p)$

P is called a *total space* or *bundle space*, Q is a *structure loop*, $\mathfrak{M}$ is a *base* or *base space* and π is a *projection*. For any $x \in \mathfrak{M}$ the inverse image $\pi^{-1}(x)$ is the F_x *fibre* over x. Any fibre is diffeomorphic to Q.

Let us consider a covering $\{U_\alpha\}$ of $\mathfrak{M}$, which can be chosen in such a way that the restriction of the fibration to each open set U_α is *trivializable* . This implies, that there exists a diffeomorphism Φ_α: $\pi^{-1}(U_\alpha) \to U_\alpha \times Q$. The set $\{U_\alpha, \Phi_\alpha\}$ is called a *local trivialization*. It is easiy to prove the following.

THEOREM 3.1: The right action on the fibre does not depend on the trivialization: if $p \in U_\alpha \cap U_\beta$, then $\tilde{R}_a p|_\alpha = \tilde{R}_a p|_\beta$.

Now consider two sets U_α and U_β of the covering with $U_\alpha \cap U_\beta \neq \varnothing$ and denote $\varphi_{\beta\alpha} = \varphi_\beta \circ \varphi_\alpha^{-1}$ the restriction of $\Phi_\beta \circ \Phi_\alpha^{-1}$ to the set $\pi^{-1}(U_\alpha \cap U_\beta)$:

$$\varphi_{\beta\alpha}: \quad \pi^{-1}(U_\alpha \cap U_\beta) \to Q.$$

¿From $\tilde{R}_a p|_\alpha = \tilde{R}_a p|_\beta$ we get $\varphi_{\beta\alpha}(q_\alpha \cdot a) = q_\beta \cdot a$, where we set $q_\alpha = \varphi_\alpha(p)$, $q_\beta = \varphi_\beta(p)$; and $p \in P$, $a \in Q$. The loop Q acts on itself on the left, therefore there exists $q_{\beta\alpha} \in Q$ such, that

$$q_{\beta\alpha} \cdot (q_\alpha \cdot a) = q_\beta \cdot a \Longrightarrow q_{\beta\alpha} = (q_\beta \cdot a)/(q_\alpha \cdot a) \equiv R^{-1}_{q_\alpha \cdot a} R_a q_\beta. \tag{9}$$

DEFINITION 3.1: The family of diffeomorphisms $\{q_{\beta\alpha}\} : U_\alpha \cap U_\beta \to Q$ is called the family of *transition functions* of the bundle $P(\mathfrak{M}, Q)$ corresponding to the covering $\{U_\alpha\}$ on $\mathfrak{M}$.

THEOREM 3.2: Transition functions $q_{\beta\alpha}$ depend on fibre.

Proof. In according to the definition we have

$$q_{\beta\alpha}(x, \tilde{R}_a p) = R^{-1}_{q_\alpha \cdot a} \circ R_a \circ R_{q_\alpha} \circ R^{-1}_{q_\alpha} q_\beta = r(a, q_\alpha) q_{\beta\alpha}(x, p),$$

where $r(a, q_\alpha) = R^{-1}_{q_\alpha \cdot a} \circ R_a \circ R_{q_\alpha}$, a right associator. Thus we find that $q_{\beta\alpha}$ is changed under the right translations and, therefore depends on the points of the fibre. □

DEFINITION 3.2: Let $\{U_\alpha, \Phi_\alpha\}$ be the local trivialization and $e \in Q$ a neutral element, then $\sigma_\alpha = \Phi^{-1}(x, e)$, $x \in U_\alpha$ is called the *local section* associated with the trivialization

Let $u \in P$ then one can write the local section as follows: $\sigma_\alpha = \tilde{R}^{-1}_{q_\alpha(u)} u$, where $q_\alpha := \varphi_\alpha(u)$.

THEOREM 3.3: The section σ_α does not depend on the choice of the point on the fibre.

Proof. Let $u, p \in P$ and $u = \tilde{R}_{q_\alpha(u)}\sigma_\alpha$, $p = \tilde{R}_{q_\alpha(p)}\sigma'_\alpha$. There exists an element $a \in Q$ such that $p = \tilde{R}_a u$. Then we have

$$p = \tilde{R}_{q_\alpha(p)}\sigma'_\alpha = \tilde{R}_a u = \tilde{R}_a \circ \tilde{R}_{q_\alpha(u)}\sigma_\alpha = \varphi_\alpha^{-1}(R_a \circ R_{q_\alpha(u)}\varphi_\alpha(\sigma_\alpha)) = \\ = \varphi_\alpha^{-1}(R_a \circ R_{q_\alpha(u)}e) == \tilde{R}_{q_\alpha \cdot a}\sigma_\alpha. \qquad (10)$$

Taking into account that $q_{\alpha(p)} = q_{\alpha(u)} \cdot a$, we obtain $R_{q_\alpha(p)}\sigma'_\alpha = R_{q_\alpha(p)}\sigma_\alpha$. This yields $\sigma_\alpha = \sigma'_\alpha$ and, hence, the section σ_α does not depend on the point of the fibre, while the dependence on the base, $\sigma_\alpha = \sigma_\alpha(x)$, can be hold. □

THEOREM 3.4: Let the point x lies in the intersection of the neighborhoods U_α and U_β, $x \in U_\alpha \cap U_\beta$, then the following formula holds

$$\sigma_\alpha(x) = \tilde{R}_{q_{\beta\alpha}}\sigma_\beta. \qquad (11)$$

Proof. Let $u = \tilde{R}_{q_\alpha}\sigma_\alpha$, rewritting this formula in the chart U_β, we obtain

$$\tilde{R}_{q_\alpha}\sigma_\alpha = \tilde{R}_{q_\beta}\sigma_\beta = \varphi_\beta^{-1}(R_{q_\beta}\varphi_\beta(\sigma_\beta)) = \varphi_\beta^{-1}(R_{q_{\beta\alpha}\cdot q_\alpha}\varphi_\beta(\sigma_\beta)) = \\ = \varphi_\beta^{-1}(R_{q_\alpha} \circ R_{q_{\beta\alpha}}e) = \varphi_\beta^{-1}(R_{q_\alpha} \circ R_{q_{\beta\alpha}}\varphi_\beta(\sigma_\beta)) = \tilde{R}_{q_\alpha} \circ \tilde{R}_{q_{\beta\alpha}}\sigma_\beta,$$

where the relation $\varphi_\beta(\sigma_\beta) = e$ has been used. Finally, we obtain that $\sigma_\alpha(x) = \tilde{R}_{q_{\beta\alpha}}\sigma_\beta$ in the intersection $U_\alpha \cap U_\beta$. □

DEFINITION 3.3: Let $P(\mathfrak{M}, Q, \pi)$ be a principal Q-bundle and $\{U_\alpha\}$ some covering of the base $\mathfrak{M}$. We say that the system $\{\mathfrak{M}, Q, \{U_\alpha\}, q_{\alpha\beta}\}$ is *principal coordinate Q-bundle*

Assume that there are two families of local sections, σ_α and σ'_α. The sections are linked by the right translations. This implies that $\exists q_\alpha \in Q : \sigma'_\alpha = R_{q_\alpha}\sigma_\alpha$. Using the set $\{\sigma_\alpha\}$, introduce new transition functions $q'_{\alpha\beta}$. Then the following transformation law holds

$$q'_{\alpha\beta} = L_{q_\alpha}^{-1} \circ R_{q_\beta} q_{\alpha\beta}.$$

3.2 Connection, curvature and Bianchi identities

Connection on the Q-bundle

Let $P(\mathfrak{M}, Q)$ be the principal Q-bundle over the manifold $\mathfrak{M}$. For any $u \in P$ a tangent space at u we denote as $T_u(P)$ and the tangent to the fiber passing through u as $\mathcal{V}_u$. We call $\mathcal{V}_u$ a *vertical subspace*. It is generated by the right translations on the fiber: $u \mapsto \tilde{R}_a u$, $a \in Q$, $u \in P$:

$$X_u = \frac{d}{dt}\tilde{R}_{a(t)}u\bigg|_{t=0}, \quad X_u \in \mathcal{V}_u,\ a \in Q,\ u \in P. \qquad (12)$$

The basic idea of connection is to compare the points in the "neighboring" fibres in a way that not dependent on a local trivialization. Let $\gamma(t) \in \mathfrak{M}$ be a smooth curve. A *horizontal lift* of γ is a curve $\tilde{\gamma}(t) \in P$ such that $\pi(\tilde{\gamma}(t)) = \gamma(t)$. Evidently, for determining $\tilde{\gamma}(t)$ it is suffucient to define at any point $u \in \tilde{\gamma}(t)$ a tangent vector $X_u : \pi_* \tilde{X}_u = X_{\pi(u)}$, where $X_{\pi(u)}$ is the tangent vector to $\gamma(t)$. A set $\{X_u\}$ is called a *horizontal subspace* $\mathcal{H}_u$.

DEFINITION 3.4: A *connection* on a principal Q-bundle is a smooth assignment to each point $p \in P$ of a horizontal subspace $\mathcal{H}_p$ of $T_p(P)$ such that:

(i)$T_p = \mathcal{V}_p \oplus \mathcal{H}_p$ (a direct sum).

(ii) The family of horizontal subspaces is invariant under the right action of the loop Q, i.e., for any $a \in Q$, $p \in P$

$$\mathcal{H}_{\tilde{R}_a p} = (\tilde{R}_a)_* \mathcal{H}_p,$$

where $\tilde{R}_{a*}$ denotes the right action of a on a tangent vector.

Applying the connection, any vector $Z \in T_p(P)$ can be presented as the following decomposition: $Z = X + Y$, where $X = horZ \in \mathcal{H}_p$ is the horizontal component of the vector Z and $Y = verZ \in \mathcal{V}_p$ is the vertical one. The diffeomorphism $\varphi(p)$ induces the map of the vertical subspace $\mathcal{V}_p$ onto the tangent space of Q, $\mathcal{V}_p \xrightarrow{\varphi_*} T_q(Q)$, $q = \varphi(p)$:

$$\tilde{V}_p = \frac{d}{dt}\tilde{R}_{a(t)}p\Big|_{t=0} \xrightarrow{\varphi_*} \hat{V}_q = \frac{d}{dt}R_{a(t)}q\Big|_{t=0}. \tag{13}$$

Note, that the vector field $\hat{V}_q$ is left quasi-invariant. Indeed, it can be written as $\hat{V}_q = (L_q)_* \hat{V}_e$, where

$$\hat{V}_e = \frac{da(t)}{dt}\Big|_{t=0} \in T_e(Q). \tag{14}$$

DEFINITION 3.5: Let $\hat{V}_e \in T_e(Q)$ be a vector in the tangent space at e. The vector field $\tilde{V}_p$ connecting with $\hat{V}_e$ by means of (13), (14) is called a *fundamental* vector field.

A notion of the fundamental vector field yields the decomposition of the tangent space on the horizontal and vertical components and leads to the following definition of the connection, which is equivalent to given above.

DEFINITION 3.6: A *connection form* in a principal Q-bundle is a vector-valued one-form taking values at $T_e(Q)$, which satisfies:

(i) $\omega(X_p) = \hat{X}_e$, where $X_p \in \mathcal{V}_p$, $\hat{X}_e \in T_e(Q)$ are determined in according to (13), (14).

(ii)$(\tilde{R}_a^* \omega) X_p = \mathrm{Ad}_a^{-1}(q)\omega(X_p)$, where $q = \varphi(p)$.

(iii) The horizontal subspace $\mathcal{H}_p$ is a kernel of ω:

$$\mathcal{H}_p = \{X \in T_p(P) : \omega(X) = 0\}.$$

The definition implies, that a connection in Q-bundle is determined by the left quasi-invariant form of Maurer-Cartan type.

With the given connection form can be associated a local one-form taking values at $T_e(Q)$. Let $\sigma : U \subset \mathfrak{M} \to \sigma(U) \subset P$, $\pi \circ \sigma = \mathrm{id}$ be a local section of a Q-bundle

$Q \to P \to \mathfrak{M}$ which is equipped with a connection one-form ω. Define the *local σ-representative* of ω to be the vector-valued one-form (taking values at $T_e(Q)$) ω^U on the open set $U \subset \mathfrak{M}$ given by $\omega^U := \sigma^*\omega$.

THEOREM 3.5: (On reconstruction of the connection form) For given one-form $\tilde{\omega}$ defined on $U \subset \mathfrak{M}$ with values at $T_e(Q)$ and the section $\sigma : U \to \pi^{-1}(U)$ there exists one and only one connection one-form ω on $\pi^{-1}(U)$ such, that $\sigma^*\omega = \tilde{\omega}$.

Proof. Let $p_0 = \sigma(x)$ and $Z \in T_{p_0}(P)$. The following decomposition is hold: $Z = X_1 + X_2$, where $X_1 := (\sigma_* \circ \pi_*)Z$ and $X_2 \in \mathcal{V}_{p_0}$, $\pi_* X_2 = 0$. Define at p_0 the one-form ω to be the vector-valued one-form given by $\omega_{p_0} = \tilde{\omega}_x(\pi_* X) + \hat{X}_2$. Continuation of the one-form ω onto all points of the fiber is realized by means of right translations, namely, $\forall p \in P$, $\exists a \in Q : p = \tilde{R}_a p_0$. This yields

$$\omega_p\big((\tilde{R}_a)_* X_{p_0}\big) = \mathsf{Ad}_a^{-1}(q_0)\omega_{p_0}(X_{p_0}), \quad q_0 = \varphi(p_0).$$

It is easy to see, that obtained one-form satisfies all conditions of the definition 3.6. □

Connection in the local trivialization

Let $\{U_\alpha, \Phi_\alpha\}$ be a local trivialization and let $\overline{\mathrm{Id}} : U_\alpha \to U_\alpha \times Q$ by $x \mapsto (x, e)$. A trivialization Φ_α defines a *canonical* section σ_α by the equation

$$\sigma_\alpha = \Phi_\alpha^{-1} \circ \overline{\mathrm{Id}}.$$

and vice versa. The canonical Ad-form we denote ω_C.

DEFINITION 3.7: Let $\omega_\alpha = \sigma_\alpha^*\omega$, where ω is the connection form. The form ω_α on U_α is called the *connection form in the local trivialization* $\{U_\alpha, \Phi_\alpha\}$.

THEOREM 3.6: At $U_\alpha \cap U_\beta$ local connection forms ω_α and ω_β, corresponding to the same connection ω on P, are related by

$$\omega_\beta = \mathsf{Ad}_{q_{\alpha\beta}}^{-1}(q_{\beta\alpha})\omega_\alpha + l_{(q_{\beta\alpha}, q_{\alpha\beta})*}\theta_{\alpha\beta}, \tag{15}$$

where $q_{\alpha\beta}$ are transition functions, and $\theta_{\alpha\beta} = q_{\alpha\beta}^*\omega_C$ denotes the pullback on $U_\alpha \cap U_\beta$ of the canonical one-form ω_C on Q. Vice versa, for any set of the local forms $\{\omega_\alpha\}$, satisfying (15), there exists the unique connection form ω on P generating this family of the local forms: $\omega_\alpha = \sigma_\alpha^*\omega$, $\forall\alpha$.

Proof. 1. The direct theorem. Let $x \in U_\alpha \cap U_\beta$. Applying (11), we obtain $\sigma_\beta(x) = \tilde{R}_{q_{\alpha\beta}}\sigma_\alpha(x)$, $\forall x \in U_\alpha \cap U_\beta$. Any vector $X \in T_x(U_\alpha \cap U_\beta)$ under the map $(\sigma_\beta(x))_*$ transform to $(\sigma_\beta)_* X \in T_{\sigma_\beta(x)}(P)$. Using the Leibniz formula [8], we get

$$(\sigma_\beta)_* = (\tilde{R}_{q_{\alpha\beta}})_*(\sigma_\alpha)_* + (L_{q_{\beta\alpha}})_*(q_{\alpha\beta})_*,$$

where $q_{\beta\alpha} := \varphi_\beta(\sigma_\alpha)$. Applying ω to both sides of this relation, we find

$$\begin{aligned}\omega_\beta(X) := \omega((\sigma_\beta)_* X) &= \omega\big((\tilde{R}_{q_{\alpha\beta}})_*(\sigma_\alpha)_* X\big) + \omega\big((L_{q_{\beta\alpha}})_*(q_{\alpha\beta})_* X\big) = \\ &= \mathsf{Ad}_{q_{\alpha\beta}}^{-1}(q_{\beta\alpha})\,\omega_\alpha(X) + l_{(q_{\beta\alpha}, q_{\alpha\beta})*}(q_{\alpha\beta}^*\,\omega_C)(X) = \\ &= \mathsf{Ad}_{q_{\alpha\beta}}^{-1}(q_{\beta\alpha})\,\omega_\alpha(X) + l_{(q_{\beta\alpha}, q_{\alpha\beta})*}\theta_{\alpha\beta}\,\omega_C(X).\end{aligned}$$

2. The inverse theorem. Let us define the one-form $\tilde{\omega}$ as follows:

$$\tilde{\omega} = \mathsf{Ad}_{q_\alpha}^{-1}(e)(\pi^*\omega_\alpha) + q_\alpha^*\omega_C, \quad q_\alpha := \varphi_\alpha(p).$$

Let $X \in T_u(P)$ be an arbitrary vector and $u = \sigma_\alpha(\pi(x))$. Decompose X in horizontal Y and vertical Z components:

$$X = Y + Z, \quad Y = (\sigma_\alpha)_*(\pi_* X), \quad \pi_* Z = 0.$$

This yields

$$\begin{aligned}\tilde{\omega}(X) &= \mathsf{Ad}_{q_\alpha}^{-1}(e)\omega_\alpha(\pi_* X) + (q_\alpha^*\omega_C)(X) = \\ &= \mathsf{Ad}_{q_\alpha}^{-1}(e)\omega((\sigma_\alpha)_*\pi_* X) + \omega_C((q_\alpha)_* X) = \\ &= (\tilde{R}_{q_\alpha}^*\sigma_\alpha^*)\omega(\pi_* Y) + \omega_C((q_\alpha)_* Z) = \\ &= \omega((\sigma_\alpha)_*\pi_* Y) + \omega_C((q_\alpha)_* Z) = \\ &= \omega(Y) + \hat{Z} = \omega(Y) + \omega(Z) = \omega(X),\end{aligned}$$

and one can see, that $\tilde{\omega} = \omega$ at any point of the section σ_α. These forms are transformed in the same way under the right translations and, therefore, coincide in $\pi^{-1}(U)$. □

COROLLARY 3.7: For arbitrary sections σ_1 and σ_2: $\sigma_2 = R_q\sigma_1$ and $\omega_1 = \sigma_1^*\omega$, $\omega_2 = \sigma_2^*\omega$, the following relation holds

$$\omega_2 = \mathsf{Ad}_q^{-1}(q_1)\omega_1 + l_{(q_1,q))*}(q^*\omega_C),$$

where $q_1 := \varphi(\sigma_1)$.

3.3 Covariant derivative. Curvature form

Let $\{x^\mu, y^i\}$ be a local coordinate system in the neighborhood $\pi^{-1}(U_\alpha)$: x^μ being coordinates in $U_\alpha \in \mathfrak{M}$ and y^i coordinates in the fiber. Locally $\pi^{-1}(U_\alpha)$ can be presented as a direct production $U_\alpha \times Q$. The connection form can written as the following: $\omega = \omega^i L_i$, $\{L_i\}$ being the basis of left quasi-invariant fields and $\{\omega^i\}$ basis of one-forms. In the coordinates $\{x^\mu, y^i\}$ this reads:

$$\omega^i = \left(\mathsf{Ad}_q^{-1}(e)\right)^i_j A^j_\mu(x)dx^\mu + \omega^i{}_j dy^j. \tag{16}$$

DEFINITION 3.8: A *covariant derivative* D_μ in the principal Q-bundle is defined as

$$D_\mu = \partial_\mu - A^i_\mu(x)\bar{L}_i,$$

where $\bar{L}_i = (R_*)^j_i\partial/\partial y^j$ are generators of the left translations.

We demonstrate, that $\omega(D_\mu) = 0$. Indeed, (16) yields

$$\omega(D_\mu) = [\left(\mathsf{Ad}_q^{-1}(e)\right)^i_j A^j_\mu(x) + \omega^i{}_p A^j_\mu(R_*)^p_j]L_i.$$

Noting that $\omega^i{}_p = (L_*^{-1})^i_p$, we obtain $\omega^i{}_p(R_*)^p_j = (\mathsf{Ad}_q^{-1}(e))^i_j$, and, hence, $\omega(D_\mu) = 0$.

Let us compute the commutator $[D_\mu, D_\nu]$. The computation yields

$$[D_\mu, D_\nu] = (\partial_\nu A^i_\mu - \partial_\mu A^i_\nu)\bar{L}_i + A^i_\mu A^j_\nu [\bar{L}_i, \bar{L}_j].$$

Introducing $[\bar{L}_i, \bar{L}_j] = C^p_{ij}(y)\bar{L}_p$, we get

$$[D_\mu, D_\nu] = -F^i_{\mu\nu}\bar{L}_i, \tag{17}$$

$$F^i_{\mu\nu} := \partial_\mu A^i_\nu - \partial_\nu A^i_\mu - A^i_\mu A^j_\nu C^i_{jp}. \tag{18}$$

DEFINITION 3.9: Let Ψ be a vector-valued r-form in the principal Q-bundle. A $(r+1)$-form $D\Psi$ defined as

$$D\Psi(X_1, X_2, \ldots, X_{r+1}) = d\Psi(horX_1, \ldots, horX_{r+1}) \tag{19}$$

is called a *covariant differential* of the form Ψ.

DEFINITION 3.10: A vector-valued two-form $\Omega(X,Y)$ defined as

$$\Omega(X,Y) = D\omega(X,Y) = d\omega(horX, horY), \tag{20}$$

where ω is a connection form, is called a *curvature form.*

LEMMA 3.8: Let X, Y be horizontal fields, then the following relation holds

$$\omega([X,Y]) = -2\Omega(X,Y).$$

Proof. Using the lemma on the exterior defferentiation [8], we obtain

$$2d\omega(X,Y) = X\omega(Y) - Y\omega(X) - \omega([X,Y]).$$

As $X, T \in \mathcal{H}_p$, then $\omega(X) = \omega(Y) = 0$. This yields $\omega([X,Y]) = -2\Omega(X,Y)$. □

COROLLARY 3.9: Curvature form can be defined as follows

$$\Omega(X,Y) = -\frac{1}{2}\omega([hor\bar{Y}, hor\bar{Y}]),$$

where $\bar{X}, \bar{Y}$ is any continuation of the vectors $X, Y \in T_p(P)$.

COROLLARY 3.10: Two-form Ω is Ad-form and transforms under the right translations as the following:

$$(\tilde{R}^*_a\Omega)(X,Y) = \mathsf{Ad}^{-1}_a(\varphi(p))\Omega|_p(X,Y). \tag{21}$$

THEOREM 3.11: Curvature form Ω satisfies the structure equation

$$\Omega = d\omega + \omega \wedge \omega. \tag{22}$$

Proof. The point is to prove (22), considering all possible pairs X, Y.

1. Let X, Y be horizontal vectors, then $\omega(X) = \omega(Y) = 0$. This yields

$$d\omega(X,Y) + \omega \wedge \omega(X,Y) = d\omega(horX, horY) = \Omega(X,Y).$$

2. Let X, Y be vertical vectors. Without loss of generality one can assume that X, Y are fundamental vector fields. This implies $\omega(X) = \hat{X}$, $\omega(Y) = \hat{Y}$, $\hat{X}, \hat{Y} \in T_e(Q)$. The computation yields

$$\begin{aligned} 2d\omega(X,Y) &= X\omega(Y) - Y\omega(X) - \omega([X,Y]) = X(\hat{Y}) - Y(\hat{X}) - \omega([X,Y]) \\ &= -[\hat{X},\hat{Y}] = -[\omega(X),\omega(Y)] \Longrightarrow d\omega(X,Y) + \frac{1}{2}[\omega(X),\omega(Y)] = 0. \end{aligned}$$

As X, Y are vertical fields, then $\Omega(X,Y) = d\omega(horX, horY) = 0$.

3. Let X be a horizontal vector field and Y be a vertical one (or fundamental vector field). From the one side, we have

$$\Omega(X,Y) = d\omega(horX, horY) = 0,$$

and from the other side,

$$d\omega(X,Y) = X\omega(Y) - Y\omega(x) - \omega([X,Y]) = 0,$$

since the commutator $[X,Y]$ is the horizontal vector field. The last statement arises from the following lemma. □

LEMMA 3.12: Let Z be a fundamental vector field and X be a horizontal vector field, then the commutator $[X, Z]$ be a horizontal vector field.

Proof. The fundamental vector field is induced by the right translations $\tilde{R}_{a(t)}$. The commutator can be defined as follows (see, e.g., [8])

$$[X, Z] = \lim_{t \to 0} \frac{1}{t}((\tilde{R}_{a(t)})_* X - X).$$

If X is a horizontal vector field, then $(\tilde{R}_{a(t)})_* X$ is a horizontal vector field, and, hence, $[X, Z]$ is a horizontal one. □

Let us compute $\Omega_{\mu\nu} = \Omega(D_\mu, D_\nu)$:

$$\Omega_{\mu\nu} = -\frac{1}{2}\omega([D_\mu, D_\nu]) = \frac{1}{2}F^i_{\mu\nu}\omega(\bar{L}_i) = \frac{1}{2}F^i_{\mu\nu}(\mathrm{Ad}_q^{-1}(e))^j_i \hat{L}_j(e), \tag{23}$$

where $\{L_j(e) = \omega(L_j)\}$ is the basis of the quasileftinvariant vector fields at $T_e(Q)$ and $q = \varphi(p)$, $p \in P$. Choosing the family of the local sections σ_α associated with the trivialization, we obtain

$$\pi^*\Omega|_\alpha = \frac{1}{2}F_{\mu\nu}dx^\mu \wedge dx^\nu, \tag{24}$$

where it is taking into account, that $\varphi_\alpha(\sigma_\alpha) = e$. Equation (23) implies, that at $\pi^{-1}(U_\alpha)$ it holds $\Omega = \mathrm{Ad}^{-1}_{q_\alpha}(e)(\pi^*\Omega|_\alpha)$, and at $\pi^{-1}(U_\beta)$ one has $\Omega = \mathrm{Ad}^{-1}_{q_\beta}(e)(\pi^*\Omega|_\beta)$. At the intersection $\pi^{-1}(U_\alpha \cap U_\beta)$ these two expressions must coincide:

$$\mathrm{Ad}^{-1}_{q_\alpha}(e)(\pi^*\Omega|_\alpha) = \mathrm{Ad}^{-1}_{q_\beta}(e)(\pi^*\Omega|_\beta).$$

Noticing that as Ω is a Ad-form we obtain

$$\Omega|_\beta = \mathrm{Ad}^{-1}_{q_{\alpha\beta}}(q_{\beta\alpha})\Omega|_\alpha. \tag{25}$$

THEOREM 3.13: Bianchi identity: $D\Omega = 0$.

Proof. It is sufficient to show that $d\Omega(X, Y, Z) = 0$, if X, Y, Z are horizontal vector fields. Applying the exterior derivative to (22), we obtain $d\Omega(X, Y, Z) = 0$, if X, Y, Z are horizontal vector fields. □

ACKNOWLEDGMENTS

I am grateful to L.V. Sabinin for helpful discussions and comments. This work was supported by CONACyT, Grant No. 1626P-E.

REFERENCES

[1] AI Mal'cev. Analytic loops. Mat. Sb. (N.S.) 36(78): 569 - 576, 1955.

[2] LV Sabinin. Differential equations of smooth loops. Proceedings of Seminar on Vector and Tensor Analysis, Moscow, 1988. 23: 133 - 146.

[3] LV Sabinin. Analytical Quasigroups and Geometry. Moscow: Peoples' Friendship University, 1991.

[4] LV Sabinin. On differential equations of smooth loops. Russian Mathematical Survey 49: 172 - 173, 1994.

[5] LV Sabinin. Smooth quasigroups and loops. Dordreht: Kluwer Academic Publishers, 1999.

[6] O Chein, H Pflugfelder and JDH Smith, eds. Quasigroups and loops: Theory and Applications. Berlin: Heldermann Verlag, 1990.

[7] AI Nesterov. Methods of nonassociative algebra in physics. Dr Sci Dissertation, Institute of Physics, Estonian Ac Sci, Tartu, Estonia, 1989.

[8] SK Kobayashi and K Nomizy. Foundations of Differential Geometry, Vol. 1,2. New York: Interscience Publisher, 1963, 1969.

Z×Z-graded Lie algebras containing a Heisenberg algebra

J. MARSHALL OSBORN Department of Mathematics, University of Wisconsin, Madison, WI 53706.

KAIMING ZHAO Institute of Systems Science, Academia Sinica, Beijing, 100080, China

ABSTRACT In this paper we study $\mathbf{Z} \times \mathbf{Z}$-graded Lie algebras $A = \sum A_{i,j}$ with $\dim A_{i,j} \leq 1$, where $A_{0,-1}$, $A_{0,0}$, $A_{0,1}$ span a Heisenberg algebra H, and $[A_{-1,0}, A_{1,0}] = 0$. We show that there are no such algebras satisfying the additional conditions (i) that A is generated by $A_{\pm1,0}$ and $A_{0,\pm1}$, (ii) $A_{0,0}$ is not in the center of A, and (iii) that $H \cap [A, A] \neq A_{0,0}$. If Condition (iii) is removed, then there are many such Lie algebras, but they do not seem interesting because of their weaker structure.

1. INTRODUCTION

This paper is one of a series in which we study algebras $A = \sum A_{i,j}$ graded by $\mathbf{Z} \times \mathbf{Z}$ where $\dim A_{i,j} \leq 1$ over a field F of characteristic 0. For each i, $j \in \mathbf{Z}$, we choose $w_{i,j}$ to be an element which spans $A_{i,j}$. We have pointed out in [OZ2] that the subalgebra generated by $w_{0,-1}$ and $w_{0,1}$ is either sl2, a Heisenberg algebra, or abelian. The same is of course true of the subalgebra generated by $w_{-1,0}$ and $w_{1,0}$. In this paper we consider the case where $w_{0,-1}$ and $w_{0,1}$ generate a 3-dimensional Heisenberg algebra, and $w_{-1,0}$ and $w_{1,0}$ commute.

We now give a formal statement of the main result that we prove in this paper.

THEOREM 1.1 *There exist no Lie algebras A with the following properties:*

(I) $A = \sum A_{i,j}$ *is graded by* $\mathbf{Z} \times \mathbf{Z}$ *with* $\dim A_{i,j} \leq 1$;

AMS Subject Classification: 17B20, 17B36, 17B65, 17B70.

(II) $A_{0,-1}$, $A_{0,0}$, $A_{0,1}$ *span a Heisenberg algebra* H;
(III) $[A_{-1,0}, A_{1,0}] = 0$;
(IV) A *is generated by* $A_{\pm1,0}$ *and* $A_{0,\pm1}$;
(V) $A_{0,0}$ *is not in the center,*
(VI) $H \cap [A, A] \neq A_{0,0}$.

In Section 2 we develop some properties involving Heisenberg modules and embeddings of the Heisenberg algebra, and in Section 3 we give the proof of Theorem 1.1. We conclude the paper by noting that if (IV) is not assumed then one can construct metabelian examples satisfying the remaining axioms using arbitrary Heisenberg modules for $\sum_{i\neq 0} A_{i,j}$.

2. ABOUT THE 3-DIMENSIONAL HEISENBERG ALGEBRA

For convenience, we shall introduce the usual Heisenberg notation here by setting $e_i = w_{0,i}$ for $i \in \{0, \pm1\}$ with $[e_{-1}, e_1] = e_0$ and e_0 in the center, and denote this Heisenberg algebra by H. We define $A_r = \oplus_{i\in\mathbf{Z}} A_{i,r}$ and $A'_r = \oplus_{i\in\mathbf{Z}} A_{r,i}$ for any $r \in \mathbf{Z}$. In a previous paper [OZ4] we established

LEMMA 2.1 *Let* $V = \sum_{i\in\mathbf{Z}} V_i$ *be an indecomposible* H*-module with* $\dim V_i \leq 1$ *and* $e_{-1}V_i \subset V_{i-1}$ *and* $e_1 V_i \subset V_{i+1}$ *for all* i. *Suppose that* $HV \neq 0$. *Then there exists an* $a \in F^*$ *so that one of the following is true:*

(a) $V = \oplus_{i\geq 0} F v_i$, $e_{-1}v_i = iv_{i-1}$, $e_1 v_i = av_{i+1}$, $e_0 v_i = av_i$;

(b) $V = \oplus_{i\leq 0} F v_i$, $e_{-1}v_i = av_{i-1}$, $e_1 v_i = iv_{i+1}$, $e_0 v_i = av_i$;

(c) $V = \oplus_{i\in\mathbf{Z}} F v_i$, *and there exists* $b \in F$ *such that* $e_{-1}v_i = (i+b)v_{i-1}$, $e_1 v_i = av_{i+1}$, $e_0 v_i = av_i$;

(d) $V = \oplus_{i\in\mathbf{Z}} F v_i$, *and there exists* $b \in F$ *such that* $e_{-1}v_i = av_{i+1}$, $e_1 v_i = (b+i)v_{i-1}$, $e_0 v_i = av_i$.

Under the duality that comes from interchanging the signs($\pm$) of subscripts of the e's and the v's we obtain that (a) and (b) are dual, and (c) and (d) are dual. If we rechoose the basis for H we can further assume that $a = 1$ in the above Lemma. In our context here, the Heisenberg algebra is embedded in the larger subalgebra A'_0, and we wish to know how A'_0 acts on the modules A'_i for $i \neq 0$.

Now we construct some $\mathbf{Z}$-graded Lie algebras containing a Heisenberg algebra. Let $Z^+ = \{0, 1, 2, \cdots\}$, and let $\mathbf{Z}^+_\infty$ denote the set of positive integers together with infinity. For any $m, n \in \mathbf{Z}^+_\infty$ with $n > 1$, the algebra $H_{m,n} = \text{span}\{e_i | i \leq -n, \text{ or } -1 \leq i \leq m\}$ is defined by

$$[e_{-1}, e_i] = ie_{i-1}, \ i \neq -1 \tag{2.1}$$

and other brackets 0. We also define the algebra $\bar{H}_{m,n} = \text{span}\{e_i | i \geq n, \text{ or } 1 \geq i \geq -m\}$ by

$$[e_1, e_i] = ie_{i+1}, \ i \neq 1 \tag{2.2}$$

and other brackets 0. We see that $\bar{H}_{m,n}$ is dual to $H_{m,n}$. We also have that $H_{m,n} \simeq \bar{H}_{m,n}$, and $H_{1,\infty}$ is the 3-dimensional Heisenberg algebra.

PROPOSITION 2.2 *Let* $C = \sum_{j\in\mathbf{Z}} C_j$ *be a* $\mathbf{Z}$*-graded Lie algebra with* $\dim C_j \leq 1$ *such that* $H = C_1 + C_0 + C_{-1}$ *is a Heisenberg algebra. Let* $V = \sum_{i\in\mathbf{Z}} V_i$ *be a nonzero* $\mathbf{Z}$*-graded* C*-module generated by* V_0 *with* $\dim V_i \leq 1$ *and* $C_0 V_i = V_i$ *for all*

$i \in \mathbf{Z}$. If C faithfully acts on V, then there exist $m, n \in \mathbf{Z}^+_\infty$ such that $C \simeq H_{m,n}$ or $\bar{H}_{m,n}$. Further there exist $s, t \in F$ and $e_i \in C_i$ satisfying (2.1) for $H_{m,n}$ and (2.2) for $\bar{H}_{m,n}$, and $v_i \in V_i$ such that one of the following is true:

(a) *$C \simeq H_{m,n}$ and $e_j v_i = v_{i+j}$ for $j \neq -1$ and $e_{-1} v_i = (s+i) v_{i-1}$, where $V_i \neq 0$ for all i, and $n \neq \infty$ if $s \in \mathbf{Z}^+$;*

(b) *$C \simeq \bar{H}_{m,n}$ and $e_j v_i = v_{i+j}$ for $j \neq 1$ and $e_1 v_i = (s+i) v_{i+1}$, where $V_i \neq 0$ for all i, and $n \neq \infty$ if $-s \in \mathbf{Z}^+$;*

(c) *$C \simeq \bar{H}_{n,m}$, for $k \geq n$,*

$$e_k v_{-s-i} = \begin{cases} \frac{(n-1)(n-2)\cdots(k-i+1)}{(-1)\cdots(-i+1)} v_{k-s-i}, & \forall\, i \in \{1, 2, \cdots k\} \\ 0, & \text{otherwise}; \end{cases}$$

$e_1 v_i = v_{i+1}$ and for $0 \geq k \geq -m$,

$$e_k v_i = (s+i)(s+i-1)\cdots(s+i+k+1) v_{i+k} \ \forall\, i \in \mathbf{Z},$$

where $-s \in \mathbf{Z}^+$ and $V_i \neq 0$ for all i;

(d) *$C \simeq H_{m,n}$, for $k \leq -n$,*

$$e_k v_{-s+i} = \begin{cases} \frac{(-n+1)(-n+2)\cdots(k+i-1)}{(1)\cdots(i-1)} v_{k-s+i}, & \forall\, i \in \{1, 2, \cdots k\} \\ 0, & \text{otherwise}; \end{cases}$$

$e_{-1} v_i = v_{i-1}$ and for $0 \leq k \leq m$,

$$e_k v_i = (s+i)(s+i+1)\cdots(s+i+k-1) v_{i+k} \ \forall\, i \in \mathbf{Z},$$

where $s \in \mathbf{Z}^+$ and $V_i \neq 0$ for all i;

(e) *$C \simeq \bar{H}_{m,\infty}$ or $H_{m,\infty}$,*

$$e_k v_i = v_{k+i}, \ \forall\, k \geq 0,\ i \geq -s.$$

$$e_k v_i = (s+i)(s+i-1)\cdots(s+i+k+1) v_{i+k} \ \forall\, 0 \leq k < m,\ i \geq -s.$$

where $s \in \mathbf{Z}^+$ and $V_i \neq 0$ only when $i \geq -s$;

(f) *$C \simeq \bar{H}_{m,\infty}$ or $H_{m,\infty}$,*

$$e_k v_i = v_{k+i}, \ \forall\, k \leq -m,\ i \leq 0.$$

$$e_k v_i = (s+i)(s+i+1)\cdots(s+i+k-1) v_{i+k} \ \forall\, 0 \geq k > -m,\ i \leq s.$$

where $-s \in \mathbf{Z}^+$ and $V_i \neq 0$ only when $i \leq s$.

Proof. When we use v_i and e_i in this proof we will always assume that $v_i \in V_i$ and $e_i \in C_i$. Consider the H-submodule U generated by V_0. By rechoosing the basis of H, from Lemma 2.1 we see that there exists $s \in F$ such that (i) $s \notin \mathbf{Z}, U = \oplus_{i \in \mathbf{Z}} F v_i$, $e_{-1} v_i = (s+i) v_{i-1}$, $e_1 v_i = v_{i+1}$, $e_0 v_i = v_i$, or (ii) $-s \in \mathbf{Z}^+, U = \oplus_{i \in \mathbf{Z}} F v_i$, $e_{-1} v_i = (s+i) v_{i-1}$, $e_1 v_i = v_{i+1}$, $e_0 v_i = v_i$, or (iii) $s \in \mathbf{Z}^+, U = \oplus_{i \geq -s} F v_i$, $e_{-1} v_i = (s+i) v_{i-1}$, $e_1 v_i = v_{i+1}$, $e_0 v_i = v_i$, or one of the dual cases to these three cases holds. We consider these three cases separately.

Case (i): $s \notin \mathbf{Z}, U = \oplus_{i \in \mathbf{Z}} F v_i$, $e_{-1} v_i = (s+i) v_{i-1}$, $e_1 v_i = v_{i+1}$, $e_0 v_i = v_i$.

In this case we can choose another basis $u_i \in V_i$ such that

$$e_{-1}u_i = u_{i-1}, \; e_1 u_i = (s+i+1)u_{i+1}, \; e_0 u_i = u_i.$$

Suppose $n > 1$ is the first integer (if it exists) such that $e_n \neq 0$.

Subcase 1: $[e_{-1}, e_n] = 0$.

By letting $e_n u_i = b_i u_{i+n}$ where $b_i \in F$, and computing $[e_{-1}, e_n]u_i = 0$, we deduce that $b_i - b_{i-1} = 0$, $\forall\, i \in \mathbf{Z}$. Since $e_n V \neq 0$, we choose e_n such that $e_n u_i = u_{i+n}$, $\forall\, i \in \mathbf{Z}$. Using this we see that $[e_1, e_n]u_i = n u_{i+n+1} \neq 0$, so we can choose $e_{n+1} = n^{-1}[e_1, e_n]$. Thus $e_{n+1}u_i = u_{i+n+1}$ for all $i \in \mathbf{Z}$. In this manner by induction on $k \geq n$, we can choose e_k such that $e_k u_i = u_{i+k}$, $\forall\, i \in \mathbf{Z}$, and

$$[e_k, e_1] = -k e_{k+1}, \; [e_k, e_{-1}] = 0.$$

Subcase 2: $[e_{-1}, e_n] \neq 0$.

We deduce that $n = 2$, and we can choose e_2 such that $[e_{-1}, e_2] = 2e_1$. By letting $e_2 v_i = b_i v_{i+2}$ where $b_i \in F$, and using $[e_{-1}, e_2]v_i = 2e_1 v_i$ we deduce that

$$(s+2+i)b_i - (s+i)b_{i-1} = 2. \tag{2.3}$$

Solving for b_i from (2.3) we get

$$b_i = \frac{b}{(s+i+1)(s+i+2)} + 1, \; \forall\, i \in \mathbf{Z}, \tag{2.4}$$

where $b \in F$. Suppose $b \neq 0$. We compute

$$[e_1, e_2]v_i = \frac{2b}{(s+i+1)(s+i+2)(s+i+3)} v_{i+3} \neq 0, \; \forall\, i \in \mathbf{Z},$$

and we choose $e_3 \in C_3$ such that $[e_1, e_2] = 2e_3$. In this manner by induction on $k \geq 3$ we can choose $e_k = (k-1)^{-1}[e_1, e_{k-1}]$ such that $[e_1, e_{k-1}] = (k-1)e_k$ and

$$e_k v_i = \frac{b}{(s+i+1)(s+i+2)\cdots(s+i+k)} v_{k+i}, \; \forall\, i \in \mathbf{Z}. \tag{2.5}$$

We know that there exists $a \in F$ such that $[e_2, e_3] = a e_5$. It follows from (2.4) and (2.5) that

$$[e_2, e_3]v_i = \frac{6b}{(s+i+1)(s+i+2)(s+i+4)(s+i+5)} v_i \neq a e_5 v_i,$$

a contradiction. Thus $b = 0$. From (2.4) we see that

$$e_2 v_i = v_{i+2}, \; \forall\, i \in \mathbf{Z}.$$

By induction on $k \geq 2$, if $[e_{-1}, e_k] \neq 0$, using a similar argument we can choose e_j for $2 \leq j \leq k$ such that $[e_{-1}, e_j] = j e_{j-1}$, $e_j v_i = v_{i+j}$ and $[e_1, e_j] = 0$.

If there exists $n' > 2$ such that $[e_{n'}, e_{-1}] = 0$, letting $e_{n'}v_i = c_i v_{i+n'}$ and by computing $0 = [e_{n'}, e_{-1}]v_i$ we get

$$(s + n' + i)c_i - (s + i)c_{i-1} = 0, \ \forall\, i \in \mathbf{Z}.$$

Thus

$$c_i = \frac{c}{(s+i+1)(s+i+2)\cdots(s+i+n')}, \ \forall\, i \in \mathbf{Z},$$

where $c \in F$. Choose $e_{n'}$ such that

$$e_{n'}v_i = \frac{1}{(s+i+1)(s+i+2)\cdots(s+i+n')} v_{i+n'}, \ \forall\, i \in \mathbf{Z}.$$

We can show inductively that, for $k \geq n'$, we can choose $e_k \in C_k$ such that

$$e_k v_i = \frac{1}{(s+i+1)(s+i+2)\cdots(s+i+k)} v_{i+k}, \ \forall\, i \in \mathbf{Z}.$$

We know that there exists $a \in F$ such that $[e_2, e_{n'}] = a e_{n'+2}$. But

$$[e_2, e_{n'}]v_i = [\frac{1}{(s+i+1)(s+i+2)\cdots(s+i+n')}$$

$$- \frac{1}{(s+i+3)(s+i+4)\cdots(s+i+n'+2)}]v_{n'+i+2} \neq a e_{n'+2} v_i,$$

a contradiction. So this case does not occur.

Thus one of the following holds: (1) for $k \geq n$, we can choose e_k such that $e_k u_i = u_{i+k}$, $\forall\, i \in \mathbf{Z}$, and $[e_k, e_1] = -k e_{k+1}$, $[e_k, e_{-1}] = 0$; (2) for $k \geq 2$, if $C_k \neq 0$, we can choose e_j for $2 \leq j \leq k$ such that $[e_{-1}, e_j] = j e_{j-1}$, $[e_1, e_j] = 0$ and $e_j v_i = v_{i+j}$ for all i.

Suppose $m < -1$ is the first integer (if it exists) such that $e_m \neq 0$. By similar arguments or by duality, we deduce that one of the following holds: (3) for $k \leq m$, we can choose e_k such that $e_k v_i = v_{i+k}$, $\forall\, i \in \mathbf{Z}$, and $[e_k, e_{-1}] = -k e_{m-k+1}$, $[e_k, e_1] = 0$; (4) For $k \leq -2$, if $C_k \neq 0$, we can choose e_j for $-2 \geq j \geq k$ such that $[e_1, e_j] = j e_{j+1}$, $[e_{-1}, e_j] = 0$ and $e_j u_i = u_{i+j}$ for all i.

The combination of (1) and (4) gives (b); The combination of (2) and (3) gives (a) with $s \notin \mathbf{Z}$. We need to remove the combination of (1) and (3) and the combination of (2) and (4).

First consider the combination of (2) and (4) with $C_{\pm 2} \neq 0$. Note that $e_2 v_i = v_{i+2}$ for all i, that $[e_1, e_{-2}] = -2e_{-1}$ and $[e_{-1}, e_{-2}] = 0$. Let $c_i \in F$ be defined by $e_{-2}v_i = c_i v_{i-2}$. Applying $[e_1, e_{-2}] = -2e_{-1}$ to v_i we deduce that $c_i - c_{i-1} = -2(s+i)$, $\forall\, i \in \mathbf{Z}$. Thus $c_i = c_0 + i(2s + i - 1)$, i.e.,

$$e_{-2}v_i = (c_0 + i(2s+i-1))v_{i-2}, \ \forall\, i \in \mathbf{Z}.$$

It is clear that there exists $a \in F$ such that $[e_{-2}, e_2] = a e_0$. We deduce that

$$[e_{-2}, e_2]v_i = [(i+2)(2s+i+1) - i(2s+i-1)]v_i$$

$$= (4i + 4s + 2)v_i \neq ae_0v_i,$$

a contradiction. Thus this combination does not occur.

Now we consider the combination of (1) and (3) with $m, n \in \mathbf{Z}$. We are going to use the basis $\{v_i\}$. Let c_i be defined by $e_nv_i = c_iv_{i+n}$. Applying $[e_{-1}, e_n] = 0$ to v_i we deduce that $(s+n+i)c_i - (s+i)c_{i-1} = 0$, $\forall\, i \in \mathbf{Z}$. Thus

$$c_i = \frac{c}{(s+i+1)(s+i+2)\cdots(s+i+n)},\ \forall\, i \in \mathbf{Z},$$

where $c \in F$. Choose e_n such that

$$e_nv_i = \frac{1}{(s+i+1)(s+i+2)\cdots(s+i+n)}v_{i+n},\ \forall\, i \in \mathbf{Z}.$$

We can show inductively that, for $k \geq n$, we can choose $e_k \in C_k$ such that

$$e_kv_i = \frac{1}{(s+i+1)(s+i+2)\cdots(s+i+k)}v_{i+k},\ \forall\, i \in \mathbf{Z}.$$

Choose $k > 1$ such that $C_{\pm k} \neq 0$. We know that there exists $a \in F$ such that $[e_k, e_{-k}] = ae_0$. But

$$[e_k, e_{-k}]v_i = [\frac{1}{(s+i)(s+i-1)\cdots(s+i-k+1)}$$

$$-\frac{1}{(s+i+1)(s+i+2)\cdots(s+i+k)}]v_i \neq ae_0v_i,$$

a contradiction. Thus this combination does not occur.

Case (ii): $-s \in \mathbf{Z}^+, U = \oplus_{i\in\mathbf{Z}}Fv_i$ and $e_{-1}v_i = (s+i)v_{i-1}$, $e_1v_i = v_{i+1}$, $e_0v_i = v_i$.

Suppose $n > 1$ is the first integer (if it exists) such that $e_n \neq 0$. Let $e_nv_i = b_iv_{i+n}$ where $b_i \in F$.

Subcase 1: $[e_{-1}, e_n] = 0$.

By computing $[e_{-1}, e_n]v_i = 0$, we deduce that $(s+n+i)b_i - (s+i)b_{i-1} = 0$, $\forall\, i \in \mathbf{Z}$. Noting that $s \in \mathbf{Z}$, we deduce that $b_i = 0$ for all $i \in \mathbf{Z}\backslash\{-s-1, -s-2, \cdots, -s-n\}$. Repeatedly applying e_{-1} to $e_nv_{-s-1} = b_{-s-1}v_{n-s-1}$ we get that

$$b_{-s-i} = \frac{(n-1)(n-2)\cdots(n-i+1)}{(-1)\cdots(-i+1)}b_{-s-1},$$

$\forall\, i \in \{1, 2, \cdots n\}$. Since $e_nV \neq 0$, we choose e_n such that $b_{-s-1} = 1$, i.e.,

$$e_nv_{-s-i} = \frac{(n-1)(n-2)\cdots(n-i+1)}{(-1)\cdots(-i+1)}v_{n-s-i} \tag{2.6}$$

$\forall\, i \in \{1, 2, \cdots n\}$. By computing $[e_1, e_n]v_{-s-i}$ for $i \in \{1, 2, \cdots n\}$, we see that $[e_1, e_n]v_{-s-i} \neq 0$, and we can choose e_{n+1} such that $e_{n+1} = n^{-1}[e_1, e_n]$ and

$$e_{n+1}v_{-s-i} = \frac{(n-1)(n-2)\cdots(n-i+2)}{(-1)\cdots(-i+1)}v_{n+1-s-i}$$

$\forall\, i \in \{1, 2, \cdots n+1\}$.

In this manner by induction on $k \geq n$, we can choose e_k such that

$$e_k v_{-s-i} = \frac{(n-1)(n-2)\cdots(k-i+1)}{(-1)\cdots(-i+1)} v_{k-s-i}$$

$\forall\, i \in \{1, 2, \cdots k\}$, and

$$[e_k, e_1] = -k e_{k+1}, \quad [e_k, e_{-1}] = 0, \ \forall\, k \in \mathbf{Z}.$$

Subcase 2: $[e_{-1}, e_n] \neq 0$.

We deduce that $n = 2$, and we can choose e_2 such that $[e_{-1}, e_2] = 2e_1$. Computing $[e_{-1}, e_2]v_i = 2e_1 v_i$ we deduce that

$$(s+2+i)b_i - (s+i)b_{i-1} = 2. \tag{2.7}$$

It is clear that $b_{-s-1} + b_{-s-2} = 2$ and $b_i = 1$ for $i \in \mathbf{Z}\backslash\{-s-1, -s-2\}$.

If $b_{-s-1} \neq b_{-s-2}$, by computing $[e_1, e_2]v_i = (b_i - b_{i+1})v_{i+3}$ we see that $[e_1, e_2] \neq 0$. Choose e_3 such that $[e_1, e_2] = 2e_3$. Then $e_3 v_i = \frac{(b_i - b_{i+1})}{2} v_{i+3}$ for all $i \in \mathbf{Z}$. From [OZ1, Lemma 1.1], it is clear that $[e_{-1}, e_3] = 0$. By induction on $k > 2$ we can show that there exists $e_j \in C_j$ for $3 \leq j \leq k$ such that $[e_1, e_{j-1}] = (j-1)e_j$ and

$$(k-1)!e_k v_i = \sum_{j=0}^{k-2} (-1)^j \binom{k-2}{j} b_{i+j} v_{i+k}.$$

We know that there exists $a \in F$ such that $[e_2, 2e_3] = a4!e_5$. But

$$[e_2, 2e_3]v_i = [(b_i - b_{i+1})b_{i+3} - (b_{i+2} - b_{i+3})b_i]v_{i+5},$$

and $a4!e_5 v_i = a(b_i - 3b_{i+1} + 3b_{i+2} - b_3)v_{i+5}$, so that

$$(b_i - b_{i+1})b_{i+3} - (b_{i+2} - b_{i+3})b_i = a(b_i - 3b_{i+1} + 3b_{i+2} - b_3), \ \forall\, i \in \mathbf{Z}.$$

Setting $i = -s-1$ we get $a = 1$, and setting $i = -s-2$ we deduce that $b_{-s-1} = 1$, a contradiction. Therefore we have

$$e_2 v_i = v_{i+2}, \ \forall\, i \in \mathbf{Z}. \tag{2.8}$$

By induction on $k \geq 2$, if $[e_{-1}, e_k] \neq 0$, using a similar argument we can show that we can choose e_j for $2 \leq j \leq k$ such that $[e_{-1}, e_j] = je_{j-1}$, $e_j v_i = v_{i+j}$ and $[e_1, e_j] = 0$.

If there exists $n' > 2$ such that $[e_{n'}, e_{-1}] = 0$, as in Subcase 1 we can choose e_k for $k \geq n'$ such that

$$e_k v_{-s-i} = \frac{(n'-1)(n'-2)\cdots(k-i+1)}{(-1)\cdots(-i+1)} v_{k-s-i} \tag{2.9}$$

$\forall\, i \in \{1, 2, \cdots k\}$, and

$$[e_k, e_1] = -k e_{k+1}, \quad [e_k, e_{-1}] = 0, \ \forall\, k \in \mathbf{Z}.$$

We know that there exists $a \in F$ such that $[e_2, e_{n'}] = ae_{n'+2}$. But

$$[e_2, e_{n'}]v_{-s-i} =$$

$$\frac{(n'-1)(n'-2)\cdots(n'-i+3)[(n'-i+2)(n'-i+1)-(-i+2)(-i+1)]}{(-1)\cdots(-i+1)}.$$

$$\cdot v_{n'+2-s-i},$$

and $ae_{n'+2}v_i = a\frac{(n'+1)n'\cdots(n'-i+3)}{(-1)\cdots(-i+1)}v_{n'+2-s-i}$. Thus

$$(n'-i+2)(n'-i+1)-(-i+2)(-i+1) = a(n'+1)n', \ \forall\, i \in \{1, 2, \cdots, n'+2\}.$$

Setting $i = 2$ we deduce that $a = 1$, and setting $i = 1$ we get $n' - 1 = n' + 1$, a contradiction. So this case does not happen.

Suppose $m < -1$ is the first integer (if it exists) such that $e_m \neq 0$. Let $e_m v_i = b_i v_{i+m}$ for all $i \in \mathbf{Z}$.

Subcase 3: $[e_1, e_m] = 0$.

Computing $[e_1, e_m]v_i = 0$, we get $b_i = b_{i+1}$ for all $i \in \mathbf{Z}$. Since $e_m V \neq 0$, we may choose e_m such that $e_m v_i = v_{i+m}$ $\forall\, i \in \mathbf{Z}$. It follows that, for all $j \leq n$, we can choose $e_j \in C_j$ such that

$$[e_j, e_{-1}] = -je_{j-1}, \ e_j v_i = v_{i+j}, \ \forall\, i \in \mathbf{Z}.$$

Subcase 4: $[e_1, e_m] \neq 0$.

We must have $m = -2$. Choose e_{-2} such that $[e_1, e_{-2}] = -2e_{-1}$. Applying this to v_i, we get $b_i - b_{i+1} = -2(s+i)$ for all $i \in \mathbf{Z}$. So there exists $b \in F$ such that $b_i = b + (s+i)(s+i-1)$, $\forall\, i \in \mathbf{Z}$, and

$$e_{-2}v_i = [b + (s+i)(s+i-1)]v_{i-2} \ \forall\, i \in \mathbf{Z}.$$

If $b \neq 0$, we see that $[e_{-1}, e_{-2}]v_i \neq 0$ for almost all i. Choose e_{-3} such that $[e_{-1}, e_{-2}] = -2e_{-3}$. Then $e_{-3}v_i = bv_{i-3}$ for all $i \in \mathbf{Z}$. By induction on $k < -2$, we can show that we can choose e_k such that $[e_{-1}, e_k] = ke_{k-1}$ and $e_k v_i = bv_{i+k}$ for all $i \in \mathbf{Z}$. There exists $a \in F$ such that $[e_2, e_3] = ae_5$. By computing the actions $[e_2, e_3]v_i = ae_5 v_i$, we get a contradiction. Thus $b = 0$, i.e.,

$$e_m v_i = (s+i)(s+i-1)v_{i+m} \ \forall\, i \in \mathbf{Z}.$$

By induction on $k \leq m$, if $[e_1, e_k] \neq 0$, then we can choose $e_j \in C_j$ for $-2 \leq j \leq k$ such that $[e_1, e_j] = je_{j+1}$ and

$$e_k v_i = (s+i)(s+i-1)\cdots(s+i+k+1)v_{i+m} \ \forall\, i \in \mathbf{Z}. \tag{2.10}$$

The combination of Subcase 1 and Subcase 4 gives (c); The combination of Subcase 2 and Subcase 3 gives (a). We need to remove the combination of Subcase 1 and Subcase 3 the combination of Subcase 2 and Subcase 4.

First consider the combination of Subcase 2 and Subcase 4 with $C_{\pm 2} \neq 0$. It is clear that there exists $a \in F$ such that $[e_{-2}, e_2] = ae_0$. We deduce that

$$[e_{-2}, e_2]v_i = [(s+i+2)(s+i+1) - (s+i)(s+i-1)]v_i$$

$$= (4i + 4s + 2)v_i \neq ae_0 v_i,$$

a contradiction. Thus this combination does not occur.

Now we consider the combination of Subcase 1 and Subcase 3 with $m, n \in \mathbf{Z}$. Choose $k \in \mathbf{Z}^+$ such that $C_{\pm k} \neq 0$. We know that there exists $a \in F$ such that $[e_k, e_{-k}] = ae_0$. But $[e_k, e_{-k}]v_{-s-k-1} = 0$, and $[e_k, e_{-k}]v_{-s} = e_0 v_{-s-k} \neq 0$, a contradiction. Thus this combination does not occur.

Case (iii): $s \in \mathbf{Z}^+, V = U = \oplus_{i \geq -s} F v_i$ and $e_{-1}v_i = (s+i)v_{i-1}$, $e_1 v_i = v_{i+1}$, $e_0 v_i = v_i$.

In this case and the next cases we shall omit some details since there are a lot of similarities to the above two cases.

Suppose $n > 1$ is the first integer (if it exists) such that $e_n \neq 0$. Let $e_n v_i = b_i v_{i+n}$ where $b_i \in F$.

Subcase 1: $[e_{-1}, e_n] = 0$.
By computing $[e_{-1}, e_n]v_i = 0$, we deduce that $(s+n+i)b_i - (s+i)b_{i-1} = 0$, $\forall\, i \geq -s$. So we can obtain that $b_i = 0$ for all $i \geq -s$. Thus $e_n V = 0$, a contradiction. So this subcase does not occur.

Subcase 2: $[e_{-1}, e_n] \neq 0$.
We deduce that $n = 2$, and we can choose e_2 such that $[e_{-1}, e_2] = 2e_1$. Computing $[e_{-1}, e_2]v_i = 2e_1 v_i$ we deduce that

$$(s+2+i)b_i - (s+i)b_{i-1} = 2, \ \forall\, i \geq -s.$$

It is clear that $b_i = 1$ for $i \geq -s$, i.e., $e_2 v_i = v_{i+2}$ for $i \geq -s$. We can show that, if $C_k \neq 0$ for $k > 2$, then we can choose e_j for $2 \leq j \leq k$ such that $[e_{-1}, e_j] = je_{j-1}$, $e_j v_i = v_{i+j}$ for $i \geq -s$ and $[e_1, e_j] = 0$.
And there does not exist $n' > 2$ such that $[e_{n'}, e_{-1}] = 0$.

Suppose $m < -1$ is the first integer (if it exists) such that $e_m \neq 0$. Let $e_m v_i = b_i v_{i+m}$ for all $i \in \mathbf{Z}$.

Subcase 3: $[e_1, e_m] = 0$.
We deduce that $e_m U = 0$, contradicting the faithfulness of C. So this case does not happen.

Subcase 4: $[e_1, e_m] \neq 0$.
We must have $m = -2$. Choose e_{-2} such that $[e_1, e_{-2}] = -2e_{-1}$. Applying this to v_i, we get $b_i - b_{i+1} = -2(s+i)$ for all $i \geq -s$. Since $b_{-s} = b_{-s+1} = 0$, we have $b_i = (s+i)(s+i-1)$, $\forall\, i \geq -s$, i.e.,

$$e_{-2}v_i = (s+i)(s+i-1)v_{i-2} \ \forall\, i \geq -s.$$

By induction on $k \leq m$, if $[e_1, e_k] \neq 0$, then we can choose $e_j \in C_j$ for $-2 \leq j \leq k$ such that $[e_1, e_j] = je_{j+1}$ and

$$e_k v_i = (s+i)(s+i-1)\cdots(s+i+k+1)v_{i+m} \ \forall\, i \geq -s.$$

The combination of Subcase 2 and Subcase 4 does not occur. So (e) holds.

Case (iv): $s \in \mathbf{Z}^+, V = \oplus_{i\in\mathbf{Z}} Fv_i$ and $e_{-1}v_i = (s+i)v_{i-1},\ e_1v_i = v_{i+1},\ e_0v_i = v_i$.

As in Case (ii), disregarding the restriction $s \notin Z^+$ there, we can get all the subcases in Case (ii). The combination of Subcase 2 and Subcase 3 gives (a) with $n \neq \infty$ since V is generated by V_0; but the combination of Subcase 1 and Subcase 4 does not happen since $V = CV_0$.

Case (v): $s, t \in \mathbf{Z}^+$ with $t > s$, $V = U \oplus \bar{U}$ where $U = \oplus_{i\geq -s} Fv_i$ and $\bar{U} = \oplus_{i\leq -t} Fv_i$, and

$$e_{-1}v_i = (s+i)v_{i-1},\ e_1v_i = v_{i+1},\ e_0v_i = v_i,\ \forall\ i \geq -s,$$

$$e_{-1}v_i = av_{i-1},\ e_1v_i = (t+i)v_{i+1},\ e_0v_i = av_i\ \forall\ i \leq -t,$$

where $a \in F^*$.

Suppose $m < -1$ is the first integer (if it exists) such that $e_m \neq 0$. Let $e_m v_i = b_i v_{i+m}$ for all $i \in \mathbf{Z}$.

Subcase 1: $[e_1, e_m] = 0$.

If $e_m v_{-s} = 0$, we easily deduce that $e_m U = 0$. Computing $[e_1, e_m]v_i = 0$ for $i \leq -t$, we get $(t+i+m)b_i = (t+i)b_{i+1}$ for all $i \leq -t$. So $b_i = 0$ for all $i \leq -t$, i.e., $e_m\bar{U} = 0$, contrary to the faithfulness of the action of C on V. Thus $e_m v_{-s} \neq 0$. Then $-m \geq t - s$. We may suppose that $e_m v_{-s} = v_{m-s} \in \bar{U}\backslash\{0\}$. By repeatedly applying e_1 to this we get

$$e_m v_{i-s} = (t+m-s)(t+m-s+1)\cdots(t+m-s+i-1)v_{m+i-s},\ \forall\ i \geq 0.$$

For $k < m$ we can choose e_k such that $[e_{-1}, e_k] = ke_{k-1}$ and

$$e_k v_{i-s} = \frac{(t+k-s)(t+k-s+1)\cdots(t+k-s+i-1)}{m(m-1)\cdots(k+1)} v_{k+i-s},\ \forall\ i \geq 0.$$

From [Lemma 1.1, OZ1], we see that $[e_1, e_k] = 0$ for all $k < m$. But the following computation

$$0 = [e_1, e_k]v_{-s} = \frac{1-(t+k-s)}{m(m-1)\cdots(k+1)} v_{k+1-s} \neq 0$$

gives a contradiction, so this subcase does not occur.

Subcase 2: $[e_1, e_m] \neq 0$.

We must have $m = -2$. Choose e_{-2} such that $[e_1, e_{-2}] = -2e_{-1}$. Applying this to v_i for $i \leq -t$, we deduce $b_i = a$ for all $i \leq -t$. And we can get that, for any $k \leq m$, if $C_k \neq 0$, then we can choose $e_j \in C_j$ for $-2 \leq j \leq k$ such that $[e_1, e_j] = je_{j+1}$, $[e_{-1}, e_j] = 0$ and $e_k v_i = av_{k+i}$ for $i \leq -t$.

If $e_{-2}v_{-s} = v_{-s-2} \in \bar{U}\backslash\{0\}$, computing $[e_1, e_{-2}]v_{-s} = 0$, we get $e_{-2}v_{1-s} = (t-s-2)v_{-s-1}$. From $[e_{-1}, e_{-2}]v_{-s} = e_{-1}e_{-2}v_{-s} = v_{-s-3}$ we know that $[e_{-1}, e_{-2}] \neq 0$, contrary to $[e_{-1}, e_{-2}] = 0$. So we must have $e_{-2}v_{-s} = 0$. Computing

$[e_1, e_{-2}]v_{-s} = 0$, we get $e_{-2}v_{1-s} = 0$. Then we can show that $e_{-2}v_i = (s+i)(s+i-1)v_{i-2}$ for $i \geq -s$. And if $e_k \neq 0$ for $k < -2$ we have

$$e_k v_i = (s+i)(s+i-1)\cdots(s+i+k+1)v_{i-2}, \ \forall\, i \geq -s.$$

In particular, we obtain that $e_k U \subset U$ and $e_k \bar{U} \subset \bar{U}$ for all $i < 0$.

Similarly we can show that $e_k U \subset U$ and $e_k \bar{U} \subset \bar{U}$ for all $i > 0$. Therefore $V \neq CV_0$, a contradiction. Therefore this case does not occur.

The cases we have discussed and their duals cover all possible cases, to complete the proof of our proposition. ∎

3. PROOF OF THEOREM 1.1

Let elements $w_{i,j}$ be chosen so that $A_{i,j} = Fw_{i,j}$ for all i, $j \in \mathbf{Z}$. Denote $A^+ = \oplus_{i>0} A'_i$ and $A^- = \oplus_{i<0} A'_i$.

LEMMA 3.1 *There exists nonzero $a \in F$ such that $[w_{0,0}, w_{i,j}] = iaw_{i,j}$ for all i, $j \in \mathbf{Z}$.*

Proof. Suppose $[w_{0,0}, w_{\pm 1,0}] = a_{\pm 1} w_{\pm 1,0}$, where $a_1, a_{-1} \in F$. If $a_1 = a_{-1} = 0$, by Condition (IV) then $w_{0,0}$ is in the center A, contrary to (V). So at least one of $a_{\pm 1}$ is not 0. Next we need to show that $a_1 = -a_{-1}$.

Suppose $a_1 \neq -a_{-1}$. Then no eigenvalues of $[A^+, A^-] \cap A'_0$ with respect to $w_{0,0}$ is 0. By (IV) the zero eigenvalue part of A'_0 must be generated by H, so is just H. Thus, $H \cap A' = A_{0,0}$, contrary to (VI). Therefore $a_1 = -a_{-1}$. ∎

By rechoosing the basis for H we may assume that $a_{-1} = -a_1 = -1$ in Lemma 3.1.

LEMMA 3.2 *(a) As H-modules, $A'_0 \simeq H_{m,n} \oplus C$ or $\bar{H}_{m,n} \oplus C$ for some $m, n \in \mathbf{Z}^+_\infty$ with $n > 1$, where C is an H-module;*

(b) $A'_{\pm 1}$ are indecomposable H-modules;

(c) at least one of $A_{0,\pm 1}$ is in $[A'_1, A'_{-1}]$.

Proof. Denote by $M_{\pm 1}$ the A'_0-submodule of $A_{\pm 1}$ generated by $w_{\pm 1,0}$ respectively. From Proposition 2.2 we see that the part of A'_0 each element of which does not annihilate M_1 or M_{-1} is $H_{m,n}$ or $\bar{H}_{m,n}$, and $M_{\pm 1}$ are indecomposable H-modules. If A'_0 has other nonzero elements other than those from $H_{m,n}$, let C denote the subspace spanned by those elements. Then $[H, C] \subset C$. Thus (a) follows.

Consider the subalgebras $B^\pm$ generated by $M_{\pm 1}$ respectively. From the fact that $[B^+, A'_{-1}] \subset B^+ + A'_0$ and $[B^-, A'_1] \subset B^- + A'_0$ it is easily to see that $A = B^+ + B^- + A'_0$. Then $M_{\pm 1} = A'_{\pm 1}$, so (b) follows.

If $A_{0,\pm 1} \notin [A'_1, A'_{-1}]$ then from the last paragraph by induction on k we can show that $A_{0,\pm 1} \notin [A'_k, A'_{-k}]$. Thus $A_{0,\pm 1} \notin [B^+, B^-]$. So $H \cap [A, A] = A_{0,0}$, contrary to (IV). Therefore (c) holds. ∎

LEMMA 3.3 *Either both A_{-1} and A_1 are degenerate, or both are nondegenerate, for the pair of operators* $\mathrm{ad}(w_{-1,0})$, $\mathrm{ad}(w_{1,0})$.

Proof. In the H-module A'_1, either $[w_{0,1}, w_{1,0}] \neq 0$ or $[w_{0,-1}, w_{1,0}] \neq 0$. By symmetry we can suppose that $[w_{0,1}, w_{1,0}] \neq 0$. Let $[w_{-1,0}, w_{1,1}] = c_1 w_{0,1}$ where $c_1 \in F$. Applying $\mathrm{ad}w_{-1,0}$ to it and using $[w_{-1,0}, w_{1,0}] = 0$ we obtain

$[[w_{0,-1}, w_{-1,0}], w_{1,1}] = -c_1 w_{0,0}$. We see that A_1 is nondegenerate for the pair $w_{-1,0}$, $w_{1,0}$ if and only if $c_1 \neq 0$. Now, if $c_1 \neq 0$, then $[w_{0,-1}, w_{-1,0}] \neq 0$ and $[w_{-1,-1}, w_{1,1}] \neq 0$. Further, if $[w_{-1,-1}, w_{1,0}] = c_2 w_{0,-1}$, then

$$[w_{-1,-1}, [w_{0,1}, w_{1,0}]] = c_2 w_{0,0}. \tag{3.1}$$

Since $[w_{0,1}, w_{1,0}] \neq 0$ and $[w_{-1,-1}, w_{1,1}] \neq 0$, we conclude that $c_2 \neq 0$, and so A_{-1} is nondegenerate also.

Suppose that A_1 is degenerate, so that $c_1 = 0$. Then $[[w_{0,-1}, w_{-1,0}], w_{1,1}] = 0$, so that either $[w_{0,-1}, w_{-1,0}] = 0$ or $[w_{-1,-1}, w_{1,1}] = 0$. The first case shows that A_{-1} is degenerate. In the latter case, we see from (3.1) that $c_2 = 0$, and hence A_{-1} is degenerate also. ∎

LEMMA 3.4 *Both A_{-1} and A_1 are nondegenerate for the pair of operators* $\mathrm{ad}(w_{-1,0})$, $\mathrm{ad}(w_{1,0})$.

Proof. Suppose both A_{-1} and A_1 are degenerate. From Lemma 3.2 we may assume that $A_0' \simeq H_{m,n} \oplus C$ for suitable $m, n \in \mathbf{Z}^+_\infty$. We also know that $A'_{\pm 1}$ are indecomposable H-modules.

Case 1: $w_{0,1}$ does not annihilate any $A_{1,i}$ and $A_{-1,i}$ for all $i \in \mathbf{Z}$.

We can choose a basis for $A_{\pm 1}$ such that

$$[w_{0,-1}, w_{1,i}] = (s+i) w_{1,i-1},\ [w_{0,1}, w_{1,i}] = w_{1,i+1},\ [w_{0,0}, w_{1,i}] = w_{1,i},$$
$$[w_{0,1}, w_{-1,i}] = -w_{-1,i+1},\ [w_{0,-1}, w_{-1,i}] = (t+i) w_{-1,i-1},$$
$$[w_{0,0}, w_{-1,i}] = -w_{-1,i}$$

where $s, t \in F$, for all nonzero $w_{1,i}$ and nonzero $w_{-1,i}$. Since A_1 is degenerate we obtain that

$$[w_{1,0}, w_{-1,1}] = 0 = [w_{-1,0}, w_{1,1}]. \tag{3.4}$$

From $[w_{0,1}, [w_{1,i}, w_{-1,-i}]] = 0$ we deduce that

$$[w_{1,i}, w_{-1,1-i}] = [w_{1,1+i}, w_{-1,-i}]$$

for all $w_{1,i} \neq 0$ and $w_{-1,-i} \neq 0$. Combining this with (3.4) we see that $w_{0,1} \notin [A_1', A_{-1}']$. From Lemma 3.2 we see that $w_{0,-1} \in [A_1', A_{-1}']$. Similarly we also have

$$[w_{1,i}, w_{-1,-1-i}] = [w_{1,-1+i}, w_{-1,-i}]$$

for all $w_{1,i-1} \neq 0$ and $w_{-1,-1-i} \neq 0$. Thus $[w_{1,i}, w_{-1,-1-i}] \neq 0$ for all $w_{1,i} \neq 0$ and $w_{-1,-1-i} \neq 0$. Since $w_{0,-1} \in [A_1', A_{-1}']$ and A_{-1} is degenerate, we obtain that, $t = s = 0$ if $A_{-1,-1} \neq 0$ and $A_{1,-1} \neq 0$. Suppose $A_{-1,-1} \neq 0$ and $A_{1,-1} = 0$, then $s = 0$ and $A_{-1,-2} \neq 0$. Thus $[w_{1,0}, w_{-1,-1}] \neq 0$, so $t = 0$. If $A_{-1,-1} = A_{1,-1} = 0$, it is clear that $t = s = 0$. Thus $t = s = 0$ in all cases. Let $\bar{A}^-$ and $\bar{A}^+$ be subalgebras generated by $\oplus_{i \geq 0} A_{-1,i}$ and $\oplus_{i \geq 0} A_{1,i}$ respectively. We can show that $\bar{A}^- + A_0' + \bar{A}^+$ is an subalgebra, contrary to the fact $A_{-1,-1} \neq 0$ and (IV). So this case does not occur.

Case 2: $w_{0,-1}$ does not annihilate any $A_{1,i}$ and $A_{-1,i}$ for all $i \in \mathbf{Z}$.

An argument identical to that in Case 1 shows that this case does not occur either.

Case 3: Now we consider the last case, i.e., $w_{0,1}$ annihilates some elements in $A'_{\pm 1}$ and $w_{0,-1}$ annihilates some elements in $A'_{\pm 1}$. Without loss of generality we assume that $w_{0,1}$ annihilates an element in A'_{-1} and $w_{0,-1}$ annihilates an element in A'_1.

We may choose the basis for $A'_{\pm 1}$ such that

$$[w_{0,-1}, w_{1,i}] = (s+i)w_{1,i-1}, \ [w_{0,1}, w_{1,i}] = w_{1,i+1}, \ [w_{0,0}, w_{1,i}] = w_{1,i},$$

$$[w_{0,1}, w_{-1,i}] = (t+i)w_{-1,i+1}, \ [w_{0,-1}, w_{-1,i}] = -w_{-1,i-1},$$

$$[w_{0,0}, w_{-1,i}] = -w_{-1,i}$$

where $s, t \in \mathbf{Z}$. Denote $B_{-1} = \oplus_{i \leq -t} A_{-1,i}$ which is an H-submodule, and $B'_{-1} = \oplus_{i > -t} A_{-1,i}$.

Subcase 1: $-t \geq 0$.

Since $A_{\pm 1}$ is degenerate we obtain that

$$[w_{1,0}, w_{-1,-1}] = 0 = [w_{-1,0}, w_{1,1}]. \tag{3.5}$$

From $[w_{0,1}, [w_{1,-i}, w_{-1,i}]] = 0$ we deduce that

$$(t+i)[w_{1,-i}, w_{-1,1+i}] = -[w_{1,1-i}, w_{-1,i}]$$

for all $i \leq -t$. Combining this with (3.5) we see that $w_{0,1} \notin [A'_1, B_{-1}]$. A similar argument shows that $w_{0,0} \notin [A'_1, B_{-1}]$ and $w_{0,-1} \notin [A'_1, B_{-1}]$. So $[A'_1, B_{-1}] \subset C \oplus_{i<0} A_{0,i}$. Let $\bar{A}^-$ be the subalgebra generated by B_{-1}. So $\bar{A}^- + A'_0 + A^+$ is a subalgebra of A. By (III) this is A. Then we see that $A^- = \bar{A}^-$. Thus $A_{0,\pm 1} \not\subset [[A'_1, A'_{-1}]$, which contradicts Lemma 3.2(c). So this subcase does not occur.

Subcase 2: $-t < 0$.

Since A_1 is degenerate we obtain that

$$[w_{1,0}, w_{-1,1}] = 0 = [w_{-1,0}, w_{1,1}]. \tag{3.6}$$

From $[w_{0,1}, [w_{1,-i}, w_{-1,i}]] = 0$ we deduce that

$$(t+i)[w_{1,-i}, w_{-1,1+i}] = -[w_{1,1-i}, w_{-1,i}]$$

for all $i > -t$. Combining this with (3.6) gives $w_{0,1} \notin [A'_1, B'_{-1}]$.

Computing $[w_{0,1}, [w_{1,t}, w_{-1,-t}]] = 0$ yields

$$[w_{1,1+t}, w_{-1,-t}] = 0. \tag{3.7}$$

From $[w_{0,1}, [w_{1,-i}, w_{-1,i}]] = 0$ for $i \leq -t$ we deduce that

$$(t+i)[w_{1,-i}, w_{-1,1+i}] = -[w_{1,1-i}, w_{-1,i}]$$

for all $i \leq -t$. Combining this with (3.7) gives $w_{0,1} \notin [A'_1, B_{-1}]$. Thus $w_{0,1} \notin [A'_1, A'_{-1}]$. From Lemma 3.2 we see that $w_{0,-1} \in [A'_1, A'_{-1}]$.

Suppose $s \neq 0$. Computing $[w_{0,1}, [w_{1,-2+t}, w_{-1,-t}]] = 0$ we deduce that

$$[w_{1,-1+t}, w_{-1,-t}] = 0. \tag{3.8}$$

From $[w_{0,1}, [w_{1,-2-i}, w_{-1,i}]] = 0$ for $i \leq -t$ we deduce that

$$(t+i)[w_{1,-2-i}, w_{-1,1+i}] = [w_{1,-1-i}, w_{-1,i}]$$

for all $i \leq -t$. So $w_{0,-1} \notin [A'_1, B_{-1}]$. Since $s \neq 0$ we know that $[w_{-1,0}, w_{1,-1}] = 0$ (from the degeneracy). Then we can derive that $w_{0,-1} \notin [A'_1, B'_{-1}]$. We obtain that $w_{0,-1}, w_{0,1} \notin [A'_1, A'_{-1}]$, contrary to Lemma 3.2. Therefore we must have $s = 0$.

Similarli if $t \neq 1$ similarly we can deduce that $w_{0,-1}, w_{0,1} \notin [A'_1, A'_{-1}]$, contrary to Lemma 3.2. Therefore we must have $t = 1$.

From the degeneracy we see that $[w_{1,0}, w_{-1,-1}] = 0$. Thus we can get $w_{0,-1} \notin [A'_1, B_{-1}]$. So $w_{0,-1} \in [A'_1, B'_{-1}]$. If $[w_{-1,0}, w_{1,-1}] = 0$, then we can derive that $w_{0,-1} \notin [A'_1, B'_{-1}]$, a contradiction. Thus $[w_{-1,0}, w_{1,-1}] \neq 0$. So $[w_{-1,1}, w_{1,-1}] \neq 0$. Since $[w_{-1,0}, w_{1,1}] = 0$, applying $\mathrm{ad}(w_{0,-1})$ to it we obtain that $[w_{-1,-1}, w_{1,1}] = 0$, a contradiction. So this subcase does not occur either.

This completes the proof of Lemma 3.4. ∎

LEMMA 3.5 *If $A_{i-1,j-1} = 0$ and if $w_{0,-1}$ and $w_{0,1}$ act degenerately on A_j, then either $[w_{0,-1}, w_{i,j}] = 0$ or $[w_{0,1}, w_{i,j-1}] = 0$.*

Proof. Suppose to the contrary that $[w_{0,-1}, w_{i,j}] \neq 0$ and $[w_{0,1}, w_{i,j-1}] \neq 0$. Then

$$[w_{-1,-1}, [w_{1,0}, w_{i,j}]] = [[w_{-1,-1}, w_{1,0}], w_{i,j}] \neq 0,$$

$$[w_{-1,0}, [w_{1,1}, w_{i,j-1}]] = [[w_{-1,0}, w_{1,1}], w_{i,j-1}] \neq 0,$$

from which we conclude that $[w_{1,0}, w_{i,j}] \neq 0$ and $[w_{-1,0}, w_{i+1,j}] \neq 0$. But this contradicts the hypothesis that $w_{-1,0}$ and $w_{1,0}$ act degenerately on A_j, to complete the proof. ∎

LEMMA 3.6 *One of the following must hold:*

(i) *A'_0 and all A'_i's are not bounded either above or below.*

(ii) *A'_0 and all A'_i's are not bounded above, the lowest nonzero element of A'_0 is $w_{0,-1}$, and the lowest nonzero element of A'_i for $i \neq 0$ is either $w_{i,-1}$ or $w_{i,-2}$.*

(iii) *A'_0 and all A'_i's are not bounded below, the highest nonzero element of A'_0 is $w_{0,1}$, and the highest nonzero element of A'_i for $i \neq 0$ is either $w_{i,1}$ or $w_{i,2}$.*

Proof. Suppose that A'_0 is bounded below. We show first that every A'_i is bounded below. If this is not true for all $i > 0$, let k be the first positive integer such that A'_k is not bounded below. Since A'_{k-1} is bounded below, there exists $\ell \in Z$ such that $A_{k-1,j} = 0$ for $j \leq \ell$. Then A_j is degenerate for $j \leq \ell$. Applying Lemma 3.5 with $j = \ell$ and $i = k$, we obtain that either $[w_{0,-1}, w_{k,\ell}] = 0$ or $[w_{0,1}, w_{k,\ell-1}] = 0$. Applying Lemma 3.5 again with $j = \ell - 1$ and $i = k$, we get the further relations $[w_{0,-1}, w_{k,\ell-1}] = 0$ or $[w_{0,1}, w_{k,\ell-2}] = 0$. But this cannot happen unless $w_{k,\ell-2} = 0$. This shows that A'_i is bounded below for $i > 0$, and by symmetry it is also true for

$i < 0$. Since $w_{0,0}$ acts as a nonzero constant on each A'_i for $i \neq 0$, each such A'_i is an irreducible Heisenberg module which is bounded below and unbounded above. If A'_0 is also bounded above, then by symmetry every A'_i is bounded above and not below. This contradiction shows that A'_0 cannot be bounded both above and below. The same argument will show that if any A'_i is bounded below, then all of them including A'_0 are bounded below.

Suppose then that A'_0 is bounded below but not above, and let $k \geq 2$ be minimal such that $w_{0,k}$ exists. If $w_{0,-1}$ is not the lowest nonzero element of A'_0, then $w_{0,-2} \neq 0$. Let ℓ be minimal so that $w_{1,\ell} \neq 0$. Then $[w_{0,k}, [w_{0,2}, w_{1,\ell}]] = 0$ and $[w_{0,2}, [w_{0,k}, w_{1,\ell}]] \neq 0$, contradicting the fact that $[w_{0,2}, w_{0,k}] = 0$. This shows that $w_{0,-2}$ doesn't exist, and so $w_{0,-1}$ is the lowest element of A'_0. If $A_{i,-3} \neq 0$ for some $i > 0$, let k be the first i positive integer for which this is true. Using Lemma 3.5, either $[w_{0,-1}, w_{k,-2}] = 0$ or $[w_{0,1}, w_{k,-3}] = 0$. But this can happen in an irreducible module bounded below only if $w_{k,-3} = 0$, to give a contradiction. Thus, $A_{i,-3} = 0$ for all i. Dually, if A'_0 is bounded above but not below, then e is the highest element of A'_0, and $A_{i,3} = 0$ for all i. ∎

LEMMA 3.7 *The cases (ii) and (iii) of Lemma 3.6 cannot occur.*

Proof. By symmetry, we may assume that A satisfies case (ii) of Lemma 3.3, and hence either $s_1 = 1$ or 2. If $s_1 = 1$, we apply $\mathrm{ad}(w_{0,-1})$ to $[w_{-1,1}, w_{1,-1}]$ to obtain $[w_{-1,0}, w_{1,-1}] = 0$, which contradicts Lemma 3.4. Thus, $s_1 = 2$, and applying $\mathrm{ad}(w_{0,-1})$ to $[w_{-1,2}, w_{1,-2}]$ gives $[w_{-1,1}, w_{1,-2}] = 0$. On the other hand, applying $\mathrm{ad}(w_{0,-1})$ to $[w_{-1,1}, w_{1,-1}]$ gives

$$0 = (s_{-1} + 1)[w_{-1,0}, w_{1,-1}] + [w_{-1,1}, w_{1,-2}] = (s_{-1} + 1)[w_{-1,0}, w_{1,-1}].$$

But the coefficient $(s_{-1}+1)$ is nonzero since $s_{-1} = 1$ or 2, so that again $[w_{-1,0}, w_{1,-1}] = 0$. This contradiction completes the proof. ∎

Proof of Theorem 1.1. In view of the preceding lemmas, we are left with the case where every A'_i including A'_0 is unbounded both above and below. By symmetry, we can assume that A'_0 is isomorphic to $H_{\infty,n} \oplus C$ (see Lemma 3.2). The actions of A'_0 on A'_{-1} and A'_1 are given by

$$\begin{aligned} &[w_{0,0}, w_{-1,i}] = -w_{-1,i}, \quad [w_{0,0}, w_{1,i}] = w_{1,i}, \\ &[w_{0,1}, w_{-1,i}] = b_{-1,i}w_{-1,i+1}, \quad [w_{0,1}, w_{1,i}] = b_{1,i}w_{1,i+1}, \\ &[w_{0,-1}, w_{-1,i}] = d_{-1,i}w_{-1,i-1}, \quad [w_{0,-1}, w_{1,i}] = d_{1,i}w_{1,i-1}, \end{aligned} \tag{3.9}$$

for all $i \in Z$, where using Lemmas 2.2 and 3.2 (and left-right symmetry) we can choose $b_{-1,i}, b_{1,i}, d_{-1,i}, d_{1,i} \in K$ in one of the following three ways:

$$b_{-1,i} = -1, \quad b_{1,i} = 1, \quad d_{-1,i} = (s_{-1} + i), \quad d_{1,i} = (s_1 + i); \tag{3.10}$$

$$b_{-1,i} = (s_{-1} + i), \quad b_{1,i} = (s_1 + i), \quad d_{-1,i} = -1, \quad d_{1,i} = 1, \tag{3.11}$$

$$b_{-1,i} = (s_{-1} + i), \quad b_{1,i} = 1, \quad d_{-1,i} = -1, \quad d_{1,i} = (s_1 + i), \tag{3.12}$$

for some $s_{\pm 1} \in F$. Then applying $\mathrm{ad}(w_{0,-1})$ on $[w_{-1,i}, w_{1,1-n-i}]$ gives

$$d_{-1,i}[w_{-1,i-1}, w_{1,1-n-i}] + d_{1,1-n-i}[w_{-1,i}, w_{1,-n-i}] = 0,$$

and applying $\mathrm{ad}(w_{0,1})$ on $[w_{-1,i-1}, w_{1,-n-i}]$ leads to

$$b_{-1,i-1}[w_{-1,i}, w_{1,-n-i}] + b_{-1,-n-i}[w_{-1,i-1}, w_{1,1-n-i}] = 0.$$

Solving the last two equations simultaneously, and noting that $w_{0,-n}$ must be in $[A'_{-1}, A_1]$, we deduce that

$$d_{-1,i}b_{-1,i-1} - d_{1,1-n-i}b_{-1,-n-i} = 0, \tag{3.13}$$

for some i's. In the three cases (3.10), (3.11), (3.12), this gives respectively

$$s_{-1} + s_1 + 1 - n = 0, \; s_{-1} + s_1 - 1 - n = 0, \; s_{-1} + s_1 - n = 0. \tag{3.14}$$

Note that $n > 1$. Now, $\mathrm{ad}(w_{0,-1})$ on $[w_{-1,i}, w_{1,-i}]$ is

$$d_{-1,i}[w_{-1,i-1}, w_{1,-i}] + d_{1,-i}[w_{-1,i}, w_{1,-i-1}] = 0,$$

and $\mathrm{ad}(w_{0,1})$ on $[w_{-1,i-1}, w_{1,-i-1}]$ is

$$b_{-1,i-1}[w_{-1,i}, w_{1,-i-1}] + b_{1,-i-1}[w_{-1,i-1}, w_{1,-i}] = 0.$$

Solving the last two equations simultaneously, and noting that $w_{0,-1}$ must be in $[A'_{-1}, A_1]$, we deduce (3.13) for some i's with $n = 1$. In the three cases (3.10), (3.11), (3.12), this gives respectively (3.14) with $n = 1$, contradicts (3.14). This contradiction shows that an Lie algebra A satisfying the hypotheses of Theorem 1.1 cannot exist. ∎

We conclude this paper with an example of a Lie algebra not satisfying (VI), but which satisfies all other conditions in Theorem 1.1. Let $A = A'_1 + A'_{-1} + A'_0$, where $A'_0 = A_{0,-1} + A_{0,0} + A_{0,1}$ is a Heisenberg algebra, and where A'_{-1} and A'_1 are A'_0 modules under the adjoint action generated by $A_{-1,0}$ and $A_{1,0}$ respectively. We let $[A'_{-1}, A'_{-1}] = 0$, $[A'_1, A'_1] = 0$, and $[A'_1, A'_{-1}] = 0$. Then it is easy to check that A has the desired properties.

REFERENCES

[B1] R. Block, New simple Lie algebras of prime characteristic, Trans. Amer. Math. Soc., 72(1958), 421-449.

[B2] R. Block, On torsion-free abelian groups and Lie algebras , Proc. Amer. Math. Soc., 9(1958), 613-620.

[OZ1] J.M. Osborn and Kaiming Zhao, A characterization of the Block algebra and its q-forms in characteristic 0, J. Alg., Vol.207(1998), 367-408.

[OZ2] J.M. Osborn and Kaiming Zhao, Kirillov's problem for Lie algebras, Proceedings of Malaga Conference, 1997.

[OZ3] J.M. Osborn and Kaiming Zhao, A characterization of certain loop algebras, preprint, 1998.

[OZ4] J.M. Osborn and Kaiming Zhao, Doubly **Z**-graded Lie algebras containing a Heisenberg algebra, Preprint, 1998.

[OZ5] J.M. Osborn and K. Zhao, Doubly **Z**-graded Lie algebras containing a Virasoro algebra, preprint, 1998.

Email addresses: osborn@math.wisc.edu, zhao@iss06.iss.ac.cn.

$A_n^{(1)}$ Toda solitons: a relation between dressing transformations and vertex operators

H. BELICH and R. PAUNOV Centro Brasileiro de Pesquisas Fisicas, Rua Dr. Xavier Sigaud 150, Rio de Janeiro, Brazil

ABSTRACT Affine Toda equations based on simple Lie algebras arise by imposing zero curvature condition on a Lax connection which belongs to the corresponding loop Lie algebra in the principal gradation. In the particular case of $A_n^{(1)}$ Toda models, we exploit the symmetry of the underlying linear problem to calculate the dressing group element which generates arbitrary N-soliton solution from the vacuum. Starting from this result we recover the vertex operator representation of the soliton tau functions.

1 INTRODUCTION

Integrable nonlinear evolution equations play important role in various topics of the mathematics and particle physics. There is a huge class of integrable equations which admit soliton solutions [1]. The last are exact localizable solutions of the equations of motion with finite values of the physical observables: momentum, energy, etc. [2]. Generically soliton solutions correspond to topologically nontrivial field configurations. Within the quantum field theory, the solitons are interpreted as nonperturbative particles which appear in the spectrum of the underlying model.

Integrable evolution equations are distinguished by the existence of zero curvature or Lax representation. Taking into account the Lax representation of the equations of motion, one can look for *gauge* transformations which preserve the form of the components of the Lax connection. It is clear that such kind of transformations, called dressing transformations, are symmetries of the underlying integrable model. The study of the dressing group which is a symmetry of the soliton (integrable) equation has been initiated by the Kyoto group [3]. In the last reference the dressing symmetry was considered on the example of the Kadomtsev-Petiashvili (KP) hierarchy. The group of dressing transformations admits a particularly simple and transparent expression when one introduces the Hirota tau functions. The last provide a bridge between integrable models and the representation theory of the affine Lie algebras [4]. It has been shown by Semenov-Tian-Shansky [5] that the dressing group is a Poisson Lie group.

The application of the dressing group to the Toda field theories in $1+1$ dimensions is due to Babelon and Bernard [6]. In this reference it was also argued that the dressing symmetry is a semiclassical analog of the quantum symmetry of the integrable model.

The N soliton solutions of the affine Toda models based on an arbitrary simple Lie algebra are obtained in [7]. In particular the group-theoretical tau functions related to the fundamental representations of the affine Lie algebra are given by

$$\frac{\tau_\Lambda(\Phi)}{\tau_\Lambda(\Phi_0)} = e^{\frac{\zeta_0-\zeta}{n+1}} \prod_{i=1}^{N} Y_i \cdot < \Lambda|(1+X_1F^{r_1}(\mu_1))\dots(1+X_NF^{r_N}(\mu_N))|\Lambda > \tag{1.1a}$$

where $X_i, i = 1, ..., N$ are numerical factors depending exponentially on the light cone coordinates, F^{r_i} are elements of the affine Lie algebra which diagonalize the adjoint action if the principal Heisenberg subalgebra [4, 7] and $|\Lambda >$ is the highest weight vector of a fundamental representation of the affine Lie algebra of highest weight Λ. Alternative expression for the tau functions is provided by using the dressing symmetry

$$\frac{\tau_\Lambda(\Phi)}{\tau_\Lambda(\Phi_0)} =< \Lambda|\hat{g}_-^{-1}(x^+,x^-)\,\hat{g}_+(x^+,x^-)|\Lambda > \tag{1.1b}$$

where $\hat{g}_-$ and $\hat{g}_+$ are triangular elements of the affine group.

In [8] it has been shown that the representations (1.1a) and (1.1b) are equivalent for the the sine-Gordon N–solitons. Our purpose is to extend this result for arbitrary $A_n^{(1)}$ Toda models. General analysis of the dressing symmetry of integrable hierarchies which admit vacuum solution has been done in [9].

In the present talk we give a summary of [10]. It is organized as follows: In Sec.1 we discuss the N-soliton solutions of the $A_n^{(1)}$ Toda equations and the related dressing problem. An explicit expression for the dressing transformation which creates solitons from the vacuum is found. The main result of Sec.1 is the Proposition 2. It states that the above mentioned dressing transformation factorizes into "monosoliton" factors. In Sec.3 we exploit the free field representations of the affine Lie algebra $A_n^{(1)}$ to demonstrate the equivalence between (1.1a) and (1.1b). Our approach is based on the observation that the solution of the dressing problem on the affine group differs from those in the loop group by a factor which is in the center of the affine group. Sec.4 is reserved for concluding remarks and discussion of possible further developments.

2 N-SOLITONS AND THE SOLUTION OF THE DRESSING PROBLEM IN THE AFFINE GROUP

The $A_n^{(1)}$ Toda equations in $1+1$ dimensions [11]

$$\partial_+\partial_-\varphi_i = m^2(e^{\varphi_i-\varphi_{i+1}} - e^{\varphi_{i-1}-\varphi_i}), \quad \partial_\pm = \frac{\partial}{\partial x^\pm}, \quad i \in \mathbb{Z}_{n+1} \tag{2.1}$$

are equivalent to the zero curvature condition

$$F_{+-} = \partial_+ A_- - \partial_- A_+ + [A_+, A_-] = 0 \tag{2.2a}$$

of a connection which belongs to the loop Lie algebra $\widetilde{sl}(n+1) = sl(n+1)\otimes\mathbb{C}[\lambda,\lambda^{-1}]$

$$A_\pm = \pm\partial_\pm\Phi + m\lambda^{\pm1}e^{\pm ad\Phi}\mathcal{E}_\pm, \quad \partial_\pm = \frac{\partial}{\partial x^\pm} \tag{2.2b}$$
$$e^{\mathrm{ad}X}Y \stackrel{\mathrm{def}}{=} e^X\, Y\, e^{-X}$$

where Φ is in the Cartan subalgebra of $sl(n+1) = A_n$. We introduce some notations which are commonly used in physiscs. Let V be a finite dimensional vector space (in our case, $V = \mathbb{C}^n$). Let $\{e_i,\ i = 1,\ldots \mathrm{dim}V\}$ be a basis of V. Then one sets $e_i = |i>$ and $e^i =< i|$ where $e^i,\ i = 1,\ldots\mathrm{dimV}$ is a basis of the dual space $V^\star$ which is dual to the basis e_i: $e^i(e_j) = \delta_{ij}$. The last identity, written in physical notations reads $< i|j >= \delta_{ij}$. The endomorphisms End$V$ of V can be identified with $V \otimes V^\star$. There is natural basis of End$V$ spanned on the elementary matrices $E^{ij} = e_i \otimes e^j = |i >< j|$. Taking into account these conventions, one can write

$$\Phi = \frac{1}{2}\sum_i \varphi_i|i >< i| \quad \text{and} \quad \mathcal{E}_\pm = \sum_{k\in\mathbb{Z}_{n+1}} |k >< k \pm 1|. \tag{2.2c}$$

The adjoint action of the element $S \in SL(n+1)$

$$S \ = \ \omega^{\frac{n}{2}} \sum_{k\in\mathbb{Z}_{n+1}} \omega^{1-k}|k >< k| \ , \ \omega = e^{\frac{2\pi i}{n+1}} \tag{2.3a}$$

defines an inner automorphism σ of order $n+1$ of the algebra A_n

$$\sigma(X) = SXS^{-1}, \quad \sigma^{n+1} = 1. \tag{2.3b}$$

Applying σ on the components of the Lax connection (2.2b) we conclude that

$$A_\pm(x,\omega\lambda) = \sigma(A_\pm(x,\lambda)) = SA_\pm(x,\lambda)S^{-1}. \tag{2.4}$$

Since (2.2b) is a flat connection, there is a covariantly constant vector $w(x,\lambda)$:

$$D_\pm w(x,\lambda) = (\partial_\pm + A_\pm)w(x,\lambda) = 0. \tag{2.5}$$

Taking into account (2.4), one concludes that $\mathcal{S}^r(w)(x,\lambda) = S^r w(x,\omega^{-r}\lambda)$ for any $r \in \mathbb{Z}_{n+1}$ is covariantly constant also. This observation allows us to construct a matrix solution of the linear system (2.5)

$$W(x,\lambda) \ = \ \left(w(x,\lambda), \omega^{\frac{n}{2}}(\mathcal{S}^{-1}w)(x,\lambda)\ldots\omega^{\frac{n^2}{2}}(\mathcal{S}^{-n}w)(x,\lambda)\right) \tag{2.6}$$

To obtain N–soliton solutions of (2.5) we introduce the expansion

$$\hat{w}(x,\lambda) = \sum_{j=0}^{N} \lambda^j w^{(j)}(x) e(x,-\lambda), \quad w^{(j)} = \sum_{k\in\mathbb{Z}_{n+1}} w_k^{(j)} |k>$$

$$e(x,\lambda) = \exp\{m(\lambda x^+ + \frac{x^-}{\lambda})\} \tag{2.7}$$

where N is an integer which will be identified with the number of solitons and $w^{(j)}(x)$, $j = 1,...,N$ are λ–independent $n+1$–dimensional vectors. We shall need the following

PROPOSITION 1 [12] *Let* $w^{(N)} = \sum_{k\in\mathbb{Z}_{n+1}} |k>$ *in the expansion (2.7). Suppose also that for each* $j = 1,...,N$ *there is an integer* $r_j = 1,\ldots,n$ *and complex numbers* μ_j, $\mu_i^{n+1} \neq \mu_j^{n+1}$ *such that*

$$(\mathcal{S}^{-r_j}\hat{w})(x,\mu_j) \quad = \quad \omega^{-\frac{r_j n}{2}} c_j \hat{w}(x,\mu_j) \tag{2.8}$$

for arbitrary constants c_j, $j = 1,\ldots,N$*. Then* $\hat{w}$ *is a solution of the system*

$$(\partial_\pm + \hat{A}_\pm(x,\lambda))\hat{w}(x,\lambda) = 0,$$
$$\hat{A}_+(x,\lambda) = 2\partial_+\Phi(x) + m\lambda\mathcal{E}_+, \quad \hat{A}_-(x,\lambda) = \frac{m}{\lambda} e^{-2ad\Phi(x)}\mathcal{E}_- \tag{2.9}$$

for certain traceless diagonal matrix Φ.

Note that $A_\pm$ and $\hat{A}_\pm$ are related by the gauge transformation $\hat{A}_\pm(x,\lambda) = g^{-1}\partial_\pm g + g^{-1}A_\pm g$ where $g = e^\Phi$. Therefore, the zero curvature condition imposed on both $A_\pm$ and $\hat{A}_\pm$ yields the $A_n^{(1)}$ Toda equations (2.1).

As a consequence from the above Proposition we see that the entries of Φ satisfy the $A_n^{(1)}$ Toda equations (2.1) Alternatively, the components $\hat{w}_k$ of $\hat{w}$ (2.7) can written as

$$\hat{w}_k(x,\lambda) \quad = \quad \prod_{j=1}^{N} (\lambda + \epsilon_{kj}(x)) e(x,-\lambda), k\in\mathbb{Z}_{n+1}$$

and from the Proposition 1 it follows that the evolution of the N-soliton system assumes the form

$$\prod_{l=1}^{N} \frac{\epsilon_{kl} + \omega^{r_j}\mu_j}{\epsilon_{kl} + \mu_j} \quad = \quad c_j \omega^{r_j(1-k)} \frac{e(\omega^{r_j}\mu_j)}{e(\mu_j)}. \tag{2.10a}$$

The components of the Toda field are given by

$$e^{-\varphi_k} \quad = \quad (-)^N \prod_{j=1}^{N} \frac{\epsilon_{kj}}{\mu_j}, \quad k = 1,\ldots,n+1 \tag{2.10b}$$

Note that $\hat{w}(x,\lambda)$ (2.9) is covariantly constant vector with respect to the connection $\hat{D}_\pm = e^{-\Phi} D_\pm e^\Phi$ where $D_\pm = \partial_\pm + A_\pm$ is given by (2.2b). Therefore, $w = e^\Phi \hat{w}$.

In accordance with the general definition, [3, 6, 9], the dressing transformations are represented by loop group elements $g(x,\lambda) \in \widetilde{SL}(n+1)$ which act on the Lax connections (2.2b) as gauge transformations $A_\pm \to A_\pm^g$

$$A_\pm^g = -\partial_\pm g g^{-1} + g A_\pm g^{-1}, \tag{2.11a}$$

such that the connection $A_\pm^g$ has the same form as the original one (2.2b) with $\Phi \to \Phi^g$. Since under gauge transformations the curvature transforms as $F_{+-} \to gF_{+-}g^{-1}$, we see that the dressing group is a symmetry of the equations of motion. Equivalently, the action of the dressing group can be described in terms of the normalized transport matrix

$$T(x,\lambda) \to T^g(x,\lambda) = g(x,\lambda)T(x,\lambda)g^{-1}(0,\lambda) \tag{2.11b}$$

where $T(x,\lambda)$ is a solution of the linear system

$$(\partial_\pm + A_\pm)T(x,\lambda) = 0, \quad T(0,\lambda) = T|_{x^+=x^-=0} = 1 \tag{2.11c}$$

The loop group $\widetilde{G}$ (in the present talk we consider $G = SL(n+1)$ only) is defined [13] as a group of smooth maps of the unit circle $S^1(|\lambda| = 1)$ into the Lie group G. It is clear that the loop group has two subgroups $\widetilde{G}_+$ and $\widetilde{G}_-$: $\widetilde{G}_+$ ($\widetilde{G}_-$) contains all the elements which admit analytic extension on λ inside (outside) the unit disc. In view of (2.2b) A_+, (A_-) is holomorphic Lie valued function for $|\lambda| < 1$, ($|\lambda| > 1$). Therefore, one can look for two solutions $g_\pm \in \widetilde{G}_\pm$ of the dressing problem (2.11a). Since g_+ and g_- produce the same result, from (2.11b) it follows that they provide a solution of the factorization problem

$$g_-^{-1}(x)g_+(x) = T(x)g_-^{-1}(0)g_+(0)T^{-1}(x) \tag{2.12}$$

Note that the transport matrix related to the vacuum solution $\Phi = 0$ (2.1), (2.2c) is $T_0(x,\lambda) = e^{-m(\lambda\mathcal{E}_+x^+ + \frac{x^-}{\lambda}\mathcal{E}_-)}$. In [10] we have shown that there is a solution of (2.11a) which interpolates between the vacuum and the N-soliton solution Φ (2.10a)-(2.10b)

$$g^{(N)}(\Phi,\{\mu\},\lambda) = e^\Phi \Gamma^{(N)}(\Phi,\{\mu\},\lambda)\Omega^{-1}$$

$$\Gamma_{kl}^{(N)}(\Phi,\{\mu\},\lambda) = \omega^{(k-1)(l-1)} \prod_{j=1}^{N} \frac{\lambda + \omega^{1-l}\epsilon_{kj}(x)}{\lambda - \omega^{1-l}\mu_j}, \quad \Omega_{ij} = \omega^{(i-1)(j-1)} \quad , \quad \lambda \in S^1 \tag{2.13a}$$

which belongs to loop group $\widetilde{SL}(n+1)$ and obeys the relation (c.f. (2.4))

$$g^{(N)}(\Phi,\{\mu\},\omega\lambda) = Sg^{(N)}(\Phi,\{\mu\},\lambda)S^{-1} \tag{2.13b}$$

To proceed, we have to introduce some Lie algebraic background [4, 7]. First of all, note that the automorphism σ (2.3a), (2.3b) defines a $\mathbb{Z}_{n+1}$-gradation in the Lie algebra $\mathcal{G} = sl(n+1)$

$$\mathcal{G} = \oplus_{k\in\mathbb{Z}_{n+1}}\mathcal{G}_k \quad , \quad \sigma(\mathcal{G}_k) = \omega^k\mathcal{G}_k \tag{2.14}$$

On the other hand, the elements

$$\mathcal{E}_i = \sum_{k \in \mathbb{Z}_{n+1}} |k><k+i| \quad , \quad i \in \mathbb{Z}_{n+1}\backslash\{0\} \tag{2.15a}$$

are mutually commuting and diagonalizable

$$\Omega^{\pm 1}\mathcal{E}_k \Omega^{\mp 1} = \omega^{\pm \frac{kn}{2}} S^{\pm k}, \quad \mathcal{E}_1 = \mathcal{E}_+, \quad \mathcal{E}_{-1} = \mathcal{E}_n = \mathcal{E}_- \tag{2.15b}$$

where Ω is defined by the last equation (2.13a). Therefore, the set of generators $\mathcal{E}_k$, $k = 1, \ldots, n$ span a new Cartan subalgebra. To complete it to a base of the Lie algebra we define the elements

$$F^i = \Omega|i+1><i|\Omega^{-1} \quad , \quad i \in \mathbb{Z}_{n+1}\backslash\{0\} \tag{2.16a}$$

and consider their grade expansions

$$F^i = \sum_{k \in \mathbb{Z}_{n+1}} F^i_k, \qquad \sigma(F^i_k) = \omega^k F^i_k \tag{2.16b}$$

The commutation relations in the new basis are

$$[F^r_i, F^s_l] = \frac{\omega^{sk} - \omega^{rl}}{n+1} F^{r+s}_{k+l}, \quad F^r_k = \begin{cases} F^r_k & r = 1, \ldots, n; \quad k \in \mathbb{Z}_{n+1} \\ \frac{1}{n+1}\mathcal{E}_k & r = 0 \quad k = 1, \ldots, n \end{cases} \tag{2.17a}$$

In the above basis the invariant non degenerate scalar product $(X, Y) = tr(X.Y)$ (the trace is taken in the defining representation) is given by

$$(F^r_k, F^s_l) = \frac{\omega^{sk}}{n+1}\delta^{(n+1)}_{k+l,0} \tag{2.17b}$$

where $\delta^{(n+1)}_{i,j}$ is the Kronecker symbol on the cyclic group.

To treat integrable evolution equations one has to extend the classical Lie algebra by introducing a spectral parameter λ (c.f. (2.2b)). The Lax connection is in the loop Lie algebra $\tilde{\mathcal{G}} = \mathbb{C}[\lambda, \lambda^{-1}] \otimes \mathcal{G}$. $\tilde{\mathcal{G}}$ is spanned on the elements $X_n = \lambda^n X$. Generic loop algebra has a central extension [4, 13] known as affine Lie algebra $\widehat{\mathcal{G}} = \tilde{\mathcal{G}} \oplus \mathbb{C}\widehat{c} \oplus \mathbb{C}\widehat{d}$

$$[X_k, Y_l] = [X, Y]_{k+l} + \frac{k}{(n+1)}\hat{c}\,\delta_{k+l,0}\,(X, Y), \quad \left[\hat{d}, X_k\right] = kX_k, \quad \left[\hat{c}, \hat{\mathcal{G}}\right] = 0. \tag{2.18a}$$

The derivation $\widehat{d} = \lambda\frac{\partial}{\partial\lambda}$ defines a $\mathbb{Z}$-gradation

$$\tilde{\mathcal{G}} = \oplus_{k \in \mathbb{Z}}\tilde{\mathcal{G}}_k, \quad \left[\widehat{d}, \widehat{\mathcal{G}}_k\right] = k\widehat{\mathcal{G}}_k \tag{2.18b}$$

Note also that setting $\widehat{c} = 0$ in (2.18a), one recovers the commutation relations in the loop algebra. The symmetry of the Lax connection (2.4) suggests the following.

DEFINITION: *The affine (loop) Lie algebra* $\widehat{sl}(n+1)$ $(\widetilde{sl}(n+1))$ *is in the principal gradation iff its elements* $X(\lambda) = \sum_{l \in \mathbb{Z}} \lambda^l X_l$ *satisfy the restriction*

$$X(\omega\lambda) = \sigma X(\lambda) \tag{2.19}$$

In view of the above definition, (2.4) is in the loop algebra $\tilde{sl}(n+1)$ in the principal gradation. The affine algebra analog of the base (2.17a) is

$$F_k^r, \quad r=1,\dots,n, \quad k\in\mathbb{Z}, \quad F_k^0, \qquad k\in\mathbb{Z}/(n+1)\mathbb{Z} \tag{2.20a}$$

$$[F_k^r, F_l^s] = \frac{\omega^{sk}-\omega^{rl}}{n+1}F_{k+l}^{r+s} + \frac{k\omega^{rl}}{(n+1)^2}\,\hat{c}\,\delta_{r+s,0}^{(n+1)}\,\delta_{k+l,0} \tag{2.20b}$$

In the above notations and in what follows we will perform a slight abuse of notations: the lower index will be used to indicate both the cyclic $\mathbb{Z}_{n+1}$ charge (2.14) as well as to label the $\mathbb{Z}$ gradation (2.18b). The elements $\mathcal{E}_k = (n+1)F_k^0$, $k\neq 0 \bmod (n+1)$ form an infinite set of decoupled harmonic oscillators

$$[\mathcal{E}_k, \mathcal{E}_l] = k\,\hat{c}\delta_{k+l,0} \quad k\neq 0 \bmod (n+1) \tag{2.20c}$$

The subalgebra spanned on $\mathcal{E}_k$ is known as the Heisenberg subalgebra in the principal gradation. In order to formulate one of the main results of [10] we have to introduce some notations. Let (2.10a) be an arbitrary N-soliton solution. To it we associate the diagonal traceless matrices

$$F_i = \frac{1}{2}\sum_{k\in\mathbb{Z}_{n+1}} f_{ki}|k><k|, \quad P_i = \sum_{k\in\mathbb{Z}_{n+1}} p_{ki}|k><k|$$

$$K(F_i) = \sum_{k\in\mathbb{Z}_{n+1}} K_k(F_i)|k><k|, \quad i=1,\dots,N \tag{2.21a}$$

where the quantities $K(F_i)$ are linear on F_i and satisfy the recurrence relations

$$K_k(F_i) - K_{k+1}(F_i) = \frac{f_{k\,i}+f_{k+1\,i}}{2} \quad , \quad k\in\mathbb{Z}_{n+1} \tag{2.21b}$$

The last of the above equations agrees with the periodicity property $K_k = K_{k+n+1}$ since $\mathrm{tr}F_i = 0$. Following [10], one imposes the relations

$$\sum_{k\in\mathbb{Z}_{n+1}} \omega^{r(1-k)}\left(\frac{\rho_{kl}(\omega^r\mu_l)}{\rho_{kl}(\mu_l)} - \omega^{-r}\frac{\rho_{k+1l}(\omega^r\mu_l)}{\rho_{k+1l}(\mu_l)}\right)\beta_k(F_l) = \delta_{r,r_l}X_lY_l$$

$$X_l = (1-\omega^{r_l})\prod_{a\neq l}\frac{\omega^{r_l}\mu_l-\mu_a}{\mu_l-\mu_a}\prod_a\frac{\mu_l+\epsilon_{1a}}{\omega^{r_l}\mu_l+\epsilon_{1a}}$$

$$Y_l = \sum_{k\in\mathbb{Z}_{n+1}}\left(1+\mu_l\frac{d}{d\lambda}\ln\frac{\rho_{kl-1}}{\rho_{k+1l-1}}(\mu_{l+1})\right)\beta(F_l) \tag{2.22a}$$

$$\beta_k(F_l) = e^{K_k(F_l)-\frac{f_{kl}}{2}}. \tag{2.22b}$$

In view of (2.21b) one has

$$e^{-f_{kl}} = \frac{\beta_k(F_l)}{\beta_{k-1}(F_l)} \tag{2.22c}$$

The functions ρ_{al}, $a\in\mathbb{Z}_{n+1}$, $l=1,\dots,N$ are components of $(n+1)$-dimensional vectors

$$\rho_l(\lambda) = \mathcal{D}^{(l\,l-1)}(\lambda)\rho_{l-1}(\lambda) \quad , \qquad \rho_{l-1}(\lambda) = \mathcal{D}^{(l-1\,l)}(\lambda)\rho_l(\lambda) \tag{2.22d}$$

where the entries of the matrices $\mathcal{D}^{(l\,l-1)}$ and $D^{(l-1\,l)}$ are

$$\mathcal{D}^{(l\,l-1)}_{ab}(\lambda) = \frac{\gamma_l \beta_b(F_l)}{(n+1)\rho_{bl-1}(\mu_l)} \sum_{q\in \mathbb{Z}_{n+1}} \omega^{q(a-b)} \frac{\lambda - \omega^q e^{f_{bl}} \mu_l}{\lambda - \omega^q \mu_l}$$

$$\mathcal{D}^{(l-1\,l)}_{ab}(\lambda) = \frac{\rho_{al-1}(\mu_l)}{(n+1)\gamma_l \beta_b(F_l)} \sum_{q\in \mathbb{Z}_{n+1}} \omega^{q(a-b)} \frac{\lambda - \omega^q e^{-f_{bl}} \mu_l}{\lambda - \omega^q \mu_l},$$

$$\gamma_l = \left(\prod_{p\in \mathbb{Z}_{n+1}} \rho_{pl-1}(\mu_l)) \right)^{\frac{1}{n+1}} \tag{2.22e}$$

The relation (2.22d) together with

$$\rho_{j0}(\lambda) = \frac{1}{n+1} \tag{2.22f}$$

determines uniquely ρ_l , $l = 1, ..., N$. Finally the entries of the diagonal matrices P_l are fixed by

$$e^{K_a(F_{l-1}) + \frac{P_{al-1}}{2}} = \frac{\rho_{al-1}(\mu_l)}{\gamma_l} \tag{2.22g}$$

Now we are in a position to formulate the following

PROPOSITION 2 [10]: *Consider the N-soliton solution (2.10a), (2.10b). Denote by* g_+ (g_-) *the analytic continuation of (2.13a) around* $\lambda = 0$, $(\lambda = \infty)$. *Then the following factorization is valid*

$$g_\pm = g_\pm(N) g_\pm(N-1) ... g_\pm(1), \qquad g_\pm(i) = e^{K(F_i)+P_i} e^{W_\pm(i)} \tag{2.23a}$$

where $K(F_i)$ *and* P_i *are fixed recursively by (2.22a)-(2.22f) and* $W_\pm(i)$ *are the loop group elements*

$$W_\pm(i) = \sum_{k\in \mathbb{Z}_{n+1}} f_{ki} W^k_\pm(\mu_i)$$

$$W^k_+(\mu) = -\sum_{r=1}^{n} \left(\frac{1-\omega^{-rk}}{1-\omega^{-r}} F^r_0 + \omega^{r(1-k)} \sum_{p¿1} \frac{F^r_p}{\mu^p} \right)$$

$$W^k_-(\mu) = \sum_{r=1}^{n} \left(-\frac{1-\omega^{r(1-k)}}{1-\omega^{-r}} F^r_0 + \omega^{r(1-k)} \sum_{p6-1} \frac{F^r_p}{\mu^p} \right). \tag{2.23b}$$

Note that $W^k_\pm(\mu)$ do not contain contributions belonging to the principal Heisenberg subalgebra. Moreover

$$W^k(\mu) = W^k_+(\mu_i) - W^k_-(\mu_i) = -\sum_{r=1}^{n} \omega^{r(1-k)} F^r(\mu), \quad F^r(\mu) = \sum_{p\in \mathbb{Z}} \frac{F^r_p}{\mu^p_i} \tag{2.24a}$$

and the element $F^r(\mu)$ diagonalize the adjoint action of the principal Heisenberg subalgebra:

$$[\mathcal{E}_k, F^r(\mu)] = (\omega^{rk} - 1)\mu^k F^r(\mu) \tag{2.24b}$$

The Proposition 2 together with (2.24a)-(2.24b) suggests a relation between the dressing symmetry and vertex operator representation of the soliton tau function [3, 7].

3 VERTEX OPERATOR REPRESENTATION FOR THE SOLITON TAU FUNCTIONS

The $A_n^{(1)}$ Toda models have a conformally invariant extension [14] known as the A_n Conformal Affine Toda (CAT) model. The A_n (CAT) equations are also integrables and admit zero curvature representation for the connection (2.2b) in the affine Lie algebra $\hat{\mathcal{G}} = \hat{Sl}(n+1) = A_n^{(1)}$. The affine algebra counterpart of Φ (2.2c) is

$$\Phi \to \Phi + \eta \hat{d} + \frac{\hat{c}}{2(n+1)}\zeta \tag{3.1}$$

where $\hat{d}$ and $\hat{c}$ stand for the derivation and the central charge respectively (c.f. (2.18a)). Combining (2.2a), (2.2b) and (3.1) one obtains the A_n CAT equations

$$\begin{aligned} &\partial_+\partial_-\varphi_i = m^2 e^{2\eta}(e^{\varphi_i-\varphi_{i+1}} - e^{\varphi_{i-1}-\varphi_i}), \quad \partial_+\partial_-\eta = 0, \\ &\partial_+\partial_-\zeta = m^2 e^{2\eta} \sum_{i\in\mathbb{Z}_{n+1}} e^{\varphi_i-\varphi_{i+1}}. \end{aligned} \tag{3.2}$$

The first of the above equations coincides with the affine Toda equations (2.1) for $\eta = 0$. According to [8], affine solitons arise after imposing the last restriction, it is also worthwhile to recall the Hirota bilinear representation [15]

$$\begin{gathered} \partial_+\tau_k\partial_-\tau_k - \tau_k\partial_+\partial_-\tau_k = m^2(\tau_{k+1}\tau_{k-1} - \tau_k^2) \\ e^{-\varphi_k} = \frac{\tau_k}{\tau_{k-1}}, \quad e^{\zeta_0-\zeta} = \prod_{k\in\mathbb{Z}_{n+1}} \tau_k, \quad k \in \mathbb{Z}_{n+1}, \quad \zeta_0 = (n+1)m^2x^+x^- \end{gathered} \tag{3.3}$$

of the system (3.2) for $\eta = 0$. Setting $\zeta = \zeta_0$ with $\varphi_k = 0$, $k \in \mathbb{Z}_{n+1}$ and $\eta = 0$ in (3.1) one ends with the vacuum solution of the A_n CAT equations (3.2)

$$\Phi_0 = \frac{m^2}{2}\hat{c}x^+x^-. \tag{3.4}$$

In [8] it was argued that the Riemann problem (2.12) is related to a similar factorization problem in the affine group. More precisely, the following relation

$$\hat{g}_\pm(x) = e^{\pm\frac{\zeta_0-\zeta}{2(n+1)}\hat{c}} g_\pm(x) \tag{3.5}$$

holds. In the above equation $g_\pm$ represent a dressing transformation in the loop group which by (2.11a) creates N-soliton solutions from the vacuum; $\hat{g}_\pm(x)$ stand for the affine group analog of $g_\pm(x)$ (2.12).

In what follows we shall need some basic facts concerning the representation theory of the Lie algebras $A_n^{(1)}$ in the principal gradation [4]. We will be interested on highest weight representations which are generated by the action of arbitrary polynomials on the negative grade elements (2.18b) on certain highest weight state $|\Lambda>$. The later is annihilated by the elements of positive grade

$$X_n|\Lambda>= 0 \ , \quad \left[\hat{d}, X_n\right] = nX_n, \quad n > 0. \tag{3.6a}$$

In the particular example of $A_n^{(1)}$, the highest weight state is characterized by (2.20a)

$$F_0^r|\Lambda>=\Lambda(F_0^r)|\Lambda>, \quad 1\leq r\leq n, \quad \hat{d}|\Lambda>=\Lambda(\hat{d})|\Lambda>, \quad \hat{c}|\Lambda>=\Lambda(\hat{c})|\Lambda> \qquad (3.6b)$$

where Λ is a linear functional on the subalgebra $\hat{\mathcal{G}}_0$ of elements of $\mathbb{Z}$–grade zero (cf. (2.18a)).

The principal Heisenberg subalgebra (2.20c) admits a Fock representation. It is built up on the Fock vacuum $|0>$

$$\mathcal{E}_k|0>=0, \quad k\geq 1, \quad <0|\mathcal{E}_k=0, \quad k\leq -1. \qquad (3.7)$$

When the value of $\hat{c}$ (2.18a) is one, all the irreducible highest weight representations of the Lie algebra $A_n^{(1)}$ are expressed in terms of the (principal) Heisenberg subalgebra. The corresponding $A_n^{(1)}$ modules are identified with the Fock space (3.7). To be more precise, we introduce the "vertex operators":

$$V^r(\mu)=\exp\{\sum_{k\leq -1}\frac{1-\omega^{-kr}}{k}\frac{\mathcal{E}_k}{\mu^k}\}\exp\{\sum_{k\geq 1}\frac{1-\omega^{-kr}}{k}\frac{\mathcal{E}_k}{\mu^k}\}$$
$$r=1,...,n \qquad (3.8)$$

The following theorem is due to Kac

THEOREM 1 [4]: *Consider the Fock space representation (3.7) of the principal Heisenberg subalgebra. For each* $l\in\mathbb{Z}_{n+1}$ *and* $r\in\mathbb{Z}_{n+1}\backslash\{0\}$ *define*

$$F^r(\mu)=\frac{\omega^{rl}}{(n+1)(\omega^r-1)}V^r(\mu) \qquad (3.9a)$$

Then

(i) (3.9a) *admits a Laurent expansion*

$$F^r(\mu)=\sum_{p\in\mathbb{Z}}\frac{F_p^r}{\mu^p} \qquad (3.9b)$$

(ii) *For each* $l\in\mathbb{Z}_{n+1}$ *the Laurent coefficients* F_p^r , $p\in\mathbb{Z}$ and $r\in\mathbb{Z}_{n+1}\backslash\{0\}$ *together with the harmonic oscillators* $F_k^0=\frac{1}{n+1}\mathcal{E}_k$, $k\in\mathbb{Z}\backslash(n+1)\mathbb{Z}$ *obey the commutation relations (2.20b).*

(iii) *For each* $l\in\mathbb{Z}_{n+1}$, *the Fock vacuum* $|0>$ *(3.7) is a highest state* $|\Lambda_l>$ *characterized by*

$$F_0^r|\Lambda_l>=\frac{\omega^{rl}}{(n+1)(\omega^r-1)}|\Lambda_l>, \ \hat{c}|\Lambda_l>=|\Lambda_l>, \ \hat{d}|\Lambda_l>=0. \qquad (3.9c)$$

The meaning of the above theorem is that there are at least $n+1$ irreducible representations of the Lie algebra $A_n^{(1)}$ which can be expressed in terms of bosonic oscillators. The value of the central charge is $\hat{c}=1$. It was shown in [4] that the representations with highest weight vector $|\Lambda_l>$ (3.9c), $l \in \mathbb{Z}_{(n+1)}$ exhaust the inequivalent fundamental representations of the affine algebra.

In what follows, it will be important to consider also the group theoretical tau functions

$$\tau_\Lambda(\Phi) =< \Lambda|e^{-2\Phi}|\Lambda >= e^{-2\Lambda(\Phi)} \tag{3.10}$$

where Φ is given by (3.1) and $|\Lambda>$ is a highest weight vector. Setting $\eta = 0$ in (3.1) and taking into account (3.9c), one gets a relation with the Hirota tau functions (3.3)

$$\tau_{\Lambda_k}(\Phi) = e^{-\frac{\zeta_0}{n+1}}\tau_k(\Phi) \tag{3.11}$$

Let us recall the following result [8, 16]

$$\frac{\tau_\Lambda(\Phi)}{\tau_\Lambda(\Phi_0)} =< \Lambda|\hat{g_-}^{-1}(x)\hat{g_+}(x)|\Lambda > \tag{3.12}$$

where $\hat{g}_\pm(x)$ (3.5) are the elements of the affine group which generate by dressing transformation a solution Φ, starting from the vacuum Φ_0 (3.4). Let us consider an N-soliton solution (2.10a), (2.10b) of the $A_n^{(1)}$ Toda model. Then the Proposition 2, (3.5) together with (3.12) suggest us to study the following affine group element

$$h(\alpha) = e^{-\alpha W_-(\phi)}e^{\alpha W_+(\phi)}, \quad W_\pm(\Phi) = \sum_{k\in\mathbb{Z}_{n+1}} \varphi_k W_\pm^k(\mu), \quad \sum_{k\in\mathbb{Z}_{n+1}} \varphi_k = 0 \tag{3.13}$$

where $W_\pm^k(\mu)$ are given by (2.23b). As a corollary of the Theorem 1 one gets the

PROPOSITION 3 *Consider an arbitrary fundamental representation of the Lie algebra $A_n^{(1)}$. Then the following commutation relations are valid*

$$W_-(\Phi)V^r(\mu) - V^r(\mu)W_+(\Phi) = \frac{1}{n+1}\sum_{s=1}^{n}\frac{\omega^{s(k+1)}}{\omega^s - 1}\hat{\varphi}_s V^{r+s}(\mu), \tag{3.14a}$$

$k \in \mathbb{Z}_{n+1}$ labels the fundamental representations of $A_n^{(1)}$; $V^r(\mu)$ are the "vertex operators" introduced by (3.8) and $\hat{\varphi}_s$, $s \in \mathbb{Z}_{n+1}$ is the discrete Fourier transform of φ_s ($\sum_s \varphi_s = 0$)

$$\hat{\varphi}_s = \sum_k \omega^{-ks}\varphi_k, \quad \varphi_k = \frac{1}{n+1}\sum_s \omega^{ks}\hat{\varphi}_s. \tag{3.14b}$$

From the above Proposition it follows that the *affine* group element (3.13) admits the following (Taylor) expansion

$$h(1) = e^{-W_-(\Phi)}\,e^{W_+(\Phi)} = \sum_{l=0}^{\infty}\frac{(-1)^l}{l!}\sum_{s_1,\dots,s_l=1}^{n}\prod_{j=1}^{l}\Psi_{s_j}^{(k)}(\Phi)V^{s_1+\dots+s_l}(\mu)$$

$$\Psi_s^{(k)}(\Phi) = \frac{1}{n+1}\frac{\omega^{s(k+1)}}{\omega^s - 1}\hat{\varphi}_s, \quad k \in \mathbb{Z}_{n+1} \tag{3.15}$$

In view of the periodicity condition $V^r = V^{r+n+1}$ (cf. (3.8)), it is opportune to introduce the commutative associative algebra $\mathcal{F}$ of (complex) dimension $n+1$. It is generated by V^s, $s \in \mathbb{Z}_{n+1}$. The multiplication is

$$V^r \star V^s = V^{r+s} \tag{3.16}$$

It is worthwhile to note that the above equation describes "fusion rules" of a class of Rational Conformal Field Theories [17]. Introducing the invariant inner product

$$< V^r, V^s >_{\mathcal{F}} = \delta^{(n+1)}_{r+s,0}, \tag{3.17}$$

$\mathcal{F}$ becomes a Fröbenius algebra. Taking into account (3.15) and (3.16) one gets

$$h(1) = \mathcal{F} \exp\{-\sum_{s=1}^{n} \Psi_s^{(k)}(\Phi) V^s\}(\mu) \tag{3.18}$$

where the symbol $\mathcal{F}$ means that the exponential is taken in the algebra (3.16).

Diagonalizing the "fusion rules" (3.16) we get

PROPOSITION 4 *The affine group element $h = h(1)$ (3.13) is given by the expression*

$$h = \frac{\hat{\beta}_0(\Phi)}{n+1} + \sum_{p \in \mathbb{Z}_{n+1}} (\omega^p - 1)\hat{\beta}_p(\Phi) F^p(\mu) \tag{3.19}$$

within any fundamental representation (3.9a)-(3.9c). The numbers $\hat{\beta}_p(\Phi)$ are the discrete Fourier transform (3.14b) of $\beta_k(\Phi)$ (2.21b), (2.22b).

In [10] we derived an exact expression for the adjoint action of the elements $g_\pm(i)$, $i = 1, \ldots, N$ defined by (2.23a), (2.23b) on the affine Lie algebra. The result is

$$g_+^{-1}(i) F^s(\omega^c \lambda) g_+(i) = \sum_{r,v \in \mathbb{Z}_{n+1}} R^{sc}_{rv}(i;\lambda)\left(F^r(\omega^v \lambda) - Z^{rv}[F_i](i;\lambda)\right), \quad |\mu_i| > |\lambda| \tag{3.20a}$$

$$R^{sc}_{rv}(i;\lambda) = P^c_v(i;\lambda)\, Q^{s+c}_{r+v}(i;\lambda). \tag{3.20b}$$

The matrices P, Q and the vector Z are expressed in terms of the quantities (2.22d) and (2.22e) as follows

$$P^c_v(i;\lambda) = \frac{\gamma_i}{n+1} \sum_{a,b \in \mathbb{Z}_{n+1}} \omega^{(a-1)v-(b-1)c} \frac{1}{\rho_{ai-1}(\mu_i)} \mathcal{D}^{(i-1\, i)}_{ab}(\lambda) \rho_{bi}(\mu_{i+1})$$

$$Q^s_r(i;\lambda) = \frac{1}{(n+1)\gamma_i} \sum_{a,b \in \mathbb{Z}_{n+1}} \omega^{(a-1)s-(b-1)r} \frac{1}{\rho_{ai}(\mu_{i+1})} \mathcal{D}^{(i\, i-1)}_{ab}(\lambda) \rho_{bi-1}(\mu_i)$$

$$Z^{sc}[F_i](i,\lambda) = \frac{\lambda \omega^c}{\gamma_i (n+1)^2 (\omega^{s+c}\lambda - \mu_i)} \sum_{a,b \in \mathbb{Z}_{n+1}} \omega^{(s+c)a-bc} \mathcal{D}^{(i\, i-1)}_{ab}(\lambda) \frac{\rho_{bi-1}(\mu_i)}{\beta_a(F_i)} +$$

$$+ \frac{\lambda \omega^c \delta^{(n+1)}_{s,0}}{(n+1)(\mu_i - \omega^{s+c}\lambda)} \tag{3.20c}$$

As a consequence of (3.20b), (3.20c), (2.22d) and taking into account the indentity

$$\frac{d\rho_{kj}}{d\lambda}(\lambda)=-\frac{1}{n+1}\sum_{l=1}^{j}\sum_{a,a',r\in\mathbb{Z}_{n+1}}\frac{\omega^{r(a'-a)}}{\lambda-\omega^r\mu_l}\mathcal{D}_{ka'}^{(j\,l-1)}(\lambda)\rho_{a'\,l-1}(\lambda)\left(\frac{\rho_{al}(\lambda)}{\gamma_l\beta_a(F_l)}-\frac{\rho_{al-1}(\lambda)}{\rho_{al-1}(\mu_l)}\right)$$

one arrives at

PROPOSITION 5 *Let $\mu_1,\ldots,\mu_N$ are radially ordered complex numbers $|\mu_1|>\ldots,>|\mu_N|$. Then the matrices $R(j;\mu_i)$ $(j<i)$ which act on $\mathbb{C}^{n+1}\otimes\mathbb{C}^{n+1}$ as well as the vectors $Z[F_j](j,\mu_i)\in\mathbb{C}^{n+1}\otimes\mathbb{C}^{n+1}$ $(j<i)$ satisfy the equations*

$$\sum_{p\in\mathbb{Z}_{n+1}}\omega^{p(1-k)}\left(R(j;\mu_{j+1})\ldots R(1;\mu_{j+1})\right)_{rv}^{p0}=\omega^{r(1-k)}\delta_{v,0}^{(n+1)}\frac{\rho_{kj}(\omega^r\mu_{j+1})}{\rho_{kj}(\mu_{j+1})}$$

$$\sum_{\substack{1\le l\le j\\ p\in\mathbb{Z}_{n+1}}}\omega^{p(1-k)}\left(R(j;\mu_{j+1})\ldots R(l;\mu_{j+1})\,Z[-F_l](l;\mu_{j+1})\right)^{p0}=-\frac{\mu_{j+1}}{n+1}\frac{d\ln\rho_{kj}}{d\lambda}(\mu_{j+1})$$

(3.21)

Combining the above Proposition with Proposition 2 and the Proposition 4 we obtain the following

THEOREM 2[10] *Let $\Phi=\frac{1}{2}\sum_{k\in\mathbb{Z}_{n+1}}\varphi_k|k><k|$ be an N-soliton solution of the $A_n^{(1)}$ Toda equations (2.10a), (2.10b), with $|\mu_1|>\ldots,>|\mu_N|$. $g_\pm$ are supposed to be same as in the Proposition 2. Then for each fundamental representation of the Lie algebra $A_n^{(1)}$, the following identities are valid*

$$\mathrm{Ad}\left(g_+^{-1}(1)\ldots g_+^{-1}(i-1)\right)g(i)=Y_i\left(1+X_iF^{r_i}(\mu_i)\right)\qquad(3.22)$$

The factors X_i, Y_i were introduced in (2.22a) and $g_\pm$ are the elements of the loop group $\widetilde{S}L(2)$ which by dressing transformation (2.11a) generate N-soliton solutions form the vacuum.

As a corollary of the above Theorem and taking into account (3.5) and (3.12) we get

$$\frac{\tau_\Lambda(\Phi)}{\tau_\Lambda(\Phi_0)}=e^{\frac{\zeta_0-\zeta}{n+1}}\prod_{i=1}^{N}Y_i\cdot<\Lambda|(1+X_1F^{r_1}(\mu_1))\ldots(1+X_NF^{r_N}(\mu_N))|\Lambda>$$

(3.23)

Therefore (1.1a) and (1.1b) are equivalent provided that

$$\prod_{i=1}^{N}Y_i=\exp\left(\frac{\zeta-\zeta_0}{n+1}\right)\qquad(3.24)$$

Note that as a consequence of (3.3), (3.11), the last identity can be equivalently written as

$$\prod_{k\in\mathbb{Z}_{n+1}}\prod_{i=1}^{N}Y_i<\Lambda|(1+X_1F^{r_1}(\mu_1))\ldots(1+X_NF^{r_N}(\mu_N))|\Lambda>=1\qquad(3.25)$$

We hope to go back to the demonstration of the above identity in a future publication.

4 REMARKS AND CONCLUSIONS

The present talk is devoted to the study of the relation between the vertex operator representation of the (soliton) tau functions and the group of dressing transformations. We recall [5, 6] that the dressing group is dual to the corresponding loop (or affine) Lie group. On the other hand, from the general analysis advanced by Leznov and Saveliev [16], it follows that

$$\tau_\Lambda^g =< \Lambda|\exp\{-mx^+\mathcal{E}_1\}\, g \exp\{mx^-\mathcal{E}_{-1}\}|\Lambda > \tag{4.1a}$$

where $\mathcal{E}_{\pm1}$ are the elements of $\mathbb{Z}$-grade ±1 of the Heisenberg subalgebra (2.20c), g is a constant element of the affine Lie group and $|\Lambda >$ is a highest weight vector and τ_Λ^g is a tau function (3.10) of the corresponding CAT equations. More precisely,

$$\exp\{-2\Lambda(\Phi^g)\} =< \Lambda|\exp\{-2\Phi^g\}|\Lambda > \tag{4.1b}$$

is a solution of the A_n CAT equations (3.2). On the other hand, it was observed in [7] that setting in (4.1a)

$$g = \exp\{\alpha_1 F^{r_1}(\mu_1)\} \ldots \exp\{\alpha_N F^{r_N}(\mu_N)\}, \quad F^r(\mu) = \sum_{p\in\mathbb{Z}_{n+1}} \frac{F_p^r}{\mu^p},$$

one recovers the N-soliton solutions of the CAT model. However, the relation with the dressing symmetry seems rather obscure. A hint of how to relate the group-algebraic approach [7, 16] with the group of dressing transformations has been suggested in [8] on the particular example of the sine-Gordon equation.

Our approach generalizes the the results of [8] for arbitrary $A_n^{(1)}$ Toda models. However, some problems remain open. First, we have not considered the relation with the Bäcklund transformations. We believe that the solution of this problem can shed a new light on the geometry of affine Toda models. For arbitrary simple Lie algebras one can generalize the Proposition 2, but it is not generally possible to obtain a matrix solution of the underlying linear problem. The reason is that the order of the automorphism (2.3a), (2.3a) is generically smaller than the dimension of any irreducible representation of the Lie algebra.

Acknowledgements

The authors are grateful to the organizers of the IV *International Conference on Non Associative Algebra and its Applications* for the kind invitation to present this talk at the University of São Paulo, Brazil. We acknowledge L. A. Ferreira for discussions on the subject, and especially for explaining to us the results of [9]. We thank G. Cuba for his collaboration on the early stage of [10]. The work of H. B. and R. P. are supported financially by CNPq-Brazil and FAPERJ-Rio de Janeiro respectively. Both the research foundations are deeply acknowledged.

REFERENCES

[1] L. Dickey, Soliton equations and Integrable systems, World-Scientific(1991).

[2] R. Rajaraman, Solitons and Instantons, North–Holland, Elsevier 1982.

[3] E. Date, M. Jimbo, M. Kashiwara and T. Miwa, Transformation Groups for Soliton Equations, in: Non–linear integrable systems, eds M. Jimbo and T. Miwa (World Scientific, Singapore, 1983); Publ. RIMS, 18, 1077 (1982).

[4] V. G. Kac, Infinite dimensional Lie algebras. Third edition. Cambridge Univ. Press. 1990.
V. G. Kac and M. Wakimoto, Exceptional Hierarchies of Soliton Equations, in Proceedings of Symposia in Pure Mathematics, vol. 49, (1989) 191.

[5] M. Semenov–Tian–Shansky, Publ. RIMS Kyoto Univ. 21, 1237 (1985).

[6] O. Babelon and D. Bernard, Phys. Lett. B260, 81 (1991) ;
Commun. Math. Phys. 149, 279 (1992).

[7] D. I. Olive, N. Turok and J. W. Underwood, Nucl. Phys. B401, 663 (1993); B409, 509 (1993)
D. I. Olive, M. V. Saveliev and J. W. Underwood, Phys. Lett. B311, 117 (1993).

[8] O. Babelon and D. Bernard, Int. J. Mod. Phys. A8, 507 (1993).

[9] L. A. Ferreira, L. Miramontes and J. S. Guillén, J. Math. Phys. 38, 882 (1997);
L. A. Ferreira, and J. S. Guillén, solv–int/9701006.

[10] H. Belich and R. Paunov, J. Math. Phys. 38, 4108 (1997),
H. Belich, G. Cuba and R. Paunov, hep–th/9712249 , J. Math. Phys., to appear.

[11] D. I. Olive and N. Turok, Nucl. Phys. B215, 470 (1983); B220, 491 (1983).

[12] E. Date, Osaka J. Math, 19, 125 (1982).

[13] A. N. Pressley and G. B. Segal, Oxford Univ. Press, 1985.

[14] O. Babelon and L. Bonora, Phys. Lett. B244, 220 (1990);
H. Aratyn, L. A. Ferreira, J. F. Gomes and A. H. Zimerman, Phys. Lett. B254, 372 (1991).

[15] T. Hollowood, Nucl. Phys. B384, 523 (1992).

[16] A. N. Leznov and M. V. Saveliev, Group Theoretical Methods for Integration of Non-Linear Dynamical Systems, Progress in Physics Series, vol. 15 (Birkhäuser–Verlag, Basel, 1992);
A. N. Leznov and M. V. Saveliev, Lett. Math. Phys. 3, 489 (1979).
A. N. Leznov, M. V. Saveliev and V. G. Smirnov, Sov. J. Theor. Math. Phys. 48, 3 (1981).

[17] E. Verlinde, Nucl. Phys. B300, 360 (1988).

Generalized quadratic rings

LUIZ ANTONIO PERESI[1] Departamento de Matemática, Universidade de São Paulo, Caixa Postal 66281, São Paulo-SP, Brazil, 05315-970.

ABSTRACT. The purpose of this paper is to survey on some results concerning polynomial identities of quadratic algebras. We give results about 2×2 matrices, Jordan algebras of degree 2, Cayley-Dickson algebras and for the class of all quadratic algebras. We communicate some new results about generalized quadratic rings, i.e., rings verifying some polynomial identities of degree ≤ 5 which are satisfied by all symmetric quadratic algebras.

1 INTRODUCTION

Polynomial identities are one of the basic tools in the structure theory of algebras. The problem of classifying algebras does not have, of course, a satisfactory answer. Thus it turns out interesting to study classes of algebras (or rings) satisfying some natural conditions like a set of polynomial identities.

Another problem consists of finding polynomial identities for a class of algebras. The main question in this direction is the following:

SPECHT'S PROBLEM. *Given a class of algebras C, is it true that any algebra $A \in C$ has a finite basis of identities (i.e., the ideal of identities of A is generated by a finite number of identities)?*

Specht [1] posed this problem for associative algebras over a field of characteristic 0. The two most outstanding partial solutions are: if an associative algebra over a field of characteristic 0 satisfies a polynomial identity of the form

$$[x_1, x_2][x_3, x_4] \ldots [x_{2r-1}, x_{2r}] = 0$$

[1]This paper was written while the author held a grant from CNPq of Brazil.

(a product of commutators) then it has a finite basis of identities (obtained by Latyshev [6], Genov [8] and Popov [9]); any matrix algebra over a field of characteristic 0 has a finite basis of identities (announced by A. V. Yakovlev in 1982 but did not publish). In 1987, Kemer [17] gave the complete solution.

THEOREM 1 *(Kemer [17]) Any associative algebra over a field of characteristic 0 has a finite basis of identities.*

This theorem proves the existence of a basis of identities but does not give the construction of the basis for a given associative algebra.

We mention also a problem of more concrete nature. Given a class of algebras $\mathcal{C}$ find the *minimal identities* (i.e., the polynomial identities of minimal degree, sometimes modulo some known identities) which holds for all algebras in C. Once the minimal identities are found it is interesting to study the variety of algebras or class of rings defined by these identities.

We introduce some notations and recall some definitions. We denote by $a \circ b$ the *Jordan product* $ab + ba$, by $[a, b]$ the *commutator* $ab - ba$, by (a, b, c) the *associator* $(ab)c - a(bc)$ and by $< a, b, c >$ the *Jordan associator* $(a \circ b) \circ c - a \circ (b \circ c)$. A nonassociative algebra is called: *flexible* if satisfies $(x, y, x) = 0$; *Jordan* if satisfies $xy = yx$ and the *Jordan law* $(x^2, y, x) = 0$; *noncommutative Jordan* if satisfies $(x, y, x) = 0$ and $(x^2, y, x) = 0$; *alternative* if satisfies the *alternative laws* $(x, x, y) = 0$ and $(y, x, x) = 0$; *Lie* if satisfies $xy = -yx$ and $(x, y, z) + (y, z, x) + (z, x, y) = 0$ (when $\text{char}(K) = 2$ we replace $xy = -yx$ by $x^2 = 0$). We say that a variety $\mathcal{M}$ of algebras is *Spechtian* if any subvariety of $\mathcal{M}$ has a finite basis of identities.

We have the following analogues of Kemer's theorem for Lie, alternative and Jordan algebras:

THEOREM 2 *(Ilt'yakov [24]) The variety of algebras generated by a finite dimensional Lie algebra over a field of characteristic 0 is Spechtian.*

THEOREM 3 *(Ilt'yakov [23]) Let A be a finitely generated alternative algebra over a field of characteristic 0 satisfying a polynomial identity. Then the variety generated by A is Spechtian.*

THEOREM 4 *(Vais and Zelmanov [19]) Any finitely generated Jordan algebra over a field of characteristic 0 satisfying a polynomial identity has a finite basis of identities.*

In this paper we survey results on polynomial identities of quadratic algebras. Also we communicate some new results about a class of rings (defined by polynomial identities) that generalize symmetric quadratic algebras.

2 QUADRATIC ALGEBRAS

Let K be a field of characteristic $\neq 2$ and let A be a nonassociative algebra with identity element 1 over K. We identify K with $K1$. The algebra A is called a *quadratic algebra* when 1, a, a^2 are linearly dependent for all a in A, i. e., for any $a \in A$ there are elements $t(a)$ (called *trace of* a) and $n(a)$ (called *norm of* a) in K such that $a^2 - t(a)\,a + n(a) = 0$. We remark that $t : A \to K$ is a linear form and $n : A \to K$ is a quadratic form (see Zevlakov et al [12], p. 37).

For a quadratic algebra A we have the direct sum $A = F \oplus V$, where $V = \{a \in A : t(a) = 0\}$. Let $a = \alpha + x$ and $b = \beta + y$ $(\alpha, \beta \in F$ and $x, y \in V)$ be arbitrary elements in A. If $xy = -f(x,y) + x.y$, where $f : V \times V \to K$ is a bilinear form and $x.y \in V$, then the product ab is given by

$$ab = (\alpha\beta - f(x,y)) + (\beta x + \alpha y + x.y).$$

The structure of quadratic division algebras was investigated by Osborn [3]. He proved that no quadratic division algebra of dimension 3 exists, and all quadratic division algebras of dimension 4 were found in terms of the structure of the quadratic forms over K.

Quadratic alternative algebras were completely classified up to isomorphism by Elduque [21]. This class of algebras includes quaternion and octonion algebras, but also many other algebras which appear in conection with some physical questions.

We shall say that a nonassociative ring A satisfies a *quadratic condition over a subring* U if, for each x in A, there are elements $u(x)$, $v(x)$, $w(x)$ in U such that $u(x)x^2 - v(x)x + w(x) = 0$. If, for every $x \in A$, at least one of $u(x)$, $v(x)$, $w(x)$ is nonzero and $v : A \to U$ is linear, we shall say that A is a *quadratic ring over* U.

The algebra A is said to be a *symmetric quadratic algebra over* K if A is a quadratic algebra over K and $t(ab) = t(ba)$ for all a and b in A. Notice that A is symmetric if and only if f is symmetric. When A is a quadratic ring over U and $v(ab) = v(ba)$ for all a and b in A, we say that A is a *symmetric quadratic ring over* U.

Let A be any quadratic algebra. For any $a \in A$ we have $a^2 = t(a)a - n(a)$ and it follows that $a^2 a = aa^2$. This means that A satisfies the identity $(a,a,a) = 0$. Furthermore, any degree 3 identity which holds for all quadratic algebras (or all symmetric quadratic algebras) over a field of characteristic $\neq 2, 3$ is a consequence of $(a,a,a) = 0$ (see Hentzel and Peresi [28]).

If A is a symmetric quadratic algebra then from the linearized form of $a^2 - t(a)a + n(a) = 0$ we obtain $[a,b] \circ [c,d] \in K$ for all $a,b,c,d \in A$. Thus A satisfies the polynomial identity

$$[[a,b] \circ [c,d], e] = 0.$$

The result remains true if we have a symmetric quadratic ring A over U such that $[U, A] = 0$.

Denote by S_n the *symmetric group* and by $(-1)^\sigma$ the *sign* of the permutation σ. Let R_a denote the linear operator defined by $R_a(b) = a \circ b$. We define the linear operator $\bar{S}_3$ by

$$\bar{S}_3 = \sum_{\sigma \in S_3} (-1)^\sigma R_{a_{\sigma(1)}} R_{a_{\sigma(2)}} R_{a_{\sigma(3)}}.$$

It is straightforward to check that

$$\bar{S}_3(a^2) - \bar{S}_3(a) \circ a = 0$$

is an identity for any quadratic algebra (see Racine [15]).

3 TWO BY TWO MATRICES

Denote by $M_n(K)$ the algebra of all $n \times n$ matrices over a field K. The Amitsur-Levitzki Theorem (see [2]) asserts that the minimal degree of a polynomial identity which holds for $M_n(K)$ is $2n$, and any multilinear homogeneous polynomial identity of degree $2n$ of this algebra is a scalar multiple of the *standard polynomial* $s_{2n}(x_1, \ldots, x_{2n})$, where

$$s_n(x_1, \ldots, x_n) = \sum_{\sigma \in S_n} (-1)^\sigma x_{\sigma(1)} \ldots x_{\sigma(n)}.$$

The algebra $M_2(K)$ is a symmetric quadratic algebra where the trace is the usual trace of a matrix and the norm is the determinant.

Rasmyslov [5] proved that the ideal of identities of $M_2(K)$ is generated by nine identities and Drensky [10] found a minimal basis of identities when $\mathrm{char}(K) = 0$.

THEOREM 5 *(Drensky [10]) If char(K) $= 0$ then the ideal of identities of $M_2(K)$ is generated by $[[x, y]^2, x] = 0$ and $s_4(x_1, \ldots, x_4) = 0$.*

We notice that, modulo $s_4(x_1, \ldots, x_4) = 0$, the identity $[[x, y]^2, x] = 0$ is equivalent to $[[a, b] \circ [c, d], e] = 0$.

Let K_{p^n} denote the finite field of characteristic p with p^n elements. We have:

THEOREM 6 *(Mal'tsev and Kuz'min [7]) The ideal of identities of the matrix algebra $M_2(K_{p^n})$ is generated by the polynomials*

$$f_1(x, y) = (x - x^q)\ (y - y^{q^2})\ (1 - [x, y]^{q-1}),$$

$$f_2(x, y) = (x - x^q) \circ (y - y^q) - ((x - x^q) \circ (y - y^q))^q,$$

where $q = p^n$.

When we consider the *Lie product* $[a, b]$ in $M_n(K)$ we obtain the Lie algebra $M_n(K)^-$. The subspace $sl_n(K)$ of all matrices with zero trace is a subalgebra of Lie of $M_n(K)^-$. Vaughan-Lee [4] proved that if K is an infinite field and $\mathrm{char}(K) = 2$ then $M_2(K)^-$ does not have a finite basis of identities. It is not known if, under the same conditions, $M_2(K)$ has or has not a finite basis of identities. When $\mathrm{char}(K) = 0$ Razmyslov [5] found 3 identities that generate the ideal of identities of $sl_2(K)$. Fillipov [11] and Vasiloviskii [20] improved this result.

THEOREM 7 *(Vasiloviskii [20]) Over an infinite field K of characteristic $\neq 2$ all identities of $sl_2(K)$ are consequences of*

$$[[[y, z], [t, x]], x] + [[[y, x], [z, x]], t] = 0.$$

An associative polynomial $f(x_1, \ldots, x_n)$ is called a *weak polynomial identity* of $sl_2(K)$ if $f(a_1, \ldots, a_n)$ vanishes in $M_2(K)$ for all $a_1, \ldots, a_n \in sl_2(K)$. When $\mathrm{char}(K) = 0$ Razmyslov [5] proved that the basis of the weak polynomial identities of $sl_2(K)$ consists of the single identitity $[x \circ y, z] = 0$. This result was improved by Koshlukov [27].

THEOREM 8 *(Koshlukov [27]) If K is an infinite field of characteristic $\neq 2$ then the weak polynomial identities of $sl_2(K)$ are consequences of*

$$[x \circ y, z] = 0.$$

For $\mathrm{char}(K) = 2$ we have:

THEOREM 9 *(Drensky [26]) If K is an infinite field of characteristic 2 then the weak polynomial identities of $sl_2(K)$ are consequences of*

$$[[x, y], z] = 0,$$

$$[x_1, x_2][x_3, x_4] + [x_1, x_3][x_2, x_4] + [x_1, x_4][x_2, x_3] = 0.$$

4 JORDAN ALGEBRA OF DEGREE 2

Let K be a field of characteristic $\neq 2$, V be a vector space of dimension n $(2 \leq n \leq \infty)$ over K, $q : V \times V \to K$ be a symmetric bilinear form and let $B_n(q, V) = K \oplus V$. We define a multiplication in $B_n(q, V)$ by setting

$$(\alpha + x)(\beta + y) = (\alpha\beta + q(x, y)) + (\beta x + \alpha y).$$

We obtain a Jordan algebra called *Jordan algebra of degree 2*. This algebra is a commutative quadratic algebra where $t(\alpha + x) = 2\alpha$ and $n(x) = \alpha^2 - q(x, x)$, and it is symmetric. Conversely, any commutative quadratic algebra A of dimension n is a Jordan algebra of degree 2, more precisely, $A = B_n(-f, V)$ where $V = \{a \in A \mid t(a) = 0\}$. The algebra $B_n(q, V)$ is simple when q is nondegenerate and is not finitely generated when $n = \infty$.

Isaev [14] gave in an explicit form a basis for the polynomial identities of $B_n(q, V)$ when q is nondegenerate and K is a finite field. This basis consists of four identities when $n = \infty$ and of some additional identities when $n < \infty$. For an infinite field of characteristic $\neq 2, 3, 5, 7$ Vasiloviskii [22] obtained a similar result. For $n = \infty$ the basis consists of two identities (one of degree 5 and the other of degree 6). For $n < \infty$ there are two additional identities.

Let A be an algebra over K and $A^{\#} = K \oplus A$ with multiplication defined by $(\alpha + a)(\beta + b) = \alpha\beta + (\alpha b + \beta a + ab)$. We say that a variety $\mathcal{M}$ of algebras over K is *unitally closed* if for any $A \in \mathcal{M}$ we have that $A^{\#} \in \mathcal{M}$. The variety $\mathcal{M}$ is called *unitally Spechtian* if any unitally closed subvariety of $\mathcal{M}$ has a finite basis of identities.

THEOREM 10 *(Il'tyakov [16]) If $\mathrm{char}(K) = 0$, $n < \infty$ and q is nondegenerate then the variety of algebras generated by $B_n(q, V)$ is unitally Spechtian.*

For the class of all Jordan algebras of degree 2 over a field of characteristic 0, Racine [15] found a basis for the minimal identities (the degree is 5) which are multilinear Jordan polynomials and are not consequences of the commutative and Jordan laws. The polynomial identities of degree ≤ 5 which hold for all Jordan algebras of degree 2 are given by

THEOREM 11 *(Hentzel and Peresi [28]) For Jordan algebras of degree 2 over a field K we have:*
(i) If $\mathrm{char}(K) \neq 2, 3$ all degree 3 polynomial identities are consequences of $ab = ba$.
(ii) If $\mathrm{char}(K) \neq 2, 3$ all degree 4 polynomial identities are consequences of $ab = ba$ and $(a^2, b, a) = 0$.
(iii) If $\mathrm{char}(K) \neq 2, 3, 5$ all degree 5 polynomial identities are consequences of $ab = ba$, $(a^2, b, a) = 0$ and $\bar{S}_3(a^2) - \bar{S}_3(a) \circ a = 0$.

5 CAYLEY-DICKSON ALGEBRAS

A *Cayley-Dickson algebra* is an eight-dimensional alternative algebra over a field K with three non-zero parameters α, β, γ in K. We denote this algebra by $C(\alpha,\beta,\gamma)$. Over a field K of characteristic $\neq 2$ it is possible to choose a basis $1, e_1, \cdots, e_7$ with the multiplication table given below (see Zhevlakov et al [12], p. 34).

	1	e_1	e_2	e_3	e_4	e_5	e_6	e_7
1	1	e_1	e_2	e_3	e_4	e_5	e_6	e_7
e_1	e_1	α	e_3	αe_2	e_5	αe_4	$-e_7$	$-\alpha e_6$
e_2	e_2	$-e_3$	β	$-\beta e_1$	e_6	e_7	βe_4	βe_5
e_3	e_3	$-\alpha e_2$	βe_1	$-\alpha\beta$	e_7	αe_6	$-\beta e_5$	$-\alpha\beta e_4$
e_4	e_4	$-e_5$	$-e_6$	$-e_7$	γ	$-\gamma e_1$	$-\gamma e_2$	$-\gamma e_3$
e_5	e_5	$-\alpha e_4$	$-e_7$	$-\alpha e_6$	γe_1	$-\alpha\gamma$	γe_3	$\alpha\gamma e_2$
e_6	e_6	e_7	$-\beta e_4$	βe_5	γe_2	$-\gamma e_3$	$-\beta\gamma$	$-\beta\gamma e_1$
e_7	e_7	αe_6	$-\beta e_5$	$\alpha\beta e_4$	γe_3	$-\alpha\gamma e_2$	$\beta\gamma e_1$	$\alpha\beta\gamma$

Any Cayley-Dickson algebra is a symmetric quadratic algebra. The trace and norm are given by $t(x) = x + \bar{x}$ and $n(x) = x\,\bar{x}$, where $x \in A \to \bar{x} \in A$ denotes the involution of the algebra. A Cayley-Dickson algebra $C(\alpha,\beta,\gamma)$ is called *split* if $n(x) = 0$ for some $x \neq 0$ in $C(\alpha,\beta,\gamma)$ or, equivalently, there are zero divisors in $C(\alpha,\beta,\gamma)$. Every split Cayley-Dickson algebra over a field K is isomorphic to the *Cayley-Dickson matrix algebra over* K deenoted by $C(K)$ (see Zhevlakov et al [12], p. 46). This algebra contains as a subalgebra $M_2(K)$. Over a finite field every Cayley-Dickson algebra is split and we have:

THEOREM 12 *(Isaev [13]) The ideal of identities of the algebra $C(K_{p^n})$ is generated by the polynomials $h_1(x)$, $f_2(x,y)$ and $h_3(x,y,z)$, where*

$$h_1(x) = (x - x^q)\ (x - x^{q^2}),$$

$$h_3(x,y,z) = g_1(x,y,z)\ g_2(y,z)\ g_3(x,y,z),$$

$$g_1(x,y,z) = (x - x^q)\ (\ (y - y^{q^2})\ (z - z^{q^2})\ ),$$

$$g_2(y,z) = (1 - y^{q^2-q})\ (1 - z^{q^2-q})\ (1 - (y \circ z)^{q-1}),$$

$$g_3(x,y,z) = (1 - (x \circ y)^{q^2-q})\ (1 - (x \circ z)^{q^2-q})\ (1 - (x \circ yz)^{q^2-q}),$$

$q = p^n$ and the arrangement of parenthesis in h_3, g_2, g_3 is arbitrary.

For fields of characteristic zero we have:

THEOREM 13 *(Il'tyakov [16]) If char$(K) = 0$ then the variety of algebras generated by a Cayley-Dickson algebra is unitally Spechtian.*

For all Cayley-Dickson algebras, the minimal identities modulo the alternative laws was found by Racine [18]. In [25] Hentzel and Peresi found a new identity of degree 6.

THEOREM 14 *(Racine [18], Hentzel and Peresi [25]) The homogeneous multilinear polynomial identities of degree ≤ 6 of the Cayley-Dickson algebras are given by:*
(i) If char(K) $\neq 2$ there is no identity of degree 2.
(ii) If char(K) $\neq 2, 3$ the identities of degree 3 and 4 are consequences of $(x, x, y) = 0$ and $(y, x, x) = 0$.
(iii) If char(K) $\neq 2, 3, 5$ the identities of degree 5 are consequences of

$$(x, x, y) = 0, \quad (y, x, x) = 0,$$

$$[[a, b] \circ [c, d], e] = 0, \quad \bar{S}_3(a^2) - \bar{S}_3(a) \circ a = 0.$$

(iv) If char(K) $\neq 2, 3, 5$ the identities of degree 6 are consequences of

$$(x, x, y) = 0, \quad (y, x, x) = 0,$$

$$[[a, b] \circ [c, d], e] = 0, \quad \bar{S}_3(a^2) - \bar{S}_3(a) \circ a = 0,$$

$$[\ \sum \{\ 24\ a(b(c(de))) + 8\ a((b, c, d)e) - 11\ (a, b, (c, d, e))\ \}\ , f\ ] = 0,$$

where $\sum$ means the alternating sum over the arguments.

6 GENERALIZED QUADRATIC RINGS

Hentzel and Peresi [28] determined a basis for the polynomial identities of degree ≤ 5 which hold for all quadratic algebras and did the same for all symmetric quadratic algebras (over a field of characteristic $\neq 2, 3, 5$). They proved that any symmetric quadratic algebra satisfies the following identities:

$$(a, a, a) = 0, \quad (a^2, a, b) - (a, a^2, b) = 0, \quad (b, a^2, a) - (b, a, a^2) = 0, \tag{1}$$

$$[(a, b, a), c] = 0, \quad ((a, b, a), c, d) = 0, \quad (c, (a, b, a), d) = 0, \tag{2}$$

$$(c, d, (a, b, a)) = 0, \quad [\,[a, b] \circ [c, d], e\,] = 0, \tag{3}$$

$$([a, b] \circ c, c, d) - ([a, b], c^2, d) = 0, \tag{4}$$

$$([a, b] \circ c, d, c) - ([a, b], d, c^2) = 0, \tag{5}$$

$$\begin{aligned} &[2a\,(a, b, b) - 2\,(a, b, ab) + 2b\,(b, a, a) \\ &-2\,(b, a, ba) + a \circ (b, a, b) + b \circ (a, b, a),\ c] = 0, \end{aligned} \tag{6}$$

$$[2a^2 \circ b^2 - (a \circ b)^2 + 2a < a, b, b > +2b < b, a, a >, c] = 0. \tag{7}$$

If we add to this set of identities the identities

$$\bar{S}_3(a^2) - \bar{S}_3(a) \circ a = 0,$$

$$\begin{aligned} -(a^2 \circ c, d, e) + (a^2 \circ e, d, c) - (e, d, a^2 \circ c) + (e, d, (a \circ c) \circ a) \\ -(c, d, (a \circ e) \circ a) + (a \circ c, d, a \circ e) - (a \circ e, d, a \circ c) \\ -((a \circ c) \circ e, d, a) + (c, d, a^2 \circ e) + ((a \circ e) \circ c, d, a) = 0, \end{aligned}$$

$$\sum_{b,c,e}\{(a, <c,d,e>, b) + (a, b, <c,d,e>)\} = 0,$$

where $\sum_{b,c,e}$ denotes the alternating sum on the variables b, c, e, we obtain the basis of identities for symmetric quadratic algebras mentioned above.

We call a ring that satisfies the identities (1)-(7) *generalized quadratic ring.* Any flexible generalized quadratic ring is noncommutative Jordan since, modulo $(a,b,a) = 0$, the identity $(a^2,a,b)-(a,a^2,b) = 0$ is equivalent to the Jordan law. The following characterization of generalized quadratic rings has been recently obtained:

THEOREM 15 *(Hentzel, Peresi and Giuliani [29]) Let A be a generalized quadratic ring of characteristic $\neq 2$. We have:*
(i) If A is simple then either A is flexible and hence noncommutative Jordan, or else A is in the variety generated by all symmetric quadratic algebras over the center of A which is a field.
(ii) If A is prime and has an identity element then either A is flexible and hence noncommutative Jordan, or else the center of A is an integral domain and A is in the variety generated by all symmetric quadratic algebras over the quotient field of the center of A.
(iii) If A is semi-prime then A is a subdirect sum with two summands: the first is a noncommutative Jordan ring (which satisfies a quadratic condition over its commutative center); the second is a not flexible symmetric quadratic ring over its center.

REFERENCES

1. W. Specht. Gesetze in rings I. Math. Z. 52:557-589, 1950.
2. A. S. Amitsur, J. Levitzki. Minimal identities for algebras. Proc. Amer. Math. Soc. 1:449-463, 1950.
3. J. M. Osborn. Quadratic division algebras. Trans. Amer. Math. Soc. 105:202-221, 1962.
4. M. R. Vaughan-Lee. Varieties of Lie algebras. Ph. D. Dissertation, Oxford University, Oxford, England, 1970.
5. Yu. P. Rasmyslov. Finite basing of the identities of a matrix algebra of second order over a field of characteristic zero. Algebra and Logic 12:47-63, 1973.
6. V. N. Latyshev. Partially ordered sets and nonmatrix identities of associative algebras. Algebra and Logic 15:34-45, 1976.
7. Yu. N. Mal'tsev, E. N. Kuz'min. A basis for the identities of the algebra of second-order matrices over a finite field. Algebra and Logic 17:18-21, 1978.
8. G. K. Genov. Some Specht varieties of associative algebras. PLISKA, Stud. Math. Bulgar. 2:30-40, 1981.
9. A. P. Popov. On the Specht property of certain varieties of associative algebras. PLISKA, Stud. Math. Bulgar. 2:41-53, 1981.

10. V. Drensky. A minimal basis of identities for a second order matrix algebra over a field of characteristic 0. Algebra and Logic 20:188-194, 1981.

11. V. T. Fillipov. Varieties of Mal'tsev algebras. Algebra and Logic 20:200-210, 1981.

12. K. A. Zhevlakov, A. M. Slin'ko, I. P. Shestakov, A. I. Shirshov. Rings That Are Nearly Associative. Academic Press, New York, 1982.

13. I. M. Isaev. Identities of a finite Cayley-Dickson algebra. Algebra and Logic 23:282-289, 1984.

14. I. M. Isaev. Identities of the algebra of a bilinear form over a finite field. In: Some questions and problems in analysis and algebra. Novosibirsk Gos. Univ., Novosibirsk, 1985, pp. 61-75.

15. M. L. Racine. Minimal identities for Jordan algebras of degree 2. Comm. Alg. 13:2493-2506, 1985.

16. A. V. Ilt'yakov. Specht ideals of identities of certain simple nonassociative algebras. Algebra and Logic 24:210-228, 1985.

17. A. R. Kemer. Finite basis property in associative algebras. Algebra and Logic 5:362-397, 1987.

18. M. L. Racine. Minimal identities of octonion algebras. J. Algebra 115:251-260, 1988.

19. A. Ya. Vais, E. I. Zelmanov. Kemer's theorem for finitely generated Jordan algebras. Izv. Vyssh. Uchebn. Zaved. Mat. 6:42-51, 1989.

20. S. Vasiloviskii. Basis of identities of a three-dimensional simple Lie algebra over an infinite field. Algebra and Logic 28:355-368, 1989.

21. A. Elduque. Quadratic alternative algebras. J. Math. Phys. 31:1-5, 1990.

22. S. Vasilovskii. A finite basis for polynomial identities of the Jordan algebras of bilinear form. Siberian Adv. Math. 1:1-43, 1991.

23. A. V. Ilt'yakov. Finiteness of basis of identities of a finitely generated alternative PI-algebra over a field of characteristic zero. Siberian Math. J. 32:948-961, 1991.

24. A. V. Ilt'yakov. On finite basis of identities of Lie algebra representations. Nova J. Alg. Geom. 1:207-259, 1992.

25. I. R. Hentzel, L. A. Peresi. Identities of Cayley-Dickson algebras. J. Algebra 188:292-309, 1997.

26. V. S. Drensky. Identities of representation of nilpotent Lie algebras. Comm. Alg. 25:2115-2127, 1997.

27. P. Koshlukov. Weak polynomial identities for the matrix algebra of order two. J. Algebra 188:610-625, 1997.

28. I. R. Hentzel, L. A. Peresi. Degree three, four and five identities of quadratic algebras. J. Algebra 206:1-16, 1998.

29. I. R. Hentzel, L. A. Peresi, O. F. Giuliani. On the variety determined by symmetric quadratic algebras. Submitted to publication, 1998.

The derivation algebra of a division composition algebra

J. M. PEREZ-IZQUIERDO * Departamento de Matemáticas y Computación, Universidad de La Rioja, 26004 Logroño, Spain.

1. INTRODUCTION

A division algebra is an algebra where the left and right multiplication operators L_x, R_x by any nonzero element x are bijective. Over the real numbers Bott, Milnor and Kervaire [1, 9] showed that the only possible dimensions of any finite-dimensional division algebra are 1, 2, 4 or 8.

In the following we will always use the word "algebra" as synonymous with "finite-dimensional algebra" and we will assume that the fields have characteristic $\neq 2, 3$.

A theorem from Frobenius [6] says that any associative real division algebra must be either the real numbers, the complex numbers or the division algebra of quaternions.

An algebra is called *alternative* if it satisfies the following identities:

$$x(xy) = x^2y \quad \text{and} \quad (yx)x = yx^2,$$

or, equivalently, if any two elements generate an associative subalgebra [13]. If we relax the hypothesis of associativity in Frobenius' theorem and we only impose alternativity then a new algebra appears, namely the octonion division algebra [13].

Since in any alternative algebra the subalgebra generated by two elements is associative, these algebras verify the identity $(xy)x = x(yx)$, so they are *flexible*. In [4] Benkart, Britten and Osborn deal with the classification of flexible real division algebras.

* The author was supported by the Spanish DGICYT (Pb 94-1311-C03-03), API-98/B15 and from the 'Programa del Personal Investigador en el Extranjero'.

In general, not many things are known about real division algebras. In [2, 3] Benkart and Osborn seek a clearer picture of this family of algebras by studying their derivation algebras. The point of view there is that, over the real numbers, studying the algebra of derivations is the same as studying the group of automorphisms, and this group gives an idea of the "symmetries" of the algebra. So, roughly speaking, by classifying the derivation algebra $\operatorname{Der} A$ [2]:

THEOREM 1. Let A be a finite-dimensional real division algebra and $\operatorname{Der} A$ its derivation algebra, then:

i) If $\dim A = 1$ or 2 then $\operatorname{Der} A = 0$.

ii) If $\dim A = 4$ then either $\operatorname{Der} A = \mathrm{su}(2)$ or $\dim \operatorname{Der} A \leq 1$.

iii) If $\dim A = 8$ then $\operatorname{Der} A$ is one of the following Lie algebras:

a) The compact algebra G_2.

b) $\mathrm{su}(3)$.

c) $\mathrm{su}(2) \oplus \mathrm{su}(2)$.

d) $\mathrm{su}(2) \oplus Z$ where Z denotes an abelian ideal of dimension 0 or 1.

e) An abelian algebra H of dimension $0, 1$ or 2.

and then analyzing the possible decompositions of A as a $\operatorname{Der} A$-module, they can obtain a first approximation of the multiplication table of A [3].

Our goal is to develop a similar program for division composition algebras over arbitrary fields of characteristic $\neq 2$ or 3. In the present paper we compute the derivation algebra of these algebras, obtaining an analogous of Theorem 1. In [10] we deal with the description and classification of these algebras.

2. COMPOSITION ALGEBRAS

DEFINITION. A composition algebra A over a field F is an algebra equipped with a quadratic form $n: A \to F$ such that:

i) $(x, y) = \frac{1}{2}(n(x+y) - n(x) - n(y))$ is nondegenerate

and

ii) $n(xy) = n(x)n(y) \ \forall x, y \in A$.

The full linearization of ii) is:

$$(xy, wz) + (xz, wy) = 2(x, y)(w, z) \qquad (1).$$

Unital composition algebras, also termed *Hurwitz algebras*, have been completely classified [7, 12]:

THEOREM 2. Any Hurwitz algebra is isomorphic to one of the following algebras:

i) the base field F,

ii) a quadratic extension of F,

iii) a generalized quaternion algebra, or

iv) a generalized octonion algebra.

These algebras can be constructed by the Cayley-Dickson process. An immediate consequence of the previous theorem is that the dimension of any Hurwitz algebra is either 1, 2, 4 or 8.

Two Hurwitz algebras are isomorphic if and only if their quadratic forms are equivalent. In case this form has nonzero isotropic vectors, then the Witt index is maximum; therefore, Hurwitz algebras of the same dimension with nonzero isotropic vectors are isomorphic. Such algebras are called *split*, and, for instance, they are the only Hurwitz algebras over algebraically closed fields. The split octonion algebra possesses a basis $\{e_1, e_2, u_1, u_2, u_3, v_1, v_2, v_3\}$ where the product is given by the following table:

	e_1	e_2	u_1	u_2	u_3	v_1	v_2	v_3
e_1	e_1	0	u_1	u_2	u_3	0	0	0
e_2	0	e_2	0	0	0	v_1	v_2	v_3
u_1	0	u_1	0	v_3	$-v_2$	$-e_1$	0	0
u_2	0	u_2	$-v_3$	0	v_1	0	$-e_1$	0
u_3	0	u_3	v_2	$-v_1$	0	0	0	$-e_1$
v_1	v_1	0	$-e_2$	0	0	0	u_3	$-u_2$
v_2	v_2	0	0	$-e_2$	0	$-u_3$	0	u_1
v_3	v_3	0	0	0	$-e_2$	u_2	$-u_1$	0

Any basis $\{e_1, e_2, u_1, u_2, u_3, v_1, v_2, v_3\}$ of a (split) octonion algebra C where the product is given by the previous table is called *canonical basis* [5]. The maps

$$\begin{array}{lll} u_1 \mapsto a u_1 & v_1 \mapsto -a v_1 & \\ u_2 \mapsto b u_2 & v_2 \mapsto -b v_2 & e_i \mapsto 0 \ \ (i = 1, 2) \\ u_3 \mapsto -(a+b) u_3 & v_3 \mapsto (a+b) v_3 & \end{array} \tag{2}$$

with $a, b \in F$ span a two-dimensional Cartan subalgebra $\mathcal{H}$ of $\operatorname{Der} C$. With respect to this subalgebra C decomposes as a sum of weight subspaces

$$C = C_0 \oplus C_{\pm\alpha_1} \oplus C_{\pm(\alpha_1+\alpha_2)} \oplus C_{\pm(2\alpha_1+\alpha_2)} \tag{3}$$

where α_1 is the short simple root of $\operatorname{Der} C = G_2$ and α_2 the long simple root.

Canonical bases arise from the Peirce decomposition of C [11]. Given two orthogonal idempotents e_1, e_2 of C set $C_{ij} = \{x \in C \mid e_1 x = ix \text{ and } xe_1 = jx\}$, with $i, j = 0, 1$. It is well known that $C_{11} = Fe_1, C_{00} = Fe_2$ as well as that C_{10} and C_{01} are dual with respect to the bilinear form $(\,,\,)$, and $\dim C_{10} = \dim C_{01} = 3$. In fact, using formulas in [11, 13] it is easy to prove:

LEMMA 3. Given x_1, x_2 and x_3 in C_{10} with $(x_3, x_1 x_2) = 1/2$ then

$$\{e_1, e_2, x_1, x_2, x_3, x_2 x_3, x_3 x_1, x_1 x_2\}$$

is a canonical basis of C.

NOTE: Given x_1 and x_2 linearly independent there always exists such an x_3.

Nonunital finite-dimensional composition algebras are quite related to Hurwitz algebras. In [8] Kaplansky proved that given any finite-dimensional composition algebra A and $a \in A$ with $n(a) \neq 0$ then the left and right multiplication operators L_u, R_u, where $u = a^2/n(a)$, are isometries, and the new algebra (A, a) with product:

$$x \circ y = R_u^{-1}(x)L_u^{-1}(y)$$

is a Hurwitz algebra with unit u^2 and the same norm (quadratic form) as A. So, the product xy of any composition algebra A can be constructed from a Hurwitz algebra $(A, \circ)$ just by

$$xy = \phi_1(x) \circ \phi_2(y) \tag{4}$$

where $\phi_i, i = 1, 2$ are appropriate isometries. Equation (4) says that any finite-dimensional composition algebra is an *isotope* of a Hurwitz algebra. Therefore, the dimension of any finite-dimensional composition algebra is 1, 2, 4 or 8. Since in (4) we only needed the operators R_u and L_u to be bijective, these are the possible dimensions of any division composition algebra too.

3. DECOMPOSITION $\bar{A} = \oplus_\alpha \bar{A}_\alpha(d)$

Given a vector space S over a field F, and $\bar{F}$ the algebraic closure of F, we will denote $\bar{F} \otimes_F S$ by $\bar{S}$. If $d: S \to S$ is a linear map then $S_\alpha = S_\alpha(d) = \{x \in S \mid \exists n \text{ s.t. } (d - \alpha I)^n x = 0\}$ is the generalized eigenspace of d with eigenvalue α. The extension of d to $\bar{S}$ will also be denoted by d. Notice that if $\alpha \in F$ then $\overline{S_\alpha(d)} = \bar{S}_\alpha(d)$. In the following A will stand for a finite-dimensional composition algebra. Our goal in this section is to describe the decomposition $\bar{A} = \oplus_\alpha \bar{A}_\alpha(d)$ for any $d \in \operatorname{Der} A$. The following results will become helpful:

LEMMA 4. Given $x \in A$, they are equivalent:

i) R_x is bijective ii) $n(x) \neq 0$ iii) L_x is bijective.

PROPOSITION 5. Let $d \in \operatorname{Der} A$, then

$$(dx, y) + (x, dy) = 0 \qquad \forall x, y \in A.$$

As an immediate consequence of this proposition we have:

LEMMA 6. The only nilpotent derivation of a division composition algebra is the trivial one.

Proof: Let d be a nilpotent derivation and $d^{k+1} = 0$ with k minimal. Take x such that $d^k x \neq 0$, so

$$0 \neq (d^k x, d^k x) = (-1)^k (d^{2k} x, x).$$

In particular, $d^{2k} \neq 0$ and $2k \leq k$, or equivalently $k = 0$. Therefore, $d = 0$. ◇

LEMMA 7. Given $d \in \operatorname{Der} A$ and $\bar{A}_\gamma = \bar{A}_\gamma(d)$ as before, then

i) $(\bar{A}_\alpha, \bar{A}_\beta) = 0$ if $\alpha \neq -\beta$.
ii) $\dim \bar{A}_\alpha = \dim \bar{A}_{-\alpha}$.
iii) $\bar{A}_\alpha \bar{A}_\beta \subseteq \bar{A}_{\alpha+\beta}$.

From now on, we will restrict ourselves to A being a division composition algebra.

In the case of $\dim A \leq 2$ the derivation algebra is trivial, and in the case of dimension four we have [5]:

THEOREM 8. Let A be a four-dimensional composition algebra. Then either

i) $\operatorname{Der} A$ is a three-dimensional simple Lie algebra and A decomposes as $1+3$ as a $\operatorname{Der} A$-module, or

ii) $\dim \operatorname{Der} A = 1$, $\operatorname{Der} A = Fd$ with semisimple action of d on A and $\dim \ker d = 2$, or

iii) $\dim \operatorname{Der} A = 1$, $\operatorname{Der} A = Fd$ with $d^2 \neq 0 = d^3$ and $\ker d$ is not totally isotropic.

NOTE: The notation $1+3$ means the sum of two irreducible modules of dimension 1 and 3 respectively.

So, in case that $\dim A = 4$ and $0 \neq d \in \operatorname{Der} A$, the previous theorem says that $\bar{A}_0(d) \neq 0$. By Lemma 7 it is a composition subalgebra of $\bar{A}$ of even dimension, and by Lemma 6 this dimension is 2. This gives the decomposition $\bar{A} = \bar{A}_0 \oplus \bar{A}_\alpha \oplus \bar{A}_{-\alpha}$. Moreover, since $\bar{A}_0 = \overline{A_0}$ and d induces a nilpotent derivation on A_0, which is a division composition subalgebra of A, it must happen $\bar{A}_0 = \ker d$, so d is diagonalizable.

In the following we will focus on division composition algebras A of dimension 8. Note that the same argument as above works to show that $\bar{A}_0 = \ker d$. The following lemma will be used frequently in this section.

LEMMA 9. Given $0 \neq a_\gamma \in \bar{A}_\gamma$ with $\gamma \neq 0$, there exists $a_{-\gamma} \in \bar{A}_{-\gamma}$ such that $(a_\gamma, a_{-\gamma}) \neq 0$ (or equivalently, $n(a_\gamma + a_{-\gamma}) \neq 0$). Moreover, for any $\bar{A}_{\gamma'} \neq 0$ we have $0 \neq (a_\gamma + a_{-\gamma})\bar{A}_{\gamma'} \subseteq \bar{A}_{\gamma'-\gamma} \oplus \bar{A}_{\gamma'+\gamma}$ and $\dim \bar{A}_{\gamma'} \leq \dim \bar{A}_{\gamma'-\gamma} + \dim \bar{A}_{\gamma'+\gamma}$. In particular, with $\gamma' = \gamma$, $\bar{A}_0 \oplus \bar{A}_{2\gamma} \neq 0 \neq \bar{A}_0 \oplus \bar{A}_{-2\gamma}$ and $\dim \bar{A}_\gamma \leq \dim \bar{A}_0 + \dim \bar{A}_{2\gamma}$.

Proof: By Lemma 7 there exists $a_{-\gamma} \in \bar{A}_{-\gamma}$ with $(a_\gamma, a_{-\gamma}) \neq 0$ and, since $n(a_\gamma) = 0 = n(a_{-\gamma})$, $n(a_\gamma + a_{-\gamma}) = 2(a_\gamma, a_{-\gamma}) + n(a_\gamma) + n(a_{-\gamma}) = 2(a_\gamma, a_{-\gamma}) \neq 0$; therefore, by Lemma 4, $L_{a_\gamma + a_{-\gamma}}$ is bijective, which, together with Lemma 7, implies the statement. ◇

PROPOSITION 10. A_0 is a composition subalgebra of dimension 2 or 4.

Proof: It is enough to prove the statement for $\bar{A}_0$. We will proceed by contradiction. Assume that $\bar{A}_0 = 0$ and take $\bar{A}_{\pm\alpha} \neq 0$. Lemma 9 tells us that $\bar{A}_{\pm 2\alpha} \neq 0$, and since $\operatorname{char} F > 3$, these spaces are different from $\bar{A}_{\pm\alpha}$. Thus, we have a first decomposition

$$\bar{A} = (\bar{A}_{\pm\alpha} \oplus \bar{A}_{\pm 2\alpha}) \oplus \bar{A}_*.$$

Let us denote the sum of all spaces $\bar{A}_{\pm i\alpha} = \bar{A}_{i\alpha} + \bar{A}_{-i\alpha}$, $i = \pm 1, \pm 2, \ldots$ by $\bar{A}(\alpha)$. It is clear that $\bar{A}(\alpha)$ is a composition subalgebra of $\bar{A}$ with dimension ≥ 4. It is equally clear that d induces a bijective derivation of $\bar{A}(\alpha)$; hence, by Theorem 8, $\dim \bar{A}(\alpha) \neq 4$. Therefore, $\dim \bar{A}(\alpha) = 8$ and $\bar{A} = \bar{A}(\alpha)$, that is

$$\bar{A} = \oplus_i \bar{A}_{\pm i\alpha}.$$

If char $F = 5$ then the previous equation turns out to be $\bar{A} = \bar{A}_{\pm\alpha} \oplus \bar{A}_{\pm 2\alpha}$. Now, with $\gamma = \pm 1, \pm 2$ in the previous lemma we obtain that $\dim \bar{A}_{i\alpha} = 2$, $i = \pm 1, \pm 2$. Take $0 \neq a_{i\alpha} \in \bar{A}_{i\alpha}$ and $a_{-i\alpha} \in \bar{A}_{-i\alpha}$ such that $(a_{i\alpha}, a_{-i\alpha}) \neq 0$. By dimensions we have that $a_{i\alpha}\bar{A}_{i\alpha} = (a_{-i\alpha} + a_{i\alpha})\bar{A}_{i\alpha} = \bar{A}_{2i\alpha}$. So $0 = n(a_{i\alpha})(\bar{A}_{i\alpha}, \bar{A}_{2i\alpha}) = (a_{i\alpha}\bar{A}_{i\alpha}, a_{i\alpha}\bar{A}_{2i\alpha}) = (\bar{A}_{2i\alpha}, a_{i\alpha}\bar{A}_{2i\alpha})$. However, since $a_{i\alpha}\bar{A}_{2i\alpha} \subseteq \bar{A}_{-2i\alpha}$, this means that $a_{i\alpha}\bar{A}_{2i\alpha} = 0$ $\forall a_{i\alpha} \in \bar{A}_{i\alpha}$; that is, $\bar{A}_{i\alpha}\bar{A}_{2i\alpha} = 0$. Similarly we can prove that $\bar{A}_{2i\alpha}\bar{A}_{i\alpha} = 0$. With $i = 1, 2$ we obtain $\bar{A}_{\alpha}\bar{A}_{2\alpha} = 0 = \bar{A}_{-\alpha}\bar{A}_{2\alpha}$, but this is impossible by the previous lemma.

Now let us assume that char $F = 7$. Here $\bar{A}$ breaks up as

$$\bar{A} = \bar{A}_{\pm\alpha} \oplus \bar{A}_{\pm 2\alpha} \oplus \bar{A}_{\pm 4\alpha}.$$

In this case, using the previous lemma with $\gamma = i\alpha$, $i = \pm 1, \pm 2, \pm 4$ we obtain that $\dim \bar{A}_{i\alpha}$ does not depend on i. Therefore, $8 = \dim \bar{A} = 6 \dim \bar{A}_{\alpha}$, which is a contradiction.

To conclude the proposition, let us suppose that char $F > 7$. Here, by Lemma 9 we have that

$$\bar{A} = \bar{A}_{\pm\alpha} \oplus \bar{A}_{\pm 2\alpha} \oplus \bar{A}_{\pm 4\alpha} \oplus \bar{A}_{\pm 8\alpha}.$$

Notice that since char $F > 7$ then $\bar{A}_{\pm 6\alpha} = 0$. Now $0 \neq (\bar{A}_{2\alpha} + \bar{A}_{-2\alpha})\bar{A}_{8\alpha} \subseteq \bar{A}_{6\alpha} + \bar{A}_{10\alpha} = \bar{A}_{10\alpha}$. Therefore $10 \in \{\pm 1, \pm 2 \pm 4 \pm 8\}$. The only possible case is $10 = -1$, or equivalently, char $F = 11$; however, with this restriction $0 \neq (\bar{A}_{8\alpha} + \bar{A}_{-8\alpha})\bar{A}_{8\alpha} \subseteq \bar{A}_{\pm 5\alpha} = \bar{A}_{-6\alpha} = 0$, which again gives a contradiction. ◇

LEMMA 11. Let $d \in \operatorname{Der} A$ such that

$$\bar{A} = \bar{A}_0 \oplus \sum_{\gamma} \bar{A}_{\gamma}$$

with $\dim \bar{A}_{\gamma} = 1$ $\forall \gamma \neq 0$ and $\dim \bar{A}_0 = 2$. Fix $e \in \bar{A}_0$ with $n(e) = 1$. Then there exists a canonical basis $\{e_1, e_2, u_1, u_2, u_3, v_1, v_2, v_3\}$ of $(\bar{A}, e)$ such that e_1 and e_2 belong to $\bar{A}_0$ and any of the spaces $\bar{A}_{\gamma}$ $(\gamma \neq 0)$ is spanned by basic elements.

Proof: Denote by $\circ$ the product in $C = (\bar{A}, e)$, and notice that $(\bar{A}_0, \circ)$ is a composition subalgebra of C. Now we can choose two orthogonal idempotents in $(\bar{A}_0, \circ)$. We can consider the Peirce decomposition $C = C_{11} \oplus C_{00} \oplus C_{10} \oplus C_{01}$ associated to these idempotents. Since d kills e_1 and e_2, in particular d kills e, d is also a derivation of C and the spaces of the Peirce decomposition are invariant under d. Therefore, these spaces are the sum of different $\bar{A}_{\gamma}$'s. By Lemma 3 we can pick up a canonical basis $\{e_1, e_2, u_1, u_2, u_3, v_1 = u_2 \circ u_3, v_2 = u_3 \circ u_1, v_3 = u_1 \circ u_2\}$ with u_1, u_2 and u_3 eigenvectors of d in C_{10}. The result follows from the fact that v_1, v_2, v_3 are also eigenvectors of d. ◇

LEMMA 12. Let $d \in \operatorname{Der} A$ such that

$$\bar{A} = \bar{A}_0 \oplus \sum_{\gamma} \bar{A}_{\gamma}$$

with $\dim \bar{A}_\gamma = 1$ $\forall \gamma \neq 0$ and $\dim \bar{A}_0 = 2$. Then $H = \{d' \in \operatorname{Der} A \mid [d, d'] = 0\}$ is a two-dimensional Cartan subalgebra of $\operatorname{Der} A$. Moreover, $\bar{H}$ decomposes $\bar{A}$ as the following sum of weight spaces:

$$\bar{A} = \bar{A}_0 \oplus \bar{A}_{\pm\alpha} \oplus \bar{A}_{\pm\beta} \oplus \bar{A}_{\pm(\alpha+\beta)}$$

with $\alpha, \beta \in \bar{H}^*$ linearly independent.

Proof: Take $e \in A_0$ with $n(e) = 1$ and consider $C = (C, \circ) = (A, e)$, and $H = \{d' \in \operatorname{Der}(A, e) \mid [d, d'] = 0\}$. Since any $d' \in H$ commutes with d it follows that every eigenspace $\bar{A}_\gamma$ is d'-invariant. Moreover, since $(\bar{A}_0, \circ)$ is a two-dimensional composition subalgebra of $(\bar{C}, \circ)$, d' kills $\bar{A}_0$. In particular, if we take a canonical basis of $\bar{C}$ as in the previous lemma, the coordenate matrix of d' relative to this basis must be diagonal. Thus, $\bar{H}$ is contained in the Cartan subalgebra $\mathcal{H}$ of $\bar{C}$ described in equation (2). Conversely, since the left and right multiplication operators by e on $\bar{A}$ commute with d, the elements u_i and v_i must be eigenvectors for these operators; so, $\mathcal{H}$ commutes with these operators and $\mathcal{H} \subseteq \bar{H}$. This shows that $\bar{H} = \mathcal{H}$. Now, the decomposition in the statement is just that in (3).

In order to check that $\bar{H}$ is a Cartan subalgebra of $\operatorname{Der} \bar{A}$ it is enough to see that if $[h, \bar{H}] \subseteq \bar{H}$ with $h \in \operatorname{Der} \bar{A}$ then $h \in \bar{H}$. This follows from the fact that in such case $[[h, \bar{H}], \bar{H}] = 0$ and since $\bar{H}$ acts diagonalizably we have $[h, \bar{H}] = 0$, so $h \in \bar{H}$. ◇

LEMMA 13. If $\bar{A} = \bar{A}_0 \oplus \bar{A}_{\pm\alpha} \oplus \bar{A}_{\pm 2\alpha}$ with $\dim \bar{A}_{2\alpha} = 1$, then
i) $\bar{A}_{2\alpha}\bar{A}_{i\alpha} = 0 = \bar{A}_{i\alpha}\bar{A}_{2\alpha}$ $i = 1, 2$
ii) $\bar{A}_{-2\alpha}\bar{A}_{-i\alpha} = 0 = \bar{A}_{-i\alpha}\bar{A}_{-2\alpha}$ $i = 1, 2$

Proof: If $\operatorname{char} F > 5$ the statement follows from Lemma 7, so let us assume that $\operatorname{char} F = 5$. We have that $\bar{A}_{2\alpha} = \bar{F}a_{2\alpha}$ and there exists $a_0 \in \bar{A}_0$ with $n(a_0) \neq 0$ such that $a_{2\alpha}a_0 = a_{2\alpha}$. Since $a_{2\alpha}A_\alpha \subseteq \bar{A}_{3\alpha} = \bar{A}_{-2\alpha}$ and $(a_{2\alpha}\bar{A}_\alpha, a_{2\alpha}) = (a_{2\alpha}\bar{A}_\alpha, a_{2\alpha}a_0) = n(a_{2\alpha})(\bar{A}_\alpha, a_0) = 0$, then we can conclude that $\bar{A}_{2\alpha}\bar{A}_\alpha = 0$. Similarly $\bar{A}_\alpha\bar{A}_{2\alpha} = 0$. Now consider $e \in \bar{A}_0$ with $n(e) = 1$. It is clear that $ea_{2\alpha} = \lambda a_{2\alpha}$ and $a_{2\alpha}e = \mu a_{2\alpha}$ for some nonzero $\lambda, \mu \in \bar{F}$. We can compute $a_{2\alpha}^2$ as follows:

$$a_{2\alpha}a_{2\alpha} = (a_{2\alpha}e) \circ (ea_{2\alpha}) = \lambda\mu a_{2\alpha} \circ a_{2\alpha} = -\lambda\mu n(a_{2\alpha})e^2 = 0,$$

where $\circ$ denotes the product of $(\bar{A}, e)$. This proves that $\bar{A}_{2\alpha}\bar{A}_{2\alpha} = 0$.

Part ii) follows in a similar way. ◇

PROPOSITION 14. Any nonzero derivation of A acts diagonalizably on $\bar{A}$, and $\bar{A}$ decomposes as a sum of eigenspaces in one of the following ways:
i) $\bar{A} = \bar{A}_0 \oplus \bar{A}_{\pm\alpha} \oplus \bar{A}_{\pm\beta} \oplus \bar{A}_{\pm(\alpha+\beta)}$, with $\alpha, \beta \in \bar{F}$
ii) $\bar{A} = \bar{A}_0 \oplus \bar{A}_{\pm\alpha}$ with $\alpha \in \bar{F}$ and $\dim \bar{A}_\alpha = \dim \bar{A}_{-\alpha} = 2$, and $\dim \bar{A}_0 = 4$.
iii) $\bar{A} = \bar{A}_0 \oplus \bar{A}_{\pm\alpha} \oplus \bar{A}_{\pm 2\alpha}$ with $\alpha \in \bar{F}$ and $\dim \bar{A}_0 = 2 = \dim \bar{A}_\alpha = \dim \bar{A}_{-\alpha}$, and $\dim \bar{A}_{2\alpha} = \dim \bar{A}_{-2\alpha} = 1$.

Proof: Let d be a derivation of A. If d acts with six different eigenvalues on $\bar{A}_0^\perp$ then, by Lemma 12, we have a decomposition as in part i). Notice also that if d has two independent eigenvalues α, β over the prime field, then $\alpha + \beta$ must be also an eigenvalue, so we are again in part i). Therefore, let us assume that all the eigenvalues of d on $\bar{A}_0^{\perp}$ are proportional (over the prime field) and that there are fewer than six of them. Under these restrictions the only possible decompositions of $\bar{A}$ are $\bar{A} = \bar{A}_0 \oplus \bar{A}_{\pm\alpha}$ and $\bar{A} = \bar{A}_0 \oplus \bar{A}_{\pm\alpha} \oplus \bar{A}_{\pm i\alpha}$. In the last case we have $(\bar{A}_{i\alpha} + \bar{A}_{-i\alpha})\bar{A}_\alpha \neq 0$, so either $1+i$ or $1-i$ belongs to $\{0, \pm 1, \pm i\}$. Changing α by 2α if necessary, we get $\bar{A} = \bar{A}_0 \oplus \bar{A}_{\pm\alpha} \oplus \bar{A}_{\pm 2\alpha}$.

Concerning the dimensions, in case $\bar{A} = \bar{A}_0 \oplus \bar{A}_{\pm\alpha}$, Lemma 9 with $\gamma = \alpha$ and $\gamma' = 0$ gives $\dim \bar{A}_0 \leq \dim \bar{A}_\alpha + \dim \bar{A}_{-\alpha}$. The same lemma with $\gamma = \alpha = \gamma'$ gives $\dim \bar{A}_\alpha + \dim \bar{A}_{-\alpha} \leq \dim \bar{A}_0$, so both sides agree and we have part ii). In regard to part iii), Lemma 9 with $\gamma = i\alpha$ and $\gamma' = 0$ says that $\dim \bar{A}_0 \leq \dim \bar{A}_{i\alpha} + \dim \bar{A}_{-i\alpha}$ $i = 1, 2$. Thus, $\dim \bar{A}_0 = 2$ and either $\bar{A}_\alpha$ or $\bar{A}_{2\alpha}$ has dimension 2. Now, we can pick a_α and $a_{-\alpha}$ with $(a_\alpha, a_{-\alpha}) \neq 0$ so that the left multiplication operator by $(a_\alpha + a_{-\alpha})$ maps $\bar{A}_0 \oplus \bar{A}_{\pm 2\alpha}$ injectively into $\bar{A}_{\pm\alpha}$. Therefore, $\dim \bar{A}_\alpha = 2 = \dim \bar{A}_{-\alpha}$ and $\dim \bar{A}_{2\alpha} = 1 = \dim \bar{A}_{-2\alpha}$.

In order to prove that d acts diagonalizably on $\bar{A}$ we should note that, since $\bar{A}_0 = \overline{\ker d}$, the only place where this may fail is on the spaces $\bar{A}_{\pm\alpha}$ of parts ii) and iii). However, in these cases $\alpha^2 \in F$ (take the trace of d^2) so $\bar{A}_{\pm\alpha}$ is the extension of the F-vector space $S = \{x \in A \mid (d^2 - \alpha^2 I)^2 x = 0\}$. If $d^2 - \alpha^2 I$ does not kill this space then we would have $x \in S$ such that:

$$0 \neq ((d^2 - \alpha^2 I)x, (d^2 - \alpha^2 I)x) = ((d^2 - \alpha^2 I)^2 x, x) = 0$$

which is not possible; therefore, d acts diagonalizably on $\bar{A}$. ◇

4. STRUCTURE OF Der A

Once that we know how d acts on $\bar{A}$, we can construct Cartan subalgebras of $L = \operatorname{Der} A$. Let $C_L(d)$ be the set $\{d' \in L \mid [d, d'] = 0\}$, where $d \in L$:

PROPOSITION 15. If $\dim C_L(d) = 1$ then $C_L(d)$ is a Cartan subalgebra of L. If $\dim C_L(d) \geq 2$ then we can find a two-dimensional Cartan subalgebra inside $C_L(d)$.

Proof: If $C_L(d) = Fd$ then for any $d' \in L$ such that $[d, d'] \in C_L(d)$ we have $[d, [d, d']] = 0$. Since d acts diagonalizably on $\bar{A}$ this means that $[d, d'] = 0$; therefore $C_L(d)$ is a Cartan subalgebra.

Let us assume now that $\dim C_L(d) \geq 2$ and consider $H = \operatorname{span}\{d, d'\} \subseteq C_L(d)$ a two-dimensional subspace. This subspace decomposes $\bar{A}$ as a sum of weight spaces. Since the dimension of H is 2 we should have 2 independent weights (see [5]) and a decomposition as follows:

$$\bar{A} = \bar{A}_0 \oplus \bar{A}_{\pm\alpha} \oplus \bar{A}_{\pm\beta} \oplus \bar{A}_{\pm(\alpha+\beta)} \tag{5}$$

where $\bar{H}^* = \operatorname{span}\{\alpha, \beta\}$. Any $d'' \in L$ with $[d'', H] \subseteq H$ verifies $[[d'', H], H] = 0$ and so $[d'', H] = 0$. Therefore d'' must leave invariant the weight spaces and so it must belong to H. ◇

We will devote the remainder of the paper to prove our main theorem:

THEOREM 16. Let A be a division composition algebra over a field of characteristic not 2 or 3, and $\operatorname{Der} A$ its derivation algebra, then:

i) If $\dim A = 1$ or 2 then $\operatorname{Der} A = 0$.

ii) If $\dim A = 4$ then either $\operatorname{Der} A$ is a Lie algebra of type A_1 or $\dim \operatorname{Der} A \leq 1$.

iii) If $\dim A = 8$ then $\operatorname{Der} A$ is one of the following Lie algebras:

   a) G_2.
   b) A_2.
   c) $A_1 \oplus A_1$.
   d) $A_1 \oplus Z$ where Z denotes an abelian ideal of dimension 0 or 1.
   e) An abelian algebra H of dimension $0, 1$ or 2.

We will begin with a particular case:

LEMMA 17. If $\dim C_L(d) = 1$ for all $d \in L$, then either L is abelian of dimension 1 or $L = A_1$.

Proof: Let us assume that L is not abelian. Since $\dim C_L(d) = 1$, d must decompose $\bar{A}$ as in parts ii) or iii) in Proposition 14. In particular,

$$\operatorname{trace}(d^2) = \begin{cases} 4\alpha^2 & \text{part ii)} \\ 12\alpha^2 & \text{part iii)} \end{cases}$$

Therefore, the bilinear form $t(d, d') = \operatorname{trace}(dd')$ is nondegenerate on L and $\bar{L}$. Moreover, since $[L, L] \neq 0$, given $d \in [L, L]$ we have $\bar{L} = \bar{F}d \oplus \sum_i \bar{L}_{i\alpha}$, so $[\bar{L}, \bar{L}] = \bar{L}$. We also have that the center of L, as well as the center of $\bar{L}$, is 0 (notice that $L = C_L(d)$ for any d in the center). So, in terms of Seligman [12] $\bar{L}$ is a classical Lie algebra and $\bar{H} = \bar{F}d$ is a classical Cartan subalgebra of $\bar{L}$ of dimension 1. From the classification of these algebras over fields of characteristic not 2 or 3 we can conclude that $\bar{L}$ is a classical Lie algebra of type A_1. ◇

So, in order to complete the proof of Theorem 16, we can assume that there exists $d \in L$ such that $\dim C_L(d) \geq 2$. Let $H = \operatorname{span}\{d, d'\}$ a Cartan subalgebra as in the proof of Proposition 15. This subalgebra allows us to use the following result [5]:

THEOREM 18. Let A be a composition algebra over an algebraically closed field of characteristic not 2 or 3. If there exists a Cartan subalgebra of $\operatorname{Der} A$ acting on A with two linearly independent weights then $\operatorname{Der} A$ is one of the following algebras:

i) G_2 ii) sl(3) iii) sl(2) $\oplus$ sl(2) iv) sl(2) $\oplus Z$ v) H

where Z is an abelian ideal of dimension 1 and H is abelian of dimension 2.

The only point that we have to check now is that if $\bar{L} = \mathrm{sl}(2) \oplus \mathrm{sl}(2)$ then $L = A_1 \oplus A_1$.

LEMMA 19. Let C be an octonion algebra and Q a quaternion subalgebra. Then, $\operatorname{Der}(C, Q) := \{d \in \operatorname{Der} C \mid dQ \subseteq Q\} = A_1 \oplus A_1$ and C is $1 + 3 + 4$ as a $\operatorname{Der}(C, Q)$-module.

Proof: We can decompose C as $C = Q \perp v_0Q$ where v_0 is an arbitrary element of $Q^\perp$ with $n(v_0) \neq 0$ that we fix. Since Q is a simple central associative algebra, any derivation $d \in \operatorname{Der} Q$ is inner, that is, $d = d_a$, $d_a: x \mapsto ax - xa$ for some $a \in Q$ and $a \perp e$ where e denotes the unit element. Given any element $b \in Q \cap e^\perp$ we can extend d to a derivation $d_{a,b}$ of C by the following formulas:

$$\begin{aligned} d_{a,b}(x) &= [a, x] \\ d_{a,b}(v_0x) &= v_0(xb + [a, x]) \qquad \forall x \in Q \end{aligned}.$$

In fact, $\operatorname{Der}(C, Q) = \operatorname{span}\{d_{a,b} \mid a, b \in Q \cap e^\perp\}$. Now it is easy to check that $\operatorname{Der}(C, Q)$ is the direct sum of two ideals, each of them being a simple Lie algebra of type A_1, namely

$$\operatorname{Der}(C, Q) = \operatorname{span}\{d_{a,a} \mid a \in Q \cap e^\perp\} \oplus \operatorname{span}\{d_{0,b} \mid b \in Q \cap e^\perp\}.$$

From the explicit formula for $\operatorname{Der}(C, Q)$ we obtain that $Q = 1+3$ as a $\operatorname{Der}(C, Q)$-module, and that any submodule of v_0Q can be written as v_0S where S is an ideal of Q, so v_0Q must be irreducible. ⋄

Our final step is

LEMMA 20. If $\bar{L} = \mathrm{sl}(2) \oplus \mathrm{sl}(2)$ then $L = A_1 \oplus A_1$.

Proof: By (5) $\bar{H}$ decomposes $\bar{A}$ as $\bar{A} = \bar{A}_0 \oplus \bar{A}_{\pm\alpha} \oplus \bar{A}_{\pm\beta} \oplus \bar{A}_{\pm(\alpha+\beta)}$ with $\alpha, \beta \in \bar{H}^*$. Take $d \in \bar{H} \cap \mathrm{sl}(2)$ and denote by $\widetilde{\mathrm{sl}(2)}$ the other copy of $\mathrm{sl}(2)$. It is clear that d decomposes $\bar{A}$ as $\bar{A} = \bar{A}_0 + \bar{A}_{\pm\alpha(d)} + \bar{A}_{\pm\beta(d)} + \bar{A}_{\pm(\alpha+\beta)(d)}$ and that, since d commutes with $\widetilde{\mathrm{sl}(2)}$, these eigenspaces are $\widetilde{\mathrm{sl}(2)}$-stable. So, not all of them are different, and hence d decomposes $\bar{A}$ as

$$\text{a) } \bar{A}_0 \oplus \bar{A}_{\pm\gamma} \quad \text{or} \quad \text{b) } \bar{A}_0 \oplus \bar{A}_{\pm\gamma} \oplus \bar{A}_{\pm 2\gamma}$$

with $\gamma \in \bar{F}$ and $\dim \bar{A}_\gamma = 2 = \dim \bar{A}_{-\gamma}$. Since $\bar{A}_\gamma$ is an $\widetilde{\mathrm{sl}(2)}$-module and $\bar{A}_{-\gamma}$ is its dual, either they are trivial or irreducible. In the first case $\widetilde{\mathrm{sl}(2)}$ kills $\bar{A}_{\pm\gamma}$, which generates $\bar{A}$ as an algebra; this forces a contradiction. Therefore, $\bar{A}_\gamma$ and $\bar{A}_{-\gamma}$ are $\widetilde{\mathrm{sl}(2)}$-irreducible.

Notice that $\bar{A}_0$ and $\bar{A}_0 \oplus \bar{A}_{\pm\delta}$ with $\delta = \alpha, \beta, \alpha + \beta$ are composition subalgebras of $\bar{A}$, so $B = (\bar{A}_{\pm\gamma})^\perp$ is a subalgebra of $\bar{A}$, and $\widetilde{\mathrm{sl}(2)}$ acts as derivations on it. In particular, B is either trivial or decomposes as $1 + 3$ as an $\widetilde{\mathrm{sl}(2)}$-module (Theorem 8).

Placing together all this information we obtain that, as an $\widetilde{\mathrm{sl}(2)}$-module, $\bar{A}$ decomposes either as $1+1+1+1+2+2$ or as $1+3+2+2$. Since $[\widetilde{\mathrm{sl}(2)}, \mathrm{sl}(2)] = 0$, in both cases $\mathrm{sl}(2)$ must preserve B, that is, B is an $\bar{L}$-submodule. Now take $e \in B$ with $\bar{L}e = 0$ and $n(e) \neq 0$. We have that $\mathrm{sl}(2) \oplus \mathrm{sl}(2) = \bar{L} \subseteq \{d \in \operatorname{Der}(\bar{A}, e) \mid dB \subseteq B\} = \mathrm{sl}(2) \oplus \mathrm{sl}(2)$ by Lemma 19, so the equality holds. By the same lemma $\bar{L}$ decomposes $\bar{A}$ as $1 + 3 + 4$. Such a decomposition is induced by a similar decomposition on A [10]. Denoting the sum of the one and three-dimensional irreducible L-submodules of A by Q we have that $B = \bar{Q}$, so Q is a subalgebra of A and $L = \{d \in \operatorname{Der}(A, e) \mid dQ \subseteq Q\} = A_1 \oplus A_1$, where $e \in Q$ is such that $Le = 0$. ⋄

REFERENCES

1. Bott, R. and Milnor, J.: "On the parallelizability of the spheres", Bull. A.M.S. 64 (1958), 87–89.

2. Benkart, G. and Osborn, J.M.: "The derivation algebra of a real divison algebra", Amer. J. Math. 103 (1981), 1135–1150.

3. Benkart, G. and Osborn, J.M.: "An investigation of real division algebras using derivations", Pacific J. Math. 96 (1981), 265–300.

4. Benkart, G., Britten, D.J. and Osborn, J.M.: "Real flexible division algebras", Can. J. Math. XXXIV, N. 3 (1982), 550–588.

5. Elduque, A. and Pérez, J.M.: "Composition algebras with large derivation algebras", J. Algebra 190 (1997), n. 2, 372–404.

6. Herstein, I.N.: Noncommutative Rings, The Carus Mathematical Monographs, 15. Mathematical Association of America, Washington, DC, 1994.

7. Jacobson, N.: "Composition algebras and their automorphisms", Rend. Circ. Mat. Palermo 7 ser. II (1958), 55–80.

8. Kaplansky, I.: "Infinite-dimensional quadratic forms admitting composition", Proc. Amer. Math. Soc. 4 (1953), 956–960.

9. Kervaire, M.: "Non-parallelizability of the n sphere for $n > 7$", Proc. Nat. Acad. Sci. 44 (1958), 280–283.

10. Pérez-Izquierdo, J.M.: "Division composition algebras through their derivation algebras", preprint.

11. Schafer, R.D.: An Introduction to Nonassociative Algebras, Academic Press. New York, 1966.

12. Seligman, G.B.: Modular Lie Algebras, Ergebnisse der Mathematik and ihrer Grenzgebiete, Band 40, Springer-Verlag, New York, 1967.

13. Zhevlakov, K.A., Slin'ko, A.M., Shestakov, I.P. and Shirshov, A.I.: Rings that are Nearly Associative, Academic Press. New York, 1982.

On the structure of the simple Moufang loop $GLL(F_2)$

M.L. MERLINI GIULIANI, Departamento de Matemática, Univ. Federal de Santa Maria, 97119-105 - Santa Maria RS, Brazil, lurdinha@ccne.ufsm.br

CÉSAR POLCINO MILIES[1], Instituto de Matemática e Estatística, Universidade de São Paulo, Caixa Postal 66.281, 05315-970 São Paulo SP, Brazil, polcino@ime.usp.br

ABSTRACT In this paper, we recall some basic facts about linear Moufang loops and then study the subloop structure of the smallest simple Moufang loop, namely, $GLL(F_2)$. We do not include detailed proofs, which are given in [4] and will be published elsewhere, but the main ideas are clearly stated.

1 INTRODUCTION

Let K be a field. Consider the set of 2×2 matrices of the form

$$\begin{bmatrix} a & \mathbf{x} \\ \mathbf{y} & b \end{bmatrix},$$

where $a, b \in K$ and $\mathbf{x}, \mathbf{y} \in K^3$, the set of ordered triples of elements of K.

We add two such matrices entrytwise and multiply them according to the following rule:

[1]The second author was partially supported by a research grant from CNPq. Proc. 300243/79-0(RN)

$$\begin{bmatrix} a_1 & \mathbf{x}_1 \\ \mathbf{y}_1 & b_1 \end{bmatrix} \begin{bmatrix} a_2 & \mathbf{x}_2 \\ \mathbf{y}_2 & b_2 \end{bmatrix} = \begin{bmatrix} a_1a_2 + \mathbf{x}_1.\mathbf{y}_2 & a_1\mathbf{x}_2 + b_2\mathbf{x}_1 - \mathbf{y}_1 \times \mathbf{y}_2 \\ a_2\mathbf{y}_1 + b_1\mathbf{y}_2 + \mathbf{x}_1 \times \mathbf{x}_2 & b_1b_2 + \mathbf{y}_1\mathbf{x}_2 \end{bmatrix},$$

where . and $\times$ denote the usual dot and cross products respectively in K^3. In this way, we obtain an alternative algebra known as *Zorn's vector matrix algebra over* K which will be denoted by $\mathbf{M(K)}$. It is also called the *split Cayley-Dickson algebra over* K.

It is well known that Zorn's algebra is simple in the sense that it contains no proper two-sided ideals and, furthermore, that any simple alternative (non-associative) algebra over K which is not a division ring is isomorphic to $\mathbf{M(K)}$.

We shall denote by $GLL(K)$ the set of all invertible elements of the algebra $\mathbf{M(K)}$. It is natural to consider loops of the form $GLL(K)$ as a non-associative analog of linear groups. Following [3] we shall call $GLL(K)$ the *general linear loop over* K. Since $\mathbf{M(K)}$ is alternative, it follows that $GLL(K)$ is a Moufang loop.

We recall that it is possible to define a determinant mapping in $\mathbf{M(K)}$ by:

$$\alpha = \begin{bmatrix} a & \mathbf{x} \\ \mathbf{y} & b \end{bmatrix} \mapsto det(\alpha) = ab - \mathbf{x}.\mathbf{y}.$$

An element $\alpha = \begin{bmatrix} a & \mathbf{x} \\ \mathbf{y} & b \end{bmatrix} \in \mathbf{M(K)}$ is invertible if and only if $det(\alpha) \neq 0$ and, in this case, its inverse is

$$\alpha^{-1} = (\det \alpha)^{-1} \begin{bmatrix} b & -\mathbf{x} \\ -\mathbf{y} & a \end{bmatrix}.$$

The set $SLL(K) = \{\alpha \in \mathbf{M(K)} \mid det(\alpha) = 1\}$ is a normal subloop of $GLL(K)$ and will be called the *special linear loop*. It is easy to see that the centre of $SLL(K)$ is $\mathcal{Z}(SLL(K)) = \{I, -I\}$. We also define the *projective special linear loop* to be the quotient loop:

$$PSLL(K) = SLL(K)/\mathcal{Z}(SLL(K)).$$

Notice that, if K is a finite field with $|K| = q$ then it can be easily shown that:

$$|GLL(F_q)| = q^3(q^4-1)(q-1) \qquad |SLL(F_q)| = q^3(q^4-1),$$
$$|PSLL(F_q)| = \tfrac{q^3(q^4-1)}{2}, \text{ if } q \neq 2^n, \quad |PSLL(F_q)| = q^3(q^4-1), \text{ if } q = 2^n.$$

We recall that L. Paige showed in [6] that $PSLL(K)$ is simple for any field K. Later M.W. Liebeck [5] showed that every finite simple Moufang loop is isomorphic to a loop of this type. If we denote by F_2 the field with two elements, we have that $GLL(F_2) = SLL(F_2)$, and $Z_0 = \{I\}$, so it follows that also $PSLL(F_2) = GLL(F_2)$. Hence, this is the smallest simple Moufang loop, and it is easy to show that $|GLL(F_2)| = 120$.

2 SOME BASIC FACTS ON $GLL(F_2)$

We first determine the order $o(\alpha)$ of an element α in $GLL(F_2)$.

LEMMA 2.1 *Let* $\alpha = \begin{bmatrix} a & \mathbf{x} \\ \mathbf{y} & b \end{bmatrix} \in GLL(F_2)$. *If* $a = b$, *then* $o(\alpha) = 2$ *and if* $a \neq b$ *then* $o(\alpha) = 3$.

Given an element $\alpha \in GLL(F_2)$, we shall denote by $\mathcal{C}(\alpha)$ its centralizer. First we show that elements of order 3 commute only with themselves or their powers.

LEMMA 2.2 *Let* $\alpha \in GLL(F_2)$ *be such that* $o(\alpha) = 3$. *Then,* $\mathcal{C}(\alpha) = \langle\alpha\rangle$.

As a consequence, we have the following.

COROLLARY 2.3 $GLL(F_2)$ *contains no cyclic subgroups of order* 6.

COROLLARY 2.4 *The* 3*-subloops of* $GLL(F_2)$ *are all cyclic of order 3.*

Hence, the only cyclic subgroups of $GLL(F_2)$ are those of order 2 or 3. These are very easy to determine.

PROPOSITION 2.5 $GLL(F_2)$ *contains* 63 *subgroups of order* 2 *and* 28 *subgroups of order* 3.

For elements of order 2, we have the following.

LEMMA 2.6 *(i) If* $o(\alpha) = 2$, *then* $|\mathcal{C}(\alpha)| \leq 32$.

(ii) If $o(\alpha) = o(\beta) = 2$, *with* $\alpha \neq \beta$, *then* $|\mathcal{C}(\alpha) \cap \mathcal{C}(\beta)| \leq 16$.

Finally, it is also possible to obtain information about the possible orders of subloops of $GLL(F_2)$.

PROPOSITION 2.7 *Let* H *be a proper subloop of* $GLL(F_2)$. *Then,* $|H| = 2^a 3^b$, *with* $0 \leq a \leq 5$ *and* $0 \leq b \leq 1$.

3 MAXIMAL (ASSOCIATIVE) SUBGROUPS

We begin this section classifying the 2-subloops of $GLL(F_2)$. Since LEMMA 2.1 shows that these are subloops of exponent 2, we recall the following, due to O. Chein [1]

LEMMA 3.1 *Every Moufang Loop of exponent* 2 *is a group and, as such, it is abelian.*

In view of this result and taking into account the facts about centralizers of elements of order 2 given in LEMMA 2.6, one can describe the lattice of 2-subloops of $GLL(F_2)$.

THEOREM 3.2 *(i) Every subgroup of order 2 is contained in at least one subgroup of order 4.*

(ii) Every subgroup of order 4 is contained in at least one subgroup of order 8.

(iii) $GLL(F_2)$ contains no subgroups of order 16.

A concrete example of a subloop of $GLL(F_2)$ isomorphic to $C_2 \times C_2 \times C_2$ can be obtained as follows. If we set:

$$\alpha = \begin{bmatrix} 1 & \mathbf{i} \\ \mathbf{j} & 1 \end{bmatrix}, \quad \beta = \begin{bmatrix} 1 & \mathbf{j} \\ \mathbf{i} & 1 \end{bmatrix}, \quad \gamma = \begin{bmatrix} 0 & \mathbf{k} \\ \mathbf{k} & 0 \end{bmatrix},$$

then

$$\alpha\beta = \begin{bmatrix} 0 & \mathbf{i}+\mathbf{j}+\mathbf{k} \\ \mathbf{i}+\mathbf{j}+\mathbf{k} & 0 \end{bmatrix} = \beta\alpha, \quad \alpha\gamma = \begin{bmatrix} 0 & \mathbf{i}+\mathbf{k} \\ \mathbf{j}+\mathbf{k} & 0 \end{bmatrix} = \gamma\alpha$$

,

$$\beta\gamma = \begin{bmatrix} 0 & \mathbf{j}+\mathbf{k} \\ \mathbf{i}+\mathbf{k} & 0 \end{bmatrix} = \gamma\beta, \quad \alpha(\beta\gamma) = \begin{bmatrix} 1 & \mathbf{i}+\mathbf{j} \\ \mathbf{i}+\mathbf{j} & 1 \end{bmatrix} = (\alpha\beta)\gamma.$$

Thus, $\quad G = \langle \alpha, \beta, \gamma \rangle \cong C_2 \times C_2 \times C_2.$

Further, we deal with groups whose order envolves both the primes 2 and 3.

THEOREM 3.3 *Let* $\alpha = \begin{bmatrix} a & \mathbf{x} \\ \mathbf{y} & a \end{bmatrix}$, $\beta = \begin{bmatrix} 1 & \mathbf{u} \\ \mathbf{v} & 0 \end{bmatrix}$, *with* $\alpha^2 = \beta^3 = 1$.

(i) If $\mathbf{x}.\mathbf{v} \neq \mathbf{y}.\mathbf{u}$, *then* $\langle \alpha, \beta \rangle \cong S_3$.

(ii) If $\mathbf{x}.\mathbf{v} = \mathbf{y}.\mathbf{u}$, *then* $\langle \alpha, \beta \rangle \cong A_4$.

LEMMA 3.4 *If G is a subgroup of $GLL(F_2)$ of order 12, then $G \cong A_4$*

A concrete example of a subloop of $GLL(F_2)$ isomorphic to A_4 can be obtained as follows. Set:

$$\alpha = \begin{bmatrix} 1 & \mathbf{i} \\ \mathbf{j} & 1 \end{bmatrix}, \quad \beta = \begin{bmatrix} 1 & \mathbf{i}+\mathbf{j} \\ \mathbf{i} & 0 \end{bmatrix}.$$

Denote $\quad \mathbf{u} = \mathbf{i}+\mathbf{j}+\mathbf{k}.$

$$\beta^2 = \begin{bmatrix} 0 & \mathbf{i}+\mathbf{j} \\ \mathbf{i} & 1 \end{bmatrix}, \qquad \alpha\beta = \begin{bmatrix} 0 & \mathbf{u} \\ \mathbf{u} & 1 \end{bmatrix},$$

$$(\alpha\beta)^2 = \begin{bmatrix} 1 & \mathbf{u} \\ \mathbf{u} & 0 \end{bmatrix}, \qquad \alpha\beta^2 = \begin{bmatrix} 1 & \mathbf{j}+\mathbf{k} \\ \mathbf{i}+\mathbf{k} & 0 \end{bmatrix},$$

$$(\alpha\beta^2)^2 = \begin{bmatrix} 0 & \mathbf{j}+\mathbf{k} \\ \mathbf{i}+\mathbf{k} & 1 \end{bmatrix}, \qquad \alpha\beta\alpha = \begin{bmatrix} 1 & \mathbf{j} \\ \mathbf{i}+\mathbf{j} & 0 \end{bmatrix},$$

$$(\alpha\beta\alpha)^2 = \begin{bmatrix} 0 & \mathbf{j} \\ \mathbf{i}+\mathbf{j} & 1 \end{bmatrix}, \qquad \beta^2\alpha\beta = \begin{bmatrix} 0 & \mathbf{i}+\mathbf{k} \\ \mathbf{k} & 0 \end{bmatrix},$$

$$\beta\alpha\beta^2 = \begin{bmatrix} 0 & \mathbf{k} \\ \mathbf{j}+\mathbf{k} & 0 \end{bmatrix}.$$

Due to the results above, we can completely describe the maximal associative subgroups of $GLL(F_2)$

COROLLARY 3.5 *A maximal (associative) subgroup of* $GLL(F_2)$ *is isomorphic to either* $C_2 \times C_2 \times C_2$, S_3 *or* A_4.

4 MAXIMAL (NONASSOCIATIVE) SUBLOOPS

We recall that Moufang loops of small order have been classifyied by O. Chein in [1] and [2]. It is well-known that Moufang loops whose order is less than 12 are associative and it is easy to see, from the classification, that the only nonassociative Moufang loop of order 12 that can be a subloop of $GLL(F_2)$ is $M_{12}(S_3, 2)$, which we denote briefly by M_{12}, and which can be defined as the loop generated by three elements

$$M_{12} = \langle \alpha, \beta, \gamma \mid \alpha^2 = \beta^3 = \gamma^2 = 1 \rangle,$$

satisfying the following relations:

$$\alpha^2 = \gamma^2 = (\alpha\beta)^2 = (\gamma\beta)^2 = (\alpha\beta^2)^2 = (\gamma\beta^2)^2 = (\alpha\gamma)^2 = [(\alpha\gamma)\beta]^2 = [(\alpha\gamma)\beta^2]^2 = 1$$

$$\alpha\gamma = \gamma\alpha, \;\; \beta\gamma = \gamma\beta^2, \;\; \beta\alpha = \alpha\beta^2, \;\; \beta(\alpha\gamma) = (\alpha\gamma)\beta^2.$$

We can describe this loop as

$$M_{12} = \{1, \alpha, \beta, \beta^2, \alpha\beta, \alpha\beta^2, \gamma, \gamma\beta, \gamma\beta^2, \alpha\gamma, (\alpha\beta)\gamma, \alpha(\beta\gamma)\}.$$

The following is a concrete example of a subloop of $GLL(F_2)$ isomorphic to M_{12}.

EXAMPLE

$$\alpha = \begin{bmatrix} 1 & \mathbf{i} \\ \mathbf{0} & 1 \end{bmatrix}, \qquad \beta = \begin{bmatrix} 0 & \mathbf{u} \\ \mathbf{u} & 1 \end{bmatrix}, \qquad \gamma = \begin{bmatrix} 1 & \mathbf{j} \\ \mathbf{0} & 1 \end{bmatrix},$$

$$\beta^2 = \begin{bmatrix} 1 & \mathbf{u} \\ \mathbf{u} & 0 \end{bmatrix}, \qquad \alpha\beta = \begin{bmatrix} 1 & \mathbf{j}+\mathbf{k} \\ \mathbf{i} & 1 \end{bmatrix}, \qquad \alpha\beta^2 = \begin{bmatrix} 0 & \mathbf{u} \\ \mathbf{i} & 0 \end{bmatrix},$$

$$\alpha\gamma = \begin{bmatrix} 1 & \mathbf{i}+\mathbf{j} \\ \mathbf{k} & 1 \end{bmatrix} = \gamma\alpha, \quad \beta\gamma = \begin{bmatrix} 0 & \mathbf{u} \\ \mathbf{j} & 0 \end{bmatrix} = \gamma\beta^2, \quad \beta^2\gamma = \begin{bmatrix} 1 & \mathbf{i}+\mathbf{k} \\ \mathbf{j} & 1 \end{bmatrix} = \gamma\beta,$$

$$(\alpha\beta)\gamma = \begin{bmatrix} 1 & \mathbf{k} \\ \mathbf{0} & 1 \end{bmatrix}, \qquad \alpha(\beta\gamma) = \begin{bmatrix} 0 & \mathbf{u} \\ \mathbf{k} & 0 \end{bmatrix}.$$

Moreover, we have the following.

PROPOSITION 4.1 *Every subgroup of $GLL(F_2)$ isomorphic to S_3 is contained in a subloop isomorphic to M_{12}.*

We shall now consider subloops of order 24.

The nonassociative Moufang loops of order 24 were described by O.Chein in [2]. They are: $M_{24}(A_4,2)$; $M_{24}(D_4,2)$; $M_{24}(G_{12},2)$; $M_{24}(G_{12},C_2 \times C_4)$; $M_{24}(G_{12},Q)$. It is easy to see that, with the possible exception of $M_{24}(A_4,2)$, which we denote briefly as M_{24}, none of these can be contained as a subloop of $GLL(F_2)$. Note that M_{24} can be constructed as

$$M_{24} = \langle L, \gamma \mid (g\gamma)^2 = 1 \text{ for all } g \in L \rangle,$$

where L is a group isomorphic to A_4.

It can be shown that $GLL(G_2)$ does have subloops isomorphic to M_{24}; more precisely, we have the following.

THEOREM 4.2 *Every subgroup of $GLL(F_2)$ isomorphic to A_4 is contained in a subloop isomorphic to M_{24}.*

Since it is also easy to see that $GLL(F_2)$ contains no subloop of order 32 or 48, we have the following.

THEOREM 4.3 *A (nonassociative) subloop L of $GLL(F_2)$ is maximal if and only if it is isomorphic to either $C_2 \times C_2 \times C_2$, M_{12} or M_{24}.*

The following diagram gives an idea of the relations between subloops of $GLL(F_2)$.

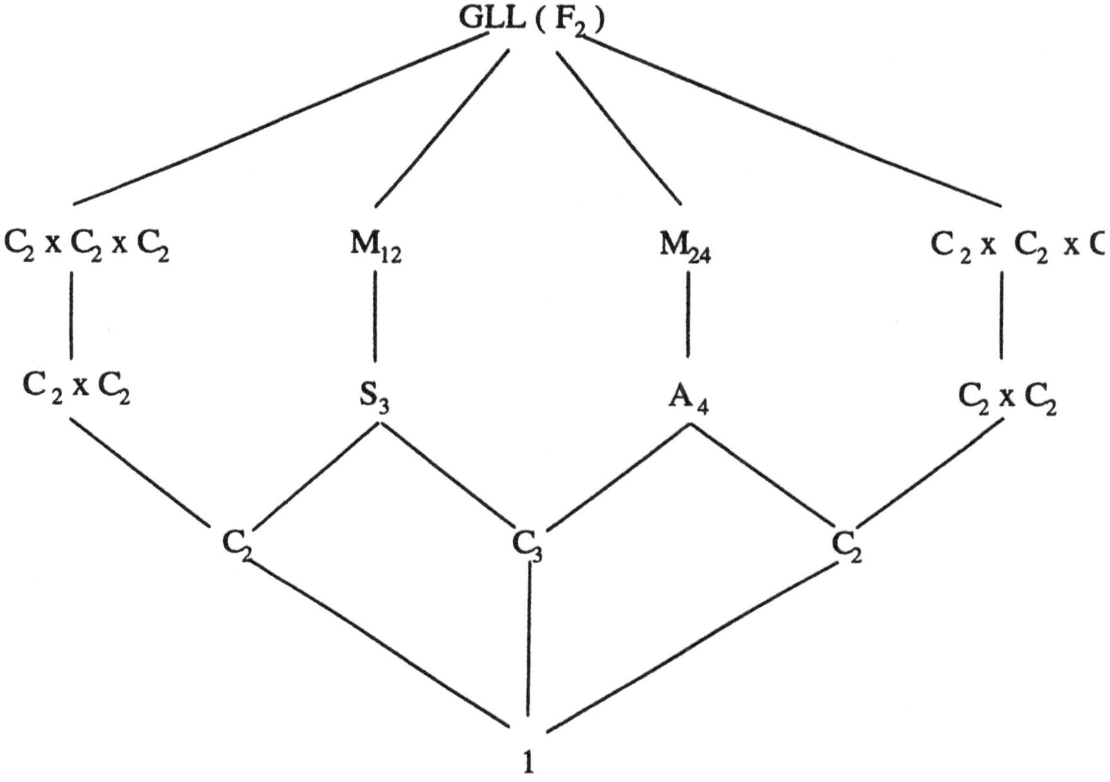

Figure 1:

REFERENCES

[1] O. Chein, Moufang Loops of Small Order I, Trans. Amer. Math. Soc. 188 (1974), 31-51.

[2] O. Chein, Moufang Loops of Small Order II, Memoirs Amer. Math. Soc. 197 (13) (1978).

[3] E.G. Goodaire, E. Jespers and C. Polcino Milies, Alternative Loop Rings, North Holland Math. Studies 184, Elsevier, Amsterdam, 1996.

[4] M.L. Merlini Giuliani and C. Polcino Milies, The smallest simple Moufang loop, preprint.

[5] M.W. Liebeck, The classification of finite simple Moufang loops, Math. Proc. Cambridge Phil. Soc., 102 (1987), 33-47.

[6] L.J. Paige, A class of simple Moufang loops, Proc. Amer. Math. Soc., 7 (1956), 471-482.

Lie groups, differential geometry, and nonlinear integrable systems

A. V. RAZUMOV Institute for High Energy Physics, 142284 Protvino, Moscow region, Russia

M. V. SAVELIEV Institute for High Energy Physics, 142284 Protvino, Moscow region, Russia

1 INTRODUCTION

The Toda systems permanently attract attention of physicists and mathematicians due to their integrability and deep links to a number of problems in the theory of differential equations, differential and algebraic geometry, and relevance to many problems of modern theoretical and mathematical physics.

In the present work the construction of the Toda-type equations in low and high dimensional spaces is reviewed and the integration procedure for these equations is discussed. We give a differential geometry formulation of the main principle of the group-algebraic approach [1, 2] to the integration of the systems under consideration, the so-called grading condition. The proof of the generalised Plücker formulae, based on the use of Toda equations is considered.

2 TODA-TYPE SYSTEMS

2.1 Zero curvature condition and grading condition

Let M be either the real manifold $\mathbb{R}^2$ or the complex manifold $\mathbb{C}$. For $M = \mathbb{R}^2$ we denote the standard coordinates by z^- and z^+. In the case of $M = \mathbb{C}$ we use the notation z^- for the standard complex coordinate and z^+ for the complex conjugate of z^-. Let G be a complex matrix Lie group and $\mathfrak{g}$ be its Lie algebra. It is well known that a connection on the trivial principal G-bundle $M \times G$ is characterised by the corresponding $\mathfrak{g}$-valued 1-form ω on M.[1] Here the connection is flat if and only if

$$d\omega + \omega \wedge \omega = 0. \tag{2.1}$$

Since this equality implies that the curvature tensor of the connection under consideration is the zero tensor, we call relation (2.1) the *zero curvature condition*. Any nonlinear integrable system of partial differential equations can be represented as the zero curvature condition for some 1-form ω of a special type and such representation is the starting point of integration procedure.

Consider first the general zero curvature condition (2.1). Since M is simply connected, the general solution to (2.1) is

$$\omega = {}^{\varphi}\omega = \varphi^{-1}d\varphi,$$

where φ is an arbitrary mapping from M to G. Furthermore, the zero curvature condition (2.1) is gauge invariant. This means that if a 1-form ω satisfies (2.1), then for any mapping $\psi : M \to G$ the 1-form

$$\omega^{\psi} = \psi^{-1}\omega\psi + \psi^{-1}d\psi \tag{2.2}$$

also satisfies (2.1). It is clear that

$$({}^{\varphi}\omega)^{\psi} = {}^{\varphi\psi}\omega,$$

therefore, any solution of (2.1) can be transformed to the trivial solution $\omega = 0$. Thus, the general zero curvature condition (2.1) considered as an equation for ω is trivial, and to obtain nontrivial equations one should impose some restrictions on the form of ω. Let us discuss a method of imposing the restrictions in question leading to the Toda-type systems.

Suppose that the Lie algebra $\mathfrak{g}$ is endowed with a $\mathbb{Z}$-gradation,

$$\mathfrak{g} = \bigoplus_{m \in \mathbb{Z}} \mathfrak{g}_m.$$

The subspace $\mathfrak{g}_0$ is a subalgebra of $\mathfrak{g}$. The corresponding connected Lie subgroup of G is denoted G_0.

Represent the 1-form ω as

$$\omega = \omega_- dz^- + \omega_+ dz^+,$$

and assume that the $\mathfrak{g}$-valued functions ω_- and ω_+ have the form

$$\omega_- = \sum_{m=-l_-}^{0} \omega_{-,m}, \qquad \omega_+ = \sum_{m=0}^{l_+} \omega_{+,m}, \tag{2.3}$$

where the functions $\omega_{\pm,m}$ take values in $\mathfrak{g}_m$ and $l_\pm$ are some positive integers. The restriction on the form of ω specified by relations (2.3) is called the *grading condition*.

[1] Following the physical slang we call ω a connection form or just a connection on M.

In terms of the functions $\omega_\pm$ the zero curvature condition reads

$$\partial_- \omega_+ - \partial_+ \omega_- + [\omega_-, \omega_+] = 0. \tag{2.4}$$

Here and below we use the notation $\partial_\pm = \partial/\partial z^\pm$. From (2.4) it follows, in particular, that

$$\partial_+ \omega_{-,-l_-} + [\omega_{+,0}, \omega_{-,-l_-}] = 0.$$

Representing $\omega_{+,0}$ in the form

$$\omega_{+,0} = \xi_+^{-1} \partial_+ \xi_+, \tag{2.5}$$

where ξ_+ is a mapping from M to G_0, we can rewrite this equation as

$$\partial_+ (\xi_+ \omega_{-,-l_-} \xi_+^{-1}) = 0.$$

Therefore, we have the following representation

$$\omega_{-,-l_-} = \xi_+^{-1} c_- \xi_+, \tag{2.6}$$

where c_- is a $\mathfrak{g}_{-l_-}$-valued function which satisfies the relation

$$\partial_+ c_- = 0. \tag{2.7}$$

Similarly, representing $\omega_{-,0}$ as

$$\omega_{-,0} = \xi_-^{-1} \partial_- \xi_-, \tag{2.8}$$

where ξ_- is a mapping from M to G_0, we obtain

$$\omega_{+,l_+} = \xi_-^{-1} c_+ \xi_-, \tag{2.9}$$

where c_+ is a $\mathfrak{g}_{l_+}$-valued function which satisfies the relation

$$\partial_- c_+ = 0. \tag{2.10}$$

Using (2.5), (2.8), (2.6) and (2.9), one can write the following representation for the components of the connection:

$$\omega_- = \xi_+^{-1} c_- \xi_+ + \sum_{m=-l_-+1}^{-1} \xi_+^{-1} \upsilon_{-,m} \xi_+ + \xi_-^{-1} \partial_- \xi_-, \tag{2.11}$$

$$\omega_+ = \xi_+^{-1} \partial_+ \xi_+ + \sum_{m=1}^{l_+-1} \xi_-^{-1} \upsilon_{+,m} \xi_- + \xi_-^{-1} c_+ \xi_-, \tag{2.12}$$

where

$$\upsilon_{-,m} = \xi_+ \omega_{-,m} \xi_+^{-1}, \qquad \upsilon_{+,m} = \xi_- \omega_{+,m} \xi_-^{-1}.$$

Note that the set of connections satisfying the grading condition is invariant with respect to gauge transformations (2.2) with the mapping ψ taking values in G_0. Hence, the zero curvature condition considered as an equation for the 1-form ω

satisfying the grading condition still have a residual gauge invariance. Here the functions $v_{\pm,m}$ and $c_\pm$ are gauge invariant, while for the mappings $\xi_\pm$ one has

$$\xi_-^\psi = \xi_-\psi, \qquad \xi_+^\psi = \xi_+\psi.$$

To obtain the explicit form of the equations which follow from the zero curvature condition it is convenient to make the gauge transformation (2.2) with $\psi = \xi_+^{-1}$. As the result we get a new $\mathfrak{g}$-valued 1-form, which we again denote by ω, with the components

$$\omega_- = c_- + \sum_{m=-l_-+1}^{-1} v_{-,m} + \gamma^{-1}\partial_-\gamma, \tag{2.13}$$

$$\omega_+ = \gamma^{-1}\left(\sum_{m=1}^{l_+-1} v_{+,m} + c_+\right)\gamma, \tag{2.14}$$

where

$$\gamma = \xi_-\xi_+^{-1}.$$

The equations which follow from the zero curvature condition for the connections with the components specified by (2.11), (2.12) and (2.13), (2.14) are identical. It is worth to note that the mapping γ is invariant with respect to the residual gauge transformations. Thus, as it should be expected, we obtain the equations for the gauge invariant quantities only. The equations for the functions c_- and c_+ are just (2.7) and (2.10). It is customary to fix these functions and to consider the equations which follow from the zero curvature condition as the equations for the mapping γ and the functions $v_{\pm,m}$. We call these equations in the general case the *Toda-type equations.*

In the case when $l_- = l_+$ and $\mathfrak{g}_m = 0$ for $-l_- < m < 0$ and for $0 < m < l_+$, the zero curvature condition for the connection with the components specified by (2.13), (2.14) gives the equation

$$\partial_+(\gamma^{-1}\partial_-\gamma) = [c_-, \gamma^{-1}c_+\gamma], \tag{2.15}$$

which we call simply a *Toda equation.* Here, when the Lie group G_0 is abelian, we say that we deal with an *abelian Toda equation*, otherwise we call the corresponding equation a *nonabelian Toda equation.* The class of nonabelian Toda equations has not been studied yet in full; see in this connection [3] and references therein.

There exists a multidimensional generalisation of the Toda-type equations [4]. Here one considers the case where either $M = \mathbb{R}^{2d}$ or $M = \mathbb{C}^d$. The standard coordinates are denoted by z^{-i} and z^{+i}. The grading condition imposed on the components of a connection

$$\omega = \sum_{i=1}^{d}(\omega_{-i}\mathrm{d}z^{-i} + \omega_{+i}\mathrm{d}z^{+i})$$

looks as

$$\omega_{-i} = \sum_{m=-l_{-i}}^{0} \omega_{-i,m}, \qquad \omega_{+i} = \sum_{m=0}^{l_{+i}} \omega_{+i,m},$$

where $l_{\pm i}$ are some positive integers. It can be show that in this case by some gauge transformation (2.2), where the mapping ψ takes values in G_0, the connection ω can be transformed to the connection with the components

$$\omega_{-i} = c_{-i} + \sum_{m=-l_{-i}+1}^{-1} v_{-i,m} + \gamma^{-1}\partial_{-i}\gamma,$$

$$\omega_{+i} = \gamma^{-1}\left(\sum_{m=1}^{l_{+i}-1} v_{+i,m} + c_{+i}\right)\gamma,$$

where the mappings $c_{\pm i}$ satisfy the relations

$$\partial_{+i}c_{-j} = 0, \qquad \partial_{-i}c_{+j} = 0$$

and

$$[c_{-i}, c_{-j}] = 0, \qquad [c_{+i}, c_{+j}] = 0.$$

The corresponding equations for the mapping γ and the functions $v_{\pm i,m}$ are called the *multidimensional Toda-type equations*. In particular, the analogue of the two dimensional Toda equation is

$$\partial_{+j}(\gamma^{-1}\partial_{-i}\gamma) = [c_{-i}, \gamma^{-1}c_{+j}\gamma],$$
$$\partial_{-i}(\gamma c_{-j}\gamma^{-1}) = \partial_{-j}(\gamma c_{-i}\gamma^{-1}), \qquad \partial_{+i}(\gamma^{-1}c_{+j}\gamma) = \partial_{+j}(\gamma^{-1}c_{+i}\gamma).$$

Concrete examples of the multidimensional Toda equations can be found in [4].

2.2 General solution

Here we discuss the integration procedure for the Toda-type equations. We begin with the investigation of the properties of a mapping φ which generates a flat connection ω satisfying the grading condition.

Introduce the following nilpotent subalgebras of $\mathfrak{g}$:

$$\mathfrak{g}_{<0} = \bigoplus_{m<0} \mathfrak{g}_m, \qquad \mathfrak{g}_{>0} = \bigoplus_{m>0} \mathfrak{g}_m,$$

and denote the corresponding connected Lie subgroups of G by $G_{<0}$ and $G_{>0}$ respectively. Suppose that $G_{<0}$, $G_{>0}$ and G_0 are closed subgroups of G and that

$$G = \overline{G_{<0}G_{>0}G_0} = \overline{G_{>0}G_{<0}G_0},$$

where bar denotes the topological closure. The above equality means that for an element a belonging to some dense subset of G, there are valid the Gauss decompositions

$$a = m_-n_+h_+ = m_+n_-h_-, \tag{2.16}$$

where $m_-, n_- \in G_{<0}$, $m_+, n_+ \in G_{>0}$ and $h_\pm \in G_0$. We assume that the Gauss decompositions (2.16) are unique. It is clear now that any mapping $\varphi : M \to G$ can be locally represented as

$$\varphi = \mu_+\nu_-\eta_-, \qquad \varphi = \mu_-\nu_+\eta_+, \tag{2.17}$$

where for some elements $a_\pm \in G$ the mappings μ_+ and μ_- take values in $a_+G_{>0}$ and $a_-G_{<0}$, the mappings ν_- and ν_+ take values in $G_{<0}$ and $G_{>0}$, and the mappings $\eta_\pm$ take values in G_0. From (2.17) it follows that

$$\mu_+^{-1}\mu_- = \nu_-\eta\nu_+^{-1}, \tag{2.18}$$

where

$$\eta = \eta_-\eta_+^{-1}.$$

One can treat (2.18) as a kind of Gauss decomposition of the mapping $\mu_+^{-1}\mu_-$. Such interpretation shows that the mappings ν_-, ν_+ and η are uniquely determined by the mappings μ_- and μ_+. Thus, the mapping φ can be reconstructed from the mappings μ_- and μ_+ and one of the mappings η_+ and η_-. This fact is very important for the integration procedure of the Toda-type equations.

Let now a mapping $\varphi : M \to G$ generate a connection which satisfies the grading condition. Using respectively the first and the second equalities of (2.17), we can write the following representation for the components of the connection ω:

$$\omega_- = \eta_-^{-1}\nu_-^{-1}(\mu_+^{-1}\partial_-\mu_+)\nu_-\eta_- + \eta_-^{-1}(\nu_-^{-1}\partial_-\nu_-)\eta_- + \eta_-^{-1}\partial_-\eta_-, \tag{2.19}$$

$$\omega_+ = \eta_+^{-1}\nu_+^{-1}(\mu_-^{-1}\partial_+\mu_-)\nu_+\eta_+ + \eta_+^{-1}(\nu_+^{-1}\partial_+\nu_+)\eta_+ + \eta_+^{-1}\partial_+\eta_+. \tag{2.20}$$

From these relations it follows that for a connection satisfying the grading condition one has

$$\partial_+\mu_- = 0, \qquad \partial_-\mu_+ = 0. \tag{2.21}$$

Further, comparing (2.20) and (2.14) one concludes that

$$\partial_+\eta_+ = 0.$$

It is convenient for our purposes to denote η_+ by γ_-. Then the above equality reads

$$\partial_+\gamma_- = 0. \tag{2.22}$$

The comparison of (2.19) and (2.13) gives

$$\eta_-^{-1}\partial_-\eta_- = \gamma^{-1}\partial_-\gamma.$$

Therefore, for some mapping $\gamma_+ : M \to G_0$ which satisfies the relation

$$\partial_-\gamma_+ = 0 \tag{2.23}$$

one has

$$\eta_- = \gamma_+\gamma.$$

From (2.17) one also obtains

$$\omega_- = \eta_+^{-1}\nu_+^{-1}(\mu_-^{-1}\partial_-\mu_-)\nu_+\eta_+ + \eta_+^{-1}(\nu_+^{-1}\partial_-\nu_+)\eta_+ + \eta_+^{-1}\partial_-\eta_+,$$

$$\omega_+ = \eta_-^{-1}\nu_-^{-1}(\mu_+^{-1}\partial_+\mu_+)\nu_-\eta_- + \eta_-^{-1}(\nu_-^{-1}\partial_+\nu_-)\eta_- + \eta_-^{-1}\partial_+\eta_-.$$

These relations imply that

$$\mu_-^{-1}\partial_-\mu_- = \gamma_-\left(c_- + \sum_{m=-l_-+1}^{-1}\lambda_{-,m}\right)\gamma_-^{-1}, \tag{2.24}$$

$$\mu_+^{-1}\partial_+\mu_+ = \gamma_+\left(\sum_{m=1}^{l_+-1}\lambda_{+,m} + c_+\right)\gamma_+^{-1} \tag{2.25}$$

for some mappings $\lambda_{\pm,m}$ taking values in $\mathfrak{g}_m$ and satisfying the relations

$$\partial_+\lambda_{-,m} = 0, \qquad \partial_-\lambda_{+,m} = 0. \tag{2.26}$$

Thus, the mappings μ_- and μ_+ are solutions of equations (2.24) and (2.25) respectively. In other words, the mappings $\mu_\pm$ are determined by the mappings $\gamma_\pm$ and $\lambda_{\pm,m}$ up to some integration constants. As we noted above, the mapping φ is determined by the mappings $\mu_\pm$ and by the mapping $\eta_+ = \gamma_-$. The mapping φ, in its turn, determines the mappings γ and $\upsilon_{\pm,m}$ which satisfy the Toda-type equations under consideration. Actually the mappings γ and $\upsilon_{\pm,m}$ can be directly expressed through the mappings $\gamma_\pm$, η and $\nu_\pm$. Indeed, for the mapping γ one has the representation

$$\gamma = \gamma_+^{-1}\eta\gamma_- \tag{2.27}$$

and relations (2.19) and (2.20) imply that

$$c_- + \sum_{m=-l_-+1}^{-1} \upsilon_{-,m} = \gamma_-^{-1}\eta^{-1}(\nu_-^{-1}\partial_-\nu_-)\eta\gamma_-, \tag{2.28}$$

$$\sum_{m=1}^{l_+-1} \upsilon_{+,m} + c_+ = \gamma_+^{-1}\eta(\nu_+^{-1}\partial_+\nu_+)\eta^{-1}\gamma_+. \tag{2.29}$$

Taking the above discussion into account, one can suggest the following integration procedure for the Toda-type equations. Let $\gamma_\pm$ be some mappings taking values in G_0 and satisfying conditions (2.22) and (2.23), and let $\lambda_{\pm,m}$ be some $\mathfrak{g}_m$-valued functions satisfying conditions (2.26). Integrate equations (2.24) and (2.25). Here we choose only the solutions which satisfy relations (2.21). Then from the Gauss decomposition (2.18) determine the mappings η and $\nu_\pm$. Finally, with the help of (2.27) and (2.28), (2.29) construct the mapping γ and the functions $\upsilon_{\pm,m}$. It can be shown that this procedure really gives a solution to the Toda-type equations under consideration. Moreover, it is clear that any solution can be obtained in this way.

In the multidimensional case one begins with mappings $\gamma_\pm : M \to G_0$ and functions $\lambda_{\pm i,m}$ taking values in $\mathfrak{g}_m$. Here it is assumed that

$$\partial_+\gamma_- = 0, \qquad \partial_-\gamma_+ = 0,$$
$$\partial_+\lambda_{-i,m} = 0, \qquad \partial_-\lambda_{+i,m} = 0.$$

Then one solves the equations

$$\mu_-^{-1}\partial_{-i}\mu_- = \gamma_-\left(c_{-i} + \sum_{m=-l_{-i}+1}^{-1} \lambda_{-i,m}\right)\gamma_-^{-1}, \tag{2.30}$$

$$\mu_+^{-1}\partial_{+i}\mu_+ = \gamma_+\left(\sum_{m=1}^{l_{+i}-1} \lambda_{+i,m} + c_{+i}\right)\gamma_+^{-1}. \tag{2.31}$$

Note that only those solutions of these equations which satisfy the conditions

$$\mu_-^{-1}\partial_{+i}\mu_- = 0, \qquad \mu_+^{-1}\partial_{-i}\mu_+ = 0$$

are used in the integration procedure. It is important that it is possible to solve equations (2.30) and (2.31) if and only if the corresponding integrability conditions are satisfied. In particular, for the case of the multidimensional Toda equations the conditions

$$\partial_{-i}(\gamma_- c_{-j} \gamma_-^{-1}) = \partial_{-j}(\gamma_- c_{-i} \gamma_-^{-1}), \qquad \partial_{+i}(\gamma_+ c_{+j} \gamma_+^{-1}) = \partial_{+j}(\gamma_+ c_{+i} \gamma_+^{-1})$$

should be satisfied. An example of multidimensional Toda equations for which the above integrability conditions can be explicitly solved is given in [4].

The next step is to use the Gauss decomposition (2.18) to obtain the mappings $\nu_\pm$ and η. Now one use (2.27) and the relations

$$c_{-i} + \sum_{m=-l_{-i}+1}^{-1} v_{-i,m} = \gamma_-^{-1} \eta^{-1} (\nu_-^{-1} \partial_{-i} \nu_-) \eta \gamma_-,$$

$$\sum_{m=1}^{l_{+i}-1} v_{+i,m} + c_{+i} = \gamma_+^{-1} \eta (\nu_+^{-1} \partial_{+i} \nu_+) \eta^{-1} \gamma_+$$

to construct the mapping γ and the functions $v_{\pm i,m}$.

2.3 Hermitian solutions

From the point of view of applications to differential geometry problems the most important solutions of the Toda-type equations are the so-called hermitian solutions.[2] These solutions are defined only for the case $M = \mathbb{C}$.

Assume that the Lie algebra $\mathfrak{g}$ is endowed with a hermitian involution which can be extended to the hermitian involution of the Lie group G. We denote the action of the involution under consideration on the elements of $\mathfrak{g}$ and G by dagger. Suppose that

$$(\mathfrak{g}_m)^\dagger = \mathfrak{g}_{-m}.$$

Hence for the subgroups $G_{<0}$, G_0 and $G_{>0}$ one has

$$(G_{<0})^\dagger = G_{>0}, \qquad (G_0)^\dagger = G_0, \qquad (G_{>0})^\dagger = G_{<0}.$$

If a mapping γ and functions $v_{\pm,m}$ is a solution of the Toda-type equations, then the connection with the components defined by (2.13) and (2.14) is flat. Let $\varphi : \mathbb{C} \to G$ be a mapping which generates the connection under consideration. Let $l_+ = l_-$ and the $\mathfrak{g}$-valued functions c_- and c_+ be connected by the relation

$$(c_-)^\dagger = -c_+.$$

In this case it can be shown that the mapping $\varphi^{\dagger -1} \gamma^\dagger$ generates the connection with the components of form (2.13) and (2.14) where instead of γ and $v_{\pm,m}$ stand the mapping γ' and the functions $v'_{\pm,m}$ given by

$$\gamma' = \gamma^\dagger, \qquad v'_{-,m} = -(v_{+,-m})^\dagger, \qquad v'_{+,m} = -(v_{-,-m})^\dagger.$$

[2] Actually in [5, 2] we use the term 'real solutions'.

Therefore, the mapping γ' and the functions $v'_{\pm,m}$ satisfy the same Toda-type equations. Thus, we can consider the solutions of the Toda-type equations having the property

$$\gamma^{\dagger} = \gamma, \qquad (v_{+,m})^{\dagger} = -v_{-,-m}. \tag{2.32}$$

For such solutions the mapping φ which generates the corresponding connection satisfies the relation

$$\varphi^{\dagger} = \gamma\varphi^{-1}a^{-1},$$

where a is some element of G such that $a^{\dagger} = a$. We call a solution of the Toda-type equation satisfying relations (2.32) a *hermitian solution*.

To obtain all such solutions one should start with the mappings $\gamma_{\pm}$ and $\lambda_{\pm,m}$ satisfying the conditions

$$\gamma_{+} = (\gamma_{-})^{\dagger-1}, \qquad \lambda_{+,m} = -(\lambda_{-,-m})^{\dagger}.$$

and choose the solutions of equations (2.24) and (2.25) for which

$$\mu_{+} = a^{-1}(\mu_{-})^{\dagger-1},$$

where a is some element of G such that $a^{\dagger} = a$.

3 DIFFERENTIAL GEOMETRY OF TODA SYSTEMS AND GENERALISED PLÜCKER FORMULAE

3.1 Geometrical meaning of grading condition

Denote by $G_{\leq 0}$ and $G_{\geq 0}$ the connected Lie subgroups of G corresponding to the subalgebras $\mathfrak{g}_{\leq 0} = \mathfrak{g}_{<0} \oplus \mathfrak{g}_0$ and $\mathfrak{g}_{\geq 0} = \mathfrak{g}_0 \oplus \mathfrak{g}_{>0}$. These Lie subgroups are closed and

$$G_{\leq 0} = G_{<0}G_0 = G_0G_{>0}, \qquad G_{\geq 0} = G_0G_{>0} = G_{>0}G_0.$$

Denote by π_- and π_+ the canonical projections from G to $F_- = G/G_{\geq 0}$ and the canonical projection from G to $F_+ = G/G_{\leq 0}$ respectively. Let ω be a flat connection on the trivial principal G-bundle $\mathbb{C} \times G$ and φ be a mapping from $\mathbb{C}$ to G which generates ω. It can be shown that the component $\omega^{(1,0)}$ takes values in $\mathfrak{g}_{\leq 0}$ and the component $\omega^{(0,1)}$ takes values in $\mathfrak{g}_{\geq 0}$ if and only if the mapping $\varphi_- = \pi_- \circ \varphi$ is holomorphic and the mapping $\varphi_+ = \pi_+ \circ \varphi$ is antiholomorphic [5, 2].

Define the following subspaces of $\mathfrak{g}$:

$$\mathfrak{m}_- = \bigoplus_{m \geq -l_-} \mathfrak{g}_m, \qquad \mathfrak{m}_+ = \bigoplus_{m \leq l_+} \mathfrak{g}_m.$$

Let p be an arbitrary point of F_- and let a be any element of G such that $\pi_-(a) = p$. Define the subspace $\mathcal{M}_{-p} \subset T_p^{(1,0)}(F_-)$ by

$$\mathcal{M}_{-p} = \pi_{-*a}(L_{a*e}(\mathfrak{m}_-)),$$

where L_a is the left shift in G by a, and e is the unit element of G. The subspaces $\mathcal{M}_{-p}$, $p \in F_-$, generate a distribution on F_- which will be denoted $\mathcal{M}_-$. In a similar way we define the distribution $\mathcal{M}_+$ on F_+.

The connection ω satisfies the grading condition if and only if the mapping φ_- is holomorphic and tangent to the distribution $\mathcal{M}_-$, and the mapping φ_+ is antiholomorphic and tangent to the distribution $\mathcal{M}_+$.

3.2 Some differential geometry related to hermitian solutions

In this section we assume that $M = \mathbb{C}$ and that $\mathfrak{g}$ is endowed with a hermitian involution which satisfies the requirements formulated in section 2.3. Let ω be a flat connection on the trivial principal G bundle $\mathbb{C} \times G \to \mathbb{C}$ which satisfies the grading condition and the relation

$$\omega^\dagger = -\omega^\chi, \tag{3.1}$$

where the mapping $\chi = \chi^\dagger$ takes values in G_0. It can be shown that with the help of gauge transformation (2.2) with ψ taking values in G_0, such a connection can be transformed to the connection with the components of form (2.13), (2.14) where the mapping γ and the functions $v_{\pm,m}$ obey relations (2.32).

In the case under consideration a mapping $\varphi : \mathbb{C} \to G$ generating the connection ω has the property

$$\varphi^\dagger = \chi^{-1}\varphi^{-1}a^{-1}, \tag{3.2}$$

where a is some element of G such that $a^\dagger = a$. From the other hand, if a mapping $\varphi : \mathbb{C} \to G$ satisfies (3.2), where $a^\dagger = a$ and $\psi^\dagger = \psi$, then it generates the connection ω which satisfies (3.1).

Define the conjugation σ in $\mathfrak{g}$ by

$$\sigma(x) = -x^\dagger$$

and the conjugation Σ in G by

$$\Sigma(a) = a^{\dagger -1}.$$

The mapping Σ generates the mappings $\Sigma_- : F_- \to F_+$ and $\Sigma_+ : F_+ \to F_-$ defined as

$$\Sigma_-(aG_{\geq 0}) = \Sigma(a)G_{\leq 0}, \qquad \Sigma_+(aG_{\leq 0}) = \Sigma(a)G_{\geq 0}.$$

These mappings are antiholomorphic and mutually inverse:

$$\Sigma_- \circ \Sigma_+ = \mathrm{id}_{F_+}, \qquad \Sigma_+ \circ \Sigma_- = \mathrm{id}_{F_-}.$$

Relation (3.2) can be written as

$$\Sigma \circ \varphi = a\varphi\chi.$$

From this relation it follows that

$$\Sigma_- \circ \varphi_- = L_a^{F_+} \circ \varphi_+, \qquad \Sigma_+ \circ \varphi_+ = L_a^{F_-} \circ \varphi_-, \tag{3.3}$$

where L^{F_-} and L^{F_+} are the standard left actions of G on F_- and F_+, respectively, and the mappings φ_- and φ_+ are defined in the previous section. It is clear that a mapping φ satisfies (3.2) if and only if the mappings φ_- and φ_+ satisfy (3.3).

Summarizing the discussion of the present section we can say that a connection ω satisfies the grading condition and relation (3.1) if and only if the mapping φ_- is holomorphic and tangent to the distribution $\mathcal{M}_-$, the mapping φ_+ is antiholomorphic and tangent to the distribution $\mathcal{M}_+$, and the mappings φ_- and φ_+ are connected by relations (3.3). Not that it suffices to satisfy only one of relations (3.3).

3.3 Some facts about semisimple Lie groups and Lie algebras

Assume that G is a semisimple Lie group of rank r. Up to the action of the group of the automorphisms of the Lie algebra $\mathfrak{g}$ all $\mathbb{Z}$-gradation of $\mathfrak{g}$ can be obtained with the help of the following procedure. Assign to the vertices of the Dynkin diagram of $\mathfrak{g}$ nonnegative integer labels q_i, $i = 1, \ldots, r$, and define

$$q = \sum_{i,j=1}^{r} h_i (k^{-1})_{ij} q_j \tag{3.4}$$

where h_i are Cartan generators and $k = (k_{ij})$ is the Cartan matrix of $\mathfrak{g}$. Now defining

$$\mathfrak{g}_m = \{x \in \mathfrak{g} \mid [q, x] = mx\},$$

we come to some $\mathbb{Z}$-gradation of $\mathfrak{g}$. The Cartan generators h_i form a basis of a Cartan subalgebra $\mathfrak{h}$ of $\mathfrak{g}$. Let Δ be the set of roots of $\mathfrak{g}$ with respect to $\mathfrak{h}$, and $\Pi = \{\alpha_1, \ldots, \alpha_r\}$ be a base of Δ. The subspace $\mathfrak{g}_m$ for $m \neq 0$ is the direct sum of the root subspaces $\mathfrak{g}^\alpha$ corresponding to the roots $\alpha = \sum_{i=1}^r n_i \alpha_i$ with $\sum_{i=1}^r n_i q_i = m$. The subspace $\mathfrak{g}_0$, besides the root subspaces corresponding to the roots $\alpha = \sum_{i=1}^r n_i \alpha_i$ with $\sum_{i=1}^r n_i q_i = 0$, includes the Cartan subalgebra $\mathfrak{h}$.

For any $\mathbb{Z}$-gradation of $\mathfrak{g}$ the subalgebras $\mathfrak{g}_{<0}$, $\mathfrak{g}_{>0}$ are parabolic subalgebras of $\mathfrak{g}$ and the subgroups $G_{<0}$, $G_{>0}$ are parabolic subgroups of G. Hence, in this case F_- and F_+ are flag manifolds of G.

Denote by x_{-i} and x_{+i}, $i = 1, \ldots, r$, the Chevalley generators of $\mathfrak{g}$. Recall that the ith fundamental representation ρ_i of $\mathfrak{g}$ is characterised by the highest weight vector v_i which satisfies the relation

$$\rho_i(h_j) v_i = \delta_{ij} v_i, \qquad j = 1, \ldots, r.$$

Since v_i is the highest weight vector, one has

$$\rho_i(x_{+j}) v_i = 0, \qquad i = 1, \ldots, r.$$

Moreover, it can be shown that

$$\rho_i(x_{-j}) v_i = 0$$

for all $j \neq i$, and that

$$\rho_i(x_{-i}) v_i \neq 0.$$

The hermitian Chevalley involution in $\mathfrak{g}$ is the antilinear antiautomorphism of $\mathfrak{g}$ defined by

$$h_i^\dagger = h_i, \qquad x_{\pm i}^\dagger = x_{\mp i}.$$

It is known that the representation space of any fundamental representation of $\mathfrak{g}$ can be endowed with a nondegenerate positive definite hermitian form (,) such that

$$\rho_i(x^\dagger) = \rho_i(x)^\dagger$$

for any $x \in \mathfrak{g}$. Note that the dagger in the left hand side of the above equality means the hermitian Chevalley involution, while the dagger in the right hand side means the hermitian conjugation with respect to the hermitian form (,).

If the Lie group G is simply connected then the fundamental representations of $\mathfrak{g}$ can be extended to representations of the Lie group G which we also denote by ρ_i. In this case the hermitian Chevalley involution of $\mathfrak{g}$ can be extended to a hermitian involution of G. Here one has

$$\rho_i(a^\dagger) = \rho_i(a)^\dagger \tag{3.5}$$

for any $a \in G$.

3.4 Toda equations for principal gradation

Consider the principal gradation of the Lie algebra $\mathfrak{g}$. This gradation is defined by the grading operator (3.4) where all q_i are equal to 2.[3] Here the subalgebra $\mathfrak{g}_0 = \mathfrak{h}$ is a Cartan subalgebra of $\mathfrak{g}$, the subalgebras $\mathfrak{g}_{\le 0}$ and $\mathfrak{g}_{\ge 0}$ are Borel subalgebras of $\mathfrak{g}$ usually denoted by $\mathfrak{b}_-$ and $\mathfrak{b}_+$. The corresponding subgroups of G are denoted by H, B_- and B_+.

With the choice $l_- = l_+ = 2$ we come to the Toda equations of form (2.15). Assume that the Lie group G is simply connected. In this case we have can use the fundamental representations of G. It is convenient for any $x \in \mathfrak{g}$ and any $a \in G$ to write xv and av instead of $\rho_i(x)v$ and $\rho_i(a)v$. Define the functions

$$\beta_i = (v_i, \gamma v_i), \qquad i = 1, \ldots, r,$$

where for any $i = 1, \ldots, r$ the vector v_i is the highest weight vector of the ith fundamental representation of the Lie group G, normalized by the condition

$$(v_i, v_i) = 1.$$

Actually, since the subspace $\mathbb{C}v_i$ is invariant with respect to the action of the group G_0, one has

$$\gamma v_i = \beta_i v_i.$$

From equation (2.15) it follows that

$$\partial_+\partial_-\beta_i - \partial_+\beta_i\,\beta_i^{-1}\partial_-\beta_i = -(v_i, c_+\gamma c_- v_i). \tag{3.6}$$

The general form of the functions c_- and c_+ is given by

$$c_- = \sum_{i=1}^r c_{-i}\, x_{-i}, \qquad c_+ = \sum_{i=1}^r c_{+i}\, x_{+i}.$$

For any $a \in G$ there is valid the following equality

$$\det\begin{pmatrix} (v_i, av_i) & (v_i, x_{+i}\, av_i) \\ (v_i, a\, x_{-i}v_i) & (v_i, x_{+i}\, a\, x_{-i}v_i) \end{pmatrix} = \prod_{\substack{j=1 \\ j\ne i}}^r (v_j, av_j)^{-k_{ij}};$$

[3]Usually the principal gradation is defined by choosing $q_i = 1$ for all $i = 1, \ldots, r$. With our choice we get the grading operator q coinciding with the Cartan element of the corresponding integral embedding of the Lie algebra $\mathfrak{sl}(2, \mathbb{C})$ into G.

see, for example, [1]. In particular, for $a \in H$, taking into account that $k_{ii} = 2$, one obtains

$$(v_i, x_{+i} a x_{-i} v_i) = (v_i, a v_i) \prod_{j=1}^{r} (v_j, a v_j)^{-k_{ij}}.$$

Using this equality and relation (3.6), we come to the following equations for the functions β_i:

$$\partial_+ (\beta_i^{-1} \partial_- \beta_i) = -c_{+i} c_{-i} \prod_{j=1}^{r} \beta_j^{-k_{ij}}.$$

3.5 Generalised Plücker formulae

In this section $M = \mathbb{C}$ and G is a complex simply connected semisimple Lie group of rank r. We assume that some Cartan subgroup H of G is fixed, and denote the corresponding Cartan and Chevalley generators of $\mathfrak{g}$ by h_i and $x_{\pm i}$. As hermitian involution and conjugation in $\mathfrak{g}$ and G we take the hermitian Chevalley involution and the corresponding conjugation.

Consider the fundamental representations of G. The stability subgroup of the subspace $\mathbb{C} v_i$ is the parabolic subgroup of G whose Lie algebra is generated by all Cartan and Chevalley generators of $\mathfrak{g}$ except the generator x_{-i}. It is customary to denote this parabolic subgroup and parabolic subalgebra by $P_{+i'}$ and $\mathfrak{p}_{+i'}$.[4]

Denote the representation space of the ith fundamental representation of G by V_i. Note that the fundamental representation ρ_i defines a left actions of G in the in the projective space $\mathbb{P}(V_i)$. Here for any $a \in G$ one has

$$a \cdot \mathrm{pr}_i(v) = \mathrm{pr}_i(\rho_i(a) v),$$

where pr_i is the canonical projection from $V_i \setminus \{0\}$ to $\mathbb{P}(V_i)$. Thus, we can say that the parabolic subgroup $P_{+i'}$ is the stability subgroup of the point $p_i \in \mathbb{P}(V_i)$ corresponding to the highest weight vector v_i. Moreover, it is possible to define the embedding $\iota_{F_{-i}}$ of the flag manifold $F_{-i} = G/P_{+i'}$ into the projective space $\mathbb{P}(V_i)$ by

$$\iota_{F_{-i}}(a P_{+i'}) = a \cdot p_i.$$

Let us endow the representation space V_i with the hermitian positive definite nondegenerate hermitian form $(\ ,\ )$ for which relation (3.5) is valid. Define the function f_i on $V_i \setminus \{0\}$ by

$$f_i(v) = \ln(v, v).$$

The Fubini–Study metric on $\mathbb{P}(V_i)$ is defined by its associated fundamental form Φ_i^{FS} which is a unique form satisfying the relation

$$\mathrm{pr}_i^* \Phi_i^{FS} = -2\sqrt{-1} \partial \bar{\partial} f_i.$$

Via the mapping $\iota_{F_{-i}}$ the Fubini–Study metric on $\mathbb{P}(V_i)$ induces a Kähler metric on F_{-i}. Hence the flag manifold F_{-i} is a Kähler manifold.

[4]These notations originate from the general construction of the parabolic subalgebras and parabolic subgroups of for semisimple Lie groups and Lie algebras which is described, for example, in [5, 2].

Let now φ_- be a holomorphic mapping from $\mathbb{C}$ to $F_- = G/B_+$, where B_+ is the Borel subgroup of G whose Lie algebra $\mathfrak{b}_+$ is generated by the Cartan generators h_i and the Chevalley generators x_{+i}. Since the parabolic subgroup $P_{+i'}$ contains B_+ there is a natural projection π_{-i} from the flag manifold F_- to the flag manifold F_{-i}, defined by

$$\pi_{-i}(aB_+) = aP_{+i'}.$$

The Kähler metric on F_{-i} via the mapping $\pi_{-i} \circ \varphi_-$ specifies a pseudo-metric on $\mathbb{C}$. The fundamental form Φ_i associated with this pseudo-metric is given by

$$\Phi_i = (\varphi_-^* \circ \pi_{-i}^* \circ \iota_{F_{-i}})\Phi_i^{FS}.$$

It can be shown [5, 2] that any local holomorphic lift $\widetilde{\varphi}_-$ of the mapping φ_- to G leads to the following representation of the fundamental form Φ_i:

$$\Phi_i = -2\sqrt{-1}\partial\bar{\partial}\ln(\widetilde{\varphi}_- v_i, \widetilde{\varphi}_- v_i).$$

Therefore, the function

$$K_i = \ln(\widetilde{\varphi}_- v_i, \widetilde{\varphi}_- v_i)$$

is a Kähler potential of the pseudo-metric on $\mathbb{C}$ which has Φ_i as its fundamental form. Choosing different lifts we get different Kähler potentials.

Suppose now that the mapping φ_- is tangent to the distribution $\mathcal{M}_-$ defined with the help of the principal gradation of $\mathfrak{g}$. Define the mapping φ_+ by

$$\varphi_+ = \Sigma_- \circ \varphi_-.$$

This mapping is antiholomorphic and tangent to the distribution $\mathcal{M}_+$.

It can be shown [5, 2] that there exists a mapping $\varphi : \mathbb{C} \to G$ such that

$$\varphi_- = \pi_- \circ \varphi, \qquad \varphi_+ = \pi_+ \circ \varphi. \tag{3.7}$$

This mapping generates the connection ω which satisfies the grading condition and relation (3.1). Performing the corresponding gauge transformation, transform the connection ω to the connection with the components

$$\omega_- = c_- + \gamma^{-1}\partial_-\gamma, \qquad \omega_+ = \gamma^{-1}c_+\gamma, \tag{3.8}$$

where $c_+ = -c_-^\dagger$ and $\gamma^\dagger = \gamma$. Actually we can assume from the very beginning that the mapping φ generates the connection with the components given by (3.8). It is connected with the fact that relations (3.7) define the mapping φ up to the multiplication from the right by a mapping taking values in H. Note that here one has

$$\varphi^{\dagger -1} = \varphi\gamma. \tag{3.9}$$

Thus, we associate with the mapping φ a hermitian solution of the Toda equation determined by the principal gradation of $\mathfrak{g}$.

Construct now the local Gauss decompositions (2.17), where having in mind (3.9) we choose

$$\mu_+ = \mu_-^{\dagger -1}, \qquad \nu_- = \nu_+^{\dagger -1}, \qquad \eta_- = \eta_+^{\dagger -1}\gamma. \tag{3.10}$$

As was shown in section 2.2, the mappings γ_- and γ_+ defined as

$$\gamma_- = \eta_+, \qquad \gamma_+ = \eta_- \gamma^{-1} \tag{3.11}$$

satisfy the relations

$$\partial_+ \gamma_- = 0, \qquad \partial_- \gamma_+ = 0.$$

Note that in our case

$$\gamma_+ = \gamma_-^{\dagger -1}.$$

Taking into account (3.10) and (3.11), we can write

$$\varphi = \mu_- \nu_+ \gamma_- = \mu_-^{\dagger -1} \nu_+^{\dagger -1} \gamma_-^{\dagger -1} \gamma.$$

Therefore, one has

$$\gamma = \gamma_-^\dagger \nu_+^\dagger \mu_-^\dagger \mu_- \nu_+ \gamma_- .$$

From this relation we obtain

$$\beta_i = (v_i, \gamma v_i) = (\mu_- \gamma_- v_i, \mu_- \gamma_- v_i).$$

It is clear that the mapping $\mu_- \gamma_-$ is a local holomorphic lift of the mapping φ_-. Hence, the functions

$$K_i = \ln \beta_i$$

are Kähler potentials of the pseudo-metrics on $\mathbb{C}$ with the associated fundamental forms Φ_i. As follows from the discussion given in section 3.4, the functions β_i satisfy the equations

$$\partial_+(\beta_i^{-1} \partial_- \beta_i) = |c_i|^2 \prod_{j=1}^{r} \beta_j^{-k_{ij}},$$

where c_i are some holomorphic functions. From these equations it follows directly [5, 2] that the Ricci curvature tensors R_i of the pseudo-metrics under consideration are connected with the corresponding fundamental forms Φ_i by the relations

$$R_i = \sum_{j=1} k_{ij} \Phi_i.$$

These relations are called the *generalised infinitesimal Plücker formulae*. The validity of these formulae was first conjectured by A. B. Givental' [6], and proved by L. E. Positsel'skii [7]. Our proof is based on the special choice of the Kähler potentials of the pseudo-metrics, and in this sense is more close to the proof of the standard Plücker formulae, see, for example, [8]. Note also that the relation between the abelian Toda fields and the Kähler potentials in question has been established for the Lie groups $\mathrm{SL}(n, \mathbb{C})$ by J.-L. Gervais and Y. Matsuo [9], and for the other classical simple groups by J.-L. Gervais and M. V. Saveliev [10], using explicit calculations in a local coordinate parametrization of the flag manifolds. The discussion above follows the lines of [5] and is valid for an arbitrary semisimple Lie group. The corresponding generalised global Plücker formulae are also valid (F. E. Burstall, personal communication; see also announcement of the proof in [11]). The case of the affine Lie groups associated with the Lie groups $\mathrm{SU}(n)$ is discussed in [12].

Acknowledgments

It is a pleasure to thank J.-L. Gervais and Yu. I. Manin for many fruitful discussions. One of us (M. S.) wishes to acknowledge the warm hospitality of the Instituto de Física Theórica, Universidade Estadual Paulista, São Paulo, Brazil, and the financial support from FAPESP during his stay there in March-July 1998. The research program of the authors was supported in part by the Russian Foundation for Basic Research under grant # 98–01–00015 and by INTAS grant # 96–690.

REFERENCES

[1] AN Leznov, MV Saveliev. Group-theoretical Methods for Integration of Nonlinear Dynamical Systems. Birkhauser: Basel, 1992.

[2] AV Razumov, MV Saveliev. Lie Algebras, Geometry, and Toda-type Systems. Cambridge University Press: Cambridge, 1997.

[3] AV Razumov, MV Saveliev. Maximally non-abelian Toda systems. Nucl Phys B494:657–686, 1997 (hep-th/9612081).

[4] AV Razumov, MV Saveliev. Multi-dimensional Toda-type systems. Theor Math Phys 112:999-1022, 1997 (hep-th/9609031).

[5] AV Razumov, MV Saveliev. Differential geometry of Toda systems. Commun Anal Geom 2:461–511, 1994 (hep-th/9311167).

[6] AB Givental', Plücker formulae and Cartan matrices. Russ Math Surv 44(3):193–194, 1989.

[7] LE Positsel'skii. Local Plücker formulas for a semisimple Lie group. Funct Anal Appl 25:291–292, 1991.

[8] P Griffiths, J Harris. Principles of Algebraic Geometry. New York: Interscience, 1978.

[9] JL Gervais, Y Matsuo. Classical A_n-W-geometry. Commun Math Phys 152:317–368, 1993 (hep-th/9203039).

[10] JL Gervais, MV Saveliev. W-geometry of the Toda systems associated with non-exceptional Lie algebras. Commun Math Phys 180:265–296, 1996 (hep-th/9312040).

[11] J Bolton, LM Woodward. Some geometrical aspects of the 2-dimensional Toda equations. In: Geometry, Topology and Physics, Proceedings of the First Brazil–USA Workshop. de Gruyter, 1997, pp 69–81.

[12] J Bolton, G Jensen, M Rigoli, LM Woodward. On conformal minimal immersions of S^2 into $\mathbb{C}P^n$. Math Ann 279:599–620, 1988.

Smooth quasigroups and loops. Recent achievements and open problems

L. V. SABININ Department of Mathematics, University of Quintana Roo, Chetumal, Quintana Roo, Mexico.

0. INTRODUCTION

During the last twenty six years in our work the foundations of Smooth quasigroups and Loops were established. Their notable relations with Geometry and Mathematical Physics were disclosed end explored (Affine connections, Riemannian Geometry, Symmetric and Reductive spaces, Special and General Relativity, Coherent states, Chiral anomalies, Poisson Mechanics and Geometry and so on).

The infinitesimal theory of smooth loops was developed as well as the Bol loops–Bol algebras theory, etc. See [1], [2], [3], [4].

The up-to-date presentation of the above theory and of its geometric applications will appear in our monograph Smooth Quasigroups and Loops" (Kluwer Academic Publishers) in this year. It contains a comprehensive Bibliography on the matter.

This promising area of research is now in the state of continuous growth and, to our mind, in the near future will replace the Lie group theory being its direct and natural generalization.

Due to the frames of our lecture we concern the smooth loops and quasigroups rather then their applications.

1. SMOOTH M-LOOPS

1.1.

DEFINITION. A loop $\langle Q, \cdot, \backslash, /, \varepsilon\rangle$ satisfying the property

$$\left.\begin{aligned} L_x \circ L_y \circ L_{\varphi x}^{-1} = L_{x\cdot(y\cdot(\varphi x)^{-1})}, \\ L_x z = x\cdot z,\ \varphi \text{ being a map } Q \xrightarrow{\varphi} Q,\ \varphi\varepsilon = \varepsilon, \end{aligned}\right\} \tag{1.1}$$

is called a left φM-loop.

We consider a smooth left φM-loop $\langle Q, \cdot, \backslash, /, \varepsilon\rangle$ locally, near ε. Then, in a fixed chart near ε, we get (1.1) in a coordinate representation:

$$\left.\begin{aligned} [x\cdot(y\cdot(\varphi x\backslash z))]^\beta = [\{x\cdot(y\cdot I\varphi x)\}\cdot z]^\beta, \\ Iw = w^{-1},\ \beta = 1,\dots n = \dim Q. \end{aligned}\right\} \tag{1.2}$$

Differentiating it by x^γ at $x = \varepsilon$ we get the *differential equation of a smooth left φM-loop*:

$$\left.\begin{aligned} a_\gamma^\beta(y\cdot z) = \varphi_\gamma^\nu a_\nu^\sigma(z)\frac{\partial(y\cdot z)^\beta}{\partial z^\sigma} + [-\varphi_\gamma^\tau A_\tau^\sigma(y) + a_\gamma^\sigma(y)]\frac{\partial(y\cdot z)^\beta}{\partial y^\sigma}, \\ \text{where } a_\gamma^\beta(z) = \frac{\partial(x\cdot z)^\beta}{\partial x^\gamma}|_{x=\varepsilon},\quad A_\gamma^\beta(z) = \frac{\partial(z\cdot x)^\beta}{\partial x^\gamma}|_{x=\varepsilon}. \end{aligned}\right\} \tag{1.3}$$

1.2.

REMARK. Note that $A_\gamma(z) = A_\gamma^\beta(z)\,\partial/\partial z^\beta$ and $a_\gamma(z) = a_\gamma^\beta(z)\,\partial/\partial z^\beta$ constitute so called the *left* and *right basic fundamentall vector fields.*

As a result we get the following proposition:

1.3.

PROPOSITION. Any smooth local left φM-loop satisfies the differential equation (1.3).

1.4.

REMARK. As it is known, any smooth system $a_\mu(x)$ $(a_\mu(\varepsilon) = (\partial_\mu)_\varepsilon)$ of vector fields uniquely defines a smooth left monoalternative loop $\langle Q, *, \varepsilon\rangle$ such that a_μ are its right basic fundamental vector fields.

The left monoalternativity means that a loop $\langle Q, *, \varepsilon\rangle$ equipped with the canonical unary operations

$$tc = \operatorname{Exp} t \operatorname{Exp}^{-1} c \quad (t \in \mathbb{R}),$$

where

$$\frac{d(\operatorname{Exp} t\zeta)^\lambda}{dt} = a_\mu^\lambda(\operatorname{Exp} t\zeta)\zeta^\mu, \quad \operatorname{Exp}(0\zeta) = \varepsilon,$$

satisfies the property

$$ta * (ua * b) = (t + u)a * b \quad (t, u \in \mathbb{R}). \tag{1.4}$$

Note, also, that in the case

$$\frac{d(tx * y)^\alpha}{dt} = a^\alpha_\beta(tx * y)(\mathrm{Exp}^{-1} x)^\beta, \quad 0x * y = y. \tag{1.5}$$

1.5.

DEFINITION. We say that a (smooth) loop $\langle Q, \cdot, \varepsilon\rangle$ is a left νPL-loop, if there exists a smooth left monoalternative loop $\langle Q, *, \varepsilon\rangle$ such that

$$\left.\begin{array}{c} P_x \circ L_y \circ P^{-1}_{\nu x} = L_z, \quad P_x y = x * y \quad L_x y = x \cdot y, \\ \nu \colon Q \to Q. \end{array}\right\} \tag{1.6}$$

The following proposition is valid.

1.6.

PROPOSITION. A local smooth loop $\langle Q, \cdot, \varepsilon\rangle$ satisfies the differential equation (1.3) if and only if it is a smooth left $\bar{\varphi}PL$-loop ($\bar{\varphi} = \mathrm{Exp} \circ \lambda \circ \mathrm{Exp}^{-1}$, $\lambda^\alpha_\beta = \varphi^\alpha_\beta$).

1.7.

DEFINITION. A left νPL-loop is said to be singular if $\nu_{*,\varepsilon} - \mathrm{Id}$ is not invertible, and non-singular otherwise.

One may prove the proposition.

1.8.

PROPOSITION. (a) Any smooth left φM-loop is a left $\bar{\varphi}PL$-loop ($\bar{\varphi} = \mathrm{Exp} \circ \varphi_{*,\varepsilon} \circ \mathrm{Exp}^{-1}$).

(b) Any non-singular left νPL-loop is a left ψM-loop, where $\psi = \tau \circ \nu \circ \tau^{-1}, \tau b = b * (\varphi b)^{-1}$, $c * c^{-1} = \varepsilon$, $\varphi = \mathrm{Exp} \circ \nu_{*,\varepsilon} \circ \mathrm{Exp}^{-1}$).

1.9.

EXAMPLES: (a) A smooth local left Bol loop ($L_x \circ L_y \circ L_x = L_{x \cdot (y \cdot x)}$) is a left φM-loop ($\varphi x = x^{-1}$) and a non-singular left φPL-loop.

Here $P_x = L_x$, since any smooth left Bol loop is left monoalternative.

(b) A smooth local loop with the property $L_x \circ L_y \circ L_x^{-1} = L_{x \cdot (y \cdot x^{-1})}$ is a left φM-loop ($\varphi = \mathrm{id}$) and a left singular φPL-loop.

1.10.

REMARK. Non-singular smooth left M-loops (PL-loops) have the direct concern to Differential Geometry (Transsymmetric spaces). Moreover, any transsymmetric space may be considered as a non-singular smooth left φM-loop with the $I\varphi$-Bruck Identity

$$\left.\begin{array}{c}[(x \cdot y) \cdot I\varphi(x \cdot y)] = x \cdot [(y \cdot I\varphi y) \cdot I\varphi x] \\ (Ix = x^{-1}).\end{array}\right\} \tag{1.7}$$

1.11

Thus it is worth to study local left PL-loops, since local left M-loops are among them.

Precisely speaking we need an infinitesimal theory of local left PL-loops. It may be achieved by the examination of the integrability conditions for (1.3). We have to proceed in this way until we get a completely integrable system.

Final relations appear to be some generalization of the Bol loop case. At this stage we omit those in our presentation since the matter needs further careful consideration.

Instead, we give a different approach using some consequences of (1.6). Let us consider

$$P(a,b) = P_{a*b}^{-1} \circ P_a \circ P_b. \tag{1.8}$$

After calculations based on (1.6), (1.8) we get

$$\left.\begin{array}{c}P(a,b) \circ L_x \circ (L^{-1}_{\tau_{I(a*b)}\tau_a\tau_b\varepsilon} \circ P(a,b))^{-1} = \\ L_{\tau_{I(a*b)}\tau_a\tau_b x}, \quad \tau_c z = c * (z * I\bar{\varphi}c). \quad Ic = c^{-1}.\end{array}\right\} \tag{1.9}$$

This identity is of the pseudo-special type (see [1], [2]) and differentiating it by a and b at ε one may obtain the relation of the type

$$\left.\begin{array}{c}[a_\lambda, [a_\mu, a_\nu]] = [a_\lambda, a_\sigma]a^\sigma_{\mu\nu} + \rho^\sigma_{\lambda,\mu\nu}a_\sigma, \\ a^{\cdot}_{\cdot\,\cdot} = const, \quad \rho^{\cdot}_{\cdot,\cdot\,\cdot} = const.\end{array}\right\} \tag{1.10}$$

Moreover, $a^\sigma_{\mu\nu}$ is expressible by φ^λ_μ, $c^\sigma_{\mu\nu} = [a_\mu, a_\nu]^\sigma(\varepsilon)$.

The relation (1.10) shows that we deal with a finite-dimensional Lie algebra $\mathfrak{g} = E + [E,E]$, $E = \{q^\lambda a_\lambda\}$ and we may use further the methods of Lie algebra theory.

1.12

In the light of the above said, it is clear, how one may introduce a proper infinitesimal object, which is an unary-binary-ternary algebra on $T_\varepsilon Q$, in the same way as for smooth pseudo-reductive loops [1], [2].

1.13. Conclusions

The above given has outlined the theory of smooth local PL-loops. We need only certain routine computations to finish that.

It will solve the problem for local non-singular left M-loops. But the problem of smooth singular M-loops is still open. The main question is to know (in infinitesimal terms) when a left singular PL-loop is a M-loop.

2. BINARY BOL LOOPS

2.1

It is quite natural to try to develop the infinitesimal theory of smooth left binary-Bol loops.

2.2.

DEFINITION. A smooth loop $\langle Q, \cdot, \varepsilon\rangle$ is said to be left binary-Bol, if any two its elements generate a left Bol subloop.

We remind that a loop $\langle P, \cdot, \varepsilon\rangle$ is a left Bol loop if $L_x \circ L_y \circ L_x = L_{x\cdot(y\cdot x)}$ $(L_x y = x \cdot y)$.

Since the expansion of the law of composition into series in normal coordinates depends only on two elements, it is obvious that, like in the Bol case, the proper infinitesimal object will be a binary-ternary multilinear algebra. The problem here is to get its characteristic identities, generalizing the Bol algebra identities. And, further, one may try to develop the structural theory of binary-Bol algebras.

3. SMOOTH LOOPS ACTIONS

3.1. DEFINITION. Let M be a smooth manifold and $\{\langle Q, \overset{x}{\cdot}, \varepsilon\rangle\}_{x\in M}$ be a system of local loops.

If $b \mapsto f_b$ $(Q \to \operatorname{Diff} M)$ is defined for b near ε and

$$xf_a f_b = xf_{(a\overset{x}{\cdot}b)} \text{ (near } \varepsilon), \quad xf_\varepsilon = x, \tag{3.1}$$

then we say that a *smooth right action* $Q \xrightarrow{f} \operatorname{Diff} M$ is given.

Analogously one may define a *smooth left action.*

3.2.

EXAMPLE. The right natural action R induced by a loop $\langle Q, \cdot, \varepsilon\rangle$, $a \overset{R}{\mapsto} R_a$ $(x \cdot a = xR_a)$.

Evidently,

$$xR_a R_b = xR_{(a\overset{x}{\cdot}b)}, \quad a \overset{x}{\cdot} b = L_x^{-1} L_{x\cdot a} b. \tag{3.2}$$

3.3

Further we use the notations $xf_a = xa$. Then (3.1) takes the form

$$(xa)b = x(a \overset{x}{\cdot} b) \tag{3.3}$$

3.4.

REMARK. If $a \overset{x}{\cdot} b = a \cdot b$ (does not depend on x) then $(xa)b = x(a \cdot b)$ and for a faithful action $a \overset{f}{\mapsto} f_a$ we get the group action.

3.5.

The above construction is rather flexible, since, for the same $a \overset{f}{\mapsto} f_a$ there may exist a number of loop systems $\{\langle Q, \overset{x}{\cdot}, \varepsilon\rangle\}_{x \in M}$. It motivates a more rigid construction.

Due to (3.3) we get $[(xa)b]c = [x(a \overset{x}{\cdot} b)]c$ and, after routine calculations,

$$x[(a \overset{x}{\cdot} b) \overset{x}{\cdot} c] = x[a \overset{x}{\cdot} (b \overset{xa}{\cdot} c)]. \tag{3.4}$$

It does not mean that

$$(a \overset{x}{\cdot} b) \overset{x}{\cdot} c = a \overset{x}{\cdot} (b \overset{xa}{\cdot} c), \tag{3.5}$$

but we may assume (3.5) as an additional requirement coming to the notion of a *strict action*.

3.6. The differential equations of a strict action

In a suitable chart, differentiating (3.3) and (3.5) by c at $c = \varepsilon$ we get

$$\left.\begin{aligned} \overset{x}{A}{}^{\beta}_{\alpha}(a)\frac{\partial (xa)^i}{\partial a^\beta} &= L^i_\alpha(xa), \quad x\varepsilon = x, \\ \Big[\frac{\partial (b \overset{x}{\cdot} a)^\alpha}{\partial a^\nu}\Big]\overset{xb}{A}{}^{\nu}_{\beta}(a) &= \overset{x}{A}{}^{\alpha}_{\beta}(b \overset{x}{\cdot} a), \quad b \overset{x}{\cdot} \varepsilon = b, \end{aligned}\right\} \tag{3.6}$$

where $\alpha, \beta, \nu = 1, \ldots, m = \dim Q, \quad i = 1, \ldots, n = \dim M$,

$$L^i_\alpha(y) = \frac{\partial (y \cdot c)^i}{\partial c^\alpha}\Big|_{c=\varepsilon}$$

(the fundamental basic fields of action),

$$\overset{x}{A}{}^{\beta}_{\alpha}(a) = \frac{\partial (a \overset{x}{\cdot} b)^\beta}{\partial b^\alpha}\Big|_{b=\varepsilon}$$

(the left fundamental basic fields of the loop $a \overset{x}{\cdot} b$).

3.7

Involving into consideration the integrability conditions for (3.6) we get

$$\left.\begin{aligned} [L_\sigma, L_\alpha]^i(x) &= c^\mu_{\sigma\alpha}(x)L^i_\mu(x), \\ \frac{\partial \overset{x}{B}{}^{\lambda}_{\mu}(a)}{\partial a^\nu} - \frac{\partial \overset{x}{B}{}^{\lambda}_{\nu}(a)}{\partial a^\mu} &= -c^\lambda_{\sigma\tau}(xa)\overset{x}{B}{}^{\sigma}_{\nu}(a)\overset{x}{B}{}^{\tau}_{\mu}(a), \end{aligned}\right\} \tag{3.7}$$

where $\overset{x}{B}{}^{\lambda}_{\sigma}$ is the inverse matrix to $\overset{x}{A}{}^{\mu}_{\nu}$.

The system (3.6), (3.7) is closed, which means that its integrability conditions are consequences of the system itself.

In the contrast with the Lie group theory we can not regard here the structure functions $c^{\lambda}_{\mu\nu}(x)$ as known, since, due to the Jacobi identities, there are some nontrivial relations of non-algebraic nature between $L^i_\alpha(x)$ and $c^{\lambda}_{\mu\nu}(x)$.

But we may regard $L^i_\alpha(x)$ and $c^{\lambda}_{\mu\nu}(x)$ as known and satisfying the nonlinear bracket relations

$$[L_\alpha, L_\beta](x) = c^{\lambda}_{\alpha\beta}(x) L_\lambda(x), \tag{3.8}$$

(see the first line of (3.7)). Then one may regard (3.7), (3.8) as a completely integrable system with unknowns

$$\varphi^\alpha(a) = (b \overset{x}{\cdot} a)^\alpha, \quad \overset{x}{A}{}^{\lambda}_{\mu}(a), \quad \psi^i(a) = (xa)^i.$$

It means that by nonlinear Lie brackets system one may (locally) restore the strict action.

The above said is only one fragment of the complete theory (similar to the restoration of a Lie group by its algebra Lie of vector fields).

Further we need some analogue of the third Lie theorem. That is not done yet (because nobody suggested the proper infinitesimal object).

3.8.

REMARK. In our opinion it is evident that the loop action theory should replace such a monstrous construction as differentiable groupoid.

3.9. The described above is only one possible approach to the loop action theory. Indeed, in the Bol loops theory it is quite natural to consider so called Nono systems ($((xa)b)a = x[(a \cdot b) \cdot a]$) as Bol actions, since for a right Bol loop $R_a \circ R_b \circ R_a = R_{(a \cdot b) \cdot a}$.

The theory of Bol action may be easily constructed in an analogy with the Bol loops theory.

After that one may generalize the situation and consider the families $\{f_a : x \to xf_a\}_{a \in Q,\, x \in M}$ of bijections with the property

$$\left.\begin{aligned} xf_af_bf_a &= xf_{F(x;a,b)}, \\ f_\varepsilon = \mathrm{id}, \quad F(x,\varepsilon,b) &= b. \end{aligned}\right\} \tag{3.10}$$

The infinitesimal theory for such a generalized Bol actions has not been considered yet.

4. THE STRUCTURE THEORY OF BOL ALGEBRAS

It has not been elaborated till now. We know that Prof. E.N. Kuz'min with collaborators (Novosibirsk, Russia) actively works in this area.

5. THE GENERALIZED GEOMETRIC ALGEBRA

One may observe that the parallelogram rule of an affine plane A in the conventional geometric algebra determines the system of loops $\{\langle A, \underset{a}{\cdot}, a\rangle\}_{a\in A}$ (defined not everywhere). Certain remarkable identities connecting these loops may be got.

Thus it is of value to consider so called loopuscular spaces, the spaces with the families of loops with some specific identities.

This matter has the direct concern to the Foundations of Geometry.

In particular, in this way the algebraic foundations of affinely connected and Riemannian spaces have been established [1], [2], [3].

It still needs a careful axiomatic consideration in analogy with the conventional geometric algebra.

In particular, the question is, how one may generalize Desargues axioms in order to get an affinely connected space.

REFERENCES

Journal article

[1] LV Sabinin. Quasigroups, Geometry and Nonlinear Geometric Algebra. Acta Applicandae Mathematicae 50: 45–66, 1998.

Book

[2] LV Sabinin. Smooth Quasigroups and Loops. Kluwer academic publishers, 1999, 223 pp.

Book chapter

[3] LV Sabinin, PO Miheev. Quasigroups and Differential Geometry In: O Chein, H Pflugfelder and JDH Smith, eds. Quasigroups and Loops: Theory and Applications. Berlin: Heldermann Verlag, 1990, pp 357–430.

Book

[4] LV Sabinin Analytic Quasigroups and Geometry. Moscow: Friendship of Nations University press, 1991, 112 pp.

On smooth left square distributive quasigroup

LIUDMILA SABININA Facultad de Ciencias, UAEM. Av. Universidad 1001, 62210 Cuernavaca, Morelos, México. e-mail: liudmila@ servm.fc.uaem.mx

ABSTRACT The aim of this paper is to describe geometric properties of smooth quasigroups defined by the Left Square Distributive Identity: $x^2(yz) = (xy)(xz)$

The following definitions of left quasigroups are equivalent and we will use each of them through the paper.

1. DEFINITION. *Let* $< Q, \cdot >$ *be a set with a binary operation such that* $a \cdot x = b$, $\forall\ a, b \in Q$, *is uniquelly solvable. (In this case* x *it is denoted by* $a \backslash b$*). Then* $< Q, \cdot >$ *is called a left quasigroup.*

2. DEFINITION. *The set* $< Q, \cdot, \backslash >$ *with two binary operation such that* $a \cdot (a \backslash b) = b$ *and* $a \backslash (a \cdot b) = b$, $\forall\ a, b \in Q$, *is called a left quasigroup.*

One can note that $< Q, \backslash, \cdot >$ is also a left quasigroup. This quasigroup is called a *parastrophe* of $< Q, \cdot, \backslash >$.

This research was supported by a PROMEP SEP Grant

3. DEFINITION. *Let $< Q, \cdot >$ be a set with a binary operation such that any left translation L_a, $\forall a \in Q$, is a bijection. ($L_a b = a \cdot b$). Then $< Q, \cdot >$ is called a left quasigroup.*

Since the left translations are bijections, one can define the group of left translations as a group generated by the set of all translations $L(Q) =< \{L_a\}_{a \in Q} >$.

Analogously, one can define right quasigroups $< K, \cdot, / >$ and a group of right translation $R(K)$ ($R_a b = b \cdot a$, $a, b \in K$).

4. DEFINITION. *Let X be a G-space. We say that X is a homogeneous space if the action on X is transitive.*

It is known that any homogeneous space can be identified with the set of left cosets G/H, where H is the stability subgroup of some point of X.

It is easy to see that the action of $L(Q)$ on the both- sided quasigroup $< Q, \cdot >$ is transitive, so $< Q, \cdot >$ is a homogeneous space.

Now we will construct a left quasigroup with the left square distributive identity

(1) $\quad x^2 \cdot (y \cdot z) = (x \cdot y) \cdot (x \cdot z)$.

Let G be a group and $\varphi: G \to G$ be its endomorphism. We consider the subgroup H of G given by $H = \{h \in G : \varphi(h) = h\}$.

We define an operation on the set of left cosets G/H by the formula $\overline{x} \circ \overline{y} = \varphi(x) \cdot x^{-1} \cdot yH$, where $\overline{x} = xH$.

It is easy to see that $< G/H, \circ >$ is a left quasigroup. Let us check up that this quasigroup has the left square distributive identity.

Indeed,

$\overline{x} \circ \overline{x} = \varphi(x)H$,

$(\overline{x} \circ \overline{x}) \circ (\overline{y} \circ \overline{z}) = \varphi(\varphi(x))\varphi(x)^{-1}\varphi(y)y^{-1}zH = \varphi^2(x)\varphi(x)^{-1}\varphi(y)y^{-1}zH$,

$(\overline{x} \circ \overline{y}) \circ (\overline{x} \circ \overline{z}) = \varphi(x)x^{-1}yH \circ \varphi(x)x^{-1}zH = \varphi[\varphi(x)x^{-1}y]y^{-1}x\varphi(x)^{-1}\varphi(x)x^{-1}zH = \varphi^2(x)\varphi(x)^{-1}\varphi(y)y^{-1}zH$.

Comparing the last two equalities we get the left square distributive identity.

Let note that we have an endomorphism of $< G/H, \circ >$, $\psi: G/H \to G/H$, given by the formula $\psi(\overline{x}) = \overline{x}^2 = \overline{\varphi(x)}$.

5. DEFINITION. *Let M be a C^k-manifold ($k \geq 2$) and $\{\sigma_x\}_{x \in M}$ be a family of local diffeomorphisms $\sigma_x: M \to M$. We consider smooth maps $e: M \to M$ given by $e(x) = \sigma_x^{-1}x$ and $\rho_y: M \to M$ given by $\rho_y x = \sigma_x y$. Then $(M, \{\sigma_x\}_{x \in M})$ is called a transsymmetric space if the following conditions hold:*

i) $\sigma_x \circ \sigma_y = \sigma_{\sigma_x y} \circ \sigma_{e(x)}$

ii) $\rho_{e(x)}$ *is a local diffeomorphism.*

Now we formulate the main result on transsymmetric spaces [1].

Let $(M, \{\sigma_x\}_{x \in M}$ be a transsymmetric space. Then there exists an affine connection ∇ on M, such that $\{\sigma_x\}$ are its local diffeomorphisms i.e., $(\sigma_x)_\star \nabla_Y Z = \nabla_{(\sigma_x)_\star} Y (\sigma_x)_\star Z$ and $\nabla T = 0$, $\nabla R = 0$ where T is the torsion and R is the curvature. This affine connection can be expressed by the formula

$(\nabla_X Y)_b = \frac{d}{dt}\{[(P^b_{x(t)})_{\star,b}]^{-1} Y_{x(t)}\}|_t = 0$,

where $x(0) = b$, $\dot{x}(0) = X_b$ and $P^b_x = \sigma_{\rho^{-1}_{e(b)} x} \circ \sigma_x^{-1}$.

6. PROPOSITION. *A smooth quasigroup* $< Q, \cdot, \backslash, / >$ *with identity* $x^2(yz) = xy \cdot xz$ *is a transsymmetric space with the family of transsymmetries* $\{L_a^{-1}\}_{a \in Q}$.

Proof. Identity (1) is equivalent to $L_x^{-1} \circ L_y^{-1} = L_{L_x^{-1}y}^{-1} \circ L_{x^2}^{-1}$, where $x^2 = (L_x^{-1})^{-1}x$. This is *i*) from the definition 5 of transsymmetric space. Since Q is a two- sided quasigroup that implies *ii*) from definition 5.

So the smooth quasigroup $< Q, \cdot, \backslash, / >$ with (1) is the manifold with the affine connection ∇ such that its tensors of torsion and curvature are covariantly constant and the left translations are automorphisms of ∇. □

7. Let us consider some particular cases:

Suppose Q is a smooth quasigroup with (1).

1. Q is a loop, i.e., there exists an element $e \in Q$, such that $e \cdot x = x \cdot e = x$.

In this case Q is an additive group of vector space. The matter is that the loop with (1) is a commutative Moufang loop [5]. Due to the Malcev-Kuzmin theory on Moufang loops and Malcev algebras any smooth commutative Moufang loop turns to be an abelian groups [4], [6].

2. Q is a TS-quasigroup. Then in Q one can construct commutative Moufang loop using the operations of quasigroup [2].
(TS-quasigroup is a quasigroup with the condition : $a \cdot b = c$ implies $b \cdot a = c$, $c \cdot a = b$, $b \cdot c = a$, $a \cdot c = b$, $c \cdot b = a$, $a, b, c \in Q$).

3. Q is a quasigroup with the identity $(yz)x^2 = (yx)(zx)$. Then Q is an isotope of commutative Moufang loop [7]. (A loop $< Q, \circ >$ is an isotope of a quasigroup $< Q, \cdot >$ if $x \circ y = R_d^{-1}x \cdot L_c^{-1}y$, for some $c, d \in Q$, and for all $x, y \in Q$, where L_c, R_d are left and right translations of $< Q, \cdot >$, respectively.)

It means, that in cases 2 and 3 there exists on Q the structure of abelian group such that the operation of this group is expressed by operations of quasigroups.

4. Q is a quasigroup with idempotent identity $x^2 = x$.
In this case Q is a left distributive quasigroup.
Smooth left distributive quasigroup is well known s-space [7]

5. There exists left neutral element ε in Q, i.e., $\varepsilon x = x$, $\forall x \in Q$ and $x^2 = \varepsilon$.
Then Q is a parastrophe of group Lie.

6. The following conditions hold:
a). Left Inverse Property (LIP), i.e., there exists a smooth map $p: Q \to Q$ such that $p(a) \cdot (a \cdot b) = b$, $\forall a, b \in Q$
b). There exists a smooth map $q: Q \to Q$ such that $R_a^{-1} = L_a^{-1} \circ q$, $\forall a \in Q$.
Then Q is an isotope of parastrophe of Moufang loop [3].

For smooth quasigroup Q with (1) the infinitesimal object is binary-ternary algebra, for cases 5 and 6 it is reduced to binary algebra.

On the above matters see also [4], [5], [6], [7].

REFERENCES

[1] L. V. Sabinin and L. L. Sabinina, On geometry of transsymetric spaces, Webs and Quasigroups (1991), 117-122.

[2] Yu.I. Manin, Cubic Forms: Algebra, Geometry, Arithmetic, North-Holland, 1986.

[3] I.A. Florja, M.I. Ursul, F-quasigroups with the inverse property, Questions of the theory of quasigroups and loops (1971). (Russian)

[4] E.N. Kuzmin, The connection between Mal'cev Algebras and analytic Moufang loops, Algebra and Logic 10 (1971), no. 1, 1-14.

[5] R.H. Bruck, A survey of Binary Systems, Springer-Verlag, Berlin, 1971.

[6] R. Jimenez, L. Sabinin and L. Sabinina, On C^1-smooth commutative Moufang Loops and Distributive Quasigroups, Webs and Quasigroups (1996-1997), 86-88.

[7] O. Chein, H. Pflugfelder, J.D.H. Smith (eds), Quasigroups and Loops: Theory and Applications, Heldermann Verlag, Berlin, 1990.

Algebraic structure of transsymmetric spaces

L. V. SBITNEVA Department of Mathematics, University of Quintana Roo, Chetumal, Quintana Roo, Mexico.

INTRODUCTION

Transsymmetric spaces have been intoduced by L. V. Sabinin as a far going generalization of Symmetric spaces and studied by L. L.Sabinina in Ph. D. Dissertation [1]. Despite of the geometric nature of the initial approach it has been clear that the whole theory may be reformulated in the terms of Smooth quasigroups and loops. In our presentation we are going to outline this algebraic approach.

0.

For the reader's convenience we give, first, the basic definitions.

DEFINITION. An algebra $\langle Q, \cdot, \backslash \rangle$ with two binary operations $(\cdot)$ and $(\backslash)$ (multiplication and left division) and with the identites

$$x \cdot (x \backslash y) = y, \quad x \backslash (x \cdot y) = y$$

is called a *left quasigroup*.

A *right quasigroup* is defined analogously.

DEFINITION. An algebra $\langle Q, \cdot, \backslash, / \rangle$ with three binary operations $(\cdot)$, $(\backslash)$ $(/)$ (multiplication, left and right divisions) and with the identites

$$x \cdot (x \backslash y) = y, \quad x \backslash (x \cdot y) = y,$$
$$(x \cdot y)/y = x, \quad (x/y) \cdot y = x$$

is called a (*two-sided*) *quasigroup*.

DEFINITION. A left quasigroup with a right neutral ε ($x \cdot \varepsilon = x$) is called a *left loop*.

Analogously a *right loop* is defined.

DEFINITION. A two sided quasigroup with neutral element ε ($\varepsilon \cdot x = x \cdot \varepsilon = x$) is called a (*two-sided*) *loop*.

As to the general theory of quasigroups, see [7]–[10].

1.

Algebraically, we deal with a *smooth left F-quasigroup* (local or global), that is, with a smooth left quasigroup $\langle Q, *, \backslash\backslash \rangle$ satisfying the identity

$$x * (y * z) = (x * y) * (Fx * z). \tag{1}$$

It is easily verified (at $y = x\backslash\backslash x$) that one may put $Fx = x\backslash\backslash x$, what is supposed further.

One trivial example demonstrates the significance of the above definition: any group with respect to the left division as the main operation is a left F-quasigroup.

Note also that the Euclidean plane Π equipped with the operation $x * y = s_x y$, where s_x is the central symmetry with respect to $x \in \Pi$, is a left quasigroup.

Let us introduce the notations

$$\sigma_x y = x * y = \rho_y x. \tag{2}$$

2.

DEFINITION. We say that a left F-quasigroup is correct if there exists ρ_{ey}^{-1}.

This is less then the existence of the right division //.

3.

REMARK. A smooth left correct F-quasigroup is nothing but a transsymmetric space.

The traditional differential-geometric definition is: *a manifold M together with a system of local diffeomorphisms (transsymmetries) $(\sigma_x)_{x\in M}$ is called a transsymmetric space if $\sigma_x \circ \sigma_y = \sigma_{\sigma_x y} \circ \sigma_{Fx}$ ($F\colon x \mapsto Fx \in M$) and $\sigma_x Fa = b$ is a locally uniquelly solvable with respect to $x \in M$ near $a \in M$.*

Note that the theory of smooth left F-quasigroups without correctness has not been developed yet despite of the obvious interest.

4.

The standard way to deal with a quasigroup is, due to L.V. Sabinin, to consider the system of its isotopic loops, so called a loopuscular structure [2].

We proceed along this line.

Let $\langle Q, *, \backslash\backslash \rangle$ be a left correct F-quasigroup. One may introduce

$$L(x, a, y) = x \underset{a}{\cdot} y = \rho_{Fa}^{-1} x * \sigma_a^{-1} y \tag{3}$$

and easily verify that $\langle Q, \underset{a}{\cdot}, a\rangle$ is a loop.

Further one may treat the family $(\langle Q, \underset{c}{\cdot}, c\rangle)_{c\in Q}$ of loops as a set Q with one ternary operation, $\langle Q, L\rangle$ (*canonical loopuscular algebra*), where $L(x,c,y) = x \underset{c}{\cdot} y$.

5.

Due to the left F-identity, our canonical loopuscular algebra possesses some remarkable properties. Moreover, the canonical loopuscular algebra allows us to restore the initial left F-quasigroup.

6.

PROPOSITION. The canonical loopuscular algebra of a left correct F-quasigroup $\langle Q, *, \backslash\backslash\rangle$ is reductive, that is

$$\begin{gathered} L_c^b L(x,a,y) = L(L_c^b x, L_c^b a, L_c^b y), \\ \text{where } L(x,a,y) = L_x^a y = \rho_{Fa}^{-1} x * \sigma_a^{-1} y. \end{gathered} \tag{4}$$

The proposition means, in particular, that Q is a homogeneous space with the group action generated by $\{L_c^b\}$.

7. GEOMETRIC COMMENT

In the smooth case one may consider $\langle Q, L\rangle$ introduced above as the reductive affinely connected space with the connection

$$\left.\begin{gathered} (\nabla_X Y)_a = \frac{d}{dt}[(L^a_{\gamma(t)})^{-1}_{*,a} Y_{\gamma(t)}]_{t=0}, \\ \gamma(0) = a, \quad \dot\gamma(0) = X_a, \end{gathered}\right\} \tag{5}$$

and prolong the investigation in the purely differential-geometric way.

8. CHARACTERISTIC IDENTITIES

Since all loops $\langle Q, \underset{c}{\cdot}, c\rangle$ are isomorphic (due to reductivity) we consider one of them, $\langle Q, \underset{\varepsilon}{\cdot}, \varepsilon\rangle$, and put $(\cdot)$ instead of $(\underset{\varepsilon}{\cdot})$. Due to the left F-identity we get:

$$\begin{gathered} L_b \circ L_x \circ L_{\varphi b}^{-1} = L_{b\cdot(x\cdot(\varphi b)^{-1})} \\ \text{(Left } \varphi M\text{-identity)}, \end{gathered} \tag{6}$$

where $\varphi \in \mathrm{End}\langle Q, \cdot, \varepsilon\rangle$, $\quad x \overset{f}{\mapsto} x \cdot (\varphi x)^{-1}$ is bijective (correctness), and

$$\begin{gathered} (x\cdot y)\cdot I\varphi(x\cdot y) = x\cdot[y\cdot(I\varphi y \cdot I\varphi x)], \quad Ix = x^{-1}) \\ \text{(Left } I\varphi\text{-Bruck identity)}. \end{gathered} \tag{7}$$

Note that any loop satisfying the left φM-identity is called a *left φM-loop*, see [8],[9]

9.

PROPOSITION. Any correct left F-quasigroup is isotopic to a correct left φM-loop, φ being its endomorphism, with the left $I\varphi$-Bruck identity. And vice versa.

(The above isotopism may be described in the explicit form.)

We call such a loop a *left correct φts-loop.*

10.

As a result, the theory of local left correct F-quasigroups is reduced to the theory of local left correct φts-loops.

The proper infinitesimal theory of left correct φts-loops has been established in the frames of Differential Geometry [3], [4] in the terms of curvature and torsion.

The main point here is based on the reductivity (Proposition 6) which implies the reduictivity of affine connection (5). Its Bianci identities give us, at a fixed point, so called a triple Lie algebra [2].

This algebra, together with some unary operation, generated by endomorphism φ, and with some identities, constitute the proper infinitesimal object. Note, that this construction has not been yet, properly published.

11.

The above approach is based on Differential Geometry and it is desirable to replace it by the direct consideration in the frames of the Smooth quasogroup theory.

For this one may use the results of the smooth φM-loop theory (see the lecture of Dr. L. V. Sabinin at this conference).

The construction is based on the characteristic differential equation of so called a left νPL-loop and on the coincidence of nonsingular νPL-loops and correct left φM-loops. In this way one has to use that φ is an endomorphism, it implies the left A-property. Finally one needs the well established infinitesimal theory of smooth left A-loops.

At present this approach is under our consideration.

12. ALGEBRAIC STRUCTURE OF PERFECT TRANSSYMMETRIC SPACES

Symmetric spaces possess some geometric properties which are not valid for the general transsymmetric spaces. This has motivated the notion of Perfect transsymmetric space. We are going, first, to present the above notion in purely algebraic terms. For this we need some preliminaries.

Let $\langle Q, \cdot, \varepsilon\rangle$ be a smooth loop and a_μ ($\mu = 1, \ldots, \dim Q$) be its right basic fundamental vector fields, $a^\alpha_\mu(y) = [\partial(x \cdot y)^\alpha / \partial x^\mu]_{x=\varepsilon}$. Then one may introduce the exponential map $\operatorname{Exp}\colon T_\varepsilon Q \to Q$ (at least locally) solving the differential equation

$$\frac{d(\operatorname{Exp} t\zeta)^\alpha}{dt} = a^\alpha_\nu(\operatorname{Exp} t\zeta)\zeta^\nu, \quad \operatorname{Exp} 0\zeta = \varepsilon, \tag{8}$$

and (locally) introduce the canonical operations $x \mapsto tx$ ($t \in \mathbb{R}$) by

$$tx = \operatorname{Exp}(t \operatorname{Exp} x)^{-1}. \tag{9}$$

The system $\langle Q, \cdot, \varepsilon, (t)_{(t \in \mathbb{R})} \rangle$ (the loop $\langle Q, \cdot, \varepsilon \rangle$ with its canonical operations) constitute so called a *canonical preodule*.

Thus, in the smooth case, instead of a loop one may consider its canonical preodule.

13.

DEFINITION. A smooth $\langle Q, \cdot, \varepsilon \rangle$ (as well as its canonical preodule) is said to be left monoalternative if

$$tx \cdot (ux \cdot y) = (t + u)x \cdot y, \quad t, u \in \mathbb{R}, \quad x, y \in Q. \tag{10}$$

14.

DEFINITION. A left monoalternative correct left φts-loop is called perfect.

15.

REMARK. Note that a transsymmetric space is perfect if and only if its correct left φts-loop is left monoalternative (that is perfect).

16.

PROPOSITION. 1. Any smooth local perfect φts-loop is a left Bol loop (that is $L_x \circ L_y \circ L_x = L_{x \cdot (y \cdot x)}$) and a left A-loop (that is $l(a, b) \circ L_x \circ [l(a, b)]^{-1} = L_{l(a,b)x}$, where $l(a, b) = L^{-1}_{(a \cdot b)} \circ L_a \circ L_b$).

2. For any perfect φts-loop $(\varphi x)^{-1} \cdot x \in N_l = \{w; w \cdot (a \cdot b) = (w \cdot a) \cdot b, \forall a, b\}$–the left nucleus.

Note that the left nucleus of a left Bol loop is its normal subloop (and a group).

17.

PROPOSITION. Let $\mathcal{Q} = \langle Q, \cdot, \varepsilon \rangle$ be a smooth perfect φts-loop , $\mathcal{N}_l = \langle N_l, \cdot, \varepsilon \rangle$ be its left nucleus. Then the factor-loop $\mathcal{Q}/\mathcal{N}_l$ is a left Bol loop satisfying the left Bruck property $(x \cdot y) \cdot (x \cdot y) = x \cdot [(y \cdot y) \cdot x]$.

18.

The infinitesimal structure of a correct left φts-loop, in particular, of a perfect one, may be described in the language of Lie algebras as a reductive pair $(\mathfrak{g}, \mathfrak{h})$ of Lie algebras, $\mathfrak{h} \subset \mathfrak{g}$, $\mathfrak{g} = \mathfrak{m} + \mathfrak{h}$, $[\mathfrak{h}, \mathfrak{m}] \subset \mathfrak{m}$, with some additional properties. See [5].

We omit here this construction because of the time frames of our talk.

19. REDUCTIVE SPACES AND LEFT CORRECT φts-LOOPS

Any left correct φts-loop may be modelled in a reductive space in the following way.

Let G be a group, Φ be its endomorphism and $H = \{x \in G;\ \Phi x = x\}$. We assume additionally that $\{x \cdot (\Phi x)^{-1}\}_{x \in G} = Q$ is a transversal, that is, for any $x \in G$, $xH \cap Q$ consists of a single element. Then $h \cdot Q \cdot h^{-1} \subset Q$ $(h \in H)$ (the reductivity of G/H), and $\langle Q, *, \varepsilon\rangle$, where $q * \tilde{q} = \pi(q \cdot \tilde{q})$, $\{\pi x\} = xH \cap Q$, is a left correct φts-loop.

Any correct left φts-loop may be constructed in such a way ([6]).

In the case of a smooth left perfect φts-loop we have, additionally, $Q = \exp \mathfrak{m}$ where $\mathfrak{m} = T_\varepsilon Q \subset \mathfrak{g} = T_\varepsilon G$.

Or, in another language, any 1-parametric group G_1 tangent to Q at ε belongs to Q, $G_1 \subset Q$.

20.

REMARK. In the above considered, G/H possesses the reductant $Q = \{x \cdot (\Phi x)^{-1}\}_{x \in G}$ and, as well, the different reductant $\{\Phi x \cdot x^{-1}\}_{x \in G}$ which coincide only if the original left F-quasigroup is left distributive $(x \cdot (y \cdot z) = (x \cdot y) \cdot (x \cdot z))$.

Using the second reductant and the projection along cosets one obtains the loop with the property $(x \cdot x) \cdot (y \cdot z) = (x \cdot y) \cdot (x \cdot z)$.

More details on this remarkable loop are contained in the talk of L.L. Sabinina at this Conference.

REFERENCES

[1] LL Sabinina On the Geometry of Smooth F-Quasigroups. PhD Dissertation, Friendship of Nations University, Moscow, 1992.

[2] LV Sabinin Methods of Nonassociative Algebra in Differential Geometry. In: S Kobayashi, K Nomizu. Foundations of Differential Geometry. Vol. 1. Moscow: Nauka press, 1981, pp 293–339.

[3] LV Sabinin, LL Sabinina. On the geometry of transsymmetric spaces. In: Webs and Quasigroups. Tver: Tver Univ. press, 1990, pp 64–68.

[4] LV Sabinin, LL Sabinina. On geometry of transsymmetric spaces. Proceedings of the International Conference "Nonlinear Geometric Algebra–89", Moscow, 1989. In: Webs and Quasigroups. Tver: Tver Univ. press, 1991, pp 117–122.

[5] LV Sbitneva, LV Sabinin, LL Sabinina. Perfect transsymmetric spaces. Publicaciones Mathematicae 54: (in print), 1999

[6] LV Sbitneva, LV Sabinin. Left F-quasigroups and Reductive spaces. Algebras, Groups and Geometries 13: 479–487, 1996.

[7] H Pflugfelder. Quasigroups and Loops: An Introduction. Berlin: Heldermann Verlag, 1990

[8] VD Belousov. Foundations of the Theory of Quasigroups and Loops. Moscow: Nauka Press, 1967 (Russian)

[9] VD Belousov. Elements of Quasigroups theory. Kishinev: Univ. Press, 1981, 115 pp. (Russian)

[10] O Chein, H Pflugfelder, JDH Smith. Quasigroups and Loops: Theory and Applications. Berlin: Heldermann Verlag, 1990.

Bernstein algebras whose lattice of ideals is distributive

C. MARTÍNEZ and J. SETÓ*. Departamento de Matemáticas. Universidad de Oviedo. 33007 Oviedo, España.

0 INTRODUCTION

Sometimes the lattice of subalgebras or the lattice of ideals of an algebra are studied in order to get information about the algebra itself. This is also the case with other algebraic structures. The lattice of subalgebras of a Bernstein algebra has been studied in [5] and [6]. Bernstein algebras with a linear lattice of ideals were characterized in [7]. Here we will study Bernstein algebras whose lattice of ideals is distributive.

A Bernstein algebra B over a field K, $chK \neq 2$, is a (non-associative) commutative algebra that possesses a non-zero algebra homomorphism $\omega : B \to K$ such that the identity $(x^2)^2 = \omega(x)^2 x^2$ holds for every $x \in B$.

It is know that every Bernstein algebra has idempotent elements. If $e \in B$ is an idempotent, then $B = Ke + Ker\omega$ and $Ker\omega = U_e + Z_e$, $U_e = \{u \in Ker\omega | eu = \frac{1}{2}u\}$, $Z_e = \{z \in Ker\omega | ez = 0\}$. Properties of Bernstein algebras can be found in [2],[3], [4] and [10]. (We will always assume $Ker\omega \neq 0$).

Here we will remind, without proofs, some of the properties of Bernstein algebras that will be used in the paper and that we can summarize in:

THEOREM 0.1. Let $B = Ke + U_e + Z_e$ be a Bernstein algebra. Then:

1. $U_e^2 \subseteq Z$, $U_e Z_e \subseteq U_e$, $Z_e^2 \subseteq U$.
2. $u^3 = 0 = u(uz) = uz^2 = (u^2)^2 = (uz)^2$, $\forall u \in U_e, \forall z \in Z_e$.
3. Dimensions of U_e and Z_e do not depend on e.

We will use some know properties of non-associative algebras that may be found

* Partially supported by DGICYT PB94-1311-C03-01 and FC-97-PBPGI97-03

in [9] and [10].

1 BASIC FACTS ABOUT IDEALS

In this section some general facts about ideals of a Bernstein algebra will be given. (See [7]). For general results about lattices we refer the reader to [1].

Let I be an ideal of B. Then either $I \subseteq Ker\omega$ or $B^2 \subseteq I$. Consequently the ideals of a Bernstein algebra B divide in two types:

a) Ideals contained in $Ker\omega$. All of them admit a decomposition $I = U_1 + Z_1$, where $U_1 \subseteq U_e$, $Z_1 \subseteq Z_e$.

b) Subspaces of B containing $B^2 = Ke + U_e + U_e^2$.

In [5] we proved that if a Bernstein algebra B has a lattice of ideals that is linear, then B is one of the following algebras:

I. $B = Ke + Ku$ with $u^2 = 0$.

II. $B = Ke + Ku + Ku^2$.

Now we will remember the definition of U_0, an ideal of every Bernstein algebra which was studied in [3].

DEFINITION 1.1. Let $B = Ke + U_e + Z_e$ be a Bernstein algebra. Then $U_0 = \{u \in U_e \mid uU = 0\} = \bigcap_{f=f^2} U_f$ is an ideal of B, clearly independent of the particular idempotent element e. Furthermore B/U_0 is a Jordan-Bernstein algebra (see [3]). In particular, if $U_0 = 0$, the Bernstein algebra B is Jordan.

Also $Z_0 = AnnB \cap Z$ is an ideal of B (possibly zero).

It can be easily proved that a minimal ideal of B (except in the trivial case $B = Ke + Z_e$) is either contained in U_0 or in Z_0.

It is clear that every subspace of Z_0 is an ideal of B. Since it is known that the lattice of subspaces of a vector space with dimension at leat 2 is not distributive, B may have a distributive lattice of ideals only if $dim(B/B^2) \leq 1$ and $dimZ_0 \leq 1$.

2 TRIVIAL AND NORMAL ALGEBRAS

In this section we will consider only trivial Bernstein algebras (that is $Ker^2\omega = 0$) and normal Bernstein algebras ($UZ = Z^2 = 0$).

In both cases algebras having distributive lattice of ideals will be characterized.

LEMMA 2.1. If $B = Ke + U_e + Z_e$ is a trivial Bernstein algebra, then B has a distributive lattice of ideals if and only if B is one of the following algebras:

1. $B = Ke + Ku + Kz$.
2. $B = Ke + Kz$.
3. $B = Ke + Ku$.

Proof:

In a trivial algebra we have $U_0 = U$ and $Z_0 = Z$. Furthermore every subspace contained in U and every subspace contained in Z are ideals. Hence $dimU \leq 1$ and $dimZ \leq 1$. So only the three possibilities mentioned above remain.

In case 1 the only proper ideals of B are:

$$B^2 = Ke + Ku, \; Ker\omega = Ku + Kz, \; Ku, \; Kz$$

In case 2 the only proper ideals of B are:

$$B^2 = Ke, \; Ker\omega = Kz.$$

In case 3 the only proper ideal of $B = B^2$ is $Ker\omega = Ku$.

Clearly the lattice of ideals is distributive in any of the three cases. •

PROPOSITION2.1. Let $B = Ke + U_e + Z_e$ be a normal Bernstein algebra whose lattice of ideals is distributive. Then either B is trivial or $B = Ke + Ku + Kz$ with $u^2 = z$.

Proof:

Let us assume that B is not trivial. Then $U^2 \neq 0$ and $dimZ = 1$, that is $U^2 = Z$ and $B = B^2$ is nuclear. Let $Z = Kz$. For every $u \in U$ the 2-dimensional subspace $K(u, z)$ is an ideal of B. If $dimU \geq 2$ and $u_1.u_2 \in U$ are linearly independents it suffices to consider $K(u_1, z), K(u_2, z)$ and $K(u_1 + u_2, z)$ to conclude that the lattice of ideals is not distributive. •

Notice that the only normal Bernstein algebra that is not trivial and has a distributive lattice of ideals has indeed a linear lattice of ideals.

3 EXCLUSIVE ALGEBRAS

In this section exclusive Bernstein algebras will be considered. These algebras satisfy $U^2 = 0$ and it is well know that such property does not depend on the particular idempotent element. In other words, $U_0 = 0$.

Since $B^2 = Ke + U$, $dim(B/B^2) = dimZ \leq 1$.

We only need to consider the non-trivial case. Consequently $dimZ = 1$. Let us suppose $Z = Kz$. We will distinguish two cases, $UZ = 0$ and $UZ \neq 0$.

THEOREM 3.1. Let $B = Ke + U_e + Z_e$ be a non-trivial exclusive Bernstein algebra. If $UZ = 0$, the lattice of ideals is distributive if and only if $B = Ke + Ku + Kz$ with $z^2 = u$.

Proof:

Let us notice that the condition $UZ = 0$ in an exclusive Bernstein algebra does not depend on the idempotent e. Since the algebra mentioned in the theorem is the only non-trivial exclusive Bernstein algebra with $UZ = 0$ and $dimU = dimZ = 1$, we only need to prove that $dimU = 1$.

Condition $UZ = 0$ implies that $0 \neq Z^2$. In this case $U_0 = U$ and every subspace of U is an ideal of B. So $dimU \leq 1$. But $0 \neq Z^2 \subseteq U$ implies $dimU = 1$. Proposition is proved. •

Now we will consider the case $UZ \neq 0$. In this case $\varphi : U \to U$ given by $\varphi(u) = uz$, that is, $\varphi = R_{z|_U}$ defines a non-zero endomorphism of U.

With the above notation we have:

THEOREM 3.2. Let $B = Ke + U + Z$ be an exclusive Bernstein algebra with $UZ \neq 0$. Then the lattice of ideals of B is distributive if and only if U is a cyclic φ-module.

Proof:

It is well know that U is a cyclic φ-module (that is, there is only one invariant factor) exactly when there are not two different elementary divisors associated to the same irreducible polynomial.

Let us consider $U = K[x]\,\omega_1 + \cdots + K[x]\,\omega_n$ the decomposition of U associated to the elementary divisors of φ, minimum polynomial$(\omega_i) = p_i(x)^{\alpha_i}$.

If we assume that (up to a new arrangement) $p_1(x) = p_2(x) = p(x)$, then we can consider elements $u_1 = p(x)^{\alpha_1 - 1}\omega_1$, $u_2 = p(x)^{\alpha_2 - 1}\omega_2$. Then those elements u_1, u_2 and $u_1 + u_2$ satisfy that $p(x)u_1 = p(x)u_2 = p(x)(u_1 + u_2) = 0$. Clearly the ideals $B(u_1), B(u_2)$ and $B(u_1 + u_2)$, generated by u_1, u_2 and $u_1 + u_2$ respectively, satisfy:

$$B(u_1) \cap B(u_2) = B(u_1) \cap B(u_1 + u_2) = B(u_2) \cap B(u_1 + u_2) = 0,$$

$$B(u_1) + B(u_1 + u_2) = B(u_2) + B(u_1 + u_2) = B(u_1) + B(u_2).$$

Clearly this is impossible in a distributive lattice. We have proved that the fact that U is a cyclic φ-module is needed for the distributivity of the lattice of ideals of B.

Notice that ideals of B belong to one of the following types:

1. $I = U_1 \subseteq U$ $\quad U_1$ a φ-invariant subspace of U and
2. $I = U_1 + Kz$ $\quad U_1$ a φ-invariant subspace of U and $z^2 \in U_1$.

Since U_1 is a φ-invariant subspace of U, U_1 is also a cyclic φ-module. This implies that

$U_1 = K[x]\,u_1^1 + \cdots + K[x]\,u_m^1$ with min.pol.$(u_i^1) = p_i(x)^{\beta_i}$, $0 \leq \beta_i \leq \alpha_i$.

(We have included the case $\beta_1 = 0$ to unify notation).

But the φ-invariant subspaces of $K[x]\,\omega_i$ form a chain:

$$0 \subseteq S_i^1 \subseteq S_i^2 \subseteq \ldots \subseteq S_i^{\alpha_i} = K[x]\omega_i$$

where S_i^j denote the only (cyclic) subspace of $K[x]\,\omega_i$ that is with minimal polynomial equal to $p_i(x)^j$.

So, given any three ideals of B, we can express them in the form:

$I_j = S_1^{j(1)} + \ldots + S_m^{j(m)} + K(\lambda_j z)$ with $\lambda = 0$ or 1, $j = 1, 2, 3$.

So $I_h \cap I_l = S_1^{r(1)} + \ldots + S_m^{r(m)} + K(\lambda_1\lambda_2 z)$ with $r(j) = min(h(i), l(i))$ and $I_h + I_l = S_1^{\delta(1)} + \ldots + S_m^{\delta(m)} + K((\lambda_1 + \lambda_2)z)$ with $\delta(i) = max(h(i), l(i))$.

It can be easily checked that the lattice of ideals of B is distributive. •

THEOREM 3.3. Given a K-vector space U and $\varphi : U \to U$ a non-zero endomorphism such that its minimal and characteristic polynomial coincide and a fixed element $u_1 \in U$, there is an exclusive Bernstein algebra $B(\varphi, U, u_1)$ of type $(dimU + 1, 1)$ and whose lattice of ideals is distributive. Furthermore, $B(\varphi, U, u_1)$ is isomorphic to $B(\varphi', U', u_1')$ if and only if $dimU = dimU'$, $|xI - \varphi| = |xI - \varphi'|$ and the minimal polynomials of u_1 and u_1' with respect to φ and φ' respectively coincide.

Proof:

Let us define $B(\varphi, U, u_1) = Ke + U + Kz$ the exclusive Bernstein algebra with $z^2 = u_1$ and $uz = \varphi(u)$ for every $u \in U$.

The rest is clear. •

Let us notice that if B is also Jordan, then $z^2 = 0$ and $(uz)z = 0$ for every $u \in U$. If the algebra is not trivial, $p.m\varphi = x^2$, so $dimU = 2$ is the only possible case for the lattice of ideals to be distributive. That is, we have the following:

COROLLARY 3.1. There is only one (up to isomorphisms) non trivial, exclusive Jordan-Bernstein algebra whose lattice of ideals is distributive. This algebra is

$$B = Ke + K(u_1, u_2) + Kz$$

with $u_1 z = u_2$ is the only nonzero product in $Ker\omega = K(u_1, u_2) + Kz$.

4 JORDAN BERNSTEIN ALGEBRAS

If $B = Ke + U + Z$ is a Jordan-Bernstein algebra then $Z^2 = 0$ and $(uz)z = 0$ for every $u \in U$, $z \in Z$ (see [2]). It is know that in such algebras the length of the lattice of ideals coincide with the dimension of the algebra, (see [8]).

So if $U_0 \neq 0$, there is a 1-dimensional ideal $Ku_0 \subseteq U_0$. Since $(u_0 z)z = 0$ it follows that $u_0 z = 0$, $\forall z \in Z$.

Using the previous facts we can prove:

THEOREM 4.1. Let $B = Ke + U + Z$ be a Jordan-Bernstein algebra. If B has a distributive lattice of ideals then B is exclusive or normal.

Proof:

Since the only non-trivial Jordan-Bernstein algebra of dimension 3 is $B = Ke + Ku + Kz$ with $z^2 = 0 = uz$, $u^2 = z$ (so normal), the result is true for Jordan-Bernstein algebra of dimension 3. We will use induction on $dimA = n \geq 4$.

Case 1. $U_0 \neq 0$. Let Ku_0 a 1-dimensional ideal of B contained in U_0. Then $u_0 U = u_0 Z = 0$. The quotient algebra $\bar{B} = B/Ku_0$ is a Jordan-Bernstein algebra whose lattice of ideals is distributive. Since $dim\bar{B} = dimB - 1$ it follows, by the induction assumption, that:

1. $\bar{B}$ is exclusive or
2. $\bar{B}$ is normal.

Clearly if $\bar{B}$ is exclusive then $U^2 = 0$ and B is also exclusive.

In the second case, $UZ \subseteq Ku_0$ and $0 = Z^2 \subseteq Ku_0$.

So for any subspace $S \subseteq Z$, the subspace $S + Ku_0$ is an ideal of B. By distributivity we conclude that $dimZ = 1$ (since we can assume that the algebra B is not trivial). That is $B = Ke + U + Kz$. If B is exclusive we have done. In other case $U^2 \neq 0$.

Let us consider I an arbitrary ideal of B. The possible form of I is:

- $I = U_1 \subseteq U_0$ with U_1 an R_z−invariant subspace of U or

- $I = U_1 + Kz$ with U_1 an R_z-invariant subspace of U.

As we did in the exclusive case, distributivity of the lattice of ideals assures that U is a cyclic R_z-module. But $(uz)z = 0\ \forall u \in U$ implies that the minimal polynomial of R_z divides to x^2. So $dimU = 2$ and there is a basis $\{u_1, u_2\}$ of U such that $u_1z = u_2$, $u_2z = 0$. If B is not exclusive (notice that $u_2^2 = 0 = u_1u_2$) then $u_1^2 = \lambda z$ with $\lambda \neq 0$. But this implies that $u_1z = \lambda^{-1}u_1.u_1^2 = 0$, a contradiction.

Case 2. $U_0 = 0$.

In this case we know that $Z_0 \neq 0$ since $Ker\omega$ is nilpotent. By the distributivity assumption $dimZ_0 = 1$, that is, $Z_0 = Kz_0$ with $z_0U = 0$.

Again let us consider $\bar{B} = B/Kz_0$. By the induction assumption, $\bar{B}$ is normal or exclusive.

If $\bar{B}$ is normal then $UZ \subseteq Kz_0$, $Z^2 = 0$, that is, $UZ = Z^2 = 0$ and B is also normal.

So let us consider the case $\bar{B}$ exclusive, that is $U^2 \subseteq Kz_0$. We can assume $U^2 = Kz_0$ (in other case B is exclusive and we have done). Since $dim(B/B^2) \leq 1$ we have $Z = Kz_0$ or $Z = K(z_0, z_1)$.

If $Z = Kz_0$, $U_1 + Z$ is an ideal of B for every subspace U_1 of U (remember $U^2 = Kz_0$ and $uz_0 = 0, \forall u \in U$). By distributivity of the lattice of ideals we conclude that $dimU = 1$. But then $U^2 = 0$, that is $U = U_0$ against the assumption.

So $Z = K(z_0, z_1) = Kz_0 + Kz_1$. Let $\varphi = R_{|z_1} : U \to U$. Every ideal of B is of the form: $U_1 + Kz_0$ with U_1 a φ- invariant subspace of U or $U_1 + Kz_0 + Kz_1$ with U_1 a φ-invariant subspace of U.

(Notice that in this case Kz_0 is the only 1-dimensional ideal of B and it is contained in every non-zero ideal of B).

Arguing as in the exclusive case and using that $\varphi^2 = 0$ we get that either $\varphi = 0$, what implies B normal, or $dimU = 2$ and there is a basis $\{u_1, u_2\}$ of U such that $u_1z = u_2$, $u_2z = 0$.

But $u_1z = u_2$ implies $u_2^2 = 0 = u_2u_1$ and so $u_2 \in U_0$ against the assumption $U_0 = 0$. •

5 THE GENERAL CASE

We know that, under the assumption of a distributive lattice of ideals, $\dim(B/B^2) \leq 1$. So we have to distinguish two cases:

1. $dimB = dimB^2$, that is, B is nuclear, and

2. $dimB = dimB^2 + 1$.

THEOREM 5.1. Let $B = Ke + U + Z$ be a nuclear Bernstein algebra whose lattice of ideals is distributive. Then B is exclusive or normal.

Proof:

Remember that if the lattice of ideals of B is a chain then B is normal.

Since $Z = U^2$ it is known that $U_0Z = 0$ and so every subspace of U_0 is an ideal of B. So $dimU_0 \leq 1$ and $dimZ_0 \leq 1$.

In a nuclear Bernstein algebra the nucleus $Ker\omega$ is nilpotent, so the case $U_0 = 0 = Z_0$ is impossible.

If $U_0 = 0$ then B is a Jordan-Bernstein algebra. Result follows from Theorem 4.1.

If $dimU_0 = 1$, the result is true for $\bar{B} = B/U_0$ by Theorem 4.1.

If $\bar{B}$ is exclusive also B is exclusive.

Let us assume $\bar{B}$ normal. Then $UZ + Z^2 \subseteq Ku_0$. If $UZ + Z^2 = 0$ then B is normal. Let us study the case $UZ + Z^2 = Ku_0$.

Every subspace $S + Ku_0$ where S is an arbitrary subspace of Z is an ideal of B. The distributivity of the lattice of ideals implies that $dimZ = 1$. If $Z = Kz$ then $Ker^2\omega = Ku_0 + Kz$. If U_1 is an arbitrary subspace of U and $u_0 \notin U_1$ then $U_1 + Ku_0 + Kz$ is an ideal of B. Again the distributivity assumption assures that $dimU = 2$. Let us assume $U = K(u_1, u_0)$. But $U^2 = Z$ and $u_0^2 = u_0u_1 = 0$ implies $u_1^2 = z$. Then $z^2 = 0 = Uz$, that is, $Z^2 = 0 = UZ$, the contradiction. •

It remains to consider the case $dimB = dimB^2 + 1$ that is $dimZ = dimU^2 + 1$.

We will prove the following:

THEOREM 5.2. Let $B = Ke + U + Z$ be a Bernstein algebra such that $dimB = dimB^2 + 1$. Then B has a distributive lattice of ideals if and only if B is normal or exclusive.

Proof:

Let us assume that B is not exclusive (in other case we have already done). If B is Jordan, theorem 4.1 gives us the wanted result. So we can assume $0 \neq U_0 \neq U$.

Let us consider $\bar{B} = B/U_0 = K\bar{e} + \bar{U} + \bar{Z}$. Since $\bar{B}$ is a Jordan-Bernstein algebra and has a distributive lattice of ideals it suffices to apply Theorem 4.1 to conclude that $\bar{B}$ is either exclusive or normal.

But $\bar{B}$ exclusive implies B exclusive.

So we only need to consider the case $\bar{B}$ is normal, i.e. $\bar{U}\bar{Z} + \bar{Z}^2 = 0$ or equivalently $UZ + Z^2 \subseteq U_0$.

This implies that for every subspace S of U satisfying $UZ + Z^2 \subseteq S \subseteq U$, the vector subspace of B, $I = S + Z$ is an ideal of B. Indeed, $US \subseteq Z$, $UI \subseteq S$, $SZ \subseteq UZ \subseteq S$ and $Z^2 \subseteq S$.

By distributivity of the lattice of ideals. $codim_U(UZ + Z^2) \leq 1$.

This fact together with $UZ + Z^2 \subseteq U_0 \subset U$ implies that $U_0 = UZ + Z^2$ and $dimU = dimU_0 + 1$.

But then we can find a basis $\{u, u_2, \ldots, u_r\}$ with $u_2, \ldots, u_r \in U_0$. Hence $U^2 = Ku^2$ and we know that under our assumption, $dimZ = dimU^2 + 1$. So $Z = K(u^2, z)$ where $z \notin U^2$.

Considering again the quotient algebra $\bar{B} = B/U_0$ we have a normal algebra $\bar{B} = K\bar{e} + K\bar{u} + K(\bar{u}^2, \bar{z})$. But it is impossible by Proposition 2.1.

The contradiction comes from the assumptions at the beginning of de proof. So B is normal or exclusive. Theorem is proved. •

Now again proposition 2.1 assures that the only normal Bernstein algebra whose lattice of ideals is distributive are trivial algebras of types $(1, 1)$ and $(2, 1)$.

So we have proved:

COROLLARY 5.2. Let $B = Ke + U + Z$ be a non trivial Bernstein algebra with a distributive lattice of ideals and $B^2 \neq B$, i.e. not nuclear, then B is exclusive.

REMARK. We also know that every non trivial exclusive algebra whose lattice of ideals is distributive is not nuclear, i.e. $dimB = dimB^2 + 1$.

In conclusion, except the three trivial Bernstein algebras of types $(1,1)$, $(2,0)$ and $(2,1)$ and the nuclear normal algebra of type $(2,1)$, Bernstein algebras with a distributive lattice of ideals are exclusive. They have been described in Propositions 2.1 and 2.2.

REFERENCES

[1] Birkhoff, G.; Lattice Theory. AMS Colloquium Publications (vol. XXV), third ed. Providence, Rhode Island, (1984).

[2] González, S. and Martínez, C.; Idempotent elements in a Bernstein algebra, J. London Math. Soc. (2) 42, 430-436, (1990).

[3] Lyubich, Y.; Mathematical structures in Population Genetics. Springer-Verlag, Berlin-Heidelberg. (1992)

[4] Martínez, C.; Free nuclear Bernstein algebras, J. of Algebra 177, 676-697, (1995).

[5] Martínez, C. and Sánchez-Nadal, J. A.; Bernstein algebras with a lattice isomorphism that doesn't preserve the nucleus, Comm. in Algebra 22, 4781-4792, (1994).

[6] Martínez, C. and Sánchez-Nadal, J. A.; Lattice isomorphism in Bernstein algebras. Mathematics and its applications 303, 269-274. Kluwer Acad. Publish. (1994).

[7] Martínez, C. and Setó, J.; Bernstein algebras whose lattice of ideals is linear. Mathematics and its applications 303, 275-278. Kluwer Acad. Publish. (1994).

[8] Setó, J.; Doctoral Dissertation. Universidad de Oviedo (1998).

[9] Shafer, R.D.; An introduction to non associative algebras. Academic Press, New York and London (1966).

[10] Wörz-Busecros, A.; Algebras in Genetics. Lectures notes in Biomathematics, vol. 36. Springer-Verlag, Berlin-Heidelberg (1980).

[11] Zhevlakov, K.A., Slin'ko, A.M., Shestakov, I.P. and Shirshov, A.I.; Ring that are nearly associative. Academic Press, New York and London (1982).

Speciality problem for Malcev algebras and Poisson Malcev algebras

IVAN P. SHESTAKOV[1] Sobolev Institute of Mathematics, Novosibirsk, 630090, Russia Departamento de Matemática, Universidade de São Paulo, Caixa Postal 66281, São Paulo-SP, Brazil, 05315-970.

ABSTRACT The problem of speciality for Malcev algebras is investigated. A Malcev algebra is called *special* if it admits an isomorphic embedding into a commutator algebra A^- for a certain alternative algebra A. It is an open question whether any Malcev algebra is special. We introduce a notion of Malcev Poisson algebra, which generalizes that of (Lie) Poisson algebra, and show that the problem of speciality for a Malcev algebra M is reduced to the existence of alternative quantization deformations for certain Malcev Poisson algebras related with M.

1 INTRODUCTION

Malcev algebras were introduced in 1955 by A.I.Malcev [11] as tangent algebras of analytic Moufang loops. They are related to alternative algebras in the same way that Lie algebras are related to associative algebras: if A is an alternative algebra, then the algebra A^-, with the multiplication $[a, b] = ab - ba$, is a Malcev one. Although the theory of Malcev algebras is quite well developed, it is still an open problem whether the analog of the Poincare—Birkhoff—Witt is true for them; that is, whether any Malcev algebra is isomorphic to a subalgebra of an algebra A^- for a suitable alternative algebra A. Some partial results on this problem were obtained in [6, 14].

In [12, 13] it was shown that the problems of speciality sometimes could be solved by methods of deformation theory. It is known that the universal enveloping algebra $U(L)$ of a Lie algebra L can be obtained by a quantization deformation of the symmetric algebra $S(L)$, which has the natural structure of Poisson algebra

[1] Supported by The Foundation "Universities of Russia" and by the Brasilian FAPESP, Proc. 98/2162-6

[4, 8, 7, 12]. We define Malcev Poisson algebras and prove that any Malcev algebra M can be embedded into a Malcev Poisson algebra $\tilde{S}(M)$, which has the universal property for homomorphisms of M into Malcev Poisson algebras. We consider next alternative quantization deformations of $\tilde{S}(M)$, and find a necessary condition for Malcev algebras to be special. It is still an open question whether this condition is sufficient and whether it is satisfied by any Malcev algebra.

2 POISSON MALCEV ALGEBRAS

Let A be an associative and commutative algebra over a (unital, associative, and commutative) ring of scalars Φ . We will call A a *general Poisson algebra*, if A is endowed with an anticommutative multiplication $\{x, y\}$ (*a Poisson bracket*), that satisfies the *Leibniz identity*

$$\{ab, c\} = a\{b, c\} + \{a, c\}b.$$

Note that it is usually assumed in the definition of Poisson algebra that the multiplication $\{a, b\}$ is a Lie one. We will call such algebras *Poisson Lie algebras*.

Let A be an anticommutative Φ-algebra, $S(A)$ be the symmetric algebra of the Φ-module A. We will denote by $[a, b]$ the multiplication in A, and by fg the multiplication in $S(A)$. Define on $S(A)$ a new multiplication $\{f, g\}$, by setting for $f = a_1a_2\cdots a_n$, $g = b_1b_2\cdots b_m$

$$\{f, g\} = \sum_{i=1}^{n}\sum_{j=1}^{m} a_1\cdots a_{i-1}b_1\cdots b_{j-1}[a_i, b_j]\, a_{i+1}\cdots a_nb_{j+1}\cdots b_m, \tag{1}$$

where $a_i, b_j \in A$, $i = 1, \ldots, n$; $j = 1, \ldots, m$.

PROPOSITION 1 *The algebra $S(A)$ is a general Poisson algebra with respect to the multiplication defined above. The algebra A is isomorphically embedded as a subalgebra into the algebra $\langle S(A), \{,\}\rangle$.*

An anticommutative algebra A is called a *Malcev algebra*, if it satisfies the identity

$$[J(x, y, z), y] = J(x, y, [y, z]), \tag{2}$$

where $J(x, y, z) = [[x, y], z]+[[y, z], x]+[[z, x], y]$ is a *Jacobian* of the elements x, y, z. These algebras that generalize the Lie algebras were introduced by A. I. Malcev in 1955 [11] (under the name *Moufang—Lie algebras*) as the tangent algebras of local analytic Moufang loops.

A general Poisson algebra A we will call a *Poisson Malcev algebra*, if it is a Malcev algebra with respect to the multiplication $\{,\}$.

It is well known (e.g, [2, 7, 12]), that for any Lie algebra L the symmetric algebra $S(L)$ with the bracket $\{,\}$ defined by (1) is a Poisson Lie algebra. Moreover, the algebra $S(L)$ has the following universal property: any (Lie) homomorphism of L into a Poisson algebra P can be extended to a unique Poisson algebra homomorphism of $S(L)$ into P. Below we will construct a similar universal algebra for homomorphisms of arbitrary Malcev algebra M into Poisson Malcev algebras. But in this case instead of the algebra $S(M)$ we have to use a certain homomorphic image.

LEMMA 1 *A (general) Poisson algebra satisfies the identities*

$$\begin{aligned} J(a^2,b,c) &= 2aJ(a,b,c), \\ 2J(a,b,c)\{a,b\} &= \{J(a^2,b,c),b\} - J(a^2,b,\{b,c\}) \\ &\quad -2a(\{J(a,b,c),b\} - J(a,b,\{b,c\})), \end{aligned}$$

where $J(a,b,c) = \{\{a,b\},c\} + \{\{b,c\},a\} + \{\{c,a\},b\}$.

COROLLARY 1 *A Poisson Malcev algebra satisfies the identity*

$$J(a,b,c)\{a,b\} = 0.$$

Fix a Malcev algebra M and consider in the general Poisson algebra $S(M)$ the (associative-commutative) ideal I generated by the elements $J(x,y,z)[x,y]$, where $x,y,z \in M$.

LEMMA 2 $\{I, S(M)\} \subseteq I$.

Denote by $\tilde{S}(M)$ the quotient algebra $S(M)/I$. By lemma 2 the bracket $\{,\}$ can also be defined on the algebra $\tilde{S}(M)$. Therefore, $\tilde{S}(M)$ has a structure of a general Poisson algebra.

THEOREM 1 *The general Poisson algebra $\tilde{S}(M)$ is a Poisson Malcev algebra that contains the algebra M as a subalgebra with respect to the bracket $\{,\}$. Moreover, any homomorphism $\varphi : M \to (P, \{,\})$ of M into a Poisson Malcev algebra P can be uniquely extended to a Poisson algebra homomorphism $\tilde{\varphi} : \tilde{S}(M) \to P$.*

REMARK 1. A passage from the algebra M to $\tilde{S}(M)$ gives new possibilities in constructing examples of Malcev algebras. Starting with the known examples of finite dimensional algebras one can in this way construct infinite dimensional Malcev algebras. A base of the algebra $\tilde{S}(M)$ one can construct from a base of M using Groebner basis methods. (e.g., [3]).

EXAMPLE 1. Let M be a 4-dimensional solvable nonnilpotent Malcev algebra over a field F with the base e, a, b, c and the nonzero products $[e,a] = a$, $[e,b] = b$, $[e,c] = -c$, $[a,b] = c$ (cf. [9]). Then the ideal I is generated by the elements c^2, ca, cb, and $\tilde{S}(M) = F[e]c + F[e,a,b]$, with the Poisson bracket defined according to (1). A base and a multiplication table for the Malcev algebra $(\tilde{S}(M), \{,\})$ can be written down quite easily.

REMARK 2. If $M = L$ is a Lie algebra then the ideal $I = 0$ and $\tilde{S}(L) = S(L)$. In particular, in this case $\tilde{S}(L)$ does not contain zero divisors and is always infinite dimensional if $L \neq 0$. In the case when M is a non-Lie Malcev algebra, the algebra $\tilde{S}(M)$ always contains zero divisors. Besides, in this case $\tilde{S}(M)$ may be finite dimensional.

EXAMPLE 2. Let $M = C_7$ be the simple 7-dimensional non-Lie Malcev algebra over a field F with the base $e_1, \ldots, e_7$ [10]. Then the ideal I is generated by the elements $e_ie_j, e_i^2 - e_j^2$, $i \neq j$, $i,j = 1,\ldots,7$, and $\tilde{S}(M) = Fe_1 + \cdots + Fe_7 + F\alpha$, where $\alpha = e_i^2$, $i = 1,\ldots,7$, $\alpha^2 = \alpha e_i = e_ie_j = 0$.

REMARK 3. All the definitions and arguments from the above can be naturally extended to superalgebras, by substitution the *supersymmetric algebra of the superspace* M for $S(M)$ (e.g., [2, 12]. In particular, using the classification of Malcev superalgebras of small dimensions in [1], one may construct new examples of Malcev superalgebras.

EXAMPLE 3. Let $M = Fx + Fm_0 + Fm_1$ be a split null extension of the unique nontrivial irreducible Malcev supermodule $V = Fm_0 + Fm_1$ (see [5]), which is defined over the 1-dimensional odd Malcev superalgebra $A = A_1 = Fx$, $x^2 = 0$, by the conditions $m_0 \cdot x = -m_0 \cdot x = m_1$, $m_1 \cdot x = x \cdot m_1 = m_0$. (In the notations of [1] it is the superalgebra $M(1, 2; 1)$). Denote $a = m_0$, $y = m_1$, then the ideal I is generated by one element a^2, and $\tilde{S}(M) = Fa + Fx + Fy + Fxy + Fax + Fay + Faxy$, where $x^2 = y^2 = 0$, $xy = -yx$, $ax = xa$, $ay = ya$.

3 ASSOCIATE GRADED POISSON MALCEV ALGEBRA OF UNIVERSAL ALTERNATIVE ENVELOPING ALGEBRA

For any alternative algebra A the algebra A^- with the multiplication $[a, b] = ab - ba$ is a Malcev algebra. A Malcev algebra M is called *special* it it admits an isomorphic embedding into an algebra A^- for a suitable alternative algebra A.

By standard arguments one can prove that for any Malcev algebra M there exist an alternative algebra $U(M)$ and a Malcev algebra homomorphism $\sigma : M \to U(M)^-$ with the following universal property: for any Malcev algebra homomorphism $\rho : M \to B^-$, where B is alternative, there exists a unique homomorphism of alternative algebras $\tilde{\rho} : U(M) \to B$ such that $\rho = \sigma \circ \tilde{\rho}$. Moreover, the algebra M is special if and only if $\sigma$is a monomorphism.

The algebra $U = U(M)$ is generated by the set $\sigma(M)$ and has the natural filtration:

$$U_1 \subseteq U_2 \subseteq \cdots, \text{ where } U_1 = \sigma(M),\ U_{k+1} = U_k + \sigma(M)^k;$$

besides, $U = \cup_i U_i$, so the filtration is exhaustive. Consider the graded algebra $gr\,U = gr\,U(M)$ associated with the filtered algebra U:

$$gr\,U = \oplus_{i=1}^{\infty} gr\,U_i,$$

where $gr\,U_1 = U_1$ and $gr\,U_i = U_i/U_{i-1}$ for $i > 1$, with the multiplication defined by the conditions: if $\bar{a} \in gr\,U_i$, $\bar{a} = a + U_{i-1}$, $\bar{b} \in gr\,U_j$, $\bar{b} = b + U_{j-1}$, then

$$\bar{a}\bar{b} = (a + U_{i-1})(b + U_{j-1}) = ab + U_{i+j-1}.$$

The algebra $gr\,U(M)$ is also generated by the set $\sigma(M)$ and is isomorphic to U as a vector space over F.

In particular, it would be sufficent for speciality of M to prove that the mapping $\sigma : x \mapsto \sigma(x)$ induces an embedding of M into $gr\,U(M)$. The algebra $gr\,U$ usually has a less complicated multiplication than U, so as a first step in construction of $U(M)$ it is natural to try to understand the structure of the graded algebra $gr\,U(M)$.

LEMMA 3 *For any $a \in U_i$, $b \in U_j$, $c \in U_k$ the following inclusions hold:*

$$[a, b] \in U_{i+j-1},$$
$$(a, b, c) \in U_{i+j+k-2}.$$

COROLLARY 2 *The associated graded algebra $gr\, U(M)$ is associative and commutative.*

Let us define on $gr\, U(M)$ one more multiplication, by setting for $\bar{a} \in gr\, U_i$, $\bar{a} = a + U_{i-1}$, $\bar{b} \in gr\, U_j$, $\bar{b} = b + U_{j-1}$,

$$\{\bar{a}, \bar{b}\} = \{(a + U_{i-1}), (b + U_{j-1})\} = [a, b] + U_{i+j-2}.$$

It follows from lemma 4 that the multiplication is correctly defined.

THEOREM 2 *The algebra $gr\, U(M)$ forms a Poisson Malcev algebra with respect to the bracket $\{,\}$. The Malcev homomorphism $m \mapsto \sigma(m)$ induces a Poisson algebra ephimorphism $\pi : \tilde{S}(M) \to gr\, U(M)$.*

We do not know whether π is an isomorphism, even in the case when M is special.

4 SPECIALITY AND DEFORMATIONS

Now we define alternative quantization deformations for Poisson Malcev algebras and show that the problem of speciality for a Malcev algebra M is closely related with the problem of existence of alternative quantization deformations for Poisson Malcev algebras $\tilde{S}(M)$ and $gr\, U(M)$.

Let A be a Poisson Malcev algebra and $A[t]$ a Φ-module of polynomials over A. Assume that $A[t]$ admits a multiplication $*$ such that

i) $\langle A[t], * \rangle$ is an alternative algebra,

ii) $a * b = ab \ (mod\ t)$,

iii) $a * b - b * a = \{a, b\}t \ (mod\ t^2)$,

iv) $t * a = a * t = at$

for any $a, b \in A$. Then we will say that A *admits an algebraic alternative quantization deformation* $*$.

THEOREM 3 *Let M be a Malcev algebra, $gr\, U(M)$ and $\tilde{S}(M)$ be the associated graded and universal Poisson Malcev algebras related to M. Then, if $\tilde{S}(M)$ admits an algebraic alternative quantization deformation $*$ such that*

$$a * b - b * a = \{a, b\}t \qquad (3)$$

for any $a, b \in M$, then M is special. On the other hand, if M is special, then $gr\, U(M)$ admits an algebraic alternative quantization deformation that satisfies (3).

By studing the obstructions to deformations of small orders for algebra $\tilde{S}(M)$, we have found the following necessary condition for speciality of Malcev algebras.

PROPOSITION 2 *If a Malcev algebra M is special then it satisfies the following quasiidentity.*

$$\text{If} \quad \sum_i J(a_i, b_i, c_i) \otimes [a_i, b_i] \otimes t_i = 0 \textit{ in } S(M) \textit{ for some } a_i, b_i, c_i, t_i \in M,$$
$$\text{then} \quad \sum_i ([[t_i, J(a_i, b_i, c_i)], [a_i, b_i]] + [[t_i, [a_i, b_i]], J(a_i, b_i, c_i)]) = 0 \textit{ in } M.$$

The known examples of Malcev algebras satisfy this quasiidentity. As a possible algebra that might not satisfy it, we suggest the algebra

$$M = alg\langle x, y, z, u, v | \, J(x, y, z) = [u, v] \rangle.$$

One has to check whether the element

$$r = J([u, v], [x, y], J(u, v, z)) + 3[[J(u, v, z), [x, y]], [u, v]]$$

is nonzero in M. It seems that this task could be solved with a computer.

REFERENCES

[1] H. Albuquerque and A. Elduque, Classification of Malcev superalgebras of small dimensions, Algebra i Logika, 35, No. 6 (1996), 629–654;
English transl.: Algebra and Logic, 35, No. 6 (1996), 351–365.

[2] F. A. Berezin, An introduction to algebra and analysis with anticommuting variables, Izdat. Moskov. Gos. Univ., Moscow, 1983; English transl., Introduction to superanalysis, Reidel, Dordrecht, 1987.

[3] D. Cox, J .Little, D. O'Shea, Ideals, varieties, and algorithms. An introduction to computational algebraic geometry and commutative algebra, Springer, 1992.

[4] W. G. Drinfeld, About constant quasiclassical solutions of the quantum Yang—Baxter equations, Dokl. Akad.Nauk SSSR, 273 (1983), 525–531.

[5] A. Elduque and I. Shestakov, Irreducible Non-Lie Modules for Malcev Superalgebras, J.Algebra, 173 (1995), 622–637.

[6] V. T. Filippov, The measure of non-Lieness for Malcev algebras, Algebra Logika, 31, No.2 (1992) 198–217;
English transl.: Algebra and Logic, 31, No.2 (1992), 126–140.

[7] J. Grabowski, Abstract Jacobi and Poisson structures. Quantization and star-products, J. Geom. Phys. 9 (1992), 45–73.

[8] S.Gutt, An explicit $*$-product in the cotangent bundle of a Lie group, Lett. Math. Phys. 7 (1983),249–258.

[9] A. T. Gainov, Binary Lie algebras of lower ranks, Algebra i Logika, 2, No. 4 (1963), 21–40.

[10] E. N. Kuz'min, I. P. Shestakov, Nonassociative Structures, VINITI, Itogi nauki i tekhniki, seria "Fundamental Branches", v.57, 179–266, Moscow, 1990; English transl. in "Encyclopaedia of Math. Sciences", v.57, Algebra VI, 199–280, ed. A. I. Kostrikin and I R. Shafarevich, Springer-Verlag, 1995.

[11] A. I. Malcev, Analytic loops (Russian), Mat. Sb. (N. S.) 36 (78), No 3 (1955), 569–575.

[12] I. P. Shestakov, Quantization of Poisson superalgebras and speciality of Jordan Poisson superalgebras, Algebra i Logika, 32, No.5 (1993), 572–585; English transl.: Algebra and Logic, 32, No.5 (1993), 309–317.

[13] I. P. Shestakov, Speciality and Deformations of Algebras, Proceedings of the International Algebraic Conference devoted to the memory of A.G.Kurosh, to appear.

[14] S. R. Sverchkov, Varieties of special algebras, Commun. Algebra, 16 (1988), 1877–1919.

Applications of Lie ring methods to group theory

PAVEL SHUMYATSKY[1] Department of Mathematics, University of Brasilia, 70910-900 Brasilia - DF, Brazil. e-mail: pavel@mat.unb.br

ABSTRACT The aim of the paper is to illustrate how a Lie-theoretic result of Zelmanov enables one to treat various problems of group theory.

1. INTRODUCTION

Lie rings were associated with p-groups in the 30s in the context of the Restricted Burnside Problem and since then Lie ring methods proved to be an important and very effective tool of group theory. In the last 10 years the sphere of use of Lie rings was amplified considerably, mainly due to Zelmanov's outstanding contribution.

It has been known for some time that the following assertions are equivalent.

1.1. Let m and n be positive integers. Then there exists a number $B(m,n)$ depending only on m and n such that the order of any m-generator finite group of exponent n is at most $B(m,n)$.

1.2. The class of locally nilpotent groups of exponent n is a variety.

1.3. The class of locally finite groups of exponent n is a variety.

1.4. Any residually finite group of exponent n is locally finite.

The Restricted Burnside Problem is exactly the question whether the first of the above assertions is true. In 1956 P. Hall and G. Higman reduced the problem to the case of prime-power exponent [11]. This shows that 1.1 is equivalent to 1.2. In [22] A. I. Kostrikin solved the Restricted Burnside Problem affirmatively for $n = p$, a prime. His proof relies on a profound study of Engel Lie algebras of positive characteristic. The reduction to Lie algebras was made earlier by W. Magnus [27] and independently by I. N. Sanov [31]. Finally, in 1989, E. Zelmanov

[1]Research supported by FAPDF and CNPq

gave a complete solution of the Restricted Burnside Problem [52], [53]. In [54] Zelmanov deduces the positive solution of the Restricted Burnside Problem from the following theorem.

THEOREM 1.5. Let L be a Lie algebra generated by $a_1, \ldots, a_m$. Suppose that L satisfies a polynomial identity and each commutator in $a_1, \ldots, a_m$ is ad-nilpotent. Then L is nilpotent.

The above theorem proved to be a clue for some other deep results on profinite and residually finite groups (see [54], [55]). In the present paper we will discuss some recently discovered group-theoretic corollaries of this theorem. Each of the principal results described in Sections 3-5 can be viewed as a development around the Restricted Burnside Problem in the sense of at least one of the assertions 1.1-1.4. Thus, Theorem 1.5 is not only one of the main tools in the solution of the Restricted Burnside Problem but also an important factor extending our understanding of the problem itself. The present paper is an attempt to make this subject accessible to a broader audience.

In the next section we describe the construction that associates a Lie algebra $L(G)$ to any group G and establish basic facts reflecting relationship between G and $L(G)$. This will be a bridge linking the above theorem of Zelmanov with group theory.

In Section 3 we consider residually finite groups in which all commutators $[x_1, x_2, \ldots, x_k]$ satisfy some restrictive condition. In particular we describe a proof of the following theorem.

THEOREM 3.1. Let k be an integer, $q = p^s$ a prime-power, and G a residually finite group such that $[x_1, x_2, \ldots, x_k]^q = 1$ for all $x_1, x_2, \ldots, x_k \in G$. Then $\gamma_k(G)$ is locally finite.

This extends the positive solution of the Restricted Burnside Problem for groups of prime-power exponent (take $k = 1$). Another result considered in Section 3 deals with residually finite groups in which the commutators $[x_1, x_2, \ldots, x_k]$ are Engel.

THEOREM 3.2. Let k, n be positive integers, and G a residually finite group such that $[x_1, x_2, \ldots, x_k]$ is n-Engel for any $x_1, x_2, \ldots, x_k \in G$. Then $\gamma_k(G)$ is locally nilpotent.

For $k = 1$ this is a well-known result of J. Wilson [49].

Section 4 is devoted to some questions on exponent of a finite group with automorphisms. The main result of the section is the following theorem.

THEOREM 4.3 (Khukhro, Shumyatsky [21]). Let q be a prime, n an integer. Suppose that a non-cyclic group A of order q^2 acts on a finite group G of coprime order in such a manner that the exponents of the centralizers $C_G(a)$ of non-trivial elements $a \in A$ divide n. Then the exponent of G is bounded in terms of n and q.

Note that the exponent of the centralizer of a single automorphism a of a finite group G has no impact over the exponent of G. Indeed, any abelian group of odd order admits a fixed-point-free automorphism of order two. Hence, we cannot bound the exponent of G solely in terms of the exponent of $C_G(a)$. In view of this the following theorem seems to be interesting.

THEOREM 4.4. Let n be a positive integer, G a finite group of odd order admitting an involutory automorphism a such that $C_G(a)$ is of exponent dividing n. Suppose that for any $x \in G$ the element $[x, a] = x^{-1}x^a$ has order dividing n. Then the exponent of G is bounded in terms of n.

In Section 5 we obtain some sufficient conditions for a periodic residually finite group to be locally finite.

THEOREM 5.3 (Shalev [36]). Let G be a periodic residually finite group having a finite 2-subgroup A such that the centralizer $C_G(A)$ is finite. Then G is locally finite.

THEOREM 5.4. Let q be a prime, G a residually finite group in which each 2-generator subgroup is finite. Suppose that G has a finite q-subgroup A such that the centralizer $C_G(A)$ is finite. Then G is locally finite.

This paper certainly is not a comprehensive survey on Lie methods in group theory. Some important areas are not even mentioned here. For the reader willing to learn more on this subject we recommend the survey [35] and the textbooks [18, Chapter VIII], [20], [47]. Throughout the paper p stands for a fixed prime. We use the term "$\{a, b, c, \ldots\}$-bounded" to mean "bounded from above by some function of $a, b, c, \ldots$".

I am grateful to A. Mann for a number of suggestions simplifying proofs in Section 3.

2. ASSOCIATING A LIE ALGEBRA TO A GROUP

Let L be a Lie algebra over a field $\mathfrak{k}$. Let k, n be positive integers and let $x_1, x_2, \ldots, x_k$,
x, y be elements of L. We define inductively

$$[x_1] = x_1; \; [x_1, x_2, \ldots, x_k] = [[x_1, x_2, \ldots, x_{k-1}], x_k]$$

and

$$[x, {}_0y] = x; \; [x, {}_ny] = [[x, {}_{n-1}y], y].$$

An element $a \in L$ is called ad-nilpotent if there exists a positive integer n such that $[x, {}_na] = 0$ for all $x \in L$. If n is the least integer with the above property then we say that a is ad-nilpotent of index n. Let $X \subseteq L$ be any subset of L. By a commutator in elements of X we mean any element of L that can be obtained as a Lie product of elements of X with some system of brackets. Denote by F the free Lie algebra over $\mathfrak{k}$ on countably many free generators $x_1, x_2, \ldots$. Let $f = f(x_1, x_2, \ldots, x_n)$ be a non-zero element of F. The algebra L is said to satisfy the identity $f \equiv 0$ if

$f(a_1, a_2, \ldots, a_n) = 0$ for any $a_1, a_2, \ldots, a_n \in L$. In this case we say that L is PI. The next theorem is straightforward from Theorem 1.5 quoted in the introduction.

THEOREM 2.1. Let L be a Lie algebra over a field $\mathfrak{k}$ generated by $a_1, a_2, \ldots, a_m$. Assume that L satisfies an identity $f \equiv 0$ and that each commutator in the generators $a_1, a_2, \ldots, a_m$ is ad-nilpotent of index at most n. Then L is nilpotent of $\{f, n, m, \mathfrak{k}\}$-bounded class.

Proof. Consider the free m-generated Lie $\mathfrak{k}$-algebra F_m on free generators $f_1, \ldots, f_m$, and let T be the ideal of F_m generated by all values of f on the elements of F_m and by all elements of the form $[g, {}_n c]$, where $g \in F_m$ and c is an arbitrary commutator in the f_i. Then F/T satisfies the hypothesis of Theorem 1.5 and hence is nilpotent of some class $u = u(m, n, f, \mathfrak{k})$. Clearly, L is the image of F/T under the homomorphism induced by the mapping $f_i \to a_i$. Hence L is nilpotent of class at most u. □

An important criterion for a Lie algebra to be PI is the following

THEOREM 2.2 (Bahturin-Linchenko-Zaicev). Let L be a Lie algebra over a field $\mathfrak{k}$. Assume that a finite group A acts on L by automorphisms in such a manner that $C_L(A)$, the subalgebra formed by fixed elements, is PI. Assume further that the characteristic of $\mathfrak{k}$ is either 0 or prime to the order of A. Then L is PI.

This theorem was first proved by Yu. A. Bahturin and M. V. Zaicev for solvable groups A [2] and later extended by V. Linchenko to the general case [25].

COROLLARY 2.3. Let F be the free Lie algebra of countable rank over $\mathfrak{k}$. Denote by F^* the set of non-zero elements of F. For any finite group A there exists a mapping

$$\phi : F^* \to F^*$$

such that if L and A are as in Theorem 2.2, and if $C_L(A)$ satisfies an identity $f \equiv 0$, then L satisfies the identity $\phi(f) \equiv 0$.

Proof. Assume that the assertion is false. Then there exists $f \in F^*$ such that for any $g \in F^*$ we can choose a Lie algebra L_g which does not satisfy the identity $g \equiv 0$ and admits an action by A with $C_{L_g}(A)$ satisfying $f \equiv 0$. Consider the direct sum $M = \oplus L_g$, where the summation is taken over all $g \in F^*$. It is easy to see that M also admits an action by A with $C_M(A)$ satisfying $f \equiv 0$. Obviously M cannot be PI, a contradiction against 2.2. □

We now turn to groups. Let G be any group. For $x, y \in G$ we use $[x, y]$ to denote the group-commutator $x^{-1}y^{-1}xy$. The long commutators $[x_1, x_2, \ldots, x_k]$ and $[x, {}_n y]$ are defined as in Lie algebras:

$$[x_1] = x_1; \; [x_1, x_2, \ldots, x_k] = [[x_1, x_2, \ldots, x_{k-1}], x_k]$$

and

$$[x, {}_0 y] = x; \; [x, {}_n y] = [[x, {}_{n-1} y], y].$$

If $x_1, x_2, \ldots, x_k$ belong to a set A then we say that $[x_1, x_2, \ldots, x_k]$ is a simple commutator of weight k in elements of A.

The following commutator identities hold in any group and can be checked manually.

LEMMA 2.4.

$$[x,y]^{-1} = [y,x]$$
$$[xy,z] = [x,z][x,z,y][y,z]$$
$$[x,yz] = [x,z][x,y][x,y,z]$$
$$[x,y^{-1},z]^y[y,z^{-1},x]^z[z,x^{-1},y]^x = 1.$$

Let us also record the Collection Formula (see [18, p. 240]). For any integer n and any subgroup H of a group G we denote by H^n the subgroup of G generated by the n-th powers of elements from H and use $\gamma_n(G)$ for the n-th term of the lower central series of G.

LEMMA 2.5. Let x, y be elements of a group G. Then

$$(xy)^{p^n} \equiv x^{p^n} y^{p^n} \ mod \ \gamma_2(G)^{p^n} \prod_{1 \le r \le n} \gamma_{p^r}(G)^{p^{n-r}}.$$

A series of subgroups

$$G = G_1 \geq G_2 \geq \ldots \qquad (*)$$

is called an N-series if it satisfies $[G_i, G_j] \leq G_{i+j}$ for all i, j. Obviously any N-series is central, i.e. $G_i/G_{i+1} \leq Z(G/G_{i+1})$ for any i. An N-series is called N_p-series if $G_i^p \leq G_{pi}$ for all i.

Generalizing constructions discovered by Magnus [27] and Zassenhaus [51], Lazard noticed in [23] that a Lie ring $L^*(G)$ can be associated to any N-series $(*)$ of a group G. He also discovered some very useful special properties of $L^*(G)$ in case $(*)$ is an N_p-series. Let us briefly describe the construction.

Given an N-series $(*)$, let $L^*(G)$ be the direct sum of the abelian groups $L_i^* = G_i/G_{i+1}$, written additively. Commutation in G induces a binary operation $[,]$ in L. For homogeneous elements $xG_{i+1} \in L_i^*, yG_{j+1} \in L_j^*$ the operation is defined by

$$[xG_{i+1}, yG_{j+1}] = [x,y]G_{i+j+1} \in L_{i+j}^*$$

and extended to arbitrary elements of $L^*(G)$ by linearity. It is easy to check that the operation is well-defined and that $L^*(G)$ with the operations $+$ and $[,]$ is a Lie ring.

The above procedure can be performed for each N-series of G. If all quotients G_i/G_{i+1} of an N-series $(*)$ have exponent p then $L^*(G)$ can be viewed as a Lie algebra over $\mathbb{F}_p$, the field with p elements. This is always the case if $(*)$ is an N_p-series. We are now concerned with the relationship between G and $L^*(G)$. For any $x \in G_i \setminus G_{i+1}$ let x^* denote the element xG_{i+1} of $L^*(G)$.

LEMMA 2.6 (Lazard, [23]). If (*) is an N_p-series then $(ad\, x^*)^p = ad\,(x^p)^*$ for any $x \in G$. Consequently, if x is of finite order t then x^* is ad-nilpotent of index at most t.

Let Fr denote the free group on free generators $x_1, x_2, \ldots$, and choose a non-trivial element $w = w(x_1, x_2, \ldots, x_s) \in Fr$. We say that a group G satisfies the identity $w \equiv 1$ if $w(g_1, g_2, \ldots, g_s) = 1$ for any $g_1, g_2, \ldots, g_s \in G$.

The following proposition can be deduced from the proof of Theorem 1 in the paper of J. Wilson and E. Zelmanov [50]

PROPOSITION 2.7. Let G be a group satisfying an identity $w \equiv 1$. Then there exists a non-zero Lie polynomial f over $\mathbb{F}_p$ depending only on p and w such that for any N_p-series (*) of G the algebra $L^*(G)$ satisfies the identity $f \equiv 0$.

In fact J. Wilson and E. Zelmanov describe in [50] an effective algorithm allowing to write f explicitely for any p and w but we do not require this. Let us just record an important special case of the above proposition.

PROPOSITION 2.8 (Higman, [16]). Let n be a p-power, G a group such that $x^n = 1$ for any $x \in G$. Then for any N_p-series (*) the algebra $L^*(G)$ satisfies the identity

$$\sum_{\pi \in S_{n-1}} [x_0, x_{\pi(1)}, x_{\pi(2)}, \ldots, x_{\pi(n-1)}] = 0. \tag{2.9}$$

In general a group G has many N_p-series so that there are many ways to associate to G a Lie algebra as described above. We will introduce now an N_p-series which is of particular importance for applications of Lie-theoretic results to group theory.

To simplify the notation we write γ_i for $\gamma_i(G)$. Set $D_i = D_i(G) = \prod_{jp^k \geq i} \gamma_j^{p^k}$. The subgroups D_i form a series $G = D_1 \geq D_2 \geq \ldots$ in the group G.

PROPOSITION 2.10 ([18, p. 250]). The series $\{D_i\}$ is an N_p-series.

We will call $\{D_i\}$ the p-dimension central series of G. It is also known as the Lazard series or the Jennings-Lazard-Zassenhaus series.

In conformity with the earlier described procedure we can associate to G a Lie algebra $DL(G) = \oplus L_i$ over $\mathbb{F}_p$ corresponding to the p-dimension central series. Here $L_i = D_i/D_{i+1}$. This algebra plays a crucial rôle in all results considered in the paper.

Let $L_p(G) = \langle L_1 \rangle$ be the subalgebra of $DL(G)$ generated by L_1. If G is finitely generated then nilpotency of $L_p(G)$ has strong impact over the structure of G. The following proposition is implicit in [55].

PROPOSITION 2.11. Let G be generated by elements $a_1, a_2, \ldots, a_m$, and assume that $L_p(G)$ is nilpotent of class at most c. Let $\rho_1, \rho_2, \ldots, \rho_s$ be the list of all simple commutators in $a_1, a_2, \ldots, a_m$ of weight $\leq c$. Then for any non-negative integer i the group G can be written as a product

$$G = \langle \rho_1 \rangle \langle \rho_2 \rangle \ldots \langle \rho_s \rangle D_{i+1}$$

of the cyclic subgroups generated by $\rho_1, \rho_2, \ldots, \rho_s$ and D_{i+1}.

Proof. We start with the following remark. For any positive integer i the subgroup D_i is generated by D_{i+1} and elements of the form $[b_1, \ldots, b_j]^{p^k}$, where $jp^k \geq i$ and $b_1, \ldots, b_j \in \{a_1, \ldots, a_m\}$. This can be shown using for example formulae 2.4 and 2.5.

The proposition will be proved by induction on i, the case $i = 0$ being trivial. Assume that $i \geq 1$ and

$$G = \langle\rho_1\rangle\langle\rho_2\rangle \ldots \langle\rho_s\rangle D_i.$$

Then any element $x \in G$ can be written in the form

$$x = \rho_1^{\alpha_1}\rho_2^{\alpha_2} \ldots \rho_s^{\alpha_s} y, \tag{2.12}$$

where $y \in D_i$. Without any loss of generality we can assume that $D_{i+1} = 1$.

By the remark made in the beginning of the proof we can write

$$y = \left(\sigma_1^{p^{k_1}}\right)^{\beta_1} \left(\sigma_2^{p^{k_2}}\right)^{\beta_2} \ldots \left(\sigma_t^{p^{k_t}}\right)^{\beta_t}, \tag{2.13}$$

where each σ_n is of the form $[b_1, \ldots, b_j]$, with $jp^{k_n} \geq i$ and $b_1, \ldots, b_j \in \{a_1, \ldots, a_m\}$.

Let $\tilde{a}_l$ denote $a_l D_2 \in L_p(G)$; $l = 1, \ldots, m$. By the hypothesis $L_p(G)$ is nilpotent of class c, that is $[\tilde{b}_1, \ldots, \tilde{b}_{c+1}] = 0$ for any $b_1, \ldots, b_{c+1} \in \{a_1, \ldots, a_m\}$. This implies that $[b_1, \ldots, b_{c+1}] \in D_{c+2}$ for any $b_1, \ldots, b_{c+1} \in \{a_1, \ldots, a_m\}$ and $\gamma_{c+1} \leq D_{c+2}$. Then, by Proposition 2.10, for any $d \geq c+1$ we have $\gamma_d \leq D_{d+1}$.

Now, if σ_n is of the form $[b_1, \ldots, b_j]$ with $j \geq c+1$ then

$$\sigma_n^{p^{k_n}} \in \gamma_j^{p^{k_n}} \leq D_{j+1}^{p^{k_n}} \leq D_{(j+1)p^{k_n}} \leq D_{i+1} = 1.$$

Hence we can assume that each σ_n is of the form $[b_1, \ldots, b_j]$ with $j \leq c$, in which case σ_n belongs to the list $\rho_1, \rho_2, \ldots, \rho_s$.

It remains to remark that by 2.10 $\sigma_n^{p^{k_n}} \in Z(G)$. Comparing now (2.12) and (2.13) we obtain that

$$x \in \langle\rho_1\rangle\langle\rho_2\rangle \ldots \langle\rho_s\rangle,$$

as required. □

The following corollary is now immediate.

COROLLARY 2.14. Assume the hypotheses of 2.6 and suppose that each ρ_j is of order at most K. Then D_i is of index at most K^s for any i. In particular, the series D_i becomes stationary after finitely many steps.

A group G is said to be residually-p if for any non-trivial element $x \in G$ there exists a normal subgroup $N \leq G$ such that $x \notin N$ and G/N is a finite p-group. M. Lazard proved in [24] that if G is a finitely generated pro-p group such that $L_p(G)$ is nilpotent then G is p-adic analytic. Since any finitely generated residually-p group can be embedded in a finitely generated pro-p group (see for example [4]), we obtain the following

PROPOSITION 2.15. If G is a finitely generated residually-p group such that $L_p(G)$ is nilpotent then G has a faithful linear representation over the field of p-adic numbers.

3. GROUPS WITH COMMUTATORS OF BOUNDED ORDER

This section is devoted to groups in which all commutators $[x_1, x_2, \ldots, x_k]$ satisfy some restrictive condition. In particular we describe a proof of the following theorem.

THEOREM 3.1. Let k be an integer, $q = p^s$ a prime-power, G a residually finite group such that $[x_1, x_2, \ldots, x_k]^q = 1$ for all $x_1, x_2, \ldots, x_k \in G$. Then $\gamma_k(G)$ is locally finite.

This generalizes the positive solution of the Restricted Burnside Problem for groups of prime-power exponent (take $k = 1$). The theorem is no longer true if the assumption that G is residually finite is dropped. Using the technique developed by A. Ol'shanskii [29] it is possible (for sufficiently big values of q) to construct a group G that satisfies the identity $[x, y]^q = 1$ and has the derived group G' non-periodic.

Another result considered in this section deals with residually finite groups in which the commutators $[x_1, x_2, \ldots, x_k]$ are Engel. An element x of a group G is called (left) n-Engel if $[g, {}_n x] = 1$ for any $g \in G$. A group G is called n-Engel if all elements of G are n-Engel. It is a long-standing problem whether any n-Engel group is locally nilpotent. In [49] J. Wilson proved that this is true if G is residually finite. His proof relies on a result of A. Shalev [32] which uses the positive solution of the Restricted Burnside Problem [52],[53]. We will prove

THEOREM 3.2. Let k, n be positive integers, G a residually finite group such that $[x_1, x_2, \ldots, x_k]$ is n-Engel for any $x_1, x_2, \ldots, x_k \in G$. Then $\gamma_k(G)$ is locally nilpotent.

We will consider first finite groups satisfying the hypotheses of 3.1 and 3.2. To prepare the use of Theorem 1.5 we will show that if G is a finite group satisfying the hypothesis of Theorem 3.1 (respectively, 3.2), then $\gamma_k(G)$ is a p-group (respectively, is nilpotent).

In the proof of the next lemma we follow advice of A. Mann and R. Solomon.

LEMMA 3.3. Let G be a finite group in which all commutators $[x_1, x_2, \ldots, x_k]$ are p-elements. Then $\gamma_k(G)$ is a p-group.

Proof. Assume that the result is false and let G be a counterexample of minimal possible order. Obviously G has no non-trivial normal p-subgroups. Let r be a prime divisor of $|G|$ distinct from p and suppose that G contains an r-subgroup R such that the quotient $N_G(R)/C_G(R)$ has an element z of order prime to r. Then z can be viewed as a non-trivial automorphism of R and therefore $[R, \underbrace{z, \ldots, z}_{k-1}] \neq 1$, [7, Theorem 5.3.6]. On the other hand, by the hypothesis, $[R, \underbrace{z, \ldots, z}_{k-1}]$ must contain p-elements, a contradiction. Therefore $N_G(R)/C_G(R)$ is an r-group for any r-subgroup R of G. Hence G possesses a normal r-complement K [7, 7.4.5]. The induction on $|G|$ shows that $\gamma_k(K)$ is a p-group. Since G has no non-trivial normal

p-subgroups, it follows that $\gamma_k(K) = 1$ and K is nilpotent. Again because G has no non-trivial normal p-subgroups, we conclude that K is a p'-group. But then so is G, a contradiction. □

LEMMA 3.4. If for any elements $y, x_1, x_2, \ldots, x_k$ of a finite group G there exists an integer n such that $[y,_n[x_1, x_2, \ldots, x_k]] = 1$ then $\gamma_k(G)$ is nilpotent.

Proof. Engel elements of any finite group generate a nilpotent subgroup [17, III, 6.14]. Since G is finite and $\gamma_k(G) = \langle [x_1, x_2, \ldots, x_k];\ x_1, x_2, \ldots, x_k \in G \rangle$, the result follows. □

The next lemma is a particular case of Lemma 2.1 in the paper of J. Wilson [49].

LEMMA 3.5. Let G be a finitely generated residually finite-nilpotent group. For each prime p let J_p be the intersection of all normal subgroups of G of finite p-power index. If G/J_p is nilpotent for each p then G is nilpotent.

Proof. Let R be the intersection of all normal subgroups N of G such that G/N is torsion-free and nilpotent. Since a torsion-free nilpotent group is residually-p for any prime p, it follows that $J_p \le R$ for any p. We therefore conclude that G/R is nilpotent. Then so is $G/[R, \underbrace{G, \ldots, G}_{k}]$ for any positive integer k. Set $R_0 = R$, $R_{k+1} = [R, \underbrace{G, \ldots, G}_{k}]$. Since G is residually nilpotent, the intersection $\cap_i R_i$ is trivial. By the choice of R the quotient R/R_1 must be periodic. This is a subgroup of a finitely generated nilpotent group G/R_1 and therefore R/R_1 is finitely generated too. Hence R/R_1 is finite, and let π be the set of primes dividing the order of R/R_1. Then R/R_k is a π-group for any k. (It is a well-known general fact: If R is any normal subgroup of a nilpotent group G, and if $R/[R, G]$ is a π-group, then R is a π-group.) It follows that $\cap_{p\in\pi} J_p \le \cap_i R_i = 1$. Since π is finite, G embeds into the direct product of finitely many nilpotent groups G/J_p, $p \in \pi$. Hence G is nilpotent. □

Proof of Theorem 3.1. By Lemma 3.3 $\gamma_k(G)$ is residually-p. To prove that $\gamma_k(G)$ is locally finite let us take arbitrarily a finite subset S of $\gamma_k(G)$ and show that S generates a finite subgroup. Since $\gamma_k(G)$ is generated by elements of the form $[x_1, x_2, \ldots, x_k]$, it follows that S is contained in some subgroup H generated by finitely many elements $a_1, a_2, \ldots, a_m$ each of which can be written in the form $[x_1, x_2, \ldots, x_k]$. Now it is important to observe that if ρ is an arbitrary commutator in $a_1, a_2, \ldots, a_m$ with some system of brackets then ρ also can be written in the form $\rho = [x_1, x_2, \ldots, x_k]$ for suitably chosen $x_1, x_2, \ldots, x_k \in G$. We want to show that H is finite. Let $L = L_p(H)$ be the Lie algebra associated with the p-dimension central series

$$H = D_1 \ge D_2 \ge \ldots$$

of H (see Section 2). Then L is generated by $\tilde{a}_i = a_i D_2$; $i = 1, 2, \ldots m$. Let b be any Lie-commutator in $\tilde{a}_1, \tilde{a}_2, \ldots, \tilde{a}_m$ and c the group-commutator in $a_1, a_2, \ldots, a_m$ having the same system of brackets as b. We have already observed that $c =$

$[x_1, x_2, \ldots, x_k]$ for some $x_1, x_2, \ldots, x_k \in G$, so we conclude that $c^q = 1$. Suppose that $c \in D_j \setminus D_{j+1}$. It is immediate from the definition of L that either $b = 0$ or $b = cD_{j+1}$. This implies (Lemma 2.6) that b is ad-nilpotent of index at most q. Further, the group H satisfies the identity $[y_1, y_2, \ldots, y_k]^q = 1$. Therefore, by 2.7, L satisfies some non-trivial polynomial identity. By Theorem 1.5 we conclude that L is nilpotent. Corollary 2.14 now shows that the index of D_i in H is bounded from above by some number that depends only on q, m and the nilpotency class of L. Since H is residually-p, it follows that the intersection of all D_i is trivial and we derive that H is finite. □

The proof of 3.2 is similar to that of 3.1.

Proof of Theorem 3.2. By Lemma 3.4 $\gamma_k(G)$ is residually finite-nilpotent. To prove that $\gamma_k(G)$ is locally nilpotent let us take arbitrarily a finite subset S of $\gamma_k(G)$ and show that S generates a nilpotent subgroup. Since $\gamma_k(G)$ is generated by elements of the form $[x_1, x_2, \ldots, x_k]$, it follows that S is contained in some subgroup H generated by finitely many elements $a_1, a_2, \ldots, a_m$ each of which can be written in the form $[x_1, x_2, \ldots, x_k]$. As in the proof of 3.1 we observe that if ρ is an arbitrary commutator in $a_1, a_2, \ldots, a_m$ with some system of brackets then ρ also can be written in the form $\rho = [x_1, x_2, \ldots, x_k]$ for suitably chosen $x_1, x_2, \ldots, x_k \in G$. We want to show that H is nilpotent. Lemma 3.5 allows us to assume that H is residually-p for some prime p. Let $L = L_p(H)$ be the Lie algebra associated with the p-dimension central series

$$H = D_1 \geq D_2 \geq \cdots$$

of H. Then L is generated by $\tilde{a}_i = a_i D_2$; $i = 1, 2, \ldots m$. Let b be any Lie-commutator in $\tilde{a}_1, \tilde{a}_2, \ldots, \tilde{a}_m$ and c be the group-commutator in $a_1, a_2, \ldots, a_m$ having the same system of brackets as b. We have already observed that $c = [x_1, x_2, \ldots, x_k]$ for some $x_1, x_2, \ldots, x_k \in G$, so we conclude that c is n-Engel. This implies that b is ad-nilpotent of index at most n. Further, the group H satisfies the identity $[y,_n [y_1, y_2, \ldots, y_k]] = 1$. Therefore, by 2.7, L satisfies some non-trivial polynomial identity. Theorem 1.5 now implies that L is nilpotent. Hence, by Proposition 2.15, H has a faithful linear representation over the field of p-adic numbers. Clearly H cannot have a free subgroup of rank two, and so by Tits' Alternative [46] H has a solvable subgroup of finite index. Lemma 3.4 shows that each finite quotient of H is solvable, so that we conclude that H is solvable too. It is now easy to deduce from a result of K. Gruenberg [9] that H is nilpotent, as required. □

We conclude this section by listing some open problems related to the results described here.

PROBLEM 1. Does there exist a group G satisfying the hypothesis of Theorem 3.1 and having $\gamma_k(G)$ of unbounded exponent?

The answer to this problem is likely to be "yes" but finding an example does not seem to be easy. Using results of B. Hartley [12] one can show that such an example cannot be finitely generated (see [42] for detail in the case $k = 2$).

PROBLEM 2. Let k, n be positive integers, G a residually finite group such that $[x_1, x_2, \ldots, x_k]^n = 1$ for any $x_1, x_2, \ldots, x_k \in G$. Is then $\gamma_k(G)$ necessarily locally finite?

4. BOUNDING THE EXPONENT OF A FINITE GROUP WITH AUTOMORPHISMS

Let q be a prime, and let A be a non-cyclic group of order q^2 acting on a finite group G. It is well-known (see [7, Theorem 6.2.4, Theorem 5.3.16]) that if G is any group of order prime to q then

$$G = \langle C_G(a); a \in A^{\#} \rangle , \tag{4.1}$$

where $A^{\#} = A \setminus \{1\}$. If, moreover, G is a p-group then we even have

$$G = \prod_{a \in A^{\#}} C_G(a). \tag{4.2}$$

Since A normalizes some Sylow p-subgroup of G for any p dividing $|G|$ ([7, Theorem 6.2.4]), it is immediate that if $|C_G(a)| \leq n$ for any $a \in A^{\#}$ then $|G| \leq n^{q+1}$ (we use that A has exactly $q+1$ cyclic subgroups). How profound is the connection between the structure of G and that of $C_G(a)$; $a \in A^{\#}$? It is known that if $C_G(a)$ is nilpotent for each $a \in A^{\#}$ then G is metanilpotent [48]. In some situations this result holds even if G is allowed to be infinite periodic [38]. In this section we will prove the following theorem.

THEOREM 4.3 (Khukhro, Shumyatsky [21]). Suppose that A is a non-cyclic group of order q^2 acting on a finite group G of coprime order, and let n be such an integer that the exponents of the centralizers $C_G(a)$ of non-trivial elements $a \in A^{\#}$ divide n. Then the exponent of G is bounded in terms of n and q.

Note that the exponent of the centralizer of a single automorphism a of a finite group G has no impact over the exponent of G. Indeed, any abelian group of odd order admits a fixed-point-free automorphism of order two. Hence, we cannot bound the exponent of G solely in terms of the exponent of $C_G(a)$. In view of this the following theorem seems to be interesting.

THEOREM 4.4. Let n be a positive integer, G a finite group of odd order admitting an involutory automorphism a such that $C_G(a)$ is of exponent dividing n. Suppose that for any $x \in G$ the element $[x, a] = x^{-1}x^a$ has order dividing n. Then the exponent of G is bounded in terms of n.

Apart from the technique described in Section 2 the proof of Theorems 4.3 and 4.4 involves using powerful p-groups. These were introduced by A. Lubotzky and A. Mann in [26]: a finite p-group G is powerful if and only if $G^p \geq [G, G]$ for $p \neq 2$ (or $G^4 \geq [G, G]$ for $p = 2$).

Powerful p-groups have many nice linear properties, of which we need the following: if a powerful p-group G is generated by elements of exponent p^e, then the exponent of G is p^e too (see [4, Lemma 2.2.5]). Combining this with 4.1, we conclude that if G is a powerful p-group satisfying the hypothesis of Theorem 4.3, then the exponent of G divides n. Thus, it is sufficient to reduce the proof of Theorem 4.3 to the case of powerful p-groups. A. Shalev was the first to discover relevance of powerful p-groups to problems on automorphisms of finite groups [33], [34]. However our situation is quite different from that considered in the papers of A. Shalev. The reason powerful p-groups emerge in the context of the Restricted Burnside Problem is the following lemma.

LEMMA 4.5. Suppose that P is a d-generator finite p-group such that the Lie algebra $L_p(P)$ is nilpotent of class c. Then P has a powerful characteristic subgroup of $\{p, c, d\}$-bounded index.

Proof. Let $\rho_1, \dots, \rho_s$ be all simple commutators of weight $\leq c$ in the generators of P; here s is a $\{d, c\}$-bounded number. Since P is a finite p-group, the nilpotency of $L_p(P)$ of class c implies that every element $g \in P$ can be written in the form $g = \rho_1^{k_1} \dots \rho_s^{k_s}$ (see Proposition 2.11). Hence $|P/P^{p^m}| \leq p^{sm}$ for any positive integer m. Let V be the intersection of the kernels of all homomorphisms of P into $GL_s(\mathbb{F}_p)$. Put $W = V$ if $p \neq 2$ (or $W = V^2$ if $p = 2$). The exponent of the Sylow p-subgroup of $GL_s(\mathbb{F}_p)$ is a $\{p, s\}$-bounded number. Then $P^{p^a} \leq W$ for some $\{p, s\}$-bounded number a, which is also $\{p, c, d\}$-bounded, since s is $\{c, d\}$-bounded. There is a $\{p, c, d\}$-bounded number $u \geq a$ such that $|P^{p^u}/P^{p^{u+1}}| \leq p^s$, for otherwise the inequality $|P/P^{p^m}| \leq p^{sm}$ would be violated for some m. Then $P^{p^u} \leq P^{p^a} \leq W$, and P^{p^u} is s-generator since $|P^{p^u}/\Phi(P^{p^u})| \leq |P^{p^u}/P^{p^{u+1}}| \leq p^s$. Now, by [4, Proposition 2.12], P^{p^u} is a powerful subgroup. The index of P^{p^u} is at most p^{us} and hence is $\{p, c, d\}$-bounded. □

We will now quote a well-known lemma which is of fundamental importance whenever one uses Lie algebra methods to study finite groups with automorphisms of coprime order.

LEMMA 4.6 ([7, 6.2.2 (iv)]). Let A be a finite group acting on a finite group G. Assume that $(|A|, |G|) = 1$ and that N is a normal A-invariant subgroup of G. Then $C_{G/N}(A) = C_G(A)N/N$.

Let H be a subgroup of a group G. Set $H_j = D_j \cap H$, where D_j is the j-th term of the p-dimension central series of G. Write

$$L(G, H) = \oplus H_j D_{j+1}/D_{j+1} \text{ and } L_p(G, H) = L_p(G) \cap L(G, H).$$

Observation 4.7. $L(G, H)$ is a subalgebra of $DL(G)$. It is isomorphic to the Lie algebra associated with the N_p-series $\{H_j\}$ of H.

Let now a finite group A act on a group G. Obviously A induces an automorphism group of every quotient D_j/D_{j+1}. This action extends to the direct sum $\oplus D_j/D_{j+1}$. Thus, A can be viewed as a group acting on $L_p(G)$ by Lie algebra automorphisms. The following remark is immediate from Lemma 4.6.

Observation 4.8. If G is finite and $(|G|, |A|) = 1$ then

$$L_p(G, C_G(A)) = C_{L_p(G)}(A).$$

The next lemma will be helpful in the proof of Theorem 4.3.

LEMMA 4.9. Suppose that L is a Lie algebra, H a subalgebra of L generated by r elements $h_1, \ldots, h_r$ such that all commutators in the h_i are ad-nilpotent in L of index t. If H is nilpotent of class u, then for some $\{r, t, u\}$-bounded number v we have $[L, \underbrace{H, \ldots, H}_{v}] = 0$.

Proof. We apply to a sufficiently long (but of $\{r, t, u\}$-bounded length) commutator

$$[l, h_{i_1}, h_{i_2}, \ldots]$$

a collecting process whose aim is to rearrange the h_i (and emerging by the Jacobi identity commutators in the h_i) in an ordered string after l, where all occurrences of a given element (h_i or a commutator in the h_i) would form an unbroken segment. Since H is nilpotent, this process terminates at a linear combination of commutators with sufficiently long segments of equal elements. All these commutators are equal to 0 because all commutators in the h_i are ad-nilpotent by the hypothesis. □

Proof of Theorem 4.3. As noted above, for every prime p dividing $|G|$ there is an A-invariant Sylow p-subgroup P of G. Since $P = \langle C_P(a); a \in A^{\#} \rangle$, such a prime p must be a divisor of n. If Theorem 4.3 is valid in the case where G is a finite p-group, the exponents of all Sylow subgroups of G are bounded in terms of q and n, which implies that the exponent of G is bounded. Thus, we may assume G to be a finite p-group (for a prime $p \neq q$). We may also assume G to be generated by q^2 elements, since every element $g \in G$ is contained in the A-invariant subgroup $\langle g^a; a \in A \rangle$.

¿From now on, in addition to the hypothesis of Theorem 4.3, G is a finite q^2-generator p-group and n is a power of p. Let $A_1, A_2, \ldots, A_{q+1}$ be the distinct cyclic subgroups of A. Set $D_j = D_j(G)$, $L = L_p(G)$, $L_j = L \cap D_j / D_{j+1}$, so that $L = \oplus L_j$. Let $L_{ij} = C_{L_j}(A_i)$. Then, by 4.2, for any j we have

$$L_j = \sum_{1 \leq i \leq q+1} L_{ij}.$$

By Lemma 4.6 for any $l \in L_{ij}$ there exists $x \in D_j \cap C_G(A_i)$ such that $l = xD_{j+1}$. By the hypothesis x is of order at most n, whence l is ad-nilpotent of index at most n (Lemma 2.6). Thus,

$$\text{any element in } L_{ij} \text{ is ad-nilpotent of index at most } n. \tag{4.10}$$

Since G is generated by q^2 elements, the $\mathbb{F}_p$-space L_1 is spanned by q^2 elements. In particular, L is generated by at most q^2 ad-nilpotent elements, each from L_{i1} for some i. But we cannot claim that every Lie commutator in these generators is again in some L_{ij} and hence is ad-nilpotent too.

To overcome this difficulty, we extend the ground field of L by a primitive qth root of unity ω, forming $\overline{L} = L \otimes \mathbb{F}_p[\omega]$. The idea is to replace L by $\overline{L}$ and to prove that $\overline{L}$ is nilpotent of $\{q, n\}$-bounded class, which will, of course, imply the same nilpotency result for L. Before that we translate the properties of L into the language of $\overline{L}$.

It is natural to identify L with the $\mathbb{F}_p$-subalgebra $L \otimes 1$ of $\overline{L}$. We note that if an element $x \in L$ is ad-nilpotent of index m, say, then the "same" element $x \otimes 1$ is ad-nilpotent in $\overline{L}$ of the same index m.

Put $\overline{L_j} = L_j \otimes \mathbb{F}_p[\omega]$; then $\overline{L} = \langle \overline{L_1} \rangle$, since $L = \langle L_1 \rangle$, and $\overline{L}$ is the direct sum of the homogeneous components $\overline{L_j}$. Since the $\mathbb{F}_p$-space L_1 is spanned by q^2 elements, so is the $\mathbb{F}_p[\omega]$-space $\overline{L_1}$.

The group A acts naturally on $\overline{L}$, and we have $\overline{L_{ij}} = C_{\overline{L_j}}(A_i)$, where $\overline{L_{ij}} = L_{ij} \otimes \mathbb{F}_p[\omega]$. Let us show that

$$\text{any element } y \in \overline{L_{ij}} \text{ is ad-nilpotent of } \{q,n\}\text{-bounded index.} \tag{4.11}$$

Since $\overline{L_{ij}} = L_{ij} \otimes \mathbb{F}_p[\omega]$, we can write

$$y = x_0 + \omega x_1 + \omega^2 x_2 + \ldots + \omega^{q-2} x_{q-2}$$

for some $x_s \in L_{ij}$, so that each of the summands $\omega^s x_s$ is ad-nilpotent of index n by 4.10. Set $H = \langle x_0, \omega x_1, \ldots, \omega^{q-2} x_{q-2} \rangle$. Note that $H \subseteq C_{\overline{L}}(A_i)$, since $\omega^s x_s \in C_{\overline{L}}(A_i)$ for all s. A commutator of weight k in the $\omega^s x_s$ has the form $\omega^t x$ for some $x \in L_{im}$, where $m = kj$. By 4.10 such an x is ad-nilpotent of index n and hence so is $\omega^t x$.

Further, combining Observations 4.7 and 4.8 with Proposition 2.8, we conclude that $C_L(A_i)$ satisfies the polynomial identity 2.9. This identity is polylinear and so it is also satisfied by $C_L(A_i) \otimes \mathbb{F}_p[\omega] = C_{\overline{L}}(A_i)$. We have already observed that $H \subseteq C_{\overline{L}}(A_i)$, whence the identity 2.9 is satisfied in H. Hence by Theorem 2.1 H is nilpotent of $\{q,n\}$-bounded class. Lemma 4.9 now says that $[L, \underbrace{H, \ldots, H}_{v}] = 0$ for some $\{q,n\}$-bounded number v. This establishes 4.11.

Since A is abelian, and the ground field is now a splitting field for A, every $\overline{L_j}$ decomposes in the direct sum of common eigenspaces for A. In particular, $\overline{L_1}$ is spanned by common eigenvectors for A, and it requires at most q^2 of them to span $\overline{L_1}$. Hence $\overline{L}$ is generated by q^2 common eigenvectors for A from $\overline{L_1}$. Every common eigenspace is contained in the centralizer $C_{\overline{L}}(A_i)$ for some $1 \leq i \leq q+1$, since A is non-cyclic. Note that any commutator in common eigenvectors is again a common eigenvector. The main advantage of extending the ground field now becomes clear: if $l_1, \ldots, l_{q^2} \in \overline{L_1}$ are common eigenvectors for A generating $\overline{L}$ then any commutator in these generators belongs to some $\overline{L_{ij}}$ and therefore, by 4.11, is ad-nilpotent of $\{q,n\}$-bounded index.

We already know that the identity 2.9 is satisfied in $C_L(A_i) \otimes \mathbb{F}_p[\omega] = C_{\overline{L}}(A_i)$. So, if f denotes the Lie polynomial in 2.9 then, by 2.3, $\overline{L}$ satisfies some identity $\phi(f) \equiv 0$ which depends only on n and q. Theorem 2.1 now shows that $\overline{L}$ (hence L) is nilpotent of $\{q,n\}$-bounded class.

By Lemma 4.5 G contains a characteristic powerful subgroup G_1 of $\{q,n\}$-bounded index. Combining the hypothesis with 4.1 we see that G_1 is generated by elements of order dividing n. It follows that the exponent of G_1 divides n [4, 2.2.5]. Therefore the exponent of G is $\{q,n\}$-bounded, as required. □

In the proof of Theorem 4.4 we use the following well-known fact.

LEMMA 4.12. Let G be a finite group of odd order with an involutory automorphism a. Then any element $x \in G$ can be written uniquely in the form $x = gh$, where $g^a = g^{-1}$ and $h \in C_G(a)$. Moreover $x^a = x^{-1}$ if and only if $x = [y, a] = y^{-1}y^a$ for some $y \in G$.

Proof of Theorem 4.4. As in Theorem 4.3 we remark that G possesses an a-invariant Sylow p-subgroup for any prime p dividing $|G|$. It is therefore sufficient to bound the exponent of a-invariant p-subgroups of G. So, without any loss of generality we may assume that G is a p-group and n is a p-power. Take an arbitrary element $x \in G$. According to 4.12 we can write $x = gh$, where $g^a = g^{-1}$ and $h \in C_G(a)$. So to prove that the order of x is n-bounded it is sufficient to prove that the order of $\langle g, h\rangle$ is n-bounded. Thus, we can assume that $G = \langle g, h\rangle$.

Let $\{D_j\}$ be the p-dimension central series of G, $L_j = D_j/D_{j+1}$, $L = DL(G)$ the Lie algebra corresponding to the series $\{D_j\}$. We can naturally view a as an automorphism of L. Set

$$L^+ = C_L(a) \text{ and } L^- = \{l \in L; l^a = -l\}.$$

Then one has:

$$[L^+, L^+] \leq L^+,\ [L^+, L^-] \leq L^-,\ [L^-, L^-] \leq L^+. \tag{4.13}$$

For a fixed j let us denote for a moment L_j by M. Since M is a-invariant, 4.12 shows that $M = M^+ \oplus M^-$, where $M^+ = L^+ \cap M$ and $M^- = L^- \cap M$. It is easy to check (using 4.6) that for any $m \in M^+$ there exists $d \in C_{D_j}(a)$ such that $dD_{j+1} = m$. By the hypothesis d is of order dividing n. Therefore, by Lemma 2.6, m is ad-nilpotent of index at most n. Similarly one concludes that any element in M^- is ad-nilpotent of index at most n.

Since $G = \langle g, h\rangle$, it follows that the subalgebra $L_p(G)$ of L is generated by $\tilde{g} = gD_2$ and $\tilde{h} = hD_2$. The assumption that $g^a = g^{-1}$ and $h^a = h$ implies that $\tilde{g} \in L^-$, $\tilde{h} \in L^+$. By 4.13 we conclude that any commutator in $\tilde{g}, \tilde{h}$ lies either in $L_j \cap L^+$ or in $L_j \cap L^-$ for some j. Combining this with the observations made in the preceding paragraph, we arrive at the conclusion that any commutator in $\tilde{g}, \tilde{h}$ is ad-nilpotent of index at most n.

Arguing like in the proof of Theorem 4.3, remark that the identity 2.9 is satisfied in $C_L(a)$. Therefore, by Theorem 2.3, L satisfies a certain polynomial identity depending only on n. Theorem 2.1 now shows that $L_p(G)$ is of n-bounded nilpotency class.

By Lemma 4.5 we derive that G contains a characteristic powerful subgroup G_1 of n-bounded index. Lemma 4.12 implies that G_1 is generated by elements of order dividing n. Therefore the exponent of G_1 divides n. The theorem follows. □

It is not clear whether the above theorem can be extended to the case of automorphism of any order prime to that of G.

PROBLEM 3. Let n be a positive integer, G a finite group admitting an automorphism a, of order prime to $|G|$, such that $C_G(a)$ is of exponent dividing n. Suppose that for any $x \in G$ the element $[x, a] = x^{-1}x^a$ has order dividing n. Is then the exponent of G bounded in terms of n and $|a|$?

5. ON CENTRALIZERS IN PERIODIC RESIDUALLY FINITE GROUPS

In this section we find some sufficient conditions for a periodic residually finite group to be locally finite. As is attested by the groups constructed in [1], [6], [8], [10], [44], in general a periodic residually finite group need not be locally finite. Our theme will be the following.

Given a periodic residually finite group G acted on by a finite group A, under what conditions on A and $C_G(A)$ does it follow that G is locally finite? Since any subgroup of G acts on G by inner automorphisms, this problem includes problems on centralizers of finite subgroups of G.

In 1972 V. Shunkov proved that if a periodic group G admits an involutory automorphism a with finite centralizer $C_G(a)$ then G contains a solvable subgroup of finite index [43]. This was strengthened later by B. Hartley and Th. Meixner who showed that G has a nilpotent subgroup of index depending only on $|C_G(a)|$ and of nilpotency class at most two [15]. Locally finite groups G having an automorphism a of arbitrary prime order p with finite centralizer $C_G(a)$ have been studied intensively in seventies and eities (see for example [13]). Khukhro showed that these groups have a nilpotent subgroup of finite index depending only on $|C_G(a)|$ and on p, and nilpotency class depending only on p [19]. In general, a very interesting direction in locally finite group theory is to classify in some sense locally finite groups G having a finite subgroup A such that $C_G(A)$ possesses some prescribed property, as for example the property to be a linear group [14].

An immediate corollary of the theorem of Shunkov is that G is locally finite. This part of the theorem, and in fact most difficult part, has no analogue for periodic groups admitting an automorphism of odd order. According to G. Deryabina and A. Ol'shanskii, for any positive integer n which has at least one odd divisor there exists an infinite group G having a non-central element of order n such that all proper subgroups of G are finite [3]. Therefore the result of Shunkov cannot be extended to periodic groups having an automorphism whose order is not a 2-power. Moreover, Obraztsov and Miller constructed for any (not necessarily distinct) odd primes p and q a finitely generated infinite residually finite periodic p-group admitting a fixed-point-free automorphism of order q [28].

On the other hand, in the recent years new means to treat the problem in the residually finite case have been found. N. R. Rocco and the author proved in [30] that if a periodic residually finite group G admits an automorphism a of order 2^s with $C_G(a)$ finite then G is locally finite. Somewhat later ([39], [40]) the author proved local finiteness of a periodic residually finite group G in the following cases:

1) G admits a 4-group A of automorphisms with $C_G(A)$ finite; or

2) G has no elements of order two and admits an involutory automorphism a such that $C_G(a)$ is abelian.

Then A. Shalev obtained in [36] some very general results on local finiteness of periodic residually finite groups acted on by a finite 2-group (see 5.2 and 5.3 bellow). These results were extended in [41] to the case when the acting group is not necessarily of 2-power order.

If A is any finite group, let $q(A)$ denote the maximal prime divisor of $|A|$. One of the results described in this section is the following theorem.

THEOREM 5.1. Let G be a residually finite group acted on by a finite solvable group A with $q = q(A)$. Assume that G has no $|A|$-torsion and $C_G(A)$ is either solvable or of finite exponent. If any $q-1$ elements of G generate a finite solvable subgroup then G is locally finite.

We should mention that for any integer $d \geq 2$ there exist infinite d-generator residually finite groups in which all $(d-1)$-generator subgroups are finite. The corresponding examples are provided by Golod's groups [6].

If under the hypothesis of 5.1 A is a 2-group then the condition imposed on G is that G is merely periodic. This important special case is due to A. Shalev [36].

THEOREM 5.2 (Shalev). Let G be a periodic residually finite group with no 2-torsion acted on by a finite 2-group A. Suppose the centralizer $C_G(A)$ is solvable, or of finite exponent. Then G is locally finite.

Other results to be described in this section are as follows.

THEOREM 5.3 (Shalev). Let G be a periodic residually finite group having a finite 2-subgroup A such that the centralizer $C_G(A)$ is finite. Then G is locally finite.

THEOREM 5.4. Let q be a prime, G a residually finite group in which each 2-generator subgroup is finite. Suppose that G has a finite q-subgroup A such that the centralizer $C_G(A)$ is finite. Then G is locally finite.

Recall that the Fitting height of a finite solvable group G is defined as the least number $h = h(G)$ such that G possesses a normal series

$$G = G_1 \geq G_2 \geq \ldots \geq G_{h+1} = 1,$$

all of whose quotients G_i/G_{i+1} are nilpotent. Thus, G is nilpotent if and only if $h(G) = 1$. In this section we will use some deep results on Fitting height of finite solvable groups.

THEOREM 5.5 (Thompson, [45]). Let G and A be finite solvable groups such that $(|G|, |A|) = 1$. Assume that A acts on G in such a manner that $h(C_G(A)) = h$. Assume further that the order of A is a product of k not necessarily distinct primes. Then $h(G)$ is $\{h, k\}$-bounded.

The following lemma is a corollary of the famous theorem of P. Hall and G. Higman on p-length of a p-solvable finite group of given exponent [11].

LEMMA 5.6 (Shalev, [36]). The Fitting height of a finite solvable group of exponent n is n-bounded.

LEMMA 5.7. Let h be a positive integer, G a residually finite-solvable group such that $h(Q) \leq h$ for any finite quotient Q of G. Then G possesses a normal series

$$G = G_1 \geq G_2 \geq \ldots \geq G_{h+1} = 1,$$

all of whose quotients are residually finite-nilpotent.

Proof. Let us use the induction on h, the case $h = 1$ being obvious. Assume that $h \geq 2$, and let H be the intersection of all finite index normal subgroups N of G such that $h(G/N) \leq h-1$. By the induction hypothesis G possesses a normal series

$$G = G_1 \geq G_2 \geq \ldots \geq G_h = H,$$

all of whose quotients are residually finite-nilpotent. Therefore, it suffices to show that H is residually nilpotent. Let x be any non-trivial element of H. Since G is residually finite, there exists a normal subgroup N of finite index in G such that $x \notin N$. By hypothesis G possesses a normal series

$$G = N_1 \geq N_2 \geq \ldots \geq N_{h+1} = N,$$

all of whose quotients are nilpotent. We note that $h(G/N) = h$ for N does not contain H. Therefore $h(G/N_h) = h - 1$ and so $H \leq N_h$. Since the quotient N_h/N is nilpotent, so is $H/H \cap N$. Thus for any non-trivial element $x \in H$ we can find a normal in G subgroup N such that $x \notin N$ and $H/N \cap H$ is finite and nilpotent. This means that H is residually finite-nilpotent, as required. □

Let G be a periodic group acted on by a finite group A. Suppose that G has no non-trivial elements of order dividing that of A, and let N be an A-invariant normal subgroup of G. Lemma 4.6 says that if G is finite then $C_{G/N}(A) = C_G(A)N/N$. We saw in the previous section that this fact is of fundamental importance for using Lie ring methods in the study of finite groups G having automorphisms of coprime order. The applicability of Lie ring methods to the study of infinite periodic groups G acted on by a finite group A depends ultimately on how successful one is in extending Lemma 4.6 to the case when G is allowed to be infinite periodic. Since in general $C_{G/N}(A) \neq C_G(A)N/N$, one has to impose additional conditions on G and A.

Given a positive integer n, a group G is said to be n-finite if any n-generator subgroup of G is finite. Thus, a group is 1-finite if and only if it is periodic. It was proved in [37] that the equality $C_{G/N}(A) = C_G(A)N/N$ holds whenever A is a 2-group. More generally, we have.

LEMMA 5.8. Let A be a finite solvable group with $q = q(A)$ acting on a $(q-1)$-finite group G with no $|A|$-torsion. Let N be a normal A-invariant subgroup of G. Then $C_{G/N}(A) = C_G(A)N/N$.

Proof. It suffices to show that any A-invariant coset xN contains an element from $C_G(A)$. Suppose first that A is of order q and let a be a generator of A. Set

$$x_0 = x^{-1}x^a,\ x_1 = x_0^a,\ \ldots,\ x_{q-1} = x_{q-2}^a.$$

Then all $x_0, x_1, \ldots, x_{q-1}$ lie in N and $x_0x_1 \ldots x_{q-1} = 1$. It follows that the subgroup $F = \langle x_0, x_1, \ldots, x_{q-1}\rangle$ is generated by at most $q-1$ elements. Hence F is finite. We note that F is A-invariant and consider the natural split extension FA.

Since G has no $|A|$-torsion, it is easy to see that A and $B = \langle ax_0^{-1}\rangle$ are Sylow q-subgroups of FA. Hence there exists an element $y \in F$ such that $B = A^y$. Therefore $xy^{-1} \in N_G(A) = C_G(A)$ and $x \in C_G(A)N$, as required.

Now let A be of non-prime order. Let D be a non-trivial proper normal subgroup of A. Since D is of order less than that of A, we can assume by induction that

$xN \cap C_G(D) \neq \emptyset$. Therefore without any loss of generality we can assume $x \in C_G(D)$. Let $H = C_G(D)$, $\bar{A} = A/C_A(H)$, $M = N \cap H$. The coset xM is obviously $\bar{A}$-invariant. Arguing by induction on $|A|$ and using that $\bar{A}$ has order less than that of A we can assume that xM contains an element $z \in C_H(\bar{A})$. Now it remains to notice that $zN = xN$ and $z \in C_G(A)$. □

LEMMA 5.9. Let q be a prime, A a finite q-group acting on a residually finite group G. Suppose that G is $(q-1)$-finite and $C_G(A)$ has no q-torsion. Then G has no q-torsion.

Proof. Suppose that the lemma is false and assume first that A is of order q. Let a be a generator of A. Since G is residually finite, we can choose a normal A-invariant subgroup N such that G/N is a finite group whose order is divisible by q. It follows that A centralizes some element xN of order q in G/N. Then $x^{-1}x^a \in N$. Set

$$x_0 = x^{-1}x^a,\ x_1 = x_0^a,\ \ldots,\ x_{q-1} = x_{q-2}^a.$$

Arguing like in the previous lemma we observe that $F = \langle x_0, x_1, \ldots, x_{q-1}\rangle$ is finite. If q divides $|F|$ then obviously A must centralize some elements of order q in F. This yields a contradiction, so assume that $|F|$ is prime to q. Then $\langle a{x_0}^{-1}\rangle$ is a Sylow q-subgroup of FA. Therefore it is a conjugate of A. Hence there exists an element $y \in F$ such that $y^{-1}ay = x^{-1}ax$. Then $z = xy^{-1} \in C_G(A)$. Since the image of z in G/N is of order q, it follows that the order of z is divided by q. Therefore $C_G(A)$ contains an element of order q, a contradiction. R Suppose now that A is of order q^n and use induction on n. Let a be an element of prime order in $Z(A)$. Set $C = C_G(a)$. If C is q-torsion free then by the previous paragraph so is G. Assume that C has non-trivial q-elements. Since A induces an automorphism group of C whose order is strictly less than that of A, the induction hypothesis implies that some of q-elements of C must lie in $C_G(A)$. □

Obviously the conclusions of Lemmas 5.8 and 5.9 remain true if we replace the assumption that G is $(q-1)$-finite by the assumption that the semidirect product GA is 2-finite. Really, keeping notation like in Lemma 5.8 let us note that if GA is 2-finite then F is finite because $F \leq \langle x, a\rangle$. Leaving other parts of proofs unchanged we reach the following results.

LEMMA 5.8′. Let A be a finite solvable group acting on a group G with no $|A|$-torsion. Suppose that GA is 2-finite. Let N be a normal A-invariant subgroup of G. Then $C_{G/N}(A) = C_G(A)N/N$.

LEMMA 5.9′. Let q be a prime, A a finite q-group acting on a residually finite group G. Suppose that GA is 2-finite and $C_G(A)$ has no q-torsion. Then G has no q-torsion.

PROPOSITION 5.10. Let p be a prime and G a periodic residually finite p-group acted on by a finite solvable group A whose order is prime to p. Suppose that $q = q(A)$ and G is $(q-1)$-finite. If $C_G(A)$ satisfies a non-trivial identity then G is locally finite.

Proof. Since any finite set of elements of G is contained in a finitely generated A-invariant subgroup, we can assume that G is finitely generated. Let $L = L_p(G)$

be the Lie algebra associated with the p-dimension central series of G. We regard A as a group acting on L. Just as in Section 4 (but using Lemma 5.8 in place of Lemma 4.6) we remark that $C_L(A) = L_p(G, C_G(A))$ and so, because $C_G(A)$ satisfies a non-trivial identity, by Proposition 2.7 $C_L(A)$ is PI. Since G is periodic, it follows from Lemma 2.6 that L has a finite set of generators in which every commutator is ad-nilpotent. Consequently, by Theorem 1.5 combined with Theorem 2.2 L is nilpotent.

Let $g_1, g_2, \ldots, g_m$ be some generating set of G, let c be the nilpotency class of L and write $\rho_1, \rho_2, \ldots, \rho_s$ for the list of all simple commutators of weight at most c in $g_1, g_2, \ldots, g_m$. Choose K to be the maximum of orders of ρ_i. Then, by 2.14, $D_j(G)$ has index at most K^s for any j. Since G is residually-p, it follows that the order of G is at most K^s. The proof is complete. □

Proof of Theorems 5.1 and 5.2. Let us note that G is residually solvable. Indeed, if $q = 2$ then G has no 2-torsion and so G is residually solvable by the Feit-Thompson Theorem [5]. If $q \geq 3$ then any two elements of G generate a finite solvable subgroup. Since any simple finite group can be generated by two elements, it follows that G is residually solvable.

Assume that $C_G(A)$ is of finite exponent. Let N be any A-invariant normal subgroup of finite index in G and $Q = G/N$. Then Q is solvable because G is residually so. By Lemma 5.8 A acts on Q in such a way that $C_Q(A)$ is of exponent at most that of $C_G(A)$. By Lemma 5.6 the Fitting height $h(C_Q(A))$ of $C_Q(A)$ is bounded in terms of the exponent of $C_Q(A)$. Theorem 5.5 now shows that the Fitting height $h(Q)$ of Q is bounded in terms of the exponent of $C_G(A)$ and $|A|$. Thus, $h(Q)$ is bounded by a number h which does not depend on Q. Therefore, by Lemma 5.7, G possesses a normal A-invariant series of length at most $h+1$ all of whose quotients are residually nilpotent. Induction on h shows that without any loss of generality G can be assumed residually nilpotent. Then G is a direct product of A-invariant p-subgroups and finiteness of G follows from Proposition 5.10.

If $C_G(A)$ is solvable and Q has the same meaning as above then $h(C_Q(A))$ is at most the derived length of $C_G(A)$ and we can repeat the argument. This completes the proof. □

Using Lemma 5.9 and 5.9′ we can in some situations derive local finiteness of G even without requiring that G has no $|A|$-torsion.

Proof of Theorems 5.3 and 5.4. Since G is residually finite, we can choose a normal subgroup H of finite index in G such that $H \cap C_G(A) = 1$. It suffices to prove that H is locally finite. Lemmas 5.9 and 5.9′ imply that H has no $|A|$-torsion. Hence, by 5.8 (or 5.8′), A acts fixed-point-freely on every A-invariant finite quotient Q of H. A well-known corollary of the classification of simple finite groups says that any finite group admitting a fixed-point-free automorphism group of coprime order is solvable. We now conclude that Q is necessarily solvable. Hence H is residually solvable. This places us in a position to apply Theorem 5.1 (or Theorem 5.2 in case q=2) and derive that G is locally finite. □

Since there is no restriction on the identity satisfied by $C_G(A)$ in Proposition 5.10, it is not unreasonable to conjecture that the following problem can be answered positively.

PROBLEM 4. Let A be a finite 2-group acting on a periodic residually finite group G which has no 2-torsion. Assume that $C_G(A)$ satisfies a non-trivial identity. Does it follow that G is locally finite?

REFERENCES

[1] S. V. Aleshin, Finite automata and the Burnside problem for periodic groups, Math. Notes 11 (1972), 199–203.

[2] Yu. A. Bahturin, M. V. Zaicev, Identities of graded algebras, J. Algebra, 205(1998), 1–12.

[3] G. S. Deryabina and A. Yu. Ol'shanskii, Subgroups of quasifinite groups, Uspekhi Math. Nauk, 41 (1986), 169–170 (in Russian).

[4] J. D. Dixon, M. P. F. du Sautoy, A. Mann, D. Segal, "Analytic pro-p groups" (London Math. Soc. Lecture Note Series 157), Cambridge Univ. Press., 1991.

[5] W. Feit, J. G. Thompson, Solvability of groups of odd order, Pacific J. Math. 13 (1963), 773-1029.

[6] E. S. Golod, On nil-algebras and residually finite groups, Izvestia Akad. Nauk SSSR, Ser Mat., 28 (1964), 273–276

[7] D. Gorenstein, "Finite Groups", Harper and Row, New York, 1968.

[8] R. I. Grigorchuk, On the Burnside problem for periodic groups, Funct. Anal. Appl., 14 (1980), 53–54

[9] K. Gruenberg, The Engel elements of a soluble group, Illinois J. Math. 3 (1959), 151-168.

[10] N. Gupta and S. Sidki, On the Burnside problem for periodic groups, Math. Z., 182 (1983), 385–386.

[11] P. Hall, G. Higman, The p-length of a p-soluble group and reduction theorems for Burnside's problem, Proc. London. Math. Soc. (3) 6 (1956), 1–42.

[12] B. Hartley, Subgroups of finte index in profinite groups, Math. Z., 168 (1979), 71–76.

[13] B. Hartely, Centralizers in locally finite groups, in "Proc. 1st Bressanone Group Theory Conference 1986", Lecture Notes in Math. 1281, Springer, 1987, pp. 36–51.

[14] B. Hartley, Simple locally finite groups, in "Finite and Locally Finite Groups", NATO ASI Series, 471, Kluwer Academic Publishers, 1995, 1–45.

[15] B. Hartley, T. Meixner, Periodic groups in which the centralizer of an involution has bounded order, Arch. Math. (Basel), 36 (1981), 211–213.

[16] G. Higman, Lie ring methods in the theory of finite nilpotent groups, in "Proc. Intern. Congr. Math. Edinburgh, 1958", Cambridge Univ. Press, 1960, 307–312.

[17] B. Huppert, "Endliche Gruppen I", Springer Verlag, Berlin, 1967.

[18] B. Huppert, N. Blackburn, "Finite Groups II", Springer Verlag, Berlin, 1982.

[19] E. I. Khukhro, Groups and Lie rings admitting an almost regular automorphism of prime order, Math. USSR Sbornik, 71 (1992), 51–63.

[20] E. I. Khukhro, "Nilpotent Groups and their Automorphisms", de Gruyter–Verlag, Berlin, 1993.

[21] E. I. Khukhro and P. V. Shumyatsky, Bounding the exponent of a finite group with automorphisms, J. Algebra, 212 (1999), 363–374.

[22] A. I. Kostrikin, On the Burnside problem, Izv. AN SSSR, Ser. Mat. 23 (1959), 3–34.

[23] M. Lazard, Sur les groupes nilpotents et les anneaux de Lie, Ann. Sci. École Norm. Supr. 71 (1954), 101-190.

[24] M. Lazard, Groupes analytiques p-adiques, Publ. Math. Inst. Hautes Études Sci., 26 (1965), 389–603.

[25] V. Linchenko, Identities of Lie algebras with actions of Hopf algebras, Commun. Algebra, 25 (1997), 3179–3187.

[26] A. Lubotzky, A. Mann, Powerful p-groups. I: finite groups, J. Algebra, 105 (1987), 484–505; II: p-adic analytic groups, ibid., 506–515.

[27] W. Magnus, A connection between the Baker–Hausdorff formula and a problem of Burnside, Ann. of Math. (2), 52 (1950), 111–126.

[28] C. F. Miller III and V. N. Obraztsov, Infinite periodic residually finite groups with all finite subgroups cyclic, in preparation.

[29] A. Yu. Ol'shanskii, "Geometry of defining relations in groups", Math. Appl. (Soviet Ser.) 70(1991).

[30] N. Rocco, P. Shumyatsky, On periodic groups having almost regular 2-elements, Proc. Edinburgh Math. Soc., 41(1998), 385–391.

[31] I. N. Sanov, Establishment of a connection between periodic groups with period a prime number and Lie rings, Izv. Akad. Nauk SSSR Ser. Mat., 16 (1952), 23–58 (Russian).

[32] A. Shalev, Characterization of p-adic analytic groups in terms of wreath products, J. Algebra, 145(1992), 204–208.

[33] A. Shalev, On almost fixed point free automorphisms, J. Algebra, 157 (1993), 271–282.

[34] A. Shalev, Automorphisms of finite groups of bounded rank, Israel J. Math., 82(1993), 395–404.

[35] A. Shalev, Finite p-groups, in "Finite and Locally Finite Groups", NATO ASI Series, 471, Kluwer Academic Publishers, 1995, 401–450.

[36] A. Shalev, Centralizers in residually finite torsion groups, Proc. Amer. Math. Soc., 126 (1998), 3495–3499.

[37] P. V. Shumyatsky, Groups with regular elementary 2-groups of automorphisms, Algebra and Logic, 27 (1988), 447–457.

[38] P. Shumyatsky, On periodic solvable groups having automorphisms with nilpotent fixed-point groups, Israel J. Math., 87 (1994), 111–116.

[39] P. Shumyatsky, On groups having a four-subgroup with finite centralizer, Quart. J. Math (Oxford), 49 (1998), 491–499.

[40] P. Shumyatsky, Nilpotency of some Lie algebras associated with p-groups, to appear in Canadian Journal of Mathematics.

[41] P. Shumyatsky, Centralizers in groups with finiteness conditions, J. Group Theory, 1 (1998), 275–282.

[42] P. Shumyatsky, On groups with commutators of bounded order, Proc. Amer. Math. Soc., 127 (1999), 2583–2586.

[43] V.P. Shunkov, On periodic groups with an almost regular involution, Algebra and Logic, 11 (1972), 260–272.

[44] V.I. Sushchansky, Periodic p-elements of permutations and the general Burnside problem, Dokl. Akad. Nauk SSSR, 247 (1979), 447–461.

[45] J.G. Thompson, Automorphisms of solvable groups, J. Algebra, 1 (1964), 259-267.

[46] J. Tits, Free subgroups in linear groups, J. Algebra, 20 (1972), 250-270.

[47] M. R. Vaughan-Lee, "The Restricted Burnside Problem", 2-nd edition, Oxford University Press, Oxford, 1993.

[48] J. N. Ward, Automorphisms of finite groups and their fixed-point groups, J. Austral. Math. Soc., 9 (1969), 467–477.

[49] J.S. Wilson, Two-generator conditions for residually finite groups, Bull. London Math. Soc., 23 (1991), 239-248.

[50] J.S. Wilson and E. Zelmanov, Identities for Lie algebras of pro-p groups, J. Pure Appl. Algebra, 81 (1992), 103-109.

[51] H. Zassenhaus, Ein Verfahren, jeder endlischen Gruppe einen Lie-Ring mit der Charakteristiki p zuzuordnen, Abh. Math. Seminar Hans. Univ. Hamburg 13(1940), 200–207

[52] E. Zelmanov, The solution of the restricted Burnside problem for groups of odd exponent, Math. USSR Izv. 36 (1991), 41-60.

[53] E. Zelmanov, The solution of the restricted Burnside problem for 2-groups , Math. Sb. 182 (1991), 568-592.

[54] E. Zelmanov, "Nil Rings and Periodic Groups", The Korean Math. Soc. Lecture Notes in Math., Seoul, 1992.

[55] E. I. Zelmanov, Lie ring methods in the theory of nilpotent groups, in "Proc. Groups'93/St. Andrews, vol. 2" (London Math. Soc. Lecture Note Ser. 212), Cambridge Univ. Press, 1995, 567–585.

Computable rings, groups and their isomorphisms

BAKHADYR KHOUSSAINOV Department of Computer Science, University of Auckland, Private Bag 92019 Auckland, New Zealand. e-mail: bmk@cs.auckland.ac.nz

ARKADII SLINKO Department of Mathematics, University of Auckland, Private Bag 92019 Auckland, New Zealand. e-mail: a.slinko@auckland.ac.nz

ABSTRACT We investigate computable isomorphism types of groups. Our main result states that for any $n \in \omega \cup \{\omega\}$ there exists a computably categorical nilpotent of class 2 group G which being expanded by a finite number of constants has exactly n computable isomorphism types. This result is based on the similar result for computable nonassociative rings.

1. INTRODUCTION. BASIC NOTIONS AND MAIN RESULTS

From algebraic point of view there is no distinction between isomorphic algebraic systems. Therefore classification of algebraic systems up to isomorphism constitutes one of the main goals of structure theories of these systems. It can be said that structure theories of algebraic systems study isomorphism types of these systems, i.e., classes of isomorphic algebraic systems. The theory of rings and the theory of groups are by no means exceptions among them. However, in effective theories of the same algebraic systems this view on isomorphism types has to undergo profound changes since isomorphism types and computable isomorphism types become different.

Computable algebraic systems such as computable groups, boolean algebras, vector spaces, lattices, have been intensively investigated in recent years [1], [2]. Intensive research efforts have been made in attempts to understand the effective content of a variety of model-theoretic and algebraic notions, results and constructions. We refer the reader to the recent surveys by Harizanov [3], Millar [4], as well as to the classic papers by Frölich and Shepherdson [5], Malcev [6] and Rabin [7], devoted to these issues. In this paper we consider computable rings and investigate relationship between isomorphism types and computable isomorphism types of these algebraic systems.

Let us recall several basic notions from the computability theory [8]. Throughout the paper ω is the set of all natural numbers. A set $X \subset \omega$ is *computable* if there is a procedure which being applied to any number n tells us if $n \in X$. A function $f: \omega \to \omega$ is *computable* if the set of pairs $X = \{(n, f(n)) \mid n \in \omega\} \subseteq \omega \times \omega$ is computable under the standard Cantor's identification of $\omega \times \omega$ and ω. A set $X \subset \omega$ is *computably enumerable* if it is the range of a computable function $f: \omega \to \omega$.

In this paper by a ring $\mathcal{R} = (R, +, \times, 0)$ we understand a nonassociative ring, i.e. a ring for which the associative law for the multilication is not assumed. While nonassociative rings are less commonly met than their associative counterpart, they do have a number of significant applications. For instance, apart from Lie rings which are widely used, nonassociative rings have been also studied in relations to projective geometry [9] and their applications to physics [10] and genetics [11].

DEFINITION 1. A ring $\mathcal{R} = (R, +, \times, 0)$ is *computable* if the set R is a computable subset of ω and the ring operations $+$ and $\times$ are computable functions from R^2 into R.

Informally, a computable ring is a ring whose elements can be enumerated and whose operations can be computed by Turing machines.

DEFINITION 2. A ring $\mathcal{R}$ is said to be *computably presentable* if its isomorphism type contains a computable ring. If $\mathcal{R}'$ is a computable ring isomorphic to $\mathcal{R}$, then $\mathcal{R}'$ is called a *computable presentation* of $\mathcal{R}$.

For example, the field of rational numbers $\mathbb{Q}$ and the ring of integers $\mathbb{Z}$ are computably presentable rings, where the standard presentations are computable presentations.

DEFINITION 3. An isomorphism $f: \mathcal{R}_1 \to \mathcal{R}_2$ from a computable ring $\mathcal{R}_1$ onto a computable ring $\mathcal{R}_2$ is said to be *computable* if f itself is a computable function. In this case we say that $\mathcal{R}_1$ is computably isomorphic to $\mathcal{R}_2$.

DEFINITION 4. The notion of computable isomorphism defines an equivalence relation on the class of all computable presentations of a given computably presentable ring $\mathcal{R}$. The classes of the partition corresponding to this equivalence relation are called *computable isomorphism types* of $\mathcal{R}$. The number of computable isomorphism types of $\mathcal{R}$ is called the *algorithmic dimension* of $\mathcal{R}$.

Thus, informally one can say that the number of computable isomorphism types of a ring is the number of its effective presentations which cannot be effectively transformed one into another. Rings of algorithmic dimension 1 are the rings with exactly one computable isomorphism type. Algebraic structures with exactly one computable isomorphism type are also called *computably categorical*. Computably categorical structures have been the subject of extensive investigations: [1], [3], [4], [12], [13], [14], [15]. The following simple proposition gives examples of rings of algorithmic dimension 1 or, equivalently, computably categorical rings.

PROPOSITION 1. *Any two computable presentations of a finitely generated ring $\mathcal{R}$ are computably isomorphic.*

Proof. Let $\mathcal{R}_1$ and $\mathcal{R}_2$ be two computable presentations of $\mathcal{R}$. Let $b_1, \ldots, b_n$ be generators of $\mathcal{R}$, with $m_1, \ldots, m_n$ and $k_1, \ldots, k_n$ being the respective images of these generators in the computable presentations $\mathcal{R}_1$ and $\mathcal{R}_2$ under respective isomorphisms. Then the partial mapping $m_i \mapsto k_i$ can be extended to a computable isomorphism from $\mathcal{R}_1$ to $\mathcal{R}_2$ in the obvious way. □

Goncharov [13], [14] and independently LaRoche and Remmel [15] studied computably categorical boolean algebras and linearly ordered sets. They proved the following two theorems.

THEOREM 1. *A boolean algebra is computably categorical if and only if the number of its atoms is finite. Moreover, every boolean algebra, which is not computably categorical, has infinitely many computable isomorphism types.*

THEOREM 2. *A linear ordering is computably categorical if and only if the number of its successive pairs, that is pairs (a, b) for which $a < b$ and the interval $[a, b]$ consists of a and b only, is finite. Moreover, every linear ordering, which is not computably categorical, has infinitely many computable isomorphism types.*

Thus, the algorithmic dimension of any Boolean algebra or linear ordering is either ω or 1. For abelian groups, as Goncharov [13], [14] showed in the theorem that follows, the same result holds, but for groups in general the situation is more complicated.

THEOREM 3. *The algorithmic dimension of any abelian group is either* 1 *or* ω. *For any natural number n there exists a (noncommutative) group of algorithmic dimension n.*

The latter result Goncharov proved by constructing and encoding a special family of sets into a metabelian group [16]. The same approach was applied in [17]. In this paper we use the same approach to show that a similar result holds also for rings. Namely, we prove:

THEOREM A. *For every $n \in \omega \cup \{\omega\}$ there exists a (noncommutative and nonassociative) ring of algorithmic dimension n.*

There are basically two reasons why the notion of a computably categorical structure has attracted a significant attention of researchers in computable algebra and model theory. The first reason is that computably categorical structures are exactly those structures which do not depend on a particular computable presentation. Thus, from the computable-model-theoretic point of view there is no distinction between two computable presentations of a computably categorical structure.

The second reason comes from model theory. The basic model-theoretic notion, which motivated the study of computably categorical structures, is the notion of countably categorical model. In classical model theory a theory T is called *(countably) categorical* if all (countable) models of T are isomorphic. A (countable) structure $\mathcal{A}$ is *(countably) categorical* if its theory $Th(\mathcal{A})$ is (countably) categorical. The analogous concept for the effective model theory deals only with computable structures and isomorphisms. It is the notion of a computably categorical structure.

Any algebraic structure $\mathcal{A}$ can be expanded by a finite number of constants $a_1, \ldots, a_n \in \mathcal{A}$. The expanded structure will be denoted $(\mathcal{A}, a_1, \ldots, a_n)$ or simply $\mathcal{A}$, when ambiguity cannot occur. Algebraically, it means that we consider n unary operations $u_i(x) = a_i$, $i = 1, 2, \ldots, n$, on $\mathcal{A}$ together and on equal rights with the existing operations in $\mathcal{A}$. From the computational point of view we also fix some algorithms which output these constants. From categorical point of view this is the change of the category under consideration, as morphisms now have to respect these new operations. For example, we call a mapping $f: \mathcal{A}_1 \rightarrow \mathcal{A}_2$ from an expanded structure $\mathcal{A}_1 = (\mathcal{A}_1, a_1, \ldots, a_n)$ onto an expanded structure $\mathcal{A}_2 = (\mathcal{A}_1, b_1, \ldots, b_n)$ an *isomorphism* if it is an isomorphism of non- expanded structures and also $f(a_i) = b_i$ for $i = 1, 2, \ldots, n$.

The notion of expansion by constants has strong connections with classical categoricity of a structure. For instance, it is an easy consequence of the classical Ryll-Nardzewski Theorem that, if the theory of an arbitrary structure $\mathcal{A}$ is countably categorical, then so is the theory of any expansion of $\mathcal{A}$ by finitely many constants. In effective model theory, computably categorical structures are the analogues of countably categorical ones. Millar [18] proved that a certain amount of decidability is enough to guarantee that the property of being computably categorical is preserved under expansions by finitely many constants. Without this assumption of partial decidability, the problem, which was known as Ash-Goncharov problem [19], remained open for some time. It was solved negatively in [20]:

THEOREM 4. *For every natural number n there exists a computably categorical graph G such that, for any $c \in G$, the expanded graph (G, c) has exactly n computable isomorphism types.*

It is the analogous problem for computable rings and computable groups that we wish to address in this paper. The second and the third of the three main theorems of this paper show that the same phenomenon can also occur not only for graphs but also in the class of rings and even in the class of nilpotent groups of class 2. This solves the Ash- Goncharov problem for rings and groups.

THEOREM B. *For every natural number n there exists a computably categorical (noncommutative and nonassociative) ring $\mathcal{R}$ such that for some $c \in \mathcal{R}$ the expanded ring $(\mathcal{R}, c)$ has exactly n types of computable isomorphisms.*

THEOREM C. *For every prime $p>2$ and for every $n \in \omega \cup \{\omega\}$ there exists a computably categorical nilpotent group $\mathcal{G}$ of class, 2 satisfying $x^p = 1$, such that for some constants $a_1, a_2, a_3 \in \mathcal{G}$ the expanded group $(\mathcal{G}, a_1, a_2, a_3)$ is of algorithmic dimension n.*

2. COMPUTABLE FAMILIES AND ENUMERATIONS

To establish Theorem A we will encode a certain (uniformly) computably enumerable family of sets of natural numbers into a ring.

DEFINITION 5. A family of nonempty sets S is called *computably enumerable* if there exists a mapping $f: \omega \rightarrow S$ such that the set of pairs $\{(i, x) \mid x \in f(i)\}$ is computably enumerable. We then call f a *(computable) enumeration* of S. If f is one-to- one we say that it is a *one-to-one* enumeration of S.

Technically, it is more convenient to view a computable enumeration of S as a procedure which produces a 2-dimensional array $\{f^i(n) \mid i, n \in \omega\}$ of finite subsets of ω according to the following rules:

(i) At stage 0 it produces empty or one element subset $f^0(0)$;

(ii) At stage k it produces subsets $f^k(0), \ldots, f^1(k-1), f^0(k)$ such that $f^{k-1}(i) \subseteq f^k(i)$, $i = 1, \ldots, k-1$, and such that

$$\mathrm{card}(f^{k-1}(0) \cup \cdots \cup f^0(k-1)) \leq \mathrm{card}(f^k(0) \cup \cdots \cup f^0(k)) + 1;$$

(iii) $\bigcup_{i \geq 0} f^i(n) = f(n)$.

We define a preordering on the set of all computable enumerations of a family S that will naturally induce an equivalence relation on this set. The equivalence classes of this relation will correspond to computable isomorphism types of the ring that we are going to construct.

DEFINITION 6. Let f and g be two computable enumerations of a family S. We say that f is *reducible* to g and denote it as $f \leq g$, if there is a computable function $\Phi: \omega \to \omega$ such that $f = g\Phi$. If $f \leq g$ and $g \leq f$ then we say that f and g are *equivalent* and denote this relation by $f \sim g$.

The equivalence classes of one-to-one enumerations are the minimal elements in the induced partial ordering. One-to-one enumerations will be needed to define a family of sets that will be encoded. Theorem A will be based on the following theorem of Goncharov [12].

THEOREM 5. *For any $n \in \omega$ there exists a family S of computably enumerable sets such that S has up to equivalence exactly n one-to-one computable enumerations.*

We now present the basic notions involved in the proof of Theorem B. We need to consider families of k-tuples of sets. We give all definitions for the case $k = 2$. We will indicate later how the case $k > 2$ can be handled.

In what follows, we use r and ℓ as the right and left projections from pairs, that is, $\ell(A, B) = A$ and $r(A, B) = B$.

DEFINITION 7. Let S be a family of pairs (A, B) of nonempty sets. A family S is called *symmetric* if $(A, B) \in S$ implies that $(B, A) \in S$. A family S is said to be *computably enumerable* if there exists a mapping $f: \omega \to S$ such that the set of triples $\{(i, x, y) \mid x \in lf(i),\ y \in rf(i)\}$ is computably enumerable. We then call f a *(computable) enumeration* of S. If f is one-to-one, we say that it is a *one-to-one* enumeration of S.

The notion of reducibility and equivalence between enumerations of a symmetric family S are exactly the same as for families of computably enumerable sets. If f is a one-to-one computable enumeration of a symmetric family S of pairs of sets then there is another computable enumeration $\tilde{f}$ of S which is a natural companion of f, namely, if $f(i) = (A_i, B_i)$, then $\tilde{f}(i) = (B_i, A_i)$.

The notion of algorithmic dimension can be also applied to a family S of pairs of sets as follows:

DEFINITION 8. If f is a one-to-one computable enumeration of a symmetric family S of pairs of sets, we say that S has algorithmic dimension 2 if f and $\tilde{f}$ are not equivalent but every computable one-to-one enumeration of S is equivalent to either f or $\tilde{f}$.

Such a family was constructed in [20]:

THEOREM 6. *There exists a computably enumerable symmetric family of algorithmic dimension* 2.

This family will be encoded into a ring in order to prove Theorem B.

3. RINGS OF A FINITE ALGORITHMIC DIMENSION.

We first show how to encode a set of natural numbers into a ring, in fact, into a field. Our reference book in relation to field theory will be [21], where all facts, not explained here, can be found. Let $F = \mathbb{Z}_\nu$ be a finite field of residues modulo ν. In the construction, that follows, ν may be an arbitrary prime number. We consider the class of all algebraic extentions of F which lie in some fixed copy $\overline{F}$ of algebraic closure of F.

Let $M = \{m_0, m_1, \ldots, m_n, \ldots\}$ be a subset of ω. If M is empty, then we assume that the field F encodes it. If M is not empty, we will put in correspondence to M the algebraic extention of F

$$F_M = F[\alpha_0, \alpha_1, \ldots, \alpha_n, \ldots],$$

where α_i is an algebraic element of order p_{m_i}, the m_ith prime. We fix these elements and always use them for our coding purposes or alternatively, from the constructive point of view, we may think that we have fixed their minimal irreducible polynomials $P_i(x) = \mathrm{Irr}_{\alpha_i}(x)$.

LEMMA 1. *The set of prime factors of orders of elements of* F_M *is exactly the set* $\{p_{m_0}, \ldots, p_{m_i}, \ldots\}$. *The set* S_i *of all elements of order* p_{m_i} *in* F_M *consists of* $\nu^{p_{m_i}}$ *elements and* $S_i = F[\alpha]$ *for every element* α *of order* p_{m_i}.

Proof. Let $a \in F_M$. Then, for some n, the element a belongs to a finite extention

$$G_n = F[\alpha_0, \alpha_1, \ldots, \alpha_n], \tag{1}$$

Since degrees in towers are multiplicative, observing the tower

$$F \subseteq F[\alpha_i] \subseteq G_n$$

we see that the degree of G_n is divisible by p_{m_i} and therefore is divisible by $p_{m_0} p_{m_1} \cdots p_{m_n}$. As the degree of α_i over $F[\alpha_0, \alpha_1, \ldots, \alpha_{i-1}]$ is less than or equal to p_{m_i} the degree of G_n cannot be greater than this product.

Now by considering the tower

$$F \subseteq F[a] \subseteq G_n$$

we see that the order of a must be a divisor of this product.

Let $q = p_{m_i}$ and α be an element of order q. Then the subfield $F[\alpha]$ has ν^q elements. It is isomorphic to the Galois field $\mathrm{GF}(\nu^q)$ of this order which, being a subfield of $\overline{F}$ is known to coincide with the set of all roots of the polynomial $x^{\nu^q} - x$ in $\overline{F}$. Therefore such subfield is unique and $F[\alpha] = F[\beta]$ for any two elements of order q. □

LEMMA 2. *The set* M *is computably enumerable if and only if the field* F_M *is computably presentable.*

Proof. Let $m_0, m_1, \ldots, m_n, \ldots$ be an effective enumeration of M. Define the field G_n as in (1). We saw in the proof of the previous lemma that the dimension of the field G_n over F is $p_{m_0} p_{m_1} \cdots p_{m_n}$.

As a vector space, G_n has a spanning set consisting of monomials

$$\alpha_0^{k_0} \alpha_1^{k_1} \ldots \alpha_n^{k_n}, \tag{2}$$

where $0 \leq k_i < p_{m_i}$. This spanning set has cardinality $p_{m_0} p_{m_1} \cdots p_{m_n}$. Therefore this spanning set is a basis. These monomials can be multiplied as usual monomials in α_i but with powers of α_i being multiplied modulo the minimal irreducible polynomial of α_i. Since the union of all such bases is a basis for F_M, this certainly gives a computable presentation of F_M.

Let now $\mathcal{A}$ be a computable presentation of F_M and $a_1, a_2, \ldots, a_n, \ldots$ be the enumeration of elements of $\mathcal{A}$ which arises from this presentation. Take $a = a_1 \in \mathcal{A}$ and consider powers $a, a^2, \ldots$ until $a^s = a^n$ for $s < n$. Then $a^{n-s} = e$ is the unit element of $\mathcal{A}$. Thus $F_1 = \{0, e, 2e, \ldots, (\nu - 1)e\}$ will be the only subfield of $\mathcal{A}$ isomorphic to F.

Since F is finite we can now constructively determine the minimal irreducible polynomial of a_1, for its degree is less than or equal to $n{-}s$. We know that $m \in M$ iff there exists an element $x \in \mathcal{A}$ such that the order of x over F_1 is p_m. The prime divisors of the degree of this polynomial, say $p_{m_1}, \ldots, p_{m_k}$, will show that there are elements of such orders, and give us the first set of elements of M to list, namely $m_1, \ldots, m_k$. Hence M is a computably enumerable set. The lemma is proved. □

LEMMA 3. *The field F_M is computably categorical.*

Proof. Let $\mathcal{A}$ and $\mathcal{B}$ be two computable presentations of F_M. Consider the subfields F_1 and F_2, of $\mathcal{A}$ and $\mathcal{B}$, respectively, isomorphic to F and constructed as in the proof of the previous lemma. The only isomorphism between them can be established by assigning one unit element to another and multiples of one unit to the corresponding multiples of another. Denote $\mathcal{A}_0 = F_1$, $\mathcal{B}_0 = F_2$ and let $\sigma_0 \colon \mathcal{A}_0 \to \mathcal{B}_0$ be the established isomorphism. Suppose that we established already an isomorphism $\sigma_i \colon \mathcal{A}_i \to \mathcal{B}_i$ between subfields $\mathcal{A}_i$ and $\mathcal{B}_i$ such that for an arbitrary prime number q either all elements of $\mathcal{A}$ and $\mathcal{B}$ of prime order q belong to $\mathcal{A}_i$ and $\mathcal{B}_i$, respectively, or none of them. This isomorphism can be readily extended to the isomorphism $f(x) \mapsto f^{\sigma_i}(x)$ of polynomial rings $\mathcal{A}_i[x]$ and $\mathcal{B}_i[x]$, which is defined as follows: if $f(x) = a_0 + a_1 x + \cdots + a_m x^m$, then $f^{\sigma_i}(x) = a_0^{\sigma_i} + a_1^{\sigma_i} x + \cdots + a_m^{\sigma_i} x^m$. Now we look for the first element α in the effective enumeration of $\mathcal{A}$ which is of prime order q over $\mathcal{A}_0 = F_1$ and which is not in $\mathcal{A}_i$. Then we find $\beta \in \mathcal{B}$ with exactly the same minimal irreducible polynomial over $\mathcal{B}_0 = F_2$ and construct an isomorphism $\sigma_{i+1} \colon \mathcal{A}_i[\alpha] \to \mathcal{B}_i[\beta]$ defining the mapping σ_{i+1} for every polynomial $f(x) \in \mathcal{A}_i[x]$ of degree less than q by the following formula

$$\sigma_{i+1}(f(\alpha)) = f^{\sigma_i}(\beta).$$

It is easy to check that σ_{i+1} is again an isomorphism. We denote then $\mathcal{A}_{i+1} = \mathcal{A}_i[\alpha]$ and $\mathcal{B}_{i+1} = \mathcal{B}_i[\beta]$. Since all elements of order q lie in $F[\alpha]$, and hence in $\mathcal{A}_{i+1}$ and $\mathcal{B}_{i+1}$, the construction can be effectively continued further. The lemma is proved. □

Let now S be a countable family of countable sets. We will list them in some order which will not be important later:

$$S = \{M_0, M_1, \ldots, M_i, \ldots\}.$$

Consider the free product of fields F_{M_i} in the variety of all (nonassociative) rings

$$A(S) = F_{M_0} \star F_{M_1} \star \cdots \star F_{M_n} \star \ldots$$

We assume that each field F_{M_i} encodes the set M_i in the way it was described earlier. Up to isomorphism, this algebra does not depend on the order in which we listed the sets of our family. In this section we will use the family of sets S, constructed in [12], which, up to equivalence, has exactly n one-to-one computable enumerations $f_1, \ldots, f_n$. We use them to construct n computable presentations $A_{f_1}(S), \ldots, A_{f_n}(S)$ of $A(S)$, such that no two of them are computably isomorphic but any other computable presentation of $A(S)$ is computably isomorphic to one of the computable algebras $A_{f_1}(S), \ldots, A_{f_n}(S)$.

We will refer to the fields F_{M_i} as to the components of $A(S)$. This algebra, as it is the free product of the components, has a basis consisting of nonassociative products

$$(a_1 a_2 \ldots a_n)_q, \qquad n \geq 1, \tag{3}$$

where elements $a_1, \ldots, a_n$ are basic monomials (2) and any two neighbouring monomials a_{i-1} and a_i, situated in the bracket $(a_{i-1}a_i)$, belong to different components. For example, in the product $(a_1(a_2((a_3a_4)(a_5a_6))))$ in each pair a_3, a_4 and a_5, a_6 the monomials must be from different components, while a_1, a_2 may be arbitrary.

We will refer to the products (3) as to the *basic products*. The multiplication table on the basis is as follows: if $(a_1a_2\ldots a_n)_p$ and $(b_1b_2\ldots b_m)_q$ are two basic products and $\max(m,n) > 1$ or $m = n = 1$ and a_1, b_1 belong to different components, then

$$(a_1a_2\ldots a_n)_p \cdot (b_1b_2\ldots b_m)_q = ((a_1a_2\ldots a_n)_p(b_1b_2\ldots b_m)_q), \tag{4}$$

If $m = n = 1$ and a_1, b_1 belong to the same component, then $a_1b_1 = \sum_{i=1}^{\ell} \beta_i c_i$, where c_i's are basic monomials of the component to which both of them belong, and

$$(a_1) \cdot (b_1) = \sum_{i=1}^{\ell} \beta_i (c_i). \tag{5}$$

Let $u = (a_1a_2\ldots a_n)_q$ be a basic product. We set $|u| = n$, and for an element $a = \sum_{i=1}^{\ell} \beta_i u_i$ of $A(S)$ we set $|a| = \max_i |u_i|$. It is clear from the multiplication table (4) and (5) that

$$|a \cdot b| = |a| + |b|, \tag{6}$$

unless $|a| = |b| = 1$ and a, b are from the same component. Unfortunately, when the free product is taken in the class of associative rings, the equation (6) is no longer valid and this is why we have to consider nonassociative products.

Let us recall that an element e of a ring $\mathcal{R}$ is called an *idempotent* if $e^2 = e$.

LEMMA 4. *The fields F_{M_i} are isomorphically imbedded in $A(S)$. The unit elements $e_0, e_1, \ldots, e_n, \ldots$ of the fields F_{M_i} are the only idempotents of $A(S)$. An element $a \in A(S)$ belongs to F_{M_i} iff $e_i a = a e_i = a$.*

Proof. It can be seen from the multiplication table that the field F_{M_i} is a subring of $A(S)$, and that the unit element e_i of it is an idempotent. Suppose that $e^2 = e$ and $e \neq 0$. Then (6) implies that $|e| = 1$. It is also clear that if e is equal to the sum of basic monomials from different components, then $|e^2| = 2$ and $e^2 \neq e$. Therefore e

belongs to one of the components. But it is a field and has a unique idempotent, namely the unit element of this field.

The field F_{M_i} is a subring of $A(S)$ and $e_i a = ae_i = a$ for all $a \in F_{M_i}$. Suppose that $e_i a = ae_i = a$. Then the multiplication table of $A(S)$ implies that a is a linear combination of basic monomials from F_{M_i} and thus is an element of F_{M_i}. of algebraic The lemma is proved. □

NOTE: The statement that e_i are the only idempotents of this free product is no longer valid if this product is associative. For example, $e = e_2e_1e_2 - e_2e_1 + e_2$ will also be an idempotent.

Now, given a one-to-one computable enumeration f let us construct a computable presentation $A_f(S)$ of $A(S)$. Let us now denote $M_i = f(i)$. Since f is computable, there exists a procedure which produces a 2-dimensional array $\{M_{in} = f^i(n) \mid i, n \in \omega\}$ of finite subsets of ω according to the following rules:

(M1) At stage 0 it produces an empty or one element subset M_{00};

(M2) At stage k it produces finite subsets $M_{k0}, \ldots, M_{1k-1}, M_{0k}$ so that $M_{k-1i} \subseteq M_{ki}$, for $i = 1, \ldots, k-1$, and such that

$$\text{card}(M_{k-10} \cup \cdots \cup M_{0k-1}) \leq \text{card}(M_{k0} \cup \cdots \cup M_{0k}) + 1;$$

(M3) $\displaystyle\bigcup_{i \geq 0} M_{in} = M_n$.

Thus using f we can construct an effective sequence of computable partial algebras

$$A(f,0),\ A(f,1),\ \ldots\ ,\ A(f,n),\ \ldots$$

such that:

(A1) $A(f,i)$ is a subalgebra of $A(f,i+1)$;

(A2) $A(f,i)$ is isomorphic to $(F_{M_{i0}} \star \cdots \star F_{M_{0i}})^{(i)}$, the latter being the subspace of $A(S)$ spanned by the basic products of degree $\leq i$, which depend only on elements from $F_{M_{0i}}, \ldots, F_{M_{i0}}$ with the addition and multiplication inherited from $A(S)$;

(A3) $A_f(S) = \displaystyle\bigcup_{k=0}^{\infty} A(f,i)$ is isomorphic to $A(S)$.

As the sets M_{ki} are finite, the fields $F_{M_{ki}}$ are finite-dimensional, hence finite, and partial algebras $A(f,i)$ are also finite-dimensional, and hence also finite. It is important to note that at stage i, when we extend $A(f,i-1)$ to $A(f,i)$ the only one idempotent will be added, namely the unit element e_i of the field F_{M_i}. In order to separate stages we start each time enumeration of additional elements with e_i.

LEMMA 5. *The ring $A_f(S)$ is computable for every computable one-to-one enumeration f of S. One-to-one enumerations f and g are equivalent, iff $A_f(S)$ and $A_g(S)$ are computably isomorphic.*

Proof. The computable presentation for $A(S)$ was constructed above. It is also straightforward that if two one-to-one enumerations f and g of S are equivalent, then the algebras $A_f(S)$ and $A_g(S)$ are computably isomorphic. On the other hand, if the algebras $A_f(S)$ and $A_g(S)$ are computably isomorphic, then for every idempotent $e_i \in A_f(S)$ we can effectively compute its image in $A_g(S)$ and compute at which stage it appears in the construction of $A_g(S)$. If it were, say the jth stage, then we set $\Phi(i) = j$. This gives us a computable function $\Phi: \omega \to \omega$ such that $f = g\Phi$. □

LEMMA 6. *Let $\mathcal{A}$ be a computable presentation of $A(S)$. Then one can construct a one-to-one computable enumeration $f = f(\mathcal{A})$ of S such that $\mathcal{A}$ and $A_f(S)$ are computably isomorphic.*

Proof. Let $\mathcal{A} = \{a_0, a_1, \dots\}$ be all elements of $\mathcal{A}$ listed in a sequence. We can effectively list all idempotents $e_0, e_1, \dots$ of $\mathcal{A}$ which form a subsequence of this sequence. By Lemma 4 these idempotents are the unit elements of the components. Let $F_i = \{e_i, 2e_i, \dots, \nu e_i\}$ be the copy of the base field F which is contained in the component F_{M_i}. An element x belongs to the component F_{M_i} iff the condition $x = xe_i = e_i x$ is satisfied. Therefore

$$f(i) = \{m \mid \exists x\ (x = xe_i = e_i x \text{ and } x \text{ is algebraic of degree } p_m \text{ over } F_i)\}$$

enumerates S and $f = f(\mathcal{A})$ is a computable enumeration of S. Moreover f is one-to-one. Clearly $\mathcal{A}$ and $A_f(S)$ are computably isomorphic. We will sketch the construction of this computable isomorphism:

Step 0: Compute the number of e_0 in the sequence, say $e_0 = a_s$, and set $M_{00} = \emptyset$ if among $a_0, \dots, a_{s-1}$ there are no elements x such that $xe_0 = e_0 x = x$ which are algebraic over F_0 of prime degree. If such an element α, say of prime degree p_m, existed, we set $M_{00} = \{m\}$ and put in correspondence σ_0 the subfield $F_0[\alpha] \subseteq \mathcal{A}$ with the field $F(M_{00}) \subseteq A(S)$.

Step i: We compute the number of e_i, say $e_i = a_t$, and look for the first element $x \in \{a_0, a_1, \dots, a_{t-1}\}$ in the sequence such that $x = xe_j = e_j x$, for one of the numbers $j = 1, 2, \dots, i$, and such that x is algebraic of a prime degree p_n over F_j. Then take

$$M_{i0} = M_{i-10},\ \dots,\ M_{i-jj} = M_{i-j-1j} \cup \{n\},\ \dots,\ M_{0i},$$

where $M_{0i} = \emptyset$ or $\{n\}$ if $j = i$. If such x is not found we leave all the sets as they were, just add an empty M_{0i}. We can now find a partial subalgebra of $\mathcal{A}$ which will be in a computable correspondence σ_i with the partial subalgebra $(F_{M_{i0}} \star \cdots \star F_{M_{0i}})^{(i)}$. The lemma is proved. □

THEOREM A. *For every positive integer n there exists a computable ring of algorithmic dimension n.*

Proof. Let S be a family of computable enumerable sets which has, up to equivalence, exactly n one-to-one computable enumerations. Such a family exists due to Theorem 5. Let $f_1, \dots, f_n$ be any n mutually non-equivalent computable one-to-one enumerations of S. We construct the algebra $A(S)$ as shown in the beginning of this section. By Lemma 5 the computable presentations $A_{f_1}(S), \dots, A_{f_n}(S)$ are not computably isomorphic. Let $\mathcal{A}$ be an arbitrary computable presentation of $A(S)$. Then by Lemma 6 $\mathcal{A}$ is computably isomorphic to a computable algebra $A_f(S)$ for some one-to-one computable enumeration $f = f(\mathcal{A})$. Since f is equivalent to one of the enumerations $f_1, \dots, f_n$, the algebra $A_f(S)$ is computably isomorphic to one of the algebras $A_{f_1}(S), \dots, A_{f_n}(S)$. The theorem is proved. □

4. COMPUTABLY CATEGORICAL RINGS AND THEIR EXPANSIONS BY CONSTANTS

In this section our task will be more difficult as we will encode a family S of pairs of sets into a ring. In order to define the algebra $A(S)$ in which the family S is encoded we have to enumerate S somehow, simply for the purpose of giving an abstract definition of this algebra. This enumeration is not assumed to be computable. As in the section 3, once $A(S)$ is defined, this enumeration will be forgotten, and we will consider how computable enumerations of S lead to computable presentations of $A(S)$. Suppose that

$$S = \{(M_0, N_0), (M_1, N_1), (M_2, N_2), \dots\}.$$

Let us consider the free product (in the variety of all nonassociative rings)

$$B(S) = F[x] \star F[y] \star (F_{M_0} \oplus F_{N_0}) \star \dots \star (F_{M_k} \oplus F_{N_k}) \star \dots ,$$

where $F[x]$ and $F[y]$ be two polynomial rings in x and y - we view these polynomials without constant terms, - and F_{M_k} and F_{N_k} denote the fields encoding M_k and N_k as was described in the previous section. Finally, we consider the quotient-algebra

$$A(S) = B(S)/R,$$

where R is the ideal of $B(S)$ generated by all sets $xF_{M_i} \cup F_{M_i}x$ and $yF_{N_i} \cup F_{N_i}y$. This algebra has also the following description. A basis of $A(S)$ can be chosen consisting of nonassociative products

$$(a_1 a_2 \dots a_n)_q, \qquad n \geq 1, \tag{7}$$

where $a_1, \dots, a_n$ belong to the standard monomial bases of the polynomial rings $F[x]$, $F[y]$, or else they are basic monomials of the fields F_{M_i} and F_{N_i}. Referring to the sets

$$F_{M_0} \oplus F_{N_0}, \ \dots, \ F_{M_i} \oplus F_{N_i}, \dots \tag{8}$$

as to the components, we stipulate that any two neighbouring monomials a_{i-1} and a_i, situated in the bracket $(a_{i-1}a_i)$, belong to different components; and, in addition, if one of the elements a_{i-1}, a_i is x, then the other cannot belong to F_{M_i}; or, similarly, if one of the elements a_{i-1}, a_i is y, then the other cannot belong to F_{N_i}.

The multiplication table on the basic products defined in (7) is as follows: if $(a_1 a_2 \dots a_n)_p$ and $(b_1 b_2 \dots b_m)_q$ are two basic products and $\max(m, n) > 1$, then

$$(a_1 a_2 \dots a_n)_p \cdot (b_1 b_2 \dots b_m)_q = ((a_1 a_2 \dots a_n)_p (b_1 b_2 \dots b_m)_q), \tag{9}$$

If $m = n = 1$ and a_1, b_1 belong to the same component, then $a_1 b_1 = \sum_{i=1}^{\ell} \beta_i c_i$, where c_i's are basic monomials of the component to which both of them belong, and we define

$$(a_1) \cdot (b_1) = \sum_{i=1}^{\ell} \beta_i (c_i). \tag{10}$$

If one of the elements a_1, b_1 is equal to x and the other belongs to the component F_{M_i}, or else if one of the elements a_1, b_1 is equal to y and the other belongs to F_{N_i}, then $a_1 \cdot b_1 = 0$. Otherwise $a_1 \cdot b_1 = (a_1 b_1)$.

Let $u = (a_1 a_2 \dots a_n)_q$ be a basic product. We set $|u| = n$, and for an element $a = \sum_{i=1}^{\ell} \beta_i u_i$ of $A(S)$ we set $|a| = \max |u_i|$. It is clear from the multiplication table (9) and (10) that

$$|a \cdot b| = |a| + |b|, \tag{11}$$

unless $|a| = |b| = 1$ and a, b are from the same component, or else one of them is x and the other is from F_{M_i}, or one of them is y and the other is from F_{N_i}.

LEMMA 7. *1) The ring $A(S)$ contains isomorphic copies of the components (8).*

2) The subset $U_M = F_{M_0} \cup F_{M_1} \cup \dots \cup F_{M_n} \cup \dots$ of $A(S)$ can be characterized as the set of all elements $a \in A(S)$ satisfying the condition that $xa = ax = 0$. The subset $U_N = F_{N_0} \cup F_{N_1} \cup \dots \cup F_{N_n} \cup \dots$ of $A(S)$ can be characterized as the set of all elements $a \in A(S)$ satisfying the condition that $ya = ay = 0$.

3) The set $E_M = \{e_0, \dots e_n, \dots\}$ of unit elements of fields $F_{M_0}, \dots, F_{M_n}, \dots$ can be characterized as the set of all idempotents $e \in A(S)$ such that $xe = ex = 0$. The set $E_N = \{f_0, \dots f_n, \dots\}$ of unit elements of fields $F_{N_0}, \dots, F_{N_n}, \dots$ can be characterized as the set of all idempotents $f \in A(S)$ such that $yf = fy = 0$.

4) The fields F_{M_i} and F_{N_i} can be characterized as the set of all elements $a \in A(S)$ satisfying the conditions $xa = ax = 0$, and $e_i a = ae_i = a$; and $ya = ay = 0$, and $f_i a = af_i = a$, respectively.

5) Two idempotents $e \in E_M$ and $f \in E_N$ are in the same component (i.e., identities of F_{M_i} and F_{N_i} for some i) iff $ef = fe = 0$.

Proof. These statements follow directly from the properties of the multiplication table of $A(S)$. □

LEMMA 8. *Let $S = \{(M_i, N_i) \mid i \in \omega\}$ be a symmetric family of pairs of sets. Then there exists an automorphism α of $A(S)$ such that $\alpha(x) = y$.*

Proof. The family S is symmetric. Therefore, together with every component $F_A \oplus F_B$, with $A \neq B$, we will have also a component $F_B \oplus F_A$; and these components are isomorphic under the isomorphism $\alpha(h_1, h_2) = (h_2, h_1)$. Also every component $F_A \oplus F_A$ has a natural automorphism defined by the same formula. Conditions $\alpha(x) = y$ and $\alpha(y) = x$ define an automorphism of $F[x] \star F[y]$. These mappings simultaneously can be extended to an isomorphism α of $A(S)$. □

Now, given a one-to-one computable enumeration f of S, let us construct a computable presentation $A_f(S)$ of $A(S)$. Let us denote $f(i) = (M_i, N_i)$. Since f is computable, there exists a procedure, which produces a 2-dimensional array $\{(M_{in}, N_{in}) \mid i, n \in \omega\}$ of pairs of finite subsets of ω, according to the following rules:

(M1) At stage 0 it produces a pair (M_{00}, N_{00}), where the subsets M_{00} and N_{00} are either both empty or contain one element each;

(M2) At stage k it produces pairs of subsets $(M_{k0}, N_{k0}), \dots, (M_{1k-1}, N_{1k-1})$, (M_{0k}, N_{0k}) so that $M_{k-1i} \subseteq M_{ki}$ and $N_{k-1i} \subseteq N_{ki}$, $i = 1, \dots, k-1$, and such that for every k

$$\mathrm{card}(M_{k-10} \cup \dots \cup M_{0k-1}) \leq \mathrm{card}(M_{k0} \cup \dots \cup M_{0k}) + 1;$$

$$\mathrm{card}(N_{k-10} \cup \dots \cup N_{0k-1}) \leq \mathrm{card}(N_{k0} \cup \dots \cup N_{0k}) + 1.$$

(M3) $\bigcup_{i\geq 0} M_{in} = M_n$, and $\bigcup_{i\geq 0} N_{in} = N_n$.

Thus, using f, we can construct an effective sequence of computable partial algebras

$$A(f,0), A(f,1), \ldots, A(f,n), \ldots$$

such that:

(A1) $A(f,i)$ is a subalgebra of $A(f,i+1)$;

(A2) $A(f,i)$ is isomorphic to

$$(F[x] \star F[y] \star (F_{M_{i0}} \oplus F_{N_{i0}}) \star \cdots \star (F_{M_{0i}} \oplus F_{N_{0i}}))^{(i)},$$

the latter being the subspace of $A(S)$ spanned by the basic products of degree $\leq i$, depending only on elements from $F[x]$, $F[y]$, $F_{M_{i0}}, \ldots, F_{M_{0i}}$, $F_{N_{i0}}, \ldots, F_{N_{0i}}$, with the addition and the multiplication inherited from $A(S)$;

(A3) $A_f(S) = \bigcup_{k=0}^{\infty} A(f,i)$ is isomorphic to $A(S)$.

As the sets M_{ki} and N_{ki} are finite, the fields $F_{M_{ki}}$ and $F_{N_{ki}}$ are finite-dimensional, hence finite. Thus, the partial algebras $A(f,i)$ are also finite-dimensional, and hence also finite. It is important to note that at stage i, when we extend $A(f,i-1)$ to $A(f,i)$ only three idempotents will be added, namely the unit element e_i of the field F_{M_i}, the unit element f_i of the field F_{N_i}, and their sum e_i+f_i. They can be distinguished by multiplying them by x and y. For example, e_i is the only idempotent out of the three with the property $xe_i = e_ix = 0$. In order to separate stages, each time, we start enumeration of additional elements with e_i followed by f_i and e_i+f_i.

LEMMA 9. *1) The ring $A_f(S)$ is computable for every computable one-to-one enumeration f of S.*

2) Let f and g be one-to-one computable enumerations. Then

(i) The expanded rings $(A_f(S), x)$ and $(A_g(S), x)$ are computably isomorphic, iff $f \sim g$;

(ii) The expanded rings $(A_f(S), x)$ and $(A_g(S), y)$ are computably isomorphic, iff $f \sim \tilde{g}$.

Proof. The computable presentation for $A(S)$ has been constructed above. The rest of the proof is similar to that of Lemma 5. □

LEMMA 10. *Let $\mathcal{A}$ be a computable presentation of $A(S)$. Then one can construct a one-to-one computable enumeration $f = f(\mathcal{A})$ of S such that $\mathcal{A}$ is computably isomorphic to both $A_f(S)$ and $A_{\tilde{f}}(S)$.*

Proof. Firstly, we can list all idempotents $a_0, a_1, \ldots, a_k, \ldots$, such that $xa_i = a_ix = 0$, in order in which they occur in the given enumeration of $\mathcal{A}$. Secondly, we find a pair for each of them listing the idempotents $b_0, b_1, \ldots, b_k, \ldots$, such that $yb_i = b_iy = 0$ and $a_ib_i = b_ia_i = 0$.

Let $F_i = \{a_i, 2a_i, \ldots, \nu a_i\}$ and $G_i = \{b_i, 2b_i, \ldots, \nu b_i\}$ be the corresponding copies of the base field F. By Lemma 7 $f(i) = (M_i, N_i)$, where

$$M_i = \{m \mid \exists z\ (z = za_i = a_iz \text{ and } z \text{ is algebraic of degree } p_m \text{ over } F_i)\}$$

$$N_i = \{n \mid \exists z\ (z = zb_i = b_iz \text{ and } z \text{ is algebraic of degree } p_n \text{ over } G_i),\}$$

enumerates S, and $f = f(\mathcal{A})$ is a computable enumeration of S. Moreover, f is one-to-one. Clearly $\mathcal{A}$ and $A_f(S)$ are computably isomorphic.

If we listed b_i first and in the order in which they occur in the given enumeration of $\mathcal{A}$, we would get the enumeration $\tilde{f}$. Thus $\mathcal{A}$ and $A_{\tilde{f}}(S)$ are also computably isomorphic. □

LEMMA 11. *Suppose that S is symmetric and its algorithmic dimension is 2. Then $A(S)$ is computably categorical.*

Proof. Let $\mathcal{A}$ and $\mathcal{B}$ be any two computable presentations of $A(S)$. Let us apply Lemma 10 now and construct one-to-one computable enumerations $f_1 = f(\mathcal{A})$ and $f_2 = f(\mathcal{B})$ of S such that $\mathcal{A}$ and $\mathcal{B}$ are computably isomorphic to $A_{f_1}(S)$ and $A_{f_2}(S)$, respectively. Since the algorithmic dimension of S is 2 we know that either f_1 is equivalent to f_2 or f_1 is equivalent to $\tilde{f}_2$. By Lemma 10 $\mathcal{A}$ and $\mathcal{B}$ are computably isomorphic. □

THEOREM B. (case $n = 2$) *There exists a computably categorical ring $\mathcal{R}$ and a constant $c \in \mathcal{R}$ such that the expanded ring $(\mathcal{R}, c)$ has exactly 2 computable isomorphism types.*

Proof. Let S be a symmetric family of pairs of sets, which algorithmic dimension is 2, with a computable enumeration f, which is not equivalent to $\tilde{f}$. Then by Lemma 11 the ring $A(S)$ is computably categorical. Let us show that the expanded ring $(A(S), x)$ has algorithmic dimension 2. The expanded rings $(A(S), x)$ and $(A(S), y)$ are isomorphic by Lemma 8 but $(A_f(S), x)$ and $(A_f(S), y)$ are not computably isomorphic; for, if they were, enumerations f and $\tilde{f}$ would be equivalent by Lemma 9.

Let now $(\mathcal{A}, z)$ be a computable presentation of $(A(S), x)$. Then either $f_{\mathcal{A}} \sim f$ or $f_{\mathcal{A}} \sim \tilde{f}$. Hence by the previous lemmata $(A(S), x)$ has exactly two computable isomorphism types. The theorem is proved. □

To conclude this section we will briefly explain the guidelines for constructing, for each $k \in \omega$, a computably categorical ring which has exactly k computable isomorphism types, when expanded by any finite number of constants. Passing from the case $n = 2$ to the general case we need no new ideas, the only effect is that the notation gets more complicated. This is why we only sketch the proof.

A natural step is to consider families of k-tuples of computably enumerable sets and define an appropriate notion of symmetry.

Let $X = (X_1, \dots, X_k)$ be a k-tuple of sets. Let us define

$$\pi X = (X_k, X_1, \dots, X_{k-1}).$$

Thus π is a map defined on the set of all k-tuples of sets.

DEFINITION 9. A family S of k-tuples of sets is called *symmetric* if $X \in S$ implies that $\pi X \in S$, that is if S is closed under π. We call the sequence X, πX, $\pi^2 X$, $\dots$, $\pi^{k-1} X$ the *orbit* of X.

It is obvious that $\pi^k X = X$. We also define $\pi^0 X = X$.

Suppose that S is a symmetric family of k-tuples. Suppose that f is a one-to-one computable enumeration of S. For each $i \leq k - 1$, we define the enumeration f_i by setting $f_i(n) = \pi^i f(n)$ for all $n \in \omega$. In particular, we see from this definition that f_0 is f.

DEFINITION 10. A symmetric family of k-tuples of computably enumerable sets has *dimension* k if there exists a one-to-one computable enumeration f of S with the following two properties:

1) The enumerations $f_0, f_1, \dots, f_{k-1}$ are pairwise inequivalent.

2) Each computable one-to-one enumeration of S is equivalent to one of the enumerations $f_0, f_1, \dots, f_{k-1}$.

In [20] it is proved that there exists a symmetric family S of k-tuples of computably enumerable sets whose dimension is k. Using this result and the ideas of the previous section we can encode S into a ring and prove the following theorem:

THEOREM B. (general case) *For every natural number k there exists a computably categorical ring $\mathcal{R}$ such that for an $c \in \mathcal{R}$, the expanded ring $(\mathcal{R}, c)$ has exactly k types of computable isomorphisms.*

Proof. (Sketch) Let S be a symmetric family of dimension k. Suppose that

$$S = \{(M_1^{(i)}, M_2^{(i)}, \dots, M_k^{(i)}) \mid i \in \omega\}.$$

The algebra $B(S)$ will be constructed as follows. Let us consider the free product (in the variety of all nonassociative rings)

$$B(S) = F[x_1] \star \dots \star F[x_k] \star B^0(S) \star \dots \star B^n(S) \star \dots,$$

where $F[x_i]$ is a polynomial ring in x_i (without constant terms), $B^i(S) = F_{M_1^{(i)}} \oplus \dots \oplus F_{M_k^{(i)}}$ and $F_{M_j^{(i)}}$ denotes the fields encoding $M_j^{(i)}$. Finally, we consider the quotient-algebra

$$A(S) = B(S)/R,$$

where R is the ideal of $B(S)$ generated by all sets $x_j F_{M_j^{(i)}} \cup F_{M_j^{(i)}} x_j$, for all $i \in \omega$.

Due to the symmetry of the family there exists an isomorphism α of $A(S)$ such that $\alpha(x_i) = x_{i+1}$ (addition modulo k).

Let f be a computable enumeration of S which satisfies properties 1) and 2) of Definition 10. As above we construct a computable presentation $A_f(S)$ of $A(S)$. In a similar way it can be shown that the ring $A(S)$ is computably categorical but its extention $(A(S), x_1)$ has algorithmic dimension k with $(A_f(S), x_1), \dots, (A_f(S), x_k)$ being k non- equivalent computable presentations of it. □

5. COMPUTABLY CATEGORICAL GROUPS AND THEIR EXPANSIONS BY CONSTANTS

In this section we start with the construction of a metabelian group $G(\mathcal{R})$ from an arbitrary nonassociative ring $\mathcal{R}$ with a unit element 1, which is due to Malcev [25]. In addition to Malcev's assumptions we will assume that the ring $\mathcal{R}$ is computable and we will observe that the group $G(\mathcal{R})$ also can be considered as a computable group. We will show that $G(\mathcal{R})$ can be expanded by constants a_1 and a_2 so that the computable isomorphism types of $\mathcal{R}$ are in one- to-one correspondence with the computable isomorphism types of the expanded group $(G(\mathcal{R}), a_1, a_2)$.

Let $\mathcal{R}$ be an arbitrary nonassociative ring of characteristic $p > 2$. In this case it can be considered as an algebra over a finite field $\mathbb{Z}_p$. We assume that $\mathcal{R}$ has a unit element 1. According to Malcev, the group $G(\mathcal{R})$ is the set of all triples (a,b,c) of elements $a,b,c \in \mathcal{R}$ with the multiplication given by the formula

$$(a,b,c)(x,y,z) = (a+x, b+y, bx+c+z). \tag{12}$$

It is easy to check that, so defined, this multiplication is associative, no matter what the ring $\mathcal{R}$ were, that the triple $e = (0,0,0)$ is the identity element for it, and that

$$(a,b,c)^{-1} = (-a,-b,ba-c). \tag{13}$$

When $\mathcal{R}$ is associative, this group can be represented as the group of upper triangular 3×3 matrices via the isomorphism

$$(a,b,c) \mapsto \begin{pmatrix} 1 & b & c \\ 0 & 1 & a \\ 0 & 0 & 1 \end{pmatrix}, \qquad a,b,c \in \mathcal{R},$$

in which form $G(\mathcal{R})$ is known as the Heisenberg group.

Clearly, if $\mathcal{R}$ is a computable ring, then under the standard enumeration of triples the group $G(\mathcal{R})$ is a computable group.

¿From the formula (12) we see that the centre $\mathcal{Z}$ of the group $G(\mathcal{R})$ consists of triples $(0,0,c)$, $c \in \mathcal{R}$, and since

$$(a,b,c)(x,y,z)(a,b,c)^{-1}(x,y,z)^{-1} = (0,0,bx-ya), \tag{14}$$

we conclude that the group $G(\mathcal{R})$ is metabelian. We will put in correspondence ϕ to $\mathcal{R}$ the expansion of $G(\mathcal{R})$ by the following two constants: $a_1 = (1,0,0)$ and $a_2 = (0,1,0)$; i.e., we set

$$\phi(\mathcal{R}) = (G(\mathcal{R}), a_1, a_2).$$

LEMMA 12. *If two computable rings $\mathcal{R}_1$ and $\mathcal{R}_2$ are computably isomorphic, then their corresponding computable expanded groups $\phi(\mathcal{R}_1)$ and $\phi(\mathcal{R}_2)$ are also computably isomorphic.*

Proof. Let $\sigma\colon \mathcal{R}_1 \to \mathcal{R}_2$ be a computable isomorphism of rings. Then it takes the identity 1_1 of $\mathcal{R}_1$ to the identity 1_2 of $\mathcal{R}_2$. It is now easy to check that the mapping

$$\bar{\sigma}((a,b,c)) = (a^\sigma, b^\sigma, c^\sigma) \tag{15}$$

is a computable isomorphism of the corresponding expanded groups. □

Let us now consider the folowing five properties for an expanded group $\mathcal{G} = (\mathcal{G}, a_1, a_2)$ introduced in the Malcev's paper [25]:

(G1) $\mathcal{G}$ is metabelian;

(G2) The subsets

$$\mathcal{C}_i = \{x \in G \mid xa_i = a_ix\}, \quad i = 1,2,$$

are abelian subgroups of $\mathcal{G}$;

(G3) The intersection of $\mathcal{C}_1$ and $\mathcal{C}_2$ is exactly the centre $\mathcal{Z}$ of the group $\mathcal{G}$, i.e., $\mathcal{C}_1 \cap \mathcal{C}_2 = \mathcal{Z}$;

(G4) For each pair $z_1, z_2 \in \mathcal{Z}$ there exists an element $h(z_1,z_2) \in \mathcal{G}$ such that

$$a_1 h(z_1,z_2) a_1^{-1} h(z_1,z_2)^{-1} = z_1, \quad a_2 h(z_1,z_2) a_2^{-1} h(z_1,z_2)^{-1} = z_2; \tag{16}$$

(G5) There exist isomorphisms $f_i: \mathcal{Z} \to \mathcal{C}_i$ of $\mathcal{Z}$ into $\mathcal{C}_i$ ($i = 1, 2$) such that for $c = a_2 a_1 a_2^{-1} a_1^{-1}$ the following conditions hold: $f_1(c) = a_1$, $f_2(c) = a_2^{-1}$, and

$$a_2 f_1(z) a_2^{-1} f_1(z)^{-1} = a_1 f_2(z) a_1^{-1} f_2(z)^{-1} = z. \tag{17}$$

THEOREM 7. *Let $\mathcal{R}$ be a computable nonassociative ring of characteristic $p > 2$ with a unit element 1. Then $G(\mathcal{R})$ is a computable group with the properties (G1)-(G5), satisfying $x^p = e$. The functions h, f_1, f_2 are computable.*

Proof. Most of these facts were proved in [25]. We should only check that, when $\mathcal{R}$ is computable, then $G(\mathcal{R})$ is computable also, and to show the computability of the three functions involved. Direct calculations show that $\mathcal{C}_1$ consists of all triples $(\alpha, 0, \gamma)$ and $\mathcal{C}_2$ consists of all triples $(0, \beta, \gamma)$, and the centre $\mathcal{Z}$ consists of all triples $(0, 0, \gamma)$, where $\alpha, \beta, \gamma \in \mathcal{R}$.

If $z_1 = (0, 0, \gamma_1)$, $z_2 = (0, 0, \gamma_2)$, then $h(z_1, z_2) = (-\gamma_2, -\gamma_1, 0)$ is a computable function satisfying (16).

If $z = (0, 0, \gamma)$, then $f_1(z) = (\gamma, 0, 0)$ and $f_2(z) = (0, -\gamma, 0)$ are computable functions satisfying (17).

These functions are clearly computable. □

LEMMA 13. *If $\mathcal{G} = (\mathcal{G}, a_1, a_2)$ is a computable expanded group with the identity $x^p = e$, p prime, satisfying (G1)-(G4), then:*

(a) The function $h(z_1, z_2)$ is computable ;

(b) The condition (G5) is also satisfied for $\mathcal{G}$ with computable isomorphisms f_1 and f_2.

Proof. (a) is clear. To prove that a computable isomorphism f_1 in (b) exists we first note that the mapping $\lambda: \mathcal{C}_1 \to \mathcal{Z}$ given by

$$\lambda: x \mapsto [a_2, x]$$

is a homomorphism of $\mathcal{C}_1$ onto $\mathcal{Z}$. Indeed,

$$[a_2, xy] = a_2 x y a_2^{-1} y^{-1} x^{-1} = a_2 x a_2^{-1} a_2 y a_2^{-1} y^{-1} x^{-1} =$$
$$a_2 x a_2^{-1} [a_2, y] x^{-1} = [a_2, x][a_2, y],$$

since $[a_2, y] \in \mathcal{Z}$. Also λ is onto because of (G4).

As the centre $\mathcal{Z}$ is an abelian group with the identity $x^p = e$ it can be considered as a vector space over $\mathbb{Z}_p$ via $k \cdot z = z^k$, $k \in \mathbb{Z}_p$. Since $\mathbb{Z}_p$ is finite, the basis $\{z_i \mid i \in \omega\}$ of $\mathcal{Z}$ as a vector space, with $z_0 = [a_2, a_1]$, can be constructively found as well as the set of elements $\{c_i \mid i \in \omega\}$ of $\mathcal{C}_1$ such that $c_0 = a_1$ and $\lambda(c_i) = z_i$.

Let $z = \sum k_i \cdot z_i = \prod z_i^{k_i}$, then we define $f_1(z) = \prod c_i^{k_i}$. Then

$$a_2 f_1(z) a_2^{-1} f_1(z)^{-1} = [a_2, f_1(z)] = \lambda(f_1(z)) = \lambda\left(\prod c_i^{k_i}\right) = \prod z_i^{k_i} = z.$$

A computable isomorphism f_2 can be constructed similarly. □

Now we will describe the second part of the construction, namely how, given the computable expanded group $\mathcal{G} = (\mathcal{G}, a_1, a_2)$ with the properties (G1)-(G4), we can construct a computable ring $\psi(\mathcal{G})$ with a unit element.

THEOREM 8. *Let $\mathcal{G} = (\mathcal{G}, a_1, a_2)$ be a computable expanded group with the identity element e and such that the properties (G1)-(G4) are satisfied. Then*

(a) The centre $\mathcal{Z}$ of this group is a computable set;

(b) Together with the following operations

$$z_1 + z_2 = z_1 z_2, \tag{18}$$
$$z_1 \times z_2 = h(e, z_2)h(z_1, e)h(e, z_2)^{-1}h(z_1, e)^{-1}, \tag{19}$$

the zero element e and the identity element $c = a_2 a_1 a_2^{-1} a_1^{-1}$, the centre $\mathcal{Z}$ forms a computable ring $\psi(\mathcal{G}) = (\mathcal{Z}, +, \times, e, c)$.

(c) The multiplication (19) does not depend on the function $h(z_1, z_2)$.

Proof. Since, according to (G3),

$$\mathcal{Z} = \{z \in \mathcal{G} \mid za_1 = a_1 z \text{ and } za_2 = a_2 z\},$$

the centre $\mathcal{Z}$ is a computable subset of ω. The ring addition (18) is clearly computable because the group multiplication is computable. The ring multiplication (19) is also computable since h is a computable function.

Let h and g be two functions satisfying (16). Let us denote $x_1 = h(z_1, e)$, $x_2 = h(e, z_2)$ and $y_1 = g(z_1, e)$, $y_2 = g(e, z_2)$. We have to prove that

$$x_2 x_1 x_2^{-1} x_1^{-1} = y_2 y_1 y_2^{-1} y_1^{-1}. \tag{20}$$

According to (16) we have $a_i y_i a_i^{-1} y_i^{-1} = a_i x_i a_i^{-1} x_i^{-1}$, from which it follows that $a_i x_i^{-1} y_i = x_i^{-1} y_i a_i$, i.e., $x_i^{-1} y_i \in \mathcal{C}_1 \cap \mathcal{C}_2 = \mathcal{Z}$. But then (20) follows, for it is equivalent to $x_1 x_2^{-1}(x_1^{-1} y_1) = (x_2^{-1} y_2) y_1 y_2^{-1}$, which is true since $x_i^{-1} y_i \in \mathcal{Z}$. We again refer to the Malcev's paper [25] for the rest of the proof. □

THEOREM 9. *(a) Let $\mathcal{R}$ be a computable ring with a unit element. Then $\psi(\phi(\mathcal{R}))$ is computably isomorphic to $\mathcal{R}$.*

(b) Let $\mathcal{G} = (\mathcal{G}, a_1, a_2)$ be a computable expanded group, satisfying $x^p = e$, with the properties (G1)-(G4). Then $\phi(\psi(\mathcal{G}))$ is computably isomorphic to $\mathcal{G}$.

Proof. (a) The mapping $(0, 0, \gamma) \mapsto \gamma$ could be verified to be a required isomorphism.

(b) As Lemma 13 shows, the condition (G5) is also true with computable functions f_1 and f_2. Now we have to follow Malcev and we have to verify only the effectiveness of his construction. Let us construct the ring $\mathcal{R} = \psi(\mathcal{G}) = (\mathcal{Z}, +, \times, e, c)$ as in part (b) of Theorem 8, and then let us form the set of triples of elements of $\mathcal{Z}$ to construct $G(\mathcal{R})$. As Malcev showed - and the condition (G5) is essential for that - the mapping

$$z = (z_1, z_2, z_3) \mapsto \tau(z) = f_1(z_1) f_2(z_2)^{-1} z_3 \tag{21}$$

is an isomorphism of $\phi(\mathcal{R})$ onto $\mathcal{G}$. Since f_1 and f_2 are computable functions τ is also computable. □

LEMMA 14. *Let $\mathcal{G} = (\mathcal{G}, a_1, a_2)$ and $\tilde{\mathcal{G}} = (\tilde{\mathcal{G}}, \tilde{a}_1, \tilde{a}_2)$ be two isomorphic computable expanded groups, both satisfying $x^p = e$, and $\mathcal{G}$ satisfies the conditions (G1)-(G4). Then*

(a) $\tilde{\mathcal{G}}$ also satisfies conditions (G1)-(G4);

(b) $\mathcal{G}$ and $\tilde{\mathcal{G}}$ are computably isomorphic to $\phi(\mathcal{R})$ and $\phi(\tilde{\mathcal{R}})$, respectively, for some isomorphic computable rings $\mathcal{R}$ and $\tilde{\mathcal{R}}$. The expanded groups $\mathcal{G}$ and $\tilde{\mathcal{G}}$ are computably isomorphic, if and only if the rings $\mathcal{R}$ and $\tilde{\mathcal{R}}$ are computably isomorphic.

Proof. (a) Let $\sigma\colon \mathcal{G} \to \tilde{\mathcal{G}}$ be the given isomorphism. Then $\sigma(\mathcal{Z}) = \tilde{\mathcal{Z}}$ and $\tilde{\mathcal{C}}_i = \sigma(\mathcal{C}_i)$. This implies that conditions (G1)-(G3) hold for $\tilde{\mathcal{G}}$. To prove (G4) we need to equip $\tilde{\mathcal{G}}$ with an appropriate function $\tilde{h}$. In order to do this we may define

$$\tilde{h}(\sigma(z_1), \sigma(z_2)) = \sigma(h(z_1, z_2)).$$

As σ is an isomorphism, the condition (G4) for $\tilde{\mathcal{G}}$ is satisfied.

(b) By Theorem 9(b) $\mathcal{G}$ and $\tilde{\mathcal{G}}$ are computably isomorphic to $\phi(\mathcal{R})$ and $\phi(\tilde{\mathcal{R}})$ for the computable rings $\mathcal{R} = \psi(\mathcal{G})$ and $\tilde{\mathcal{R}} = \psi(\tilde{\mathcal{G}})$. It is clear from the way $\tilde{h}$ was introduced that these two rings are isomorphic. Also, if σ was computable isomorphism, then $\mathcal{R}$ and $\tilde{\mathcal{R}}$ are computably isomorphic. On the other hand, if $\mathcal{R}$ and $\tilde{\mathcal{R}}$ are computably isomorphic, then $\mathcal{G}$ and $\tilde{\mathcal{G}}$ are computably isomorphic because by the same Theorem they are computably isomorphic to $\phi(\mathcal{R})$ and $\phi(\tilde{\mathcal{R}})$, respectively. □

THEOREM 10. *Let $\mathcal{R}$ be a computable ring with identity of characteristic $p>2$. Then, if $S_{\mathcal{R}} = \{\mathcal{R}_\alpha \mid \alpha \in I\}$ is the full set of representatives of the computable isomorphism types of $\mathcal{R}$, then $S_{\mathcal{G}} = \{\phi(\mathcal{R}_\alpha) \mid \alpha \in I\}$ is the full set of representatives of the computable isomorphism types of $\mathcal{G} = \phi(\mathcal{R})$.*

Proof. Since $\mathcal{G} = \phi(\mathcal{R})$ satisfies (G1)-(G4) by Theorem 7, then by Lemma 12 (b) all groups, which are isomorphic to $\mathcal{G}$, are in $S_{\mathcal{G}}$. The rest of the theorem follows from Lemma 14. □

COROLLARY 1. *Let $\mathcal{R}$ be a computable ring with identity of characteristic $p>2$. Then the algorithmic dimension of the expanded group $\phi(\mathcal{R}) = (G(\mathcal{R}), a_1, a_2)$ is equal to the algorithmic dimension of $\mathcal{R}$. In particular $\mathcal{R}$ is computably categorical if and only if $\phi(\mathcal{R})$ is computably categorical.*

THEOREM 11. *Let $\mathcal{R}$ be a computable ring with identity of characteristic $p>2$ and $r \in \mathcal{R}$. Let us consider $\phi(\mathcal{R}) = (G(\mathcal{R}), a_1, a_2)$ and define $a_3(r) = (0, 0, r) \in \phi(\mathcal{R})$. Then the algorithmic dimension of the expanded group*

$$(\phi(\mathcal{R}), a_3(r)) = (G(\mathcal{R}), a_1, a_2, a_3(r))$$

is equal to the algorithmic dimension of the expanded ring $(\mathcal{R}, r)$.

Proof. Suppose that $(\mathcal{R}_\alpha, r_\alpha)$ and $(\mathcal{R}, r)$ are two (computably) isomorphic expanded rings and let $\sigma\colon \mathcal{R} \to \mathcal{R}_\alpha$ be a (computable) isomorphism such that $\sigma(r) = r_\alpha$. Then the (computable) isomorphism $\bar{\sigma}$ of $\phi(\mathcal{R})$ onto $\phi(\mathcal{R}_\alpha)$, defined by (15), takes $a_3(r)$ to $a_3(r_\alpha)$, i.e. $(\phi(\mathcal{R}_\alpha), a_3(r_\alpha))$ and $(\phi(\mathcal{R}), a_3(r))$ are (computably) isomorphic.

Suppose that $(\phi(\mathcal{R}_\alpha), a_3(r_\alpha))$ and $(\phi(\mathcal{R}_\beta), a_3(r_\beta))$ are (computably) isomorphic and $\mu\colon \phi(\mathcal{R}_\alpha) \to \phi(\mathcal{R}_\beta)$ is the corresponding (computable) isomorphism such that $\mu(a_3(r_\alpha)) = a_3(r_\beta)$. Then this (computable) isomorphism induces a (computable) isomorphism $\bar{\mu}$ of

$$\bar{\mu}\colon \psi(\phi(\mathcal{R}_\alpha)) \to \psi(\phi(\mathcal{R}_\beta))$$

Due to (21) after identifying $\psi(\phi(\mathcal{R}_\alpha))$ with $\mathcal{R}_\alpha$ and $\psi(\phi(\mathcal{R}_\beta))$ with $\mathcal{R}_\beta$ we will see that $\bar{\mu}(r_\alpha) = r_\beta$, i.e. $(\mathcal{R}_\alpha, r_\alpha)$ and $(\mathcal{R}_\beta, r_\beta)$ are (computably) isomorphic. □

THEOREM C. *For every prime $p > 2$ and for every $n \in \omega \cup \{\omega\}$ there exists a computably categorical nilpotent group $\mathcal{G}$ of class 2, satisfying $x^p = 1$, such that for some constants $a_1, a_2, a_3 \in \mathcal{G}$ the expanded group $(\mathcal{G}, a_1, a_2, a_3)$ is of algorithmic dimension n.*

Proof. Let us consider a ring $\mathcal{R}$ and its element $r \in \mathcal{R}$ which satisfy Theorem B. Let us construct $\phi(\mathcal{R}) = (G(\mathcal{R}), a_1, a_2)$. Then according to Corollary 1 $\phi(\mathcal{R})$, and hence $G(\mathcal{R})$, are computably categorical. At the same time by Theorem 11 $(\phi(\mathcal{R}), a_3) = (G(\mathcal{R}), a_1, a_2, a_3)$, where $a_3 = a_3(r)$, is of algorithmic dimension n. The theorem is proved. □

REFERENCES

[1] J.N. Crossley, ed. Aspects of Effective Algebra. Proceedings of a Conference at Monash University, Australia, 1979, Upside Down A Book, Yarra Glen, 1981.

[2] J. Remmel. Recursive Boolean Algebras. In: Handbook of Boolean Algebras, v.3. North Holland, 1990.

[3] V. Harizanov. Pure Recursive Model Theory. In: Yu. Ershov, S. Goncharov, A. Nerode, J. Remmel eds. Handbook of Recursive Mathematics, to appear.

[4] T. Millar. Abstract Recursive Model Theory. In: E. Griffor, ed. Handbook of Recursion Theory. North-Holland, to appear.

[5] A. Frölich and JC Shepherdson. Effective Procedures in Field Theory. Phil Trans Roy Soc London, ser. A, 248: 407–432, 1956.

[6] A.I. Malcev. Constructive Algebras. Uspekhi Matem. Nauk. 16:3–60, 1961.

[7] M.O. Rabin. Computable Algebra, General Theory and Theory of Computable Fields. Trans Amer Math Soc 95:341-360, 1960.

[8] R. Soare. Recursively Enumerable Sets and Degrees. Berlin: Springer, 1987.

[9] M. Hall Jr. The theory of groups. New York, N.Y: The MacMillan Co, 1959.

[10] P. Jordan, J. von Neumann, E. Wigner. On an algebraic generalization of the quantum mechanical formalism. Ann Math 36: 29–64, 1934.

[11] I.M.H. Etherington. Genetic algebras. Proc Roy Soc Edinburgh 59: 242–258, 1939.

[12] S.S. Goncharov. The Problem of the Number of Non-Autoequivalent Constructivizations. Algebra i Logika 19:58–61, 1980.

[13] S.S. Goncharov. Autostability of Models and Abelian Groups. Algebra i Logika 19: 23–44, 1980.

[14] S.S. Goncharov, VD Dzgoev. Autostability of Models. Algebra i Logika 19:45–58, 1980.

[15] J. Remmel. Recursive Isomorphisms Types of Recursive Boolean Algebras. J Symb Logic 46: 572–594, 1981.

[16] S.S. Goncharov. Groups with a Finite Number of Constructivizations. Soviet Math Dokl. 23:621–639, 1981.

[17] S.S. Goncharov, AV Molokov, NS Romanovsky. Nilpotent Groups of Finite Algorithmis Dimension. Siberian Math Journal 30:82–88, 1989.

[18] T. Millar. Recursive Categoricity and Persistence. J Symb Logic 51: 430–434, 1986.

[19] Yu. Ershov and S.S. Goncharov, eds. Logic Notebook: Problems in Mathematical Logic. Novosibirsk: Novosibirsk University, 1986.

[20] P. Cholak, S. Goncharov, B. Khoussainov, R. Shore. Computably Categorical Structures and Expansions by Constants, submitted.

[21] S. Lang. Algebra. Addison-Wesley Pub. Co., Advanced Book Program, 1984.

[22] B. Khoussainov, R. Shore. Categoricity and Scott Families. In: Eds: D.S. Bridges ...[at al.]. Combinatorics, Complexity and Logic: Proceedings of DMTCS'96, Springer, 1996, pp 299–307.

[23] B. Khoussainov, R. Shore, Computable Isomorphisms, Degree Spectra of Relations and Scott Families. Ann. of Pure and Applied Logic, to appear.

[24] V. Harizanov. Turing Degree of Nonzero Member in Two Element Degree Spectrum. Ann of Pure and Applied Logic 60:1–30, 1993.

[25] A.I. Malcev. On one correspondence between rings and groups. Matem Sbornik 50:257–266, 1960.

Quantum V-algebras [1]

J. F. GOMES, F. E. MENDONÇA DA SILVEIRA, A. H. ZIMERMAN *Instituto de Física Teórica - UNESP, Rua Pamplona, 145 , 01405-900, Sao Paulo - SP, Brazil.* jfg@axp.ift.unesp.br, eugenio@axp.ift.unesp.br, zimerman@axp.ift.unesp.br

G. M. SOTKOV[2] *Departamento de Campos e Partículas - CBPF Rua Dr. Xavier Sigaud, 150-Urca 22290-180, Rio de Janeiro-RJ, Brazil.* sotkov@cbpfsu1.cat.cbpf.br

ABSTRACT The problem of the classification of the extensions of the Virasoro algebra is discussed. It is shown that all H-reduced $\hat{\mathcal{G}}_r$-current algebras belong to one of the following basic algebraic structures: local quadratic W-algebras, rational U-algebras, nonlocal V-algebras, nonlocal quadratic WV-algebras and rational nonlocal UV-algebras. The main new features of the quantum V-algebras and their heighest weight representations are demonstrated on the example of the quantum $V_3^{(1,1)}$-algebra.

1 INTRODUCTION

The concept of algebras and groups of symmetries (finite and infinite, Lie and non-Lie etc) is, by no means, the key stone of all the field and string theories of unification of the interactions. An impressive example of the computational power of the algebraic methods, however, is provided by the theory of the *second order phase transitions*, in two dimensions. It turns out [1] that the complete nonperturbative description of the critical behaviour of a class of 2-D statistical mechanics models is given by the highest weight ($h.w.$) unitary representations $\{c(m), \Delta_{p,q}(m)\}$ of the

[1] Talk given by G.M. Sotkov

[2] On leave of absence from the Institute for Nuclear Research and Nuclear Energy, Bulgarian Academy of Sciences, 1784, Sofia

Virasoro algebra

$$[L_n, L_l] = (n-l)L_{n+l} + \frac{c}{12}n(n^2-1)\delta_{n+l,0}, \qquad n,l = 0, \pm1, \pm2, ... \tag{1}$$

In words, all the physical data of the critical model - the exact values of the critical exponents, the partition function, all the correlation functions etc - are encoded in the representation theory of the algebra (1). The exact formulation of the above statement is as follows:

• **Physical data** (critical *RSOS* models on 2-*D* planar lattice [2]): For each fixed $m = 3, 4, 5, ...$, the $m-th$ *RSOS* model is defined by attaching to each site $\vec{i}$ a height l_i of length $l_i = 1, 2, ..., m$ under the condition that the maximal length difference of the nearest neighbours ($n.n.$) heights is one, i. e., $|l_{\vec{i}} - l_{\vec{i+1}}| = 1$. The only $n.n.$'s interact and the energy of a given configuration is

$$H = -\sum_{<ij>} J_{ij} l_i l_j + h \sum_i l_i.$$

The partition function $\mathcal{Z}(T,h) = Tr_{\mathcal{H}} \exp\left(-\frac{H}{kT}\right)$ ($Tr_{\mathcal{H}}$ denotes a sum over all allowed height configurations), found in [2] shows that, at a certain critical temperature $T = T_c(m)$, the $m-th$ *RSOS* exhibits a second order phase transition. This means that, at $\tau = \frac{T-T_c}{T_c} \to 0$, all thermodynamical characteristics of the model have *power-like singularities*:

$$C_V \sim \tau^{-\alpha}, \quad M \sim \tau^{-\beta}, \quad X \sim \tau^{-\gamma}, ...$$

The critical exponents $\alpha(m)$, $\beta(m)$, $\gamma(m)$, ..., turns out to be certain nonnegative rational numbers [2]. For example, the $m = 3$ odd lattice ($l_i = 1, 3$) model is equivalent to the Ising model and it has $\alpha = 0$, $\beta = \frac{1}{8}$; the $m = 5$ ($l_i = 1, 3, 5$) describes the 3-state Potts model etc.

• **Mathematical data** ([3]): For each fixed c, the $h.\ w.$ states $|\Delta, c>$ of the Virasoro algebra (1) are defined by requiring

$$L_0|\Delta, c> = \Delta|\Delta, c>, \quad L_n|\Delta, c> = 0, \quad n > 0. \tag{2}$$

The $h.\ w.$ unitary representations[3] of (1) are given by

$$\Delta_{p,q}(m) = \frac{[(m+1)p - mq]^2 - 1}{4m(m+1)}, \qquad c(m) = 1 - \frac{6}{m(m+1)}, \tag{3}$$

where $1 \leq p \leq m-1$, $1 \leq q \leq p$, $m = 3, 4, ...$

• **Identification** ([1], [4], [5]): The *scale invariance* of 2-*D* statistical models, at the critical point $T = T_c$, is shown to be a part of a larger group of conformal transformations $(z, \bar{z}) \to (f(z), \bar{f}(\bar{z}))$, which governes the critical behaviour of these models in the continuous (thermodynamical) limit. Therefore, the critical *RSOS* models can be described in terms of certain *conformal invariant 2-D field theories* (CFT's) ($l_{\vec{i}} \equiv l^{(m)}_{i_1 i_2} \to l^{(m)}(z, \bar{z})$). The symmetries of these CFT's are generated by

[3] The unitary condition was found in ref [6].

the two components $T(z)$ and $\bar{T}(\bar{z})$ of the conserved traceless stress-tensor $T_{\mu\nu}$. Its short distance operator-product expansion (OPE) is completely determined by the symmetry

$$T(z_1)T(z_2) = \frac{c/2}{z_{12}^4} + \frac{2T(z_2)}{z_{12}^2} + \frac{\partial_2 T(z_2)}{z_{12}} + O(1) \tag{4}$$

and the same for $\bar{T}(\bar{z})$. Introducing the corresponding conserved charges $L_n = \oint T(z) z^{n+1} dz$, where $n = 0, \pm 1, \pm 2, \dots$ (and the charge $\bar{L}_n$ for $\bar{T}(\bar{z})$), and substituting them in (4), we realize that the algebra of the charges L_n's (and $\bar{L}_n'$s) contains two (mutualy commuting) *Virasoro algebras* (1). As a consequence, the (Hilbert) space of states of such quantum CFT can be constructed as a tensor product of two *h. w.* representation spaces (2), (with $c(m) = \bar{c}(m)$): $|\Delta, \bar{\Delta}, c >= P(|\Delta, c > \otimes |\bar{\Delta}, c >)$, where P is denoting an appropriate projection on the subspace of states in $\mathcal{H}_{\Delta,c} \otimes \mathcal{H}_{\bar{\Delta},c}$, satisfying certain physical conditions - crossing symmetry, semi-locality etc - see refs [1] and [4]. To each *h. w.* state, one can make, in correspondence, a *primary* field $\phi_{\Delta,\bar{\Delta}}(z, \bar{z})$ of *spin* $s = \Delta - \bar{\Delta}$ and *dimension* $d = \Delta + \bar{\Delta}$ such that $|\Delta, \bar{\Delta}, c >= \phi_{\Delta,\bar{\Delta}}(0,0)|0>$. One of the most important properties of the primary fields $\phi_{\Delta_{p,q}} \equiv \phi_{p,q}$, from the Kac-table (3), is that, together with the standard conformal Ward identities

$$T(z_1)\phi_{p,q}(z_2)|0>= \left(\frac{\Delta_{p,q}}{z_{12}^2}\phi_{p,q}(z_2) + \frac{1}{z_{12}}\partial_2\phi_{p,q}(z_2) + O(1) \right) |0>, \tag{5}$$

it has to satisfy the so-called *null vector conditions*, which say for $p = 2$, $q = 1$ appears to be in the form

$$\left\{ L_{-1}^2 - \frac{2}{3}(1 + \Delta_{21}(m)L_{-2} \right\} |\Delta_{21}, c >= 0. \tag{6}$$

Eqns. (5) and (6) allow us to calculate the structure constants of the OPE's $\phi_{p_1q_1}(z_1)\phi_{p_2q_2}(z_2)$, as well as the exact 4-point (and n-point) correlation functions $< \phi_{p_1q_1}(1)\phi_{p_2q_2}(2) \cdot \dots \cdot \phi_{p_nq_n}(n) >$. Finally, the identification with the $RSOS$ models is done by comparing the ABF-exponents $\alpha, \beta, \gamma, \dots$[2], with the Kac-dimensions (3). For the Ising model ($m = 3$, $c(m) = \frac{1}{2}$), we have $\alpha = 0$, $\beta = \frac{1}{8}$ and $\Delta_{21}(3) = \frac{1-\alpha}{2-\alpha} = \frac{1}{2}$, $\Delta_{22}(3) = \frac{\beta}{2-\alpha} = \frac{1}{16}$.

• **The problem of classification of 2-D universality classes.** The purely algebraic description of the critical $RSOS$ models address the question whether one can find appropriate infinite algebras, which representation theories provide the exact solutions for all known 2-D critical statistical models having second order phase transition. The algebras, we are looking for, have to contain the Virasoro algebra (1) as a subalgebra. The following three examples of extended Virasoro algebras are to ilustrate the main features of the *new algebraic structures* one need to introduce in the description of the universality classes, in two dimensions.

EXAMPLE 1.1. $N = 1$ **supersymmetric Virasoro algebra**[9]: An infinite Lie super-algebra, containing together with the bosonic Virasoro generators L_n, an infinite set of fermionic ones G_s ($s \in \frac{1}{2}Z$)

$$[L_n, G_s] = \left(\frac{n}{2} - s\right) G_{n+s}, \qquad [G_s, G_t]_+ = 2L_{s+t} + \frac{c}{2}\left(s^2 - \frac{1}{4}\right)\delta_{s+t,0}, \tag{7}$$

where $[\,,\,]_+$ denotes the *anticommutator* of G_t's. The *h. w.* representations of (7)[3]

$$c(m) = \frac{3}{2}\left(1 - \frac{8}{m(m+2)}\right), \qquad \Delta_{pq}(m) = \frac{[(m+2)p - mq]^2 - 2}{12m(m+2)}, \tag{8}$$

where $m = 4, 5, \dots$ and $1 \leq p \leq m-2$, $1 \leq q \leq p$, give rise to a family of superconformal minimal models [7], [8] which describe the critical behaviour of $k = 2$ generalized $RSOS$ models. The difference with the standard ($k = 1$) $RSOS$ is that now the allowed *maximal* length difference between *n. n.* heights $l_{\bar{i}}$ and $l_{i\bar{+}1}$ is $k = 2$. As it is evident from (7), the $k = 2$ critical $RSOS$ possess symmetry larger than the conformal one. The stress-tensor $(T, \bar{T})$ and the new spin $\frac{3}{2}$ supercurrent $(G, \bar{G})$, $G_n = \oint z^{n+\frac{1}{2}} G(z)dz$, generate 2-$D$ superconformal transformations.

The critical $k - RSOS$ models for $k = 3, 4, \dots$ require *fractional* spin $\frac{l}{k}$ extensions of the Virasoro algebra [10, 11], [12]. Our next example represents the main features of such "parafermionic type" algebras.

EXAMPLE 1.2. Z_N **Parafermionic algebra** [12], [13]: The Z_N generalizations of the Ising (Z_2) and Potts (Z_3) models are lattice spin models, where each site (i) is occupied by a "spin variable" $\sigma(i)$ that takes values $\sigma^l_{(i)} = \exp\left(\frac{2\pi i l}{N}\right)$ in the discrete group Z_N. To describe (multi) critical behaviour of these models, one has to consider, together with $T(z)$, a set of $N-1$ new conserved currents $\psi_l^+(z) = \psi_{N-l}^-(z)$, where $l = 1, 2, \dots, N-1$, of spins $s_l = \frac{l(N-l)}{N}$, with OPE's in the form [12], [13]

$$\begin{aligned} \psi_1^\pm(z_1)\psi_1^\pm(z_2) &= c_{11} z_{12}^{-\frac{2}{N}} \left(\psi_2^\pm(z_2) + O(z_{12})\right), \\ \psi_1^+(z_1)\psi_1^-(z_2) &= z_{12}^{\frac{2}{N}} \left(\frac{1}{z_{12}^2} + \frac{N+2}{N} T(z_2) + O(z_{12})\right), \end{aligned} \tag{9}$$

where $c_{11} = \sqrt{\frac{2(N-1)}{N}}$.Introducing the parafermionic (PF) conserved charges in (9)

$$\begin{aligned} A^\pm_{\frac{1\mp l}{N}+n}\phi_l(0) &= \oint dz \psi_1^\pm(z) z^{\mp\frac{l}{N}+n}\phi_l(0), \\ \psi_1^\pm(z\exp(2\pi i))\phi_l(0) &= \exp\left(\frac{2\pi i l}{N}\right)\psi_1^\pm(z)\phi_l(0), \end{aligned}$$

we derive the Z_N PF-extension of the Virasoro algebra (2)

$$\begin{aligned} &\sum_{p=0}^{\infty} C^p_{(\frac{2}{N})}\left(A^\pm_{\frac{3\mp l}{N}-p+m} A^\pm_{\frac{1\mp l}{N}+p+n} - A^\pm_{\frac{3\mp l}{N}-p+n} A^\pm_{\frac{1\mp l}{N}+p+m}\right) = 0, \\ &\sum_{p=0}^{\infty} C^p_{(-\frac{2}{N})}\left(A^+_{m-\frac{1+l}{N}-p} A^-_{\frac{1+l}{N}+n+p-1} + A^-_{n-\frac{1-l}{N}-p-1} A^+_{\frac{1-l}{N}+p+m}\right) \\ &= \frac{N+2}{N} L_{m+n-1} + \frac{1}{2}\left(n - 1 + \frac{l}{N}\right)\left(\frac{l}{N} + m - 2\right)\delta_{m+n-1,0} \end{aligned} \tag{10}$$

of central charge $c(N) = 2\frac{N-1}{N+2}$, where $N = 2, 3, \dots$, and structure constants $C^p_{(r)} = \frac{\Gamma(p-r)}{p!\Gamma(-r)}$. The *h. w.* representations of this infinite associative algebra, found in ref

[12], are of dimensions $\Delta_l = \frac{l(N-l)}{2N(N+2)}$ and Z_N charge - $l = 1, 2, ..., N-1$ for the (*order parameter*) fields $\sigma_l(z, \bar{z})$; and $\Delta_j = \frac{j(j+1)}{N+2}$ where $j = 1, 2, ... \leq [\frac{N}{2}]$ are the dimensions for the Z_N neutral (*energy operator*) fields $\epsilon_j(z, \bar{z})$. It is important to note that the *origin* of the fact that in the PF-algebra (10), the Lie commutator $[a, b] = ab - ba$ is replaced by an infinite sum of bilinears $A^{\pm}A^{\mp}$ is in the *branch cut singularities* $z^{\pm\frac{2}{N}}$ $(N \geq 3)$ in the OPE's (9). These types of singularities are a consequence of the *fractal spins* $s_1^{\pm} = 1 - \frac{1}{N}$ of the PF-currents $\psi_1^{\pm}$. Observe that for $N = 2$ (and for half-integer spins, in general), the OPE's have odd poles z_{12}^{-1} (or z_{12}^{-2s-1}) singularities, which lead to anticommutators $[a, b]_+ = ab + ba$. For integer spins $s = 1, 2, 3, ...$, the leading singularities in the OPE's are even poles z_{12}^{-2s} (as in eqn. (4)) and they give rise to the standard Lie commutators.

EXAMPLE 1.3. **W_3-Zamolodchikov algebra** [15], [14]: The most important property of the spin 3 extension of the Virasoro algebra (generated by $T(z)$ and $W(z)$ of spin $s_W = 3$) is that the commutator of the charges W_n of the spin 3-current $W_n = \oint W(z) z^{n+2} dz$ is quadratic in the Virasoro generators L_n's

$$[W_n, W_l] = (n-l)[d(n,l)L_{n+l} + b\Lambda_{n+l}] + \frac{c}{360}n(n^2-4)(n^2-1)\delta_{n+l,0}, \tag{11}$$

where

$$\Lambda_n = \sum_{k=-\infty}^{\infty} : L_k L_{n-k} : + \frac{1}{5} f_n L_n, \quad f_{2s} = 1 - s^2, \quad f_{2s+1} = (1-s)(2+s)$$

and

$$d(n,l) = \frac{1}{6}\left[\frac{2}{5}(n+l+2)(n+l+3) - (n+2)(l+2)\right], \qquad b = \frac{16}{22+5c}.$$

The *h. w.* states $|\Delta, w, c>$ of this *non-Lie associative* algebra are defined by

$$L_0|\Delta, w, c> = \Delta|\Delta, w, c>, \qquad W_0|\Delta, w, c> = w|\Delta, w, c>,$$

$$L_n|\Delta, w, c> = W_n|\Delta, w, c> = 0, \quad n > 0.$$

Its *h. w.* unitary representations $c(m) = 2\left(1 - \frac{12}{m(m+1)}\right)$, where $m = 4, 5, ...$, and $\Delta_{p_i q_i}(m)$, $w_{p_i q_i}(m)$ $(i = 1, 2)$ found in ref [14] give rise to a family of Z_3 symmetric CFT's that provide the exact solutions for a new class of critical statistical models. The simplest representative $m = 4$ of this class is again the critical 3-states Potts model.

The above examples of three different associative extensions of the Virasoro algebra, (7), (10), (11), suggest the following *organization* of the list of all known infinite algebras (and their 2-D CFT's):

(i) *Lie-algebraic extensions*: Conformal current (affine) $\hat{\mathcal{G}}_r$-algebras [16]:

$$[J_n^a, J_l^b] = i f^{abc} J_{n+l}^c + kn\delta^{ab}\delta_{n+l,0}, \tag{12}$$

where $n, l = 0, \pm 1, \pm 2, ...$; $a, b = 1, 2, ..., dim\ \mathcal{G}$; f^{abc} are the structure constants of an arbitrary (finite dimensional) semisimple Lie algebra $\mathcal{G}_r$; k is called the level of

$\hat{\mathcal{G}}_r$. Its generators are the conserved charges of the spin $s = 1$ chiral current $J^a(z) = \sum_{n=-\infty}^{\infty} z^{-n-1} J_n^a$, which also satisfy $[L_n, J_l^a] = -lJ_{n+l}^a$. The *h. w.* representations of (12) (and its CFT's) were constructed in refs [17, 18].

(ii) *Lie-superalgebraic extensions*: The $N = 1$ superVirasoro algebra (7); $N = 2, 3, 4$ superconformal algebras [19, 20, 21]; the affine $\hat{\mathcal{G}}_r$-superalgebras, where $\mathcal{G}_r$ is an arbitrary rank r finite dimensional superalgebra; $N = 1$ superconformal current $\hat{\mathcal{G}}_r$-algebras [23, 22], with generators J_n^a and ψ_n^a determined by (7), (12) and $[J_n^a, \psi_l^b] = f^{abc}\psi_{n+l}^c$, $[\psi_n^a, \psi_l^b]_+ = k\delta^{ab}\delta_{n+l,0}$.

(iii) *PF-extensions*: The Z_N (and D_{2N})-PF algebra (10) and its (p, M)-generalizations [12], by considering PF currents of spins $s_l = p\frac{l(N-l)}{N} + M_l$; Gepner's $\mathcal{G}_r$-parafermions [24].

(iv) *Quadratic W-algebras*: The W_n-algebras [25], [27], [26] generated by the charges of the spin $s = 2, 3, ..., n$-currents, and the more general $W\mathcal{G}_n$ [25]; the $W_n^{(l)}$-algebras [28], [29], [35]; the supersymmetric W_n-algebras etc.

To complete our table of extended Virasoro algebras, we have to add the family of the recently discovered classical Poisson brackets *nonlocal and nonlinear (quadratic)* V-algebras [31], [32], [33], [34].

(v) *V-algebras*: The simplest example is given by $V_3^{(1,1)} \equiv VA_2^{(1,1)}$-algebra [33], [34], generated by one local spin 2 $T(\sigma)$ (the stress-tensor) and two spin $\frac{3}{2}$-non local currents $V^\pm(\sigma)$:

$$\{T(\sigma), V^\pm(\sigma')\} = \frac{3}{2}V^\pm(\sigma')\partial_{\sigma'}\delta(\sigma - \sigma') + \partial_{\sigma'}V^\pm(\sigma')\delta(\sigma - \sigma'),$$

$$\{V^\pm(\sigma), V^\mp(\sigma')\} = \pm\frac{2}{k}\delta''(\sigma-\sigma') \mp \frac{2}{k}T(\sigma')\delta(\sigma-\sigma') + \frac{3}{2k^2}V^\pm(\sigma)V^\mp(\sigma')\epsilon(\sigma - \sigma'),$$

$$\{V^\pm(\sigma), V^\pm(\sigma')\} = -\frac{3}{2k^2}V^\pm(\sigma)V^\pm(\sigma')\epsilon(\sigma - \sigma'). \tag{13}$$

where $\epsilon(\sigma) = sign\ \sigma$. The $V_3^{(1,1)}$ is the first member of the $VA_n^{(1,1)}$-family of V-algebras, spanned by two non-local currents $V_{(n)}^\pm$ of spins $s = \frac{n+1}{2}$ and $n-1$ local currents W_{n-l+2} of spins $s_l = n-l+2$, where $l = 1, 2, ..., n$. The Bilal's VB_2-algebra [31] is quite similar to (13), but $V^\pm$-currents have spin $s^\pm = 2$ in this case.

Our main purpose, in what follows, is the construction of the *quantum* V_{n+1}-algebras and their minimal conformal models (i. e., their *h. w.* representations). The most important result is that the classical spins $s^{cl} = \frac{n+1}{2}$, of the nonlocal currents $V^\pm$, gets renormalized, i. e., $s^{qu} = \frac{n+1}{2}\left(1 - \frac{1}{2k+n+1}\right)$ and their algebra shares the main patterns of the PF-algebras. While the quantum local currents W_{n-l+2} manifest properties similar to the W_n-algebras. Therefore, the quantum $V_{n+1}^{(1,1)}$-algebras represent an appropriate *unification* of the features of the Z_{2k+3} PF-algebra with the W_{n+1}-one.

2 CONSTRAINED $\mathcal{G}_r$-CURRENT ALGEBRAS

The list of the five known families of extended Virasoro algebras we have made, however, does *not* solve the problem of the *classification of 2-D universality classes* (i. e., all allowed critical behaviours in two dimensions). We need a method of exhausting all the possible extensions of the Virasoro algebra. The hint is coming from the fact that all the considered algebras[4] - the Virasoro-one, the PF-, the W_n- and V_n-ones - can be obtained by imposing a specific set of constraints on the currents of certain $\hat{\mathcal{G}}_r$-current algebras ($SL(2,R)$, for (1) and (10), and $SL(3,R)$, for (11) and (13) etc):

$$\{J^a(\sigma), J^b(\sigma')\} = \imath f^{abc} J^c(\sigma')\delta(\sigma - \sigma') + k\partial_{\sigma'}\delta(\sigma - \sigma'). \tag{14}$$

It suggests that the desired classification of the extended Virasoro algebras can be reached by the methods of the Hamiltonian reduction [29], [36], [30], i. e., by considering all consistent sets of constraints on the currents $J^a(\sigma) \in \hat{\mathcal{G}}_r$

$$\hat{J}(\sigma) = g^{-1}\partial g = \sum_{\text{allroots}} J_{\{\alpha\}} E_{\{\alpha\}} + \sum_{i=1}^{r} J_i \frac{\vec{\alpha}_i \cdot \vec{H}}{\alpha_i^2}$$

where $E_{\{\alpha\}}$, $h_i = \frac{\vec{\alpha}_i \cdot \vec{H}}{\alpha_i^2}$ are the generators of the finite Lie algebra $\mathcal{G}_r$. Therefore, the question now is *whether and how* one can *classify* all constraints to be imposed on $\hat{J}(\sigma)$.

We start with few selected examples of constrained $SL(n,R)$ ($n = 2,3,4$) algebras, which demonstrate the way the *algebraic structure* of the reduced algebras depends on the specific choice of the constraints.

EXAMPLE 2.1. $SL(2,R)$ **reductions.**

(1a) $A_1/\mathcal{N}_+ \equiv$ *Virasoro algebra*: Take $J_\alpha = 1$ as a constraint and $J_1(\equiv J_0) = 0$ as its gauge fixing (i. e., J_1 is the canonically conjugated momentum of J_α, since $\{J_\alpha(\sigma), J_1(\sigma')\} = -J_\alpha(\sigma)\delta(\sigma - \sigma') \approx \delta(\sigma - \sigma')$).[5] Under these conditions, the classical Poisson bracket (PB) algebra of the remaining current $J_{-\alpha} \equiv T$ can be derived from eqn. (11), by calculating the corresponding Dirac brackets

$$\{T(\sigma), T(\sigma')\}_D = \frac{k^2}{2}\delta'''(\sigma - \sigma') - 2T(\sigma')\delta'(\sigma - \sigma') + \partial_{\sigma'}T(\sigma')\delta(\sigma - \sigma'), \tag{15}$$

which is nothing, but the *classical* PB's Virasoro algebra. Another form of the Dirac method, proposed by Polyakov [29] consists in imposing the constraints and gauge fixing conditions on the infinitesimal $\hat{\mathcal{G}}_r$-gauge transformations

$$\delta_\epsilon J^a(\sigma) = f^{abc} J^c(\sigma)\epsilon^b(\sigma) + \frac{k}{2}\partial_\sigma \epsilon^a. \tag{16}$$

Next, we solve the $\delta_\epsilon J_\alpha = \delta J_1 = 0$ equations for the redundant gauge parameters ϵ_1 and ϵ_α

$$\epsilon_1 = \frac{k}{2}\partial\epsilon, \quad \epsilon_\alpha = -\frac{k^2}{2}\partial^2\epsilon + J_{-\alpha}\epsilon, \quad \epsilon \equiv \epsilon_{-\alpha},$$

[4] The supersymmetric extensions arise from the constrained superconformal current algebras

[5] An equivalent explanation of the $J_1 = 0$ condition (which is not a constraint) is that, due to the residual gauge transformation $h = \exp(\beta(z))E_{-\alpha}$, $J' = h^{-1}Jh + kh^{-1}\partial h$, which leaves invariant the constraint $J_\alpha = 1$, one can make $J_1' = 0$, by an appropriate choice of $\beta(z)$.

and substituting them in $\delta J_{-\alpha}$, we find

$$\delta_\epsilon J_{-\alpha} = -\frac{k^2}{2}\partial^3\epsilon + 2J_{-\alpha}\partial\epsilon + \partial J_{-\alpha}\epsilon,$$

i. e., the functional form of eqn. (15).

(1b) $A_1/U(1) \equiv$ *classical PF-algebra*: Take $J_0(\equiv J_1) = 0$ as a constraint (no residual gauge transformations exists). In this case, as a consequence of eqn. (14), we have $\{J_0(\sigma), J_0(\sigma')\} = \partial_{\sigma'}\delta(\sigma - \sigma')$ $(k = 2)$. To find the Dirac brackets of the $J'_{\pm\alpha}$s, we have to invert the ∂_σ-operator, i. e., $\partial_\sigma(\partial_{\sigma'}^{-1}) = \delta(\sigma - \sigma')$, hence $\partial_{\sigma'}^{-1} = \frac{1}{2}\epsilon(\sigma - \sigma'))$ and thus to introduce *nonlocal* $\epsilon(\sigma)$-terms in the $J_{\pm\alpha} \equiv V^{\pm}$-algebra

$$\begin{aligned}
\{V^{\pm}(\sigma), V^{\pm}(\sigma')\}_D &= -V^{\pm}(\sigma)V^{\pm}(\sigma')\epsilon(\sigma - \sigma'),\\
\{V^{+}(\sigma), V^{-}(\sigma')\}_D &= \partial_{\sigma'}\delta(\sigma - \sigma') + V^{+}(\sigma)V^{-}(\sigma')\epsilon(\sigma - \sigma').
\end{aligned} \tag{17}$$

Following the Polyakov method, we get

$$\epsilon_0(\sigma) = -\int \epsilon(\sigma - \sigma')(V^{+}(\sigma')\epsilon^{-}(\sigma') - V^{-}(\sigma')\epsilon^{+}(\sigma'))d\sigma',$$

and pluging it back in the $\delta_{\epsilon^\pm}V^{\pm}$-transformations, we arrive at eqn. (17). The reason to call this *nonlocal PB* algebra as a *classical parafermionic* one is that an appropriate $N \to \infty$ limit of the (quantum) PF OPE' (9) reproduces exactly eqn. (17), as we will demonstrate, in detail, in the next section.

EXAMPLE 2.2. **Constrained $SL(3, R)$-algebras**.

(2a) $A_2/\mathcal{N}_+ \equiv W_3$*-algebra*: In this case, $\mathcal{N}_+ = \{E_{\alpha_1}, E_{\alpha_2}, E_{\alpha_1+\alpha_2}\}$ and $J_{\alpha_i} = 1$, $J_{\alpha_1+\alpha_2} = 0$ are the constraints; $J_i = J_{-\alpha_1} = 0$ are the gauge fixing conditions, in Drinfeld-Sokolov gauge. The classical W_3-algebra, generated by one spin 2 $T(z) \equiv J_{-\alpha_2}(z)$ and one spin 3 $W_3(z) = J_{-\alpha_1-\alpha_2} - \frac{1}{2}\partial J_{-\alpha_2}$ currents, has the form $(k = 2)$

$$\{T(\sigma), W_3(\sigma')\} = 3W_3(\sigma')\partial_{\sigma'}\delta(\sigma - \sigma') + 2\partial_{\sigma'}W_3(\sigma')\delta(\sigma - \sigma'),$$

$$\begin{aligned}
\{W_3(\sigma), W_3(\sigma')\} =& -4\delta^{(v)}(\sigma-\sigma') + 5T(\sigma')\delta'''(\sigma-\sigma') - \frac{15}{2}\partial_{\sigma'}T(\sigma')\delta''(\sigma-\sigma')\\
&-\left(T^2(\sigma') - \frac{9}{2}\partial^2_{\sigma'}T(\sigma')\right)\delta'(\sigma-\sigma') + \partial_{\sigma'}\left(\frac{1}{2}T^2(\sigma') - \partial^2_{\sigma'}T(\sigma')\right)\delta(\sigma-\sigma').
\end{aligned} \tag{18}$$

(2b) $A_2/\mathcal{N}_+^1 \otimes U(1) \equiv V_3^{(1,1)}$*-algebra*: Take $\mathcal{N}_+^{(1)} = \{E_{\alpha_2}, E_{\alpha_1+\alpha_2}\}$, $U(1) = \{\vec{\lambda}_1 \cdot \vec{H}\}$; $J_{\alpha_2} = 1$, $J_{\alpha_1+\alpha_2} = 0$, $\sum_{i=1}^{2}\lambda_1^{(i)}J_i = 0$ are the constraints and $J_{-\alpha_1} = \sum_{i=1}^{2}\alpha_2^i J_i = 0$ are the gauge fixing conditions. The remaining currents $V^+ = J_{\alpha_1}$, $V^- = J_{-\alpha_1-\alpha_2}$, of spin $\frac{3}{2}$ (nonlocal), and $T = J_{-\alpha_2}$, of spin 2 (local), generate the following nonlocal $V_3^{(1,1)}$-algebra $(k = 2)$ (13). We have to mention that if one *relaxes* the $U(1)$-constraint $J = \vec{\lambda}_1 \cdot \vec{J} = 0$, then the *local* currents $V^{\pm}$ $(s = \frac{3}{2})$, T $(s = 2)$ and J $(s = 1)$ span the well known local quadratic (in J) Bershadsky-Polyakov $A_2/\mathcal{N}_+^{(1)} \equiv W_3^{(2)}$-algebra [28].

(2c) $A_2/\mathcal{N}_+^{(2)} \otimes U(1) \otimes U(1) \equiv V_3$ *-algebra*: In this case, $\mathcal{N}_+^{(2)} = \{E_{\alpha_1+\alpha_2}\}$, the constraints are $J_i = 0$, $J_{\alpha_1+\alpha_2} = 0$ and the gauge fixing (*g. f.*) conditions are $J_{-\alpha_1-\alpha_2} = 0$, $J_i = 0$. The *nonlocal* V_3-algebra, of the four spin 1 currents $V_i^{\pm} = J_{\pm\alpha_i}$ $(i = 1, 2)$, has the form [37]

$$\{V_i^{\pm}(\sigma), V_j^{\pm}(\sigma')\} = \frac{1}{2k^2}[V_i^{\pm}(\sigma)V_j^{\pm}(\sigma') + V_i^{\pm}(\sigma')V_j^{\pm}(\sigma)]\epsilon(\sigma - \sigma'),$$

$$\begin{aligned}\{V_i^{+}(\sigma), V_j^{-}(\sigma')\} &= \delta_{ij}\partial_{\sigma'}\delta(\sigma - \sigma') - \frac{1}{2k^2}[V_i^{+}(\sigma)V_j^{-}(\sigma') \\ &+ \delta_{ij}\sum_{s=1}^{2} V_s^{-}(\sigma)V_s^{+}(\sigma')]\epsilon(\sigma - \sigma').\end{aligned} \tag{19}$$

The stress-tensor $T(\sigma) = \frac{1}{2}\sum_{s=1}^{2} V_s^{+}(\sigma)V_s^{-}(\sigma)$ satisfies the standard Virasoro algebra (15), but without a central term.

(2d) $A_2/\mathcal{N}_+^{(2)} \otimes U(1) \equiv V_3^{(2)}$ *-algebra*: $\mathcal{N}_+^{(2)}$ is the same as in (2c), $U(1) = (\lambda_1 - \lambda_2)^i J_i$; the constraints are $J_{\alpha_1+\alpha_2} = 1$, $\sum_{i=1}^{2}(\lambda_1 - \lambda_2)^i J_i = 0$ and the *g. f.* conditions are $(\alpha_1 + \alpha_2)^i J_i = 0$. The $V_3^{(2)}$-algebra, of the local spin 2 stress-tensor $T = J_{-\alpha_1-\alpha_2} - \frac{1}{2}(J_{\alpha_1}J_{-\alpha_1} + J_{\alpha_2}J_{-\alpha_2})$, and four nonlocal currents $V_1^+ = J_{\alpha_1}$ $(s_1^+ = \frac{1}{2})$, $V_1^- = J_{-\alpha_1} - 2\partial J_{\alpha_2}$ $(s_1^- = \frac{3}{2})$, $V_2^- = J_{\alpha_2}$ $(s_2^- = \frac{1}{2})$ and $V_2^+ = J_{-\alpha_2} + 2\partial J_{\alpha_1}$ $(s_2^+ = \frac{3}{2})$, takes the form [37]

$$\{V_i^{\pm}(\sigma), V_j^{\pm}(\sigma')\} = (i - j)[V^{\pm}_{\frac{1}{2}(i+j\mp 1)}(\sigma)]^2\delta(\sigma - \sigma') + \frac{3}{8}V_i^{\pm}(\sigma)V_j^{\pm}(\sigma')\epsilon(\sigma - \sigma'),$$

$$\{V_i^{-}(\sigma), V_i^{+}(\sigma')\} = 2V_1^{-}(\sigma)V_2^{+}(\sigma)\delta(\sigma - \sigma') - \frac{3}{8}V_i^{-}(\sigma)V_i^{+}(\sigma')\epsilon(\sigma - \sigma'),$$

$$\begin{aligned}\{V_1^{-}(\sigma), V_2^{+}(\sigma')\} &= -4\partial^2_{\sigma'}\delta(\sigma - \sigma') - \frac{3}{8}V_1^{-}(\sigma)V_2^{+}(\sigma')\epsilon(\sigma - \sigma') \\ &+ 3[V_1^{+}(\sigma)V_2^{-}(\sigma') + V_1^{+}(\sigma')V_2^{-}(\sigma)]\partial_{\sigma'}\delta(\sigma - \sigma') \\ &+ \{T(\sigma) + \frac{3}{2}[V_1^{+}(\sigma)V_1^{-}(\sigma) + V_2^{+}(\sigma)V_2^{-}(\sigma)]\}\delta(\sigma - \sigma'),\end{aligned}$$

$$\{V_2^{-}(\sigma), V_1^{+}(\sigma')\} = \delta(\sigma - \sigma') - \frac{3}{8}V_2^{-}(\sigma)V_1^{+}(\sigma')\epsilon(\sigma - \sigma'). \tag{20}$$

Thus, $V_3^{(2)}$ is an example of *nonlocal quadratic* (non-Lie) algebra.

(2e) $A_2/\mathcal{N}_+^{(1)} \equiv W_3^{(1,1)}$*-algebra*: In this case, the constraints are $J_{\alpha_2} = J_{\alpha_1+\alpha_2} = 1$ and the *g. f.* conditions are $\alpha_2^i J_i = (\alpha_1 + \alpha_2)^i J_i = 0$. The algebra $W_3^{(1,1)}$, of the local currents $J^{\pm} = J_{\pm\alpha_1}$ $(s^{\pm} = 1)$ and $T_2 = J_{-\alpha_2}$ $(s_2 = 2)$, $T_{12} = J_{-\alpha_1-\alpha_2}$ $(s_{12} = 2)$, appears to be a local quadratic algebra [37], of W-type.

EXAMPLE 2.3. **Constrained $SL(4, R)$-algebras.**

(3a) $A_3/\mathcal{N}_+ \equiv W_4$*-algebra:* $\mathcal{N}_+ = \{E_{[\alpha]} : [\alpha]$are all positive roots$\}$, the constraints are $J_{\alpha_i} = 1$ $(i = 1,2,3)$, $J_{\alpha_1+\alpha_2} = J_{\alpha_2+\alpha_3} = 0$, $J_{\alpha_1+\alpha_2+\alpha_3} = 0$ and the *g. f.* conditions are $J_i = 0$, $J_{-\alpha_1} = 0$, $J_{-\alpha_2} = 0$, $J_{-\alpha_1-\alpha_2} = 0$. The algebra of the remaining currents $T = J_{-\alpha_3}$, $W_3 = J_{-\alpha_2-\alpha_3}$ and $W_4 = J_{-\alpha_1-\alpha_2-\alpha_3}$ is the standard quadratic W_4-algebra [25], [26], [27].

(3b) $A_3/\mathcal{N}_+^{(1,1)} \otimes U(1) \equiv V_4^{(1,1)}$ *-algebra:* $\mathcal{N}_+^{(1,1)} = \{E_{[\alpha]_1}\}$, where $[\alpha]_1$ are all positive roots, but α_1, $U(1) = \lambda_1 \cdot H$, the constraints are $J_{\alpha_2} = J_{\alpha_3} = 1$, $J_{\alpha_1+\alpha_2} = J_{\alpha_2+\alpha_3} = 0$, $J_{\alpha_1+\alpha_2+\alpha_3} = 0$, $\lambda_1^i J_i = 0$ and the *g. f.* conditions are $\alpha_2^i J_i = \alpha_3^i J_i = 0$, $J_{-\alpha_2} = J_{-\alpha_1-\alpha_2} = 0$, $J_{-\alpha_1} = 0$. The spin 2 currents $V^+ = J_{\alpha_1}$ and $V^- = J_{-\alpha_1-\alpha_2-\alpha_3}$ are nonlocal and $W_3 = J_{-\alpha_2-\alpha_3}$ $(s_3 = 3)$, $T = J_{-\alpha_3}$ $(s_T = 2)$ are local ones. Their algebra is a nonlocal extension of the W_3-one (18)

$$\begin{aligned}\{V^+(\sigma), V^-(\sigma')\} &= \partial^3_{\sigma'}\delta(\sigma-\sigma') - T(\sigma')\partial_{\sigma'}\delta(\sigma-\sigma') - W_3(\sigma')\delta(\sigma-\sigma') \\ &+ \frac{1}{3}V^+(\sigma)V^-(\sigma')\epsilon(\sigma-\sigma'),\end{aligned}$$

$$\begin{aligned}\{W_3(\sigma), V^\pm(\sigma')\} &= \mp\frac{10}{3}V^\pm(\sigma')\partial^2_{\sigma'}\delta(\sigma-\sigma') \mp 5\partial_{\sigma'}V^\pm(\sigma')\partial_{\sigma'}\delta(\sigma-\sigma') \\ &\pm \frac{1}{3}[T(\sigma')V^\pm(\sigma') - 6\partial^2_{\sigma'}V^\pm(\sigma')]\delta(\sigma-\sigma'),\end{aligned}$$

$$\{V^\pm(\sigma), V^\pm(\sigma')\} = -\frac{1}{3}V^\pm(\sigma)V^\pm(\sigma')\epsilon(\sigma-\sigma'), \tag{21}$$

and the remaining PB $\{W_3(\sigma), W_3(\sigma')\}$ has the same form as in (18), but T^2, in the quadratic terms, is replaced by $T^2 + 6V^+V^-$.

(3c) $A_3/\mathcal{N}_+^{(1,2)} \equiv U_4^{(1,2)}$ *-algebra:* $\mathcal{N}_+^{(1,2)} = \{E_{[\alpha]_2}\}$, where $[\alpha]_2$ are all positive roots, but α_2, the constraints are $J_{\alpha_1} = J_{\alpha_3} = 1$, $J_{-\alpha_1-\alpha_2} = J_{-\alpha_2-\alpha_3} = 0$, $J_{-\alpha_1-\alpha_2-\alpha_3} = 0$ and the *g. f.* conditions are $\alpha_1^i J_i = 0$, $\alpha_3^i J_i = 0$, $J_{\alpha_1+\alpha_2} = J_{\alpha_2} = 0$, $J_{\alpha_2+\alpha_3} = 0$. The *nonlocal quadratic* $U_4^{(1,2)}$-algebra is generated by one spin 1 current $J = \lambda_2^i J_i$, three local spin 2 currents $V^+ = J_{\alpha_2}$, $V^- = J_{-\alpha_1-\alpha_2-\alpha_3}$, $T = J_{-\alpha_1} + J_{-\alpha_3} + 4J^2$ and one nonlocal spin 2 current $U = J_{-\alpha_3} - J_{-\alpha_1}$ [37]

$$\{U(\sigma), J(\sigma')\} = \{V^\pm(\sigma), V^\pm(\sigma')\} = 0, \qquad \{U(\sigma), V^\pm(\sigma')\} = \frac{1}{2}V^\pm(\sigma)U(\sigma')\epsilon(\sigma-\sigma'),$$

$$\{J(\sigma), V^\pm(\sigma')\} = \mp\frac{1}{4}V^\pm(\sigma)\delta(\sigma-\sigma'), \qquad \{J(\sigma), J(\sigma')\} = \frac{1}{8}\partial_\sigma\delta(\sigma-\sigma'),$$

$$\begin{aligned}\{V^\pm(\sigma), V^\mp(\sigma')\} &= -\frac{1}{2}\partial^3_\sigma\delta(\sigma-\sigma') + T(\sigma)\partial_\sigma\delta(\sigma-\sigma') + \frac{1}{2}\partial_\sigma T(\sigma)\delta(\sigma-\sigma') \\ &- \frac{1}{4}U(\sigma)U(\sigma')\epsilon(\sigma-\sigma'),\end{aligned}$$

$$\begin{aligned}\{U(\sigma), U(\sigma')\} &= -\partial^3_\sigma\delta(\sigma-\sigma') + 2T(\sigma)\partial_\sigma\delta(\sigma-\sigma') + \partial_\sigma T(\sigma)\delta(\sigma-\sigma') \\ &- [V^+(\sigma)V^-(\sigma') + V^+(\sigma')V^-(\sigma)]\epsilon(\sigma-\sigma'),\end{aligned} \tag{22}$$

where $\mathcal{T} = T - 4J^2$. As it is shown in ref [37], by choosing another set of gauge fixing conditions, the nonlocal $U_4^{(1,2)}$-algebra takes the form of the *rational* ($\frac{1}{U}$-terms) local algebras of ref [35]. If one further imposes $J = \lambda_2^i J_i = 0$ as a new constraint, the J-reduced $U_4^{(1,2)}$-algebra (i. e., $U_4^{(1,2)}/U(1)$) coincides with the nonlocal $V_4^{(1,2)}$-algebra (see sec. 7 of ref [33]). The main difference with $U_4^{(1,2)}$ is that the spin 2 currents $V^{\pm}$ become *nonlocal*, in the $V_4^{(1,2)}$-case.

The analysis of the above examples of H-reduced $SL(n,R)$-current algebras allows to conclude that they all fit into the following *basic algebraic structures*:

(A) W-algebras (quadratic): **(2a)**, **(2e)**, **(3a)** and $W_n^{(2)}$ of ref [28], $W_S^{\mathcal{G}}$ of ref [35];

(B) U-algebras (rational or nonlocal): **(3c)**;

(C) V-albegras (nonlocal or PF-type): **(1b)**, **(2b)**, **(2c)**;

and the following mixtures of **(A)** and **(B)** with **(C)**:

(D) WV-algebras (nonlocal (PF) quadratic): **(2d)** and **(3b)** (and all $V_{n+1}^{(1,1)}$-algebras of sec. 3 of ref [33]);

(E) UV-algebras (nonlocal (PF) rational): $V_4^{(1,2)} = U_4^{(1,2)}/U(1)$.

This observation adresses *the question* about the algebraic conditions that a given set of constraints (and gauge fixing conditions) $\{H\} \in \mathcal{G}$ should satisfy in order to lead to one of the above mentioned algebraic structures (U, W, V, UV, UW). To answer this question, as well as whether other families of algebras can exist, we need an *efficient method* for describing all unequivalent (and irreducible) sets of first class constraints one can impose on the currents of a given affine algebra $\hat{\mathcal{G}}$. Given a Lie algebra $\mathcal{G}_r$, by introducing a grading operator[6] $Q_r^{(s)} = \sum_{n=1}^{r} s_n \frac{2\vec{\lambda}_n \cdot \vec{H}}{\alpha_n^2}$ we provide it with a specific graded structure[7]

$$\mathcal{G}_r = \oplus_i \mathcal{G}_{\pm i}^{(s)}, \quad [Q_r^{(s)}, \mathcal{G}_{\pm i}^{(s)}] = \pm i \mathcal{G}_{\pm i}^{(s)}, \quad [\mathcal{G}_i, \mathcal{G}_j] \subset \mathcal{G}_{i+j}.$$

For each fixed $l = 1, 2, ..., r$ (and $Q_r^{(s)}$), define the nilpotent subalgebra $\mathcal{N}_+^{(l,s)} = \oplus_{i=l} \mathcal{G}_i$ and choose a generic element $\epsilon_+^{(l)} \in \mathcal{G}_l$, i. e., $\epsilon_+^{(l)} = \sum_{\alpha \in [\alpha]_l} \mu_\alpha E_{[\alpha]_l}$ where $E_{[\alpha]_l}$ are all the step operators of grade l and μ_α are arbitrary constants. Next, we consider the $\epsilon_+^{(l)}$-invariant subalgebras of $\mathcal{G}_0^{(l)} = \oplus_{i=0}^{l-1} \mathcal{G}_i$ and $\mathcal{G}_-^{(l)} = \oplus_{i=l}^{r} \mathcal{G}_{-i}$

$$\mathcal{K}_\epsilon^0(s,l) = \mathrm{kerad}\epsilon_+^{(l)} \cap \mathcal{G}_0^{(l)} = \{g_0^0 \in \mathcal{G}_0^{(l)} : [\epsilon_+^{(l)}, g_0^0] = 0\},$$

$$\mathcal{K}_\epsilon^-(s,l) = \mathrm{kerad}\epsilon_+^{(l)} \cap \mathcal{G}_-^{(l)} = \{g^- \in \mathcal{G}_-^{(l)} : [\epsilon_+^{(l)}, g^-] = 0\}.$$

Finally, we define the "constraint" subalgebra as $\mathcal{H}_c^{(l,s)}(\epsilon_+^{(l)}) = \mathcal{N}_+^{(l)}(\epsilon) \oplus \mathcal{H}_0^{(l)}$, where $\mathcal{H}_0^{(l)} \subset \mathcal{K}_\epsilon^0(s,l)$ denotes those subalgebras of $\mathcal{K}_\epsilon^0$, which elements (i. e., the

[6] $\vec{\lambda}_n$ are the fundamental weights of $\mathcal{G}_r$, $\vec{\alpha}_n$ its simple roots, $\vec{H}$ its Cartan subalgebra and s_n are nonnegative integers.

[7] The nonequivalent graded structures $\mathcal{G}_r$ can have (i. e., the set of the allowed $Q_r^{(s)}$), are given by the Kac theorem [38]; this method was introduced in ref [39], in the construction of the conformal non-Abelian Toda models.

currents belonging to $\mathcal{H}_0^{(l)}$) are constrained to zero; $\mathcal{N}_+^{(l)}(\epsilon)$ caries the information about the constraints we are imposing on the currents from $\mathcal{N}_+^{(l)}$, namely all the elements of the subalgebra $\mathcal{N}_+^{(l+1)} \subset \mathcal{N}_+^{(l)}$ are zero and the elements of $\mathcal{G}_l$, which are constrained to be constants $\mu_\alpha \neq 0$, are collected in $\epsilon_+^{(l)} = \sum_\alpha \mu_\alpha E_\alpha$ (all the remaining $\mathcal{G}_l$ elements are zero). In this language, *the problem of the classification* of the allowed set of constraints $\mathcal{H}_c^{(l,s)}(\epsilon)$ reads as follows: for each fixed $Q_r^{(s)}$ and fixed grade l (say $l = 1$), to make a list of all the nonequivalent choises of the $\epsilon_+^{(l)\prime}$s. One can further organizes the different sets of $\epsilon_+^{(l)\prime}$s (l and $Q_r^{(s)}$ fixed) in *families* $(\mathcal{K}_\epsilon^0(s,l),\ \mathcal{K}_\epsilon^-(s,l))$, according to their *invariant subalgebras*. For example, the family $(\lambda_i \cdot H, \emptyset)$ is characterized by the conditions: (a) $\mu_\alpha = 0$, for the $\vec{\alpha^i} \cdot \vec{\lambda_i} \neq 0$ $(\alpha \in [\alpha]_l)$, and (b) $i = 1$ or $i = r$ (for the $l = 1$ case), in order to have $\mathcal{K}_\epsilon^-(s,1) = \emptyset$. We call *equivalent* the sets of constraints (and gauge fixing conditions) which can be obtained from each other by certain discrete transformations from the Weyl group of $\mathcal{G}_r$. As it is shown in sec. 8 of ref [33] (for the grade $l = 1$), they give rise to the same $\mathcal{H}_c^{(l,s)}(\epsilon)$-reduced $\mathcal{G}_r$-algebra. Therefore, it is sufficient to consider only one representative of such "Weyl families" of constraints. The problem of the *irreducibility* is more delicate. Depending on our choice of $\epsilon_+^{(l)\prime}$s, it might happens that the $\mathcal{G}_r/\mathcal{H}_c^{(l,s)}(\epsilon)$-algebra splits into two (or more) mutualy commuting algebras [37]. This is the case when one takes, for example $\mu_\alpha = 0$, for all E_α that contains the simple root α_i (i. e., E_{α_i}, $E_{\alpha_i+\alpha_{i+1}}$, $E_{\alpha_{i-1}+\alpha_i}$, $E_{\alpha_{i-1}+\alpha_i+\alpha_{i+1}}$ etc).

The organization of the constraints in the families $(\mathcal{K}_\epsilon^0(s,l),\ \mathcal{K}_\epsilon^-(s,l))$ simplifies the derivation of the $\mathcal{G}_r^{(l,s)}(\epsilon, \mathcal{H}_0^{(l)}) = \mathcal{G}_r/\mathcal{H}_c^{(l,s)}(\epsilon)$-algebras (i. e., the calculation of the corresponding Dirac brackets). Depending on the algebraic data $\{\mathcal{G}_r, Q^{(s)}, l, \epsilon_+^{(l)}\}$, which defines $\mathcal{H}_c^{(l,s)}(\epsilon)$, one can *classify* all the $\mathcal{G}_r^{(l,s)}(\epsilon, \mathcal{H}_0^{(l)})$-algebras in the following $\{\mathcal{H}_0^{(l)}, \mathcal{K}_\epsilon^-(s,l)\}$-families of algebras:

THEOREM. Given $\mathcal{G}_r$ and the graded structure $(Q_r^{(s)},\ l,\ \epsilon_+^{(l)})$, which define the constraints subalgebra $\mathcal{H}_c^{(l,s)}(\epsilon) \subset \mathcal{G}_r$. Each $\mathcal{H}_c^{(l,s)}(\epsilon)$-reduced $\mathcal{G}_r$-current algebra $\mathcal{G}_r^{(l,s)}(\epsilon, \mathcal{H}_0^{(l)})$ belongs to one of the following five types of extended Virasoro algebras:

(1) W-algebras, when $\mathcal{H}_0^{(l)} = \emptyset$ (or $\mathcal{H}_0^{(l)} \neq \emptyset$ but $[\mathcal{H}_0^{(l)}, \mathcal{G}_0^\pm] = 0$) and $\mathcal{K}_\epsilon^- = \emptyset$ ($\mathcal{G}_0^\pm$ are the $\pm$ step operators of grade 0);

(2) U-algebras, when $\mathcal{H}_0^{(l)} = \emptyset$ (or $\mathcal{H}_0^{(l)} \neq \emptyset$ but $[\mathcal{H}_0^{(l)}, \mathcal{G}_0^\pm] = 0$) and $\mathcal{K}_\epsilon^- \neq \emptyset$; dim $\mathcal{K}_\epsilon^-$ is the number of the nonlocal currents (or of the "rational currents" of ref [35]);

(3) V-algebras, when $\mathcal{H}_0^{(l)} \neq \emptyset$ and $\mathcal{H}_0^{(l)} = U(1)^r$ or $U(1)^{r-1} = \{\oplus_{i=2}^r \lambda_i \cdot H\}$ or $\{\oplus_{i=1}^{r-1} \lambda_i \cdot H\}$, $\mathcal{K}_\epsilon^- = \emptyset$; the case $\epsilon_+^{(l)} = 0$, $\mathcal{H}_0^{(l)} = U(1)^r$, which also lead to V-algebras, has to be treated separately (see ref [37]);

(4) VW-algebras, when $\mathcal{H}_0^{(l)} \neq \{\emptyset,\ U(1)^r,\ U(1)^{r-1}\}$ and $\mathcal{K}_\epsilon^- = \emptyset$; the case $[\mathcal{H}_0^{(l)}, \mathcal{G}_0^\pm] = 0$ ($\mathcal{G}_0^\pm$ are the $\pm$ step operators of grade zero) has to be excluded, since it gives rise to W-algebras;

(5) VU-algebras, when $\mathcal{H}_0^{(l)} \neq \{\emptyset,\ U(1)^r,\ \oplus_{i=2}^r \lambda_i \cdot H,\ \oplus_{i=1}^{r-1} \lambda_i \cdot H\}$, and $\mathcal{K}_\epsilon^- \neq \emptyset$; in this case, again $[\mathcal{H}_0^l, \mathcal{G}_0^\pm] \neq 0$.

The algebraic conditions that separate W- from the U-algebras are given in ref [35]. The equivalence of the *rational* U-algebras, to certain *nonlocal* algebras, and the explicit form of the gauge transformations, from the "rational" gauge fixing conditions to "nonlocal" gauge fixing conditions, is demonstrated in ref [37]. The proof of this theorem, for the generic $Q_r^{(s)}$ grade one ($l = 1$) case [37], is based on the analysis of the properties of the inverse matrix Δ_{ij}^{-1} of the constraints and the gauge fixing conditions. The *origin of the nonlocal terms* in the V- , VW- and VU-algebras, are the $\mathcal{H}_0^{(l)}$ constraints and their gauge fixing's. Their PB's are always in the form $\{J_i(\sigma), J_j(\sigma')\} = k\delta_{ij}\partial_\sigma\delta(\sigma-\sigma')$ or $\{J_{-\alpha_i}(\sigma), J_{\alpha_i}(\sigma')\} = k\partial_\sigma\delta(\sigma-\sigma')$. Their contributions to Δ_{ij}^{-1} are the nonlocal $\epsilon(\sigma-\sigma')$-terms.

The explicit form of each $\mathcal{G}_r^{(l,s)}(\epsilon, \mathcal{H}_0^{(l)})$- algebra (from a given class U, V, VW etc) indeed depends on the algebra $\mathcal{G}_r$ and on the choice of $\epsilon_+^{(l)}$ and $\mathcal{H}_0^{(l)}$ as one can see from our Examples 1, 2 and 3. The full algebraic structure (all explicit PB's) is known in the case of the W_n-algebras [25] and of the simplest A_r-family of VW-algebras ($V_{r+1}^{(1,1)}$-algebras [33]) defined by $Q = \sum_{i=l}^{r} \lambda_i \cdot H$, $l = 1$, $\epsilon_+^{(1)} = \sum_{i=2}^{r} E_{\alpha_i}$, $\mathcal{H}_0^{(1)} = \{\lambda_1 \cdot H\}$. Various examples of the U- V- and VU-algebras ($V_{(n,m)}$) have been constructed by Bilal [32], by calculating the second Gelfand-Dikii brackets, associated with certain matrix differential operators.

3 QUANTUM V-ALGEBRAS

The classification of all the classical (PB's) extensions of the Virasoro algebra is an important step forward the classification of the universality classes in two dimensions. The complete solution of this problem requires, however, the knowledge of the exact critical exponents, i. e., we need to know the *h. w.* representation of the corresponding *quantum* W-, U-, V- and VW-, VU-algebras. The quantization of the classical W-algebras is a rather well understood problem. It consists in replacing the currents functions T, W_n by currents operators $\hat{T}$, $\hat{W}_n$, acting on some Hilbert space, and their PB's $\{a, b\}$ by the commutators $-\frac{i}{\hbar}[a, b]$. The only changes that occur in this procedure are the new (*quantum corrections*) coefficients in front of the central term $\delta^{(s)}(\sigma)$ and those of the quadratic terms. Another option is to start with the quantum current algebra $\mathcal{G}_r$ and to implement the operators constraints $\mathcal{H}_c^{(l,s)}(\epsilon)$ on it, following the methods of the *quantum Hamiltonian reduction* [36]. The advantage of this method is that it provides a simple way of deriving the W-algebra *h. w.* representations from the *h. w.* representations of the $\mathcal{G}_r$-current algebra. The specific nonlocal terms $V^+(\sigma)V^-(\sigma')\epsilon(\sigma-\sigma')$ that appears in the V- (and VW-, VU-) algebras, as well as the nonlocal nature of the part of the currents ($V_i^{\pm}$), are the main obstacle to the construction of the corresponding *quantum* V-algebras. It turns out [33], [34] that their quantization require deep changes in the classical algebraic structure (13), (19), (20), namely: (a) renormalization of the bare spins of the nonlocal currents (say for $V_{n+1}^{(1,1)}$, $s_{cl}^{\pm} = \frac{n+1}{2}$ goes to $s_q^{\pm} = \frac{n+1}{2}\left(1 - \frac{1}{2k+n+1}\right)$); (b) the quantum counterpart of the PB's of the $V^{\pm}$'s charges appears to be specific PF-type commutators, similar to eqn. (10);
(c) breaking of the global $U(1)$ symmetry, to some discrete group Z_{2k+n}.

The fact that all the complications in the quantization of the classical V- and VW-algebras are coming from the $\mathcal{H}_0^{(l)}$-constraints suggests the following *strategy*: relax the $\mathcal{H}_0^{(l)}$-constraints (i. e., leave the currents $\lambda_a^i J_i \in \mathcal{H}_0^{(l)}$ unconstrained) and consider the corresponding local "*intermidiate*" W-algebra, generated by the V- (or

VW-) algebra currents (which are all local now) and the additional spin one $\mathcal{H}_0^{(l)}$-currents. Since all the currents are local, the quantization of this algebra is similar to the one of the W_{n+1}- or $W_{n+1}^{(l)}$-algebras [25], [36]. The problem we address here is the following: *Given the quantum W-algebra and its h. w. representations, to derive the quantum $V = W/\mathcal{H}_0^{(l)}$-algebra and its h. w. representations by implementing the (operator) constraint $\mathcal{H}_0^{(l)} \approx 0$.* The method we are going to use is an appropriate generalization of the derivation of the Z_N parafermionic algebra [12] from the affine $SU(2)$-one (or $SL(2,R)$, for the noncompact PF's), by imposing the constraint $J_3(z) \approx 0$.

EXAMPLE 3.1. **Quantization of the PF-algebra.** Following the arguments of ref [12], we define the quantum (compact) V_2-algebras as $V_2 = \{SU(2)_k, J_3(z) = 0\}$. Therefore, the V_2-generators $\psi^{\pm}$ have to represent the $J_3 = \sqrt{\frac{k}{2}}\partial\phi$ -independent part of the $\hat{SU}(2)_k$-ones, i. e.,

$$J^{\pm} = \psi^{\pm}\exp(\mp\alpha\phi), \quad T = T_V + \frac{1}{2}(\partial\phi)^2, \quad J_3(z_1)\psi^{\pm}(z_2) = O(z_{12}),$$

$$\phi(z_1)\phi(z_2) = -\ln(z_{12}) + O(z_{12}). \tag{23}$$

Taking into account the $SU(2)$ OPE's

$$J_3(z_1)J^{\pm}(z_2) = \pm\frac{i}{z_{12}}J^{\pm}(z_2) + O(z_{12}) \tag{24}$$

and eqn. (23), we find $\alpha = i\sqrt{\frac{2}{k}}$ and, as a consequence, the spins of $\psi^{\pm}$ are $s^{\pm} = 1 - \frac{1}{k}$ (we have used that $s_{J^{\pm}} = 1$). Finally, eqns. (23) and (24) lead to the following V_2-algebra OPE's

$$\begin{aligned}
\psi^{\pm}(z_1)\psi^{\pm}(z_2) &= z_{12}^{-\frac{2}{k}}\psi^{\pm}_{(2)}(z_2) + O(z_{12}),\\
\psi^{+}(z_1)\psi^{-}(z_2) &= z_{12}^{\frac{2}{k}}\left(\frac{k}{z_{12}^2} + (k+2)T_V + O(z_{12})\right),
\end{aligned} \tag{25}$$

which are nothing, but the PF-algebra OPE's (9), with $k = N$ and $\psi_1^{\pm} = \frac{1}{\sqrt{k}}\psi^{\pm}$. Although the V_2-algebra (25) is, by construction, the quantum version of the classical PB's PF-algebra (17), the discrepancy between the spins $s^{\pm} = 1 - \frac{1}{k}$ and $s_{V^{\pm}} = 1$ requires a more precise definition of the relation of algebras (25) and (17). The exact *statement* is as follows: let $V^{\pm} = \frac{1}{k}\psi^{\pm}$ and the $V^{\pm}$ PB's are defined as a certain limit of the OPE's (25):

$$\{V^a(z_1), V^b(z_2)\} = \lim_{k\to\infty}\frac{k}{2\pi i}[V^a(z_1)V^b(z_2) - V^a(z_2)V^b(z_1)] \tag{26}$$

$(a, b = \pm)$. Then, the $k \to \infty$ limit of the OPE's (25) reproduces the PB's (17). The proof is straightforward. Applying twice the OPE's (25), we find

$$z_{12}^{\frac{2}{k}}\{V^{\pm}(z_1)V^{\pm}(z_2) - V^{\pm}(z_2)V^{\pm}(z_1)e^{-\frac{2\pi i}{k}\epsilon}(z_{12})\} = \frac{1}{k^2}O(z_{12}),$$

$$z_{12}^{-\frac{2}{k}}\{V^-(z_1)V^+(z_2) \quad - \quad V^+(z_2)V^-(z_1)e^{\frac{2\pi\imath}{k}\epsilon}(z_{12})\} = \frac{k+2}{k^2}O(z_{12})$$
$$+ \quad \frac{1}{k}\left(\frac{1}{z_{12}^2+\imath 0} - \frac{1}{z_{21}^2+\imath 0}\right), \tag{27}$$

where the identity $\imath\pi\epsilon(z_{12}) \equiv \ln\frac{z_{12}+\imath 0}{z_{21}+\imath 0}$ has been used. The $k \to \infty$ limit of eqns. (27) reproduces exactly the classical PF-algebra[8] (17). The conclusion is that the nonlocal PB's-algebra (17) is a semiclassical limit ($k \to \infty$) of the PF's OPE's (25). As we have seen, the quantization requires renormalization of the spins $s_q = s_{cl} - \frac{1}{k}$ of the nonlocal currents $V^{\pm}$. Therefore, the PB's (17) have to be replaced by the PF-commutators (10) and for k-positive integers, the classical global $Z_2 \otimes U(1)$-symmetry is broken to $Z_2 \times Z_k$, in the quantum theory.

The structure of the classical $V_3^{(1,1)}$ PB's algebra (13) is quite similar to the PF-one (17). An importante difference is that in its derivation from the classical $SL(3,R)$ (see our example 2b), one has to impose, together with the $\mathcal{H}_0^1$-type (PF) constraint $\vec{\lambda}_1 \cdot \vec{J} = 0$, two more constraints, on the nilpotent subalgebra $\mathcal{N}_+^{(1)}$: $J_{\alpha_2} = 1$ and $J_{\alpha_1+\alpha_2} = 0$. In order to demonstrate how this type of (purely W-) constraints are treated, in the frameworks of the quantum Hamiltonian reduction, we consider the simplest example of such reduction: the $\mathcal{N}_+$-reduced $SL(2,R)$ ($J_\alpha = 1$) which gives rise to the Virasoro algebra (example 1a).

EXAMPLE 3.2. **Virasoro algebra *h. w.* representation from the $SL(2,R)$-ones [36].** The implementation of the constraint $J_\alpha = 1$ as an operator identity on the $SL(2,R)_k$-space of states $\mathcal{H}_{A_1}^{(k)}$ requires an introduction of a pair of fermionic ghosts ($b(z)$, $c(z)$) of spins (0,1) and of the larger space of states $\mathcal{H}_{A_1}^{(k)} \otimes \mathcal{H}_{b,c}$. The reduced representation space of the constrained system $\{A_1/\mathcal{N}_+\}$ can be defined by means of the $BRST$ operator

$$Q_{BRST} = \oint [J_\alpha(z) - 1]c(z)dz, \quad Q_{BRST}^2 = 0,$$

as Q_{BRST}-invariant states-$|\psi\ > \in \mathcal{H}_{A_1}^{(k)} \otimes \mathcal{H}_{b,c}$ ($Q_{BRST}|\psi\ >= 0$), which are not Q_{BRST}-exact, i. e., $|\psi\ > \neq Q_{BRST}|*\ >$. The statement is that this $BRST$-cohomology $H_{Q_{BRST}}(\mathcal{H}_{A_1}^{(k)} \otimes \mathcal{H}_{b,c}) = \ker Q/\mathrm{Im} Q$ is isomorphic to the irreducible representation space $\mathcal{H}_{Vir}^{(k)} (\equiv \mathcal{H}_{A_1}/\mathcal{N}_+)$ of the Virasoro algebra [36]. To make the constraints condition $J_\alpha = 1$ consistent with the conformal invariance, we have to improve the $SL(2,R)$-Sugawara stress-tensor, in such a way that $s_{imp}(J_\alpha) = 0$

$$T_{impr} = \frac{1}{k+2} : J^a(z)J^a(z) : + \partial J_3.$$

Therefore, the new central charge is $c_{impr} = \frac{3k}{k+2} - 6k$. Taking into account the contribution $c_{gh} = -2$, of the ghost stress-tensor $T_{bc} = (\partial b)c$, we find that the total central charge is $c_{tot} = 13 - 6\left(\frac{1}{k+2} + k + 2\right)$. Since the dimension of the $\hat{SL}(2,R)_k$ representation of weight $\vec{\Lambda}$ is

$$\Delta_\Lambda = \frac{1}{2(k+2)}\vec{\Lambda} \cdot (\vec{\Lambda} + 2\vec{\alpha})$$

[8]The noncompact case $SL(2,R)/U(1)$ corresponds to the change $\phi \to \imath\phi$, which turns out to be equivalent to the $k \to -k$ one, in the OPE's, spins etc.

the improved dimensions are found to be

$$\Delta_{\Lambda}^{impr} = \frac{1}{2(k+2)}\vec{\Lambda}\cdot(\vec{\Lambda}+2\vec{\alpha}) - \vec{\alpha}\cdot\vec{\Lambda}. \tag{28}$$

An important observation of ref [36] is that the $h.\ w.$ states of the reduced space $\mathcal{H}_{Vir}^{(k)}$ are of levels $k+2=\frac{m}{m+1}$, where $m=3,4,\ldots$, and weights

$$\vec{\Lambda} = [(1-p)(k+2)-(1-q)]\vec{\alpha},$$

with $1\le p\le m-1$, $1\le q\le p$. Therefore, $c_{tot}=1-\frac{6}{m(m+1)}$ and $\Delta_{\Lambda}^{impr}=\Delta_{p,q}$, i. e., the $\mathcal{H}_{Vir}^{(k)} = H_{Q_{BRST}}(\mathcal{H}_{A_1}^{(k)}\otimes\mathcal{H}_{b,c})$ (for the above values of the levels, and the A_1-weight $\vec{\Lambda}$) coincides with the space of the $h.\ w.$ unitary representations (3) of the Virasoro algebra.

EXAMPLE 3.3. **Quantum $V_3^{(1,1)}$-algebra**. As it was pointed out in ref [33], [34], the intermidiate $W_3^{(1,1)}$-algebra is Weyl equivalent (w_{α_1}) to the Bershadsky-Polyakov algebra $W_3^{(2)}$. The improved stress-tensor is given by

$$T^{impr} = \frac{1}{k+3} : J^q(z)J^q(z) : -(\lambda_2 - \frac{1}{2}\lambda_1)^i\partial J_i \tag{29}$$

and we have to introduce the following two pair of ghosts: (b,c) and (ϕ,ϕ^+), of spins (0,1) and $(\frac{1}{2},\frac{1}{2})$. Constructions, similar to the ones in Example 3.2, allow to derive the $W_3^{(1,1)}$ central charge

$$c_{W_3^{(1,1)}} = \tfrac{8k}{k+3} - 6k - 1$$

and the dimensions $\Delta_{\vec{r},\vec{s}}$ and $U(1)$ charges $q_{r,s}$ of its $h.\ w.$ representations (NS-sector) [28] read as

$$\begin{aligned}\Delta^{W}_{\vec{r},\vec{s}} &= \frac{1}{2(k+3)}\vec{\Lambda}_{r,s}\cdot(\vec{\Lambda}_{r,s}+2\vec{\beta}) - \vec{\beta}\cdot\vec{\Lambda}_{r,s},\\ q_{\vec{r},\vec{s}} &= \frac{1}{3}\left[2\frac{p}{q}(r_1-r_2)-(s_1-s_2)\right],\end{aligned} \tag{30}$$

where $\beta=\lambda_2-\frac{1}{2}\lambda_1$, with $\vec{\Lambda}_{r,s}$ representing the weights of the following specific level $k+3=2\frac{p}{q}$ representations of $SL(3,R)_k$

$$\vec{\Lambda}_{r,s} = \sum_{i=1}^{2}\vec{\lambda}_i[(1-r_i)(k+3)-(1-s_i)],$$

where $1\le s_i\le 2p-1$, $1\le r_i\le q$. The quantum $W_3^{(1,1)}$-algebra is generated by one spin $s=1$ $J(z)$, two spin $s=\frac{3}{2}$ $G^{\pm}(z)$ and one spin $s=2$ $T(z)$ currents, with OPE's [28]

$$J(z_1)J(z_2) = \frac{2k+3}{3z_{12}^2} + O(z_{12}), \qquad J(z_1)G^{\pm}(z_2) = \pm\frac{1}{z_{12}}G^{\pm}(z_2) + O(z_{12}),$$

$$
\begin{aligned}
G^{+}(z_1)G^{-}(z_2) &= \frac{(k+1)(2k+3)}{z_{12}^3} + 3\frac{k+1}{z_{12}^2}J(z_2) \\
&+ \frac{1}{z_{12}}\left[3J^2(z_2) - (k+3)T(z_2) + 3\frac{k+1}{2}\partial J(z_2)\right] + O(z_{12}).
\end{aligned}
$$

$$
G^{\pm}(z_1)G^{\pm}(z_2) = O(z_{12}), \tag{31}
$$

According to the definition of the $V_3^{(1,1)}$-algebra $V_3^{(1,1)} = \{W_3^{(1,1)}; J = 0\}$, its generators $V^{\pm}(z)$ and T_V have to commute with $J(z)$, i. e.,

$$
J(z_1)V^{\pm}(z_2) = J(z_1)T_V(z_2) = O(z_{12}). \tag{32}
$$

Therefore, $V^{\pm}$, T_V are related to the $J = \sqrt{\frac{2k+3}{3}}\partial\phi$-independent parts of the $W_3^{(1,1)}$-currents

$$
G^{\pm} = V^{\pm}\exp(\pm a\phi),\ T_W = T_V + \frac{1}{2}(\partial\phi)^2,\ \phi(z_1)\phi(z_2) = \ln(z_{12}) + O(z_{12}). \tag{33}
$$

As a consequence of eqns. (32) and (33), we get $a = \sqrt{\frac{3}{2k+3}}$, and for the spins of the quantum currents $V^{\pm}$ $\left(s_{cl} = \frac{3}{2}\right)$, we obtain $s_q^{\pm} = \frac{3}{2}\left(1 - \frac{1}{2k+3}\right)$. The $W_3^{(1,1)}$$-OPE$'s (31), and eqn. (33), lead to the following OPE's for $V^{\pm}$ and T_V $(k \neq -3, -\frac{3}{2}, -1)$

$$
\begin{aligned}
&V^{+}(z_1)V^{-}(z_2) = z_{12}^{\frac{3}{2k+3}}\left[\frac{(2k+3)(k+1)}{z_{12}^3} - \frac{k+3}{z_{12}}T_V(z_2) + O(z_{12})\right], \\
&T_V(z_1)V^{\pm}(z_2) = \frac{s_q^{\pm}}{z_{12}^2}V^{\pm}(z_2) + \frac{1}{z_{12}}\partial V^{\pm}(z_2) + O(z_{12}), \\
&V^{\pm}(z_1)V^{\pm}(z_2) = z_{12}^{-\frac{3}{2k+3}}V_{(2)}^{\pm}(z_2) + O(z_{12}),
\end{aligned} \tag{34}
$$

which define the *quantum* $V_3^{(1,1)}$-algebra. The $T_V(1)T_V(2)$ OPE has the standard form (4), of the Virasoro OPE's, with central charge $c_V = -6\frac{(k+1)^2}{k+3}$. The $V_3^{(1,1)}$-algebra (34) has a structure similar to the PF-one (9), and for $L = 2k+3$ positive integers $(L > 3)$, the OPE's (34) involves more currents $V_l^{\pm}$ $(l = 1, 2, ..., L-1)$ of spins $s_l^{\pm} = \frac{3l}{2L}(L - l)$. Introducing the (Laurent) mode expansion for the currents[9] $V^{\pm}$

$$
V^{\pm}(z)\phi_s^{\eta}(0) = \sum_{m=-\infty}^{\infty} z^{\pm\frac{3s}{2L}+m-1\mp\eta}V^{\pm}_{-m\pm\eta-\frac{1}{2}+3\frac{1\mp s}{2L}}\phi_s^{\eta}(0),
$$

we derive, from (34), the following "commutation relations" (of PF-type) for the $V_3^{(1,1)}(L)$-algebra $(|L| > 3)$

$$
\sum_{p=0}^{\infty} C^{p}_{(-\frac{3}{L})}\left(V^{+}_{-3\frac{s+1}{2L}+m-p-\eta+\frac{1}{2}}V^{-}_{3\frac{s+1}{2L}+n+p+\eta-\frac{1}{2}} + V^{-}_{-3\frac{1-s}{2L}+n-p+\eta-\frac{1}{2}}V^{+}_{3\frac{1-s}{2L}+m+p-\eta+\frac{1}{2}}\right)
$$

[9] $\phi_s^{\eta}(0)$ denotes certain Ramond ($\eta = \frac{1}{2}$, s-odd) and Neveu-Schwartz ($\eta = 0$, s-even) fields, where $s = 1, 2, ..., L-1$.

$$= \frac{1}{2}(L+3)\left[-L_{m+n} + \frac{(L-1)L}{2(L+3)}\left(\frac{3s}{2L}+n+\eta\right)\left(\frac{3s}{2L}+n+\eta-1\right)\delta_{m+n,0}\right] \quad (35)$$

where $C^{p}_{(M)} = \frac{\Gamma(p-M)}{p!\Gamma(-M)}$, with $m, n = 0, \pm 1, \pm 2, \ldots$, and

$$\sum_{p=0}^{\infty} C^{p}_{(\frac{3}{L})}\left(V^{\pm}_{3\frac{3\mp\epsilon}{2L}-p+m+\eta-\frac{1}{2}} V^{\pm}_{3\frac{1\mp\epsilon}{2L}+p+n+\eta-\frac{1}{2}} - V^{\pm}_{3\frac{3\mp\epsilon}{2L}-p+n+\eta-\frac{1}{2}} V^{\pm}_{3\frac{1\mp\epsilon}{2L}+p+m+\eta-\frac{1}{2}}\right) = 0 \quad (36)$$

In the particular cases, when $L = 2, 3$, the OPE's $V^{\pm}V^{\pm}$ have also a pole, which makes eqn. (36) nonvalid. The simplest example of such $V_3^{(1,1)}$-algebra, for $L = 2$, is spanned by $V^{\pm}$ of $s^{\pm} = \frac{3}{4}$ and T_V, only. The relations (36) are now replaced by

$$\sum_{p=0}^{\infty} C^{p}_{(\frac{1}{2})}\left(V^{-}_{-p+m+\eta-\frac{3}{4}} V^{-}_{p+n+\eta-\frac{5}{4}} + V^{-}_{-p+n+\eta-\frac{3}{4}} V^{-}_{p+m+\eta-\frac{5}{4}}\right) = \delta_{m+n+2\eta,0},$$

and by the similar one, for $V^{+}V^{+}$. As in the PF-case, one can easily verify that certain limits of the OPE's (34) reproduces the classical PB's $V_3^{(1,1)}$-algebra (13).

The relations (33), between $W_3^{(1,1)}$ and $V_3^{(1,1)}$ currents, lead to the following form for the $W_3^{(1,1)}$-vertex operators $\phi^{W}_{(r_i,s_i)}(z)$ in terms of the $V_3^{(1,1)}$-ones $\phi^{V}_{(r_i,s_i)}(z)$ and the free field ϕ

$$\phi^{W}_{(r_i,s_i)} = \phi^{V}_{(r_i,s_i)} \exp\left[q_{(r_i,s_i)}\sqrt{\frac{3}{L}}\phi\right]. \quad (37)$$

The construction (37) is a consequence of eqns. (33), of the following OPE's

$$\begin{aligned} T^{W}(z_1)\phi^{W}(z_1) &= \frac{\Delta^{W}_{r,s}}{z_{12}^2}\phi^{W}_{(r,s)}(z_2) + \frac{1}{z_{12}}\partial\phi^{W}_{(r,s)}(z_2) + O(z_{12}), \\ J(z_1)\phi^{W}(z_2) &= \frac{q_{r,s}}{z_{12}}\phi^{W}_{(r,s)}(z_2) + O(z_{12}), \end{aligned}$$

and of the fact that $\phi^{V}_{(r,s)}$ are J-neutral, i. e., $J(z_1)\phi^{V}_{(r,s)}(z_2) = O(z_{12})$. Finally, we realize that the dimensions $\Delta^{V}_{(r,s)}$ of the $V_3^{(1,1)}$ primary fields $\phi^{V}_{(r,s)}$ are related to the $\phi^{W}_{(r,s)}$ dimensions and charges, given by eqns. (30), as follows

$$\Delta^{V}_{(r,s)} = \Delta^{W}_{(r,s)} - \frac{3}{2L}q^{2}_{(r,s)}. \quad (38)$$

Taking into account the explicit values of $\Delta^{W}_{(r,s)}$ and $q_{(r,s)}$ (30), for the class of "completely degenerate" *h. w.* representations of $W_3^{(1,1)}$ $(L+3 = 4\frac{p}{q})$, we derive the dimensions of the *h. w.* representations of $V_3^{(1,1)}$.

The main purpose of our discussion about the quantization of the classical (PB's) nonlocal $V_3^{(1,1)}$-algebra (13) is to point out the *differences* with the quantization of the W_3- and $W_3^{(1,1)}$-algebras, and the *similarities* with the PF-algebra. The origin of all these complications is the renormalization of the spins of the nonlocal currents $V^{\pm}$, $s^{\pm}_{q} = s^{\pm}_{cl} - \frac{3}{2L}$, which makes the singularities of the $V_3^{(1,1)}-OPE$'s (34)

L-dependent. For certain values of L, this requires to introduce new currents $V_l^{\pm}$ and $W_p^{\pm}$ (see ref [33]), in order to close the OPE-algebra. The typical PF-feature is the replacing of the Lie commutators, with an infinite sum of bilinears of generators, as in eqns. (35) and (36). One might wonder whether the $V_{n+1}^{(1,1)}$-algebras (defined in ref [33]), exhibit similar features. Our preliminary results show that the renormalization of the spins of the nonlocal currents $V_{(n+1)}^{\pm}$ is a commun property of all $V_{n+1}^{(1,1)\prime}$s

$$s_n^{\pm}(q) = \frac{n+1}{2}\left(1 - \frac{1}{2k+n+1}\right).$$

As usual, the spins of the local currents W_{l+1} remain unchanged. All this indicates that quantum $V_{n+1}^{(1,1)}$-algebras share many properties of $V_3^{(1,1)}$. The construction of the *h. w.* representations of these algebras, as well as the quantization of the V- and WV-algebras of other types, say as in (19) and in (20), is an interesting open problem. The same is valid for the simplest $U_4^{(1,2)}$-algebra, for the UV-algebra $V_4^{(1,2)}$, of ref [33], and for the various explicit examples of U- and UV-algebras, given in ref [37].

It is important to note, in conclusion, that the classification of the *classical extensions* of the Virasoro algebra, described in this paper, *does not solve* the problem of the classification of universality classes in two dimensions. The complete solution of this challenge problem requires the construction of the *h. w.* representations of the corresponding quantum W-, U-, V (and WV-, UV-)-algebras. We consider the above discussed quantization of the $V_3^{(1,1)}$-algebras as a demonstration that *relatively simple tools*, for the realization of this program, *do exist*.

Acknowledgments F. E. Mendonça da Silveira thanks FAPESP, for financial support. G. M. Sotkov thanks FAPESP, IFT-UNESP and DCP-CBPF, for financial support and hospitalily. This work has been partially supported by CNPq.

REFERENCES

[1] A. Belavin, A. Polyakov and A. B. Zamolodchikov, Nucl. Phys. B241 (1984) 333;

[2] G. E. Andrews, R. J. Baxter and P. J. Forrester, J. Stat. Phys. 35 (1984) 193;

[3] V. G. Kac, Lecture Notes in Physics 94 (1979) 441;

[4] Vl. S. Dotsenko, Nucl. Phys. B235 (1984) 54;

[5] D. A. Huse, Phys. Rev. B30 (1984) 3908;

[6] D. Friedan, Z. Qiu and S. Shenker, Phys. Rev. Lett. 52 (1984) 1575; Comm. Math. Phys. 107 (1986) 535;

[7] D. Friedan, Z. Qiu and S. Shenker, Phys. Lett. B151 (1985) 37;

[8] M. A. Bershadsky, V. G. Knizhnik and M. G. Teitelman, Phys. Lett. B151 (1985) 31;

[9] A. Neveu and J. H. Schwarz, Nucl. Phys. B31 (1971) 86;

[10] D. Kastor, E. Martinec and Z. Qiu, Phys. Lett. B200 (1988) 434;

[11] F. Ravanini, Mod. Phys. Lett. A3 (1988) 397;

[12] A. B. Zamolodchikov and V. A. Fateev, Sov. Phys. JETP 62 (1985) 215;

[13] A. B. Zamolodchikov and V. A. Fateev, Sov. Phys. JETP 63 (1985) 913;

[14] A. B. Zamolodchikov and V. A. Fateev, Nucl. Phys. B280 (1987) 644;

[15] A. B. Zamolodchikov, Theor. Math. Phys. 65 (1986) 1205;

[16] V. Kac, Funct. Anal. Appl. 1 (1967) 328; R. V. Moody, Bull. Amer. Math. Soc. 73 (1967) 217;

[17] V. G. Knizhnik and A. B. Zamolodchikov, Nucl. Phys. B247 (1984) 83;

[18] D. Gepner and E. Witten, Nucl. Phys. B278 (1986) 493;

[19] M. Ademollo et all, Phys. Lett. B62 (1976) 105; Nucl. Phys. B111 (1976) 77;

[20] W. Bucher, D. Friedan and A. Kent, Phys. Lett. B172 (1986) 316;

[21] P. Di Vecchia, J. L. Peterson and M. Yu, Phys. Lett. B172 (1986) 211;

[22] P. Di Vecchia, V. G. Knizhnik, J. L. Peterson and P. Rossi, Nucl. Phys. B253 (1985) 701;

[23] V. G. Kac and I. Todorov, Comm. Math. Phys. 102 (1985) 175;

[24] D. Gepner, Nucl. Phys. B290 (1987) 10;

[25] V. A. Fateev and S. L. Lukyanov, Int. J. Mod. Phys. A3 (1988) 507;

[26] A. Bilal and J. L. Gervais, Nucl. Phys. B314 (1989) 646;

[27] F. A. Bais, P. Bouwkneght, K. Schoutens and M. Surridge, Nucl. Phys. B304 (1988) 371;

[28] M. Bershadsky, Comm. Math. Phys. 139 (1991) 71;

[29] A. Polyakov, Int. J. Mod. Phys. A5 (1990) 833;

[30] J. Balog, L. Fehér, L. O'Raifeartaigh, P. Forgács and A. Wipf, Ann. of Phys. 203 (1990) 76;

[31] A. Bilal, Nucl. Phys. B422 (1994) 258;

[32] A. Bilal, Comm. Math. Phys. 170 (1995) 117;

[33] J. F. Gomes, G. M. Sotkov and A. H. Zimerman, Phys. Lett. 435B (1998) 49;

[34] J. F. Gomes, G. M. Sotkov and A. H. Zimerman, "Parafermionic Reductions of WZW Model", hep-th/9803234;

[35] L. Fehér, L. O'Raifeartaigh, P. Ruelle, I. Tsutsui and A. Wipf, Phys. Rep. 222 N°1 (1992) 1;

[36] M. Bershadsky and H. Ooguri, Comm. Math. Phys. 126 (1989) 49;

[37] J. F. Gomes, F. E. Mendonça da Silveira, G. M. Sotkov and A. H. Zimerman, "Symmetries of Non-Abelian Toda Models", in preparation;

[38] V. G. Kac, Funct. Anal. Appl. 3 (1969) 252;

[39] A. N. Leznov and M. V. Saveliev, Comm. Math. Phys. 74 (1980) 111; Comm. Math. Phys. 89 (1983) 59.

Conditions d'existence d'une loi de Hardy-Weinberg périodique dans des populations sélectivement neutres

C. MALLOL* Departamento de Matemática y stadistica, Universidad de La Frontera, Casilla 54-D, Temuco, Chile.

R. VARRO Département de Mathématiques et d'Informatique Appliquées, Université de Montpellier III, B.P. 5043, 34032 Montpellier Cedex 1, France.

Résumé. Dans ce travail nous recherchons des conditions pour que dans une population panmictique subdivisée en sous-populations, composées d'individus diploïdes, soumises à la mutation des gènes et à la migration, la structure génique d'un gène autosomique sélectivement neutre présente une succession cyclique d'états d'équilibre. Nous étudions ce problème d'un point de vue algèbrique : on associe à la population une algèbre non associative (algèbre de mutation), on obtient une condition nécessaire et suffisante pour qu'elle soit de Bernstein périodique. De plus en couplant ce résultat à un théorème d'isomorphisme dans les algèbres de mutation nous obtenons une classification complète des algèbres de mutation qui sont de Bernstein périodiques.

1 INTRODUCTION.

Dans le but d'étudier d'un point de vue algèbrique la mutation des gènes on a introduit en [1] une nouvelle classe d'algèbres non associatives : les algèbres de mutation. Dans ce qui suit nous allons montrer que ce type d'algèbre permet aussi de prendre en compte le phénomène de la migration.

Considérons des types géniques $a_1, \ldots, a_m$ non soumis à la sélection naturelle, présents dans une population panmictique subdivisée suivant C colonies (ou sous-populations, ou dèmes, ou niches) à l'intérieur desquelles ont lieu des mutations génétiques et entre lesquelles se produisent des migrations; à chaque génération se succèdent d'abord la mutation puis la migration. Pour tout $1 \leq p, q \leq C$, $1 \leq i, j \leq m$, notons :

$\mu_{p,ij}$ le taux de mutation par génération dans la colonie p du type a_j vers le type a_i,

---

* Subventionné par Fondecyt N° 1970890 et N° 1980548.

$m_{pq,i}$ la fréquence de migration des individus de type a_i de la colonie q vers la colonie p. On a donc $0 \le \mu_{p,ij}$, $0 \le m_{pq,i}$, $\sum_{1\le i\le m} \mu_{p,ij} = 1$ et $\sum_{1\le p\le C} m_{pq,i} = 1$.

Associons à cette population le R-espace A de base $\{e_{11},..., e_{1m},..., e_{C1},..., e_{Cm}\}$ et dans cette base la matrice $M = (M_{pq})_{1\le p,q\le C} diag(M_1,..., M_C)$ où on a posé $M_{pq} =$ diag$(m_{pq,1},..., m_{pq,m})$ et $M_k = (\mu_{k,ij})_{1\le i,j\le m}$; en particulier quand les taux de migrations ne dépendent pas des types géniques on a $M = (M_0 \otimes I_m) diag(M_1,..., M_C)$ où $M_0 = (m_{pq})_{1\le p,q\le C}$ (cf. [2]).

Munissons A de la structure d'algèbre définie par : $e_{pi}e_{qj} = \frac{1}{2}M(e_{pi} + e_{qj})$.

Soit $x(t) = \sum_{1\le p\le C,\ 1\le i\le m} x_{pi}(t)e_{pi}$, $\sum_{p,i} x_{pi}(t) = 1$, une distribution en fréquences des types géniques à la génération t. Alors à la génération t+1, la distribution des fréquences est $x(t+1) = x(t)^2 = \sum_{q,j} x_{qj}(t)M(e_{qj}) = \sum_{p,i} (\sum_{q,j} m_{pq,i}\mu_{q,ij}x_{qj}(t))e_{pi}$, et la fréquence du type a_i dans la colonie p à la génération t+1 est $x_{pi}(t+1) = \sum_{1\le q\le C, 1\le j\le m} m_{pq,i}\mu_{q,ij}x_{qj}(t)$.

2 ISOMORPHISME DANS LES ALGÈBRES DE MUTATION.

Dans ce paragraphe, K désigne un corps commutatif infini tel que car(K) $\neq 2$ et m est un entier ≥ 2.

Nous rappelons la définition d'une algèbre de mutation.

DÉFINITION 2.1. Une K-algèbre A est de *mutation* s'il existe une application linéaire $M : A \to A$ et une base $(e_i)_{1\le i\le m}$, appelée de mutation, telle que pour tout $1 \le i, j \le m$ on a : $M(e_i) = \sum_{1\le k\le m} \mu_{ki}e_k$, $\sum_{1\le k\le m} \mu_{ki} = 1$ et $e_ie_j = \frac{1}{2}M(e_i + e_j)$ ◊

Une telle algèbre, notée (A, M), est commutative, en général non associative, pondérée par $\omega(e_i) = 1$, donc $\omega \circ M = \omega$ et on a $xy = \frac{1}{2}(\omega(y)M(x) + \omega(x)M(y))$ pour tout x et y dans A .

La matrice associée à l'application linéaire M dans une base de mutation est appelée matrice de mutation de l'algèbre. On note Mut$_m$(K) (resp. GMut$_m$(K)) l'ensemble des matrices de mutation (resp. de mutation inversibles).

Nous allons montrer que deux algèbres de mutation sont isomorphes si et seulement si leurs matrices de mutation sont conjuguées dans GL$_m$(K), pour cela nous aurons besoin des deux lemmes suivants.

LEMME 2.2. *L'ensemble* GMut$_m$(K) *est un sous-groupe de* GL$_m$(K) *isomorphe au groupe affine* GA$_{m-1}$(K).

Démonstration. On vérifie sans problème que si M et M' sont dans Mut$_m$(K) alors $MM' \in$ Mut$_m$(K). Soit $M = (\mu_{ij})_{1\le i,j\le m} \in$ GMut$_m$(K) et $M^{-1} = (\nu_{ij})_{1\le i,j\le m}$, de $MM^{-1} = I$ on déduit que $\sum_{1\le k\le m} \mu_{ik}\nu_{kj} = \delta_{ij}$, en particulier pour tout j on a $1 = \sum_{1\le i,k\le m} \mu_{ik}\nu_{kj} = \sum_{1\le k\le m} \nu_{kj}$, c'est-à-dire que $M^{-1} \in$ Mut$_m$(K).

On rappelle que le groupe affine d'ordre n est défini par :

$$\mathrm{GA}_{n+1}(\mathrm{K}) = \{ \begin{pmatrix} 1 & 0 \\ a & U \end{pmatrix} / \ a \in \mathrm{M}_{n,1}(\mathrm{K}),\ U \in \mathrm{GL}_n(\mathrm{K}) \}$$

Soit $M \in$ GMut$_m$(K), dans une base de mutation B = $\{e_1,..., e_m\}$ on a $M(e_i) =$

$\sum_{1\leq j\leq m}\mu_{ji}e_j$ avec $\sum_{1\leq j\leq m}\mu_{ji} = 1$. Dans la base B' = $\{e_1, e_2 - e_1, \ldots, e_m - e_1\}$ on trouve $M(e_1) = e_1 + \sum_{2\leq j\leq m}\mu_{j1}(e_j - e_1)$ et $M(e_i - e_1) = \sum_{2\leq j\leq m}(\mu_{ji} - \mu_{j1})(e_j - e_1)$. Comme $M \in \mathrm{GMut}_m(\mathrm{K})$ la matrice $(\mu_{ji} - \mu_{j1})_{2\leq i,j\leq m} \in \mathrm{GL}_{m-1}(\mathrm{K})$. Donc M est semblable à une matrice de $\mathrm{GA}_{m-1}(\mathrm{K})$ et si P est la matrice de passage de la base B à la base B', l'application qui à M associe $P^{-1}MP$ définit un isomorphisme de groupes de $\mathrm{GMut}_m(\mathrm{K})$ dans $\mathrm{GA}_{m-1}(\mathrm{K})$. ◊

LEMME 2.3. *Soit* L *un corp commutatif,* K ⊂ L. *Si* $M, M' \in \mathrm{M}_{m+1}(\mathrm{K})$ *sont conjuguées dans* $\mathrm{GA}_m(\mathrm{L})$ *alors M et M' sont conjuguées dans* $\mathrm{GA}_m(\mathrm{K})$.

Démonstration. Soit $\mathrm{T}_m(\mathrm{K}) = \{ \begin{pmatrix}\alpha & 0\\ a & U\end{pmatrix} / \ \alpha \in \mathrm{K},\ a \in \mathrm{M}_{m-1,1}(\mathrm{K}),\ U \in \mathrm{M}_{m-1}(\mathrm{K}) \}$ et E l'ensemble des matrices $P \in \mathrm{T}_m(\mathrm{K})$ vérifiant $PM = M'P$ (∗). Les éléments des matrices P vérifiant (∗) sont solution d'un système linéaire à coefficients dans K, par conséquent le L-espace vectoriel E admet une base $\{P_1, \ldots, P_r\}$ formée de matrices à coefficients dans K. Comme par hypothèse la fonction polynômiale $(\alpha_1, \ldots, \alpha_r) \to \det(\alpha_1 P_1 + \ldots + \alpha_r P_r)$ n'est pas identiquement nulle sur L^r et que K est infini, il existe donc $(\lambda_1, \ldots, \lambda_r) \in \mathrm{K}^r$ tel que $P = \lambda_1 P_1 + \ldots + \lambda_r P_r \in \mathrm{E}$, $\det P \neq 0$; par conséquent on a $P = \begin{pmatrix}\alpha & 0\\ a & U\end{pmatrix}$, avec $\alpha \neq 0$, $U \in \mathrm{GL}_m(\mathrm{K})$, par suite $\alpha^{-1}P \in \mathrm{GA}_m(\mathrm{K})$, d'où le résultat ◊

THÉORÈME 2.4. *Soient (A, M)* et *(A', M') des algèbres de mutation de dimension m. Les énoncés suivants sont équivalents :*

(a) *(A, M) et (A', M') sont isomorphes,*
(b) *M et M' sont conjuguées dans* $\mathrm{GMut}_m(\mathrm{K})$,
(c) *M et M' sont conjuguées dans* $\mathrm{GL}_m(\mathrm{K})$.

Démonstration.

(a)⟺(b) Soient $(e_i)_{1\leq i\leq m}$ et $(e'_i)_{1\leq i\leq m}$ les bases de mutation, ω et ω' les pondérations respectives de (A,M) et (A', M'). Si f est un isomorphisme de A sur A', par unicité de la pondération de A on a $\omega' \circ f = \omega$ d'où $\omega'(f(e_i)) = 1$, donc $f \in \mathrm{GMut}_m(\mathrm{K})$ et de $f(e_i^2) = f(e_i)^2$ on déduit que $f \circ M = M' \circ f$. Réciproquement, si M et M' sont conjuguées dans $\mathrm{GMut}_m(\mathrm{K})$, il existe $P \in \mathrm{GMut}_m(\mathrm{K})$, $P = (\alpha_{ij})_{i,j}$ et $PM = M'P$. En posant $f(e_i) = \sum_{1\leq j\leq m}\alpha_{ji}e_j$ pour tout $1 \leq i \leq m$, on trouve $f(e_i^2) = (f \circ M)(e_i) = (M' \circ f)(e_i) = f(e_i)^2$ car $(\omega' \circ f)(e_i) = \sum_{1\leq j\leq m}\alpha_{ji} = 1$ et $f(e_i)f(e_j) = \frac{1}{2}(f(e_i)^2 + f(e_j)^2) = \frac{1}{2}f(e_i^2 + e_j^2) = f(e_ie_j)$

L'implication (b)⟹(c) est triviale. Montrons (c)⟹(b). Compte tenu de 2.2 on peut prendre M et M' dans $\mathrm{GA}_{m-1}(\mathrm{K})$. Soit L une clôture algèbrique de K, d'après 2.3 il suffit donc de montrer que si M et M' sont semblables dans $\mathrm{GL}_m(\mathrm{L})$ alors M, M' sont semblables dans $\mathrm{GA}_{m-1}(\mathrm{L})$.

On a $M = \begin{pmatrix}1 & 0\\ x & B\end{pmatrix}$ et $M' = \begin{pmatrix}1 & 0\\ x' & B'\end{pmatrix}$. De la conjugaison entre M et M' dans $\mathrm{GL}_m(\mathrm{L})$ il résulte que B et B' sont conjuguées dans $\mathrm{GL}_{m-1}(\mathrm{L})$. Donc B et B' ont la même forme de Jordan J et il existe P, Q dans $\mathrm{GL}_{m-1}(\mathrm{L})$ tels que $B = PJP^{-1}$ et $B' = QJQ^{-1}$. Or, on a :

$$\begin{pmatrix}1 & 0\\ 0 & P^{-1}\end{pmatrix}\begin{pmatrix}1 & 0\\ x & B\end{pmatrix}\begin{pmatrix}1 & 0\\ 0 & P\end{pmatrix} = \begin{pmatrix}1 & 0\\ P^{-1}x & J\end{pmatrix} \text{et}$$

$$\begin{pmatrix}1 & 0\\ 0 & Q^{-1}\end{pmatrix}\begin{pmatrix}1 & 0\\ x' & B'\end{pmatrix}\begin{pmatrix}1 & 0\\ 0 & Q\end{pmatrix} = \begin{pmatrix}1 & 0\\ Q^{-1}x' & J\end{pmatrix}.$$

On est ramené à montrer que les matrices $\begin{pmatrix}1 & 0\\ P^{-1}x & J\end{pmatrix}$ et $\begin{pmatrix}1 & 0\\ Q^{-1}x' & J\end{pmatrix}$ sont conjuguées dans $GA_{m-1}(L)$.

Remarquons ensuite que si $\begin{pmatrix}1 & 0\\ a & U\end{pmatrix}$, $\begin{pmatrix}1 & 0\\ a' & U'\end{pmatrix}$ sont conjuguées dans $GA_p(L)$ et si $\begin{pmatrix}1 & 0\\ b & V\end{pmatrix}$, $\begin{pmatrix}1 & 0\\ b' & V'\end{pmatrix}$ sont conjuguées dans $GA_q(L)$ alors $\begin{pmatrix}1 & 0 & 0\\ a & U & 0\\ b & 0 & V\end{pmatrix}$ et $\begin{pmatrix}1 & 0 & 0\\ a' & U' & 0\\ b' & 0 & V'\end{pmatrix}$ sont conjuguées dans $GA_{p+q}(L)$. En effet, de

$$\begin{pmatrix}1 & 0\\ a & U\end{pmatrix}\begin{pmatrix}1 & 0\\ z & P\end{pmatrix} = \begin{pmatrix}1 & 0\\ z & P\end{pmatrix}\begin{pmatrix}1 & 0\\ a' & U'\end{pmatrix} \text{ et } \begin{pmatrix}1 & 0\\ b & V\end{pmatrix}\begin{pmatrix}1 & 0\\ w & Q\end{pmatrix} = \begin{pmatrix}1 & 0\\ w & Q\end{pmatrix}\begin{pmatrix}1 & 0\\ b' & V'\end{pmatrix}$$

on en déduit que

$$\begin{pmatrix}1 & 0 & 0\\ a & U & 0\\ b & 0 & V\end{pmatrix}\begin{pmatrix}1 & 0 & 0\\ z & P & 0\\ w & 0 & Q\end{pmatrix} = \begin{pmatrix}1 & 0 & 0\\ z & P & 0\\ w & 0 & Q\end{pmatrix}\begin{pmatrix}1 & 0 & 0\\ a' & U' & 0\\ b' & 0 & V'\end{pmatrix}.$$

Alors compte tenu de l'écriture de J et de la remarque ci-dessus, il suffit de montrer que dans les trois cas suivants, si les matrices $\begin{pmatrix}1 & 0\\ a & U\end{pmatrix}$ et $\begin{pmatrix}1 & 0\\ a' & U'\end{pmatrix}$ sont semblables dans $GL_p(L)$ alors elles sont semblables dans $GA_{p-1}(L)$.

(i) Si 1 n'est pas valeur propre de U. On a :

$$\begin{pmatrix}1 & 0\\ y & I\end{pmatrix}\begin{pmatrix}1 & 0\\ a & U\end{pmatrix}\begin{pmatrix}1 & 0\\ -y & I\end{pmatrix} = \begin{pmatrix}1 & 0\\ a+(I-U)y & U\end{pmatrix} \text{ et } \begin{pmatrix}1 & 0\\ z & I\end{pmatrix}\begin{pmatrix}1 & 0\\ a' & U\end{pmatrix}\begin{pmatrix}1 & 0\\ -z & I\end{pmatrix} = \begin{pmatrix}1 & 0\\ a'+(I-U)z & U\end{pmatrix}$$

or $I-U \in GL_{p-1}(L)$, par conséquent on peut choisir y et z pour que les matrices $\begin{pmatrix}1 & 0\\ a & U\end{pmatrix}$ et $\begin{pmatrix}1 & 0\\ a' & U\end{pmatrix}$ soient semblables dans $GA_{p-1}(L)$ à la matrice $\begin{pmatrix}1 & 0\\ 0 & U\end{pmatrix}$.

(ii) Si $U = I$.

De la conjugaison dans $GL_p(L)$ entre $\begin{pmatrix}1 & 0\\ a & U\end{pmatrix}$ et $\begin{pmatrix}1 & 0\\ a' & U\end{pmatrix}$ on déduit que: $a = 0 \Leftrightarrow a' = 0$, donc il existe $C \in GL_{p-1}(L)$ tel que $a = Ca'$. Il en résulte que $\begin{pmatrix}1 & 0\\ a & U\end{pmatrix}\begin{pmatrix}1 & 0\\ 0 & C\end{pmatrix} = \begin{pmatrix}1 & 0\\ 0 & C\end{pmatrix}\begin{pmatrix}1 & 0\\ a' & U\end{pmatrix}$.

(iii) Si $U = I + N$, N étant la matrice de Jordan nilpotente d'ordre p-1.

Soit $P = \begin{pmatrix}a & z'\\ z & W\end{pmatrix} \in GL_p(L)$ tel que $P\begin{pmatrix}0 & 0\\ a & N\end{pmatrix} = \begin{pmatrix}0 & 0\\ a' & N\end{pmatrix}P$ (*), où $a = {}^t(\alpha_1, \ldots, \alpha_{p-1})$ et $a' = {}^t(\beta_1, \ldots, \beta_{p-1})$. Si $z' = 0$ le résultat est établi. Si $z' \neq 0$, de (*) il vient $z'a = 0$ et $z'N = 0$, d'où $\alpha_{p-1} = 0$. Mais alors en élevant la relation (*) à la puissance p-1 on obtient

$\begin{pmatrix} 0 & 0 \\ N^{p-2}a' & 0 \end{pmatrix} = 0$, d'où $\beta_{p-1} = 0$. Par suite il existe $w \in M_{p-1,1}(L)$ tel que $Nw = a - a'$ et donc on a $\begin{pmatrix} 1 & 0 \\ w & I \end{pmatrix}\begin{pmatrix} 0 & 0 \\ a & N \end{pmatrix} = \begin{pmatrix} 0 & 0 \\ a' & N \end{pmatrix}\begin{pmatrix} 1 & 0 \\ w & I \end{pmatrix}$. ◊

3 ALGÈBRES DE MUTATION ET DE BERNSTEIN PÉRIODIQUES.

Dans ce paragraphe K désigne un corps commutatif infini de caractéristique nulle.

Soit A une K-algèbre, pour $x \in A$ on définit $x^{[1]} = x$, $x^{[k+1]} = x^{[k]}x^{[k]}$, $k \geq 1$.

DÉFINITION 3.1. Une algèbre pondérée (A, ω) est de *Bernstein d'ordre n et de période p*, en abrégé une *$B(n, p)$-algèbre*, si pour tout x dans A on a :

$$x^{[n+p+1]} = \omega(x)^{2^n(2^p-1)}x^{[n+1]}$$

avec $(p, n) \in \mathbb{N}^* \times \mathbb{N}$ minimal pour l'ordre lexicographique. ◊

Pour des résultats concernant les $B(n, p)$-algèbres on se référera à [3] et [4]. En particulier les $B(n,1)$-algèbres sont les *algèbres de Bernstein d'ordre n* introduites dans [5] et étudiées en [6], [7] et [8]; les $B(0,2)$-algèbres sont les *algèbres d'Etherington-Walcher* ([9], [10], [11], [12]). D'un point de vue génétique, si A est une algèbre stochastique ([13], [14]) associée à une population et $x \in A$ une distribution de fréquences des types génétiques présents dans la population à la génération initiale, donc $\omega(x) = 1$. Alors $x^{[k+1]}$ donne la distribution des fréquences à la k-ième génération. Si de plus A est une $B(n,p)$-algèbre on a $x^{[n+p+1]} = x^{[n+1]}$, autrement dit à partir de la n-ième génération la distribution en fréquences atteint un état d'équilibre qui se répéte avec la périodicité p.

Un élément $e \in A$, $e \neq 0$ est un *idempotent d'ordre s*, si on a $e^{[s+1]} = e$ et $s = \#\{e, \ldots, e^{[s]}\}$. On note $Ip(A; s)$ l'ensemble des idempotents d'ordre s de A.

Soit L un corps commutatif tel que $K \subset L$ et $A_L = L \otimes_K A$ l'extension de la K-algèbre A au corps de base L. Il est clair que si A est pondérée par $\omega : A \to K$ alors A_L est pondérée par $\omega(\alpha \otimes x) = \alpha\omega(x)$ et que si A est une K-algèbre de mutation, de matrice de mutation M alors la L-algèbre A_L est aussi de mutation pour la même application M.

PROPOSITION 3.2. *Soit* L *un corps commutatif tel que* $K \subset L$, (A, M) *une* K-*algèbre de mutation de Bernstein d'ordre n et de période p, alors on a* :

(a) $M^n(M^p - I) = 0$.

(b) $Ip(A; d) \neq \emptyset$ *pour tout diviseur d de p.*

(c) (A_L, M) *est une* $B(n, p)$-*algèbre.*

Démonstration. (a) Soit $x \in A$, de la relation $x^2 = \omega(x)M(x)$ et de $\omega \circ M = \omega$ il découle que $x^{[k+1]} = \omega(x)^{2^k-1}M^k(x)$, alors pour tout entier $m \geq 0$ et $q \geq 1$ on a :

$$x^{[m+q+1]} = \omega(x)^{2^m(2^q-1)}x^{[m+1]} \Leftrightarrow \omega(x)^{2^{m+q}-1}(M^{m+q} - M^m)(x) = 0. \qquad (3.2.1)$$

Si (A, M) est une $B(n, p)$-algèbre on a donc $\omega(x)^{2^{n+p}-1}(M^{n+p} - M^n)(x) = 0$ pour tout

$x \in A$, et du fait que pour tout $y \notin Ker(\omega)$ et $z \in Ker(\omega)$ on a $(M^{n+p} - M^n)(y+z) = 0$ on déduit que $M^{n+p} = M^n$.

(b) Le corps K étant infini on a $Ip(A; p) \neq \emptyset$ (cf. [3]). Soit $e \in Ip(A; p)$ et d un diviseur de p, $p = ds$. Posons $c = (d\omega(e))^{-1}\Sigma_{0 \le k \le s-1} M^{kd}(e)$, on vérifie sans problème que $c^{[d+1]} = c$ et que d est le plus petit entier vérifiant cette identité.

(c) Soit $\alpha, \beta \in$ L, x, y dans A tels que $\omega(x) \neq 0$, $\omega(y) = 0$. On note, comme il est d'usage, αx pour $\alpha \otimes x$. De $(\alpha x + \beta y)^2 = \alpha\omega(x)M(\alpha x + \beta y)$ et $y^2 = 0$, il découle que :

$$(\alpha x + \beta y)^{[q+1]} = \omega(\alpha x)^{2^q} M^q(\alpha x + \beta y), \text{ pour tout } q \ge 1.$$

D'où : $(\alpha x + \beta y)^{[n+p+1]} = \omega(\alpha x)^{2^{n+p}} M^{n+p}(\alpha x + \beta y) = \omega(\alpha x)^{2^{n+p}} M^n(\alpha x + \beta y) = \omega(\alpha x)^{2^n(2^p-1)}(\alpha x + \beta y)^{[n+1]}$. ◊

Nous pouvons maintenant caractériser les algèbres de mutation qui sont de Bernstein périodiques.

Dans ce qui suit, on note $U_p(\mathrm{K}) = \{ x \in \mathrm{K}; x^p = 1 \}$.

THÉORÈME 3.3. *Soit (A, M) une* K*-algèbre de mutation,* L *une clôture algèbrique de* K*, alors sont équivalents:*

(i) *(A, M) est une $B(n, p)$-algèbre,*

(ii) *le pôlynôme minimal de $M : A_{\mathrm{L}} \to A_{\mathrm{L}}$ est $X^n(X-1)(X-\xi_1)\ldots(X-\xi_s)$ où $\xi_1, \ldots, \xi_s \in U_p(\mathrm{L})-\{1\}$ sont deux à deux distincts et tels que le groupe $\langle 1, \xi_1, \ldots, \xi_s\rangle$ soit d'ordre p.*

Démonstration. (i)⇒(ii) On se place dans A_{L}. De 3.2 (a) et de $1 \in Spec(M)$, on déduit que le polynôme minimal de M est de la forme $X^m(X-1)(X-\xi_1)\ldots(X-\xi_s)$ avec $\xi_1, \ldots, \xi_s \in U_p(\mathrm{L})$. Or, on ne peut pas avoir $m < n$ car sinon on aurait $M^m(M^p - I) = 0$ ce qui, d'après la relation (3.2.1) entrainerait $x^{[m+p+1]} = \omega(x)^{2^m(2^p-1)}x^{[m+1]}$ contredisant la minimalité de (p, n). Soient q l'ordre du groupe $\langle 1, \xi_1, \ldots, \xi_s\rangle$, $y \in KerM^n$ et $x_i \in Ker(M - \xi_i I)$ $(1 \le i \le s)$. Alors $M^n y = 0$ entraine que $\omega(y) = 0$ et de $x_i^2 = \omega(x_i)M(x_i)$ on déduit que $(1 - \xi_i)\omega(x_i)^2 = 0$ donc $\omega(x_i) = 0$ pour $1 \le i \le s$; par conséquent il existe nécessairement $x_0 \in Ker(M - I)$ tel que $\omega(x_0) \neq 0$. Si on pose $x = \sum_{1 \le i \le s} x_i + y$, on montre sans difficulté que $x^{[m+1]} = \omega(x)^{2^m-1}M^m(x)$ d'où $x^{[n+q+1]} = \omega(x)^{2^{n+q}-1}M^{n+q}(x) = \omega(x)^{2^n(2^q-1)}x^{[n+1]}$ c'est-à-dire que A_{L} est une $B(n, q)$-algèbre, contradiction.

(ii)⇒(i) On a $M^n(M^p - I) = 0$. On ne peut pas avoir $M^n(M^q - I) = 0$ pour un entier q < p, sinon $\xi_1, \ldots, \xi_s \in U_q(\mathrm{L})$ et le groupe $\langle 1, \xi_1, \ldots, \xi_s\rangle$ serait d'ordre < p, par conséquent on a $M^n(M^p - I) = 0$ avec (p, n) minimal, ce qui, d'après l'assertion (3.2.1) entraine que A est une $B(n, p)$-algèbre. ◊

COROLLAIRE 3.4. *Si (A, M) est une $B(n, p)$-algèbre alors $dim(A) \ge n + 2$.* ◊

Dans ce qui suit on note M_σ la matrice de permutation associée à une permutation $\sigma \in S_d$, on a $M_\sigma \in \mathrm{GMut}_d(\mathrm{K})$.

THÉORÈME 3.5. *Si $U_p(\mathrm{K}) \subset \{-1; 1\}$, alors, à isomorphisme près, les algèbres de mutation de dimension d qui sont des $B(0, p)$*-algèbres *sont les algèbres (A, M_σ) avec*

$\sigma \in S_d$ d'ordre p.

Démonstration. On voit de suite que pour une permutation $\sigma \in S_d$ d'ordre p on a $(M_\sigma)^p = I$ et donc que (A, M_σ) est une $B(0, p)$-algèbre.

Réciproquement, soit (A, M) une $B(0, p)$-algèbre de dimension d. Soit $p = p_1^{\alpha_1} \ldots p_q^{\alpha_q}$ la décomposition de p en facteurs premiers, pour simplifier les notations on pose dans la suite $r_k = p_k^{\alpha_k}$. Soit L une clôture algèbrique de K et pour tout $1 \le k \le q$, soit $\xi_k \in$ L tel que $\langle \xi_k \rangle = U_{r_k}(\mathrm{L})$, alors le polynôme minimal de M est $(X-1)\dfrac{(X^{r_1}-1)\ldots(X^{r_q}-1)}{(X-1)^q}$ et donc la matrice M est conjuguée dans $\mathrm{GL}_d(\mathrm{L})$ à la matrice blocs diagonale $diag(D(\xi_1),\ldots, D(\xi_q))$ où $D(\xi_k) = diag(1, \xi_k, \ldots, \xi_k^{r_k-1})$, or $D(\xi_k)$ est conjuguée dans $\mathrm{GL}_{r_k}(\mathrm{L})$ à la matrice d'un cycle d'ordre r_k. Il existe donc $\sigma \in S_d$ d'ordre p telle que les matrices M et M_σ sont conjuguées dans $\mathrm{GL}_d(\mathrm{L})$ donc dans $\mathrm{GL}_d(\mathrm{K})$, alors d'après le théorème 2.2 on a que (A, M) et (A, M_σ) sont isomorphes. ◊

NOTATIONS 3.6. A tout entiers m, $n = n_0, n_1, \ldots, n_s \ge 1$ et à toute permutation σ d'ordre p opérant sur $\{1, \ldots, m\}$, on associe la matrice M définie sur une base $\{e_1,\ldots,e_m\} \cup \bigcup_{0\le k\le s}\{e_{k,i}\}_{1\le i\le n_k}$ par :

$$M(e_i) = e_{\sigma(i)},\ 1 \le i \le m,$$
$$M(e_{k,1}) = e_1,\ 0 \le k \le s,$$
$$M(e_{k,i}) = e_{k,i-1},\ 2 \le i \le n_s,\ 0 \le k \le s.$$

Il est clair que M est une matrice de mutation, et on note $A(\sigma, n, n_1,\ldots, n_s)$ l'algèbre de mutation associée à cette matrice. ◊

THÉORÈME 3.7. *Soit (A, M) une K-algèbre de mutation de dimension d, si on a $U_p(\mathrm{K}) \subset \{-1; 1\}$, alors les énoncés suivants sont équivalents :*

(i) *A est une $B(n, p)$-algèbre;*

(ii) *il existe des entiers m, $n_1, \ldots, n_s \ge 1$; $n + m + n_1 + \ldots + n_s = d$ et $\sigma \in S_m$ d'ordre p tels que A soit isomorphe à l'algèbre $A(\sigma, n, n_1,\ldots, n_s)$.*

Démonstration. L'implication (ii)⇒(i) découle du théorème 3.3, du fait que le polynôme caractéristique de la matrice de mutation de l'algèbre $A(\sigma, n, n_1,\ldots, n_s)$ est $X^n(X^p-1)$. Quant à (i)⇒(ii), si (A, M) est une $B(n, p)$-algèbre, en utilisant les mêmes notations que dans la preuve de 3.5, il résulte du théorème 3.3 et de l'hypothèse sur le corps K, que le polynôme minimal de M est $X^n(X-1)\dfrac{(X^{r_1}-1)\ldots(X^{r_q}-1)}{(X-1)^q}$ Par conséquent M est conjuguée dans $\mathrm{GL}_d(\mathrm{L})$ à une matrice blocs diagonale de la forme $diag(J_1,\ldots, J_s, D(\xi_1),\ldots, D(\xi_q))$ où $J_1,\ldots, J_s$ sont des blocs de Jordan nilpotents d'ordres respectifs n, $n_1, \ldots, n_s$ et $D(\xi_k) = diag(1, \xi_k, \ldots, \xi_k^{r_k-1})$. Soit σ la permutation associée à la matrice $diag(D(\xi_1),\ldots, D(\xi_q))$, alors par 2.2 les algèbres (A, M) et $A(\sigma, n, n_1,\ldots, n_s)$ sont isomorphes. ◊

Du résultat ci-dessus découle immédiatement le

COROLLAIRE 3.8. *Soit (A, M) une K-algèbre de mutation* ($où U_p(K) \subset \{-1; 1\}$), *alors : (A, M) est une B(n, p)-algèbre si et seulement si il existe* $e \in Ip(A; 1)$, *B et C des sous-espaces de A tels que* $A = Ke \oplus B \oplus C$, $Ke \oplus B$ *est une B(0, p)-algèbre et* $Ke \oplus C$ *est une algèbre quasi-constante d'ordre n.* ◊

REMARQUE 3.9. Revenons au modèle exposé dans l'introdution de ce travail. Il découle du théorème 3.7 que, dans le cas particulier où *(A, M)* est une R-algèbre de mutation avec *M* une matrice stochastique, alors une condition nécessaire pour que *(A, M)* soit une *B(n, p)*-algèbre est que la trace de *M* soit un entier.

Nous remercions chaleureusement le referee pour sa lecture attentive de ce travail.

BIBLIOGRAPHIE

[1] C Mallol, R Varro. Les algèbres de mutation. Non-Associative Algebras and its Applications. Kluwer Acad. Publ., Dordrecht, 1994, pp. 245-250.

[2] P Holgate. Interaction between migration and breeding studied by means of genetic algebras. J Appl Prob 5: 1-8, 1968.

[3] R Varro. Introduction aux algèbres de Bernstein périodiques (cas Moufang, idempotents, caractéristique 2). Non-Associative Algebras and its Applications. Kluwer Acad. Publ., Dordrecht, 1994, pp.384-388.

[4] R Varro. Décompositions de Peirce des algèbres de Bernstein périodiques. Nova J Algebra Geom 3: 157-164, 1994.

[5] VM Abraham. Linearizing quadratic transformation in genetic algebras. Proc London Math Soc 40: 346-363, 1980.

[6] C Mallol. A propos des algèbres de Bernstein. Thèse d'Etat, Université de Montpellier II, France, 1989.

[7] R Hentzel, LA Peresi, P. Holgate. On k-th order Bernstein algebras and stability at the k+1 generation in polyploïds. J Math Applied in Medicine and Biology 7: 33-40, 1990.

[8] S Gonzalez, JC Gutierrez, C Martinez. On Bernstein algebras of n-th order. Non-Associative Algebras and its Applications. Kluwer Acad. Publ., Dordrecht, 1994, pp. 158-163.

[9] IMH Etherington. Commutative train algebras of rank 2 and 3. J London Math Soc 15: 136-149, 1940.

[10] S. Walcher. Algebras which satisfy a train equation for the first three plenary powers. Arch. Math. 56: 547-551, 1991.

[11] C Mallol, R Varro. A propos des algèbres vérifiant $x^{[3]} = \omega(x)^3 x$. Linear Algebra Appl.225: 187-194, 1995.

[12] C Mallol, R Varro. Dérivations dans les algèbres d'Etherington-Walcher. Nova J Math Game Theory Algebra 4: 129-146, 1996.

[13] ML Reed. Algebraic structure of genetic inheritance. Bull Amer Math Soc 3: 107-130, 1997.

[14] A Wörz-Busekros. Algebras in genetics. Lect Notes in Biomathematics 36, Berlin: Springer-Verlag, 1980.

Finite semifields three-dimensional over the left nuclei

G. P. WENE Division of Mathematics and Statisitics, The University of Texas at San Antonio, San Antonio, Texas

1. INTRODUCTION

The term *semifield* will refer to a not necessarily associative division ring. Semifields are used in the study of projective planes (see, for example, the book by Dembowski[1]). The projective planes of *Lens-Barlotti type V.1* are those translation planes (with translation line uv) which are also dual translation planes (with translation point v). These are precisely the planes which can be coordinatized using semifields (Dembowski[1]).

Petit [7] [8] presents a construction for semifields based on results contained in Ore's papers on rings of noncommutative polynomials [5][6] (see also Jacobson[3]). We will review Petit's construction, giving a proof of his existence criteria (no proof is given in his papers), and give explicit construction for semifields of characteristic 2, 3, 5 and 7.

2. THE TWISTED POLYNOMIALS RINGS

Let A be the set $\{\sum_{i=0}^{n} \alpha_i X^i \mid \alpha_i \in F, n = 1, 2, \ldots\}$ where F is the finite field $GF(p^n)$ of p^n elements. The set A can be given the structure of a (left) vector space over F by defining addition in terms of addition of the corresponding coefficients and scalar multiplication by $\beta \in F$ by

$$\beta \sum_{i=0}^{n} \alpha_i X^i = \sum_{i=0}^{n} \beta \alpha_i X^i .$$

This paper is in final form and no version of it will be submitted for publication elsewhere.

We want a multiplication in A that is associative and satisfies both distributive properties and such that the degree of a product shall be equal to the sum of the degrees of the factors. By Ore [5] we need only define the product $X \cdot a$; this gives

$$X \cdot a = a^{\sigma}X + a^{\delta}$$

where a^{σ} , $a^{\sigma} \neq 0$, and a^{δ} are elements of F. Ore calls the elements a^{σ} and a^{δ} the *conjugate* and *σ−derivation* of a respectively. We will use the notation in Jacobson[3] for these rings: $A = F[X;\sigma,\delta]$; each ring A will be a left principle ideal domain.

The insistence that $1X = X1 = X$ and associativity forces the derivation operator δ to satisfy the following properties:

(i) $1^{\delta} = 0$,

(ii) $(a+b)^{\delta} = a^{\delta} + b^{\delta}$,

(iii) $(ab)^{\delta} = a^{\sigma}b^{\delta} + a^{\delta}b$.

Since the field F is the finite field $GF(p^r)$, we can determine all conjugates σ and all σ−derivations δ.

LEMMA 2.1 If σ and $\delta$are a conjugate and corresponding σ−derivation, respectively, of the field $GF(p^r)$ then $a^{\sigma} = a^{p^i}$ for some i, where $0 \leq i < r$, and $a^{\delta} = \gamma_0\,(a - a^{\sigma})$ for some constant γ_0 in $GF(p^n)$.

Proof. Since $(a+b)^{\delta} = a^{\delta} + b^{\delta}$, we must have

$$a^{\delta} = \sum_{i=0}^{n-1} \gamma_i a^{p^i},$$

where each $\gamma_i \in GF(p^n)$.

The relation $(ax)^{\delta} = a^{\sigma}x^{\delta} + a^{\delta}x$ gives $\sum_{i=0}^{n-1} \gamma_i a^{p^i} x^{p^i} = a^{\sigma} \sum_{i=0}^{n-1} \gamma_i x^{p^i} + a^{\delta}x$. Equating the powers of x on each side gives $\gamma_0 ax = a^{\sigma}\gamma_0 x + a^{\delta}x$ for $i = 0$ and $\gamma_i a^{p^i} x^{p^i} = a^{\sigma}\gamma_i x^{p^i}$ for all $i, 1 \leq i < n$. ■

The conjugate map σ is an automorphism of the field F and the discussion on pages one through five of Jacobson[3] apply. The ring $A = F[X;\sigma,\delta]$ will be a principle ideal domain. Since the σ−derivation δ is given by $a^{\delta} = \gamma_0\,(a - a^{\sigma})$ these rings are the *twisted polynomial rings* of Jacobson. Each ring $A = F[X;\sigma,\delta]$ is isomorphic to a ring $A = F[X;\sigma,\widehat{\delta}]$ where $\widehat{\delta}$ is the zero derivation.

DEFINITION 2.1. By $A = F[X;\sigma]$, where $a^{\sigma} = a^{p^i}$ for some $0 \leq i < n$, we will mean the ring of polynomials in X with coefficients in the field F with multiplication defined by $X \cdot a = a^{p^i}X$ for all a in F. The ring $A = F[X;\sigma]$ will be commutative if and only if σ is the identity automorphism. We will use F to denote both the field F and the subring of A isomorphic to the field F.

Any automorphism T of the ring $A = F[X;\sigma]$, when restricted to F must be an automorphism of F. The automorphism T is completely determined by it action on F and the image of X under T.

THEOREM 2.2. The noncommutative ring $A = GF(p^n)\,[X;\sigma]$, $\sigma \neq 1$, has $n(p^n - 1)$ automorphisms.

Proof. Let $* : F \to F$ be an automorphism of F and let T be an automorphism of A that extends $*$ where $X^T = rX + s, r, s \in F$, $r \neq 0$.

$(Xa)^T = (a^\sigma X)^T = a^{\sigma *}\,(rX + s) = a^{\sigma *} rX + a^{\sigma *} s$

$(Xa)^T = X^T a^* = (rX + s)\,a^* = ra^{*\sigma} X + sa^*$.

Since the automorphisms σ and $*$ commute, $s = 0$. ■

3. THE SEMIFIELDS

Let $A = F\,[X;\sigma]$ and let $M(X)$ be an irreducible element of A of degree m. Let A_m denote the left vector space over F of all polynomials in A of degree no more than $m-1$. Define a multiplication $\star$ in A_m by

$$A(X) \star B(X) = R(X)$$

where $R(X)$ is the unique polynomial satisfying $= A(X)B(X) = Q(X)M(X) + R(X)$ where the degree of $R(X)$ is less than the degree of $M(X)$. There is an Eucledian algorithm for the ring A; the fact that the σ in $X \cdot a = a^\sigma X + a^\delta$ is an automorphism of F insures both left and right factorization (see Ore [5]). We will denote this not necessarily associative semifield by $A_m/M(X)$. If $a^\sigma = a$, the semifield will be a finite field; since we are interested in the case in which the semifield $\Delta = A_m/M(X)$ is not associative, we will insist in all cases $\sigma \neq 1$ and that the degree of $M(X)$ is greater than 1. This construction will produce the Hughes-Kleinfeld[2] semifields, the Sandler[9] semifields and other, new, semifields.

EXAMPLE 3.1. Let F be the field $GF(4) = \{0, 1, \omega, \omega^2 (= 1 + \omega)\}$. The polynomial $M(X) = X^2 + \omega$ is irreducible in the ring $A = GF(4)[X;\sigma]$ where $a^\sigma = a^2$. The elements of the semifield $\Delta = A/M(X)$ are the elements of the set

$$\{\alpha_0 + \alpha_1 X \mid \alpha_0, \alpha_1 \in GF(4)\}.$$

The multiplication of Δ is given in terms of the addition and multiplication of $GF(4)$:

$$(\alpha_0 + \alpha_1 X)\,(\beta_0 + \beta_1 X) = (\alpha_0\beta_0 + \omega\alpha_1\beta_1^2) + (\alpha_0\beta_1 + \alpha_1\beta_0^2)X.$$

EXAMPLE 3.2. Let F be the field $GF(9) \approx GF(3)(m)$ where $m^2 = 2$. The polynomial $M(X) = X^3 + 2X + 2$ is irreducible in the ring $A = GF(9)[X;\sigma]$ where $a^\sigma = a^3$. The elements of the semifield $\Delta = A_3/M(X)$ are the elements of the set

$$\{\alpha_0 + \alpha_1 X + \alpha_2 X^2 \mid \alpha_0, \alpha_1, \alpha_2 \in GF(9)\}.$$

The multiplication of Δ is given in terms of the addition and multiplication of $GF(9)$:

$$(\alpha_0 + \alpha_1 X + \alpha_2 X^2)\,(\beta_0 + \beta_1 X + \beta_2 X^2) = (\alpha_0\beta_0 + \alpha_1\beta =_2^3 + \alpha_2\beta_1) +$$

$$(\alpha_0\beta_1 + \alpha_1\beta_0^3 + \alpha_1\beta_2^3 + \alpha_2\beta_1 + \alpha_2\beta_2)\,X + (\alpha_0\beta_2 + \alpha_1\beta_1^3 + \alpha_2\beta_0 + \alpha_2\beta_3)X^2.$$

Each of the semifields $\Delta = A_m/M(X)$ will have a basis $\{1, X, X^2, \cdots, X^{m-1}\}$ over the field F. Let a be any element of Δ. The *right multiplication* R_a of Δ which is determined by a is defined by

$$R_a : x \to xa \text{ for all } x \text{ in } \Delta.$$

Clearly R_a is a linear operator on Δ. *Left multiplication* L_a is defined similarly. Let $M(R_a)$ denote the matrix of R_a relative to the basis $\{1, X, X^2, \cdots, X^{m-1}\}$.

THEOREM 3.1. Let $A = GF(p^n)[X;\sigma]$ where $a^\sigma = a^{p^i}$ for some $i \in \{1, 2, \cdots, n\}$. Let

$$x = \sum_{i=0}^{m-1} a_i X^i$$

be an element of the semifield $\Delta = A_m/M(X)$. The determinant of $M(R_x)$ is a homogeneous polynomial of degree

$$\sum_{j=0}^{m-1} p^{jn}$$

in the variables a_i.

Proof. Define constants $C^l_{i,j}$, for $0 \leq i, j, l \leq n-1$ by

$$X^i(a_j X^j) = a_j^{\sigma^i}(X^i X^j) = a_j^{\sigma^i} \sum_{l=0}^{m-1} C^l_{i,j} X^l.$$

The (i, j) th entry of the matrix $M(R_x)$ will be $a_j^{\sigma^i} \sum_{l=o}^{n-1} C^l_{i,j}$. The determinant of $M(R_x)$ will be a homogeneous polynomial of degree $\sum_{j=0}^{m-1} p^{jn}$ in the variables a_j. ■

Let $N_i(A_m/M(X))$, $i = l, m, r$, denote the left, middle and right nuclei of the semifield $A_m/M(X)$, respectively.

THEOREM 3.2. Let K denote the subfield of F fixed by σ: $K = \{k \in F \mid k^\sigma = k\}$. If the coefficients of $M(X)$ are elements of K then F is contained in both $N_l(A_m/M(X))$ and $N_m(A_m/M(X))$; the right nucleus $N_r(A_m/M(X))$, is the field consisting of all elements $\sum_{i=0}^{n-1} k_i X^i$, each $k_i \in K$.

Proof. We note that

$$(a, bX^i, cX^j) = (abX^i)(cX^j) - a\left(bc^{\sigma^i}X^{i+j}\right) = abc^{\sigma^i}X^{i+j} - abc^{\sigma^i}X^{i+j} = 0$$

and

$$\begin{aligned}(aX^i, b, cX^j) &= \left(ab^{\sigma^i}X^i\right)(cX^j) - (aX^i)(bcX^j) \\ (aX^i, b, cX^j) &= ab^{\sigma^i}c^{\sigma=^i}X^{i+j} - a(bc)^{\sigma^i}X^{i+j} = 0\end{aligned}$$

for all a, b, c in F and all $0 \leq i, j < n$. Hence $F \subset N_l \cap N_m$.

The semifield contains the field consisting of all elements $\sum_{i=0}^{n-1} k_i X^i$, each $k_i \in K$. This field is isomorphic to the field $K[X]/M(X)$. We will show that $X \in N_r$.

$$\begin{aligned}(aX^i, bX^j, X) &= \left(ab^{\sigma^i}X^{i+j}\right)X - (aX^i)(bX^{j+1}) \\ (aX^i, bX^j, X) &= ab^{\sigma^i}X^{i+j+1} - ab^{\sigma^i}X^{i+j+1} = 0.\end{aligned}$$

Hence the elements of K and X generate the right nucleus. If the right nucleus contained any other elements, it would not be commutative. ■

The above result is the best we can do. Example 1.1 has $F = GF(4)$ as the left and middle nuclei; the results of Sandler[9] show that the field F may be properly contained in $N_l \cap N_m$.

THEOREM 3.3. Let K denote the subfield of $F = GF(p^r)$ fixed by σ. If the coefficients of $M(X)$ are elements of K, and L, $K \subset L \subset F$, is a field and B is the ring $L[X;\sigma]$, then the semifield $A_m/M(X)$ contains a subsemifield isomorphic to $B_m/M(X)$.

Proof. If the polynomial $M(X)$ is irreducible over F, it is irreducible over any subfield of F. ■

REMARK 3.1 By Theorem 2.3, the set $\left\{\sum_{i=0}^{n-1} \alpha_i X^i \mid \alpha_i \in L\right\}$ of $A_m/M(X)$ is a subsemifield of $A_m/M(X)$ for all fields L, such that $K \subset L \subset F$. The set $\left\{\sum_{i=0}^{n-1} \alpha_i X^i \mid \alpha_i \in L\right\}$ is a field under the operations of $A_m/M(X)$.

4. SEMIFIELDS DEFINED BY QUADRATIC POLYNOMIALS

Let F be the field $GF(p^n)$ with nonidentity automorphism σ and let Δ be the vector space F^2. Define a multiplication $\star$ in Δ by

$$(a,b)\star(c,d) = (a,b)\begin{pmatrix} c & d \\ \alpha\left(c - c^{p^m}\right) + \gamma d^{p^m} & c^{p^m} + \alpha d + \beta d^{p^m} \end{pmatrix}.$$

The matrix has determinant $c^{p^m+1} + \alpha c^{p^m} d + \beta c d^{p^m} - \gamma d^{p^m+1}$ which will be zero if and only if $c = d = 0$. If the determinant is zero but neither c nor d is zero, the polynomial $X^{p^m+1} + \alpha X^{p^m} + \beta X - \gamma$ has a root in $GF(p^n)$. Writing the elements of Δ as the set $\{a + bX \mid a, b \in F\}$, we have

$$\begin{aligned}(a+bX)+(c+bX) &= (a+c)+(b+d)X\\ (a+bX)\star(c+bX) &= ac + b(\alpha(c-c^\sigma)+\gamma d^\sigma) + (ad + bc^\sigma + \alpha bd + \beta bd^\sigma)X.\end{aligned}$$

We readily see that $Xa = a^\sigma X + \alpha\,(a - a^\sigma)$ and $X^2 - (\alpha+\beta)X - \gamma$ define a ring of twisted polynomials and an irreducible polynomial over that ring.

The polynomial $X^{p^m+1} + \alpha X^\sigma + \beta X - \delta$ has no roots in $GF(p^n)$ if and only if it has no roots in in the field $GF(p^{nd})$ for all n prime to $p^m + 1$ where $a^\sigma = a^{p^m}$. We have now proved

THEOREM 4.1. The polynomial $X^{p^m+1} + \alpha X^p + \beta X - \gamma$ has no roots in $GF(p^n)$ if and only if the polynomial $X^2 - (\alpha+\beta)X - \gamma$ is irreducible in $GF(p^{nd})[X;\sigma,\delta]$ where $a^\delta = \alpha\,(a - a^\sigma)$ and $(r, p^m + 1) = 1$.

This construction yeilds, with $\alpha = 0$, the Hughes-Kleinfeld[2] semifields, i.e., those semifields Δ that are quadratic extensions of a finite field F, where F is contained in the left and right nuclei of Δ. If $\alpha = \beta = 0$ and $\sigma^2 = I$, I the identity automorphism, the field F is contained in the nucleus. Since the ring $F[X;\sigma,\delta]$ is isomorphic to the ring $F[X;\sigma,0]$, using $\alpha \neq 0$ does not lead to the construction of any additional semifields. It is now relatively easy to determine isomorphism classes; we give without proof

THEOREM 4.2. Semifields constructed from the ring $A = F[X;\sigma]$ by the polynomials $P(X) = X^2 + \beta X + \gamma$ and $Q(X) = X^2 + \widehat{\beta} X + \widehat{\gamma}$ are isomorphic if and only if there is an automorphism $*$ of F and a nonzero element $s \in F$ such that $\widehat{\beta} = s^\sigma \beta$ and $\widehat{\gamma} = s^{\sigma+1}\gamma$.

REMARK 4.1. The determinant of $M(R_{(c,d)})$ is a homogeneous function of the variables c and d of degree $p^m + 1$ where $a^\sigma = a^{p^m}$.

5. SEMIFIELDS DEFINED BY CUBIC POLYNOMIALS

Petit's method can be used to construct semifields three dimensional over their left nuclei that are not algebras three dimensional over their nuclei. These semifields are, therefore, new.

Each polynomial $M(X) = X^3+\alpha_2X^2+\alpha_1X+\alpha_0$, $\alpha_i \in GF(p^n)$, irreducible over the ring $A = GF(p^n)\,[X;\sigma]$, can be used to construct a semifield. Our discussion will not be exhaustive; we will insist that each element α_i, $i = 0, 1, 2$, is an element of $GF(p)$. We determine necessary and sufficient conditions that the polynomial $M(X)$ be irreducible over the ring $A = GF(p^n)\,[X;\sigma]$. Our result shows that $M(X)$ irreducible over the ring A implies that any one of four polynomials must be irreducible over $= GF(p)$, Petit shows that only one needs be irreducible.We use the result from Lidl and Niederreiter[4] to establish an infinite family of semifields.

THEOREM 5.1. The polynomial $M(X) = X^3 + \alpha_2X^2 + \alpha_1X + \alpha_0$, $\alpha_i \in GF(p)$, is irreducible over the ring $A = GF(p^n)\,[X;\sigma]$ where $a^\sigma = a^{p^d}$, if and only if n is prime to $p^{2d} + p^d + 1$ and one of the polynomials

(i) $X^{p^{2d}+p^d+1} - \alpha_2X^{p^{2d}+p^d} + \alpha_1X^{p^{2d}} - \alpha_0$,

(ii) $X^{p^{2d}+p^d+1} - \alpha_2X^{p^d+1} + \alpha_1X - \alpha_0$,

(iii) $X^{p^{2d}+p^d+1} - \alpha_1X^{p^d+1} + \alpha_0\alpha_2X - \alpha_0^2$ and

(iv) $X^{p^{2d}+p^d+1} - \alpha_1X^{p^{2d}+p^d} + \alpha_0\alpha_2X^{p^{2d}} - \alpha_0^2$

is irreducible over $GF(p)$. If one of these polynomials is irreducible over $GF(p)$, then all are irreducible over $GF(p)$.

Proof. If $X^3 + \alpha_2X^2 + \alpha_1X + \alpha_0 = (X + a)\left(X^2 + bX + c\right)$, we must have

$$\begin{aligned} a + b^{p^d} &= \alpha_2 \\ ab + c^{p^d} &= \alpha_1 \\ ac &= \alpha_0. \end{aligned}$$

Elimanating b and c from these equations gives

$$a^{p^{2d}+p^d+1} - \alpha_2a^{p^{2d}+p^d} + \alpha_1a^{p^{2d}} - \alpha_0 = 0.$$

Solving for c gives

$$c^{p^{2d}+p^d+1} - \alpha_1c^{p^d+1} + \alpha_0\alpha_2c - \alpha_0^2 = 0.$$

If $X^3 + \alpha_2 X^2 + \alpha_1 X + \alpha_0 = (X^2 + aX + b)(X + c)$, we must have

$$\begin{aligned} a + c^{p^{2d}} &= \alpha_2 \\ ac^{p^d} + b &= \alpha_1 \\ bc &= \alpha_0. \end{aligned}$$

Elimanating a and b from these equations gives

$$c^{p^{2d}+p^d+1} - \alpha_2 c^{p^d+1} + \alpha_1 c - \alpha_0 = 0.$$

Solving for b will give us

$$b^{p^{2d}+p^d+1} - \alpha_1 b^{p^{2d}+p^d} + \alpha_0\alpha_2 b^{p^{2d}} - \alpha_0^2 = 0.$$

By Theorem 1.2.9 of Jacobson's book, only one of these need be irreducible for the polynomial $X^3 + \alpha_2 X^2 + \alpha_1 X + \alpha_0$ to be irreducible in the ring A. If one is irreducible, then each is irreducible. By the Corollary, page 98 of the book by Lidl and Niederreiter[4], an irreducible polynomial over $GF(q)$ of degree n remains irreducible over $GF(q^k)$ if and only if k and n are relatively prime. ∎

REMARK 5.1. A polynomial may be irreducible over $GF(p)$ and may fail to be irreducible over $GF(p^n)[X;\sigma]$ where $a^\sigma = a^p$ for all $n \geq 2$. The polynomial $Q(X) = X^3 + X + 1$ is irreducible over $GF(7)$ but the polyomial $X^{57} + X^{49} + 6$ is factorable over $GF(7)$.

REMARK 5.2. We will insist, unless otherwise stated, that $a^\sigma = a^p$, where the field F is $GF(p^n)$.We can now construct semifields 3-dimensional over their left nuclei as follows: Find a third degree polynomial irreducible over $GF(p)$, use the above theorem to compute the corresponding polynomial of degree $p^2 + p + 1$, verify that this polynomial is irreducible over $GF(p)$. The number of degree three polynomials irreducible over $GF(p)$ is $\frac{1}{3}(p^3 - p)$. We find the following degree three polynomials irreducible over $GF(p)$ to be irreducible over the rings $A = GF(p^n)[X;\sigma]$ for all for all n prime to $p^2 + p + 1$:

All two of the degree three polynomials irreducilbe over $GF(2)$: $X^3 + X^2 + 1$ and $X^3 + X + 1$.

All eight of the degree three polynomials irreducilbe over $GF(3)$.

All 40 degree three polynomials irreducilbe over $GF(5)$.

There are 112 degree three polynomials irreducilbe over $GF(7)$. I have found only three that remain irreducible over the ring $A = GF(7^n)[X; a^\sigma = a^7]$; they are $X^3 + 3X + 2$, $X^3 + X + 2$ and $X^3 + X^2 + 3$.

A short FORTRAN program was used to verify that the 13-degree polynomials are irreducible over $GF(3)$. A more ambitious FORTRAN program, implimenting Berlekamp's algorithm (page 130-134 of Lidl and Niederreiter[4]), was used to show the irreducibility of the 31-degree and 57-degree polynomials for characteristics 5 and 7 respectively.

EXAMPLE 5.1. Let $A = GF(2^n)[X;\sigma]$ where $a^\sigma = a^2$ the polynomial $M(X) = X^3 + X + 1$ is irreducible over $GF(2)$. The polynomial $M(X)$ defines a semifield for all n prime to 7. By Theorem 5.1, the polynomial $M(X) = X^3 + X + 1$ determines a semifield if and only if the polynomial $X^7 + X + 1$ is irreducible over $GF(2)$. This polynomial is irreducible over $GF(2)$.

COROLLARY 5.2. Let $M(X) = X^3 + \mu_2 X^2 + \mu_1 X + \mu_0$ be irreducible over the ring $GF(p^n)[X;\sigma]$ where $a^\sigma = a^p$. Let Δ denote the corresponding semifield. The left and middle nuclei are equal and isomorphic to $GF(p^n)$; the right nucleus is the set $\left\{\sum_0^2 \alpha_i X^i \mid \alpha_i \in GF(p)\right\}$ and is isomorphic to $GF(p^3)$.

Proof. This is a corollary to Theorem 3.2. ■

THEOREM 5.3. Let $M(X) = X^3 + \mu_2 X^2 + \mu_1 X + \mu_0$ be irreducible over the ring $A = GF(p^n)[X;\sigma]$ where $a^\sigma = a^p$. Let Δ denote the corresponding semifield. If $T : \Delta \to \Delta$ is an automorphism of S, then $(aX + b)^T = a^* rX + b^*$ where $*$ is an automorphism of the field $GF(p^n)$ and r is a nonzero element of $GF(p^n)$. There is a one-to-one correspondence between automorphisms of Δ and automorphisms of the ring A.
Proof. Apply the results of Section 4 to these semifields.. ■

We can determine isomorphism classes using the follow:

THEOREM 5.4. The polynomials $M(X) = X^3 + \alpha_2 X^2 + \alpha_1 X + \alpha_0$ and $P(X) = X^3 + \beta_2 X^2 + \beta_1 X + \beta_0$ determine isomorphic semifields over the ring $GF(p^n)[X; a^\sigma = a^p]$ if and only if there exists some a in $GF(p^n)$ and

1. $\alpha_0 = \beta_0 a^{P^2+p+1}$
2. $\alpha_1 = \beta_1 a^{p^2+p}$
3. $\alpha_2 = \beta_2 a^{p^2}$.

If one of α_1 and α_2 is nonzero, then $\alpha \in GF(p)$.

Proof. Let A be the ring $GF(p^n)[X; a^\sigma = a^p]$ and let $M(X) = X^3 + \alpha_2 X^2 + \alpha_1 X + \alpha_0$ be irreducible in A. Let B be the ring $GF(p^n)[Y; a^\sigma = a^p]$ and $P(Y) = Y^3 + \beta_2 Y^2 + \beta_1 Y + \beta_0$ be irreducible in B. Let S denote the semifield $A_3/M(x)$ and $\overline{S}$ denote the semifield $B_3/P(y)$. Let $T : S \to \overline{S}$ be an isomorphism of the semifields. Since T must map the left nucleus of S to the left nucleus of $\overline{S}$, T, when restricted to the left nucleus of S, induces an automorphism $*$ of the field $GF(p^n)$. The isomorphism T will now be fully determined once we know the image of X under T.

Let

$$X^T = \sum_{i=0}^{2} \delta i Y^i.$$

Then for any element $a \in GF(p^n)$,

$$\begin{aligned}
(Xa)^T &= (a^p X)^T \\
(Xa)^T &= (a^p)^* \left(\sum_{i=0}^{2} \delta_i Y^i\right) \\
(Xa)^T &= \sum_{i=0}^{2} (a^p)^* \delta_i Y^i \\
X^T a^T &= X^T a^T \\
X^T a^T &= \left(\sum_{i=0}^{2} \delta_i Y^i\right) a^* \\
X^T a^T &= \sum_{i=0}^{2} (a^*)^{p^i} \delta_i Y^i
\end{aligned}$$

Equating coefficients shows that δ_i is nonzero if and only if $i = 1$.
If $X^T = aY$ for some nonzero element $a \in GF(p^n)$, then

$$\begin{aligned}
(X^2)^T &= (aY)^2 = a^{p+1} Y^2 \text{ and} \\
(X^3)^T &= a^{p^2+p+1} Y^3 \\
(X^3)^T &= -a^{p^2+p+1} \left(\beta_2 Y^2 + \beta_1 Y + \beta_0\right)
\end{aligned}$$

But $(X^3)^T = -\left(\alpha_2 X^2 + \alpha_1 X + \alpha_0\right)^T = -\left(\alpha_2^* a^{p+1} Y^2 + \alpha_1^* a Y + \alpha_0^*\right)$. Equating components gives us the relations 1, 2 and 3 above. The result now follows. ■

EXAMPLE 5.2 Let $A = GF(2^6)\left[X; a^\sigma = a^2\right]$. The polynomial $X^3 + X + 1$ is irreducible over A. Then $N_l = N_m = GF(2^6)1$ and $N_r = \left\{\sum_0^2 \alpha_i X^i \mid \alpha_i \in GF(p)\right\}$ which is isomorphic to $GF(2^3)$.

The left nucleus is isomorphic to the field $GF(2^6) = F_4$and contains subfields isomorphic to $GF(2^3) = F_3, GF(2^2) = F_2$ and $GF(2) = F_1$.

The semifields determined by the elements $\left\{\sum_0^2 \alpha_i X^i \mid \alpha_i \in F_k\right\}$ for $k = 1, 2, 3$ and 4 are not associative except for $k = 1$.

Since the polynomial $X^3 + X + 1$ is fixed by all automorphisms of A mapping $X + a$ to $X + a^{p^i}, i = 0, 1, \ldots, n-1$, there are at least 6 automorphisms of this semifield.

6. CONCLUSION

The construction is very sensitive to the automorphism σ used to define the noncommutative ring A. The test (Theorem 5.1) developed above show that the polynomial $X^3 + X + 1$ defines semifields for all $A = GF(2^n)[X; \sigma]$ where $a^\sigma = a^2$ and n prime to 7; the test may fail for $a^\sigma = a^{p^i}, 2 \le i \le n-1$ and yet the polynomial still defines a semifield over some $GF(2^n)$ for $n \ge 3$.

EXAMPLE 6.1. Our test shows that $X^3 + X + 1$ determines the polynomial $X^{21} + X + 1$ that is factorable over $GF(2)$. This polynomial is irreducible in the ring $A = GF(2^3)[X; \sigma]$ where $a^\sigma = a^4$.

If $X^3+X+1=(X+a)\left(X^2+bX+c\right)$ then $a+b^4=0$, $ab+c^4=1$ and $ac=1$.

$$\left(a+b^4\right)^2=a^2+b^8=a^2+b=0.$$

We now substitute into the middle equation to get

$$a^7+a+1=0.$$

Since we are working in $GF(2^3)$, $a^7=1$, which gives a contradiction.

Results concerning factoring polynomials in the ring $GF(p^n)[X;\sigma]$ appear quite limited; Theorem 1.3.16 [3] gives necessary and sufficient conditions that the polynomial X^r-u, where r is a prime, is irreducible in the ring $GF(p^r)[X;\sigma]$; u cannot be a norm over $GF(p)$.

REFERENCES

[1] P. Dembowski. Finite Geometries. Berlin: Springer-Verlag, 1997, 375p.

[2] D. R. Hughes and E. Kleinfeld. Semi-nuclear extensions of Galois fields,.Amer. J. Math. 82 (1960). 389-392.

[3] N. Jacobson. Finite-Dimensional Division Algebras over Fields. Berlin: Springer-Verlag, 1996, 278p.

[4] R. Lidl and H. Niederreiter. Introduction to Finite Fields and Their Applications. Cambridge: Cambridge University, 1986, 407p.

[5] O. Ore. Theory of non-commutative polynomials. Ann. of Math. Series 2 34 (1933). 480-508.

[6] O. Ore. On a special class of polynomials. Trans. Amer. Math. Soc. 35 (1934). 559-584.

[7] J. Petit. Quasi-corps généralisant un type d'anneau quotient. C. R. Acad. Sc. Paris. t. 265. p. 708-711 (27 novembre 1967).

[8] J. Petit. Sur les quasi-corps distributifs à base monogéne. C. R. Acad. Sc. Paris. t. 266. p. 402-404 (12 févier 1968).

[9] R. Sandler. Autotopisms groups of some finite nonassociative algebras. Amer. J. Math. 84 (1962). 239-264.

Supplementing radicals of nonassociative rings

RICHARD WIEGANDT Mathematical Institute, HAS, PO Box 127, H-1364 Budapest, Hungary. wiegandt@ math-inst.hu

ABSTRACT The notion of supplementing radical is redefined for Plotkin radicals of nonassociative rings (algebras, Ω-groups, near-rings, etc) and investigated. If the essential closure of a Plotkin radical class is contained in a unique minimal Kurosh-Amitsur semisimple class, then the supplementing radical exists, is unique, and coincides with the one of the usual definition. Any hereditary Plotkin radical has a unique supplementing radical. If in the universal class considered, every semisimple class is hereditary, then every Plotkin radical has a unique supplementing radical. Supplementing radicals are given as upper radicals of subdirectly irreducible rings. Results are applied to Andrunakievich s-varieties, in particular to alternative algebras.

1. INTRODUCTION AND PRELIMINARIES

The notion of a supplementing radical (called also as complementary radical) was introduced by Andrunakievich [1] who described the supplementing radicals of hereditary Kurosh-Amitsur radicals of associative rings. We give a new, weaker definition for supplementing radicals which is in general not equivalent to the one of Andrunakievich, but in many cases, including the cases of associative rings and Andrunakievich s-varieties, the two definitions are equivalent.

The author gratefully acknowledges the kind hospitality and financial support of the Johannes Kepler Universität, Linz. Research was supported also by the Hungarian National Foundation for Scientific Research Grant # T23434

It is the purpose of this note to study the existence and unicity of a radical supplementing to an arbitrary Plotkin radical of nonassociative rings. In our investigations we exploit ideas on essential covers used first successfully by Chinese authors [6], [7], [19] (see also [17]) in this context for Kurosh-Amitsur radicals of associative rings. Hereditary Plotkin radicals have always unique supplementing radicals and alternative results to those of [8] are proved. To an arbitrary Plotkin radical a supplementing radical exists if and only if the essential cover of the considered Plotkin radical class is contained in a minimal Kurosh-Amitsur semisimple class. If there exists a unique such semisimple class, then also the supplementing radical is unique and is the supplementing radical in the sense of Andrunakievich. This is always the case in a universal class having hereditary semisimple classes. Also the representation of supplementing radicals as an upper radical of subdirectly irreducible rings is treated. The results are then applied to Andrunakievich s-varieties, in particular to alternative algebras.

A class $\mathcal{V}$ of nonassociative rings (R-algebras, or Ω-groups, in particular near-rings, etc) is called a *universal class*, if $\mathcal{V}$ is closed under taking ideals and homomorphic images. Members of the universal class $\mathcal{V}$ will be briefly referred to as *rings*. A mapping γ which assigns to every $A \in \mathcal{V}$ an ideal $\gamma(A)$ of A is called a *preradical*, if $f(\gamma(A)) \subseteq \gamma(f(A))$ for every homomorphism $f: A \mapsto f(A)$. A preradical γ is said to be

complete *if* $I \triangleleft A \in \mathcal{V}$ and $\gamma(I) = I$ imply $I \subseteq \gamma(A)$,
idempotent *if* $\gamma(\gamma(A)) = \gamma(A)$ for all $A \in \mathcal{V}$,
hereditary *if* $I \cap \gamma(A) \subseteq \gamma(I)$ for all $I \triangleleft A \in \mathcal{V}$.

A preradical γ is called a *Hoehnke radical*, if $\gamma(A/\gamma(A)) = 0$ for all $A \in \mathcal{V}$. Complete and idempotent preradicals (Hoehnke radicals) are referred to as *Plotkin radicals*, (*Kurosh-Amitsur radicals*, respectively).

To each preradical γ we associate two classes, the *radical class*

$$\mathbb{R}_\gamma = \{A \in \mathcal{V} \mid \gamma(A) = A\}$$

and the *semisimple class*

$$\mathbb{S}_\gamma = \{A \in \mathcal{V} \mid \gamma(A) = 0\}.$$

Plotkin radicals are uniquely determined by their radical classes, and Hoehnke radicals by their semisimple classes. In particular, there is a one-to-one correspondence between Kurosh-Amitsur radicals, their radical classes and their semisimple classes. For every preradical γ the radical class $\mathbb{R}_\gamma$ is homomorphically closed.

A subclass $\mathbb{C}$ of $\mathcal{V}$ is *hereditary*, if $I \triangleleft A \in \mathbb{C}$ implies $I \in \mathbb{C}$. The radical class of a hereditary preradical is hereditary, and a Plotkin radical with hereditary radical class is itself hereditary.

Next, we introduce operators acting on subclasses of $\mathcal{V}$. The *upper radical operator* $\mathcal{U}$ is defined by

$$\mathcal{U}\mathbb{C} = \{A \in \mathcal{V} \mid \text{ if } B \text{ is a homomorphic image of } A \text{ and } B \in \mathbb{C}, \text{ then } B = 0\}.$$

If $\mathbb{C}$ is a *regular* class, (that is, every nonzero homomorphic image of an ideal of a ring in $\mathbb{C}$ has a nonzero homomorphic image in $\mathbb{C}$), then $\mathcal{U}\mathbb{C}$ is a Kurosh-Amitsur radical class. The *semisimple operator* $\mathcal{S}$ is defined by

$$\mathcal{S}\mathbb{C} = \{A \in \mathcal{V} \mid \text{ if } I \triangleleft A \text{ and } I \in \mathbb{C} \text{ then } I = 0\}.$$

If $\mathbb{C}$ is the radical class of an idempotent preradical, then $\mathcal{S}\mathbb{C}$ is its semisimple class.

An ideal I of A is *essential* in A, denoted by $I \triangleleft \cdot A$, if $I \cap K \neq 0$ whenever $0 \neq K \triangleleft A$. The *essential cover operator* $\mathcal{E}$ is defined by

$$\mathcal{E}\mathbb{C} = \{A \in \mathcal{V} \mid \text{there exists an } I \triangleleft \cdot A \text{ such that } I \in \mathbb{C}\}.$$

A subring C of A is called an *accessible subring* of A, if there exists a finite chain $C = C_1 \subseteq \ldots \subseteq C_n = A$ of subrings of A such that $C_i \triangleleft C_{i+1}$ for $1 \leq i \leq n-1$, $n > 1$ and $C = A$ for $n = 1$. The *hereditary closure operator* $\mathcal{H}$ is defined by

$$\mathcal{H}\mathbb{C} = \{A \in \mathcal{V} \mid A \text{ is an accessible subring of a ring in } \mathbb{C}\}.$$

The *largest homomorphically closed subclass* $h\mathbb{C}$ of a class $\mathbb{C}$ is the class

$$h\mathbb{C} = \{A \in \mathbb{C} \mid \text{every homomorphic image of } A \text{ is in } \mathbb{C}\}.$$

2. SUPPLEMENTING RADICALS

Let $\mathcal{V}$ be a universal class of rings, and γ a preradical in $\mathcal{V}$.

DEFINITION 2.1 (P. Fuchs [10]). A Kurosh-Amitsur radical $\overline{\gamma}$, (if exists), is called a *supplementing radical* to γ if

(i) $\gamma(A) \cap \overline{\gamma}(A) = 0$ for all $A \in \mathcal{V}$,

(ii) if χ is any Kurosh-Amitsur radical such that $\gamma(A) \cap \chi(A) = 0$ and $\overline{\gamma}(A) \subseteq \chi(A)$ for all $A \in \mathcal{V}$, then $\overline{\gamma}(A) = \chi(A)$, (that is, $\overline{\gamma}$ is a maximal Kurosh-Amitsur radical with property (i)).

PROPOSITION 2.2. *a) If γ is a complete preradical and χ any preradical with $\gamma(A) \cap \chi(A) = 0$ for all $A \in \mathcal{V}$, then $\mathcal{E}\mathbb{R}_\gamma \subseteq \mathbb{S}_\chi$.*

b) If γ is an idempotent preradical and χ a Hoehnke radical such that $\mathcal{E}\mathbb{R}_\gamma \subseteq \mathbb{S}_\chi$, then $\gamma(A) \cap \chi(A) = 0$ for all $A \in \mathcal{V}$.

Proof. a) Let $A \in \mathcal{E}\mathbb{R}_\gamma$. Then there exists an essential ideal I of A such that $I \in \mathbb{R}_\gamma$. Since γ is complete, it follows that $I \subseteq \gamma(A)$. Consequently

$$I \cap \chi(A) \subseteq \gamma(A) \cap \chi(A) = 0,$$

which implies that $\chi(A) = 0$, since I is essential in A. Thus $A \in \mathbb{S}_\chi$.

b) Let $A \in \mathcal{V}$ and K an ideal of A maximal with respect to $\gamma(A) \cap K = 0$. Since γ is idempotent, we have that $\gamma(A) \in \mathbb{R}_\gamma$. Further,

$$\gamma(A) \cong \gamma(A)/(\gamma(A) \cap K) \cong (\gamma(A) + K)/K \triangleleft A/K,$$

and by the maximality of K also $(\gamma(A)+K)/K \triangleleft \cdot A/K$ holds. Hence $A/K \in \mathcal{E}\mathbb{R}_\gamma \subseteq \mathbb{S}_\chi$. Taking into account that χ is a Hoehnke radical, we conclude that $\chi(A) \subseteq K$. Thus,

$$\gamma(A) \cap \chi(A) \subseteq \gamma(A) \cap K = 0. \qquad \square$$

THEOREM 2.3. *Let γ be a Plotkin radical in $\mathcal{V}$. If the class $\mathcal{E}\mathbb{R}_\gamma$ is contained in a minimal Kurosh-Amitsur semisimple class $\mathbb{S}$, then the radical $\overline{\gamma}$ corresponding to $\mathbb{S}$ is a supplementing radical to γ.*

Let γ be a complete preradical. If γ has a supplementing radical $\overline{\gamma}$, then $\mathbb{S}_{\overline{\gamma}}$ is a minimal Kurosh-Amitsur semisimple class containing $\mathcal{E}\mathbb{R}_\gamma$.

Proof. From $\mathcal{E}\mathbb{R}_\gamma \subseteq \mathbb{S} = \mathbb{S}_{\overline{\gamma}}$ we get that $\gamma(A)\cap\overline{\gamma}(A) = 0$ for all $A \in \mathcal{V}$ by Proposition 2.2.b. Now, let χ be any Kurosh-Amitsur radical such that $\gamma(A) \cap \chi(A) = 0$ and $\overline{\gamma}(A) \subseteq \chi(A)$ for all $A \in \mathcal{V}$. Applying Proposition 2.2.a, we obtain that $\mathbb{S}_\chi \subseteq \mathbb{S}_{\overline{\gamma}} = \mathbb{S}$. Hence $\mathcal{E}\mathbb{R}_\gamma \subseteq \mathbb{S}_\chi \subseteq \mathbb{S}_{\overline{\gamma}}$ follows. Further, the minimality of $\mathbb{S}_{\overline{\gamma}}$ infers $\mathbb{S}_\chi = \mathbb{S}_\gamma$, whence $\chi = \overline{\gamma}$.

For the second statement, apply Proposition 2.2.a, and get $\mathcal{E}\mathbb{R}_\gamma \subseteq \mathbb{R}_{\overline{\gamma}}$. If χ is any Kurosh-Amitsur radical with semisimple class $\mathbb{S}_\chi$ and $\mathcal{S}\mathbb{R}_\gamma \subseteq \mathbb{S}_\chi \subseteq \mathbb{S}_{\overline{\gamma}}$, then $\mathbb{R}_{\overline{\gamma}} \subseteq \mathbb{R}_\chi$. Hence $\overline{\gamma}(A) \subseteq \chi(A)$ for all $A \in \mathcal{V}$. Proposition 2.2.b infers that $\gamma(A) \cap \chi(A) = 0$ for all $A \in \mathcal{V}$. Since $\overline{\gamma}$ is a supplementing radical to γ, it follows that $\overline{\gamma}(A) = \chi(A)$ for all $A \in \mathcal{V}$, whence $\mathbb{S}_\chi = \mathbb{S}_{\overline{\gamma}}$. Thus $\mathbb{S}_{\overline{\gamma}}$ is minimal. □

THEOREM 2.4. *For a Plotkin radical γ in $\mathcal{V}$ the following conditions are equivalent:*

(i) *γ has a unique supplementing radical $\overline{\gamma}$,*
(ii) *$\mathcal{E}\mathbb{R}_\gamma$ is contained in a unique minimal Kurosh-Amitsur semisimple class $\mathbb{S}$,*
(iii) *there exists a Kurosh-Amitsur radical γ^* such that $\gamma(A)\cap\gamma^*(A) = 0$ for all $A \in \mathcal{V}$ and for any Kurosh-Amitsur radical χ with $\gamma(A) \cap \chi(A) = 0$ for all $A \in \mathcal{V}$ it folows $\chi(A) \subseteq \gamma^*(A)$.*

Proof. (i) ⇒ (ii) Obvious by Theorem 2.3.

(ii) ⇒ (iii) Let γ^* be the radical belonging to the semisimple class $\mathbb{S}$. By Proposition 2.2.b it follows that $\gamma(A) \cap \gamma^*(A) = 0$ for all $A \in \mathcal{V}$. Next, let χ be a Kurosh-Amitsur radical such that $\gamma(A) \cap \chi(A) = 0$ for all $A \in \mathcal{V}$, and let $A \in \mathcal{E}\mathbb{R}_\gamma$. Then there exists an essential ideal I of A such that $I = \gamma(I)$. Since γ is complete, we have that $I \subseteq \gamma(A)$. Hence

$$I \cap \chi(A) \subseteq \gamma(A) \cap \chi(A) = 0,$$

and so the essentiality of I in A yields $\chi(A) = 0$, that is, $A \in \mathbb{S}_\chi$. Thus, we have got that $\mathcal{E}\mathbb{R}_\gamma \subseteq \mathbb{S}_\chi$, and hence the minimality of $\mathbb{S}$ yields $\mathbb{S} \subseteq \mathbb{S}_\chi$. Consequently $\mathbb{R}_\chi \subseteq \mathbb{R}_{\gamma^*}$, and therefore $\chi(A) \subseteq \gamma^*(A)$ for all $A \in \mathcal{V}$.

(iii) ⇒ (i) Trivial with $\gamma^* = \overline{\gamma}$. □

3. HEREDITARY PLOTKIN RADICALS

PROPOSITION 3.1 ([9] Corollary 4.2). *Let γ be a Plotkin radical. The corresponding radical class $\mathbb{R}_\gamma$ is hereditary if and only if $\gamma(A) \cap I \subseteq \gamma(I)$ for all $I \triangleleft A \in \mathcal{V}$.* □

By the one-to-one corrrespondence between Plotkin radicals and their radical classes we may call a Plotkin radical *hereditary*, if its radical class is hereditary.

PROPOSITION 3.2 ([5] Proposition 1.4 (iv). *If $\mathbb{C}$ is a hereditary subclass of $\mathcal{V}$, then its essential cover $\mathcal{E}\mathbb{C}$ is a regular class, and hence $\mathcal{U}\mathcal{E}\mathbb{C}$ is a Kurosh-Amitsur radical class.* □

Proof. Let $A \in \mathcal{E}\mathbb{C}$. Then there exists an essential ideal K of A such that $K \in \mathbb{C}$. If $0 \neq I \triangleleft A$, then $0 \neq I \cap K \in \mathbb{C}$ by the hereditariness of $\mathbb{C}$. Put $L = I \cap K$, and choose an ideal M of A which is maximal with respect to $L \cap M = 0$. Then, as is well-known,

$$0 \neq L \cong (L+M)/M \triangleleft \cdot I/M.$$

Hence $0 \neq I/M \in \mathcal{E}\mathbb{C}$, and the regularity of the class $\mathbb{C}$ has been established.

Now we are in the position to prove that hereditary Plotkin radicals satisfy condition (iii) of Theorem 2.4.

THEOREM 3.3. *Let γ be a Plotkin radical with radical class $\mathbb{R}_\gamma$. If the class $\mathcal{E}\mathbb{R}_\gamma$ is regular, then $\mathcal{U}\mathcal{E}\mathbb{R}_\gamma$ is a Kurosh-Amitsur radical class. Let γ^* denote the radical assignment corresponding to $\mathcal{U}\mathcal{E}\mathbb{R}_\gamma$. Then*

(i) $\gamma(A) \cap \gamma^*(A) = 0$ *for all* $A \in \mathcal{V}$,

(ii) *if χ is any idempotent preradical such that $\gamma(A) \cap \chi(A) = 0$ then $\chi(A) \subseteq \gamma^*(A)$ for all $A \in \mathcal{V}$.*

Proof. Only (i) and (ii) needs proof. For proving (i), let $A \in \mathcal{V}$ be any ring. Applying Zorn's Lemma we may choose an ideal K of A such that $\gamma(A) \cap K = 0$ and K is maximal with respect to this property. Then, as is well-known,

$$\gamma(A) \cong (\gamma(A)+K)/K \triangleleft \cdot A/K.$$

Since γ is a Plotkin radical (completeness), we have also

$$(\gamma(A)+K)/K \subseteq \gamma(A/K) \triangleleft \cdot A/K.$$

Thus $A/K \in \mathcal{E}\mathbb{R}_\gamma$ by the idempotence of γ, and by the regularity of $\mathcal{E}\mathbb{R}_\gamma$ we have $\mathcal{E}\mathbb{R}_\gamma \subseteq \mathcal{S}\mathcal{U}\mathcal{E}\mathbb{R}_\gamma$. Hence $A/K \in \mathcal{S}\mathcal{U}\mathcal{E}\mathbb{R}_\gamma$ and this means that A/K is semisimple with respect to the radical γ^*. Thus $\gamma^*(A) \subseteq K$ which implies

$$\gamma(A) \cap \gamma^*(A) \subseteq \gamma(A) \cap K = 0.$$

For proving (ii), suppose that $\chi(A) \not\subseteq \gamma^*(A)$ for some $A \in \mathcal{V}$, although $\gamma(A) \cap \chi(A) = 0$ for all $A \in \mathcal{V}$. Denoting by $\mathbb{R}_\chi$ the radical class of χ, we have

$$0 \neq (\chi(A)+\gamma^*(A))/\gamma^*(A) \cong \chi(A)/(\chi(A) \cap \gamma^*(A)) \in \mathbb{R}_\chi,$$

because χ is idempotent and $\mathbb{R}_\chi$ is homomorphically closed.

On the other hand,

$$0 \neq (\chi(A)+\gamma^*(A))/\gamma^*(A) \triangleleft A/\gamma^*(A) \in \mathcal{S}\mathcal{U}\mathcal{E}\mathbb{R}_\gamma,$$

and the semisimple class $\mathcal{S}\mathcal{U}\mathcal{E}\mathbb{R}_\gamma$ is regular. Hence $(\chi(A)+\gamma^*(A))/\gamma^*(A)$ has a nonzero homomorphic image B in $\mathbb{R}_\chi \cap \mathcal{S}\mathcal{U}\mathcal{E}\mathbb{R}_\gamma$. Since the class $\mathcal{E}\mathbb{R}_\gamma$ is regular, we have $\mathcal{U}\mathcal{E}\mathbb{R}_\gamma = \mathcal{U}\mathcal{S}\mathcal{U}\mathcal{E}\mathbb{R}_\gamma$. Thus $0 \neq B \in \mathcal{S}\mathcal{U}\mathcal{E}\mathbb{R}_\gamma$ implies that $B \notin \mathcal{U}\mathcal{E}\mathbb{R}_\gamma$, and so B has a nonzero homomorphic image C in $\mathcal{E}\mathbb{R}_\gamma \cap \mathbb{R}_\chi$. Hence by the completeness of γ we have that $0 \neq \gamma(C) \triangleleft \cdot C$, and by $C \in \mathbb{R}_\chi$ it follows that $\chi(C) = C$. So we arrived at

$$0 \neq \gamma(C) = \gamma(C) \cap C = \gamma(C) \cap \chi(C) = 0,$$

a contradiction. Thus $\chi(A) \subseteq \gamma^*(A)$ for all $A \in \mathcal{V}$. □

Chen [7] called a Kurosh-Amitsur radical γ *quasi-hereditary*, if $\mathcal{E}\mathbb{R}_\gamma$ is a regular class. Chen's definition makes sense also for Plotkin radicals.

COROLLARY 3.4. *If γ is a quasi-hereditary Plotkin radical, then γ has a unique supplementing radical $\overline{\gamma}$ which is the radical corresponding to the Kurosh-Amitsur radical class $\mathcal{U}\mathcal{E}\mathbb{R}_\gamma$.*

Proof. Proposition 3.2 and Theorem 3.3 yield that γ satisfies condition (iii) of Theorem 2.4. Theorem 3.3 infers also the representation of $\mathbb{R}_{\overline{\gamma}} = \mathcal{U}\mathcal{E}\mathbb{R}_\gamma$. □

Theorem 3.3 extends Chen's [7] Theorem 3.1 to Plotkin radicals which provides also another representation of $\overline{\gamma}$.

COROLLARY 3.5. *If γ is a quasi-hereditary Plotkin radical, then the unique supplementing radical $\overline{\gamma}$ is given by its corresponding radical class $\mathbb{R}_{\overline{\gamma}} = h\mathbb{S}_\gamma$.* □

EXAMPLE 3.6. The supplementing radical $\overline{\gamma}$ of a hereditary Plotkin radical need not be hereditary. Let $\mathcal{V}$ be the class of all associative rings, $\gamma(A)$ the maximal torsion ideal of A. The torsion radical γ is a hereditary Kurosh-Amitsur radical, but its supplementing radical $\overline{\gamma}$ is not hereditary, (see Gardner [12]).

Two preradicals γ and δ with radical classes $\mathbb{R}_\gamma$ and $\mathbb{R}_\delta$ may fulfill any of the following requirements:

(a) $\mathbb{R}_\gamma \cap \mathbb{R}_\delta = 0$,

(b) $\gamma(A) \cap \delta(A) = 0$ for all $A \in \mathcal{V}$.

Condition (b) implies trivially (a).

PROPOSITION 3.7. *If γ and δ are idempotent preradicals with hereditary radical classes $\mathbb{R}_\gamma$ and $\mathbb{R}_\delta$, then condition* (a) *implies* (b), *whence* (a) *and* (b) *are equivalent.*

Proof. Suppose that (b) is not satisfied. Then there exists an $A \in \mathcal{V}$ such that $0 \neq \gamma(A) \cap \delta(A) \triangleleft A$. Since γ is idempotent, we have

$$\gamma(A) \cap \delta(A) \triangleleft \gamma(A) \in \mathbb{R}_\gamma.$$

Hence the hereditariness of $\mathbb{R}_\gamma$ yields $\gamma(A) \cap \delta(A) \in \mathbb{R}_\gamma$. Analogous reasoning infers $\gamma(A) \cap \delta(A) \in \mathbb{R}_\delta$. Thus we conclude that $\mathbb{R}_\gamma \cap \mathbb{R}_\delta \neq 0$. □

The class of all Plotkin radical classes in $\mathcal{V}$ can be partially ordered by inclusion. In this way we get a lattice, the *lattice of all Plotkin radical classes in $\mathcal{V}$*. Recall that in a lattice L with least element 0 the *pseudocomplement* (if exists) of an element $a \in L$ is the unique largest element $a^* \in L$ such that $a^* \wedge a = 0$.

COROLLARY 3.8. *If both γ and its supplementing radical $\overline{\gamma}$ are hereditary, then the radical class $\mathbb{R}_{\overline{\gamma}}$ is the pseudocomplement of $\mathbb{R}_\gamma$ in the lattice of hereditary Plotkin radicals.*

Proof. Obvious by Theorem 3.3 and Proposition 3.7. □

Remark. The radical class $\mathbb{R}_{\overline{\gamma}}$ need not be the pseudocomplement of γ in the lattice of all Plotkin radical classes: there may exist a nonhereditary Plotkin radical χ which satisfies only condition (a) but not (b) and hence $\chi(A) \not\subseteq \overline{\gamma}(A)$ for some $A \in \mathcal{V}$.

4. HEREDITARY SEMISIMPLE CLASSES

Theorem 2.4.(ii) arises the question as whether every Plotkin radical class is contained in a unique minimal Kurosh-Amitsur semisimple class. The answer is no. The following example shows that supplementing radicals are not unique.

EXAMPLE 4.1 (P. Fuchs [10]). Let R be a nonassociative ring with exactly one proper ideal I where I has exactly one proper ideal J such that J is simple and $I/J \not\cong J$, $I/J \not\cong R/I$, $R/I \not\cong J$. Then $\mathbb{R}_\gamma = \{0, R, R/I\}$ turns out to be a Kurosh-Amitsur radical class in the universal class $\mathcal{V} = \{0, R, I, J, R/I, I/J\}$ such that $\mathcal{E}\mathbb{R}_\gamma = \mathbb{R}_\gamma$. One checks that $\mathbb{S}_1 = \{0, R, R/I, I, J\}$ and $\mathbb{S}_2 = \{0, R, R/I, R/J\}$ are minimal Kurosh-Amitsur semisimple classes containing $\mathcal{E}\mathbb{R}_\gamma = \mathbb{R}_\gamma$, hence by Theorem 2.3 *both of the radicals γ_1 and γ_2 corresponding to $\mathbb{S}_1$ and $\mathbb{S}_2$, are supplementing radicals to γ.*

Henceforth in this section we shall assume that in $\mathcal{V}$ every Kurosh-Amitsur semisimple class is hereditary. It is well known that beside universal classes of associative rings many other classes of nonassociative rings do satisfy this requirement. Examples are normal varieties in the sense of Terlikowska–Osłowska [14], [15] (see also Puczyłowski [13]) and universal classes of nonassociative rings or algebras satisfying the following two conditions due to Beidar [2]:

(B1) if $K \triangleleft I \triangleleft A$ then to every element $a \in A$ there exists an integer $m > 1$ such that $aK^{(m)} + K^{(m)}a \subseteq K$,
(B2) if $K \triangleleft I \triangleleft A$ then to every element $a \in A$ there exists an integer $n > 1$ such that $((aK + Ka + K)_I)^{(n)} \subseteq K$, where $K^{(0)} = K$ and $K^{(k-1)}K^{(k-1)} = K^{(n)}$ for $k > 0$ and $(X)_I$ denotes the ideal of I generated by the subset X. Autodistributive rings, Γ-rings and Nobusawa's rings form normal varieties, further, Andrunakievich s-varieties, Jordan R-algebras with 1/2 and (γ, δ)-algebras with 1/6 satisfy Beidar's conditions (B1) and (B2), (see also Veldsman [16]).

PROPOSITION 4.2. *Every subclass $\mathbb{C} \subseteq \mathcal{V}$ is contained in a unique smallest Kurosh-Amitsur semisimple class, namely in $\mathcal{SUHC}$.*

Proof. Let $\mathbb{S}$ be any Kurosh-Amitsur semisimple class containing $\mathbb{C}$, (such a class $\mathbb{S}$ exists, for instance $\mathbb{S} = \mathcal{V}$). Then $\mathbb{S}$, being hereditary, contains also $\mathcal{H}\mathbb{C}$. Hence $\mathcal{UH}\mathbb{C}$ is the largest Kurosh-Amitsur radical class disjoint to $\mathcal{H}\mathbb{C} \setminus \{0\}$, and $\mathcal{SUH}\mathbb{C}$ is the unique smallest Kurosh-Amitsur semisimple class containing $\mathcal{H}\mathbb{C}$. □

COROLLARY 4.3. *In a universal class $\mathcal{V}$, where every Kurosh-Amitsur semisimple class is hereditary, every Plotkin radical γ has a unique supplementing radical $\overline{\gamma}$ which is a Kurosh-Amitsur radical with radical class $\mathbb{R}_{\overline{\gamma}} = \mathcal{UHE}\mathbb{R}_\gamma$.*

Proof. The statement is a direct consequence of Theorem 2.4 and Proposition 4.2. □

Remark. Supplementing radicals (called also complementary radicals) of associative rings are defined by condition (iii) of Theorem 2.4. Corollary 4.3 shows that the two definitions of supplementing radicals are equivalent whenever in the universal class $\mathcal{V}$ every Kurosh-Amitsur semisimple class is hereditary.

5. UPPER RADICALS OF SUBDIRECTLY IRREDUCIBLE RINGS

In this section we shall need that every ring of the universal class $\mathcal{V}$ can be decomposed into a subdirect sum of subdirectly irreducible rings of $\mathcal{V}$. Obviously, $\mathcal{V}$ fulfils always this requirement: each ring $A \in \mathcal{V}$ decomposes into a subdirect sum of subdirectly irreducible rings and the subdirectly irreducible components are homomorphic images of A.

In the sequel the letter s will stand for the class of all subdirectly irrreducible rings of $\mathcal{V}$, and $H(A)$ will denote the *heart*

$$H(A) = \cap(I \triangleleft A \mid I \neq 0)$$

of the ring A.

PROPOSITION 5.1. *Let $\mathbb{C}$ be a subclass of $\mathcal{V}$. Then*

a) $s = \mathcal{E}s$,
b) $\mathcal{E}(s \cap \mathbb{C}) \subseteq s \cap \mathcal{E}\mathbb{C}$,
c) *the class $s(\mathbb{C}) = \{A \in s \mid H(A) \in \mathbb{C}\}$ is contained in the class $s \cap \mathcal{E}\mathbb{C}$,*
d) *if $\mathbb{C}$ is hereditary then $s(\mathbb{C}) = s \cap \mathcal{E}\mathbb{C}$.*

Proof. a) Let $A \in \mathcal{E}s$. Then there exists a ring B such that $B \triangleleft \cdot A$ and $B \in s$. If $0 \neq I \triangleleft A$ then $0 \neq B \cap I \triangleleft B$, and so $H(B) \subseteq B \cap I \subseteq I$. Thus

$$0 \neq H(B) \subseteq \cap(I \triangleleft A \mid I \neq 0) = H(A),$$

hence A is subdirectly irreducible.

b) Trivial by a).
c) Trivial.
d) Let $A \in s \cap \mathcal{E}\mathbb{C}$. Then there exists a ring $B \in \mathbb{C}$ such that $H(A) \subseteq B \triangleleft A$. By the hereditariness of $\mathbb{C}$ it follows that $H(A) \in \mathbb{C}$, that is, $A \in s(\mathbb{C})$. □

PROPOSITION 5.2. *If γ is any preradical with radical class $\mathbb{R}_\gamma$, then $\mathcal{UE}\mathbb{R}_\gamma = \mathcal{U}(s \cap \mathcal{E}\mathbb{R}_\gamma)$.*

Proof. The containment $s \cap \mathcal{E}\mathbb{R}_\gamma \subseteq \mathcal{E}\mathbb{R}_\gamma$ yields $\mathcal{UE}\mathbb{R}_\gamma \subseteq \mathcal{U}(s \cap \mathcal{E}\mathbb{R}_\gamma)$. Let A be any ring not in $\mathcal{UE}\mathbb{R}_\gamma$. Then A has a nonzero homomorphic image B in $\mathcal{E}\mathbb{R}_\gamma$, and so there exists an essential ideal K of B such that $K \in \mathbb{R}_\gamma$. Let $B = \sum_{\text{subdirect}} (B_\lambda \mid B_\lambda \in s)$ be any subdirect decomposition of B into subdirectly irreducible rings B_λ with projections $\pi_\lambda(B) = B_\lambda$. Since $K \neq 0$, there must exist an index λ such that $\pi_\lambda(K) \neq 0$. B_λ is subdirectly irreducible, so $\pi_\lambda(K) \triangleleft \cdot B$. Further, $K \in \mathbb{R}_\gamma$ implies $\pi_\lambda(K) \in \mathbb{R}_\gamma$. Hence $B_\lambda \in s \cap \mathcal{E}\mathbb{R}_\gamma$. Thus A has a nonzero homomorphic image B_λ in $s \cap \mathcal{E}\mathbb{R}_\gamma$, and therefore $A \notin \mathcal{U}(s \cap \mathcal{E}\mathbb{R}_\gamma)$. This proves $\mathcal{U}(s \cap \mathcal{E}\mathbb{R}_\gamma) \subseteq \mathcal{U}\mathcal{E}\mathbb{R}_\gamma$. □

As an immediate consequence we recover [8] Theorem 2.5.

COROLLARY 5.3. *If γ is a hereditary Plotkin radical then its supplementing radical class $\mathbb{R}_{\overline{\gamma}}$ is given by*

$$\mathbb{R}_{\overline{\gamma}} = \mathcal{U}(s \cap \mathcal{E}\mathbb{R}_\gamma) = \mathcal{U}s(\mathbb{R}_\gamma).$$

Proof. Obvious by Corollary 3.4 and Propositions 5.1.d and 5.2. □

THEOREM 5.4. *Assume that in the universal class $\mathcal{V}$ every semisimple class is hereditary. If γ is any Plotkin radical, then the radical class of its supplementing radical $\overline{\gamma}$ is $\mathbb{R}_{\overline{\gamma}} = \mathcal{U}\mathcal{H}(s \cap \mathcal{E}\mathbb{R}_\gamma)$.*

Proof. By Corollary 4.3 we know that $\overline{\gamma}$ exists and is uniquely determined, moreover, $\mathbb{R}_{\overline{\gamma}} = \mathcal{U}\mathcal{H}\mathcal{E}\mathbb{R}_\gamma$. We are going to prove that the corresponding semisimple classes coincide: $\mathcal{S}\mathcal{U}\mathcal{H}\mathcal{E}\mathbb{R}_\gamma = \mathcal{S}\mathcal{U}\mathcal{H}(s \cap \mathcal{E}\mathbb{R}_\gamma)$.

Since $\mathcal{H}(s\cap\mathcal{E}\mathbb{R}_\gamma) \subseteq \mathcal{H}\mathcal{E}\mathbb{R}_\gamma$, we have the inclusion $\mathcal{S}\mathcal{U}\mathcal{H}(s\cap\mathcal{E}\mathbb{R}) \subseteq \mathcal{S}\mathcal{U}\mathcal{H}\mathbb{R}_\gamma$. So we have to prove the opposite inclusion. Let $A \in \mathcal{E}\mathbb{R}_\gamma$ be an arbitrary ring. Now there exists an essential ideal K of A with $K \in \mathbb{R}_\gamma$. Since γ is complete, $K \subseteq \gamma(A)$ follows and so $\gamma(A) \triangleleft \cdot A$. Let A be a subdirect sum of rings A/K_λ, $\cap K_\lambda = 0$, $\lambda \in \Lambda$ such that each A/K_λ is subdirectly irreducible. Since A/K_λ is subdirectly irreducible, either $\gamma(A) \subseteq K_\lambda$ or $(\gamma(A) + K_\lambda)/K_\lambda \triangleleft \cdot A/K_\lambda$. The index set Λ decomposes into the union of two disjoint subsets Ξ and Θ such that $\gamma(A) \subseteq K_\lambda$ for all $\lambda \in \Xi$ and $(\gamma(A) + K_\lambda)/K_\lambda \triangleleft \cdot A/K_\lambda$ for all $\lambda \in \Theta$. Hence we have

$$\gamma(A) \cap \Big(\bigcap_{\lambda\in\Theta} K_\lambda\Big) \subseteq \Big(\bigcap_{\lambda\in\Xi} K_\lambda\Big) \cap \Big(\bigcap_{\lambda\in\Theta} K_\lambda\Big) = \bigcap_{\lambda\in\Lambda} K_\lambda = 0,$$

and so by $\gamma(A) \triangleleft \cdot A$ it follows that $\bigcap_{\lambda\in\Theta} K_\lambda = 0$. Thus A is a subdirect sum of the subdirectly irreducible rings A/K_λ, $\lambda \in \Theta$. In view of

$$\gamma(A)/(\gamma(A) \cap K_\lambda) \cong (\gamma(A) + K_\lambda)/K_\lambda \triangleleft \cdot A/K_\lambda$$

and $\gamma(A)/(\gamma(A) \cap K_\lambda) \in \mathbb{R}_\gamma$ (because γ is idempotent) for all $\lambda \in \Theta$ we get that

$$(\gamma(A) + K_\lambda)/K_\lambda \subseteq \gamma(A/K_\lambda) \triangleleft \cdot A/K_\lambda$$

(because γ is complete). This shows that $A/K_\lambda \in s\cap\mathcal{E}\mathbb{R}_\gamma$ (because γ is idempotent). Since $s \cap \mathcal{E}\mathbb{R}_\gamma \subseteq \mathcal{H}(s \cap \mathcal{E}\mathbb{R}_\gamma)$ and the semisimple class $\mathcal{S}\mathcal{U}\mathcal{H}(s \cap \mathcal{E}\mathbb{R}_\gamma)$ is closed under taking subdirect sums, we conclude that $A \in \mathcal{S}\mathcal{U}\mathcal{H}(s \cap \mathcal{E}\mathbb{R}_\gamma)$. Thus $\mathcal{E}\mathbb{R}_\gamma \subseteq \mathcal{S}\mathcal{U}\mathcal{H}(s \cap \mathcal{E}\mathbb{R}_\gamma)$ has been established. Hence $\mathcal{S}\mathcal{U}\mathcal{H}\mathcal{E}\mathbb{R}_\gamma \subseteq \mathcal{S}\mathcal{U}\mathcal{H}(\mathcal{S}\mathcal{U}\mathcal{H}(s \cap \mathcal{E}\mathbb{R}_\gamma)) =$

$\mathcal{SUH}(s \cap \mathcal{E}\mathbb{R}_\gamma)$, and therefore the corresponding semisimple classes coincide, as claimed. Hence also $\mathcal{UHE}\mathbb{R}_\gamma = \mathcal{UH}(s \cap \mathcal{E}\mathbb{R}_\gamma)$ holds. □

6. SPECIAL CASES

A detailed discussion of supplementing radicals of hereditary Plotkin radicals in Andrunakievich s-varieties as well as in the variety of abstract affine near-rings can be found in [8]. In the case of *arbitrary* Plotkin radicals of Andrunakievich s-varieties we can add something more.

PROPOSITION 6.1. *Let γ be a Plotkin radical. If every member of the class $s \cap \mathcal{E}\mathbb{R}_\gamma$ has an idempotent heart, then $\mathcal{H}(s \cap \mathcal{E}\mathbb{R}_\gamma) \subseteq s \cap \mathcal{EH}\mathbb{R}_\gamma$. Moreover,*

$$s \cap \mathcal{EH}\mathbb{R}_\gamma = \{A \in s \mid H(A) \in \mathcal{H}\mathbb{R}_\gamma\}.$$

Proof. Let $B \triangleleft A \in s$ and $H(A)$ be idempotent. Then we know from [8] Proposition 3.1 that also B is subdirectly irreducible and $H(B) = H(A)$. By induction we get that every accessible subring C of A is subdirectly irreducible with heart $H(A)$. Further, if in addition also $A \in \mathcal{EH}\mathbb{R}_\gamma$, then there exists an essential ideal I of A such that $I \in \mathcal{H}\mathbb{R}_\gamma$. Since also I is subdirectly irreducible with heart $H(A)$, we get that $H(A) \in \mathcal{H}\mathbb{R}_\gamma$. Hence by $H(A) \triangleleft \cdot C$ we conclude $C \in \mathcal{EH}\mathbb{R}_\gamma$. Thus $C \in s \cap \mathcal{EH}\mathbb{R}_\gamma$ holds proving $\mathcal{H}(s \cap \mathcal{E}\mathbb{R}_\gamma) \subseteq s \cap \mathcal{EH}\mathbb{R}_\gamma$.

A subdirectly irreducible ring A is in $\mathcal{EH}\mathbb{R}_\gamma$ if and only if there exists an essential ideal B of A such that $B \in \mathcal{H}\mathbb{R}_\gamma$. Observe that B is subdirectly irreducible with heart $H(A)$. $B \in \mathcal{H}\mathbb{R}_\gamma$ if and only if $H(A)$ as well as B are accessible subrings of a ring $C \in \mathbb{R}_\gamma$. □

COROLLARY 6.2. *If a Plotkin radical γ fulfils the requirement of Proposition 6.1, then its supplementing radical $\overline{\gamma}$ is the upper radical of a class of subdirectly irreducible rings with hearts in $\mathcal{H}\mathbb{R}_\gamma$.* □

In the rest of this section we shall work in the variety *Alt* of alternative R-algebras over a commutative ring R with 1/3. As proved by Hentzel [12], *Alt* is an Andrunakievich 2-variety of index 2, so all the previous considerations apply to the universal class *Alt*.

In [4] we considered the subclass

$$\{A \in Alt \mid A/\beta(A) \text{ is associative}\},$$

where β is the Baer (prime) radical in *Alt*, and proved that this class is a hereditary Kurosh-Amitsur radical class. So, denoting the corresponding radical assignment by u, this class will be denoted by $\mathbb{R}_u$. We shall use the notations introduced earlier in a consistent way.

THEOREM 6.3. *The supplementing radical $\overline{u}$ to the hereditary Kurosh-Amitsur radical u is given by the radical class*

$$\mathbb{R}_{\overline{u}} = \mathcal{U}\mathcal{E}\mathbb{R}_u = h\mathbb{S}_u = \mathcal{U}(s \cap \mathcal{E}\mathbb{R}_u) = \mathcal{U}s(\mathbb{R}_u).$$

The class $s(\mathbb{R}_{\overline{u}})$ is given by

$$\begin{aligned} s(\mathbb{R}_{\overline{u}}) &= \{A \in s \cap Alt \mid H(A) \quad \text{is nonassociative}\} = \\ s \cap \mathcal{E}\mathbb{R}_{\overline{u}} &= \{A \in Alt \mid A \quad \text{is a Cayley-Dickson algebra}\}, \end{aligned}$$

and the supplementing radical $\overline{\overline{u}}$ to $\overline{u}$ is the upper radical of all Cayley-Dickson algebras.

Proof. The representations of $\mathbb{R}_{\overline{u}}$ are direct consequences of Corollaries 3.4, 3.5 and 5.3.

Let A be a subdirectly irreducible algebra in Alt. Since Alt is an Andrunakievich 2-variety, the heart $H(A)$ of A is either a simple prime algebra or $(H(A))^2 = 0$. Simple prime associative algebras as well as zero-algebras are in $\mathbb{R}_u$, but simple prime nonassociative algebras are in $\mathbb{R}_{\overline{u}}$. Thus $H(A)$ is in $\mathbb{R}_{\overline{u}}$ if and only if $H(A)$ is nonassociative. As is well-known (see [18] §8, Theorem 3 on p. 172), every subdirectly irreducible prime nonassociative algebra in Alt is a Cayley-Dickson algebra over its center.

The containment $s(\mathbb{R}_{\overline{u}}) \subseteq s \cap \mathcal{E}\mathbb{R}_{\overline{u}}$ is Proposition 5.1.c. To prove the opposite containment, let $A \in s \cap \mathcal{E}\mathbb{R}_{\overline{u}}$ be any algebra, and assume that $A \notin s(\mathbb{R}_{\overline{u}})$. Then $H(A)$ is associative and

$$0 \neq H(A) \subseteq u(A) \triangleleft \cdot A.$$

Further, by $A \in \mathcal{E}\mathbb{R}_{\overline{u}}$ there exists an ideal I of A with $\overline{u}(I) = I$. Hence, being in an Andrunakievich 2-variety,

$$0 \neq \overline{u}(I) \subseteq \overline{u}(A) \triangleleft A,$$

and therefore

$$u(A) \cap \overline{u}(A) \neq 0,$$

a contradiction. Thus $A \in s(\mathbb{R}_{\overline{u}})$ and $s \cap \mathcal{E}\mathbb{R}_{\overline{u}} = s(\mathbb{R}_{\overline{u}})$.

An application of Corollary 5.3 yields $\mathbb{R}_{\overline{u}} = \mathcal{U}s(\mathbb{R}_{\overline{u}})$. □

Assigning to every algebra $A \in Alt$ the associator ideal $\mathcal{D}(A)$, (that is, the ideal generated by all associators (x, y, z) with $x, y, z \in A$), is a complete Hoehnke radical which is not idempotent, as shown in [4]. We recall Theorem 3 of [3]: if $A \in Alt$ is a semiprime algebra with d.c.c. on principal right ideals, then

$$A = \mathcal{U}(A) \boxplus \mathcal{D}(A),$$

where $\mathcal{U}(A)$ is the maximal nuclear ideal of A and $\mathcal{D}(A)$ is a discrete direct sum of Cayley-Dickson algebras. Notice that since $\beta(A) = 0$, $\mathcal{U}(A) = u(A)$ is associative, and obviously $\mathcal{D}(A) = \overline{u}(A)$. Hence we arrived at

COROLLARY 6.4. *Every semiprime algebra $A \in Alt$ with d.c.c. on ideals decomposes into a direct sum*

$$A = u(A) \boxplus \overline{u}(A). \qquad \square$$

Note that also the structure of $u(A)$ is known, it is a discrete direct sum of simple rings of linear transformations of finite rank on vector spaces (cf. [3]).

Although the assignment $A \mapsto \mathcal{D}(A)$ is only a complete Hoehnke radical which is not idempotent, its radical class

$$\mathbb{R}_{\mathcal{D}} = \{A \in Alt \mid \mathcal{D}(A) = A\}$$

is the radical class of a Kurosh-Amitsur radical Δ, as it has been proved in [4]. Thus the supplementing radical $\overline{\Delta}$ to Δ is determined by the radical class

$$\mathbb{R}_{\overline{\Delta}} = \mathcal{U}\mathcal{H}\mathcal{E}\mathbb{R}_{\mathcal{D}} = \mathcal{U}\mathcal{H}(s \cap \mathcal{E}\mathbb{R}_{\mathcal{D}})$$

in view of Corollary 4.3 and Theorem 5.4. Moreover, for every semiprime algebra $A \in Alt$ with d.c.c. on principal right ideals the decomposition $A = \mathcal{U}(A) \boxplus \mathcal{D}(A)$ infers

COROLLARY 6.5. $A = \overline{\Delta}(A) \boxplus \Delta(A)$. $\square$

Acknowledgement. The author is indebted to Peter Fuchs for many fruitful discussions and advise, in particular for Definition 2.1 and Example 4.1.

REFERENCES

[1] V. A. Andrunakievich, Radicals of associative rings I, Mat. Sb., 44(86) (1958), 179–212 (Russian), English translation: Amer. Math. Soc. Transl., (2) 52 (1966), 95–128.

[2] K. I. Beidar, Semisimple classes of algebras and the lower radical (in Russian), Mat. Issled. Kishinev, 105 (1988), 13–29.

[3] K. I. Beidar and R. Wiegandt, Splitting theorems for nonassociative rings, Publ. Math. Debrecen, 38 (1991), 121–143.

[4] K. I. Beidar and R. Wiegandt, Radical assignments and radical classes, Preprint, 1998.

[5] G. F. Birkenmeier, Radicals whose esential covers are semisimple classes, Comm. in Algebra, 22 (1994), 6239–6258.

[6] Cai Chuanren, Dual radicals and F. A. Szász' Problem 21, Acta Math. Sinica, 32 (1989), 394–400 (Chinese).

[7] Chen Shencan, Supplementing dual and quasi-hereditary radicals, Northeast. Math. J., 8 (1992), 357–363.

[8] B. de la Rosa, Y. Fong and R. Wiegandt, Complementary radicals revisited, Acta Math. Hungar., 65 (1994), 253–264.

[9] B. de la Rosa, S. Veldsman and R. Wiegandt, On the theory of Plotkin radicals, Chinese J. Math., 21 (1993), 33–54.

[10] P. Fuchs, private communication.

[11] B. J. Gardner, Strong semi-simplicity, Period. Math. Hungar., 24 (1992), 23–35.

[12] R. Hentzel, The Andrunakievich lemma for alternative rings, Algebras, Groups and Geometries, 6 (1989), 55–64.

[13] E. R. Puczyłowski, On general theory of radicals, Algebra Univ., 30 (1993), 53–60.

[14] B. Terlikowska–Osłowska, Category with a self-dual set of axioms, Bull. Acad. Polon. Sci., 25 (1977), 1207–1214.

[15] B. Terlikowska–Osłowska, Radical and semisimple classes of objects in categories with self-dual set of axioms, Bull. Acad. Polon. Sci., 26 (1978), 7–13.

[16] S. Veldsman, Sufficient conditions for a well-behaved Kurosh-Amitsur radical theory, Proc. Edinburgh Math. Soc., 32 (1989), 377–394.

[17] R. Wiegandt, A note on supplementing radicals, Northeast. Math. J., 11 (1995), 476–482.

[18] K. A. Zhevlakov, A. M. Slin'ko, I. P. Shestakov, A. I. Shirshov, Rings that are nearly associative, Academic Press, 1982.

[19] Zhou Yiqiang, On the supplementing radical of a radical class, Natur. Sci. J. Hunan Normal Univ., 12 (1989), 197–199.